Sustainability

Sustainability

Fundamentals and Applications

Edited by

Rao Y. Surampalli
Global Institute for Energy, Environment and Sustainability, Lenexa, KS, USA

Tian C. Zhang
Civil and Environmental Engineering, University of Nebraska-Lincoln, Omaha, NE, USA

Manish Kumar Goyal
Department of Civil Engineering, Indian Institute of Technology, Indore, India

Satinder K. Brar
Department of Civil Engineering, Lassonde School of Engineering, York University, Toronto, Canada

R. D. Tyagi
Institut National de la Recherche Scientifique, University of Quebec, Quebec, Canada

WILEY

This edition first published 2020

Registered Offices
John Wiley & Sons, Inc., 111 River Street, Hoboken, NJ 07030, USA
John Wiley & Sons Ltd, The Atrium, Southern Gate, Chichester, West Sussex, PO19 8SQ, UK

Editorial Office
9600 Garsington Road, Oxford, OX4 2DQ, UK

For details of our global editorial offices, customer services, and more information about Wiley products visit us at www.wiley.com.

Wiley also publishes its books in a variety of electronic formats and by print-on-demand. Some content that appears in standard print versions of this book may not be available in other formats.

Library of Congress Cataloging-in-Publication Data

Names: Surampalli, Rao Y., editor.
Title: Sustainability : fundamentals and applications / edited by Rao Y. Surampalli, Global Institute for Energy, Environment, and Sustainability, USA [and 4 others].
Other titles: Sustainability (John Wiley & Sons)
Description: First edition. | Hoboken, N.J. : John Wiley & Sons, Inc., 2020. | Includes bibliographical references and index.
Identifiers: LCCN 2019051966 (print) | LCCN 2019051967 (ebook) | ISBN 9781119433965 (hardback) | ISBN 9781119433897 (adobe pdf) | ISBN 9781119434030 (epub)
Subjects: LCSH: Sustainable development.
Classification: LCC HC79.E5 .S8655 2020 (print) | LCC HC79.E5 (ebook) | DDC 338.9/27–dc23
LC record available at https://lccn.loc.gov/2019051966
LC ebook record available at https://lccn.loc.gov/2019051967

Cover Design: Wiley
Cover Image: © franckreporter/Getty Images

Set in 9.5/12.5pt STIXTwoText by SPi Global, Chennai, India

10 9 8 7 6 5 4 3 2 1

Contents

Editor Biographies

Dr. Rao Y. Surampalli, PE, F.AAAS, Dist. M.ASCE, received his MS and PhD degrees in Environmental Engineering from Oklahoma State University and Iowa State University, respectively. He is a registered professional engineer in the branches of civil and environmental engineering, and also a Board Certified Environmental Engineer (BCEE) of the American Academy of Environmental Engineers (AAEE) and a Diplomate of the American Academy of Water Resources Engineers (DWRE). His expertise is in bioconversion of wastes into value-added products (biodiesel, bioplastics, biofertilizers, etc.); occurrence, fate/transport and removal of emerging contaminants (e.g. pharmaceuticals, antibiotics and endocrine-disrupting chemicals) in the environment; water and wastewater treatment; solid and hazardous waste management; water resources management; recovery and reuse of resources and energy; environmental toxicity and risk of nanomaterials; bioremediation of contaminated sites and groundwater; control of greenhouse gas emissions; and climate change mitigation and adaptation. He is an Adjunct Professor at seven universities and Distinguished/Honorary Professor at six universities. Currently, he serves, or has served on 72 national and international committees, review panels or advisory boards including the American Society of Civil Engineers (ASCE) National Committee on Energy, Environment and Water Policy. He is a Distinguished Engineering Alumnus of both the Oklahoma State and Iowa State Universities, an elected Fellow of the Water Environment Federation and International Water Association, an elected Fellow of the American Association for the advancement of Science, an elected Member of the European Academy of Sciences and Arts, an elected Member of the Russian Academy of Engineering, and a Distinguished Member of the ASCE. He is also Editor-in-Chief of the ASCE *Journal of Hazardous, Toxic and Radioactive Waste*, and Springer *Journal of Nanotechnology for Environmental Engineering*, and past Vice-Chair of the Editorial Board of *Water Environment Research Journal*. He has authored over 600 technical publications in journals and conference proceedings, including over 300 refereed journal articles, 15 patents, 24 books and 151 book chapters.

Dr. Tian C. Zhang, PE. F.ASCE, F.AAAS, is a Professor in the Department of Civil Engineering at the University of Nebraska-Lincoln (UNL), USA. He received his PhD in environmental engineering from the University of Cincinnati in 1994 and joined the UNL faculty in August 1994. Professor Zhang teaches courses related to water/wastewater treatment, remediation of hazardous wastes and non-point pollution control. Professor Zhang's research involves fundamentals and applications of nanotechnology and conventional technology for water, wastewater, and stormwater treatment and management, remediation of contaminated environments and detection/control of emerging contaminants in the environment. Professor Zhang has published more than 145 peer-reviewed journal papers, 81 book chapters and 16 books since 1994. Professor Zhang is a Diplomate, Water Resources Engineer (D.WRE) of the American Academy of Water Resources Engineers, a Board Certified Environmental Engineer (BCEE) of the American Academy of Environmental Engineers, an elected Fellow of the American Society of Civil Engineers (F. ASCE), an elected Fellow of the American Association for the advancement of Science (F. AAAS) and an elected member of the European Academy of Sciences and Arts (EASA). Professor Zhang is an Associate Editor of *Journal of Environmental Engineering* (since 2007), *Journal of Hazardous, Toxic and Radioactive Waste* (since 2006) and *Water Environment Research* (since 2008). He has been a registered professional engineer in Nebraska, USA since 2000.

Dr. Manish Kumar Goyal is presently working as an Associate Professor in the Department of Civil Engineering, Indian Institute of Technology, Indore, India. Earlier, he worked at McGill University, Montreal, Canada, and Nanyang Technological University (NTU), Singapore. Dr. Goyal is actively involved in teaching and research in different areas such as water resources modeling, snowmelt hydrology, climate change, and extreme events analysis and soil carbon sequestration. He has authored more than 80 technical publications, including 15 book chapters, 50 refereed (peer-reviewed) journal articles, and 20 national and international conferences. He was awarded the Japan Society for the Promotion of Science (JSPS) Fellowship Program to pursue research at the University of Tokyo, Tokyo, Japan. Earlier, he was awarded the Young Scientist Award from Asia-Pacific Economic Cooperation (APEC) Climate Center, Busan, Republic of Korea, and he was also a recipient of the Canadian Commonwealth Scholarship award to pursue research at the University of Waterloo, Waterloo, Canada. Dr. Goyal is a member of the editorial board of about half a dozen journals. He has been a peer reviewer for more than 20 international journals including *Journal of Hydrology* and *International Journal of Climatology*.

Dr. Satinder K. Brar is James and Joanne Love Chair in Environmental Engineering at the Lassonde School of Engineering at York University in Toronto, Canada. She graduated with a Master's in organic chemistry from the National Chemical Laboratory, Pune, India, with a Master's in technology in environmental sciences and engineering from the Indian Institute of Technology, Bombay, Mumbai, India, and a PhD in environmental biotechnology from Institut National de la Recherché Scientifique (INRS), Québec, Canada. Dr. Brar is a recipient of the American Society of Civil Engineers (ASCE) State-of-the-Art of Civil Engineering Award (2007) for her article titled, "Bioremediation of Hazardous Wastes – A Review," which was published in the *Practice Periodical of Hazardous, Toxic & Radioactive Waste Management* – Special issue on Bioremediation. She has also received the Rudolf gold medal (2008) for the originality of the article published in *Practice Periodical of Hazardous, Toxic & Radioactive Waste Management*. Her research interests lie in the development of finished products (formulations) of wastewater and wastewater sludge-based value-added bioproducts, such as enzymes, organic acids, platform chemicals, biocontrol agents, biopesticides, butanol and biohydrogen. She is also interested in the fate of endocrine disrupter compounds, pharmaceuticals, nanoparticles and other toxic organic compounds during value addition of wastewater and wastewater sludge, in turn, finding suitable biological detoxification technologies. She is on the editorial board of *Brazilian Archives of Biology and Technology Journal* and associate editor of two internationally reputed journals. She has won several accolades through her professional career through awards, such as outstanding young scientist in India and several others. She has more than 300 research publications, which include 4 books, 40 book chapters, 200 original research papers, 50 research communications in international and national conferences, and has registered 2 patents to her credit.

R. D. Tyagi is an internationally recognized Professor at Institut National de la Recherché Scientifique – Eau, Terre, et Environnement (INRS-ETE), University of Québec, Canada. He holds the position of Canada Research Chair (senior) on "Bioconversion of wastewater and wastewater sludge to value added products." Professor Tyagi is a member of the Hall of Excellence of the University of Québec (Canada) and is also a member of the European Academy of Sciences and Arts; Academician. He conducts research on hazardous/solids waste management, water/wastewater treatment, sludge treatment/disposal, and bioconversion of wastewater and wastewater sludge into value-added products. He has developed the novel technologies of simultaneous sewage sludge digestion and metal leaching, bioconversion of wastewater sludge (biosolids) into *Bacillus thuringiensis*-based biopesticides, bioplastics, biofertilizers and biocontrol agents. He is a recipient of the Outstanding Scientist Award of the International Bioprocessing Association (an International Forum on Industrial Bioprocesses) for significant contributions in the area of environmental biotechnology. He is also a recipient of the American Society of Civil Engineers (ASCE) State-of-the-Art of Civil Engineering Award, the ASCE Rudolph Hering Medal, the ASCE Wesley Horner Medal and the ASCE Best Practice Oriented Paper Award. He received the Superior Award, Excellence in Environmental Engineering Award of the American Academy of Environmental Engineers in 2015. He also received the 2010 Global Honor Award (Applied Research) of the International Water Association (IWA), the 2010 Grand Prize (University Research) of the American Academy of Environmental Engineering, the Life-Time Achievement Award (Specialist Medal, 2007) for outstanding research contributions in wastewater sludge management by the IWA, and the Excellence Award of the Natural Sciences and Engineering Research Council of Canada for industry–university collaborative research. Dr. Tyagi has published/presented over 600 papers in refereed journals, conferences proceedings and is the author of 13 books, 80 book chapters, 10 research reports and 9 patents.

List of Contributors

Alicia Kyoungjin An
School of Energy and Environment
City University of Hong Kong
Hong Kong
China

Antonio Avalos-Ramírez
INRS
Université du Québec
Québec
Canada

Adani Azhoni
Department of Civil Engineering
National Institute of Technology
Karnataka
Surathkal, Mangalore
India

Jaeho Bae
INHA University
Incheon
Republic of Korea

K.K. Bara
Department of Microbiology
Guru Nanak Dev University
Amritsar
India

Puspendu Bhunia
School of Infrastructure
Indian Institute of Technology
Bhubaneswar
Bhubaneswar
Odisha
India

Yann Le Bihan
Department of Civil Engineering
Lassonde School of Engineering
York University
Toronto
Canada

Lalit Borana
Discipline of Civil Engineering
Indian Institute of Technology Indore
Indore
India

Satinder K. Brar
Department of Civil Engineering
Lassonde School of Engineering
York University
Toronto
Canada

Bhupinder Singh Chadha
Department of Microbiology
Guru Nanak Dev University
Amritsar
India

Yu Chen
State Key Laboratory of Hydraulics and Mountain River Engineering
Sichuan University
Chengdu
China

Jew Das
Discipline of Civil Engineering
Indian Institute of Technology Indore
Indore
India

Bhaskar Jyoti Deka
School of Energy and Environment
City University of Hong Kong
Hong Kong
China

Suchit Deshmukh
Discipline of Mechanical Engineering
Indian Institute of Technology Indore
Indore
India

Rosa Galvez
Centre de Recherche Industrielle du Québec (CRIQ)
Québec
Canada

Herath Gamini
Professor of Economics
School of Business
Monash University
Malaysia

Manish Kumar Goyal
Discipline of Civil Engineering
Indian Institute of Technology Indore
Indore
India

Jiaxin Guo
School of Energy and Environment
City University of Hong Kong
Hong Kong
China

Akash Kumar Gupta
School of Infrastructure
Indian Institute of Technology
Bhubaneswar
India

Shivam Gupta
Central Agricultural University
Imphal
India

Krishnamoorthy Hegde
INRS
Université du Québec
Québec
Canada

Gilbert Hinge
Department of Civil Engineering
Indian Institute of Technology
Guwahati
Guwahati
India

Srinidhi Jha
Discipline of Civil Engineering
Indian Institute of Technology Indore
Indore
India

Rajwinder Kaur
INRS
Université du Québec
Québec
Canada

Rehan Khan
Discipline of Mechanical Engineering
Indian Institute of Technology Indore
Indore
India

L.R. Kumar
INRS
Université du Québec
Québec
Canada

Pratik Kumar
INRS
Université du Québec
Québec
Canada

Ritunesh Kumar
Discipline of Mechanical Engineering
Indian Institute of Technology Indore
Indore
India

Eunseok Lee
INHA University
Incheon
Republic of Korea

Pengzhi Lin
State Key Laboratory of Hydraulics and Mountain River Engineering
Sichuan University
Chengdu
China

F. Luo
Huazhong University of Science and Technology
Wuhan
China

Dalila Larios Martinez
INRS
Université du Québec
Québec
Canada

Neeraj Mishra
Center for Wage Employment
National Institute of Rural Development and Panchayati Raj
Hyderabad
India

Mitra Naghdi
INRS
Université du Québec
Québec
Canada

Carlos S. Osorio-González
INRS
Université du Québec
Québec
Canada

Preetika Kuknur Pachapur
INRS
Université du Québec
Québec
Canada

Vinayak Laxman Pachapur
INRS
CRIQ - Research Center Industrial Du Québec
Québec
Canada

A.N. Pathak
Amity University
Jaipur
India

Yong Qi
Civil and Environmental Engineering Department
University of Nebraska-Lincoln
Omaha
NE
USA

S.K. Ram
INRS
Université du Québec
Québec
Canada

Prangya R. Rout
INHA University
Incheon
Republic of Korea

Joseph Sebastian
INRS
Université du Québec
Québec
Canada

Ashutosh Sharma
Department of Civil Engineering
Indian Institute of Technology Guwahati
Guwahati
India

Pritee Sharma
Discipline of Economics
School of Humanities and Social Sciences
Indian Institute of Technology Indore
Indore
India

Kanak Singh
Discipline of Economics
School of Humanities and Social Sciences
Indian Institute of Technology Indore
Indore
India

Preetinder Singh
Ludhiana Beverages
Pvt Ltd
Jaspalon
Doraha (Ludhiana)
India

Rajneesh Singh
School of Infrastructure
Indian Institute of Technology
Bhubaneswar
India

Niranjan Suralikerimath
INRS
Université du Québec
Québec
Canada

Rao Y. Surampalli
Global Institute for Energy, Environment, and Sustainability
Lenexa
KS
USA

Anita Talan
INRS
Université du Québec
Québec
Canada

R.D. Tyagi
INRS
Université du Québec
Québec
Canada

Akshaya K. Verma
Department of Civil Engineering
Siksha 'O' Anusandhan
Bhubaneswar
India

Mausam Verma
INRS
Université du Québec
Québec
Canada

Bhoomika Yadav
INRS
Université du Québec
Québec
Canada

Tian C. Zhang
Civil and Environmental Engineering
Department
University of Nebraska-Lincoln
Omaha
NE
USA

Preface

The book arose from the need to develop a compendium of contemporary issues, problems and their possible solutions in the field of sustainability. Sustainability is the way to live in harmony with the environment, and it becomes quite crucial in the present times, when economic development is being achieved at the expense of environmental quality across the globe. The exploitation of natural resources such as intensification of industrial activities, mining of resources and rapid consumption of fossil fuels lead to considerable changes in the earth's ecosystem, thereby leading toward threats such as insufficiency of vital natural resources, irreversible environmental impact and a future decrease in ecosystem productivity. The present regime of economic development at the cost of the environment is going to influence human life and the ecosystem around it, as evident in the impacts of human-induced climate change illustrated by several studies. The primary concerns are health and safety, agricultural production, water availability and socio-economic vulnerability. Looking at the adverse impacts, it is necessary to integrate sustainability into economic policies from a regional level to a global level to pave a path toward a sustainable future. Accordingly, this book is an attempt to compile, direct and publicize the most relevant information about sustainability, its fundamentals and applications.

The book is put together in a context: (i) fundamentals and framework of sustainability, (ii) its dimensions and different aspects, and (iii) applications. It puts forward all the relevant topics in the form of 29 chapters; it begins by introducing sustainability, and quantifying its dimensions by analyzing its interactions with environmental, social and ethical aspects. Finally, the book looks into the applicability of fundamental concepts of sustainability at regional scales via the use of modern-day tools such as remote sensing, machine learning and artificial intelligence. Additionally, the book covers several case studies addressing a variety of issues relevant to sustainable development.

This book is intended to be of interest to students, researchers, engineers, scientists, and policymakers; it provides state-of-the-art insights into technological development in accordance with elementary scientific research to understand the past, embrace the present and sustain the future.

We express our deep gratitude and appreciation for the hard work and dedication of all contributing authors in the making of this book. The content and opinions expressed in each chapter of this book are those of the authors and should not be necessarily interpreted as the opinion of their affiliated organizations.

The editors acknowledge the hard work and patience of all authors who have contributed to this book.

Rao Y. Surampalli
Tian C. Zhang
Manish Kumar Goyal
Satinder K. Brar
R. D. Tyagi

Part I

Fundamentals and Framework

1

Introduction to Sustainability and Sustainable Development

Prangya R. Rout[1], Akshaya K. Verma[2], Puspendu Bhunia[3], Rao Y. Surampalli[4], Tian C. Zhang[4], R.D. Tyagi[5], S.K. Brar[5], and M.K. Goyal[6]

[1] *Department of Environmental Engineering, Inha University, Incheon, Republic of Korea*
[2] *Department of Civil Engineering, Siksha 'O' Anusandhan, Bhubaneswar, India*
[3] *School of Infrastructure, Indian Institute of Technology Bhubaneswar, Bhubaneswar, India*
[4] *Civil Engineering Department, University of Nebraska-Lincoln, Omaha, USA*
[5] *INRS, Université du Québec, Québec, Canada*
[6] *Civil Engineering, Indian Institute of Technology, Indore, India*

1.1 Background and Definition

In the present-day world, which is more and more global, people have taken note of global issues. Sustainability is one of these vital issues and has progressively gained prominence in practice and in educational deliberations around the world over the past several decades (Rodrigo et al. 2017). Our contemporary practices, including the increased exploitation of non-renewable energy sources and unrestrained waste generation in all possible ways, have appeared as huge risks to the sustainable development concept (Adenike 2018). Although the idea of sustainable development has been accepted universally, it has hardly progressed from a concept to practical application since its inception (Yan et al. 2018). It is often said that "development" as intended today has moved, not necessarily progressed, through various forms and fashions during the post-Second World War era (Bell and Morse 2013). However, global development-mediated environmental deterioration and communal pressures have put civilizations under severe stress in the past couple of decades (Pedersen 2018). The global community now has a bigger accountability to implement sustainable actions so that natural resources may be preserved for forthcoming generations (Adenike 2018). The need for sustainable development was universally approved in September 2015 by attendees of the United Nations General Assembly; as an outcome, 17 Sustainable Development Goals were projected, prioritizing the role of education as the key strategy to stimulate sustainability (Fatima and Carolina 2017).

According to the Brundtland Report, "sustainable development is development that meets the needs of the present without compromising the ability of future generations to meet their own needs" (WCED 1987). The International Institute for Environment and

Sustainability: Fundamentals and Applications, First Edition.
Edited by Rao Y. Surampalli, Tian C. Zhang, Manish Kumar Goyal, Satinder K. Brar, and R. D. Tyagi

Development (IIED) redefined it in a broader sense as "the objective of integrating economic activity with environmental integrity, social concerns and effective governance systems, while maximizing the contribution to the wellbeing of the existing generation, fairly sharing the cost and benefits, without compromising the potential for the upcoming generations to meet their needs" (IIED 2002; Fatima and Carolina 2017). In other words, sustainability is an advancement to safeguard the forthcoming and existing potential to satisfy vital human requirements within the ecological and resource limits of the planet earth (Blum et al. 2017). Therefore, sustainability aims to retain the environmental carrying capacity. The terms "sustainable development" and "sustainability" are often used interchangeably in the literature; however, some authors argue that sustainability denotes a progression, while sustainable development denotes the end state. For example, from the ecological point of view, sustainability implies the environmental aspect of sustainable development (Holden et al. 2014). However, a differentiation from Robinson (2004) states that sustainable development is related to government and governing by incremental approach, while sustainability is associated with non-government organizations (NGOs) and academic environmentalists and focuses on human beings' capability to continue to live within environmental limitations (Robinson 2004; Fatima and Carolina 2017).

1.2 Basic Concepts and Issues

The concept of sustainable development or sustainability found its roots in the 1987 Brundtland Commission Report *Our Common Future*, in which the earliest attempt had been made to correlate environmental stability with that of the matters of economic development. Sustainable development is a multifaceted and normative concept that evolved during the 1970s as an ecologically centered concept but gradually transmuted into a complete socio-economic-centered concept by the 2000s. The concept contains the philosophies of equality and mutual dependence, among not only the generations but also the nations and peoples of the earth. The concept also incorporates futurity, interdisciplinarity, participation, learning and adaptation for the development of socio-cultural, socio-economical and natural environments, which are crucial for the wellbeing of the human race and of nature. To understand the concept better, we have to look into the two main characteristics as follows:

(a) "Sustainability is a *people-centered* and *conservation-based* concept that implies the development of the standard of human life by respecting nature's capacity to afford life-support facilities and resources."
(b) "Sustainable development is a *normative concept* that exemplifies standards of decision and action to be respected as 'the society' strives for satisfying its needs of survival and well-being."

Recently, sustainable development issues have caused the global community to think beyond the customary classified action of environmental, economic and social concerns and paved the way for an advanced holistic and integrated means of development. The thematic illustration in Figure 1.1 pertinently portrays the integrations and collaborations among the environmental, social and economic domains of sustainability.

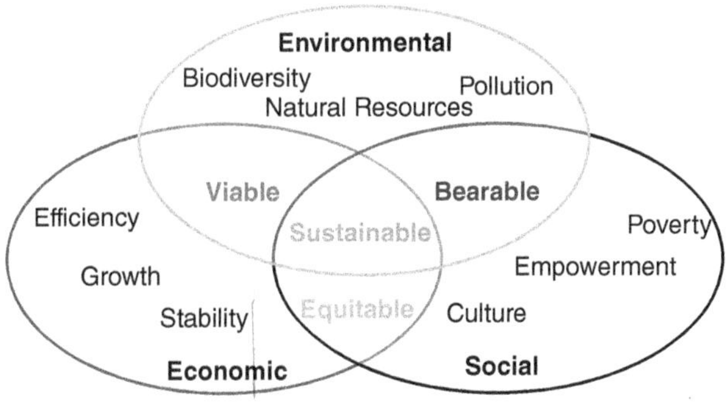

Figure 1.1 Illustration of sustainable development and the three pillars of sustainability.

The social conception of sustainable development aims at maintaining the steadiness of communal and cultural domains (Munasinghe and McNeely 1995). Contemporary society requires integration of heterogeneity and a policymaking framework for social sustainability. The sustainable development concept from the economic point of view is founded on the idea of maximum income generation and, at the same time, conserving a decent capital, which ultimately yield benefits (Maler 1990). The environmental concept of sustainable development focuses on the survivability of biological systems, which are vital to the universal steadiness of the whole ecosystem. Efforts should be made to accelerate the distinctive capability of such biological systems to acclimatize to change, instead of maintenance of a stagnant state (Weaver and Rotmans 2006). In this way, sustainable development demands the integration of environmental management, economic growth and social development as reciprocally supportive and interdependent pillars for persistent development. There is a need for multi-stakeholder approaches including governments, non-governmental bodies, higher learning organizations, and research and development units to deal with development matters.

There are conceptual/definitional issues associated with sustainability/sustainable development. Many authors describe sustainability/sustainable development as a confusing and contested concept. Sustainability and sustainable development are often criticized as complex concepts and not satisfactorily applicable for practical purposes. Some authors have commented on the existence of several meanings for sustainability/sustainable development, as the conceptualization of these terms comes from diverse viewpoints such as environment, sociology, economics, etc., which emphasizes specific components rather than covering the entire context. However, Weaver and Rotmans (2006) argued for a context-oriented definition of sustainable development that is suitable for varied participants within a specific realm of application instead of trying to formulate a "generic" description of sustainability (Munasinghe and Shearer 1995). Critics also point out that if something is to remain sustainable (the same), how it can develop (White 2013)? Conversely, Costanza and Patten (1995) noted that much of the criticism is pointed in the wrong direction because the critics are (i) failing to justify interrelationships between time and space; and (ii) unable to understand that the actual complications are connected to prediction instead of description (Costanza and Patten 1995). In brief, the arguments give rise to two important issues: the first issue is "What should be sustained?" and "What kind of development do we need?," and the second is the viability issue of "What can be sustained?" as well as "What types of systems can we expect?" (Hediger 2000). The first issue involves assessment of society's goals pertaining to economic, social and environmental objectives, whereas the second issue necessitates knowing how the diverse systems work together and how the interacting systems can be managed effectively (Hediger 2000). In contrast, the contested concept of sustainability issues gives scope for better understandings of sustainability/sustainable development coming from intricate and interdisciplinary viewpoints to develop methods to address present-day matters. Therefore, regardless of the inherent ambiguity in the concept of sustainability/sustainable development, it is now recognized as an irreducible universal concept where economic, social and environmental issues are inter-reliant dimensions that must be approached within an integrated framework.

Sustainability or sustainable development covers wide-ranging social, economic and environmental issues aside from the conceptual/definitional issues discussed in the previous paragraph. A few of the issues and actions to mitigate the same are presented in Table 1.1. Since its inception, sustainability has barely transformed from a concept into practice due to the existence of numerous inherent issues with the three prime systems. The basic structure of sustainability can be mapped successfully if we can somehow correctly articulate and arrive at a consensus on these issues. However, addressing the issues associated with the individual components in isolation, without taking care of the issues of other components, essentially ends in failure. Each component is individualistically vital; thus, addressing the issues associated with each component is unavoidably obligatory to

Table 1.1 Important sustainability issues and key actions.

Sustainability issues	Key actions
Ecological balance	Stabilization of the ecosystem and enrichment of biodiversity by safeguarding prevailing natural resources and by bringing together new habitats/life forms
Health and safety	Ensuring a harmless and healthier environment to live in
Pollution	Reduction in the level of pollution load in the environment by adopting green technologies
Climate change	Improvement of the resilience of the environment toward different aspects of climatic change
Social issues	Reduction in crime and adverse effects on society throughout the lifetime of the development
Waste	Reduction in the generation of waste by recycling in accordance with the waste hierarchy
Water	Reduction in the use of fresh water by reusing treated wastewater and by incorporating water-efficient technologies and appliances
CO_2 emission	Reduction in CO_2 emissions by adopting energy-competent design and utilizing low-carbon technologies
Economic issues	Ensuring economic growth, stability and efficiency by maintaining productivity

eradicate elements of unsustainability from human civilization, as the components are closely associated (Robinson and Tinker 1995). Concisely, it is a prerequisite to maintain a well-defined balance between the requirements and interests of the three pillars of sustainability, e.g. environment, economy and society, and to avoid antagonism between them, in order to achieve the targeted sustainability/Sustainable Development Goals.

1.3 Evolution of Sustainability and Sustainable Development

The terms "sustainable" and "sustainability" appeared in the *Oxford English Dictionary* for the first time in the late twentieth century. However, very similar terms have been used in German (*Nachhaltigkeit*, meaning "lastingness"), French (*durabilité* meaning "durable") and Dutch (duurzaamheid/duurzaam meaning "durable" or "sustainable") for centuries (Du-Pisani 2006). The term "sustainable" originated from the Latin word "sustinere," meaning maintain, defend, bear, etc. (Castiglioni and Mariotti 1981; Bolis et al. 2014). Back in 1713, the concept was acknowledged in forestry, after Hans Carl von Carlowitz used the German term of sustainability ("Nachhaltigkeit") for the first time in his book *Sylvicultura oeconomica* (Carlowitz 1732), mentioning "one should harvest only the same amount of wood which equals to trees planted." Emphasizing the significance of the right management of wood stock, Carlowitz suggested that equilibrium should be maintained between reforestation and harvesting old trees so that there would be no scarcity of timber (Bolis et al. 2014; Carlowitz 1732). The concern related to population explosion-driven increased consumption of natural resources and energy also started evolving in the late eighteenth century, after Thomas Malthus highlighted the fact that the "population of planet earth is not sustainable, as it increases exponentially in relation to the available resources" in his book entitled *An Essay on the Principle of Population* (Lampridi and Melliou 2015). The attention shifted to natural resources in the nineteenth century since alarms were raised that coal, the most important source of energy deposits, might be exhausted if not maintained sustainably. Sustainable management of a resource is only possible if its use does not exceed its reproductive capacity (Castiglioni and Mariotti 1981; Carlowitz 1732). The Brundtland Commission conducted the first official conference on sustainability in 1987. The term sustainability gained visibility only after its implementation in the environmental domain, particularly in connection to environmental protection and conservation (Bolis et al. 2014; Lampridi and Melliou 2015). Furthermore, the economic aspect of sustainability was explored, since careless environmental management like overexploitation of agriculture, fisheries, etc. resulted in reduced use of resources (Bolis et al. 2014). Negligent actions have grave consequences, compromising economic development; therefore, sustainability from an economic viewpoint is a core issue.

In addition to the environmental and economic aspects of sustainable development, the social side is also very significant as well. The way the environment is managed has a direct effect on the economy, and the economy affects the social lives of people; hence, the idea of social sustainability. In fact, the concept of sustainability gained global acceptance only after realization of the fact that resilient socio-economic influences are mostly driven by environmental alterations (Bolis et al. 2014). Therefore, the origin of the sustainability concept can be traced back to the olden days; however, factors like population growth, the danger of depletion of natural resources and fear of compromise in living standards of present and upcoming generations made possible the global adoption of the sustainability concept (Du-Pisani 2006).

1.4 Challenges and Solutions

Attaining sustainable development necessitates global actions contributing to economic and social progress and strengthening environmental protection as well. Several interactive economic, social and environmental factors are not only affecting sustainable development but also endangering the existence of planet earth. The world is facing pronounced challenges in the major domains of sustainability, e.g. social, economic and environmental. The prevailing global sustainable development challenges are population growth, poverty, debt, disease, food insecurity, malnutrition, unemployment, income inequity within and among countries, unsustainable consumption and production patterns, natural resource depletion, excessive generation of waste, global warming, environmental degradation, weak institutional capacities, poor and inadequate infrastructure, armed conflict, etc. To confront these challenges, sustainable development strategies need to be action-oriented and inclusive to take care of the needs of the poorest, diminish inequality, encourage the safeguarding of natural resources, ensure economic domination, etc. It is difficult to cover comprehensively the entire range of sustainable challenges as mentioned above. However, highlighting a few of the imperative challenges and solutions in Table 1.2 may be helpful in justifying the theme of this chapter.

1.5 Adaptation and Resilience

Adaptation and resilience help organisms in their survival in an ecological niche. Adaptation may be referred to as a dynamic evolutionary process that ensures the evolutionary fitness of the organisms in their environment. As a result, numerous changes may appear related to morphology, physiology or behavior. Resilience is the characteristics required to adapt and is a preferred way to achieve ecosystem sustainability. It represents the level that is required by the system for self-organization, learning and adaptation (Walker et al. 2004). Humans depend on ecosystems for their survival and thereby continuously affecting them. Resilience is a characteristic of such associated socio-ecological systems. The system is more likely able to tolerate abrupt disturbances without collapse by enhancing the resilience. One of the major goals of adaptation and resilience is to reduce the vulnerability of ecological communities to the natural disasters and climate-related hazards and to speed their recovery (Pielke et al. 2007). The major attributes of resilience include adaptive capacity, persistence and transformability. These attributes are required for the system to resist collapse and maintain important ecological functions. The dynamics of adaptation can be well understood by the adaptive cycle, which represents the natural pattern of changes in an ecosystem. It com-

Table 1.2 Important sustainable development challenges and solutions.

Sustainability challenges	Solutions
Poverty eradication	Continual, comprehensive and evenhanded economic progress in developing nations is a crucial prerequisite for eliminating poverty and hunger Social security schemes to reduce disparity and social barring are indispensable for eradicating poverty
Sustainable agriculture	Increasing access to financial services, land lease, efficient irrigation facilities, reuse of treated wastewater, water collection and distribution systems, training, education, healthcare, social services by agricultural producers Strengthening investments in sustainable agriculture practices like rural infrastructure development; increase in storage capacities; minimizing post-harvest food wastage; expansion of agricultural cooperatives, strengthening of urban–rural linkages, research development on sustainable agricultural technologies, etc. Taking care of other necessary factors like conservation of land and water, maintenance of biodiversity, resistance to natural calamities, adaptation to changing environmental conditions, protection of ecosystems, etc. to promote and support sustainable agriculture
Food security and nutrition	Improving food security by ensuring availability of adequate healthy food for present and future generations Focusing on sustainable aquaculture, crops, livestock, forestry, fisheries, etc. for food security and nutrition Finding the reason behind excessive foodstuff price volatility and emphasizing actions to mitigate price-related issues
Population	Systematic consideration of population trends and proper planning to address demographic change-related challenges including migration Providing access to reproductive health; emphasizing modern methods of family planning, sexual health, reduction in maternal and child mortality, and protection of all human rights of women, men and youth
Health	Strengthening health systems for the prevention and treatment of non-communicable ailments, such as diabetes, cancer, respiratory diseases, cardiovascular disorder, etc. and communicable diseases like AIDS, influenza, polio, malaria, tuberculosis, etc. Improving health infrastructure through increased health financing, training and development of the health workforce; ensuring availability of reasonably priced medical facilities including medicines of good quality
Social protection	Providing social safety for every member of the community, nurturing progress, flexibility and social integrity Promotion and protection of the human rights and fundamental freedoms of all migrants and avoiding tactics that might worsen their vulnerability
Employment	Implementing policies to create new jobs and providing employment opportunities to young people Approving macroeconomic plans that promote justifiable economic growth by increasing employment prospects Ensuring occupational safety, healthcare, social security, fundamental rights at work, education, skill development for employees

(Continued)

Table 1.2 (Continued)

Sustainability challenges	Solutions
Education	Ensuring access to primary education for all including the rural community, indigenous people, persons with disabilities, ethnic minorities, etc.
	Reforming the education system to prepare curricula linked to sustainability; use of information technologies to enhance learning outcomes; enhanced teacher training to prepare students for careers in sustainability-related fields
	Participation of higher educational institutions to focus on development of innovative programs, professional training and lifelong learning, entrepreneurship and business skills training, etc.
Climate change	Participation in appropriate global response to accelerate the reduction of global greenhouse gas emissions; minimizing the adverse impacts of climate changes like extreme weather events, persistent drought, rise in sea level, ocean acidification, coastal erosion, etc.
Water and sanitation	Facilitating access to basic sanitation and safe drinking water for all by developing integrated water resource management; maintaining the balance between water supply and demand and investing in infrastructure for water and sanitation services
	Adopting processes to lessen water pollution, water losses, increase in water quality, improving wastewater treatment and ensuring efficient use of water
Energy	Improving energy efficiency by increasing the share of renewable energy; using energy efficient and low-emission technologies; relying on cleaner fossil fuel technologies; using traditional energy resources in a manageable manner
	Ensuring access to sustainable modern energy services for all in a reliable, affordable and environmentally acceptable manner by mobilizing financial resources
Waste management	Commitment to reduce, reuse and recycle waste (three Rs), focusing on increasing energy recovery from waste, exploring waste as resources
	Adopting lifecycle approaches and implementing policies for resource efficiency and environmental friendly waste management
	Assessing the risks posed by hazardous waste to human beings and the environment; minimizing human and environmental exposure to hazardous wastes, developing safer alternatives to hazardous chemicals in products; emphasizing all possible measures to prevent the unreliable management of hazardous waste and its illegal dumping.
Gender equality	Commitment to offer equal education, basic services and economic opportunities to women and men, and addressing women's sexual and reproductive health
	Undertaking administrative restructuring to give women equal rights to men in accessing ownership, economic resources, credit and inheritance of land and other forms of property
	Prioritizing actions to encourage gender equality and women's empowerment in all segments of our societies; removing the barriers to their equal participation in decision making at all levels
Land degradation	Regional, national and global coordinated action to monitor land degradation and restore degraded lands in arid, semi-arid and dry sub-humid areas
	Establishment of forecasting and early warning systems related to land degradation, desertification, dust storms and sandstorms, and installing indicators for monitoring and assessing the extent of land degradation and desertification

Table 1.2 (Continued)

Sustainability challenges	Solutions
Biodiversity	Guaranteeing maintenance of biodiversity by boosted habitat connectivity and ensuring ecosystem resilience
	Adoption of traditional knowledge and innovations through the age-old practices of Aboriginal communities that guarantee conservation and sustainable use of biodiversity
	Emphasizing the role of access and benefit-sharing arising from the utilization of genetic resources in contributing to the conservation and sustainability of biodiversity
Disaster risk reduction	Integration of early warning systems in circumventing loss of life; lessening economic losses and social damage
	Providing technical assistance, technology transfer, capacity building and training programs, comprehensive hazard and risk assessments, sharing of reliable geospatial information and knowledge to reduce the level of disaster-linked risks effectively
	Establishing coordination and interdependence among disaster risk reduction, recovery and long-term development planning actions for better risk reduction, resilience intensification and smooth transition among rescue, relief and development
Human settlements	Facilitating an all-inclusive methodology to human settlements and urban development that makes available affordable housing, and prioritizes slum upgrading and urban regeneration
	Augmenting the quality of human settlements by guaranteeing good living conditions; housing, mobility, proper working conditions and access to all basic services fir both urban and rural inhabitants
	Strengthening existing cooperation mechanisms; adequate and predictable financial contributions and additional application tools to improve the harmonized execution of urban development planning

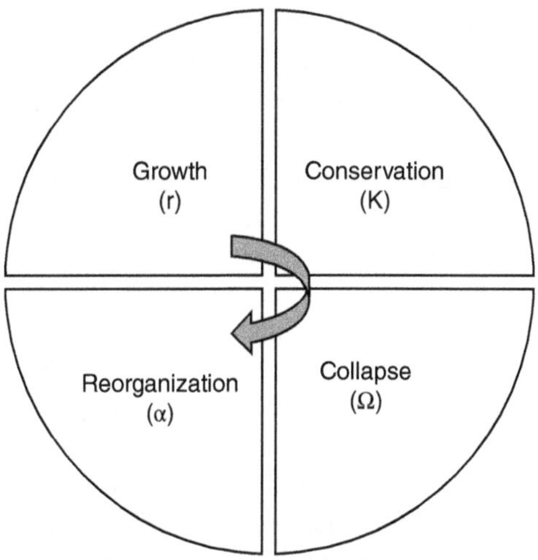

Figure 1.2 Four distinct stages of the adaptive cycle.

prises four distinct phases: (i) growth (r); (ii) conservation (K); (iii) collapse (Ω); and (iv) reorganization (α). Figure 1.2 shows the schematic representation of the adaptive cycle showing transitioning between the four phases (r, K, Ω, α) within a specific regime (Holling 1986).

Two major transitions are exhibited in the adaptive cycle: (i) from growth to conservation (slow and long), and (ii) from collapse to reorganization (rapid and short). As a result, the disturbed communities of organisms undergo a spontaneous renewal by establishment of the patterns and redistribution of resources throughout the system. During the "r" phase of the cycle, the recently disturbed system starts absorbing the available resources and experiences a hasty growth. Due to the unstable connections of the components with the system, the overall control is weak, and success depends on the ability to adapt to a wide variety of conditions. The growth rate slows down during the "K" phase due to the accumulation and sequestration of resources, leading to the increase in connectedness of the components with the system. During the "Ω" phase, a recent disturbance creates a tremor in the system; connections are destroyed, which affects the stability and functionality of the system. As a result, accumulated resources are released and the system collapses. The system is widely open for reorganization during the "α" phase. Loss of resources is minimized so that they can be available for the "r" phase. The ability of reorganization of the disturbed ecosystem can be enhanced by increasing the adaptive capacity. Adaptive capacity is generally associated with genetic diversity, biological diversity and the heterogeneity of landscape mosaics (Peterson et al. 1998). Folke et al. (2002) identified the numerous critical factors required to deal with the dynamics of natural resources: (i) learning to live with change and uncertainty; (ii) fostering diversity for resilience; (iii) gathering knowledge for learning; and (iv) generating prospects to self-organize leading to socio-ecological sustainability (Folke et al. 2002). Several aspects have been proposed in the literature to represent the different forms of resilience. These forms are (i) "stable resilience" relative to the capacity to resist uncontrolled vulnerability just after a disturbance (Angeon and Bates 2015); (ii) "progressive resilience" focusing on an anticipatory and holistic approach aiming toward long-term goals (Vale 2014); and (iii) "climatic resilience" focusing on adapting to and mitigating the effects of climate change (Leichenko 2011). Recently, Sanchez et al. (2016) proposed the concept of "sustainable resilience" or "enduring resilience" (Sanchez et al. 2016). The concept was based on the integration of technological networks with social and ecological networks, including short- to long-term mitigation. Sanchez et al. (2017) also stated that reactive, short-term, narrow resilience policies are not suitable to deal with the variability associated with long-term development (Sanchez et al. 2017).

Adaptation is focused on actions and often in a local context, while resilience is the bottom-up and top-down momentum to produce sustainability as the outcome. Resilient strategies should be specific and focused so that communities are able to curb and recover from different vulnerabilities (Gwimbi 2009). Stakeholders at each level have important roles to play in achieving adaptation and resilience sustainably. Some of the adaptation actions are: (i) changing land use; (ii) upgrading the infrastructure design; (iii) adjusting life styles; (iv) enforcing policies and regulations; and (v) increasing awareness to understand the climate risks.

1.6 Economic, Ecological, Social, Technological and Systems Perspectives

The economic, ecological, social, technological and systems perspectives of sustainability need to be understood in order to link the events of functioning of natural components like ecosystems, biodiversity, etc. to the structure and operation of the related human components like social systems, economy, etc. (Cabezas et al. 2003).

1.6.1 Economic Aspect

All economic activity is dependent upon renewable and non-renewable natural resources provided by the environment. The rate of exhaustion of natural resource stocks is mostly governed by the shortage of the stock, ultimately influencing its price. Even renewable resources can be depleted due to high extraction rates without considering the regeneration limits of the resource. The identification and implementation of feedback loops between natural and economic systems can affect sustainable resource availability and economic productivity (Settle et al. 2002). For instance, more concessional tariffs cause firms to extract natural resources unsustainably, thereby leaving a reduced resource base lfor forthcoming generations, whereas resources with several substitutable sources are used sustainably, as economists suggest switching from one resource to another when scarcity-driven prices are rising.

1.6.2 Ecological Aspect

The ecological aspect of sustainability integrally necessitates the ability of ecosystems to withstand environmental changes and retain their capacity to function normally in changed environmental conditions, as well as genetic diversity-based enhanced evolutionary features of the constituent species to adapt to the environmental changes (Cabezas et al. 2003). Species diversity and interactions among them ranging from competition to symbiosis play a vital role in maintaining the stability and productivity of the ecological environments. However, anthropogenic activity-mediated loss of biodiversity due to overharvesting and natural habitat damage ultimately affects the evolutionary capacity of the ecosystems by weakening their ability to acclimatize to changing environments. The characteristic features of complex ecosystems are primarily governed by their internal dynamics like species population and external disturbances including natural and anthropogenic sources. The extent of disturbance an ecosystem tolerates before exhibiting change in characteristic features depends on the system's resilience to disturbance. However, ecosystems usually go through many changes naturally and impart variations in functions and services availed by the human societies under these transition states (Scheffer et al. 2001; Portela and Rademacher 2001). The maintenance and restoration of biodiversity in ecosystems are highly essential for realization of sustainable ecology.

1.6.3 Social Aspect

Social systems must reflect the interaction between humans and the biological systems as an indicator of sustainability attainment. For example, the domestication of animals and plants led to proliferation of the human population by ensuring surplus of food, collection and dissemination of wealth, and merger of smaller territories into larger ones. Principally, the domestication-mediated controlled food web contributed to human domination and resource richness. Therefore, domesticated food sources are significantly affected by society's economy whereas the non-domesticated sources are mostly affected by the legal and political systems of society. Thus, human society affects sustainability by manipulating the food web in favor of domesticated species and devoting significant proportions of the resource pool to social infrastructure (Cabezas et al. 2003).

1.6.4 Technological Aspect

Technologies facilitate the availability of new resources or enhanced access to already available resources for humanity from the environment, thereby playing a key role in sustainability. However, the procedures involved in the use of the technology or manufacture of the technology itself more often have negative environmental impacts. Therefore, new technologies must be evaluated by taking into account their socio-economic benefits and environmental ill impacts. From the sustainability point of view, technology development should ensure improved efficiency and better use of resources. However, efficient and lower environmental impact technologies very often do not lead to lower or better use of resource criteria, and thus, may not contribute to the attainment of sustainability. In fact, sometimes, the availability of technologies results in increased pollution and resource use just because of the non-co-incidence of economic, social and political systems with sustainability goals. Therefore, not only the technologies but also the social, economic and political systems within which they operate must be fully understood for realization of sustainability (Cabezas et al. 2003).

1.6.5 Systems Aspect

Sustainability typically is not a feature of the parts of the system, but of the system as a whole, since the operating constituent parts of the system sustain each other, and a system cannot be sustainable unless subsystems operate in a coordinated manner. Any system is sustainable within definite boundaries, and not sustainable outside those boundary conditions. Two types of deviations from sustainable conditions named self-correcting and non-self-correcting indicate the probable development of catastrophic conditions. To understand the system's prospective sustainability is a very challenging task since it is very difficult to comprehend the entire system, which usually includes a broad range of disciplines, and most principles are discipline-specific. Therefore, principles of universal evaluation methods are needed that are applicable to the entire range of systems and subsystems in order to realize a system's prospective sustainability (Cabezas et al. 2003).

1.6.6 Integrated Aspect

Technological development is a double-edged sword that can cause several environmental problems and can solve them as well. Ancient technologies are still in place and dominating in major sectors. However, new technologies are emerging with improved output that are environmentally sustainable. There are no limits to technological development for the efficient utilization of available natural resources. Consequently, harmful waste emissions will be diminished, and non-renewable resources will last longer and ensure the future of ecological services. The availability of natural resources is one of the major constraints to accelerating economic growth, but scientific and technological advancement may overcome the related crisis to a certain extent. The failure in ecosystem services requires certain corrective interventions by policymakers. The loss of ecosystem services may be due to environmental pollution, which affects the per capita income, and consequently economic growth. As economic growth (income per capita) increases, environmental degradation rises up to a point; after that, environmental quality starts improving (Figure 1.3).

The environmental Kuznets curve demonstrates the degradation and subsequent improvement of the environment in the process of the development of the economy. During the early stages of economic growth, environmental degradation is prominent due to the uncompromising habits of people toward the development. However, after attaining a level of economic development, the stakeholders become more sensitive and are prepared to bear the cost of the improvement of environmental quality. Consequently, institutional innovations become important to restore ecological services and improve productivity for public welfare (Grossman and Krueger 1995). Economic development and energy development are closely related. Fossil fuels fulfill most of the commercial energy demand; as a result the emissions released during the burning cause global environmental problems (White and Walsh 2008) and the process is believed to be unsustainable. Adverse environmental impacts can be minimized by reducing dependency on fossil fuels, increasing energy use efficiency, and exploring clean and green energy options. Green energy offers a better alternative, but its potential for implementation as a viable option to meet the commercial energy demand is still unexplored (Pan 2002). The Earth Summit at Rio in 1992 identified four major pillars, which are essential to provide good quality of life and a healthy environment (Earth Summit 1992).

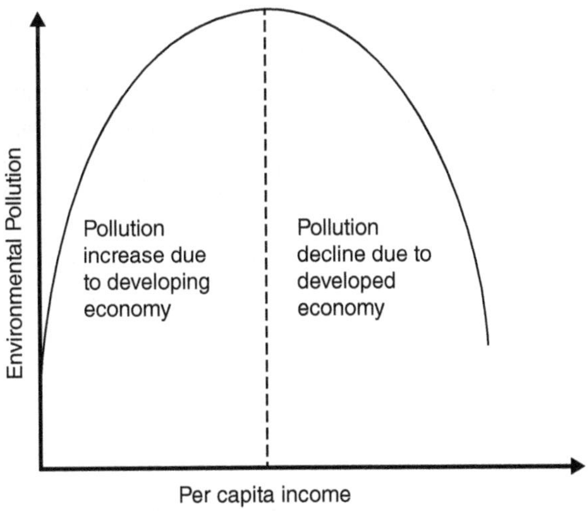

Figure 1.3 The environmental Kuznets curve.

Table 1.3 Paradigms of sustainable development.

Sustainability attributes	Criteria
Economy	Growth; development; productivity; trickle down
Environment	Ecosystem integrity; carrying capacity; biodiversity
Society	Equity; empowerment; accessibility; participation; sharing; cultural identity; institutional stability

Source: Adapted from Kahn (1995).

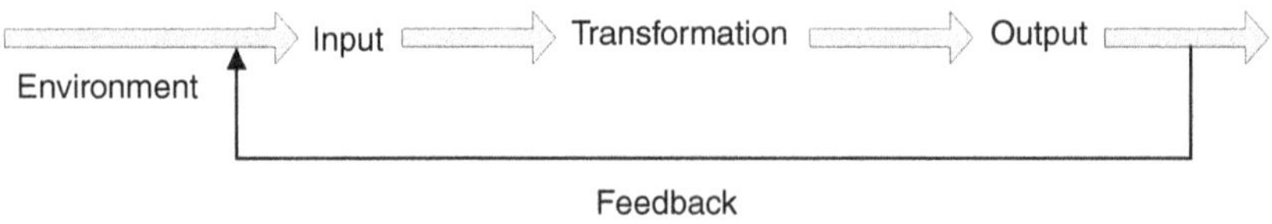

Figure 1.4 Conceptual system model. Source: Adapted from Littlejohn and Cameron (1999).

- Promote efficient utilization of energy and overcome the dependency.
- Conserve and minimize the impact on the environment.
- Improve the national economic development by the use of technology.
- Improve the quality of life for society.

Kahn (1995) reported the paradigm of sustainable development as per *Agenda 21* and identified three major attributes, namely economic sustainability, environmental sustainability and social sustainability (Table 1.3) (Kahn 1995).

Economic sustainability encompasses the criteria of growth, development and productivity that lead to conventional development. It focuses on the provisions of physical inputs into the production system (Goodland 1995). Environmental sustainability assimilates the criteria related to ecosystem services and presumes that natural capital should be conserved to maintain the economic inputs by reducing the rate of extraction of natural resources. Social sustainability includes the numerous criteria that require the conservation of the environment through sustainable economic growth. One of the major aims of social sustainability is poverty reduction (Kori and Gondo 2012). The wellbeing of society and the way society utilizes the natural resources are interrelated and can be explained using a conceptual model. Littlejohn and Cameron (1999) adopted a system conceptual model employing different attributes of system sustainability for the purpose (Figure 1.4) (Littlejohn and Cameron 1999).

The systems perspective helps to understand the positional relationship of any development in the context of larger environmental and social systems. This also explains the dependency of outputs from the system based on inputs and other related attributes.

1.7 Conclusions

The concept of sustainability or sustainable development is primarily about human welfare and their relationships with nature in a framework where nature–society imbalances can intimidate economic and social steadiness. As analyzed in this chapter, the sustainable development concept cannot be defined with a universal definition due to its changing nature over the years. It is fundamentally a multifaceted normative concept that integrates social, ecological and economic dimensions. The global community is facing pronounced challenges in the major domains of sustainability due to population growth, depletion of non-renewable resources of energy, environmental degradation, poverty, excessive generation of waste, etc. A few imperative challenges and the probable solutions have been covered in this chapter. The adaptation and resilience aspect of sustainability is discussed and summarized; adaptation is focused on local actions while resilience produces sustainability as the outcome. Finally, to link the processes of ecosystem functioning to the structure and operation of social systems, the economics, ecological, social, technological and systems perspectives of sustainability are reviewed and discussed. It is concluded in this chapter that the attainment of sustainable development necessitates a holistic approach that ensures coordinated operations among the social, economic and environmental domains.

References

Adenike, A.A. (2018). The role of microorganisms in achieving the Sustainable Development Goals. *Journal of Cleaner Production* 182: 139–155.

Angeon, V. and Bates, S. (2015). Reviewing composite vulnerability and resilience indexes: a sustainable approach and application. *World Development* 72: 140–162.

Bell, S. and Morse, S. (2013). The brave new frontier of sustainability: where are we? In: *Measuring Sustainability: Learning from Doing*, 1–27. London, Routledge.

Blum, C., Bunke, D., Hungsberg, M. et al. (2017). The concept of sustainable chemistry: key drivers for the transition towards sustainable development. *Sustainable Chemistry and Pharmacy* 5: 94–104.

Bolis, I., Morioka, S.N., and Sznelwar, L.I. (2014). When sustainable development risks losing its meaning. Delimiting the concept with a comprehensive literature review and a conceptual model. *Journal of Cleaner Production* 83: 7–20.

Cabezas, H., Pawlowski, C.W., Mayer, A.L., and Hoagland, N.T. (2003). Sustainability: ecological, social, economic, technological, and systems perspectives. *Clean Technology and Environmental Policy* 5: 167–180.

Carlowitz, H.C. (1732). *Economy of silvicuture: or message and instructions to maintain the nature in wild arboriculture (Sylvicultura oeconomica, oder haußwirthliche Nachricht und naturmäßige Anweisung zur wilden Baum-Zucht)*. Leipzig, Germany: Friedrich Brauns Erben.

Castiglioni, L. and Mariotti, S. (1981). *Latin Language Vocabulary (Vocabolario della lingua latina)*. Turin, Italy: Loescher.

Costanza, R. and Patten, B.C. (1995). Defining and predicting sustainability. *Ecological Economics* 15: 193–196.

Du-Pisani, J.A. (2006). Sustainable development – historical roots of the concept. *Environmental Sciences* 3 (2): 83–96.

Earth Summit (1992). Agenda 21. *Journal of IAEM* 19 (2): iii–viii.

Fatima, A.D. and Carolina, M. (2017). Interdisciplinarity: practical approach to advancing education for sustainability and for the Sustainable Development Goals. *The International Journal of Management Education* 15: 73–83.

Folke, C., Colding, J., and Berkes, F. (2002). Building resilience for adaptive capacity in social-ecological systems. In: *Navigating Social-Ecological Systems: Building Resilience for Complexity and Change*, 1–30. Cambridge, UK: Cambridge University Press.

Goodland, R. (1995). The concept of environmental sustainability. *Annual Review of Ecology and Systematics* 26: 1–26.

Grossman, G.M. and Krueger, A.B. (1995). Economic growth and the environment. *Quarterly Journal of Economics* 110: 353–378.

Gwimbi, P. (2009). Linking rural community livelihoods to resilience building in flood risk reduction in Zimbabwe. *JÀMBÁ: Journal of Disaster Risk Studies* 2 (1): 71–79.

Hediger, W. (2000). Sustainable development and social welfare. *Ecological Economics* 32: 481–492.

Holden, E., Linnerud, K., and Banister, D. (2014). Sustainable development: Our Common Future revisited. *Global Environmental Change* 26: 130–139.

Holling, C.S. (1986). Resilience of ecosystems; local surprise and global change. In: *Sustainable Development of the Biosphere*, 228–269. Cambridge, UK: Cambridge University Press.

IIED (2002). *Breaking New Ground: Mining, Minerals and Sustainable Development*. London: International Institute of Environment and Development (IIED) with support from the World Business Council on Sustainable Development http://pubs.iied.org/pdfs/9084IIED.pdf.

Kahn, M. (1995). Concepts, definitions, and key issues in sustainable development: the outlook for the future. Proceedings of the 1995 International Sustainable Development Research Conference, Manchester, UK (March 27–28, 1995).

Kori, E. and Gondo, T. (2012). Environmental sustainability: reality, fantasy or fallacy? Second International Conference on Environment and BioScience. *IPCBEE* 44: 105–109.

Lampridi, M. and Melliou, C. (2015). The birth and evolution of sustainable development. *Introduction to Sustainable Development* 1: 29–34.

Leichenko, R. (2011). Climate change and urban resilience. *Current Opinion in Environmental Sustainability* 3 (3): 164–168.

Littlejohn, A.H. and Cameron, S. (1999). Supporting strategic cultural change: The Strathclyde Learning Technology initiative as a model. *Association of Learning Technologies Journal* 7 (3): 64–75.

Maler, K.G. (1990). Economic theory and environmental degradation; a survey of some problems. *Revista de Analisis Economico* 5: 7–17.

Munasinghe, M. and McNeely, J. (1995). *Protected Area Economics and Policy*. Geneva/ Washington, DC: World Conservation Union (IUCM) and World Bank.

Munasinghe, M. and Shearer, W. (1995). *Defining and Measuring Sustainability: The Biogeophysical Foundations*. Tokyo/Washington, DC: United Nations University and World Bank.

Pan, J. (2002). Rural energy patterns in China: a preliminary assessment from available data sources. *Program on Energy and Sustainable Development Working Paper* 12: 1–27.

Pedersen, C.S. (2018). The UN Sustainable Development Goals (SDGs) are a great gift to business. *Procedia CIRP* 69: 21–24.

Peterson, G.D., Allen, C.K., and Holling, C.S. (1998). Ecological resilience, biodiversity and scale. *Ecosystems* 1: 6–18.
Pielke, R., Prins, G., Rayner, S., and Sarewitz, D. (2007). Climate change 2007: lifting the taboo on adaptation. *Nature* 445: 597–598.
Portela, R. and Rademacher, I. (2001). A dynamic model of patterns of deforestation and their effect on the ability of the Brazilian Amazonia to provide ecosystem services. *Ecological Modeling* 143: 115–146.
Robinson, J. (2004). Squaring the circle? Some thoughts on the idea of sustainable development. *Ecological Economics* 48 (4): 369–384.
Robinson, J. and Tinker, J. (1995). Reconciling ecological, economic and social imperatives: toward an analytical framework. *SDRI Discussion Papers Series*. Vancouver, Canada: Sustainable Development Institute.
Rodrigo, G.G.C., Raquel de, F.D., Lisiane, V.M. et al. (2017). Towards sustainable development through the perspective of eco-efficiency – a systematic literature review. *Journal of Cleaner Production* 165: 890–1004.
Sanchez, A.X., van der Heijden, J., Osmond, P. and Prasad, D. (2016). Urban sustainable resilience values: driving resilience policy that endures. CIB World Building Congress Proceedings, Tampere, Finland (May 30 – June 3, 2016).
Sanchez, A.X., Osmond, P., and van der Heijden, J. (2017). Are some forms of resilience more sustainable than others? *Procedia Engineering* 180: 881–889.
Scheffer, M., Carpenter, S., Foley, J.A. et al. (2001). Catastrophic shifts in ecosystems. *Nature* 413: 591–596.
Settle, C., Crocker, T.D., and Shogren, J.F. (2002). On the joint determination of biological and ecological systems. *Ecological Economics* 42: 301–311.
Vale, L.J. (2014). The politics of resilient cities: whose resilience and whose city? *Building Research and Information* 42 (2): 191–201.
Walker, B., Holling, C.S., Carpenter, S.R., and Kinzig, A. (2004). Adaptability and transformability in social-ecological systems. *Ecology and Society* 9: 5.
WCED (World Commission on Environment and Development) (1987). *Our Common Future*. New York: Oxford University Press.
Weaver, P.M. and Rotmans, J. (2006). Integrated sustainability assessment: what is it, why do it and how? *International Journal of Innovation and Sustainable Development* 1 (4): 284–303.
White, M.A. (2013). Sustainability: I know it when I see it. *Ecological Economics* 86: 213–217.
White, S. and Walsh, J. (2008). Greener pathways: jobs and workforce development in the clean energy economy: a report. Center on Wisconsin Strategy, the Workforce Alliance and the Apollo Alliance. https://www.greenforall.org/greener_pathways_jobs_and_workforce_development_in_the_clean_energy_economy.
Yan, Y., Wang, C., Quan, Y. et al. (2018). Urban sustainable development efficiency towards the balance between nature and human well-being: connotation, measurement, and assessment. *Journal of Cleaner Production* 178: 67–75.

2

The Need, Role and Significance of Sustainability

Anita Talan[1], *A.N. Pathak*[2], *and R.D. Tyagi*[1]

[1]*INRS Eau, Terre et Environnment, Québec, Canada*
[2]*Amity University, Jaipur, Rajasthan, India*

2.1 Introduction

In 1980 the concepts of sustainable development and the three pillars of sustainability were defined to identify the level of sustainability to be achieved or already achieved. Sustainability is needed to overcome the challenges faced in development and growth, such as economic and financial crises, and helps us to define economic sustainability. The progress of society or communities and even the growth of an individual can be defined on the basis of sustainable development. Initially, the concepts of various domains of sustainability were mostly qualitative notions, but over time there has been an evolution. Hence, precise specifications need to be defined in quantitative terms for attaining sustainability. In addition, a wide range of indicators is required to understand the need, role and significance of sustainability.

This chapter more precisely defines the different approaches to attaining sustainability and its need and significance by considering the three pillars of sustainability. To achieve sustainability, it is necessary to focus on different aspects of the goals of sustainability, which are discussed in depth in later sections. The chapter also discusses methods to create and maintain development without causing any harm to the interdependent pillars of sustainability.

2.2 Three Pillars of Sustainability

The World Summit on Sustainable Development (2005) pins down three crucial areas that are factors in the growth of the social science and philosophy of sustainable development. These crucial areas are three 'pillars' of sustainability (environment, social and economic) that form the basis to manage the important issues that are faced globally. The Brundtland Commission defined sustainability in the context of meeting the needs of the existing

Sustainability: Fundamentals and Applications, First Edition.
Edited by Rao Y. Surampalli, Tian C. Zhang, Manish Kumar Goyal, Satinder K. Brar, and R. D. Tyagi

generation without compromising the capacity of future generations to meet their own needs. Therefore, we must think about the prospective necessities of the future while making our decisions about the present (Mauerhofer 2013).

Sustainable development is also defined as an integrative theory that reflects all dimensions of community, ecological and financial aspects. These three domains are known as pillars of sustainability, which are responsible for human and capital development to achieve a sustainable quality of life (QOL) for society and its residents (Schoolman et al. 2012). Attempts have been made to balance the three main pillars of sustainability. Sometimes these are criticized as they include different values, such as diversity, costs, profits, justice, wellbeing and traditions, etc., which are indirectly proportional to each other (Hansmann et al. 2012). The provocative benefits of different investors repeatedly conflict within a single dimension of sustainability, such as communal clashes or commercial conflicts, clashes over conservational matters or preferences. Therefore, harmonizing with the interest of one pillar is sometimes more important than to balance aspects of all three pillars. The balancing of aspects of the three pillars of sustainability is not always responsible for the highly intricate interdependencies between communal activities and natural ecology systems (Schoolman et al. 2012). Many studies suggest using different perspectives to determine the interactions between the focused aims of sustainable development while using the three pillars of sustainability. The three pillars of sustainability have equal impact on each other positively as well as negatively, and the combined positive effects represent an important task in terms of sustainability-oriented decision making.

The meanings of sustainable development and sustainability are not identical, but their fundamental sense is similar. As per the Rio Declaration: "Human beings are at the centre of concerns for sustainable development. They are entitled to a healthy and productive life in harmony with nature." The message from the Brundtland Commission and the Rio Declaration could be summarized as the following points for all three pillars of sustainability (Figure 2.1):

- Originally, the idea of sustainable development was a very realistic and anthropocentric one, focusing on human beings and their QOL. The base of sustainability is defined by human needs. According to Maslow's hierarchy of needs human beings are always inspired by unsatisfied and unfulfilled necessities. The rudimentary necessities of human beings should be fully satisfied before higher needs can be fulfilled (Moldan et al. 2012). To maintain the QOL it is necessary to fulfill all the universal needs of the person before a person can act selflessly. This selfless behavior is considered as an important condition for attaining sustainable development. Thus, people striving for sustainable development should focus on the essential elements to maintain their wellbeing, such as the basic necessities for a good and healthy life, social relationships, and freedom of selection and action.
- Moreover, human life must be meaningful and healthy and in synchronization with the environment. This is required to create a balance between the three pillars of sustainability. Human life is neither self-regulating nor quarantined, but it is part of a complex ecosystem that is dependent on relationships and interdependencies. Since humans are extremely dependent on the environment, if they don't maintain harmony with the environment, the other two pillars of sustainability might be impacted creating an unbalanced situation (McCormick et al. 2013).

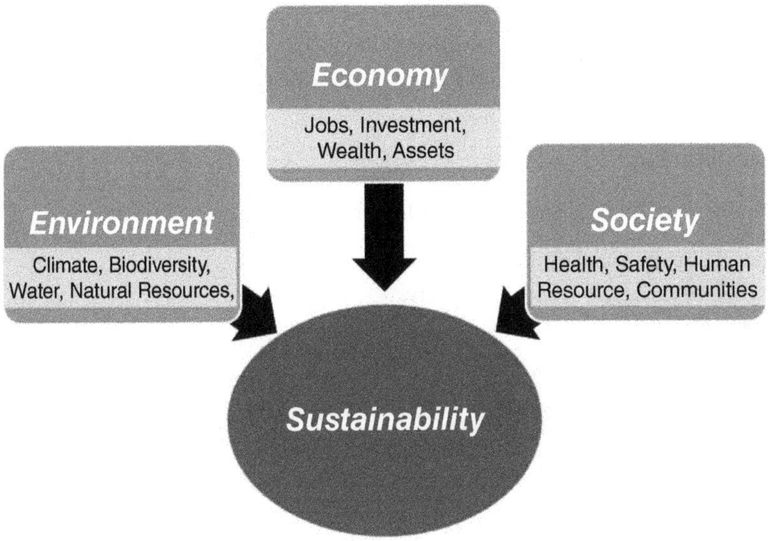

Figure 2.1 Three pillars of sustainability.

- Furthermore, the important feature of sustainability is to have a sustained environment. Changes and new formulations should take existing and future generations into account (McCormick et al. 2013). The time period given to achieve sustainability should be based on the normal human life expectancy of the existing generation.

2.2.1 Economic Development

Economic sustainability is defined as the ability of a society to support economic growth. It focuses on decisions being made in the most equitable way possible while considering the other aspects of sustainability. Therefore, there are three basic economic components to achieve economic sustainability globally or even locally and these are as follows:

- Developing new markets and opportunities for selling things.
- Minimizing energy utilization and raw material inputs.
- Cost reduction through efficiency and improvements.

While these components are implemented worldwide, economic development is the most problematic issue of sustainability as different factions of society disagree on political ideology and are not economically balanced, which directly affects the growth of the corporate sector, professions and employment. Economic sustainability emphasizes people's indulgence in business, and several other associations that follow sustainability guidelines to all extents. This participation also inspires and supports employees, and cumulative effects can be seen in the areas where groups of people are required to achieve certain tasks as one person working alone can rarely achieve much. (Poveda 2017). The stock market of supply and demand is entrepreneurial. The present lifestyle demands a lot of resources, which in turn leads to exploitation of the environment. Therefore, to conserve resources and for the sake of the environment, getting what is needed or what we consume under control is an important issue. Economic development is fulfilling the needs of the people without compromising QOL, especially in unindustrialized parts of the world (Duhn 2012).

2.2.2 Social Development

The basic components to achieve social sustainability are:

- Health and safety of every social domain.
- Extra benefits to disadvantaged groups of society.
- Impacts on local communities.
- Good QOL.

Social sustainability is affected by amenities, infrastructure, social and cultural life, voice and influence, and space for urbanization to grow (Dempsey et al. 2011). Nobel Prize winner Amartya Sen gives the following dimensions (Figure. 2.2) for social sustainability, which helped in the development of the theory of "social choice" in the early 1970s. These dimensions are also helpful in achieving global sustainability.

This pillar has many characteristics. Awareness of social equity and legislature to ensure the safety and security of people, protecting them from environmental hazards and other destructive activities of the corporate sector and other associations, are the most important features of social development. In developed countries, the government keeps strong checks and has legislation programs in place to ensure the safety and protection of people's health and wellbeing. It is also about maintaining the basic necessities of life without compromising QOL. Presently, sustainable housing and improvements in housing construction using sustainable materials are also part of social development. The basic elemental part of this pillar is education, which is encouraging every domain of society to participate in environmental sustainability. It is about teaching people the effects of environmental protection and warning them of the dangers imposed if goals are not achieved. In knowledge- and understanding-based economies, education plays a significant role to uplift the lives of people and is an important factor in determining the QOL, and thus, achieving good communities where people can live safely and fulfill their basic needs. Level of education

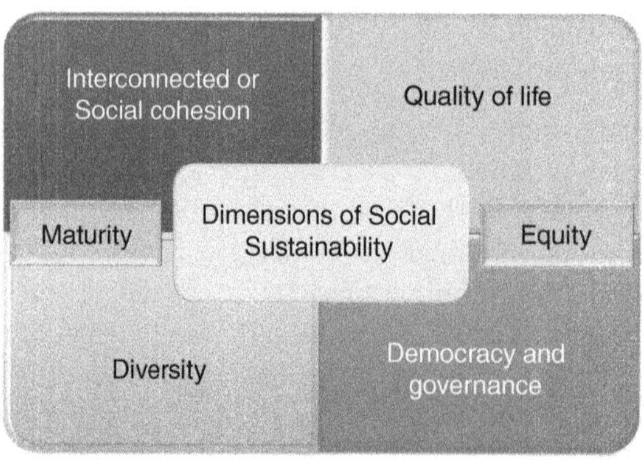

Figure 2.2 Dimensions of social sustainability (Martin et al. 2002).

is directly proportional to the level of job that an individual has. Individuals with good skills, competencies and knowledge usually have a broad range of jobs and opportunities for growth. The feeling of being responsible for the uplifting of society and benefiting the environment is the basic factor in social development (Singh et al. 2012).

2.2.3 Environmental Protection

Everyone is aware of the need to protect the environment or natural resources; people even know what needs to be done to protect the environment in terms of recycling, reducing and/or reusing resources. Various steps have been taken to prevent pollution by reducing the release of carbon. Incentives have been given to people using renewable energy as a power source at home and in industry. Environmental safety and security is the third pillar of sustainability, but it is a principal concern for future generations. It is mostly concerned with education about and protection of our ecosystems, air and water quality, reliability and sustainability of natural resources, and paying attention to the components that hinder environmental sustainability. Its major concern is working toward a green future and, according to the Environmental Protection Agency (EPA), biotechnology and developing technologies are the key factors to attain this sustainability. Steps have been taken to protect the future environment from the negative impacts of technology (Krueger et al. 2012).

The major components of environmental protection are:

- Waste reduction and lower production of effluent and emissions.
- Elimination of toxic substances.
- Reduction of impact on human and animal health.
- Increased use of renewable resources.

The new theory of environmental sustainability focuses on the development of science and technology, with six main focused areas as follows:

- *Environment systems*: this area includes climatic changes, weather risk management and control, mitigation and adaption.
- *Formation of communities and societies*: urban development, formation of new cities and providing public services such as transport, healthcare, etc.
- *Energy systems*: this area is more concerned with fulfilling the energy requirements of the people. It also covers energy use, conservation of energy, producing renewable forms of energy such as bioenergy, and creating ways to use energy efficiently.
- *Terrestrial systems*: it is necessary to focus on this area as humans are completely dependent on the cycles or ecosystems of land covering all natural and human-made managed bionetworks. Forests, food cycles, biodiversity and environmental services all are important for the uplifting of the lives of humankind.
- *Aquatic systems*: the terrestrial and aquatic ecosystems are all interdependent. Hence, it is equally important to consider all water ecosystems, fisheries, water streams and aquatic biodiversity.
- *Carbon and nitrogen cycles*: to maintain the ecosystems the balance of carbon and nitrogen is required to support the development and growth of flora and fauna. It covers all causes and basins, feedbacks and interlinking to other ecological systems.

The Organisation for Economic Co-operation and Development OECD Environmental Strategy for the First Decade of the 21st Century in 2001 defines four specific principles for environmental sustainability:

1) *Regeneration*: the efficient use of renewable resources to be maintained by not permitting their usage to exceed the rates of natural renewal.
2) *Substitutability*: the efficient use of non-renewable resources and alternatives to be used to preserve the non-renewable resource base for future generations. The limited use of these resources is required to enhance other forms of wealth.
3) *Assimilation*: the assimilative capacity of the environment should not be exceeded by releasing hazardous or polluting substances into the environment. It is necessary to control the release of these polluting substances, which are hindering efforts toward environmental sustainability and, as a result, decreasing global sustainability (Figure 2.3).
4) *Avoiding irreversibility*: to increase the efficiency of the process it is necessary to have reversible processes. Hence, avoiding irreversibility is required to achieve development of all three pillars of sustainability.

These four OECD criteria led to five interlinked goals to improve the nominal cost and effective conservation policies in reference to the sustainability of the environment:

1) To maintain the integration of ecosystems by using natural resources efficiently and managing the natural resource base.
2) To decouple environmental pressures from economic growth.
3) To analyze the indicators to quantify development and use the information for improved decision making.
4) To improve QOL by analyzing the social and environmental interface.
5) To improve governance and cooperation in society via the effects of global environmental interdependence.

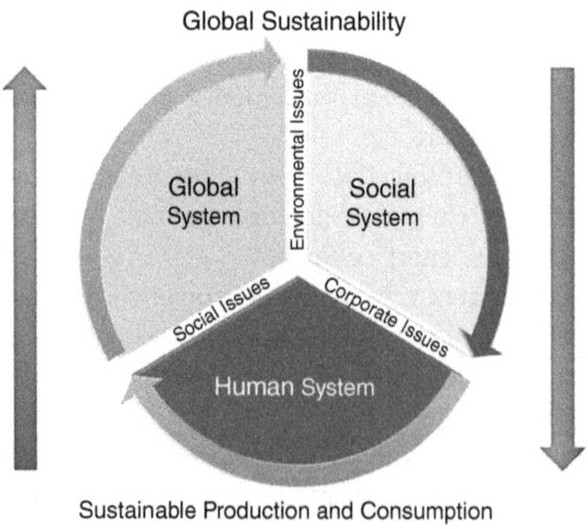

Figure 2.3 The relationships between the human system, social system and global system and their effects toward achieving global sustainability.

Furthermore, to expand the list of basic principles with the aim of environmental sustainability, other objectives have been also added (Moldan et al. 2012):

- To understand the non-linear evolution of complex ecosystems.
- Long-term perspectives without defining a designated time limit.
- Taking all the ecosystem feedback into account, especially positive feedback.
- To understand the different scales to measure the effects of indicators.
- The ability to react to a changing situation or environment, i.e. flexibility.
- To analyze the importance and significant role of local conditions in sustainability.
- To respect cultural and natural biodiversity for the growth of communities.

2.3 Primary Goals of Sustainability

To achieve the goals of sustainability the professional network of sustainable development thinks, acts and works altogether at a global level. In 2012, the United Nations Conference on Sustainability came up with Millennium Development Goals (MDGs) aiming toward poverty reduction globally and much more. The Sustainable Development Goals (SDGs) eventually led to a to-do list among other things (Griggs et al. 2013):

- Complete removal of hunger and poverty.
- Achieving gender equality.
- Sustainable economic growth promoting employability and increasing economic index.
- Promoting healthcare services.
- Better standards of education.
- A sustainable environment to deal with the effects of weather change, pollution and other ecological factors.
- Sanitation.
- QOL.

The SDGs cover a wide range of social and economic issues including poverty, hunger, health, education and many more. The SDGs are also known as "Transforming our World: the 2030 Agenda for Sustainable Development" or "Agenda 30." The goals were developed to replace the MDGs, which were ended in 2015. Unlike the MDGs, the SDG framework does not differentiate between "developed" and "developing nations," but is applied to all countries globally. The 2030 Agenda includes 17 SDGs, which list 169 targets to be achieved globally with a vision for a better sustainable world (Griggs et al. 2013); the list of objectives is given in Table 2.1 (Griggs et al. 2013).

To achieve goals of sustainability, strategies must be designed to deal with the problems that hinder growth and development globally. This requires implementation of new policies and change in existing policies at the global level with respect to their impact on local and global development. The main motive of these global strategies is to save the environment and to have economic development. These strategies are based on the theory of sustainable development, which is composed of various growth indicators, as follows.

Table 2.1 Objectives for a better sustained world under the 2030 Agenda.

Objective	Aim of the objective
Eradication of poverty	End poverty in all its forms everywhere
No hunger	End hunger, achieve food security and improved nutrition, and promote sustainable agriculture
Health and wellbeing of people	Ensure healthy lives and promote wellbeing for all at all ages
Quality and excellence education	Ensure inclusive and equitable quality education and promote lifelong learning opportunities for all
Gender equality	Achieve gender equality and empower all women and girls
Clean water and sanitation	Ensure availability and sustainable management of water and sanitation for all
Affordable and clean energy	Ensure access to affordable, reliable and sustainable and modern energy for all
Decent work and economic growth	Promote sustained, inclusive and sustainable economic growth, full and productive employment and decent work for all
Industry, innovation and infrastructure	Build resilient infrastructure, promote inclusive and sustainable industrialization, and foster innovation
Reduced inequalities	Reduce income inequality within and among countries
Sustainable cities and communities	Make cities and human settlements inclusive, safe, resilient and sustainable
Responsible consumption and production	Ensure sustainable consumption and production patterns
Climate change	Take urgent action to combat climate change and its impacts by regulating emissions and promoting developments in renewable energy
Life below water	Conserve and sustainably use the oceans, seas and marine resources for sustainable development
Life on land	Protect, restore and promote sustainable use of terrestrial ecosystems, sustainably manage forests, combat desertification, halt and reverse land degradation, and halt biodiversity loss
Peace, justice and strong institutions	Promote peaceful and inclusive societies for sustainable development, provide access to justice for all and build effective, accountable and inclusive institutions at all levels
Partnerships to achieve the defined goals	Strengthen the means of implementation and revitalize the global partnership for sustainable development

2.3.1 Growth Revival

People living in absolute poverty at the global level are incapable of satisfying their basic necessities, and thus, are a major hindrance to achieving sustainability. Poverty is responsible for reducing the communal capacity to use resources sustainably. Such poverty is mostly found in developing and underdeveloped nations. Therefore, it is essential to reverse stagnant or declining economies to increase growth trends to eliminate poverty globally.

From country to country the growth rate varies but a certain minimum is needed to have an observable impact on absolute poverty (Letaifa and Rabeau 2013). With the given current population growth rates, the required overall national income growth rate depends on whether the economy is developed or developing, e.g. a ~5% increase is required in developing economies, 5.5% in Latin America, and 6% in Africa and West Asia.

Growth must be revived in developing countries as it provides a direct link between economic growth, poverty alleviation and environmental conditions. Developing nations form a co-dependent global economy, but the potential of these countries also depends on the growth rate patterns of developed nations. A minimum 3–4% growth is required by international financial organizations to play an important role in the expansion of global economy. These minimum progress rates are environmentally maintainable if developed countries can take the initiative for improvements in processes and activities (Purcarea et al. 2013).

Efficiency can be increased by reducing raw material and energy, and if developed nations have less market demand from developing nations. However, if developing nations focus on poverty elimination, their domestic demand will increase, leading to a direct increase in the economic growth of the nation. Hence, internal stimulus to developing world growth should lead to sustainability. A balance between the exports and imports of developing and developed countries plays a major role in growth revival of sustainable development among the nations, which defines a need for reorientation in global monetary associations for sustainable development.

2.3.2 Quality of Growth

Sustainable development is more about quality of growth than growth. QOL is the state of wellbeing of individuals, societies and communities, focusing on every negative and positive feature of people's lives. It is about the satisfaction of all basic human necessities, including health, family, education, employment, religious principles, economy, wealth and natural resources (Li et al. 2018). From the global development point of view, QOL has a wide range of contexts including areas of global development such as international relations, health services, employment and legislature (Dempsey et al. 2011).

It requires a sustainable change in growth by reducing the use of raw material and energy. To maintain environmental sustainability, economic distribution and to condense economic crises, these changes are required globally. The economic development of any nation is dependent on the stock of capital that sustains it. For example, the income from a wheat farm is conventionally measured in terms of product extracted minus the cost of extraction. Various other expenses in regenerating the farm are not considered, unless money is spent on such work. Thus, profits do not take full account of the losses in future expenses, and this type of accounting takes place in exploitation of other available natural resources. Therefore, every economy of the world must take full account of growth measurement by improvements or decline in the natural resource base. Fair distribution of income is an important aspect of growth and development as given in the previous example. Speedy progress of an economy united with unusually high wages for a chosen few is worse than slower progress shared with redeployment favoring poor people. In the long run, to achieve sustainability small-scale industries are more easily sustained as they do not put pressure on natural resources. Economic sustainability is unsustainable if it increases the chances of crises. However, this susceptibility may be condensed by using recent expertise and knowledge that directly lower manufacturing threats, reduce market fluctuations, and create capital reserves, food stock and other currencies. A progressive way that is less vulnerable to risks is more sustainable for growth than other ways (Aras and Crowther 2008).

Human wellbeing is required to attain sustainability. Economic and non-economic variables are required to grow in quality for sustainable development. Work must be done to remove disabilities from the disadvantaged social domain. Observable changes in the quality of growth require altered approaches to growth to understand the impact of all efforts made. Sometimes, for the purposes of sustainability financially attractive activities should be rejected. Economic and social sustainability are (and must be) equally reinforced. Money and resources spent on education and health directly raise human productivity. Economic sustainability speeds up the progress of social sustainability by giving new opportunities and employability to disadvantaged groups of society.

2.3.3 Essential Needs for Basic Necessities

The fulfillment of basic human needs is one of the main aims of dynamic productive activities in sustainable development. Sustainability needs proper distributions of goods and services for every social domain to prevent major environmental consequences. The principal development challenge is to fulfill the basic necessities of the increasing global population. The most basic need of livelihood is employment to meet the minimum requirement for food. Food grains and starchy roots are the primary sources, but nowadays the focus is on the availability of proteins. The required growth rates in calorie and protein production seem achievable. However, the increase in the rate of food production should not be based on economically disturbed production guidelines and compromised food security.

Energy is another most important essential human need. To address the increasing demand for energy, the energy consumption pattern should be changed universally. The most urgent requirement is to fulfill the energy needs of poor people, who depend mainly on fuel wood. A person living in poverty cuts wood faster than it grows, creating an imbalance in nature. Therefore, actions are required to reduce fuel wood consumption to preserve the ecological base (Dempsey et al. 2011).

To achieve sustainability, the other basic needs of communities such as housing, water, sanitation and health are also equally important. Scarcities in these needs may also cause ecological pressure. In communities where poverty prevails, failure to meet the basic necessities is one of the main reasons for the increasing number of communicable diseases and their spread (e.g. gastro-intestinal manifestations, malaria, etc.). The rise in population and their movement to cities for the chance of a better livelihood threaten to make these problems worse over time. Thus, people taking initiatives for sustainable development must find ways to provide basic necessities to the developing world community using low-cost technologies.

2.3.4 Population Sustainability

The sustainability of growth and development is directly linked to the dynamics of population progression. The concern is not just about an increase in the global population size but about the levels of energy and materials available for the population. An increase in the population size of a country using high levels of materials and energy is a burden on the environment compared with a population increase in a poor country. Sustainable development can easily be achieved at a stabilized population size dependent on the productive ability of the environment. In developed countries, the rate of population growth is below 1%, while in developing countries the rate is very high, creating inconsistency within the communities. Poor people are unable to migrate to new lands, which is also a problem in population sustainability. Hence, the challenge now is to control population growth by decreasing the population rate, particularly in developing countries (Brammer et al. 2012).

Economic and social development leads to a decline in birth rate in industrial-oriented countries. Urbanization, gender equity and high-paying jobs all play important roles in developed countries. Similar activities should be recognized and encouraged in developing countries in order to have a sustainable population. Emphasis should be placed on female education, the health sector and fair distribution of basic needs in every social domain. Developing countries must increase awareness of the importance of reducing the fertility rate, family planning strategies and the right to self-determination to prevent the productive potential to maintain and support its population. Even distribution of the population between rural and urban areas will help to maintain the stabilized population (Brammer et al. 2012).

Developing countries are developing at a considerably faster pace than the authorities' capacity to cope. Because of this, problems such as shortages of water, energy, housing and food persist, and people are exposed to pollution and to natural and industrial hazards. Thus, more manageable cities can be achieved from low growth rates of population, which will overcome these problems. Urbanization is necessary for sustainable development but the challenge lies in maintaining QOL. Therefore, to reduce the pressure on developed cities it is necessary to develop smaller urban communities. More positive approaches and services are required to have a stable population.

2.3.5 Conservation and Enhancement of Resources

If the needs of present and future generations are to be met on a sustainable basis, then renewable sources of energy in the environment must be preserved and enhanced. Sustainable changes are required in approaches and policies to stand up to the high energy demands of global society while increasing consumption is needed for the development of developing countries in order to provide the basic necessities of to their populations. However, the conservation of nature is not merely a step toward developmental goal, but is a part of human beings' moral obligation to coming generations. A lack of resource alternatives always increases pressure on the environment. Accepted developmental policies have widened the methods of earning a sustainable livelihood, particularly for poor people. As per today's needs agricultural resources have to be conserved globally as, in many nations, agricultural land has been reduced to a great extent and other natural resources are also overexploited.

Conservation of agricultural basins is essential to meet the basic food requirements of growing populations. Scientific assessment of land should be carried out before using it for agriculture and forestry. The rate of depletion of soil, fisheries or forest resources should not surpass the degree of renewal (Kellert et al. 2000).

The pressure on agriculture land can be minimized by increasing the output rate. However, sometimes temporary developments in agriculture productivity lead to environmental pressure such as loss of high-grade genes, the presence of pesticides in fruits and vegetables, nitrate pollution of groundwater, and acidity and alkalization of irrigated lands. Increased agriculture productivity in developing as well as developed nations is mostly based on efficient use of water resources and agrochemicals, and extended use of organic manures and non-chemical methods to regulate pests. For fisheries and forestry, if people were to rely on available stocks only, such stocks could meet the demand in the near future as the population is rising every year. Hence, methods are needed to increase the production of fish, wood and other forest products under controlled conditions.

Global development is always dependent on the base of natural energy resources and the environmental capacity to absorb the by-products produced by using these natural resources for producing energy. Proper use of renewable resources is necessary to overcome the supply and emission problems. Developed parts of the world should check their energy consumption as their high energy demands are polluting the environment. Most of the problems arise in the use of renewable resources. Therefore, sustainability requires a focus on the conserving and efficient use of energy. Before the 1960s, studies assumed the rate of consumption increased exponentially but did not anticipate any problems until the coming centuries, which meant that the use of renewable fuels was a more distant possibility. With technological development, it has been proposed that scarcity of resources can be overcome by using the resources efficiently, recycling and substitution. Apart from this, a more immediate need is to modify the pattern of world trading to allow exporters to have a high share and to improve the supply in developing countries (Van Assche et al. 2017).

The prevention of toxic waste in the environment will always be an important task in conservation of natural resources. With increasing use of pesticides and fertilizers, municipal sewage, fuel burning, chemical usage and other industrial activities put a high pressure on air and water quality. Each of these factors contributes to environmental pollution and increases the environmental pressure, especially in developing parts of the world. Globally there is a necessity to take steps to prevent environmental pollution from creating difficulties, as clean-up processes after an event are very expensive. Countries should follow approaches such as promoting low-waste technologies, enforcing emission standards, and anticipating the impacts of products, technologies and wastes to prevent pollution and to conserve natural resources.

2.3.6 Sustainability of Science and Technology

The achievement of the SDGs requires the reorientation of technology to create a balance between humans and nature. The potential of new technological innovations needs to be enhanced to achieve an effective response to the challenges faced by developing nations in achieving sustainability. In addition, technological advancement must be modified to be extremely responsive to ecological factors. The technologies of developed nations should not simply be accepted and adapted to the environmental and socio-economic situations of emerging nations. The world of research and development is facing many issues in developing countries such as control of diseases or arid land agriculture. In spite of known issues, not enough is done to adapt recent innovations in energy conservation, information technology, material sciences and biotechnology to the needs of the population in developing countries.

Therefore, for sustainability of science and technology the gaps need to be filled by increasing research, development, strategy and allowance capacities for poor people. Technological research should focus on manufactured goods and developmental innovations with commercial significance. Technologies must produce socially important goods such as pollution control or increased product life. Government policies should make it mandatory for institutions or organizations to work on sensitive areas that are technologically weak (Kates et al. 2001).

The development of science and technology raises questions of risk management for advances in nuclear reactors, electricity distribution, communication networks and mass transportation. These systems are well connected by networks that make them resistant to small disturbances but more vulnerable to unexpected disturbance, which can lead to a finite threshold. Applying knowledgeable analyses of vulnerabilities and failures to technology design, manufacturing processes and possibility plans during operations can make the consequences less catastrophic. The best risk analysis has not been applied continuously across systems and technologies, which is a hindrance to sustainability. Thus, there is a need for development of new technologies to prevent accidents and damage to support safety and control. Technological failures have a negative impact on environment sustainability and, thus, should be taken into account and minimized. Potential environmental impacts of new technologies need to be assessed before they are widely used. Adverse consequences of implementation of new technologies must be evaluated beforehand and, if serious, the decision should be reviewed or deferred. A similar focus is also required for natural systems to prevent damage (Kates et al. 2001).

Another aspect of sustainability of science and literature refers to topographical and institutional distribution as well as its disciplinary composition. For science and technology advancement and development we count the publications and citations and the interface of policy and society for the social sciences. It is important not to underestimate the contributions of developed nations (e.g. American countries) as these nations are fundamental to societal challenges in sustainable development; however, issues that apply in certain parts of the world may not be present in other geographic conditions. Therefore, it is necessary to compile the sources and their potential contribution to have a complete view of the sustainability of science and technology with respect to geographic conditions. Scientific advances take place in universities, laboratories and policy-driven scientific organizations across the world. The large and diverse set of contributions in science and technology across the world leads to challenges in terms of conceptual unification but also provides vast opportunities for developments in the field to have interdisciplinary and worldwide effects.

Development in different geographical areas plays an important role both internationally and regionally by creating new scientific insights and enabling societal sustainability. It is essential that the literature is available in local languages in order for the linkages in development to be understood. Defining the field of science is not easy as it is somewhat subjective, and various methods have been proposed to do this in the past few years even though difficulties remain. Thus, new concepts and methods are required from science to build and analyze sustainability.

2.3.7 Merging Economics with Environment

One of the major concepts in the developmental pathway is the necessity to merge commercial and environmental thoughts in decision making. This can only be achieved by changing attitudes and objectives at each level stepping toward sustainability. In terms of global sustainability this integration is necessary. In terms of individual gains, however, the two concepts are seen as incompatible, and an unawareness of consequences is the basis of today's decision making. In merging economic and environmental ideas it is vital to stress the importance of inter-industrial linkages as opposed to an isolated sector. These interconnections create pathways for development and patterns of interdependence of economics and the environment. The independent sectors consider their impacts as side effects on other sectors while achieving their own objectives. Hence, many of the development and ecological problems that persist in sustainability have their root cause(s) in disintegration of duties toward nature. Sustainability needs such disintegration to be overcome by merging ecology and economics. Legal and institutional policies need to be changed to force common interest toward nature and sustainable development. Independently, law cannot enforce people's interests, so educated communities need to come up with knowledge and support that will positively affect ecology. People should fight for the proper management of their natural resources promoting initiatives toward development and community strengthening. Both the public and private sectors need to change their attitudes, procedures and policies in order to benefit other sectors and ecology (Hopwood et al. 2005).

Ecological development and its regulation is not just about safety regulations, pollution control and zoning laws; objectives of environmental sustainability are also based on taxation, investment laws and procedures, science and technology choice, foreign exchange and various other components of the developmental pathway. This integration of economics with the environment should match globally as direct linkage between resources and their consumption is there in the international market. Economic interactions through tourism, trade, investments and finance will grow and advance economic and environment interdependence. Hence in the future, sustainability requires the merging of economics and ecology globally (Dempsey et al. 2011).

2.4 Significance of Sustainability

Everyone wants a better place to live and grow. People want better housing, better educational institutions, high-paying jobs, increased employability, proper sanitation and disposable waste, safer streets and easy access to their needs. Living in a community with different types of people can sometime create problems; these problems can be grouped into three major areas. People living in a community needs a better environment (e.g. green spaces, play areas, pollution-free areas, decent housing, nice gardens, and all these resources should be renewable over generations); a better economy (with increased employability, high incentives, reasonable prices); and better social conditions (e.g. good leisure facilities, social activities including sports, arts and many more, friendly neighbors) (Nazli Nik Ahmad and Salat Ahmed Haraf 2013). All the above issues can be resolved through sustainability. Sustainability is needed to have a stable relationship between human activities and the environment so that future generations can enjoy a good QOL. A sustainable environment plays an important role in achieving global sustainability as our planet is the main resource for fulfilling basic human needs, providing air, water and soil. All other needs are directly and indirectly dependent on these resources.

Earth is the main provider of basic necessities to humans but humanity is at the threat of extinction if better care of the environment is not taken to live more sustainably. Environmental sustainability can be achieved by reducing the toxic effects of human societies on the environment, condensing the use of natural resource bases and using sustainable resources. The increased rate of population growth places pressure on the environment as well as on the natural resource base that is so important for the survival of humans and other living creatures on this earth. Fossil fuel consumption seems to be of secondary importance in comparison to the importance of access to clean water, soil and air. The consumption of natural resources to maintain the modern lifestyle is increasing at an alarming rate (Wiese et al. 2012). This will create problems for the survival of future generations because there will be no resources left at a certain point. The best solution to this problem is sustainability and recycling and, hence, achieving global sustainability (Figure 2.3).

The main aim of sustainable development is to efficiently use the renewable and non-renewable resources of the planet that are eco-friendly and inexpensive in nature. These renewable sources are in harmony with nature as they can be recycled and reused over time, reducing the negative impact on humans as well as on the environment. This helps to reduce the pressure on natural resources and to utilize the available stock of the natural basin efficiently. Sustainable development requires a reorganized modern lifestyle, economy, social activities, transportation and additional services that have negative impacts on environmental sustainability. Sustainability can only be achieved with the sacrifices of every single individual of the community. Small changes in the lifestyle and proper utilization of available renewable sources can help in achieving this goal. This will also create business opportunities and employment. Every step that contributes to sustainable development is important, no matter how unimportant it seems. The government has taken steps to conserve natural resources with the help of authorized amendments and affiliates for various assignments that are responsible for reducing the detrimental impact of human actions on atmosphere (Lang et al. 2012).

2.5 Challenges Toward Sustainability

Sustainability is facing many challenges in eradicating of poverty from the world. Continued implementation of present developmental strategies is not sufficient to achieve the goals of sustainability. Various dimensions of sustainability remain uneven, and the crisis in the international market has slowed down the progress and accelerated the stress on natural resources, resulting in an increase in the prices of basic necessities. There are sectors to which sustainability has to respond, such as urbanization, globalization, community diversity and economic index. These sectors influence sustainability by putting down tremendous challenges. The movement of people in search of better living to urban areas is a factor behind urbanization, globalization and financial decline, exposing nations to the risk of food, healthcare, energy, terrestrial and aquatic crises as well as ecological degradation. Furthermore, ecological degradation has reached a precarious stage. Thus, change at the individual, community, national and international level is required.

2.5.1 Closing/Bridging Development Gap

Meeting our present needs is essential and obvious, but sustainability is also required to fulfill the needs of future generations. If sustainability is to think about this, the focus should be on removal of the "development gap," which is defined as the vast differences in the services provided to different social domains in education, sanitation, employability, healthcare, housing and food. Poverty is not about possession but about the decline in a complex combination of social and political aspects that blocks sustainable development. No individual person or country can transform but substantial changes are required globally to fight poverty. The United Nations (UN), OECD and other sustainability organizations are struggling to bring sustainability to every part of the world. Other financial organizations like the World Bank are seeking ways to incorporate sustainability (Lang et al. 2012).

In 2000 at the UN Millennium Summit, eight MDGs were agreed by 192 UN Member States to save humanity:

- To eradicate extreme poverty and hunger.
- To achieve universal primary education.
- To promote gender equality and empower women.
- To reduce child mortality.
- To improve maternal health.
- To combat HIV/AIDS, malaria and other diseases.
- To ensure environmental sustainability.
- To develop a global partnership for development.

2.5.2 Forward Steps for the Global World

In spite of the approaches taken and policies implemented for sustainability, the development gap remains the same. Irresponsibility and lack of coordination are the reasons for the development gap. Nations have begun approaches to aid population, trade and investments, and other economic policies in order to obtain lasting development results. Consistent policies are required to make sure that the economic objectives of the countries are coherent and do not weaken each other; for example, subsidies given to the domestic market do not prevent gains in the international market and business investments do not conflict with development goals. People should understand the sensitivity of interdependent growth, sector linkages and how things relate to each other in development. The goal is to cooperate with the steps taken for development in developed countries and to prevent bad habits from being established in developing nations. Sustainability has to put poor as well as rich countries together on the path of development, maintaining cooperation and coordination in their policies and approaches (Kyburz-Graber et al. 2006).

2.5.3 Sustainable Toolkit for Companies

Sustainability requires the toolkit for companies. This toolkit is a part of corporate social responsibility (CSR) policies, which are defined as global private business self-regulation policies. This is a system of corporate community where self-regulation is incorporated into a corporate model focused on the triple bottom line of societies, earth and revenue (Saleh et al. 2011). CSR policies comprise three basic policies:

1. *Environmental policies such as those of the International Organization for Standardization (ISO).* ISO is an international organization promoting global standardization by setting the standards for specifications and requirements for materials, products, processes, procedures, formats, information and quality management. The ISO is an autonomous and non-governmental body, the associates of which are the standard society of 163 countries that runs on the common standards of nations. The standards are used for manufactured products, technology, food safety, agriculture and healthcare. These standards are used to create safe, reliable and good-quality products and services. These standards expand business by increasing the production rate while minimizing errors and waste, which facilitates the product to enter into the international market, thereby assisting in global trade development. The main objectives of these standards are to safeguard the consumers and the final users of the products and to assert that ISO-certified products obey the regulations that are set internationally.
2. *Corporate policies such as the Global Compact.* Corporate policies are initiatives that have been executed to inspire corporate sectors to follow policies leading to corporate sustainability. The United Nations Global Compact (UNGC) is an initiative taken to encourage businesses globally to attain corporate sustainability. It is a principle-founded agenda for trades around the world, and has ten codes of social and corporate issues. Under this framework corporations are brought under the UN agencies, workforce and public. The UNGC is the world's largest commercial sustainability agenda based on ten standards (Cetindamar 2007):

- Human rights:

Standard 1: the corporate world should uphold and respect the protection of globally declared human rights.
Standard 2: the corporate world should ensure it is not complicit in human rights exploitations.

- Labor standards:

Standard 3: the corporate world should uphold the freedom of association and the effective recognition of the right to collective bargaining.
Standard 4: corporate world businesses should be able to eradicate all forms of involuntary and enforced labor.
Standard 5: the real elimination of child labor.
Standard 6: the elimination of judgment in occupation.

- Environment:

Standard 7: the corporate world should support a cautionary attitude to ecological challenges.

Standard 8: the corporate world should accept initiatives to encourage ecological concern.

Standard 9: they should inspire progress and distribution of ecological friendly technologies.

- Anti-corruption:

Standard 10: the corporate world should work in combination to fight against corruption in all forms including blackmail and enticement.

3. *Product and services policies such as lifecycle assessment (LCA) (Figure 2.4) of products.* The LCA, also known as lifecycle analysis or eco-balance, is a method to measure all the ecological influences related to the process of the production of goods from raw material extraction to its disposal after use. It helps to know about environmental concerns and effects created by product formation in the environment. The LCA is a standardized methodology that gives reliability and transparency. As per ISO 14040 and 14044 the LCA is composed of four phases (Anex and Lifset 2014):

   - *Goal and scope*: in this phase, the need to conduct the LCA is defined. The subjective choices such as the reason for LCA execution, definition and lifecycle of the product, and description of the system boundaries are described in this phase.
   - *Inventory analysis*: in this phase, all the ecological contributions and productivities related to the goods are studied, such as the use of sustainable raw material and green energy, the release of toxic pollutants and waste streams. This phase gives complete information about the effects of the product and its services on the environment.
   - *Impact assessment* (IA): this step allows the taking of better business decisions. After inventory analysis of all environmental impacts, the evaluation is done and decisions are made that best suit the company. The issues and impacts are translated into environmental themes such as climate change or human health (Bond and Pope 2012).

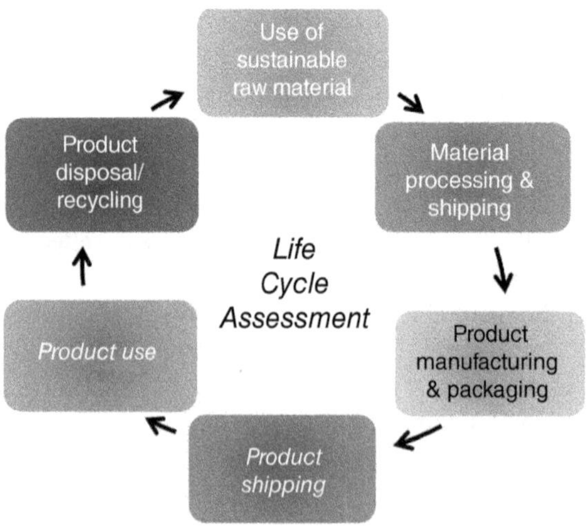

Figure 2.4 The lifecycle assessment of products.

- *Interpretation*: during this phase all the conclusions are checked so that they are all well substantiated. Various checks are conducted to test whether conclusions are supported by data and procedures used. This enables the results to be improved significantly.

2.6 Sustainable Future

A sustainable future depends on the conservation and preservation of resources by recycling, reducing and reusing. But the billions of people living on the planet and producing waste are continuously degrading the ecology and its natural resources, which hinders the conservation and preservation of resources. This problem of waste generation provides a great opportunity for transforming waste raw materials into unlimited efficient clean energy using modern technology. To have a sustainable future every dimension of sustainability should be taken care of, with all interlinked impacts on people and the environment being addressed. Social and economic developments are the foremost requirement in a development pathway eradicating extreme poverty. To have a sustainable future every sector needs development with clear specifications and its benefits to the social domain made clear.

2.7 Conclusions

The following are the important conclusions from the chapter:

1) In the broadest sense, sustainability is based on three pillars of sustainability and strategies followed for attaining sustainability encourage coordination between humans and the pillars of sustainability (environment, social and economic).
2) To define the progress of a nation in terms of sustainability, all parameters such as evolution, geographic distribution, disciplinary composition and collaboration and their individual and combined impacts on the environment, social and economic pillars need to be studied.
3) In order to attain sustainability, there is a requirement for a communal system with social equity and understanding of relationships to provide a way out for the pressures arising from disharmony in society; a production system that can preserve natural resources and the environment; a technical system that can constantly give novel answers to the problems arising in the development pathway; and a global system that can accelerate the sustainability of the corporate world in terms of trade and finance (Rimanoczy and Pearson 2010). Therefore, it is necessary to have in-depth knowledge of all the parameters that have an effect on sustainability, and by studying the challenges, to find methods to overcome these challenges and mitigate their effects on a sustainable future.
4) There are several theories on sustainability and concepts that should become an integral part of government agendas and policies, corporate policies and organization goals to achieve sustainability. Developmental policies have to be aware of resource availability, cost distributions and benefits. This development implies a concern with providing an equal distribution of resources within the generations (Koff and Maganda 2016). Therefore, it is necessary to change the vision of every nation to attain sustainable development.

Acknowledgments

We are appreciative to Ms. Rajwinder Kaur for her modification of earlier drafts to improve the chapter.

References

Anex, R. and Lifset, R. (2014). Life cycle assessment. *Journal of Industrial Ecology* 18 (3): 321–323.

Aras, G. and Crowther, D. (2008). Governance and sustainability: an investigation into the relationship between corporate governance and corporate sustainability. *Management Decision* 46 (3): 433–448.

Bond, A. and Pope, J. (2012). The State of the Art of Impact Assessment in 2012. Impact Assessment and Project Appraisal Special issue.

Brammer, S., Jackson, G., and Matten, D. (2012). Corporate social responsibility and institutional theory: new perspectives on private governance. *Socio-Economic Review* 10 (1): 3–28.

Cetindamar, D. (2007). Corporate social responsibility practices and environmentally responsible behavior: the case of the United Nations Global Compact. *Journal of Business Ethics* 76 (2): 163–176.

Dempsey, N., Bramley, G., Power, S., and Brown, C. (2011). The social dimension of sustainable development: defining urban social sustainability. *Sustainable Development* 19 (5): 289–300.

Duhn, I. (2012). Making 'place'for ecological sustainability in early childhood education. *Environmental Education Research* 18 (1): 19–29.

Griggs, D., Stafford-Smith, M., Gaffney, O. et al. (2013). Policy: Sustainable Development Goals for people and planet. *Nature* 495 (7441): 305–307.

Hansmann, R., Mieg, H.A., and Frischknecht, P. (2012). Principal sustainability components: empirical analysis of synergies between the three pillars of sustainability. *International Journal of Sustainable Development and World Ecology* 19 (5): 451–459.

Hopwood, B., Mellor, M., and O'Brien, G. (2005). Sustainable development: mapping different approaches. *Sustainable Development* 13 (1): 38–52.

Kates, R.W., Clark, W.C., Corell, R. et al. (2001). Sustainability science. *Science* 292 (5517): 641–642.

Kellert, S.R., Mehta, J.N., Ebbin, S.A., and Lichtenfeld, L.L. (2000). Community natural resource management: promise, rhetoric, and reality. *Society and Natural Resources* 13 (8): 705–715.

Koff, H. and Maganda, C. (2016). The EU and the human right to water and sanitation: normative coherence as the key to transformative development. *The European Journal of Development Research* 28 (1): 91–110.

Krueger, T., Page, T., Hubacek, K. et al. (2012). The role of expert opinion in environmental modelling. *Environmental Modelling and Software* 36: 4–18.

Kyburz-Graber, R., Hofer, K., and Wolfensberger, B. (2006). Studies on a socio-ecological approach to environmental education: a contribution to a critical position in the education for sustainable development discourse. *Environmental Education Research* 12 (1): 101–114.

Lang, D.J., Wiek, A., Bergmann, M. et al. (2012). Transdisciplinary research in sustainability science: practice, principles, and challenges. *Sustainability Science* 7 (1): 25–43.

Letaifa, S.B. and Rabeau, Y. (2013). Too close to collaborate? How geographic proximity could impede entrepreneurship and innovation. *Journal of Business Research* 66 (10): 2071–2078.

Li, N., Chan, D., Mao, Q. et al. (2018). Urban sustainability education: challenges and pedagogical experiments. *Habitat International* 71: 70–80.

Martin, D., Hondros, J. and Scambary, B. (2002). Enhancing Indigenous social sustainability through agreements with resource developers. Australian Research Council Linkage Project: Indigenous Community Organisations and Miners: Partnering Sustainable Development. Australian National University, Canberra.

Mauerhofer, V. (2013). The "Governance-Check": assessing the sustainability of public spatial decision-making structures. *Land Use Policy* 30 (1): 328–336.

McCormick, K., Anderberg, S., Coenen, L., and Neij, L. (2013). Advancing sustainable urban transformation. *Journal of Cleaner Production* 50: 1–11.

Moldan, B., Janoušková, S., and Hák, T. (2012). How to understand and measure environmental sustainability: indicators and targets. *Ecological Indicators* 17: 4–13.

Nazli Nik Ahmad, N. and Salat Ahmed Haraf, A. (2013). Environmental disclosures of Malaysian property development companies: towards legitimacy or accountability? *Social Responsibility Journal* 9 (2): 241–258.

Poveda, C.A. (2017). The theory of dimensional balance of needs. *International Journal of Sustainable Development and World Ecology* 24 (2): 97–119.

Purcarea, I., del Mar Benavides Espinosa, M., and Apetrei, A. (2013). Innovation and knowledge creation: perspectives on the SMEs sector. *Management Decision* 51 (5): 1096–1107.

Rimanoczy, I. and Pearson, T. (2010). Role of HR in the new world of sustainability. *Industrial and Commercial Training* 42 (1): 11–17.

Saleh, M., Zulkifli, N., and Muhamad, R. (2011). Looking for evidence of the relationship between corporate social responsibility and corporate financial performance in an emerging market. *Asia-Pacific Journal of Business Administration* 3 (2): 165–190.

Schoolman, E.D., Guest, J.S., Bush, K.F., and Bell, A.R. (2012). How interdisciplinary is sustainability research? Analyzing the structure of an emerging scientific field. *Sustainability Science* 7 (1): 67–80.

Singh, S.J., Krausmann, F., Gingrich, S. et al. (2012). India's biophysical economy, 1961–2008. Sustainability in a national and global context. *Ecological Economics* 76: 60–69.

Van Assche, K., Beunen, R., Duineveld, M., and Gruezmacher, M. (2017). Power/knowledge and natural resource management: Foucaultian foundations in the analysis of adaptive governance. *Journal of Environmental Policy and Planning* 19 (3): 308–322.

Wiese, A., Kellner, J., Lietke, B. et al. (2012). Sustainability in retailing – a summative content analysis. *International Journal of Retail and Distribution Management* 40 (4): 318–335.

3

Sustainable Development: Dimensions, Intersections and Knowledge Platform

Pritee Sharma and Kanak Singh

Discipline of Economics, School of Humanities and Social Sciences, Indian Institute of Technology Indore, Indore, India

3.1 Introduction

Sustainability is an open, dynamic and evolving idea that aims to encompass various possibilities of interdependencies, independencies and interpretations among various sectors, stakeholders, actions and contexts across space and time. The definition of sustainable development was created as an attempt to reconcile the tension between environmental and developmental concerns that were at the heart of global policymaking. The concept equally engages with the economic, social and environmental domains, and has been widely incorporated into policy. For example, many countries have developed ideas for sustainable development. Initially, the concept of sustainable development was central to environmental law and policy, yet its application continues to be the focus of much debate, with many different versions of the concept (Redclift 2005; Beckerman 1994).

A key issue in the concept of sustainability is its intergenerational equity, which refers to fairness across generations. According to the widely cited Brundtland Report (Brundtland 1987), with sustainable development future generations should be able to meet their own needs. This implies an ethical dimension: essentially, we need to pass on to forthcoming generations a fair share of resources. Furthermore, the definition is based on the assumption that we can identify the impact of current activities on the future availability of resources; yet this is highly uncertain (Bell et al. 2013).

Problems with definition have restored the use of sustainable development as a working legal principle. The concept has been discussed and developed by international agreements, applied in courts of law and referred to in provisions that are imposed on public and private bodies. It is not an enforceable law but referred to in policymaking documents as "soft laws."

According to the Rio Declaration, every nation has a portion of responsibility toward sustainable development. Also, it is important to note the variation in the needs of developing and developed nations on environmental grounds (United Nations 1992). However, sustainability science has a narrower scope than sustainable development. The

Sustainability: Fundamentals and Applications, First Edition.
Edited by Rao Y. Surampalli, Tian C. Zhang, Manish Kumar Goyal, Satinder K. Brar, and R. D. Tyagi

natural and engineering sciences offer tremendous prospects for contribution to a more sustainable future. It is their responsibility because they produce not only solutions but also problems. The idea that science and technology be accompanied by societal priorities is much needed. This gives sustainability science its context and rationale. Sustainability science:

- Urges us to uplift the grounded principle fundamentals and the relationship of the human and environment domains and suggest bridges within ecosystem and social sciences and technology and governance.
- Serves as a platform to discuss various perspectives on and possibilities for sustainable development along with various contexts for its design, implementation and evaluation in respective scenarios.

Today, our understanding of earth is rapidly increasing in three dimensions of space, time and structure. The tools of complex system science are disclosing new and interesting dynamics that have proven roles to play in sustainable development, orientation, design and implementation processes. For example, it can guide a more sustainable use of renewable resources, which leads to an important link between resource efficiency and economic growth.

We live in mental and social space. Our interactions with the physical world and with other living beings are mediated through biology, demography and the mind (self-awareness). Explorations in different sciences like biology, economics, sociology, psychology and others predict probabilistic terms of human interactions among complex systems. Scientific exploration represents another link between ends and means, which is much needed for sustainable development.

The objective of this chapter is to confer an overall understanding of sustainability. This chapter aims to give a new outlook, one that enable us to uncover the deep conceptualizations of sustainable development and mobilizes the understanding of later chapters in this book. The chapter briefly discusses the economic, social and environmental dimensions of sustainability. The chapter also covers sustainable development indicators with intersections of the three main dimensions of sustainability. The last part of the chapter discusses the current knowledge systems for achieving sustainability and also future avenues of the same.

3.2 Understanding Complex Systems

The definition of a complex system can be stated as one system with multiple independent entities with their respective properties. These entities are interrelated. As a bunch they exhibit one or maybe more "predictive" behavior. It is a relatively new idea with an edge over traditional concepts like medicine, science or finance. The problem-solving mechanism is hard to solve because the factors involved in a complex system may or may not share explanations. This is very much visible in addressing socio-ecological conflicts resulting from human activities, which often may have unexpected outcomes (Meadows 1999). According to Ladyman et al. (2013) features of complex systems include the following:

- Their dynamics and development are hyper-sensitive.
- The independent interacting components are often coupled, with multiple interactive stages among them.
- The system is a developing and evolving idea over time. The dynamics depend on multiple pathways. Examining its history is critical in evaluation.
- An termination of one component or stage can lead to chains of collapse, and, further, could be catastrophic for the entire system.
- Each entity of a complex system may itself be a complex system. Hence, the entities are nested.
- Complex systems have the scope to provide both dipping and growing feedback loops.

Complex systems could be of different types like the complex adaptive systems (CASs) as they are adaptive via learning from history (e.g. the stock market, the brain, the immune system and manufacturing businesses). Another form of complex system is chaotic systems. Chaos theory mentions the nature of changing systems, which have sensitivity toward original circumstances (Gleick 1987): an effect commonly known as the "butterfly effect." They are a set of small non-linear interactions like fractals, which offer a mathematical understanding of cloud formation.

Market behavior is an appropriate example of a complex system. Another example of a complex system is the earth system.

In the context of sustainable development, it is critical to understand the earth system in different contexts and fragments. This kind of systems thinking gives us an optimistic approach in the development of sustainability, which is most likely to be achieved in practice.

3.2.1 The Limits to Growth

Over four decades ago, Meadows et al. (1972) wrote *The Limits to Growth*. It presented a kind of "system model" that linked the global economy with the environment. Of the many conclusions made, the most prominent one was that the economic growth curve and development will face limitations due to the finite natural resources on planet earth. The global model presented in *The Limits to Growth* was developed to have an outlook on the future scenarios in the growth cycle with various inferences (Cole 1999).

The criticisms of this research were rooted in the fact that it was misunderstood. As in the case of system models, this model was never intended to make forecasts or be predictive. In fact, it was one of the first attempts to understand and explore the behavior of the world's economic system. *The Limits to Growth* can be seen as an early call for the transition to sustainability. Meadows et al. (2004) suggested that economic decline is likely, unless measures are taken to limit consumption and population growth and to increase the efficiency with which materials and energy are used. Turner (2008), in a quest to compare the original model projections with what has actually happened in reality since 1972, found that data describing trends over the past 30 years are in sync with the key features of the "original model" resulting in the downfall of various blanket systems by the middle of the twenty-first century. It also gave some important insights into understanding and controlling global pollution.

Atkisson (2010) also evaluated *The Limits to Growth* model and added that the authors missed some of the efficiency gains that have occurred since then in history. According to Atkisson, the real contribution of *The Limits to Growth* is the awareness and alarm it raised regarding the dynamics of the earth system. Davidson (2000) suggests that environmental degradation is one possible outcome of economic growth if the saturation points are not taken care of, rather than ecological collapse. To acknowledge these limit points, it is important to understand the consumption patterns along with their deterministic driving factors.

3.2.2 Global Consumption

It is clear that currently human are consuming more than the earth system can support as per the ecological footprint (EF) analysis. An interesting way to look at it would be to think about how much the current global population consumes. A useful overview of global consumption patterns and trends was provided by the World Business Council for Sustainable Development (WBCSD) in 2008 (WBCSD 2008), which conveys the following points:

- World gross domestic product (GDP) is projected to grow by 325% between 2007 and 2050 as the result of an increase in population and strong GDP growth.
- China's GDP will overtake the USA's GDP; and India's GDP will overtake that of Japan by 2025.
- Middle-class consumers will triple by 2030, which will place 80% of the world population in the middle-class domain.
- Low-income consumers (people who currently earn less than US$3.35/day) will have a global combined spending power of approximately US$5 trillion.
- Food will dictate low-income household budget.
- Economic development itself will lead to a change in consumption patterns.

3.2.3 Population and Environment: The IPAT Equation

The IPAT equation helps in understanding different factors that impact rthe environment(Ehrlich and Holdern 1971) . It suggests that environmental impacts (*I*) of the human population are the result of three variables: population (*P*), affluence (*A*) and technology (*T*).

$$I = P \times A \times T \tag{3.1}$$

This equation gives an approximate framework to assess the impacts of human population increase, with multiple aspects other than population number. The conceptual factor of this formula is strong, but the relative criticality among different parameters in the equation needs more clarity. Hence, in 1974, Holdren and Ehrlich (1974) stated the same equation with slight modifications.

$$\begin{gathered} Environmental\ Degradation = population \times consumption\ per\ capita \times \\ damage\ per\ unit\ of\ consumption \end{gathered} \tag{3.2}$$

This new change predominantly emphasizes the role of consumption, rather than just wealth capacity in defining human impact. It has added a more precise outlook on human impact. The use of amended versions of Eq. (3.1) has suggested that, through economic development, there may be a reduction in the consumption of energy or of goods per unit of GDP, a process called "dematerialization" (Ausubel and Waggoner 2008).

3.2.4 Notions of Strong and Weak Sustainability

Sustainability has two common notions to address the variations present between and within societies in an economic development context. It is important to consider these notions because a blanket approach to attaining sustainable solutions is not possible. These two notions are weak sustainability and strong sustainability; both of these notions are contextual and can be valid concurrently.

Weak sustainability is maintaining the complete stock capital value.

$$Weak\ sustainability = K + H + SC + N \geq X^{*} \tag{3.3}$$

where K = manmade capital; H = human capital; SC = social capital; N = natural capital; and X^* = pre-determined threshold level of all the forms of capital expressed in monetary terms. Therefore, in the case of weak sustainable development, a reduction in natural capital can be substituted by manmade capital, but this substitution has a limit. It is important that a minimum limit of each capital is achieved with an overall sense of development. In weak sustainability, these trade-offs among human, social and economic capital are unrestricted or have very few limits.

On the other hand, in strong sustainability these trade-offs are strictly restricted. According to the notion of strong sustainability natural and human-made capital are counterparts that are not supposed to be substituted. Thus, environmental degradation is the result of waste generation from economic cycles.

$$Strong\ sustainbility = K \geq K^{*}, H \geq H^{*}, SC \geq SC^{*}, and\, N \geq N^{*} \tag{3.4}$$

Daly (1990) suggested the following principles for strong sustainable development:

- Renewable resources should be encouraged for particular stocks.
- Non-renewable resources are exhausted over time in economic cycles.
- Pollution emissions should be maintained.

3.3 Dimensions of Sustainable Development: Economic Dimension

There are many ways to define, achieve and measure sustainable development. All these various concepts are basically supported by the three fundamental dimensions and the evolving flow between them. Sustainable development is defined as specific intersections of these dimensions. They could be socio-economic, biophysical or psychological. The 2030 Agenda for Sustainable Development was developed by the United Nations (UN) General Assembly. Sustainable development is about economic, social and environmental justice and equity. Another example to understand sustainability could be the model(s) of human population and the economic subsystem within the finite biosphere. It basically depicts the interference of humankind with the ecosystem. Figure 3.1 shows the role of solar energy inflow and the earth and the environment as a sink for biogeochemical cycles including mobilization of resources.

As shown in Figure 3.2, there are three main dimensions that sustainability seeks to integrate: economic, environmental and social (including sociopolitical). This is also known as the "triple bottom line" model, which was officially endorsed by the UN in 2007, also referred to as "people, profit and planet."

A circular economy has a main role in addressing sustainable development. As such it fails to recognize the co-evolutionary nature of economic, social and ecological change (Mulder and Van Den Bergh 2001). Co-evolution describes the dynamics of relationships within systems. Co-evolution takes place when at the minimum level one of the feedback loops is changed by an activity that occurs within the system. This in turn starts a further chain of ongoing and reciprocal steps of change.

3.3.1 Neoclassical Thought

Economics can be defined as a social science that helps to explain the economic aspects of human societies, or how societies decide what, how and for whom to produce (Begg et al. 2011; Parkin and Koorey 2012). Mainstream or neoclassical economics are based on three assumptions:

- Stakeholders have the independence for choices among outcomes.
- Individuals lead to utility maximization (via consuming goods), and industries are supposed to maximize profit.
- On the basis of information, people have the independence to take decisions.

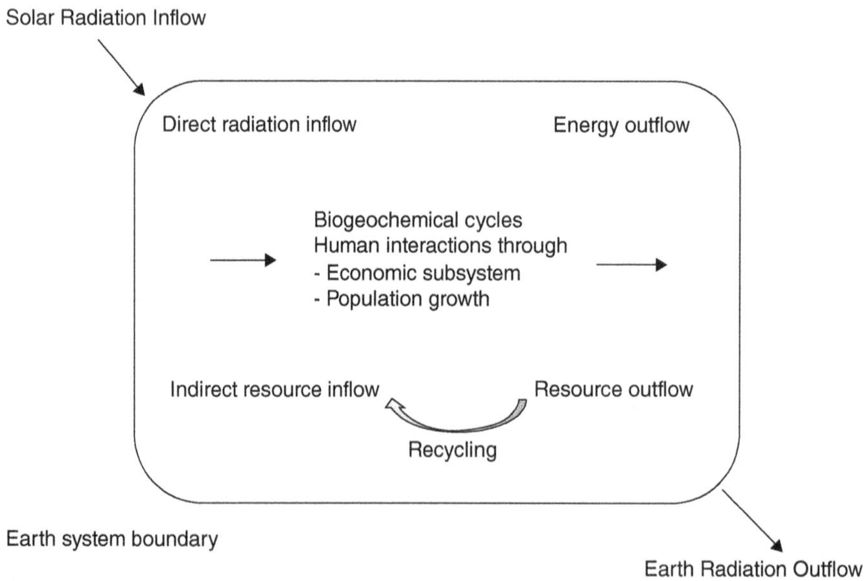

Figure 3.1 Conceptual model of human interactions within an earth system framework. *Source:* de Vries 2013.

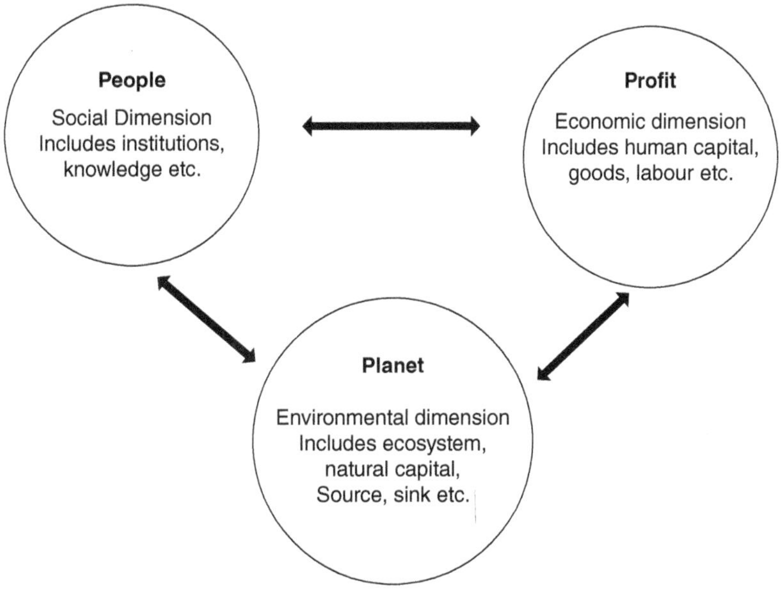

Figure 3.2 A depiction of the relationship between the dimensions of sustainable development. Source: Lawn 2006.

One of the main logics behind neoclassic economics is the superiority of the market economy to other forms of economic organization based on higher efficiency. Traditionally, GDP has been used to measure economic activity, but there are many criticisms of this method, such as the following:

- There is no account of unpaid domestic work, hence subsistence-based factors are not included.
- Environmental and social breakdowns are never subtracted from GDP.
- There is no measure of inequality in terms of sharing livelihoods and quality. This is one of the critical problems in the Sustainable Development Goals (SDGs), which we will discuss in later sections of the chapter.
- There is no account of stocks (quantity measured at a specific point in time) but flows (rate) are accounted for.

According to Aldred (2009), there is a dominant focus on economic growth in traditional economics, which infuses into government policies. Aldred also criticized the increasing monetization of various important aspects of human life. For example, instead of looking at climate change mitigation as a cost–benefit analysis, it should have a political and ethical outlook based on its major implications for future generations. In order to combat the above criticism, ecological economics has gained popularity as an upcoming academic discipline. Neoclassical economics had two subdivisions, namely, environmental economics and natural resource economics, which state the dynamics between natural resources and economics and take the environment as a subset of the human economy. However, ecological economics views economics as an economic activity taking place in the environmental context. Ecological economics therefore considers the economy to be a subsystem of nature, whereas neoclassical approaches take the opposing view (Common and Stagl 2005). Some important characteristics of ecological economics include:

- There is significance given to preserving natural capital as natural capital cannot be substituted by human-made capital.
- There is an assumption that economic growth will be steady because of the carrying capacity of the earth's ecosystem.
- There is a relative scale of economy to the ecosystem it depends on.
- There is a clear sign of interest in sustainable development and its concepts, such as equitable distribution, improvement of human wellbeing and inclusion of socio-ecological factors in decision making.

3.3.2 Green Economics

On the other hand, green economics has emerged from the grass roots level. According to this new concept there is a need for a big shift from the capitalist nature of economics. Rio+20 was one of the major international events that ignited the quest for a green economy. The main guidelines for a green economy are still evolving, although they were reviewed further by the UN and were further presented as a set of principles of green economy for future decision making. The concept of a green economy exhibits the following characteristics:

- A green economy is a way toward sustainable development.
- It should be able to create green jobs and decent work scenarios.
- It is resource- and energy efficient.
- It should take account of planetary boundaries, and ecosystem capacity.
- It should use integrated decision making.
- The indicator system used should be beyond the scope of traditional GDP.
- The notion should be fair and equitable for the entire planet and future generations.
- It should preserve biodiversity and natural capital.
- It should address poverty reduction, essential services, social protection, wellbeing and livelihoods.
- It should facilitate fair governance.
- It should be able to internalize externalities.

There is another set of principles for a green economy, which has deeper green motives with more room for efficiency and intersections. It is interesting to compare them (see below) with the above-mentioned principles:

- The dominance of use value, quality and intrinsic value to support need-based generation with maximum efficiency.
- To focus on natural flows within planetary boundaries rather than worrying about a nature dominant scenario.
- The by-product from one production should become the resource for another production.
- Popularize energy efficiency and multifunctionality.

- Promote a sustainable economic scale.
- Integrate a range of forms of organizations as diversity in marketplace globally.
- Belief in self-reliance, self-organization and self-design.
- Maximum participation and direct democracy.
- Scope of enhancement for continuous development and human creativity.
- Integration of spatial and landscape design along with built environment.

The focus on a green economy at Rio+20 can be viewed in terms of an attempt to apply new thinking or a new impetus of sustainable development specifically by linking it to the concept of green growth (Morrow 2012). The main components are economic, environmental and social, which are often represented by three pillars or three interlocking circles (Kates et al. 2005; Dresner 2008). The green economy is widely seen as a means of supporting the achievement of sustainability (Allen and Clouth 2012), or as enabling its components (UNCTAD 2010).

According to David Huberman (International Union for the Conservation of Nature [IUCN]) (Huberman 2010), the green economy is a new tool to make the economic system more sustainable by doing things differently. Hence, it contributes as an economic dimension toward sustainable development. The green economy needs monitoring to establish itself. This can be achieved by the designing of specific indicators explicitly for the green economy. Such measures are critically needed in the present world. Below are some indicators for green growth via four interrelated groups:

- Description of nature asset base.
- Accounting for quality of life (QOL) in an environmental context.
- Accounting for environmental and resource productivity of production and consumption.
- Description of policy reciprocation and economic avenues.

Some already developed alternative indicators are the Genuine Progress Indicator (GPI), the Happy Planet Index (HPI), the Index of Sustainable Economic Welfare (ISEW) and the Global Green Economy Index (CGEI).

As another aggregate indicator system, EF is already being used in Taiwan for the development of the green economy. There are interventions required to support the green economy through appropriate policies, institutions and governance (see Figure 3.3).

According to Barbier and Markandya (2012), policy interventions are imperative for the success of sustainable development via the green economy policy. These interventions should include improvements in environmental valuation and their integration policy and strategy development. Encouragement for property rights, market-based instruments, ecosystem restoration, participation from local communities and good governance is also crucial.

Interestingly, there is a stage known as "ecological modernization," which supports progress of the green economy, whereby institutions like corporations and markets are enabled but resource efficiency is stressed, e.g.the products are manufactured using less energy and fewer raw materials. This gives an important contribution of resource efficiency toward a green economy, which in turn facilitates SDGs.

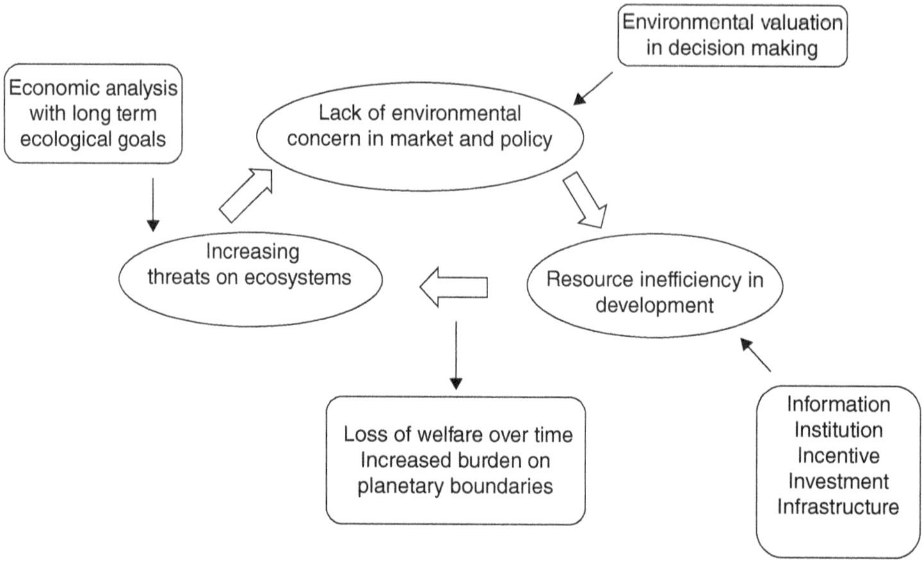

Figure 3.3 Interventions that are required to support the transition to a green economy. *Source:* Newton and Cantarello 2014.

3.3.3 Resource Efficiency and Policy Pathways

Resource efficiency is defined in terms of quantification as the difference in resource intensity over time. Resource intensity measures the amounts of resources (in terms of domestic material consumption, material footprint, energy or water) used per unit of GDP (UN 2017).

$$\text{Resource intensity} = \frac{\text{Resource inputs used}}{\text{Dollar unit of output produced}} \tag{3.5}$$

$$\text{Resource efficiency} = \text{More goods and wellbeing} + \text{Fewer resources used} \tag{3.6}$$

According to the UN (2017), there is empirical evidence for prominent links among resource efficiency gains and employment creation, increments in the human development index (HDI), energy management and climate change mitigation. HDI principally consists of three aspects: life expectancy, education per capita income and access to water and sanitation. Resource efficiency can be measured with the help of data and information. According to the nature and scope of analysis the various indicators can be derived to measure resource efficiency. The base points of these derivations are mostly the SDGs-2030.

According to the UN (2017), two indicators for measuring resource efficiency are: material consumption and material footprint. The domestic material consumption of an economy is its material resource for production purposes, which can be divided into four categories: biomass, fossil fuels, metal ores and non-metallic minerals. This indicator counts the amount of material resource extracted from nature plus imports minus exports.

$$\text{Domestic material consumption is material use of production} \tag{3.7}$$

$$\text{DMC} = \text{Extraction} + \text{Imports} - \text{Exports}$$

where DMC, domestic material consumption. Material footprint is an indicator that further complements DMC (UN 2017).

Nine policy pathways have emerged as good approaches that countries can adopt to promote resource efficiency. These policy pathways have some successful examples from Asia and the Pacific region. Their nature of applicability may vary from the macro level (with pathway numbers 1–5) to the sector level (with pathways 6–9) (UN 2017).

3.4 Dimensions of Sustainable Development: Environmental Dimension

There is a remarkable difference in various concepts and terms like biodiversity, biosphere, ecosystem, ecosystem function and ecosystem services, which are important to understand for a clear grasp on sustainable development. "Biodiversity" is unfortunately a term that has been defined in a variety of ways by different authors, often leading to confusion about precisely what is being discussed (DeLong 1996). According to the Convention on Biological Diversity (CBD), biodiversity is the vast range of variability and inconsistency among all the species and all the possible sources of life within the span of ecological complexes. Therefore, biodiversity has three major dimensions: genetic, species and ecosystem diversity. Biosphere, on the other hand, is the zone around the planet earth where life can exist.

Ecosystem functioning can be described as the cumulative vital activities of flora and fauna along with mutual participation of microbes. It also includes the effect of these interactions on the physiochemical state of the environment. Therefore, ecosystem functioning is contextual to its biophysical and chemical characteristics. Ecosystem function refers to both pools of material (organic and inorganic) and the rate of processes in an ecosystem.

What do we see when we look at a tree? Is it just a tree in the landscape that gives esthetic satisfaction or is it a canopy that obstructs rain and reduces erosion? Is it a large foliar area that intercepts particulate matter and cleans the air that we breathe or timber with a specific economic value? Or is it the carbon sink that will diminish future climate change? Ecosystem services provide answers to these questions from various perspectives. The ecosystem service method is based on the assumption that there's a link between human wellbeing and the state of the environment and the services that it can provide. This assumption is fair and objective in nature, which may help us make some remarkable decisions that are optimal and more sustainable unlike the ones the human race has made historically.

3.4.1 Ecosystems and Ecosystem Services

Ecosystems and ecosystem services are all of these things that we acquire from nature and that accord to our wellbeing. The definition of ecosystem services employed by the MEA (Millennium Ecosystem Assessment) (WRI 2005) is: "ecosystem services are the benefits people obtain from ecosystems." It is a fairly new concept, and few people are trained to see these services. Even fewer people know how to integrate them into important decision-making processes, e.g. where we should conduct economic development or landscape planning.

It is important that emphasis is placed on the decisions that strike the most favorable balance between jobs, food production, water quality protection and the other services that nature provides.

All over the world, people are making decisions every day that affect natural resources, and people cannot help but have an impact. After all, the natural world is the source of all economic activity that drives human societies. We struggle to make laws and choices that minimize the impact on the natural world, but the traditional dialogue of conservation versus development has led to a healthy conversation about how best to manage natural resource use and human development. The aim of the SDGs is to improve the ecosystem condition and human condition together. This can be done by making it easy for managers, decision makers and policymakers to incorporate the values of natural capital into all kinds of resource-related decisions. It is important to notice that biodiversity conservation and ecosystem services have some key differences.

The units of focus for biodiversity conservation are the ecosystems; for example, biodiversity is connected with a specific type of ecosystem with its unique properties, whereas in ecosystem services the units of focus are the interactions between ecosystems and the people dependent on them. Secondly, the levels of interactions within ecosystem that support biodiversity are measured in biodiversity conservation, whereas in the case of ecosystem services the interactions among ecosystems, their landscape context and the spatial connection to people is observed. Also, in terms of valuation, biodiversity conservation can be calculated via supply only, whereas valuation of ecosystem services can be done by considering supply, service and value; e.g. how much service is flowing from an ecosystem, who is the beneficiary and what is the value of that flow of service. In a decision-framing context, the focal point in biodiversity conservation is always to conserve or restore to maximize biodiversity. On the other hand, thinking in terms of ecosystem services the decision-framing process highlights healthy ecosystems as an entry point for human wellbeing and economic prosperity.

Ecosystems need to be managed in a very thoughtful and sustainable way, so that they continue to support humanity by ecosystem-based management and resilience-based governance. The MEA specified the major four categories for ecosystem services that have direct or indirect contributions to human welfare (WRI 2005):

- Provisioning services – this would include components like foods from wildlife, crops.
- Regulating services – climate regulation via carbon storage and the hydro cycle.
- Cultural services – e.g. recreational benefits.
- Supporting services – like soil formation, photosynthesis.

All these categories, along with their components, are interlinked to biodiversity. Maintaining natural capital stocks grants the possibility of a sustainable exchange of ecosystem services in future with endorsement of human welfare (TEEB 2010).

3.4.2 Mapping of Ecosystem Services

According to the European Union (EU) (Science for Environmental Policy 2015), for the effective implementation of the ecosystem services approach, it is important that all decision makers and stakeholders clearly follow the trade-offs and synergies between different ecosystem services and biodiversity. Ecosystem mapping is an important tool as well as the first step in the process. The following points are explained through ecosystem mapping:

1) How ecosystem and biodiversity are directly proportional to each other.
2) What the drivers in ecosystem services are and how they perform within the system (Malinga et al. 2015).
3) The synergies and trade-offs between multiple ecosystems.
4) The costs and benefits of maximizing ecosystem services (Schägner et al. 2013).
5) How supply and demand vary spatially.

The main characteristics of synergies and trade-offs in ecosystems are that synergies are services that rise or fall together. For example, when river corridors are restored, the ability of the ecosystem to provide the service of water purification can be increased. At the same time, the restored areas can also help to mitigate the impacts of high flows, increasing the protection of property and buildings by the service of flood mitigation. Trade-offs occur when prioritizing one service or group of services results in the loss of some other valuable service such as when a community decides to expand agricultural production (taking advantage of the service of food provision provided by those lands), but experiences a decline in water quality as a result of losing some of the natural system's ability to clean the water as it flows across the landscape. It is critical that decision makers take maximum initiative to investigate and then highlight the importance of biodiversity and ecosystem services (BESs) in human wellbeing and the socio-economic domain.

One of such assessments could be cost–benefit analysis of ecosystem services and their intercepts in sections of societies and localities at all levels over time. Another important step should be acknowledgment of human-induced impacts on nature and its accountability in the public domain. This step can serve as a necessary outcome of biodiversity assessment (TEEB 2010).

Deciding how much of one service to "trade off" to get improvement in another service is a decision that must be made locally. It is based on the values and priorities of each community. Ecosystem service approaches and tools do not make this decision for people. Instead, they can help to provide the best scientific information so that society can make informed decisions about its resources and future.

3.4.3 Ecosystem Services Assessment Frameworks and Tools

InVest, the Integrated Tool to Value Ecosystem Services and their trade-offs, is an open access tool based on land cover maps for use with GIS (Geographic Information System) software. Other tools take a variety of approaches; e.g. SolVES (Social Values for Ecosystem Services tool) (Science for Environment Policy 2015). There are many choices when it comes to frameworks and tools that can be used to bring ecosystem services information into natural resource management decisions. Some important issues to consider when determining which tools are appropriate for specific context are:

- Identify the goal: What is the decision the project is trying to inform? What critical information should be provided to help ensure a better outcome?
- Identify the audience: Who will receive this information and what motivates them? What critical information can be provided that will resonate with this audience?

- Stock of resources: Like time, data, capacity and stakeholders.
- Identify the right tools: What tools are needed to produce critical information and in the appropriate form(s)? What other communication strategies are needed to get the information to the right people? Are results being communicated using the appropriate channels?

There are many other frameworks and decision-support tools available besides InVEST. In summary, these different options vary in complexity as well as the data and expertise required to properly apply them.

3.4.4 Challenges in Biodiversity and Ecosystem Services

According to TEEB (2010), the main challenges in BESs are as follows:

- The invisibility of many of nature's services.
- Monetarily valuing ecosystem services and biodiversity.
- Difficulty in measuring the insurance/resilience value of well-functioning ecosystems and its integration in total economic evaluation.
- The unclear relationship between poverty and biodiversity.
- The lack of policies for reforming and redirecting subsidies, which has a negative side effect on natural capital.
- Ecological infrastructure and climate change.

Though the challenges in BESs are very controversial as well as complex, a study conducted by Ruckelshaus et al. (2013) suggests six lessons that were derived during application and quantification of BESs with various new approaches in 20 pilot demonstrations. The main purpose of these lessons is to bridge the gap between scientific knowledge and real-world decision making. The main point to be acknowledged is that the BES approach can be successful if it is considered as a continuously evolving policy-driven idea. This will need contribution from all stakeholders, technology, knowledge systems and governance. It is important to bridge all these factors for result-oriented actions (Ruckelshaus et al. 2013).

3.5 Dimensions of Sustainable Development: Social Dimension

The social dimension has popularly been recognized as the most challenging "pillar" of sustainable development, especially when it comes to its analytical, practical and theoretical underpinnings. The concept of social justice is at the core of the SDGs. Social justice is about creating a society based on principles of equality, solidarity and human rights like equality of opportunity, life chances and life styles.

The following needs that are important for each individual are included in the concept of the basic needs of society's individuals (Manfred et al. 1989):

1) existence;
2) protection;
3) devotion;

4) understanding and self-awareness;
5) participation;
6) recreation;
7) individuality;
8) freedom.

The eight needs included in the category of "basic needs," in essence, make up a related, functionally compatible set of needs, which determines human behavior at a given age and in a given situation (Vanags et al. 2012). If these concepts are achieved in the practical world then the world will become fair for the most vulnerable members of society like the poor, women,and children, although fair is a relative term and so is social justice. This is the reason that social dimensions in SDGs face huge challenges in terms of coordinating implications and injustice among various factors. The concept of planetary boundaries, envisioning the human impacts on the earth system, has received major recognition. Recently, attempts have been made to combine this idea with the complementary concept of social boundaries (Raworth 2012). This concept by Raworth (2012) is based on the theory that sustainable development can be achieved ensuring that all classes of people share the resources and rights that they need like food, water, healthcare, energy, education and employment in an equitable manner (see Figure 3.4). In addition, the resources and rights should be sustainable in nature and be available for future generations.

According to Raworth (2012), the social foundation comprises broadly all human deprivation. There is an environmental foundation beyond which there are dimensions responsible for environmental degradation. Between the two boundaries there is a space that represents an environmentally safe and socially just space for humanity. In this space all concepts of sustainable development like the green economy, sustainable social, economic and environmental development can co-exist. How far is humanity from this

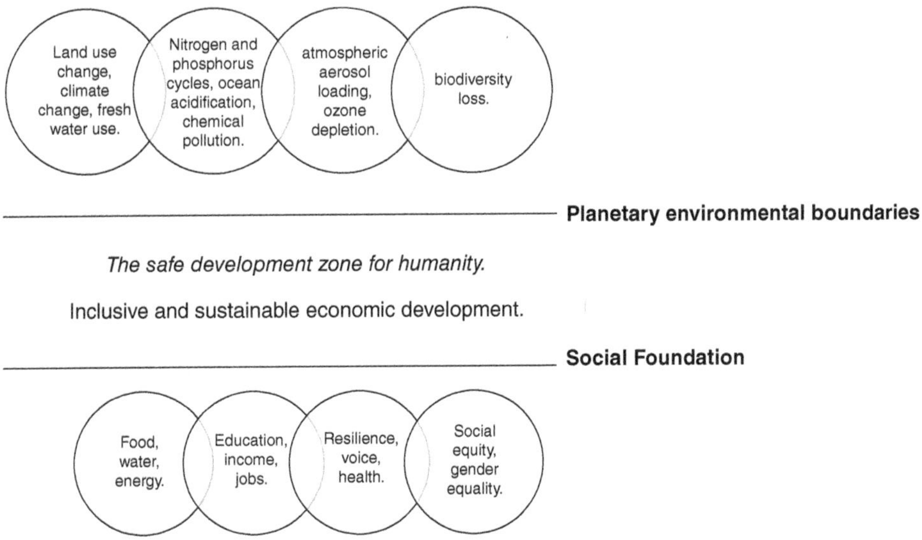

Figure 3.4 Conceptual diagram of sustainable development within planetary environmental boundaries and social boundaries. *Source:* Raworth2012.

space, especially from a social perspective? This is the further step that can help us to assess the level of work to be done and its scope and direction.

3.6 Mapping Social Development Through Sustainable Development Goals (SDGs)

The following is a brief discussion of the social development scenario including the main objectives, the policy imperatives needed, the role of race and ethnicity in employment opportunities and barriers to social inclusion.

Agenda 2030 aims to leave no one behind on the road toward successful application of its targets. This demands a contextual pace of target implementation in society. It is important that the Agenda's goals and targets are achieved faster among the most disadvantaged social groups. Without quicker improvements among those who are lagging further behind, the systematic disparities described in the Report on the World Social Situation 2016 (UN 2016) will not decline. Though the data needed to monitor progress in all goals and targets for each group that is disadvantaged or at risk are not systematically available yet, the existing data explain the complexity of establishing whether some people are being left behind. Much depends on contexts and on the indicators used to assess progress. This section discuss recent trends in SDG Indicator 2.2.1, the prevalence of stunting (having a low height for age) among children under the age of 5 years in the ethnic groups in three developing countries (Ghana, Mali and Peru).

According to the UN (2016), Ghana has really improved in the case of child health over the past two decades. According to the data from the UN (2016) the state of children in three major ethnic groups that were much behind further in terms of stunting from 1998, improved potentially from 1998 to 2008. Stunting declined by 4.2% annually among these groups but only by 0.9% in total. After regular efforts, the same three ethnic groups acknowledged improvement between 2008 and 2014. In the case of Mali, stunting declined at a much slower rate among children in the three most backward and unprivileged ethnic groups than that among the rest of the population, meaning that children in these groups were among the worst cases in that period relatively. In Peru, there are tremendous inequalities in child healthcare scenarios. The cases of stunting were high, almost double among children in the poorest indigenous group (the Quechua people) than the children of Spanish-speaking households in both the years 2000 and 2012. Nevertheless, the stunting rate declined by more than 20% among Quechua and Aymara children due to increased international and government efforts like reversion of marginalization of communities in remote Andean regions with increased budgets for quality and coverage of healthcare services (Tam et al. 2016). Thus, on the basis of this indicator alone, development was inclusive of minority ethnic groups in Peru during this period (UN 2017).

3.6.1 Policy Imperatives for "Leaving No One Behind": Social Inclusion

With its dominant and central objective to leave no one behind, the historic and ambitious 2030 Agenda recognizes that development will only be sustainable if it is inclusive. Promoting inclusion is fundamental to achieving a sustainable future (UN 2017). No single set of policies or a blanket approach is applicable across all countries and in all contexts and dimensions to challenge exclusion and boost inclusion. Instead, governments should bring an intense equity focus to policymaking, keeping in mind the following imperatives.

- The first imperative is to settle on a universal approach to social policy, integrated by special or specific measures contextually.
- The second imperative suggests that policymakers must integrate broader social protection systems integrating all small indicators (UN 2017).
- Policies that are specifically designed to handle discrimination should be encouraged.
- There should be a more systematic flow of data to ensure that no one is left behind.
- Inclusive institutions should be promoted.
- Policy coherence should be emphasized by aligning macro-economic factors with social goals.

3.6.2 Barriers to Social Inclusion and Gender Equality

Social capital is the resource created when people come together out of a shared purpose or goal that goes beyond individual benefits (Putnam 2000). However, there is a long and challenging way ahead toward creation of strong social capital. It is a common feature of societies to focus on differences based on race, sex, gender, ethnic origin and other characteristics that have no relation to people's achievements or to overall wellbeing of the society. The Report on the World Social Situation 2016 argued that discrimination is one of the key drivers of social exclusion (UN 2017). The core problem in the world today is differential approach in societies. The report also suggests that such discrimination leads to social exclusion although the institutional barriers faced by disadvantageous groups are easy to detect. The informal barriers are frequently more deeply embedded in societies making them difficult to recognize.

Goal number five of Agenda 30 aims to remove gender inequality to bring in strong socio-economic and environmental progress. It is a widely accepted concept that gender equality along with empowerment of women is very important for the achievement of the SDGs and for the world's future. Also, the importance of women's roles in decision making regarding climate change has been widely recognized at the global level. The underpinning reasons for and results of unsustainable growth and gender inequality are deeply interlinked and rooted in the dominant economic models. It is important to remember that economic growth cannot take place without considering unpaid work.

A successful case study to be mentioned in this regard is the Australian National Women's Alliance ("the Alliances"). The Alliances is networks of issues and sector-based groups, representing over 180 women's organizations. Collectively, their roles are: (i) bringing together women's organizations and individuals from across Australia to share information, identify issues that affect them, engage women in decision making to make their voices heard and identify solutions, especially with women from marginalized and disadvantaged groups; and (ii) actively engaging with the Australian Government on policy issues to enable a more informed and fundamental dialogue between women and government that is representative in nature. The major action plan includes gender policy analysis and gender budgeting, sex- and age-disaggregated data, and intersectionality.

3.7 Sustainable Development Indicators

To implement actions to reach a goal, indicators are important. One single indicator for sustainable development as a goal and performance measure is not possible. It will fall short in collaborating all views, diversities, issues and actions. Also, it is critical to have subjective and value-laden weight addition for each respective case. Therefore, a sustainable development indicator system (SDIS) is a necessary platform to work on. Any SDIS should cover the following main features:

- Its construction should be interdisciplinary in approach including various aspects from philosophy to physics.
- It should be possible to quantify all the elements present in the SDIS. For example, physical stocks and flows to describe ecosystems, monetary units to describe ecosystems.
- These quantifications should have provision for conversion into units per capita or into units per land area to make the system more informative.
- SDIS is measured on different scales, times and aggregation levels.
- The SDIS should be scientifically acceptable and politically applicable.
- It should have simple and transparent submodels within it to provide accurate context. The quantitative should be accompanied by the qualitative.
- It should be an ongoing, participatory process among the groups of stakeholders involved. The main objective is to widen the circle toward QOL in worldview.

QOL has taken center stage in sustainable development with the underlying intention to improve the lives of human beings. There is a growing amount of information available on social and economic trends across nations. This information in the form of data is used to construct an emphasis on QOL as the essential ingredient of sustainable development. Example of such indices would be infant mortality rate, life expectancy, energy used per person, household size and the literacy rate for women.

One of the earliest examples of a QOL index is "quality life standard" (QLS) introduced in 1973. QOL was defined as a measure of world system performance. The QLS was computed by taking into account four indices derived from materials standard for living, crowding, food and pollution. Its value over time was calculated by model simulation. In 2000, the World Bank decided some goals for the period of 1990–2015 for various issues, such as enrolling all children in primary school and making progress toward gender equality (World Bank Group 2000). The last seven objectives were concerned with the environment, e.g. to reverse the loss of environment resources by 2015. Each of these goals has their own parameters and indicators ranging from the local to the global scale.

The other important aggregate indicator is the HDI, proposed in the 1990s as an alternative to GDP. HDI is a geometric mean of a life expectancy index, an education index and an income index, and thus takes into account health, education and standard of living aspects of QOL. There is another out-of-the-box attempt to include another form of capital other than economic/financial capital in the economic equation, which is known as the Genuine Savings Rate (GSR) for countries, and was proposed by the World Bank in 1997. The main idea behind it is that natural capital is destroyed for short-term gains and is not compensated by other forms of capital. It suggests that economic growth is sustainable when the total amounts of capitals are sustained.

Table 3.1 Sustainable development aggregate indicators.

Indicator	Rationale	Assumption
Genuine Savings Rate (GSR)	Economic growth is only sustainable only when the total amount of capital is sustained	Thus, various forms of capitals are substitutable
Index of Sustainable Economic Welfare (ISEW)	GDP is not an overall measure for wellbeing in global context because most affluent countries are already beyond optimum income	Undesirable and harmful activities like pollution abatement and car accidents should be subtracted from the official GDP figures
Ecological footprint (EF)	The state of the planet is directly linked to individuals	Environmental impact can be expressed in terms of equivalent space used in the activity, product or project
Sustainable Net Benefit Index (SNBI)	Alternative index and presentation of items present and used in its calculation	The welfare-related items are sorted into separate "uncancelled benefit" and "uncancelled cost" accounts

Another aggregate indicator is GPI, which later evolved into ISEW. The main assumption of ISEW is that GDP cannot be the overall measure of wellbeing because it takes many undesirable activities like pollution abatement or military exercise to be positive, although ISEW is considered as a hypothesis for which more considerations are required in weighing up different factors .

Another aggregate indicator is EF. EF is a more environmentally rooted indicator. It calculates environmental impact in terms of space used in the activity. It is severely criticized by scientists for its lack of scientific consistency and neglect of externalities other than negative environmental ones (Opschoor 2000). The calculation of both indexes involves the extraction from the national accounts of the various transactions deemed relevant to human wellbeing (Eckersley 1998).

Sustainable Net Benefit Index (SNBI) is quite similar to ISEW and GPI. In this system, welfare-related items are sorted into separate "uncancelled benefit" and "uncancelled cost accounts." The total of the uncancelled cost account is subtracted from the uncancelled benefit account to obtain the SNBI. This approach is advantageous as it is consistent with the concept of income and capital. It also allows comparison of benefits and cost of a growing macro-economy. Table 3.1 is a quick summary of the main sustainable development aggregate indicators.

3.8 Exploring Knowledge Systems for Sustainability

Knowledge can be defined as "a body of propositions that are adhered to, whether formally or informally, and are routinely used to claim truth" (Díaz et al. 2015, p.13). Knowledge systems comprise different actors, agents, methods and institutions that together become responsible for the overall process of production, transfer and implication of knowledge. Hence, knowledge is deeply interlinked with agents, operations, tools, methodologies, scientific components and institutions in becoming a constant evolving, dynamic entity. The knowledge-based resources (modern as well as inherited) and principles of governing these resources are the basis of any knowledge management system (Van Kerkhoff and Szlezák 2016).

If Western science-based knowledge is compared with traditional and local knowledge domains, then the latter provide many different ways of learning, which are inspired by nature and society through deep and steady observation framed by distinct worldviews with their respective sets of strengths and limitations. It is a known fact that the world is deeply rooted in industrialization; thus, the issue is more than recognizing local and traditional knowledge for sustainable development but also finding collaborative methods to include all this knowledge for sustainable development (Tengö et al. 2017).

The bridging of different knowledge systems is extremely important. This involves the conception of various multifaceted forms of knowledge exchange modes and then focusing on the exacting dimensions of the system like the actors, institutions and processes (Figure 3.5).

Figure 3.5 shows that the agents or actors, institutions and processes are the bedrock of the five steps that are essential for fruitful conduction of knowledge across various systems. "Mobilize" broadly means to clearly communicate knowledge in the simplest forms that can be shared among all with ease. "Translate" means synergies between the knowledge systems, basically to facilitate collective awareness of the shared knowledge. "Negotiate" means collective appraisal of the similarities and distinctiveness and the mutual conflicts throughout knowledge contributions. "Synthesize" means configuring a widely accepted notion of common knowledge that manages to exploit the virtues of each specific knowledge system rather than coming up with one cumulative knowledge system. It is important to stress the fact that knowledge should be in a user-friendly format for decision makers,

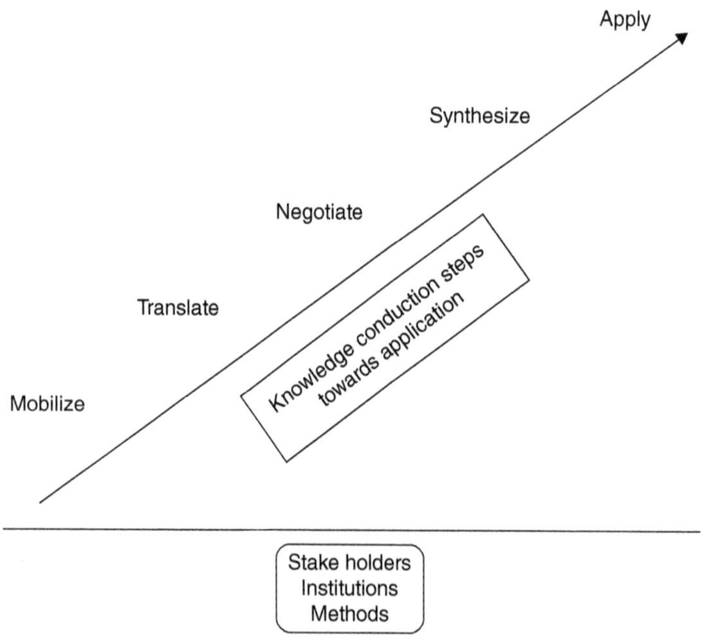

Figure 3.5 Five conduction steps for knowledge exchange in a system. *Source:* Tengö et al. 2017.

and all the agents and actors at different scales in space and time, with the provision of feedback into respective knowledge systems.

This concept is predominantly referred to in "Weaving knowledge systems in IPBES, CBD and beyond – lessons learned for sustainability" (Tengö et al. 2017). The framework explained in the study conducted by Tengö et al. (2017) is based on two examples from global science policy arenas: (i) the Intergovernmental Science-Policy Platform on Biodiversity and Ecosystem services (IPBES) - thematic assessment of pollinations, pollination and food productions and its piloting of bridging indigenous and local knowledge into assessments; and (ii) the CBD – the plan of action on customary sustainable use of biodiversity. This framework can have a universal approach for different arenas and hence is explained in a broader context here.

The very basic essentiality in achieving and facilitating this aim requires continuous generation and sharing of this concept in the form of knowledge. The underpinning stages involved in the generation and validation and distribution of knowledge also need a detailed review.

It is important to acknowledge the knowledge domain as more dynamic than science itself. Knowledge systems comprise not just the scientific facts but agents, methods and institutions, which in a combined fashion produce and transfer knowledge in a usable form. When knowledge systems are applied to the social goal of sustainability, it becomes an overall action plan for sustainable development (Van Kerkhof and Szlezák 2010). These various roles of technology, stakeholders and agencies are worth researching and exploring in the quest to achieve sustainable development.

It is also established that these interrelationships inside the knowledge system are dynamic in nature and undergo rapid changes. In these scenarios, which are fluid in nature, it is important that the role of science in connection to socio-ecological contexts is improved. This becomes the robust knowledge foundation that helps in decision making and actions to solve unsustainability.

3.8.1 Multiple Evidence-Based Approach for Creation of Knowledge Systems

According to Tengö et al. (2014) the multiple evidence base (MEB) is an approach that gives parallels whereby indigenous, local and scientific knowledge systems are viewed to generate different expressions of knowledge, which has the potential to generate new insights and innovations through complementarities (Tengö et al. 2014). The starting point is that the challenges of achieving sustainability require radical and deliberate changes in knowledge systems (Cornell et al. 2013). The term "knowledge democracy" highlights the relationship between science and the rest of society. Furthermore, sustainability scientists have a responsibility to collaborate openly with all actors (Cornell et al. 2013).

The bridging of these gaps needs intense collaboration between different science communities, local knowledge domains, social domains and all others with suitable contributions for engaging in solutions and operations to address the complex problems of sustainability. It requires an alertness and eagerness on the part of the science community to acknowledge the environmental, social and political nature of responding to global change in changing scenarios. It is also important to create fluidity in knowledge arenas where there is acknowledgment and a place for both scientific and traditional knowledge.

There is an urgent need for new skills, concepts and tools for the co-existence of multiple knowledge systems. Social learning is the most critical step between knowledge and action. Civil society, endogenous stakeholders, scientists, naturalists and policymakers should be able to acknowledge each other and their respective parts in achieving global sustainability.

There are research institutes across the world successfully operating with a hybrid character. Examples include ETH in Zurich, Barefoot College in Rajasthan and Lund University.

To add value, projects should emerge from open and interactive knowledge systems, where various actors co-produce knowledge, innovating and adding methods that are notable and compelling for themselves and all stakeholders in their own range of action.

3.8.2 Knowledge Systems for Sustainability in Future Scenarios

As mentioned earlier, sustainability needs multidimensional and collaborative action as it is an evolving concept that is a CAS. By CAS we mean that these systems have the capacity to learn and evolve from experiences. For example, the climate system is so complex and unique that comparative studies are not favorable options and controlled experiments are also difficult. Therefore, a predefined set of knowledge systems is not an option in the case of sustainability. Hence, there is a strong constant need to strengthen the current knowledge system in the context of sustainability. New technologies like lifecycle metrics, modeling land use and cover change, sustainable impact assessment and decision-support systems have emerged as efficient ways to address the SDGs. The increasing effectiveness of these tools in collaboration with experiences and the prior knowledge arena will lead to a future knowledge platform for sustainability.

To understand the above-mentioned fate of knowledge systems, let us take few examples. Lifecycle-based sustainability metrics is one such tool in the knowledge system that results in an environmental profile of the resource, measuring environmental performances for every material stage of its respective life. Lifecycle metrics and indicators will continue to evolve in the decades ahead and, in this process, will provide more exact meaning to sustainability.

Rapid developments in the field of artificial intelligence and software applications have given a fresh dimension to stakeholder engagements and interactive formulation through running games and models. These kinds of intersections have the potential to involve sciences, governance and management in coming up with unique knowledge systems. One such example is ComMod (companion modeling), which helps in building multi-agent stimulations. It is a knowledge bridge and mode of communication between scientists, policymakers and stakeholders.

Thus, it can be confidently predicted that future knowledge systems will be a mix and match of these tools with interesting twists and turns of experiences, indigenous knowledge and ecosystem intersections. A few of the above-mentioned technical tools to address sustainability are discussed in detail in further chapters.

3.9 Conclusion

Sustainable development is an open and dynamic concept with various possibilities for intersections among dimensions like socio-economic, political and environmental scenarios. There are many ways to define, achieve and measure sustainable development. The notions of strong and weak sustainability are important to consider because a blanket approach to attaining sustainable solutions is not possible. The circular economy has a main role to play in addressing sustainable development. It is important that emphasis is placed on the decisions that strike the most favorable balance among jobs, food production, water quality protection and the other services that nature provides. The concept of social justice is at the core of the SDGs. QOL has become a central issue in sustainable development, with underlying the intention to improve the lives of human beings. An SDIS is a necessary platform to work on. The very basic necessity for achieving and facilitating this aim is continuous generation and sharing of this concept in the form of knowledge. There is an urgent need for new skills, concepts and tools to enable the co-existence of multiple knowledge systems.

References

Aldred, J. (2009). *The Skeptical Economist: Revealing the Ethics inside Economics.* London: Earthscan.

Allen, C. and Clouth, S. (2012). *A Guidebook to the Green Economy. Issue 1: Green Economy, Green Growth and Low Carbon Development: History, Definitions and a Guide to Recent Publications,* UN Division for Sustainable Development. New York: Department of Economic and Social Affairs (UNDESA).

Atkisson, A. (2010). *Believing Cassandra: How to Be an Optimistic in a Pessimist's World.* London: Earthscan.

Ausubel, J.H. and Waggoner, P.E. (2008). Dematerialization: variety, caution, and persistence. *Proceedings of the National Academy of Sciences of the United States of America* 105: 12774–12779.

Barbier, E.B. and Markandya, A. (2012). *A New Blueprint for a Green Economy.* London: Routledge.

Beckerman, W. (1994). Sustainable development: is it a useful concept? *Environmental Values* 3 (3): 191–209.

Begg, D., Vernasca, G., Fischer, S., and Dornbusch, R. (2011). *Economics,* 10e. Maidenhead, UK: McGraw-Hill.

Bell, A.R., Osgood, D.E., Cook, B.I. et al. (2013). Palioclimatic histories improve access and sustainability in index insurance programs. *Global Environment Change A* 23: 774–781.

de Vries, B.J.M. (2013). *Sustainability Science,* 1e. Cambridge, UK: Cambridge University Press.

Brundtland Report (1987). *Our Common Future.* Report of the World Commission on Environment and Development. Oxford, UK: Oxford University Press.

Cole, M.A. (1999). Limits to growth, sustainable development Kuznets curves: an examination of the environmental impact of economic development. *Sustainable Development* 7: 87–97.

Common, M. and Stagl, S. (2005). *Ecological Economics: An Introduction*. Cambridge, UK: Cambridge University Press.

Cornell, S., Berkhout, F., Tuinstra, W. et al. (2013). Opening up knowledge systems for better responses to global environmental change. *Environmental Science & Policy* 2013 (28): 60–70.

Daly, H.E. (1990). Towards some operational principles of sustainable development. *Ecological Economics* 2: 1–6.

Davidson, C. (2000). Economic growth and the environment: alternatives to the limits paradigm. *Bioscience* 50: 433–440.

DeLong, D. (1996). Defining biodiversity. *Wildlife Society Bulletin (1973–2006)* 24 (4): 738–749.

Díaz, S., Demissew, S., Carabias, J. et al. (2015). The IPBES conceptual framework – connecting nature and people. *Current Opinion in Environmental Sustainability* 14: 1–16.

Dresner, S. (2008). *The Principles of Sustainability*. London: Earthscan.

Eckersley, R. (1998). Redefining progress: shaping the future to human needs. *Family Matters* (51): 6–12.

Ehrlich, P.R. and Holdren, J.P. (1971). Impact of population growth. *Science*. American Association for the Advancement of Science 171 (3977): 1212–1217.

Gaston, K.J. and Spicer, J.I. (2004). *Biodiversity: An Introduction*, 2e. Oxford, UK: Blackwell Publishing.

Gleick, J. (1987). *Chaos: The Making of a New Science*. Harrisonburg, VA: R.R. Donneley & Sons Company.

Holdren, J.P. and Ehrlich, P.R. (1974). Human population and the global environment. *American Scientist* 62 (3): 282–292.

Huberman, D. (2010). *A Guide Book for IUCN's Thematic Pragramme Area on Greening the World Economy (TPA5)*. Gland, Switzerland: IUCN.

Kates, R.W., Parris, T.M., and Leiserowitz, A.A. (2005). What is sustainable development? Goals, indicators, values and practice. *Environment* 47: 8–21.

Ladyman, J., Lambert, J., and Wiesner, K. (2013). What is a complex system? *European Journal for Philosophy of Science* 292: 641–642.

Malinga, R., Gordon, L., Jewitt, G., and Lindborg, R. (2015). Mapping ecosystem services across scales and continents – a review. *Ecosystem Services* 13: 57–63.

Manfred, A.M.-N., Elizalde, A., and Hopenhayn, M. (1989). Development and human needs. In: *Human Scale Development: Conception, Application and Further Reflections* (eds. M. Max-Neef and M. Hopenhayn), 18. New York: Apex.

Meadows, D. (1999). *Leverage Points: Places to Intervene in a System*. Hartland, VT: The Sustainability Institute.

Meadows, D.H., Meadows, D.D., Randers, J., and Behrens, W.W. (1972). *The Limits to Growth*. New York: Universe Books.

Meadows, D.H., Randers, J., and Meadows, D.L. (2004). *The Limits to Growth: The 30-Year Update*. London: Earthscan.

Millennium Ecosystem Assessment (2005). *Ecosystems and Human Well-Being: Biodiversity Synthesis*. Washington, DC: World Resources Institute.

Morrow, K. (2012). Rio+20, the green economy and re-orienting sustainable development. *Environmental Law Review* 14: 279–297.

Mulder, P. and Van Den Bergh, J.C. (2001). Evolutionary economic theories of sustainable development. *Growth and Change* 32: 110–134.

Newton, A.C. and Cantarello, E. (2014). *An Introduction to Green Economy*, 1e. New York: Routledge.

Opschoor, H. (2000). The ecological footprint: measuring rod or metaphor. *Ecological Economics* 32: 363–365.

Parkin, J. and Koorey, G. (2012). Network planning and infrastructure design. In: *Cycling and Sustainability* (ed. J. Parkin). Bingley, UK: Emerald Publishing.

Lawn, P.A. (2006). *Sustainable Development Indicators in Ecological Economics*. Cheltenham, UK: Edward Elgar Publishing.

Putnam, R.D. (2000). Bowling alone: America's declining social capital. In: *Culture and Politics* (eds. L. Crothers and C. Lockhart), 223–234. New York: Palgrave Macmillan.

Raworth, K. (2012). A safe and just space for humanity. Can we live within the doughnut? Oxfam Discussion Papers. Oxford, UK: Oxfam.

Redclift, M. (2005). Sustainable development (1987–2005): an oxymoron comes of age. Special issue:. *Critical Perspectives on Sustainable Development* 13 (4): 212–227.

Ruckelshaus, M., McKenzie, E., Tallis, H. et al. (2013). Notes from the field: lessons learned from using ecosystem service approaches to inform real world decisions. *Ecological Economics* 115: 11–21.

Schägner, J.P., Brander, L., Maes, J., and Hartje, V. (2013). Mapping ecosystem services' values: current practice and future prospects. *Ecosystem Services* 4: 33–46.

Science for Environment Policy (2015). Ecosystem Services and the Environment. In-depth Report 11 produced for the European Commission, DG Environment by the Science Communication Unit, UWE, Bristol, UK. http://ec.europa.eu/science-environment-policy (accessed November 2019).

Tam, Y., Huicho, L., Huayanay-Espinoza, C.A. et al. (2016). Remaining missed opportunities of child survival in Peru: modelling mortality impact of universal and equitable coverage of proven interventions. *BMC Public Health* 16: 1048. https://doi.org/10.1186/s12889-016-3668-7.

TEEB (2010). *The Economics of Ecosystems and Biodiversity Ecological and Economic Foundations* (ed. P. Kumar). London: Earthscan.

Tengö, M., Brondizio, E.S., Elmqvist, T. et al. (2014). Connecting diverse knowledge systems for enhanced ecosystem governance: the multiple evidence base approach. *Ambio* 43: 579–591.

Tengö, M., Hill, R., Malmer, P. et al. (2017). Weaving knowledge systems in IPBES, CBD and beyond: lessons learned for sustainability. *Current Opinion in Environmental Sustainability* 26: 17–25.

Turner, G. (2008). A comparison of *The Limits to Growth* with thirty years of reality: socio-economics and the environment in discussion (SEED). CSIRO Working Paper Series Number 2008–09. Canberra: CSIRO.

UNCTAD (2010). *The Green Economy: Trade and Sustainable Development Implications. Background Note Prepared by the UNCTAD Secretariat*. New York: United Nations.

United Nations (1992). UN Conference on Environment and Development, Rio de Janeiro (June 3–14). https://sustainabledevelopment.un.org/content/documents/Agenda21.pdf (accessed November 2019).

United Nations (2016). Report on the World Social Situation: Leaving No One Behind – The Imperative of Inclusive Development. New York: UN DESA. . https://www.un.org/development/desa/publications/report-of-the-world-social-situation-2016.html.

United Nations (2017). Analyzing Resource Efficiency Transitions in Asia and the Pacific Greening of Economic Growth Series.10 ST/ESCAP/2807. New York: UN ESCAP. https://www.unescap.org/publications/analysing-resource-efficiency-transitions-asia-and-pacific (accessed November 2019).

Van Kerkhoff, L. and Szlezák, N.A. (2010). Knowledge Systems for Sustainable Development Special Feature Sackler Colloquium: the role of innovative global institutions in linking knowledge and action. *Proceedings of the National Academy of Sciences of the United States of America* https://doi.org/10.1073/pnas.0900541107.

Van Kerkhoff, L. and Szlezák, N.A. (2016). The role of innovative global institutions in linking knowledge and action. *Proceedings of the National Academy of Sciences of the United States of America* 113: 4603–4608.

Vanags, J., Geipele, I., Mote, G. and Jirgena, H. (2012). Sustainable development: social dimension and limitation of material needs request. The 7th International Scientific Conference "Business and Management 2012," Vilnius, Lithuania (May 10–11), 768–774.

WBCSD (2008). *Sustainable Consumption Facts and Trends from a Business Perspective.* Geneva: World Business Council for Sustainable Development (WBCSD).

World Bank Group (2000). *A Better World for all - Progress towards the International Development Goals.* Washington, DC: World Bank.

4

Measurement of Sustainability

Rajneesh Singh[1], Akash Kumar Gupta[1], Puspendu Bhunia[1], Rao Y. Surampalli[2], Tian C. Zhang[2], Pengzhi Lin[3], and Yu Chen[3]

[1] *School of Infrastructure, Indian Institute of Technology, Bhubaneswar, Odisha, India*
[2] *Civil & Environmental Engineering Department, University of Nebraska-Lincoln, Omaha, NE, USA*
[3] *State Key Laboratory of Hydraulics and Mountain River Engineering, Sichuan University, Chengdu, China*

4.1 Introduction

Humans, being one of the dominant species on this planet, are continually influencing the fate of its ecosystem. Humans can engineer the ecosystem in several ways, e.g. extracting minerals, participating in the extinction of species, creation of new species by genetic interference, polluting and overburdening the world, etc. Furthermore, with the increase in population, the interference of humans has gone beyond the extraction of minerals and favoring beneficial species. Humans have now begun controlling the whole ecosystem in every possible manner, which further factors into the transformation of the entire biosphere, even at the fundamental level (Bell and Morse 2003; Kareiva et al. 2007; Kidd 1992). The population of humans, measured at around 1 billion in the late eighteenth century, had reached a mammoth 7.6 billion in 2017 and is slated to reach 11.184 billion in 2100 (Gupta 2018). The population explosion in the past 100 years has surpassed the total increase during the past 10 000 years by about seven times. It is astonishing that 5 million humans took 10 000 years to reach 1 billion, while 1 billion people took a mere 130 years to reach 7.6 billion. In the past century, between 1955 and 1975, the population growth of humans was observed to be up to 2.05% (Daily and Ehrlich 1996). The increased acceleration of inhabitants of an ecosystem has led to degradation in the environment and generates numerous socio-economic problems. The results are now visible in terms of loss of biodiversity, climate change, overexploitation of resources, socio-economic inequity and instabilities.

The above results are mostly due to the developmental activities carried out by humans to maintain their own sophisticated, luxury lifestyle. For example, to connect hilly villages, roads have been built by destroying the flora and fauna and obstructing the natural waterways. If we have a closer look at the above situation, we could conclude that now humans too are left with no choice but to alter their behavior. Thus, it is not astonishing to find that sustainability has become a common buzzword in academia and public discourse. Owing

Sustainability: Fundamentals and Applications, First Edition.
Edited by Rao Y. Surampalli, Tian C. Zhang, Manish Kumar Goyal, Satinder K. Brar, and R. D. Tyagi

to the above, sustainable development has been placed among the issues needing immediate attention (Daily and Ehrlich 1996; Goodland and Ledec 1987). However, in order to ensure sustainability, it has to be possible to measure it.

The most widely applied and reliable way of measuring sustainability is with sustainability indicators and indices (SIIs), which have primarily been chosen to ensure ultimate success while stepping toward sustainable development (Daily and Ehrlich 1996; Mihelcic et al. 2003). Initially, the measurement of sustainability was limited to environmental and ecological conditions (Goodland 1995); later it was extended to the economic and social contexts as well. However, remarkable development was achieved after the Earth Summit organized in Rio de Janeiro in 1992 (Neumayer 2000; Stockhammer et al. 1997). At this summit, a set of fundamental principles and a work plan were suggested for the prevalence of sustainable development. To be specific, the summit held in Rio stressed the need to develop indicators and indices for the measurement of sustainability. About 170 countries gave their consent for the above-proposed action plan. Subsequently, various non-governmental organizations (NGOs), legislating agencies, local communities and academic scholars also involved themselves in the planning and execution of the proposed work plan, aimed at gauging the trajectory of sustainable socio-economic growth and environmental conditions.

Nowadays, numerous indicators and indices have already been developed and are being applied in ensuring sustainable growth at global, national and local levels. For instance, according to a recent report from the United Nations Environment programme, the total number of sustainable goal indicators is 232 (Merino-Saum et al., 2018). However, due to disparities in views, there is no such methodology that is universally acceptable (Hardi and Zdan 1997; UNCED 1992).

The indicators can be defined in many ways, but there general consensus on the definition "an operational representation of an attribute (quality, characteristic, property) of the system," while data are actual measurements or observations of the values of indicators (Niemi and McDonald 2004). This chapter is an attempt to establish an understanding of the SIIs, which is imperative for the safeguarding of sustainability. The chapter specifically focuses on the measures for choosing an indicator to measure sustainability.

4.2 Types and Choice of Indicators

In order to apply a suitable indicator, the categorizing of the indicators is mandatory. The categorization of indicators also enables us to foresee the areas needing special attention and the remedies to the problems getting in the way of ensuring sustainability. The indicators are mostly in two categories: namely (i) traditional indicators, and (ii) sustainable indicators (Heink and Kowarik 2010). The traditional indicators are a measure of change from only a segment of a community (e.g. stockholder profits, asthma rates and water quality) and should be applied to the segments that are completely independent of/unaffected by interference from other communities/segments (Kidd 1992). In contrast, the sustainability indicators explain the dependency of a particular segment over the other segments, which are tightly interconnected (Figure 4.1). For further explanation, if we look closely at Figure 4.1, we can see that the available natural resources of our world, serving as a base for other ongoing happenings, are raw materials for industries that offer jobs. While the

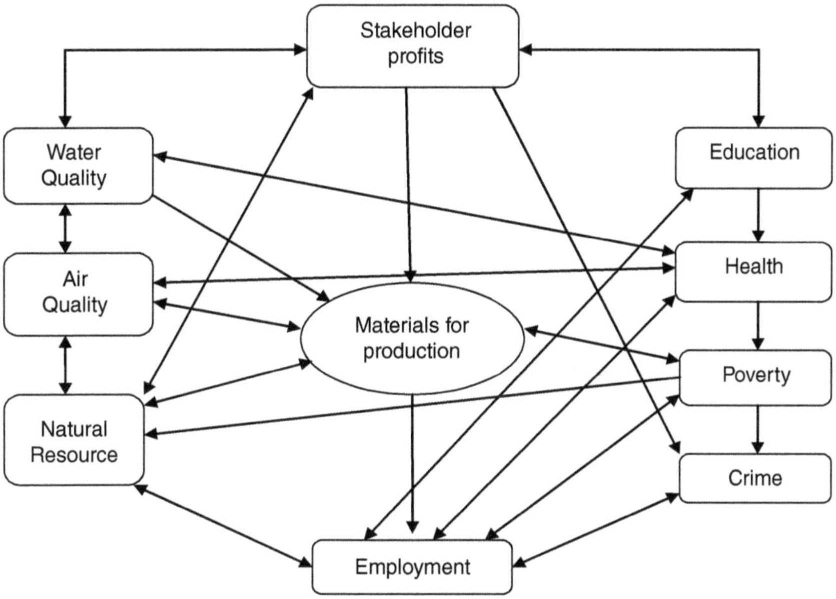

Figure 4.1 The interconnectivity among sustainable indicators.

jobs affect the poverty rate and contribute to the dynamism in stock markets, the other entities of Figure 4.1, such as air and water quality, which have a significant impact on employee health, and may affect their willingness to work and productivity. Medical expenses and cash crunches may also have an impact on the productivity of a company, which in turn adds dynamism in the stock markets.

Accordingly, sustainable indicators for the measurement of sustainability should have an integrated approach toward the multidimensional analysis of any condition. Otherwise, the application of traditional indicators in such a multidimensional analysis may not be beneficial. For example, despite being associated with the societal and ecological health of the country, gross domestic product (GDP) chiefly ensures only economic wellbeing (Labuschagne et al. 2005). There have been many instances when GDP has gone up despite worsening health conditions of the residents of the country. The above argument disqualifies GDP from being used as a sustainable indicator. In contrast, an indicator named the Index of Sustainable Economic Welfare (ISEW) gives an even clearer picture of the growth in the economy along with the impact of ongoing harmful activities. The ISEW is regarded as a correction to GDP as it considers the harmful impacts of the ongoing economic activity. For example, ISEW considers the damage caused by air pollution generated from the industrial activities contributing to GDP. It also calculates the consequences of the weakening of the ozone layer in the stratosphere.

After comparative analysis of both types of indicators (Table 4.1), it can be concluded that sustainable indicators could be of great use in establishing a healthy and vibrant community (Heink and Kowarik 2010). They monitor the health of economic, social and environmental scenarios of a community along with a parallel analysis of the ongoing negative trends. Sustainable indicators further give an impetus toward addressing the associated problems, ensuring an all-round sustainability.

Table 4.1 Traditional vs sustainable indicators.

Traditional indicators	Sustainability indicators	Emphasis of sustainability indicators
Economic indicators		
Per capita income against average US resident's income	Number of hours of paid employment at the average wage needed for fulfilling the fundamental requirements	What wage can buy defines basic needs in terms of sustainable consumption
Unemployment rate, the number of companies and the number of jobs	Variety and significance of local job base. Number and variability in the capacity companies. Number and variety of industries. Diversity in expertise needed for different unit operational chores	Reliance of the employment market and its fluidity in times of economic change
Economical enormity measured using GNP and GDP	Circulation of wages paid in the indigenous market and foreign investment in the indigenous economy for availing the natural renewable resources and workforce. Dependence of indigenous economy on the renewable indigenous resources	Local financial resilience
Environmental indicators		
Level of water and air pollution in the surroundings	Application of non-eco-friendly materials in the process of development. Scaling of pollution caused by the operation of transportation vehicles	Gauging of pollution caused by miscellaneous chores
Solid waste generated	Categorizing the output into durable, repairable or readily recyclable or compostable	Conservative and cyclical use of materials
Cost of fuel	Accounting of renewable and non-renewable energy used and their ratio	Sustainable resource usage
Social indicators		
SAT and other standardized test scores	Number of students trained for jobs available in indigenous economy. Students going to college and serving society	Matching job skills and training to meet the needs of the local economy
Count of registered voters	Count of voters participating in elections and community development-related meetings	Participation in democratic process, ability to participate in the democratic process

4.3 Framework of Lifecycle-Based Sustainability Metrics

A general framework has been proposed for sustainable lifecycle assessment. As shown in Figure 4.2, the first step should be the effective division of the product lifecycle into different stages (Dreyer et al. 2006; McCool and Stankey 2004). The division of the lifecycle into different stages facilitates a close examination of the lifecycle of a system. The second step involves the identification of parameters to be used in the lifecycle and corresponding modeling of the lifecycle of each stage. Next, from every stage, different sustainability

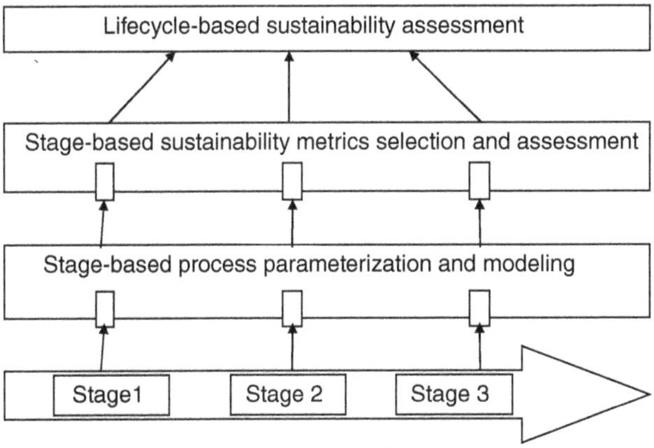

Figure 4.2 Steps involved in framing of the assessment of the lifecycle-based sustainability.

assessing indices are identified by assessing their performance in assessing the particular segment of the already divided lifecycle. Finally, lifecycle-based sustainability is assumed to be achieved on the basis of the characterization done at every stage.

4.4 Technological Aspects of Supply Chain and Process Sustainability

To begin with the prevalence of sustainability in the supply chains one need to set up guidelines, which should further be provided to suppliers and internal colleagues. If the suppliers comply with the guidelines, this will be enough to avoid any social or environmental harm. However, setting up the codes of conduct is a cumbersome duty, as it comprises meeting the expectations of suppliers as well as consumers. Owing to the above-mentioned, for most companies, the code of conduct is merely an extension of their guidelines for commercial activities, rather than focusing on minimizing social and environmental harm caused by the unsustainable supply chain (Dreyer et al. 2006).

In developing code of conduct for sustainable supply chain management, many international standards are consulted. However, setting up of a code of conduct for widespread application is not feasible, and could lead to a reduction in the required standards and incompatibility (Carter and Rogers 2008). In such conditions, companies are allowed to write their own codes, and the same may be followed by other companies working in the same dimension.

For the collective societal fundamentals, establishments may consult with the UN Declaration of Human Rights and International Labor Organization core conventions and recommendations, which may assist in meeting with the expectations of an even broader range of stakeholders. The above may address other issues pertaining to work, employment, social security, social policy and human rights. In addition, periodical review of the code of conduct is also mandatory. The codes on environmental concern may vary from industry to industry and may involve dialogue and active collaboration with other government and private agencies (Hussain et al. 2015). The key steps involved in developing a code of conduct for the sustainable supply chain management are: (i) having a dialogue with and consent from the stockholders; (ii) identifying the expectations of the suppliers on the basis of existing fundamentals and avoiding the forming of new standards (avoid undermining international law and regulations); (iii) consulting with the cross-functional teams, in particular the people who are involved in the chain; and (iv) address the requirements of suppliers and consumers.

4.5 Sustainable Economy Indices

GDP is the widely applied indicator for ensuring economic sustainability. It represents the monetary equivalent of the worth of the entire output and services offered within a definite interval. GDP is an indicator of the economic performance of a country. Being the primary accounting agenda and scale for gauging the economic status of a country, GDP serves as a dominant response instrument and drives the countrywide strategy (Bell and Morse 2003). However, GDP has limitations of its own, as it only emphasizes the final output and does not include sustainable conservation of the existing assets and resources (Clarkson 1995). GDP does not include the depletion of or degradation in natural or human-built resources and social capital on which the economy actually depends. Owing to the above, GDP nowadays is considered an outdated tool for measuring economic sustainability (Klein 2009). Therefore, new indices have been introduced that give a holistic view to address the growth of the community by including economic, societal and ecological aspects of the country. Currently, the economic indicators that are often applied and have a more sustainable approach are the ISEW and genuine progress indicator (GPI) (Klein 2009; Lawn 2003). The objective of these indicators is to display economic progress by including factors that affect the process of growth and the capacity to withstand it in the future.

4.5.1 Index of Sustainable Economic Welfare

This concept was coined by Herman Daly and John Cobb (Moldan et al. 2012). It was formulated by keeping in view the problems of internal steadiness and transparency of the structure. ISEW estimates economic wellbeing or at least its transformation over time by considering GDP, societal changes, damage to environment and income disparity. Calculation of ISEW comprises three steps (Moldan et al. 2012): (i) the consumption base, measuring the probable consumption against the actual extent of production; (ii) the estimation of subtractable items (defensive costs), which includes economic activities that do not necessarily contribute to economic welfare but are necessary for maintaining a standard of welfare; and (iii) the result, estimated by subtracting the subtractable items from the consumption base, is weighed through an index applied to take account of the income disparity and availability of labor. ISEW gives new dimensions to economic wellbeing as it combines different facets of society and the environment. It also points out the limitations associated with any project being evaluated on the basis of a single equivalent (economic) as it also includes more parameters based on sustainability for the measurement of welfare (Lawn 2003).

4.5.2 Genuine Progress Indicator (GPI)

Like the ISEW, GPI is an attempt to encapsulate environmental and social aspects of the economic welfare of any country. GPI is an improved gauging of wellbeing compared withGDP as it also associates itself with the ecological value of every production unit (Klein 2009). Comparatively GDP and GPI are analogous to gross profit and net profit, where the latter portrays a better picture of the growth of the country at a particular interval of time. Thus, GPI could be zero if poverty and environmental losses exceed the economic gains in production of goods or services. This really gives a clear picture of the sustainable welfare of any community.

4.6 Environmental Indicator for Manufacturing Competitiveness

Of late, the increasing consumer base and growing demand have led companies to consider much more their impact on the environment and society (Singh et al. 2009). Due to the above, manufacturers have now begun to look for devices or techniques to measure sustainability. However, despite having a good number of indicators, it is still very hard for the manufacturers to choose an indicator due to sustainability being a wide and relatively new topic (Feng and Joung 2009). Also, due to the numerous numbers of indicators, it is necessary to categorize them. To address the above, the National Institute of Science and Technology has categorized the existing indicators. After the categorization, a subcategorization of the indicators has also been performed. For effective categorization of indicators, the five dimensions of sustainability (environmental stewardship, economic growth, social wellbeing, technological advancement and performance management) were used and each of these was further broken down into subcategories (Figure 4.3). The subcategories are as follows.

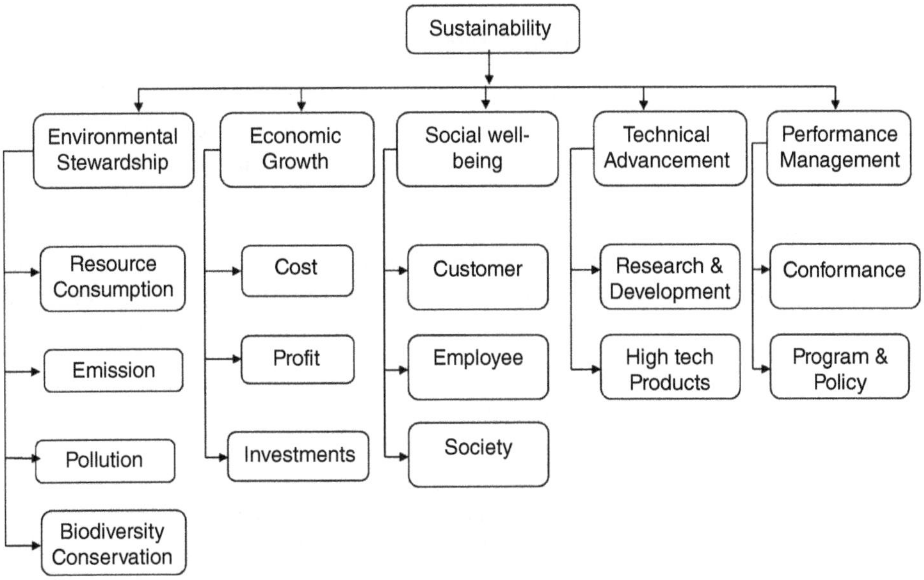

Figure 4.3 Categorization of indicators to induce manufacturer competitiveness.

4.6.1 Environmental Stewardship Indicators

The subcategories of environmental stewardship indicators include effect of emissions, exploitation of reserves, pollution and natural habitat preservation. This category contains the highest number of indicators, and it can be linked to the 2005 Environmental Sustainability Index (ESI), 2008 Environmental Performance Index (EPI), United Nations Commission on Sustainable Development (UN CSD), etc.

4.6.2 Economic Growth Indicators

Indicators to measure associated profit and costs a and eco-friendly investment can be considered for sustainable and economic manufacturing. These indicators also address investments, promoting monetary and community development of the society.

4.6.3 Social Wellbeing Indicators

Indicators addressing the overall wellbeing of employees, customers and the community living around the manufacturing plants are part of this category. The wellbeing of a community is reported to involve programs being run through corporate social responsibility (CSR).

4.6.4 Technological Advancement Indicators

These are made for assessing the new techniques and technology being applied and the dedication of R&D toward sustainable manufacturing. Examples include the purchase, incorporation and selling of high tech newly designed products and their patents.

4.6.5 Performance Management Indicators

This category of indicators is different as it is a collective and indirect measure of the other dimensions of sustainability. It has two components: conformance indicators being utilized in evaluating an organization in terms of its determination to meet the guidelines for sustainable manufacturing; and assessing the policies and programs of an organization to meet the requirements for sustainable manufacturing.

Due to this categorization, companies may now choose indicators and identify the focus areas needing special attention. In addition, a total of 11 indicators have been identified (Feng and Joung 2009) for sustainable manufacturing, which are listed below:

1) Global Report Initiative (GRI);
2) Dow Jones Sustainability Index (DJSI);
3) 2005 Environmental Sustainability Indicator (ESI);
4) Environment Performance Index (EPfI);
5) United Nations Indicators of Sustainable Development (UN CSD);
6) Organisation for Economic Co-operation and Development (OECD) Core Environmental Indicators (CEIs);
7) Ford Product Sustainability Index (Ford PSI);

8) International Organization of Standardization (ISO) Environmental Performance Evaluation (EPE) standard (ISO14031);
9) Environmental Pressure Indicator for European Union (EPrI);
10) Japan National Institute of Science and Technology Policy (NISTEP);
11) European Environmental Agency Core Set of Indicators (EEA CSI).

4.7 Monitoring and Evaluation Processes

Figure 4.4 shows the ten steps for monitoring and evaluation processes (Dreyer et al. 2006; Kusek and Rist, 2004). These steps are guidelines to ensure sustainability, and are discussed in the following.

4.7.1 Assessment of Readiness

Readiness assessment is done to check the willingness of the organization seeking sustainability and its capability to carry out the required modifications in its process or code of conduct. In general, it serves as a base for finalizing the framework. In conducting the readiness assessment, one is required to identify the driving need for the facilitation of monitoring and evaluation (Dreyer et al. 2006). In addition, taking help from the previously conducted readiness assessment may also be considered.

4.7.2 Agreeing on Outcomes to Monitor and Evaluate

In this step, we aim to address the significance of outcomes, related issues, the role of stakeholders, and setting up and finalizing the outcomes (Heink and Kowarik 2010). In attempting the above, strategic priorities should be used to attain the desired outcomes.

4.7.3 Selecting Key Performance Indicators to Monitor Outcomes

To analyze the success of our planned objective(s), the indicators are chosen. The selection of indicators is meant to carry out the qualitative and quantitative analysis of sustainability.

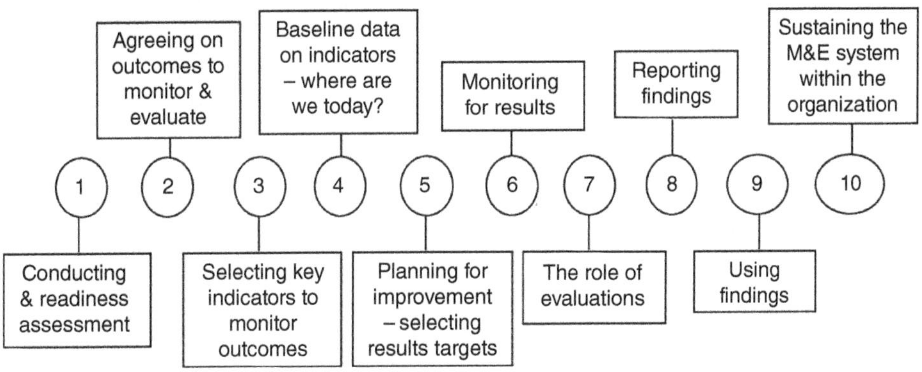

Figure 4.4 Steps involved in the monitoring and evaluation.

This step is among the most vital paramcters. The selected indicator(s) must be transparent, pertinent, economic, sufficient and monitorable. The step may even involve the construction of new indicators to meet the specific needs.

4.7.4 Baseline Data on Indicators – Where Are We Today?

Before the monitoring and evaluation begin, the planners are required to recognize and define the baseline, which will help in the comparison of the initial situation and that after progress. This would also be of help in setting up the models and conducting pilot-scale studies. The collection of data might increase if the number of indicators is high. This might involve identification of the source of data, data collection techniques, rate of data collection, a person to analyze and report the data, difficulty in data collection and recruitment of the person using the developed data.

4.7.5 Planning for Improvement – Selecting Results Targets

This step sets up the targets to be achieved and also defines the factors to consider while choosing indicators for ensuring sustainability. The above-mentioned planning also consists of considering the previous examples and setting up a framework for the measurement of overall performance. The factors to be kept in consideration are the existing capacity, budgets, personnel, funding and the existing organization policies.

4.7.6 Monitoring for Results

This step is performed to consider the types and level of monitoring and to identify the need for and principles of monitoring and analysis of the obtained data (Bell and Morse 2008). The monitoring is generally of two types: implementation of the code of conduct and tracking of the performance to analyze the effectiveness of means and strategies applied, while the monitoring of the result is carried out for performance assessment.

4.7.7 The Role of Evaluations

Up to this point, the steps were mostly centered around the monitoring of sustainability and no significant stress was placed on the evaluation of the process. In this step, the causes of change and identification of the basis of attribution will be assessed. The evaluation is done with the intention to incorporate the findings from earlier data to aid the decision-making process.

4.7.8 Reporting Findings

This step is performed to monitor and evaluate the obtained findings. It enables the dissemination of the findings to the right audience and assesses the after-effect of a bad performance. It is performed to evaluate, convince, educate, explore, investigate and also to document. This step involves the stakeholders' participation in order for them to better understand the process and thus lend their support to it. It is also utilized in preparing for a presentation of the data obtained to the authorities.

4.7.9 Using Findings

The reported findings will now be utilized in this step. It will also involve the strategies adopted for sharing the generated information and exploring the other benefits of the findings (feedback, knowledge and learning). This may also be utilized in the justification of new budgets or in the increment/reduction in raw material inflow.

4.7.10 Sustaining the Monitoring and Evaluation System Within the Organization

This step examines the importance of incentives and disincentives in monitoring and evaluation systems, as well as analyzing possible hurdles in attaining a sustainable result. It also examines the six critical components of any monitoring and evaluation system; namely, demand, transparent roles and tasks, dependable and trustworthy information, accountability, capability and encouragements.

4.8 Conclusion

This chapter is an attempt toward developing an understanding and a systematic approach to the gauging of sustainability. Moreover, the chapter conclusively deals with the steps involved in the monitoring and evaluation of sustainability. The chapter also discusses aspects such as the sustainability in a supply chain and among manufacturers. This chapter briefly describes the choice of suitable indicators suited to one's own need. The analysis of the indicators reveals that sustainable indicators are better than the traditional indicators. With the establishment and selection of suitable indicators, the sustainability of any societal, economic and environmental process can be evaluated and monitored.

References

Bell, S. and Morse, S. (2003). *Measuring Sustainability: Learning by Doing*. London: Earthscan.

Bell, S. and Morse, S. (2008). *Sustainability Indicators: Measuring the Immeasurable?* 2e. London: Earthscan.

Carter, C.R. and Rogers, D.S. (2008). A framework of sustainable supply chain management: moving toward new theory. *International Journal of Physical Distribution and Logistics Management* 38: 360–387.

Clarkson, M.B.E. (1995). A stakeholder framework for analyzing and evaluating corporate social performance. *The Academy of Management Review* 20: 42–56.

Daily, G.C. and Ehrlich, P.R. (1996). Socioeconomic equity, sustainability, and Earth's carrying capacity. *Ecological Applications* 6: 991–1001.

Dreyer, L.C., Hauschild, M., and Schierbeck, J. (2006). A framework for social life cycle impact assessment. *The International Journal of Life Cycle Assessment* 11: 88–97.

Feng, S. and Joung, C. (2009). An overview of proposed measurement infrastructure for sustainable manufacturing. Proceedings of the 7th Global Conference on Sustainable Engineering, Chennai, India (December 2–4).

Goodland, R. (1995). The concept of environmental sustainability. *Annual Review of Ecology and Systematics* 26: 1–24.

Goodland, R. and Ledec, G. (1987). Neoclassical economics and principles of sustainable development. *Ecological Modelling* 38: 19–46.

Gupta, J. (2018). The puzzle of the global commons or the tragedy of inequality: revisiting Hardin. *Journal of Environment: Science and Policy for Sustainable Development* 61: 16–25.

Hardi, P. and Zdan, T. (1997). *Assessing Sustainable Development: Principles in Practice.* Winnipeg, Canada: International Institute for Sustainable Development https://www.iisd.org/pdf/bellagio.pdf.

Heink, U. and Kowarik, I. (2010). What are indicators? On the definition of indicators in ecology and environmental planning. *Ecological Indicators* 10: 584–593.

Hussain, M., Khan, M., and Al-Aomar, R. (2015). A framework for supply chain sustainability in service industry with CFA. *Renewable Sustainable Energy Reviews* 55: 1301–1312.

Kareiva, P., Watts, S., McDonald, R., and Boucher, T. (2007). Domesticated nature: shaping landscapes and ecosystems for human welfare. *Science* 316: 1866–1869.

Kidd, C.V. (1992). The evolution of sustainability. *Journal of Agricultural and Environmental Ethics* 5: 1–26.

Klein, J. (2009). The rock builder. *Time* 173: 22–24.

Kusek, J.Z. and Rist, R.C. (2004). Ten steps to a results-based monitoring and evaluation. Washington, DC: World Bank. https://www.oecd.org/dac/peerreviews/World%20bank%202004%2010_Steps_to_a_Results_Based_ME_System.pdf.

Labuschagne, C., Brent, A.C., and van Erck, R.P.G. (2005). Assessing the sustainability performances of industries. *Journal of Cleaner Production* 13: 373–385.

Lawn, P.A. (2003). A theoretical foundation to support ISEW, GPI and other related other related indexes. *Ecological Economics* 44: 105–118.

McCool, S. and Stankey, G. (2004). Indicators of sustainability: challenges and opportunities at the interface of science and policy. *Environmental Management* 33: 294–305.

Merino-Saum, A., Baldi, M.G., Gunderson, I., and Oberle, B. (2018). Articulating natural resources and sustainable development goals through green economy indicators: A systematic analysis. *Resources, Conservation and Recycling* 139: 90–103.

Mihelcic, J., Crittenden, J., Small, M. et al. (2003). Sustainability science and engineering: the emergence of a new Metadiscipline. *Environmental Science and Technology* 37: 5314–5234.

Moldan, B., Janouskova, S., and Hak, T. (2012). How to understand and measure environmental sustainability: indicators and targets. *Ecological Indicators* 17: 4–13.

Neumayer, E. (2000). On the methodology of ISEW, GPI and related measures: some constructive suggestions and some doubt on the "threshold" hypothesis. *Ecological Economics* 34: 347–361.

Niemi, G.J. and McDonald, M.E. (2004). Application of ecological indicators. *Annual Review of Ecology, Evolution, and Systematics* 358: 89–111.

Singh, R., Murty, H., Gupta, S., and Dikshit, A. (2009). An overview of sustainability assessment methodologies. *Ecological Indicators* 9: 189–212.

Stockhammer, E., Hochreiter, H., Obermayr, B., and Steiner, K. (1997). Index of sustainable economical welfare (ISEW) as an alternative to GDP in measuring economic welfare. *Ecological Economics* 21: 19–34.

UNCED (1992). Rio Declaration on Environment and Development. Report of the United Nations Conference on Environment and Development (August 12), A/CONF.151/26. New York: UNCED.

5

Sustainable Impact Assessment

L.R. Kumar, Anita Talan and R.D. Tyagi

INRS, Université du Québec, Québec, Canada

5.1 Introduction

An increasing growth of the interest in "sustainability" or "sustainable development" over the past 20 years and the challenges perceived by scholars have led to the redesigning of the concept of "impact assessment" (IA). Designed in the later 1960s and early 1970s, the main focus of IA was on assessing the environmental influences of a proposed project. However, to incorporate sustainable development along with social and economic impacts of a project, the term "impact assessment" has been redefined in the past few decades (Bond and Pope 2012). Sustainable impact assessment (SIA) is a tool that can guide policy and decision makers in considering the needs of future generations while framing a policy or taking any decision (Devuyst 2001). The objective of SIA is to ensure that the "plans and activities" of any proposed project make a substantial contribution to sustainable development (Bond and Pope 2012). This chapter includes defining SIA, its advantages and contribution to society, how it's different from other types of IA, and how it is conducted and reviewed, followed by some case studies reported in the literature.

5.2 Types of Impact Assessment (IA)

5.2.1 Environment Impact Assessment (EIA)

EIA is a predictive tool that will manage the environmental impacts of any proposed project. Previously, the main concern of EIA was to solve the problem of clean drinking water (Bishop and Prosser 2001). It acts as a policy formulation and management tool for both planning and decision making. It was designed to be a preventive tool against soil and water pollution (Miles 2010). Different countries have different laws and regulations guiding EIA processes, and few developing countries do not have EIA as management tool. Recently, in some countries, EIA has proven to be more effective than other tools like

Sustainability: Fundamentals and Applications, First Edition.
Edited by Rao Y. Surampalli, Tian C. Zhang, Manish Kumar Goyal, Satinder K. Brar, and R. D. Tyagi

cost–benefit analysis and risk assessment (RA) (Saidi 2010). As a result of increased population and industrialization, many developing countries face challenges with land and water pollution and/or scarcity. If EIAs are properly conducted, management and allocation of land and water resources will be better, avoiding the possibilities of pollution and even scarcity. However, EIA is facing some of the following challenges in some developing countries.

- *Limited resources*: In developing countries, the major concerns in conducting EIA are limited resources and lack of technical knowledge. People who know nothing or very little are employed to conduct EIAs. This lack of expertise eventually leads to poor-quality reports that hardly contribute to the required assessment (Chanchitpricha and Bond 2013).
- *Data reliability*: Many developing countries have poor data collection methods; thus, the data available are often incomplete or unreliable. Absence of proper data collection and acquisition limits the use of EIA, which eventually leads to expensive data acquisition costs.
- *Ethical challenges*: The main aim of an EIA is protection of the environment from pollution/damage. Sometimes, this can be seen as a hindrance to socio-economic development. For instance, building a new hospital in a developing country is a good initiative. However, this could be considered as negative for the environment because it may impact the nearby lake and aquatic life residing in it. These situations make it difficult for policy- and decision makers to view EIA as a tool for sustainable development.

5.2.2 Strategic Environment Assessment (SEA)

EIA tends to focus on environmental protection, while SEA focuses on environmental objectives and policy alternatives and aims to achieve them (Wen-feng and Hills 2000). EIA follows a well-defined procedure with well-defined stages, including a "beginning" and an "end." EIA involves public participation at one stage which mostly takes the form of a public hearing. While SEA may have public participation at several stages in some cases, in other cases public participation is not even necessary. Private developers and environmental authorities are the main actors in EIA, although in some developed countries, there are only planning authorities not environmental authorities. In most of the countries, the EIA process is controlled by the Environment Ministry.

EIA predicts the negative impacts of a proposed project, whereas SEA does not predict impacts for strategic actions. It just focuses on effects, issues and implications. SEA is thus more concerned with the direction of the change rather than its magnitude. For example, while constructing a transportation infrastructure for a country, the increased accessibility of the rural population to health centers can be considered. However, it is difficult to predict the changes in mortality and morbidity rate as a result of construction. Conducting EIA requires a lot of detailing and predicts the outcomes with a great deal of precision. It focuses on a narrower range of issues, while SEA focuses on broader issues, broader territories, a broader range of stakeholders, but cannot analyze or predict impacts with the same level of precision (Table 5.1) (Wen-feng and Hills 2000).

The SIA concept has been developed from project-level assessments such as EIA, SEA and policy assessments such as regulatory impact assessment (RIA). SIA is closely

Table 5.1 Difference between environmental impact assessment (EIA) and strategic environment assessment (SEA).

Parameter	EIA	SEA
Process	Linear	Iterative
Use	Used in above 100 countries	Institutionalized in not many countries
Application	Early stages	Not fixed
Screening	Projects are listed	Decided case by case
Assessment process	Consolidation of technical expertise and local issues	Consolidation of political issues, stakeholder discussion and expert judgment
Public participation	General public is involved	Representative bodies are involved
Assessment	More quantitative	More qualitative
Quality review	Focus is on quality of information	Focus is on stakeholder process and quality of information
Decision making	Compares against norms and standards	Compares alternatives against policy objectives
Monitoring	Focusing on measuring actual impacts. Narrow but high level of detail	Focus on plan implementation. Broader but low level of detail

Source: Adapted from Wen-feng and Hills (2000).

related to EIA when applied to projects and SEA when applied to policies, plans and programs (PPPs).

EIA applied to project proposals has limitations; for example, it is applied at a late stage of decision making and achieves limited success in evaluating alternatives, while SEA has been used over the past decade as a tool for evaluating the impacts of decisions on the environment at much higher levels. However, there is no fixed approach for conducting SEA. There are several definitions of SEA based on different ideas and authors; SEA has been further classified into EIA-driven SEA and objective-led SEA. The EIA-driven SEA approach is a reactive, ex-post process (conducted after policy formulation) that aims to evaluate the environmental impacts of a policy, plan or program, identify potential modifications to improve the environmental outcomes and to evaluate the acceptability of the impacts. However, there is little scope for policy alternatives in this model as it is applied at a late stage of the decision-making process (Morrison-Saunders et al. 2014).

"Objective-led SEA" is a proactive, ex-ante process (conducted during the early stages of policy formulation) that is used during the early stages of policy formulation rather than evaluating them after the fact. A clear and well-defined set of environmental objectives is a prerequisite for objective-led SEA. The defined objectives must be consistent with different levels of decision making. Ideally, assessment carried out at higher levels of decision making should establish appropriate objectives at lower levels of decision making. However, the process is rarely streamlined in practice (Morrison-Saunders et al. 2014).

5.2.3 Health Impact Assessment (HIA)

According to the World Health Organization (WHO), HIA is defined as "a combination of procedures, methods and tools for analyzing effects of a policy, program, or project on the population health and distribution of these effects within a population" (WHO 1999). It is conducted at the policy formulation stage and takes into consideration the distribution of impacts on vulnerable sections in accordance with the recommendations made by WHO (1999). HIA may have several challenges, such as:

- the increased level of uncertainty and difficulty in predicting health outcomes (Huang 2012);
- tools are rarely enforced by the law;
- people outside the heath sector are not familiar with health concepts and literature (Hirono et al. 2017).

5.2.4 Risk Assessment (RA)

RA is used for evaluating the health effects of the exposure of individuals or populations to hazardous materials and situations on a factual basis (Mindell and Joffe 2003). RA is conducted during manufacturing (extraction, etc.) and after use (consumption, disposal, etc.). The approach is based on identification of basic characteristics and uses of products that may cause risks. RA's exclusive focus is on adverse effects of exposure to toxins (Hirono et al. 2017). EA involves several challenges as follows:

- exposure to animals is well-controlled, but there are not many studies on human exposure;
- the high degree of uncertainty surrounding the factual basis of RA leads to frequent disagreements within the expert community about interpreting and evaluating these data;
- RA is in technical terms but acceptability of risk is a political judgment (Huang 2012).

Table 5.2 compares different types of IAs reported in the literature. In general, every IA has its limitation that its focus is on evaluating impacts related to one particular aspect. There is a requirement for an integrated and holistic assessment including economic, environmental and social aspects.

5.3 Sustainable Impact Assessment (SIA)

The basis of sustainability assessment is attaining sustainable development. Sustainable development was first described by the Brundtland Commission in 1987 as "development that meets the needs of the present without compromising with the needs of future generations" (Laedre et al. 2015). The Brundtland Commission defined sustainability as per a two-pillar model reflecting environment and development concerns.

Since then, many alternative definitions and diverse interpretations have been made. Currently, sustainability is based on the concept of "three pillars" or the "triple bottom line" (TBL), which accounts for both social and economic factors in development, emphasizing that "material gains are not the only measures or preservers of human well-being" Tuomasjukka et al. 2017. SIA is still in an early and developing phase. Moreover, SIA does

Table 5.2 Comparison of different types of impact assessments.

Parameter	Health impact assessment	Environmental impact assessment	Strategic environment assessment	Risk assessment
Analysis level	Policy, program of project	Project	Policy, program or plan	Hazardous substance
Values	Democracy; equity; sustainable development (SD)	Sustainability	SD	Scientific rigor
Concern	Population health	Environmental impacts		Concerns about adverse effects of exposure
Place in policy formulation process	Policy formulation stage	Post policy formulation	Policy formulation stage	Both pre and post
Data used	Quantitative and qualitative	More quantitative	More qualitative	Quantitative; animal data and statistical exposition
Public involvement	Depending on type	Must	Wherever possible	Not important

Source: Mindell and Joffe (2003); Hirono et al. (2017); Huang (2012).

not have a common definition. As shown in Figure 5.1, the term SIA can be defined as "integrated assessment, which accounts for economic, environmental and social impacts of any project proposal" (Tuomasjukka et al. 2017).

However, some scholars argue that the term "integration" should be more than the sum of separate environmental, social and economic assessments. Laedre et al. (2015) remind us that "the principle that the sum of the individual parts does not equalize to the integrated

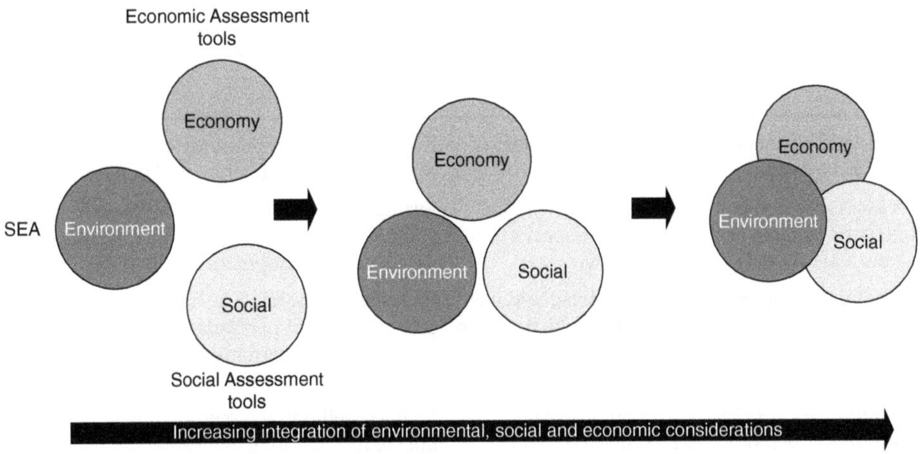

Figure 5.1 Integration of environmental, social and economic considerations. *Source:* Adapted from Morrison-Saunders et al. (2014).

assessment" and suggest that "integration means new relationships are established, which are relevant to individual entities that have specific characteristics but in combination act in a different way." The objective of integrated assessment is articulated by Post et al. (1998). Integrated assessment defines the term from the perspective of an identified problem or proposed project – the relations between the human communities, their economic organization and their actual resource base. SIA qualifies, quantifies and values the impacts of any project on the three (economic, social and natural) subsystems and their intersystem relations. SIA attempts to identify beneficial interventions and to fully expose unavoidable trade-offs.

Therefore, both EIA-driven and objectives-led SIA should not only consider the environmental, social and economic impacts of proposals, but should also examine the interrelations between these aspects (Morrison-Saunders et al. 2014).

SIA has the following characteristics (Laedre et al. 2015):

- *A participative assessment*: Politicians, civil servants, stakeholders and civil society can provide their inputs for policymaking.
- *An integrated assessment*: SIA broadens the traditional analysis into social, economic and environmental analyses of policies with the impact of each component (pillar) against the others being weighted.
- *Ex-ante process*: SIA is conducted during the early stages of policy formulation.

The common difference between SIA and other forms of IA is their objective or function: SIA is conducted in the early stages of the decision-making process. SIA considers different alternatives in order to strengthen positive outcomes of the proposed policy while alleviating the side-impacts of the policy, which eventually creates a positive overall impact of the policy. A distinctive characteristic of SIA is that it assess whether the policy contributes to sustainable development and does not assess based on the policy goal (Laedre et al. 2015). The differences between the main objectives of SIA and any other IA are highlighted in Table 5.3.

Table 5.3 Difference between sustainable impact assessment (SIA) and any other impact assessment (IA).

SIA	Any other IA
Evaluates a policy proposal during early stages of policy formulation	Evaluates a policy proposal after the policy formulation or just before decision is taken
SIA assesses whether a proposed policy will contribute to sustainable development or not; does not evaluate the policy against its own goals	Evaluates the policy proposal against its own goals; does not account for contributing to sustainable development or not
Promotes development of policy alternatives and supportive accompanying measures, which seek to emphasize and promote policy benefits while alleviating potential negative impacts	Identifies potential negative and positive consequences of particular policy; does not suggest policy alternatives
SIA follows well-defined methodologies and procedures; the procedure is transparent and creates a learning environment	Follows well-defined methodologies but does not create a learning environment

Source: Adapted from Laedre et al. (2015).

5.4 Advantages of Conducting Sustainable Impact Assessment

SIA is an important tool that assists in decision- or policymaking. SIA is not a substitute for any political decision or judgment as it does not provide clearcut outcomes or recommendations. However, SIA provides important input on impact, effects and consequences of the proposed project, and, thus, helps decision makers to take the correct decision for the sake of society, the economy and the environment. SIA is also considered as an effective and valuable communication tool through which policymakers bring in the important information and analysis to put in front of government heads. IA provides principles and minimum standards for policymaking (Liu et al. 2014). Advantages of conducting SIA are highlighted in Figure 5.2. Other advantages of conducting SIA are:

- *Better planning and implementation*: A well-analyzed project minimizes the negative environment and health-related impacts; it also helps to avoid costs associated with damage compensation.
- *Savings in foreseen cost*: SIA is a tool that helps in saving cost incurred at late project implementation. It employs an "anticipate and avoid" method instead of a "react and fix" method.
- *Legal and regulatory backing*: SIA has well-defined rules and guidelines. Involvement of stakeholders has helped to make SIA an effective management tool.

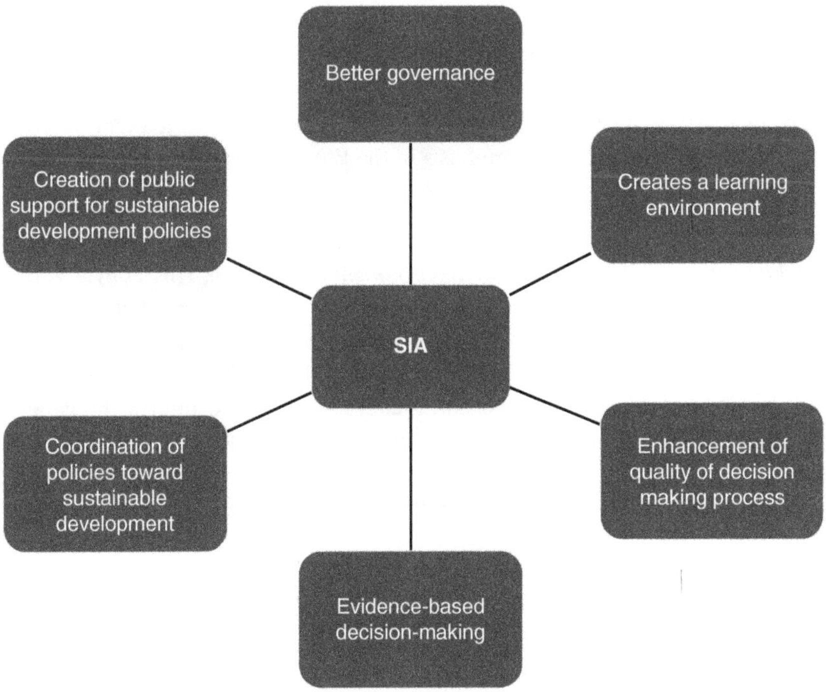

Figure 5.2 Advantages of sustainable impact assessment.

- *A learning environment*: Information is fed back into the political decision-making system creating a learning environment.
- *Transparency*: Stakeholder participation allows for greater transparency during the SIA conduction process and creates more sustainable and consensual policy solutions. SIA takes full account for accountability at different levels, procedures and methodologies used, and reasons for chosen mitigation options and solutions.
- *Long-term impact on society*: SIA not only considers the impacts on the environment, society and economics, but also interprets interrelations between the three spheres and considers both short-term and long-term impact on human life, ecology, the environment and economics, leading to sustainable development for society. *SIA uses a variety of tools and methodologies to capture less monetary aspects of sustainability.*

5.5 Recent Evolution of Impact Assessment

In the US the National Environment Policy Act (NEPA) was introduced in 1969–1970, which was the elaborated form of an environmental impact statement (EIS) affecting the quality of the environment. The regulatory panel of environmental quality applied PPPs in addition to projects. Thus, NEPA established both SEA and EIA. Compared to strategic-level SEA, project-level SEA was quickly developed and accepted worldwide. In the early years of EIA, attention was given to single project impacts but later on the limitations of EIA came to notice, such as EIA's inability to assess collective data or impacts or to consider a wide range of items under single consideration.

With all the limitations observed, more attention was given to the concept and procedure of SEA. Generally, SEA is around 15 years behind EIA in Europe and in other parts of the world. The European Union EIA Directive was accepted in 1985 but its SEA directive was approved very late (2001). The high importance of SEA has facilitated the rapid development of SEA globally, with a particular emphasis in the European region and other developed nations. In the European region three events enabled the rapid development of SEA:

- *The Aarhus Convention (Convention on Access to Information, Public Participation in Decision-making and Access to Justice in Environmental Matters)*: This convention is broadly adopted within Europe for the United Nations Economic Commission. The main aim of this convention is to give access to information, enable public involvement in decision making and to have justice in ecological matters. It is said that SEA must be followed to make it mandatory for the decision makers or parties to inform the general public about sustainability and its impacts for which decision making is necessary.
- *The European Union Directive*: The rules and regulations of this directive are only applied to specific actions and alternatives to set the framework for future projects of certain types only.
- *The International SEA Protocol*: This was an international agreement developed based on conventional EIA.

Global non-government organizations (NGOs) also play a role in developing effective SIA protocols with strong participation from the general public. Three main factors are required for an IA; namely, the limitations, the need to develop apolicy or project, and sustainability.

The procedures developed for sustainability brought many positive changes. Decision making evolves with time to achieve sustainable development of social, economic and ecological aspects. Developed as well as developing nations are concerned about SIA evolution for the betterment of society and the future. Various objects of every dimension are combined to give the cumulative effects for conducting SIA.

5.5.1 Integration of Different Impact Assessments for Meaningful Remarks and Conclusions

Integrated assessment is defined as a structured process to deal with all the complex issues using scientific knowledge from various aspects and/or stakeholder knowledge that is available to policy makers (Geneletti 2014). It is an iterative, sustained process where integrated understanding from the scientific and stakeholder community are communicated to the decision makers; the experiences and learning effects of decision makers form the inputs for scientific and social assessment. Although stakeholder participation is not a condition, non-scientific knowledge, social values and community preferences are essential to conduct integrated assessment. The engagement of these values is important to improve the quality of the integrated assessment procedure by allowing a broader range of options and perspectives in policymaking.

Integrated assessment follows two methods, and two main classes of integrated assessment are identified: (i) *analytical methods*, which are based on model analysis, scenario analysis and risk analysis; and (ii) *participatory methods*, which rely on approaches based on social sciences. In highly complex cases, both analytical and participatory methods are used for assessment. Rotmans and van Asselt (1999) differentiated two main types of existing models of integrated assessment: (i) *biosphere-oriented models*, which represent process-oriented interpretation of geophysical and biophysical processes and feedback do not represent the socio-economic system; and (ii) *macroeconomic-oriented models*, which represent the decision-analytic procedure of complex problems based on economic concepts and environmental dynamics. The main goal of the integrated assessment models is to combine scientific knowledge from various fields in an analytical framework. Complete integration of all the disciplines or areas is still a challenging research objective to be achieved.

5.6 Different Approaches of SIA

5.6.1 EIA-Driven Integrated Assessment

EIA-driven integrated assessment is a method for sustainability assessment that has originated from years of international experience with the traditional approach of conducting EIA. Similar to EIA, this method is promoter-driven and reactive, and is always instigated in response to planning and proclamation of a new proposal. As well as ecological impacts, its goal is also to identify the economic and social impacts of a project or proposal, and approaches to reduce potential negative impacts. During any policy- or decision making, all possible impacts are estimated with baseline conditions to determine the acceptance for the proposal. There is a potential for trade-offs among social, environmental and economic factors by bringing together the TBL impacts (Morrison-Saunders et al. 2014). To have sustainable development, EIA-driven integrated assessment always seeks to ensure that impacts should not have unacceptable negative effects. The acceptance for any new proposal or policy is based on the fact that it should not have a less sustainable outcome than the current status without the proposal. Previously, this approach was defined as "direction to target," where the exact position of a sustainable state for that particular proposal is unknown (Morrison-Saunders et al. 2014). In terms of proposal outcomes, it is presumed that if impacts occur as anticipated and reduction measures are implemented as planned for negative outcomes, then the limits to accept the proposal will be met and sustainability will be achieved for that proposal or policy.

5.6.2 Objectives-Led Integrated Assessment

The objective-led integrated assessment is based on objectives-led SEA, which basically works for policy analysis processes and is extended to include the three pillars of sustainability (Morrison-Saunders et al. 2014). It is for achieving well-defined social, economic and environmental objectives by estimating the extent of impacts when compared with baseline conditions by implementation of a proposal that directly or indirectly contributes to the objectives. It can be applied to both strategic-level and project-level proposals. Like objectives-led SEA, the objectives-led integrated assessment also aims to provide maximum contribution to attaining the objective of sustainability. This assessment is an integral part of the process to choose between alternative options to develop the best proposal to meet TBL objectives (Lederwasch and Mukheibir 2013). It is not a process for evaluating or refining the selected option after the development of the proposal. A positive approach to objectives-led integrated assessment requires approval of a broad set of social, economic and ecological objectives before developing the proposal. For a government-assisted assessment decision makers have to define the government objectives relevant to the proposal or policy developed. This condition develops a boundary within which the proposal upholder is able to develop the proposal to meet its own outcomes making a complete process that is government-driven as well as proponent-driven. However, this approach is only feasible for developing government projects, but not to be applied to private projects or policies. This method and decision making are aligned toward maximum achievement of TBL objectives. If the proper methodology is followed, then it is assumed that the project will attain sustainability defined by social, economic and environmental objectives in terms of contribution to sustainable development. Gibson in 2001 said that sustainability is about achieving positive change rather than minimizing negative effects; by this definition objective-led integrated assessment has more potential to contribute to sustainability than EIA-driven integrated assessment (Gibson 2001; Kardos 2012).

5.6.3 New Conception of Sustainability Assessment

To control natural processes and the scale of human activities, sustainability assessment is required to be integrated into urban planning. In this context, indicator-based sustainability assessment tools are key tools to provide information to support policy- and decision making. Indicators are needed to monitor the implementation of developed policies and to provide the feedback needed to achieve the accepted state(s) of sustainability. To conduct the assessment procedure the major problem is gathering reliable and accessible data.

Therefore, today sustainability problems are measured using indicators in scientific areas, government organizations and policymakers. Indicators are potent tools for evaluating the impacts of social, environmental and economic issues, and making policy for achieving sustainability. The choice of indicator depends on various factors indirectly affecting sustainability, which makes this selection subjective in nature (Agol et al. 2014).

To develop an effective indicator framework the measurement of impacts is one of the basic needs to develop the best decision or policy. Recently, decision makers and associated workforce developed sustainable development indicator (SDI) frameworks for a wide range of units. However, challenges persist owing to poor quality of data available or collected. Moreover, further studies are providing potential approaches to obtaining reliable and accessible data at different scales of items. Indicator-based assessment is important in policy- or decision making in providing information to researchers and practitioners, and important in SIA in taking action for sustainability (Kwofie et al. 2015). Indicators help decision makers to understand different ways to achieve an objective and suggest ways to think, categorize, measure and act. This framework provides focus, purpose, direction, clarity and attention as well as exposing the limitations of a policy development. The ecological, social and economic aspects of society are linked with each other; thus, instead of the "one problem, one indicator approach" the SDI's common indicator framework provides a platform for growing communities and their sustainability. Large numbers of tools are developed based on indicators to measure sustainable development, but this development is a very challenging job. Quantitative measurement of sustainability requires various forms of information including objectives, assessment criteria, indices, indicators and performance parameters. The first SIA framework developed was STRESS (Stress Response Environmental Statistical System) to guide ecological data analysis and indicator development. This framework was developed in the late 1970s by Statistics Canada based on ecological behavior, which distinguished pressure, state and response of ecosystem. Later, in the 1980s the PSR (Pressure-State-Response) model was developed based on STRESS by the Organisation for Economic Co-operation and Development (OECD) (Dizdaroglu 2017).

5.7 Determining Criteria for Sustainability

Criteria for sustainability are used to decide whether the environmental, social and economic goals or objectives of any policy/project have been met or not. Two types of approach have been defined for determining sustainability criteria: (i) the "bottom-up" approach in which objectives are defined in relation to baseline conditions; and (ii) the "top-down" approach to the development of assessment for sustainability criteria. It begins with the concept of sustainability as a state to which society aspires, and then moves on to define this state in terms of sustainability criteria (Morrison-Saunders et al. 2014).

The principles used for defining sustainability will clearly depend upon the prevailing conception of sustainability. In order to provide an example of principle-based criteria, Table 5.4 presents the sustainability principles that have been developed for Western Australia and the criteria for sustainability assessment that have been derived from the

Table 5.4 Principles and criteria for sustainability according to Western Australian Society.

Principles	Criteria
• *Equity and human rights*: Sustainability recognizes that environment needs to be created where all people can express their full potential and lead productive lives	• Increase access, equity and human rights
• *Biodiversity and ecological integrity*: Sustainability recognizes that all life has intrinsic value and is interconnected and biodiversity, ecological integrity is part of irreplaceable life support systems upon which earth depends	• Improves biodiversity and ecological integrity and builds life support systems
• Long-term economic health	• Provides both short- and long-term economic gain
• *Net benefit from development*: Sustainability means overall development including environmental, social and economic benefit for future generations	• Provides *net social–economic benefit*
• *Settlement efficiency and quality of life*: Sustainability recognizes need to reduce demand on material and wastes while simultaneously improving the quality of life (health, housing, employment, community)	• Improves the quality of life
• *Community, regions, "sense of place" and heritage*: Sustainability recognizes the significance and diversity of community and regions and the critical acceptance of "sense of place" and heritage	• Builds up community and regions, "sense of place" and heritage protection
• *Planning of common good*: Sustainability recognizes that planning for the common good requires equal distribution of public resources so that ecosystem balance is maintained and a shared resource is available to all	• Increases "common good" resources
• *Hope, vision, symbolic and iterative change*: Sustainability recognizes that applying these principles as part of broad strategic vision for earth can generate hope in the future, and thus it will involve symbolic change that is part of many successive steps in future	• Brings change and a *sense of hope for the future* as it is linked to a broader strategic vision
• *Precaution*: Sustainability requires serious or irreversible damage to environmental, economic or social capital, designing for surprise and managing for adaptation	• Ensures that there are *acceptable levels of risk* for the worst-case scenarios

Source: Adapted from Newman and Rowe (2003).

principles (Newman and Rowe 2003). Clearly the criteria listed in Table 5.4 are generic and not sufficient to define the basis of an assessment for the sustainability process.

5.8 Procedure to Follow Impact Assessment

IA is the process to determine the consequences of present and proposed policies in future. This process is used to ensure the sustainability of projects and policies. It is used to carry out projects and processes in a cost-effective way. Before a policy proposal the concerned department needs to carry out the IA. Similar steps are followed while planning IA but its purpose and orientation differs every time. The steps to be followed by the department responsible for SIA include:

- IA planning;
- impact analysis;
- discussion with affected stakeholders and public;
- collaboration with affected organizations;
- report presentation and summary of findings;
- forwarding results to decision-making team;
- IA report ready to publish.

All the analytical steps are part of the impact analysis, which is mainly related to problems such as its definition, policy objectives, policy options, impacts, comparison and recommendation, and its monitoring. The complete methodology to conduct SIA is highlighted in Figure 5.3.

5.8.1 Statement of Objective

The main objective of IA is to provide quantitative and qualitative decision parameters that will guide decision makers in taking decisions. The goal is to determine all positive and negative effects associated with a policy, project, investment or program to achieve competing objectives. IA is important to measure the impact magnitude. It is related to the importance that society places on the availability of resources for future generations and the importance of the impacts to stability and sustainability. To determine the significance of impacts on society, the environment and the economy is one of the most difficult aspect of IA procedure. Magnitude quantification is an objective and scientifically based process. So,

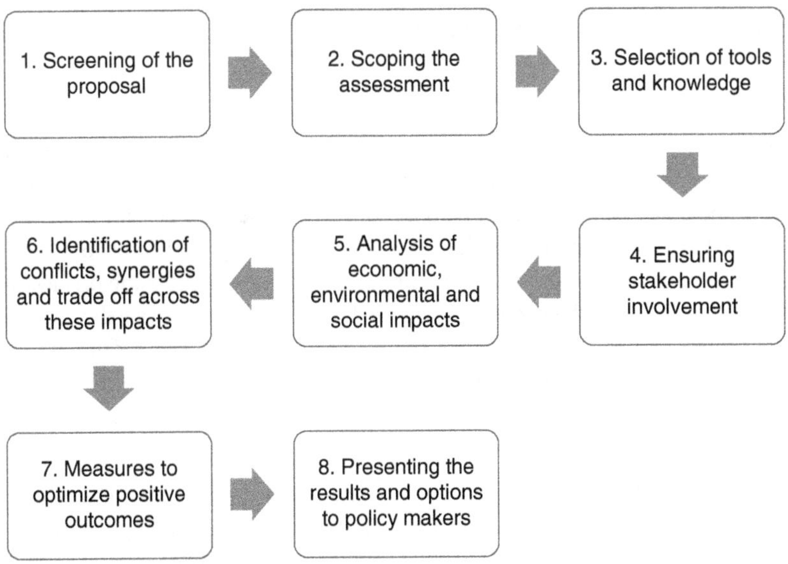

Figure 5.3 Sequential steps of conducting sustainable impact assessment (SIA).

final decision making is determined on the basis of magnitude of impacts. Decision makers consider all consequences of alternatives in making the final decision. Significant impacts are adjusted as per possible project features and policies. Factors such as area influenced, resources affected, persistence of impacts, resource sensitivity and regulatory status of resources are considered in determining the significance of impacts during IA procedure (Karvonen et al. 2017).

5.8.2 Possibilities for Achieving Objectives

To conduct the SIA procedure, accepted and alternative impacts have been considered and to measure the impacts is one of the important tasks of IA. Achieving IA is a difficult task considering all the parameters affecting sustainability. To achieve the objective there are different possibilities depending on the defined project and policies. An objective can only be achieved by making small changes at the social and economic level, community behavior, quality of life, community and people relationships, and enhancement in knowledge and skills of every domain of the society. These small changes are the driving force to attain global sustainability on the basis of generated reports of SIA. Impact indicators are equally important to determine the success rate of an element of the service to decide on collection of evidence and in choosing impact indicators. Once the evidence is collected, it is used to enhance service sustainability (Karvonen et al. 2017). During the achievement of the complete process, certain steps have to be taken:

- Encouraging participants with a clear and manageable focus that can be achieved with the available resources.
- Choices made based on research evidence.
- Aligning themes for assessment with broader objectives and priorities.
- Following approaches that are responsible for changes.
- Recognizing qualitative and quantitative characters.
- Making judgments based on baseline impact evidence.

5.8.3 Proposed Actions and Alternatives

The quality of an accepted and developed proposal always depends on the quality of the competing alternatives. The alternatives are all the options, choices or actions that could achieve the ends of the objective. From the SIA point of view, these ends are not just goals of an organization but also include the broader set of goals of society, economic and environmental quality. The most important action or alternative of assessment is the one available to fit the choice set index and the center of analysis. For alternative analysis the actions should be rigorous, objective and detailed, once the alternatives have been developed. In choosing the alternatives, the regulations provide explicit information and guidance on alternative selection, which even considers the option of a "no action" alternative. The "no action" alternative has two explanations:

1) *No change*: The condition is the same as the current condition or ongoing activity; e.g., the operations of a hydroelectric project under the terms of the existing license.
2) *No activity*: No action is taken or no project is started, e.g. the decision not to build a hydroelectric project.

Moreover, the "no action" alternative is considered to provide baseline conditions, which are used to evaluate and compare the feasibility of other alternatives. The "no action" alternative does not mean that there would be no impact. Available alternatives should ideally be considered as alternative approaches to achieving the objectives of the action rather than alternative designs only. An alternative approach is an altered method to accomplish the objectives being functionally different, while alternative design is a functionally similar method. For example, an alternative approach in road construction would be the extension of public transit, while an alternative design would be a different orientation of the highway (Steinemann 2001).

5.8.4 Environmental Characterization Report

Environmental sustainability is a global requirement that affects all developed and developing nations individually or collectively (Mekuriaw and Teffera 2013). Environment safety and its preservation have become one of the major issues worldwide. For the sake of the environment, governments have set certain environmental policies, laws and regulations. The administrative framework is designed to assess the environmental impact before accepting any development and investment activity in the nation (EPA 2012). The Environmental Protection Agency (EPA) has been asked to manage and predict impacts that may be caused by new investment activity, ongoing modification and activity termination to promote sustainability. Thus, a policy proposal is not complete without assessment of its impacts on the environment. Hence, an environmental characterization report is very necessary to assess the feasibility of a new policy to be followed within the nation to attain sustainability through SIA.

5.8.5 Identification of Impact and Analysis of Magnitude and Importance of Impact

Impacts or effects can be described in qualitative, quantitative and economic terms when valid estimates are possible. To attain sustainability, it is better to express all the impacts in economic terms, which makes it easier to compare different impacts as the units of expression are the same. However, not all impacts can be quantified in economic terms, and in this case the problem is to describe and quantify the impacts in their own terms. Before impact analysis, the impact needs to be identified as in some cases the level of unpredictability may be too high or too low to quantify precisely. Thus, in these cases the ranges of probable values should be given. During IA, the risk treatment and associated uncertainties with precise costs and benefits should be included. Sometimes in assessing the impacts one has to deal with collective uncertainties due to incomplete knowledge of processes or even of human behavior. There are some cases while performing IA procedure that are not quantified, such as cases involving ethical issues; in these cases qualitative assessment should be considered for policymaking.

Assessing impact magnitude is important to assess trade-offs, cost effectiveness, cost-benefit and multicriteria analysis (MCA). To determine the costs and benefits of a policy proposal or project the magnitude of an impact should be estimated. Analytical approaches and quantitative tools with process methods are important components for IA. Using quantitative variables derived from analytical tools to determine the impact magnitude would be of no use without the right methods or tools to bring IA to an action. Impact magnitude can only be achieved correctly by using methodologies of social sciences concerned with all aspects of social dimension, and better communication tools based on the social learning process to manage global links and sustainability. Impacts are important to improve the quality and clarity of the policymaking process. Their main aim is to contribute to an effective and efficient regulatory environment and coherent implementation of decisions for sustainable development. Impacts form the basis for the communication that explains the process of IA to government and integrates direct and indirect effects of a proposed measure.

5.8.6 Assessment of Impact

The SIA is an analytical tool to assess impacts that can help in decision-making process for projects, plans, policies and programs. To ensure an objective determination of impact significance, it is best to establish significance levels prior to determining impact magnitude. Predetermined significance criteria allow an objective comparison of alternatives and a determination of the anticipated effectiveness of mitigations. It is often useful to establish these criteria in consultation with regulatory or resource management agencies. All regulatory criteria that exist in the environment serve as points of reference against which impacts can be measured. Values of criteria exceeding standard limits are considered significant and impacts are not considered significant if no change has occurred. However, there are many projects and policies for which there are no regulatory criteria and, in such cases, establishing criteria for insignificant and significant impacts relies on professional judements that are well defined in IA. For each resource the criteria often need to be established separately. Some agencies of developed countries have certain significant criteria that are applied to all resource bases. For example, the criteria that are used by the US Nuclear Regulatory Commission. On the basis of impact significance, impact can be categorized into three categories: (i) *small* – these are those impacts whose effects are not detectable in the environment and are so small in size that they will not destabilize or put any pressure on resources; (ii) *moderate* – impacts are sufficient to create noticeable effects on resources but do not disturb resource stability; and (iii) *large* – these are those impacts that destabilize the resource completely, giving clear noticeable environmental effects.

5.8.7 Different Methodologies for Sustainable Impact Assessment

SIA can be divided into two categories based on the conception: process-oriented and outcome-oriented. Process-oriented SIA focuses more on values than on facts (Paredis et al. 2006). Therefore, SIA is conducted through participative mechanisms and non-monetary methods. In contrast, an outcome-oriented SIA will probably focus more on facts and therefore on quantitative modeling (Table 5.5).

Table 5.5 Different conceptions of sustainable impact assessment (SIA).

	SIA	
Parameter	**Outcome-oriented**	**Process-oriented**
Factual based	Quantitative modeling	Soft, qualitative models
Values based	Cost–benefit analysis	Non-monetary multicriteria methods, deliberate democracy

Source: Adapted from Paredis et al. (2006).

5.9 ToSIA – Software Tool for Sustainable Impact Assessment

ToSIA is a software tool that analyzes sustainability impacts in the forest-based sector. The forest-based sector provides significant amounts of income and employment to European citizens. It also plays a central role in developing a more sustainable economy as it utilizes a renewable resource. Many socio-economic factors and policies affecting the sector are currently changing at a rapid pace. Changes in the industry's competitiveness are contributed by fluctuating energy prices, policies aiming at a strongly increased share of renewable energy use and trade regulations. At the same time, societal demand for goods and services from forests is also increasing (Stenger et al. 2009). ToSIA is beneficial for policymakers, industry, other stakeholders, consultants and researchers. The tool has been developed in a European Commission-funded project (Päivinen 2011). ToSIA was applied in the scenario where forest resources were transformed into valuable products and services. ToSIA addresses three sustainability dimensions: environmental, economic and social. The approach for conducting SIA with (forest wood chains) FWCs starts from the forest and ends with end-of-life of wood products as depicted in Figure 5.4.

5.9.1 Modeling Forest Wood Chains

Modeling FWCs includes the following steps:

1) *The goal and scope defining*: The goal of an SIA study in FWCs is the identification of alternatives for conversion of forest wood to value-added products. The forest management alternatives can be tree plantation or natural regeneration. Moreover, the system boundary needs to be defined before calculations in terms of location.
2) *Performing material flow calculations*: ToSIA will calculate material flows within the system boundaries (described in the next subsection). The material entering into the system is defined as input and the material going out of the system is defined as output/export.
3) *Specifying the structures of the analyzed FWCs*: This means that all processes within FWCs and end products need to be well defined. The important components in an FWC are processes, products and product shares (Päivinen 2011). Examples of FWC processes

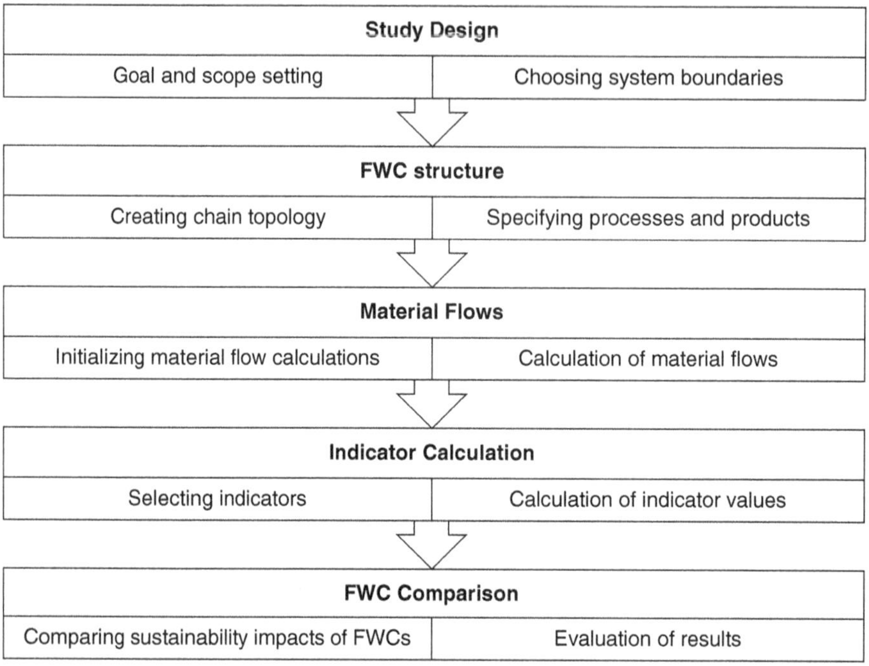

Figure 5.4 ToSIA work flow for conduction of sustainable impact assessment (SIA) for forest wood chains (FWCs).

are stand regeneration, harvesting, transport, sawing, pulping, paper-making, printing, etc. (discussed in the next subsection).

5.9.2 Material Flow Calculations

Material flows are calculated for both input and output streams at every process. Output material flows are calculated based on process efficiency. Material flows are captured in two measurement units: (i) organic carbon content within the wood (tons); and (ii) forest area (hectares) (Päivinen 2011). The parameters like forest area, the harvesting amounts or the number of products consumed are initialized in the software for material flow calculations. An example of material flow calculations is provided in Figure 5.5.

5.9.3 Sustainability Indicator Calculations

ToSIA can calculate any indicator that is linked to FWC processes via the flow amount. Relative indicator values are provided per unit of input material flow into a process. Sustainability indicators for the FWCs are production costs, employment, energy use, greenhouse gas (GHG) emissions, etc. as provided in Table 5.6 (Päivinen 2011). GHG emissions are used to quantify environmental impacts of products.

In ToSIA, sustainability indicator values for a process are calculated by multiplying the input material flow of the process with each of the processes' relative indicator values. After calculating the indicator values for processes, the values can be summed for the

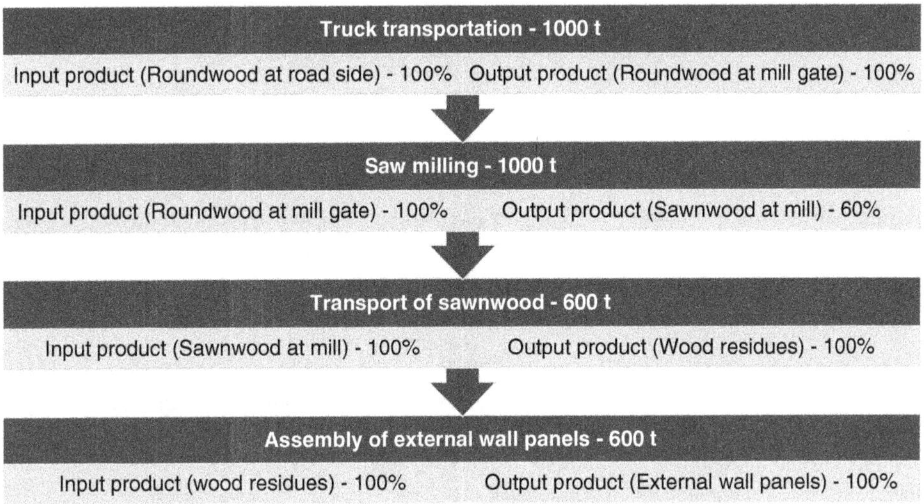

Figure 5.5 Material flow calculations for forest wood chains (FWCs).

Table 5.6 Indicators for sustainability in forest wood chains (FWCs).

Specific indicator used	Definition	Unit
Labor costs	Average labor costs	€
Resource use	Total amount of wood harvested	kg
Number of persons employed	Absolute numbers per city population	—
Non-fatal accidents	Absolute numbers per 1000 employees	—
Energy use	Fuel used directly in the process	MJ
Greenhouse gas emissions and carbon stock	Carbon stock in whole-tree biomass excluding stump and roots	CO_2 equiv.

Source: Adapted from Päivinen (2011).

whole FWC. For most of the indicators, the aggregation can be done by summing up the indicator values of individual processes. ToSIA also provides alternatives for a particular situation. For example, in this case, two forest management alternatives – natural regeneration (FM1) and planting (FM2) – have been assessed for conversion of forest woods to value-added products in terms of sustainability. The results of sustainability indication revealed FM1 provides more employment and income for the studied region. Moreover, FM1 also has positive impacts for climate protection because of larger carbon storage and slightly smaller energy use.

ToSIA provides a complete analysis of how new policies, changes in market conditions or new technologies will affect the sustainability of entire FWCs. As ToSIA uses a data-driven approach, the reliability of the sustainability assessment depends on both completeness of the data and assumptions related to the data. Future applications of the tool will contribute to identification of strategies that will improve the sustainability of the sector.

However, multicriteria and cost–benefit evaluation methods are currently under development and will soon be integrated into ToSIA (Päivinen 2011).

5.9.4 Other Tools for Evaluating Sustainability

Material flow analysis (MFA), indicator selection feature and simulation modeling are integrated into the ToSIA tool for FWCs. However, there are individual tools for assessing sustainability for different types of projects/policies. Table 5.7 discusses different tools for sustainability assessments with their orientation, advantages and disadvantages (Karvonen et al. 2017).

Table 5.7 Different tools for evaluation of sustainability with their advantages and disadvantages.

Tool	Main orientation	Advantages	Disadvantages
Cost–benefit analysis	Economic	Assess monetized impacts of a policy formulation	Sometimes values are subjective, e.g. we can't predict a valued for odor reduction
Input–output (IO) methods methods	Economic	IO analysis is reliable and well documented	Market needs to be well defined and known for higher certainty levels
			New products may cause uncertainty
Lifecycle analysis (LCA) methods	Environment	All process inputs and emissions of a product during lifecycle are considered during LCA	Datasets outdate fast and large datasets are difficult to handle
		Standardized methods are available	LCA demands large datasets
Material flow analysis (MFA)	Environment	Main focus is on material requirement for the production of a specific product	Helpful only when system boundaries are defined well
Multicriteria analysis (MCA)	Any	Number of approaches available	Unsustainable alternatives need to be excluded beforehand
		Enables thorough evaluation and balancing between alternatives with respect to indicators	Preferences have to be provided through questionnaire
			Stakeholders' lack of knowledge may corrupt the evaluation

Source: Adapted from Karvonen et al. (2017).

5.10 Case Studies for Use of Sustainable Impact Assessment

SIA is an accepted form of strategic impact assessment, and has been used worldwide. It is a well-accepted method of ex-ante policy assessment in the European countries. SIA procedure was used as an application to measure international trade policies and was first developed in 1999 by the Institute for Development Policy and Management (IDPM). Since then SIA has been accepted and used in conducting the agendas of international trade negotiations. It has been carried out for assessment of various social, environmental and other dimensions of sustainability. Two major case studies of SEA are studied, one is of international trade negotiations, and the other is for watershed programs on agro-ecosystems affecting social and environment sustainability, respectively. Studies have been done to compare agro-ecosystems with and without watershed programs to assess the sustainability impacts.

5.10.1 Sustainable Impact Assessment in International Trade Negotiations

Before SIA, strategic impact assessments were followed for trade agreements and were initially started by the controlling authorities of the US and Canada for the North American Free Trade Agreement (NAFTA). As per accepted policies, these assessments provide information on environmental consequences largely confined to their own countries (Kardos 2012). On the basis of these accepted assessment procedures, the policy framework has been developed for conducting SEA in international trade negotiations (Cooper 2002). This framework is being implemented in the Canadian government negotiation on the Free Trade Agreement of the Americas (FTAA). The SIA of trade agreements followed by the North American government aims to give complete understanding of impacts on environment across countries with all economic and social considerations. This complete information provided on environmental problems enables negotiators to make safer and more stable trading. The European Commission in 1999 adopted wider approaches to assessing the impact of trade agreements on social, economic and environmental sustainability in their own countries as well as integrated sustainability with other nations involved in trading. To know about the economic analysis of trade policy the concerned organizations have to hold discussions with stakeholders and affected general public. For outside nation impacts, the SIA may highlight affected areas and odds that arise due to external trading and corresponding external partners in the short- as well as the long term. SIA is required to check integrated sustainability that is highly dependent on trading policies. The European Commission has defined the objectives of its SIA studies (EC 2002) as a means of incorporating sustainability into European trade policy:

- By analyzing the problems of international trade negotiations with respect to sustainability.
- By giving complete information to negotiators on all possible consequences of trade policy.
- By giving guidelines to increase the commercial activity inside as well as outside nations to give positive environmental impact.

The European Commission determined that the main objective of SIA is not to assess the desirability of further liberalization. Instead, it is the implementation of the SIA that contributes to show the power of a nation to make liberalization a part of sustainable development and to maximize the positive impacts of liberalization on sustainability in the long term. As per various studies done on impact analysis of trade negotiations, the European Commission recognizes that the results of SIA must be integrated into international trade negotiations to achieve political and technical development of a country. For example, if IA indicates a "red light" the government will modify its negotiating position to improve the result. The authorities may publish its result, but decision making always carries a degree of confidentiality. The SIA studies for international trade negotiations done by every nation remain confidential, and only a few exact results are published with a broad sense so as not to reveal their own position to their trading partners. This confidentiality adds a complexity to the SIA procedure, and the analyses must be independent of a nation's own negotiating position.

There are complexities in integrating SIA into decision making. However, to a certain extent they are a smaller measure than they are for international trade negotiations. The SIA provides a broader global aim of sustainability by ensuring that negotiated agreements and policies are supportive in nature and are not destructive to the environment, social and economic sustainability. The country's main aim behind carrying out the SIA is its own development, but its main purpose is to support parallel trading partners to strengthen the role of world trading in attaining global sustainability. The SIA process of international trade negotiations also strengthens international relations by adjusting new rules and regulations and trade development by achieving developmental goals. Beyond this, SIA programs aim to frame broader international actions on trade and sustainability problems through the World Trade Organization, United Nations initiatives, and other national, international and NGOs working for the development of trade to push sustainability one step ahead.

5.10.2 Watershed Program Case Study

Concerns about soil degradation and poorly managed water resources have led to the development of watershed programs throughout the world (Firdaus et al. 2014). The growing number of watershed projects is an approach to rural development and management of natural resources. However, until now little has been known about their impacts. It is a step toward attaining sustainability of a country's land and water resources. Having behind them the authority of high political parties, these sustainable projects contribute to equity and quality of life, while the wellbeing and relationships of the rural community seem to be more unpredictable (Nasrabadi et al. 2013). Therefore, from every aspect research is needed to ensure agro-ecosystem sustainability by implementing new projects and policies. Research is needed to obtain spatially and temporally acceptable indicators to work on watershed programs for sustainable development (Veisi et al. 2012). Researchers have defined nine indicators that represent economic, social and ecological impacts of watershed programs (Veisi et al. 2012).

Here, we will talk about the watershed program in Iran (Firdaus et al. 2014). A causal relative method was used to study the sustainability impacts of watershed program at Gonbad chai watershed. This watershed benefits two villages – Gonbad chai and Tahonabad of Iran. Therefore, these two villages are included in this case study. Some other villages, Bagche, Sabz-abad and Gorgoz, were also included as they share the same watershed basin and have social characteristics that are relatively close to the main villages. To obtain the data, interviews were conducted using a set of questionnaires containing open and closed questions. To assess the impact, sample size was determined using a formula (Cochran 1977). In total, 69 samples were taken from an agro-ecosystem with watershed and 67 samples without watershed. The social impact was characterized using three social criteria (Firdaus et al. 2014): quality of life consisting of seven items; public services, with six items; and social capital and social structure, with six items. The environmental impact was characterized into five ecological criteria: crop management, with ten items; biodiversity, including five items; soil health, with nine items; hydrological processes, with two items; and energy, with five items.

The economic impact was taken as the marginal to cost–benefit ratio. The data weighting was done using conjoint analysis (Khoshbakht et al. 2012) and the data were standardized. T-test was used to determine the differences between the two sets agro-ecosystems with and without watershed. The impact analysis of the watershed program for the economic criteria of the cost–benefit ratio showed a significant difference, with $T = 7.05$; $P > 0.001$ between the two agro-ecosystems, having a value of $X = 65.34$ for the watershed agro-ecosystem and $X = 49.15$ for the one without a watershed. The *T*-test signified that the cost–benefit ratio was increased in an agro-ecosystem with a watershed program. These data were congruent with recent developmental approaches in rural Iran (Athari et al. 2017).

Impacts of a watershed program for the objects of social criteria include:

1) *Public services*: Equal access to public services is required to ensure social stability and to improve production while conserving natural resources. On the basis of survey and data analysis there was a significant difference between the agro-ecosystems with ($X = 47.87$) and without ($X = 38.53$) watershed ($T = 4.79$; $P > 0.001$; Khoshbakht et al. 2012). The agro-ecosystem with a watershed program showed increased access to public services between the villages and urban areas.
2) *Quality of life*: The *T*-test values ($T = 1.59$; $P > 0.05$) showed no differences in quality of life between the agro-ecosystems with ($X = 43.07$) and without ($X = 40.81$) a watershed program. This implies that people in villages with watershed programs believed there was an improvement in their quality of life (Khoshbakht et al. 2012). However, this result varies from place to place and from community to community (Nasrabadi et al. 2013).
3) *Social capital*: The comparative analysis of social capital showed a significant difference in the *T*-test results ($T = 2.65$; $P < 0.01$) between the agro-ecosystems with ($X = 72.69$) and without ($X = 78.75$) a watershed program (Khoshbakht et al. 2012). Based on the analysis it was clearly defined that social capital of the agro-ecosystem with a watershed program had declined significantly.

Impacts of a watershed program for the objects of environmental criteria include:

1) *Crop management:* Using a watershed program, there were improvements in farm management practices. This is an impact of increased knowledge and education of agriculture experts. The analysis showed significant difference ($T = 2.35$; $P < 0.05$) between the

agro-ecosystems with (X = 46.06) and without (X = 43.90) a watershed program (Khoshbakht et al. 2012).

2) *Biodiversity:* The analysis for crop management clearly defined that the watershed altered the use of land favoring crop production. The T-test value of T = 2.35 and $P<0.05$ showed a noticeable difference, which also confirmed the findings of other studies of Kuppannan and Devarajulu (2009).
3) *Soil health*: The study clearly showed that watershed management program implementation protects from soil degradation. It increases soil fertility and the organic matter content of the soil. The same finding was also seen in Kuppannan and Devarajulu (2009) and the results of both the studies agree.
4) *Hydrological processes*: Watershed projects are always implemented to promote water harvesting and conservation techniques (Sharma and Ruud 2003). With respect to hydrological processes, the analysis ($T = 5.10$; $P<0.001$) showed significant results with ($X = 54.27$) and without ($X = 44.13$) a watershed agro-ecosystem. Even this study confirmed the results of studies of Nasrabadi et al. (2013).
5) *Energy*: The agro-ecosystem with a watershed program showed improved energy indicators directly affecting yield. Increased energy indicators signify an increasing yield and lower energy requirement for water pumping systems (Khoshbakht et al. 2012).

An AMOEBA diagram (Astier et al. 2012) was used to achieve integration and synthesis of the results. The AMOEBA diagram fits for qualitative items to show the efficacy of an object for each indicator and provides a simple graphical comparison to depict advantages and limitations of watershed programs under evaluation. Comparative analysis showed that watershed program installation has some negative impacts on some terms of social criteria while having increased social capital and structure on the other side. This study also revealed that the project had a positive impact on public services, economic conditions, crop management practices, conservation of natural resources and conservation of biodiversity with water resources. This case study showed the benefits and impacts of watershed programs in an agro-ecosystem are significantly greater than those in a system without a watershed project. Based on this study, it can be judged that watershed programs raise the stability of an agro-ecosystem by improving environmentally sustainable development indicators. These projects enable improved quality of life and a better living for farmers.

5.11 Conclusions

1) SIA is an integrated assessment, which accounts for economic, environmental and social impacts of any project. It not only assesses the economic, environmental and social impacts, but also evaluates the interrelation between the three spheres. SIA takes account of monetary and non-monetary aspects of a project that can have long-term impacts on society. However, SIA is conducted in developed countries and is yet to be implemented in developing countries.
2) SIA is conducted at the early stages of policy formulation where stakeholders, public servants and civil society take part in it, and information is fed back into the political decision-making system, creating a learning environment for everyone. SIA considers

different alternatives in order to strengthen positive outcomes and mitigate the negative impacts of a proposed policy.

3) Although there are well-defined methodologies for conducting SIA, there is no single approach to conducting SIAs. An SIA can be conducted in two ways based on conception – process-oriented SIA, which is based on qualitative models (value based), and outcome-oriented SIA conducted through quantitative modeling (factual based).
4) Stakeholder participation is important in conducting SIAs. Stakeholders' lack of knowledge about the policy alternatives can corrupt the evaluation of a SIA.
5) As per Western Australian society, the criteria for sustainability are generic and not sufficient to form the basis of sustainability criteria. The indicators for sustainability depend specifically on the project: if the project is related to production of value-added products from existing resources, indicators are energy use, non-fatal accidents, employment given to people, production costs and GHG emission. For a watershed program, the main indicators are soil health, water conservation, energy use, quality of life and social capital.
6) There are various tools and software available for assessing SIA. ToSIA is one of them, and uses a data-driven approach where completeness and reliability of data are crucial for the whole analysis. All the uncertainties and assumptions related to data must be properly fed to the software. Since SIA is dependent on data acquisition and a large amount of data need to be fed for evaluation, highly qualified technical experts should be employed to conduct SIA successfully.
7) The software for assessing sustainability should have different features: MFA, product lifecycle analysis (LCA), cost–benefit evaluation, input–output calculations and MCA. A lot of individual software exists, such as SuperPro designer (for cost evaluation and LCA), but software has to be conceptualized in a user-friendly way including features like selection of sustainability indicators. ToSIA is a software program for SIA where simulation modeling, material flow calculations and selection of indicators are present, but MCA and cost–benefit evaluation features are under development and need to be incorporated into the software.
8) Public awareness and public participation are of the utmost importance in conducting SIA. It is important to spread awareness about SIA and local issues related to society and environment in developing countries to increase public participation in the decision-making process for successful implementation of SIA.

Acknowledgment

The authors would like to acknowledge Mr Tilak Raj (Alumni, Indian Institute of Technology, Delhi, India) for his valuable suggestions in writing the chapter.

References

Agol, D., Latawiec, A.E., and Strassburg, B.B. (2014). Evaluating impacts of development and conservation projects using sustainability indicators: opportunities and challenges. *Environmental Impact Assessment Review* 48: 1–9.

Astier, M., García-Barrios, L., Galván-Miyoshi, Y. et al. (2012). Assessing the sustainability of small farmer natural resource management systems. A critical analysis of the MESMIS program (1995–2010). *Ecology and Society* 17 (3): 1–20.

Athari, Z., Rad, G.P., Abbasi, E. et al. (2017). Designing a model for integrated watershed management in Iran. *Water Policy* 19 (6): 1143–1159.

Bishop, V. and Prosser, R. (2001). *Water Resources: Process and Management*. London: Collins Educational.

Bond, A. and Pope, J. (2012). *The State of the Art of Impact Assessment in 2012*. Abingdon, UK: Taylor & Francis.

Chanchitpricha, C. and Bond, A. (2013). Conceptualising the effectiveness of impact assessment processes. *Environmental Impact Assessment Review* 43: 65–72.

Cochran, W.G. (1977). *Sampling Techniques*, 3e. Chichester, UK: John Wiley & Sons.

Cooper, J. (2002). Evaluating public participation in the environmental assessment of trade negotiations. Report prepared for the Department of Foreign Affairs and International Trade, December.

Devuyst, D. (2001). Introduction to sustainability assessment at the local level. In: *How Green is the City* (eds. D. Devuyst, L. Hens and W. De Lannoy), 1–41. New York: Columbia University Press.

Dizdaroglu, D. (2017). The role of indicator-based sustainability assessment in policy and the decision-making process: a review and outlook. *Sustainability* 9 (6): 1018.

EC (2002). Sustainability Impact Assessment (SIA) Presentation of the Sustainability Impact Assessment (SIA) of the European Commission. Directorate General for Trade, European Commission, Brussels. http://europa.eu.int/comm/trade/issues/global/sia/index_en.htm (accessed November 2019).

EPA (2012). National Report of Ethiopia, the United Nations Conference on Sustainable Development (Rio+20). Federal Democratic Republic of Ethiopia, Addis Ababa.

Firdaus, R., Nakagoshi, N., and Idris, A. (2014). Sustainability assessment of humid tropical watershed: a case of Batang Merao watershed, Indonesia. *Procedia Environmental Sciences* 20: 722–731.

Geneletti, D. (2014). Integration of impact assessment types improves consideration of alternatives. *Impact Assessment and Project Appraisal* 32 (1): 17–18.

Gibson, R.B. (2001). *Specification of Sustainability-Based Environmental Assessment Decision Criteria and Implications for Determining "Significance" in Environmental Assessment*. Ottawa: Canadian Environmental Assessment Agency.

Hirono, K.T., Raab, K., Wendel, A. et al. (2017). The challenges and opportunities of peer review in health impact assessment. *Chronicles of Health Impact Assessment* 2 (1): 37–48.

Huang, Z. (2012). Health impact assessment in China: emergence, progress and challenges. *Environmental Impact Assessment Review* 32 (1): 45–49.

Kardos, M. (2012). The reflection of good governance in sustainable development strategies. *Procedia-Social and Behavioral Sciences* 58: 1166–1173.

Karvonen, J., Halder, P., Kangas, J., and Leskinen, P. (2017). Indicators and tools for assessing sustainability impacts of the forest bioeconomy. *Forest Ecosystems* 4 (1): 2.

Khoshbakht, K., Veisi, H. and Rezai, M.E. (2012). Sustainability impact assessment of watershed programs. *Proceedings of the International Conference on Applied Life Sciences*. London: InTech.

Kuppannan, P. and Devarajulu, S.K. (2009). Impacts of watershed development programmes: experiences and evidences from Tamil Nadu. *Agricultural Economics Research Review* 22: 387–396.

Kwofie, T.E., Adinyira, E., and Fugar, F. (2015). Nature of communication ineffectiveness inherent in the procurement systems on mass housing projects. *Journal of Construction Engineering* 2015: 914520.

Laedre, O., Haavaldsen, T., Bohne, R.A. et al. (2015). Determining sustainability impact assessment indicators. *Impact Assessment and Project Appraisal* 33 (2): 98–107.

Lederwasch, A. and Mukheibir, P. (2013). The triple bottom line and progress toward ecological sustainable development: Australia's coal mining industry as a case study. *Resources* 2 (1): 26–38.

Liu, L., Liu, J., and Zhang, Z. (2014). Environmental justice and sustainability impact assessment: in search of solutions to ethnic conflicts caused by coal mining in Inner Mongolia, China. *Sustainability* 6 (12): 8756–8774.

Mekuriaw, A. and Teffera, B. (2013). The role of environmental impact assessment for sustainable development. *Proceedings of the IAIA13 Conference Proceedings' Impact Assessment the Next Generation 33rd Annual Meeting of the International Association for Impact Assessment*, 13-16. https://www.researchgate.net/publication/317617739_The_role_of_Environmental_Impact_Assessment_for_sustainable_development (accessed November 2019).

Miles, T. (2010). Environmental impact assessment handbook – a practical guide for planners, developers and communities. *Proceedings of the Institution of Civil Engineers – Engineering Sustainability* 163 (3): 177.

Mindell, J. and Joffe, M. (2003). Health impact assessment in relation to other forms of impact assessment. *Journal of Public Health* 25 (2): 107–112.

Morrison-Saunders, A., Pope, J., Bond, A., and Retief, F. (2014). Towards sustainability assessment follow-up. *Environmental Impact Assessment Review* 45: 38–45.

Nasrabadi, A., Karami, E., and Ahmadvand, M. (2013). Determinants of participation in watershed development projects in Khorasan, Iran. *Journal of Agricultural Science and Technology* 15 (6): 1085–1094.

Newman, P. and Rowe, M. (2003). *Hope for the Future: the Western Australian State Sustainability Strategy*. Perth, Australia: Western Australian Government.

Päivinen, M.R., Lindner, M., Werhahn-Mees, W. et al. (2011). Conducting sustainability impact assessments of forestry-wood chains: examples of ToSIA applications. *European Journal of Forest Research* 131: 21–34.

Paredis, E., Heyerick, A., Doom, R., Bauler, T., Zaccai, E., Waktare, M., Bonifazi, A., Boulanger, P.-M., Lussis, B. and Varone, F. (2006). Methodology and feasibility of sustainability impact assessment. Case: federal policy processes. Brussels: Belgian Science Policy. https://www.academia.edu/993847/Methodology_and_feasibility_of_sustainability_impact_assessment._Case_federal_policy_processes (accessed November 2019).

Post, R.A., Kolhoff, A.J., and Velthuyse, B.J. (1998). Towards integration of assessments, with reference to integrated water management projects in third world countries. *Impact Assessment and Project Appraisal* 16 (1): 49–53.

Rotmans, J. and van Asselt, M. (1999). Integrated assessment modelling. In: *Climate Change: An Integrated Perspective* (eds. P. Martens and J. Rotmans), 239–275. London: Springer.

Saidi, T.A. (2010). Environmental impact assessment as a policy tool for integrating environmental concerns in development. AISA Policy Brief 19. Pretoria, South Africa: AISA. www.ai.org.za/wp-content/uploads/downloads/2011/11/No-19.-Environmental-Impact-Assessment-as-a-Policy-Tool-for-Integrating-Environmental-Concerns-in-development.pdf (accessed November 2019).

Sharma, S. and Ruud, A. (2003). On the path to sustainability: integrating social dimensions into the research and practice of environmental management. *Business Strategy and the Environment* 12 (4): 205–214.

Steinemann, A. (2001). Improving alternatives for environmental impact assessment. *Environmental Impact Assessment Review* 21 (1): 3–21.

Stenger, A., Harou, P., and Navrud, S. (2009). Valuing environmental goods and services derived from the forests. *Journal of Forest Economics* 15 (1): 1–14.

Tuomasjukka, D., Athanassiadis, D., and Vis, M. (2017). Threefold sustainability impact assessment method comparison for renewable energy value chains. *International Journal of Forest Engineering* 28 (2): 116–122.

Veisi, H., Khoshbakht, K., and Sabahi, H. (2012). A participatory assessment of sustainability of the agro-ecosystem of Absard, Iran. *International Journal of Agricultural Sustainability* 10 (4): 1–17.

Wen-feng, M. and Hills, P. (2000). Environmental impact assessment, strategic environmental assessment and sustainable development. *China Environmental Science* S1 http://en.cnki.com.cn/Article_en/CJFDTotal-ZGHJ2000S1034.htm.

World Health Organization (WHO) (1999). Health impact assessment: main concepts and suggested approach: a Gothenburg consensus paper. Conference proceedings. Geneva: WHO. www.impactsante.ch/pdf/HIA_Gothenburg_consensus_paper_1999 (accessed November 2019).

6

Analytical Tools and Methodologies to Evaluate Sustainable Development Goals of the United Nations with Special Reference to Asia

Gamini Herath

School of Business, Monash University, Malaysia

6.1 Introduction

Sustainable development (SD) is commonly defined as increasing human welfare in a way that meets the needs of the current generation without threatening the ability of future generations to meet their own needs (WCED 1987). SD has gained significance since the 1970s due to severe depletion of land, soils, water, biodiversity and climate change threatening the livelihoods of millions of poverty-stricken people around the world.

Deforestation and industrialization generated excessive greenhouse gases (GHGs) leading to climate change with frequent occurrence of extreme events (IPCC 2001; Meigh et al. 1998). Heavy logging in the Malaysian state of Sarawak led to a loss of around 30% of the forest area between 1962 and 1985. The heaviest logging of forest between 1981 and 1990 was in Southeast Asia. Expansion of industry increased GHGs contributing to climate change, which has affected the frequency and intensity of floods and droughts, and the sea level rises threatening people who rely on fisheries and aquaculture for their livelihoods. Climate change will be a serious threat to many small nations in Asia such as the Maldives.

Excessive abstraction of groundwater in India, China, Bangladesh and Pakistan reduced water quality and caused a drop in the water table (Shah et al. 2008). Chemical intensive agriculture has enhanced land and soil erosion and water contamination (Alauddin and Quiggin 2008; Herath 2012). Thus SD is critically important for the world to ensure food, water and energy security.

The aim of this chapter is to discuss selected methodologies and tools that are appropriate to understand and evaluate SD with special reference to the Sustainable Development Goals (SDGs) of the United Nations (UN) adopted in 2016. These tools would help policymakers to optimize the synergies among the goals and indicators. The tools are important because SDGs are interconnected and the environmental, social and economic goals must be integrated to enhance synergies.

Sustainability: Fundamentals and Applications, First Edition.
Edited by Rao Y. Surampalli, Tian C. Zhang, Manish Kumar Goyal, Satinder K. Brar, and R. D. Tyagi

6.2 Sustainable Development Goals of the United Nations

SD is based on the intersection of economic, environmental and social dimensions. The UN recently proposed the SDGs 2030 agenda with 17 goals as shown in Table 6.1. Some of these goals are extremely important, such as eradicating poverty (SDG 1), reducing income inequality (SDG 10) and mitigating climate change (SDG 13). These goals have a total of 169 targets, which provide a normative framework to catalyze action (Sachs et al. 2016).

Policymakers will face many headwinds in implementing the SDGs. Unreliable and non-standardized data, weak political will and institutional capacity have made rapid responses to the SDGs problematic. Governance problems, lack of technology and policy incoherence have become major obstacles (Herath 2017). The synergies must be maximized by adopting policies that could mutually support each other through correct coordination to realize the co-benefits. However, for most governments the challenge is the complexity of the SDGs and the lack of robust and easily usable tools for evaluation of SD (Jungcurt 2016). New approaches and tools are indispensable for planning and learning by policymakers to achieve policy coherence for SD.

6.3 Planning and Decision Making Methodologies for the Sustainable Development Goals

The Bertlemanns Stiftung along with the sustainable development solutions network (SDSN) jointly published the SDG Index to rank and track SDG achievements (Sachs et al. 2016).

Table 6.1 The Sustainable Development Goals.

Goal No.	Goal
1.	Eliminate poverty everywhere.
2.	End hunger, improve food security and nutrition.
3.	Promote health and well-being for all.
4.	Ensure equitable quality education and lifelong learning.
5.	Promote gender equality.
6.	Ensure sustainable management of water and sanitation for all.
7.	Ensure access to sustainable and modern energy for all.
8.	Promote sustainable economic growth, and productive employment.
9.	Build infrastructure, promote inclusive and sustainable industrialization.
10.	Reduce inequality within and among countries.
11.	Make cities and human settlements inclusive, safe, resilient and sustainable.
12.	Ensure sustainable consumption and production patterns.
13.	Take action to mitigate climate change and its impacts.
14.	Conserve and sustainably use the oceans, seas and marine resources.
15.	Protect, terrestrial ecosystems, combat desertification, and halt and reverse land degradation.
16.	Promote inclusive societies for sustainable development, provide access to justice for all.
17.	Strengthen global partnership for sustainable development.

Source: UN (2016).

The approach includes a total of 79 SDG indicators selected on the basis of data availability, i.e. at least for 80% of all countries with a population greater than 1 million. A total of 149 countries of the UN members were included in the SDG Index. The Index uses four steps: assessing normality; rescaling between 0 and 100 (higher value representing better performance) by defining the upper and lower bounds of the distribution; aggregating across the indicators for a given goal; and aggregating all the goals using equal weighting to obtain the overall SDG Index for a country. Equation (6.1) below shows how the min–max normalization is done by rescaling the raw data:

$$x^* = \{x - \text{lower}(x)\} / \{\text{upper}(x) - \text{lower}(x)\} \times 100 \tag{6.1}$$

where x is the value of the raw data. The upper and lower denotes bounds for best and worst performance. The SDG Index is developed by aggregating the rescaled variables, with first aggregation at the individual SDG level (Sachs et al. 2016). For each of the goals under the SDGs, the index was calculated as the arithmetic mean of the normalized values of all the priority indicators within the goal. In calculating the average, equal weights were assigned to each indicator whole number. The overall index was then calculated as the arithmetic mean of the goal scores across all the 17 SDGs using again equal weight for every goal score.

The SDG Index has been used to evaluate the performance of different countries. For example, Japan ranked 18th out of 149 countries with an index of 75 according to the 2016 global SDG Index (Sachs et al. 2016). Japan is already a high-level performer in SDG1, SDG3, SDG4, SDG6, SDG7, SDG9, SDG10 and SDG11 in absolute terms with scores over 80. Japan falls behind Sweden (the highest ranked country) for most of the SDGs. India is ranked 116 out of 157 nations in 2016 on a global index. Many of the richest countries performed poorly and were obstacles for the implementation process for poorer countries.

Wackernagel et al. (2017) argued that countries with a higher SDG Index had high ecological footprints per person, and the SDGs vastly underperform on sustainability. The developed countries achieved economic wellbeing involving exploitation of natural resources and carbon-intense industrialization. They concluded that the SDGs need to more robustly embrace resource constraints and climate change.

6.4 Tools to Tackle Interlinked Goals and Policy Incoherence

6.4.1 Network Mapping Technique

Nilsson et al. (2016a) developed a mapping approach to SDG interactions using a 7-point scale. Network mapping indicates an unequally knit network with some goals linked to many other goals, and others having fewer links with the rest of the goal set. Network mapping techniques are very useful for SDGs where interactions among variables are important. This technique emphasizes the linkages among goals because present policies of most governments often undermine the internal consistency necessary to maximize the synergies. This method creates a map of the relationship of one SDG to the other SDGs (Le Blanc 2015). For example SDG 1, SDG 8, SDG 10 and SDG12 are each linked to 12 other goals. In contrast, SDG 7, SDG 9 and SDG 14 are linked only to three other goals. Elucidation of such links helps minimize trade-offs and maximize synergies.

6.4.2 The Nexus Approach

SDGs require adaptations to issues such as climate change, which has received increased attention in recent years. Adaptation requires integrated effort to reduce trade-offs but enhances existing synergies and resource use efficiency. The nexus concept takes cognizance of the intrinsic relationships among some fundamental developmental variables. For example, in food production, elements such as water, energy and climate are intrinsically interrelated, which is relevant to the SDGs. The limited sectoral approaches adopted in the past pay less attention to interdependencies among various sectors. The nexus approach facilitates adaptation strategies because it identifies interdependencies leading to coherent policy that increases synergies and reduces trade-offs. The climate–food–water–energy nexus is popular and has been studied in detail (Rasul 2016; Houghton 2014). This nexus recognizes that the food, energy, water and climate are interrelated and puts pressure on ecosystems. These interdependencies can create trade-offs and synergies in moving toward the SDG 1 (reducing poverty), SDG 2 (eliminating hunger), SDG 6 (water and sanitation), SDG 13 (climate change) and SDG 4 (education). The nexus approach has the richness to accommodate and provide the balance necessary to manage the economic, environmental and social dimensions necessary for SD.

6.4.3 Scoring Approaches

There is an important relationship between science and policymaking for SDGs and they support each other in understanding the implementational issues of SDGs. In other words, science and policymaking must be integrated in decision making. This may occur at various decision-making levels but the basic principle is critical. The International Council of Science (ICS) developed a framework to use interactions among the SDGs and their targets using a scoring system (Nilsson et al. 2016a). The scores reflect adverse and positive relationships among goals. A perfectly supportive relationship exists where achieving one goal is linked to the gains in another (Table 6.2) (Nilsson et al. 2016a). The scoring system can be used to support decision making by helping policymakers to identify policy options that minimize negative interactions and enhance positive ones.

This scale focuses on the degree of interdependence between two goals or two targets, which can vary. When one goal reinforces the other, the scale is positive. The scale ranges from +3, when the goals or targets reinforce each other, to −3, which applies when goals and targets mutually exclude each other. This initial framework is commonly used and many SDGs have been evaluated (e.g. SDG 2, SDG 3, SDG 7 and SDG 14) (Table 6.2). The scoring approach facilitates systematic understanding of options. The positive scores with plus values represent opportunities for maximizing synergies, and the negative scores show trade-offs. However, users are cautioned due to a number of issues that can affect interdependence. Some of the main issues are irreversibilities (land conversion, species extinction), unidirectional interaction among the goals and uncertainty about the strength of the interaction. These are discussed in detail in Jungcurt (2016) and are not presented here.

Table 6.2 Scoring the influence of one Sustainable Development Goal or target on another.

Nature of Interaction	Explanation	Score
Indivisible	Inextricably linked to the achievement of another goal	+3
Reinforcing	Aids the achievement of another goal	+2
Enabling	Creates the conditions that further another goal	+1
Consistent	No significant positive or negative interactions	0
Constraining	Limit options on another goal	−1
Counteracting	Clashes with another goal	−2
Canceling	Makes it impossible to reach another goal	−3

Source: Nilsson et al. (2016a).

6.4.4 Stakeholders Forum Classification of Type and Nature of Sustainable Development Goal Interlinkages

The stakeholder approach developed by Newcastle University, Australia elaborates the interlinkages by defining eight types of interactions grouped into three categories (Coopman et al. 2016): supporting, enabling/disenabling and relying. A score for any interlinkage indicates the strength of the linkage. A score of 0 as shown in Table 6.3 shows negative or no linkage, and a score of 1 – 3 shows an increasingly positive interlinkage (Jungcurt 2016). Jungcurt (2016) showed that the linkage of SDG 12 with other goals and indicators is medium or strong. He also identified the missing linkages between targets 12 and targets 11 on reducing inequalities where linkages can support better integration of policymaking.

Table 6.3 Classification of type and nature of Sustainable Development Goal interlinkages.

Category	Type and definition	Score
Supporting	Commonly supporting: both targets contribute to the same objective	1
	Mutually supporting: target A's objective is achieved by target B's means of implementation and vice versa	2
Enabling	Disenabling: implementing target B may hinder or reverse target A	0
	Indirect enabling: implementing target B indirectly enables achievement of target A	1
	Direct enabling in both directions: implementing target B indirectly enables achievement of target A and implementing target A directly enables achievement of target B	3
Relying	Partial reliance: target B is a subcategory of target A	1
	Full reliance: target B's implementation is necessary for, but not intrinsic to target A's achievement	2

Source: Jungcurt (2016), Coopman et al. (2016).

6.4.5 Integrated Sustainable Development Goals (iSDGs) Planning Model

The iSDG is a simulation tool to clarify interconnections among the SDGs and their targets. The model simulates the general trends and development that affect SDG implementation. Here they analyze the effects of policies after implementation. The Millennium Institute developed the planning model (iSDG) to simulate trends toward achieving the SDGs and explore potential impacts of alternative scenarios. iSDG provides a dashboard that displays the performance of all 17 SDGs, which policymakers can use to test the impacts of policies before they are designed or the effects of policies after implementation. After applying an intervention, users can see how the country's progress toward that goal, and all other goals, changes. This interface allows users to visualize how combinations of interventions can lead to positive synergies. The model offers a range of interventions and sees how these contribute toward the other SDGs. This model can assess the synergies and trade-offs and can minimize the dissonance. The model covers all 17 goals and 79 indicators (Jungcurt 2016).

6.5 Carrying Capacity (CC)

6.5.1 Social Carrying Capacity (SCC)

The CC concept highlights that higher intensity of use of resources can lead to adverse consequences (Prato 2001). Peng et al. (2005) shows that in his population model, CC is depicted as a logistic equation (Wolf 1984). The predator–prey model is given in differential Eq. (6.2):

$$dN / dt = rN(K - N / K), \tag{6.2}$$

where N is population size; t is time; r is the rate of population growth per capita and K is the upper asymptote or maximal value of N, which is the CC.

The SCC conceptualizes the level of use for various activities. In tourism, it is an area that can withstand impact while providing a sustained quality of the experience (Wagar 1964). SCC is based on a subjective decision-making tool incorporating real impacts and impacts perceived by tourists and the general population. Saveriades (2000) formulates the SCC as the maximum level of use that is acceptable without undue damage to the quality of the visitors' experience and impact on the residents. Shelby and Heberlain (1984) developed an SCC model that focuses on alarm signals indicating the necessity to reduce the negative impacts of tourism.

The SCC was estimated using 6-point Likert scale statements in 2014 with KILIM Geopark in Malaysia as their main tourism destination using seven variables (Kayat and Radzi 2012). Using a self-administered questionnaire survey, information resources, visitation data and purpose of visit, etc. were evaluated. The perception was evaluated using safety, hospitality and tourism facility information (Kayat and Radzi 2012). The method initially involves examination of the relationship between variables, and then studies the relationship between the key factors and the selected independent variables (age, gender, nationality, education level, occupation and types of visitors). Correlation analysis presents 16 weak relationships and six medium strengths relationships, with all relationships significant at the 0.01 level (two-tailed).

Cumberbatch and Moses (2011) investigated SCC in beach management in Barbados using information from locals and tourists from beaches during the period of 2005–2009.

The findings show distinct demographic use patterns with respect to time of use, activities and amenity preferences. It also shows that the current beach management plans failed to recognize the diversity of the users, which may lead to conflicts.

The SCC is socially determined, rather than biologically fixed (Seidel and Tisdell 1999). Tourism CC is the maximum number of visitors (day visitors and tourists) that can be contained in a tourist area (abundance CC). SCC is a dynamic indicator and can be affected by many factors that are specific to the particular destination.

6.5.2 Limits of Acceptable Change (LAC)

A modification of the SCC is the LAC planning and management decision-making tool based on a new understanding of CC (Lindberg et al. 1997). LAC can also be used to assess sustainability because it identifies environmental quality indicators and tourism impacts. This information helps to identify thresholds that the conservation goals should meet. The model involves nine steps given in Table 6.4.

LAC assesses the impact of an activity on the destination to determine the change that will be tolerated and what actions need to be implemented to ensure that limits are not violated. LAC is a revised model of CC within an integrative framework (Stankey et al. 1985). LAC identifies in general how much of the area can be used and how much change can be allowed (Stankey et al. 1985). LAC emphasizes that use levels should be kept within acceptable ecological and social levels to limit adverse impacts for a given area (Prosser 1986).

LAC is useful to prevent damage to wilderness areas because wilderness conditions require better management. Many studies of wilderness show that recreation-related uses are the major sources of damage to natural areas, which can be social or biophysical.

6.6 Evalution Models Under Uncertainty

6.6.1 The Precautionary Principle (PP)

The PP is relevant to SD because the adverse impacts on the environment or human health can have serious consequences. The PP requires action even though "complete" scientific

Table 6.4 Steps in limits to acceptable change.

Step number	Characteristics of step
1.	Identify areas of concern and issues
2.	Define and describe opportunity classes
3.	Select indicators of resource and social conditions
4.	Inventory resource and social conditions
5.	Specify standards for resources and social indicators
6.	Identify alternative opportunity class allocations
7.	Identify management actions for each alternative
8.	Evaluation and selection of an alternative
9.	Implement actions and monitor conditions

Source: Prosser (1986).

evidence of damage is not available or in instances of scientific uncertainty (Prato and Herath 2007, 2010). An important challenge is that ecosystem processes are not completely understood, but avoiding irreversible losses of species and/or preserving biodiversity are critical for SD.

The basic argument is that critical environmental policy decisions cannot be postponed until all scientific information is available (Francis 1996). The PP focuses conservation policy away from the current generation to future generations and moves the burden of proof to the resource developers and users, hence aligning with SD (Wills 1997).

The safe minimum standard (SMS) can be adopted as a precautionary model. It emphasizes the protection of the environment wherever thresholds of irreversible damage are likely to be exceeded. The basic features are presented in Table 6.5 (Ready and Bishop 1991; Herath 2002). A two-person zero sum game is presented, with nature and society being the players. First, the cost of implementing SMS is viewed as an insurance premium paid against potential future losses. Society has two strategies: (i) no timber harvesting or SMS, which involves leaving the area intact if the damage due to timber harvesting (in terms of species extinction) is in excess of the SMS; and (ii) timber harvesting, which allows timber harvesting to proceed, but poses a risk of species extinction (Herath 2002). Table 6.5 represents the values in terms of losses to identify the strategy that minimizes the maximum possible loss. When there is no timber harvesting and a disease outbreak occurs, but the disease is cured due to the presence of the species, losses are zero. When there is no timber harvesting and no disease outbreak occurs, the losses are zero. If timber harvesting occurs and there is no outbreak of the disease, then the total net gain of logging is (expressed as losses) is $-B_d$. If timber harvesting occurs and the disease breaks out, then the net cost to society is the loss of the species (L) less the benefits from logging. This net cost is $L-B_d$, assuming that the benefits from species preservation are greater than the benefits from logging, which is positive. In this game, the maximum possible loss is zero without timber harvesting or SMS, and the maximum possible loss is $L-B_d$ with timber harvesting. If society benefits are greater than the benefits from logging, the value is positive. In this game, the maximum possible loss is zero.

Berrens et al. (1998) used the SMS to examine the importance of ensuring minimum environmental flows in the Colorado River. The study estimated a present value of US$1.29 billion in regional benefits of restoring environmental flows during the study period. SMS is useful when there are critical minimum populations where, if it drops below this population, population growth becomes negative, and the population becomes the SMS.

Table 6.5 Social loss matrix for timber harvesting strategies.

Strategy	State 1: disease outbreak	State 2: no disease outbreak	Max. loss
No timber harvesting (safe minimum standard)	0	0	0
Timber harvesting	$L-B_d$	$-B_d$	$L-B_d$

Source: Ready and Bishop (1991), Herath (2002).

In eutrophication, if the accumulated nutrient levels in a lake cross the threshold, this may trigger algal blooms – an example of the SMS (Randall 2011).

6.7 Adaptive Management

Adaptive management is used by researchers to generate guidelines for management of natural resources (Prato 2006). Adaptive management develops optimal management capacity by maintaining ecological resilience and institutional flexibility, which enables stakeholders to respond when conditions change. The aim is to manage within a range of acceptable outcomes while avoiding irreversible outcomes.

Tourism areas, for example, are regularly monitored to assess the development-intended outcomes, which form the basis of adaptive management. Monitoring will reveal the adverse impacts on biodiversity because planned processes might not yield the expected outcomes. Adaptive management aims to develop a flexible management plan when necessary. Plans sometimes do not work because of the complex and dynamic nature of ecosystems and the absence of complete knowledge. Ecosystem processes have time lags like complex social systems, which are often unpredictable. Plans must be flexible because rapid intervention may be required to avoid negative tourism impacts. It might be necessary to reduce visitor numbers to avoid serious damage to recreational areas.

6.8 Multicriteria Decision Analysis (MCDA)

SD is a multidimensional concept, which can deal with sustainability and uncertainty to solve complex decision problems (Boggia and Cortina 2010). MCDA produces an aggregate MCDA index to rank alternative decisions (Singh et al. 2012). Many applications are reported on the use of MCDA to evaluate sustainability (Bond et al. 2012), sustainable management and planning of forests (Kangas et al. 1993), as well as catchment basins (Prato and Herath, 2007). MCDA can be combined with geographic information system (GIS) in strategic sustainable planning applications (Ferretti and Pomarico 2012). Multiattribute value theory (MAVT), multiattribute utility theory (MAUT) and the analytic hierarchy process (AHP) are the most common approaches within this school.

6.8.1 The Analytic Hierarchy Process

The theoretical foundations of AHP were developed by Saaty (1977, 1980). AHP aggregates the separate criteria into an integrated criterion. When applying the AHP, the preferences of the decision elements are compared in a pairwise manner with regard to the element preceding them in the hierarchy. If two criteria are of equal importance, a value of 1 is given in the comparison, whereas a value of 9 indicates the absolute importance of one criterion over the other.

Pairwise comparison data can be analyzed using either regression analysis or an eigenvalue technique.[1] The pairwise comparisons made by the respondents can be synthesized into pairwise comparison matrices, one of which is given in Eq. (6.3).

$$A = \begin{bmatrix} a_{11} & a_{12} & \cdot & \cdot & \cdot & a_{1n} \\ a_{21} & a_{22} & \cdot & \cdot & \cdot & a_{2n} \\ \cdot & \cdot & \cdot & \cdot & \cdot & \cdot \\ \cdot & \cdot & \cdot & \cdot & \cdot & \cdot \\ \cdot & \cdot & \cdot & \cdot & \cdot & \cdot \\ a_{n1} & a_{n2} & \cdot & \cdot & \cdot & a_{nn} \end{bmatrix}. \tag{6.3}$$

where $\boldsymbol{a_{ij}}$ represents the pairwise comparisons rating for attributes i and j.

$$AW = nW \tag{6.4}$$

where A is the $n \times n$ comparison matrix and $W = (w_1, w_2, \ldots w_n)^T$. Saaty (1977) proposed the following definition:

$$AW = \lambda_{max} W \tag{6.5}$$

where λ_{max} is the maximum eigenvalue (Perron root) of matrix A. Saaty (1977, 1980) proved that the largest eigenvalue λ_{max} is always greater than or equal to n (number of rows or columns).

AHP was used to assess sustainability of water supply and sanitation services (WSSs) in Algeria (Boukhari et al 2016) using 6 dimensions, 12 criteria and 50 indicators at three different levels of hierarchy. Pairwise comparisons were obtained from 12 experts in different fields. The results show that the most important were the economic (38.3%) and technical (26.1%) dimensions. AHP was used for sustainability assessment of coastal beaches in China by Tian et al. (2017) who used 14 indicators derived from the economic, social and ecosystem dimensions for a wind power project on a coastal beach of Yancheng. Results show that the wind power farms on coastal beaches are not sustainable because the wind power farms generated some negative impacts.

AHP was used to rank and prioritize urban sustainability indicators for Malaysia (Noor 2013). The study indicates that the environmental dimension was the most important followed by the economic, social and institutional dimensions. The significance of prioritizing the environment arose due to the major sustainability issue of the environment in Malaysia.

Herath (2004) used AHP to evaluate the Wonga wetlands in Victoria, Australia. He evaluated three alternative wetland management plans with a diverse range of flora and fauna and recreational opportunities, and high-quality water in the Murray. The decision problem was to choose the best forest management plan. Jalil (2010) focused on SD and proper management of household waste in Malaysia. A methodology based on descriptive analysis has been used to discuss how household waste in Malaysia could be converted into vermicompost. Azri and Romle (2015) used AHP to examine the green city concept, which could assure sustainability for present and future generations in Malaysia.

---

1 Crawford and William (1985) showed how pairwise comparisons data can be analysed using regression analysis. The evidence suggests that both eigenvalue and regression techniques yield similar results.

Riparian buffer strips are used to protect watercourses from farmland activities (Qureshi and Harrison 2001). AHP was used to study the environmental, social and economic impacts of riparian revegetation options for the Scheu Creek, a highly degraded small catchment in north Queensland. Any revegetation strategy would have considerable social benefits. Typically, legislation designed to protect natural resources cannot force farmers to implement a revegetation option (Qureshi and Harrison 2003). In this study, it was noted that no single optimal policy could be identified by multicriteria analysis (MCA).

6.9 Multiattribute Utility Theory

MAUT is based on expected utility theory where the preferences for individual attributes have been specified and the decision maker can then establish preferences *between* the attributes by specifying the weights in a multiattribute utility function of the form:

$$U(x_1,x_2,\ldots,x_n) = k_1U(x_1) + k_2U(x_2) + \cdots + k_nU(x_n). \tag{6.6}$$

where the k's are the weights applied to the utility of each attribute, and the sum of the k's $= 1$.

Keeney and Raiffa (1976) described ways to check for utility independence to separate the individual utility functions. The theorem, which follows from utility independence, is as follows. If each Y_i is utility independent of $\bar{Y}_i$, $i = 1, \ldots, n$, then the utility function is either additive:

$$U(Y_1,\ldots,Y_n) = \sum_{i=1}^{n} k_iU_i(Y_i) \tag{6.7}$$

or multiplicative

$$1 + KU(Y_1,\ldots,Y_n) = \prod_{i=1}^{n}\left[1 + Kk_iU_i(Y_i)\right] \tag{6.8}$$

where U and U_i are utility functions scaled from zero to one, the k_i's are scaling constants with $0 < k_i < 1$, and $K > -1$ is a non-zero scaling constant. If U is multiplicative

$$\sum_{i=1}^{n} k_i \neq 1,$$

and if additive

$$\sum_{i=1}^{n} k_i = 1$$

6.9.1 Applications of MAUT

An application is the use of AHP and MAUT to optimize the selection of a design alternative given the objectives: net-zero, Leadership in Energy and Environmental Design (LEED) and cost (Doczy and AbdelRazig 2017). Here the decision maker can specify a project's objectives and use weights to prioritize the goals for two building projects to select the best alternative.

Table 6.6 Overall index score for goal 2 for five hypothetical countries.

	Countries				
Index	**A**	**B**	**C**	**D**	**E**
Goal 2 (eliminate hunger)	0.359	0.383	0.409	0.983	0.916

Source: Talukder and Hipel (2016).

A theoretical illustration of using MAUT using hypothetical data for five countries was developed by Talukder and Hipel (2016). The aim was to develop an index and dashboard for SDG 2. Here the set of indicators are selected for (SDG 2: end hunger). Hypothetical data are developed for the targets and the hypothetical countries and, using MAUT, an index is developed for each target useful for the UN's 17 SDGs to monitor progress and ensure accountability to achieve the SDGs by 2030 (Talukder and Hipel 2016). The MAUT technique of WebHIPRE (Hierarchical Preferences), an Internet-based software program, can aggregate the indicators' scores and weightings generating indices on a 0–1 scale. The utility functions were assumed to be additive and linear.

The overall index scores for SDG 2 for the hypothetical countries are 0.359, 0.383, 0.409, 0.983 and 0.91 for countries A, B, C, D and E, respectively (Table 6.6). The overall score of the index for SDG 2 is highest in country "D" and lowest in "A." Higher scores imply better performance related to SDG 2; in Web-HIPRE it is useful to compare the performance on the goals, targets and indicators among countries. The methodology is easy to use and can calculate the overall score considering all indicators. This method has the capacity to study performance related to the goals and targets of SDGs and hence is very useful.

Prato and Herath (2007) used MAUT in a different context by evaluating five farming systems in the Goodwater Creek catchment in northeast Missouri, USA. The farmers in the watershed practice intensive use of fertilizer and herbicides with adverse impacts on drinking water quality and aquatic ecosystems. Atrazine concentration in the Goodwater Creek has exceeded the maximum contaminant levels for drinking water of 3 ppm established by the Environmental Protection Agency (Prato and Herath 2007). The five farming systems were ranked using an additive utility function. Criteria weights were obtained from a survey of 20 farmers in the Goodwater catchment. Using the final utility scores, the best farming system was that with corn–soybean rotation, reduced tillage, medium fertilizer application and banded pesticide application.

MAUT was used to examine ecosystem-based fisheries to develop socio-economic indicators in Alaska (Seung and Zhang 2011). The study incorporated several socio-economic indicators in developing a the MAUT approach for a fishery. The study indicates that there is inadequate appreciation of the importance of socio-economic indicators and MAUT can be used to improve decision making.

6.9.2 Multicriteria Decision Analysis and Valuation of Natural Resources

A further development of the MAUT is the choice modeling (CM) approach, which has been used to examine willing to pay (WTP) for water restrictions and water service and waste water service attributes in Australia (Henscher et al. 2005, 2006). They found that consumers are WTP up to $18 to improve the appearance of the urban environment. They would pay only $10.00 to prevent a 10% reduction in water use (1997 dollars). Henscher et al (2006) found that residential consumers are WTP $23.79 to move from a situation with continuous restrictions to a situation with no water restrictions. Another study of the WTP for water and waste water attributes in Sydney showed that customers' WTP is smaller when they face more frequent interruptions and longer periods of interruption (Henscher et al. 2005).

6.9.3 Integrating Environmental Values with the Analytic Hierarchy Process

Economists have developed many approaches to establish monetary values for non-market resources (Herath 2002; Bennet et al. 2000). The contingent valuation (CV) method and the travel cost method are widely used to estimate the total economic value (TV), which includes both use and non-use values (Herath 1999). Separating use and non-use values is important because of its implications for the management of wetlands. The AHP, which is a MCDA approach, can be used to separate use and non-use values from TV. Wattage and Mardle (2005) used the CV method to assess the TV of the Muthurajawela Marsh and the Negombo Lagoon (MMNL) area in Sri Lanka, and to split the TV into use and non-use values using the AHP. This approach is expected to contribute to resolving conflicts between wetland conversion for development and conservation.

The AHP model was used to decompose the TV into use and non-use values. A pairwise comparison survey was undertaken using standard AHP practices to make pairwise comparisons of use and non-use values at each level of the tree. The aggregated preferences are 0.552 for use values and 0.448 for non-use values in conservation of wetlands, which suggests that option value should be split equally between use and non-use values. This study demonstrates two different uses of the AHP in evaluating and managing a wetland aquatic system. AHP can be used to determine stakeholders' preferences and their values, which are useful in evaluating policies for sustainably managing wetlands.

6.9.4 Combined Models of Multicriteria Decision Analysis and Geographic Information System

A recent model named the GeoUmbriaSUIT, illustrates how MCDA can be combined with GIS in a spatial MCDA leading to the creation of a multicriteria spatial decision support system (MCSDSS). It combines the data management of GIS with the judgment capabilities of MCDA. A MCSDSS has three parts (Malczewski 2010), namely a geographical component, an MCDA model-based system and an interface. How much the MCDA and GIS parts are integrated and the presence of a unique interface determines the level of integration of the system. However, the only type of integration that allows the use of the same database and interface and enables access to both the MCDA and GIS tools at any time during the analysis is the complete MCSDSS (Malczewski 2010).

The model GeoUmbriaSUIT was applied to Malta, to compare the sustainability of different territorial areas. The three dimensions of sustainability are represented by a specific set of indicators proposed as the first MCSDSS for complete integration of MCDA and GIS, especially developed for evaluating sustainability. The study used a weighted summation algorithm, calculated on two sets of indicators, environmental and socio-economic, to obtain an index of sustainability to compare different territorial areas.

Another study analyzed the importance of sustainability in tourism using a methodology referred to as the PM4SD (Project Management for Sustainable Development). The UN SDGs refer to tourism as a target. The study reviews the European and global sustainable tourism policies and definitions of sustainable tourism (Garrod and Fyall 1998). Some definitions are less oriented to SD, while other definitions argue that market forces are inimical to the achievement sustainability, and hence require alternative governance systems (Cardia and Jones 2017).

6.10 Conclusions

This chapter presents a selected portfolio of tools to evaluate sustainability issues with special reference to the evaluation of the SDGs. Many other methods not discussed here are being developed by academics and researchers. Development of these models proceeds continuously but their application lags behind and there is a paucity of use, especially to evaluate the SDGs in developing countries. Hybrid models, where different aspects of models are combined, are appearing on the scene and these are more versatile. The uses are location-specific and more flexible models and rules that are fit for a diversity of situations and more general use are essential. The promotion of these models by highlighting their potential is the only way to ensure their wider use for the SDGs and their further development.

References

Alauddin, M. and Quiggin, J. (2008). Agricultural intensification, irrigation and the environment in South Asia: issues and policy options. *Ecological Economics* 65: 111–124.

Azri, M.A.R. and Romle, B.A.R. (2015). Sustainable development: development for a sustainable future. A case of Putrajaya Green City. *Australian Journal of Basic and Applied Sciences* 9: 30–34.

Bennet, J., Rolfe, J., and Louviere, J. (2000). Choice modelling and its potential application to tropical rainforest preservation. *Ecological Economics* 35: 289–302.

Berrens, R.P., Brookshir, D.S., McKee, M., and Schmidt, C. (1998). Implementing the safe minimum standard approach: two case studies from the US Endangered Species Act. *Land Economics* 74: 147–171.

Boggia, A. and Cortina, C. (2010). Measuring sustainable development using a multi-criteria model: a case study. *Journal of Environmental Management* 91: 2301–2306.

Bond, A., Morrison-Saunders, A., and Pope, J. (2012). Sustainability assessment: the state of the art. *Impact Assessment and Project Appraisal* 30: 53–62.

Boukhari, S., Djebbar, Y., Amarchi, H., and Sohani, A. (2016). Using AHP to sustainability of water supply and sanitation services in Algeria. *Water Science and Technology: Water Supply* 18: 1282–1293.

Cardia, G. and Jones, A. (2017). PM4SD as a methodological framework for sustainable tourism. In: *Tourism, Culture and Heritage in a Smart Economy*. Springer Proceedings in Business and Economics (eds. V. Katsoni, A. Upadhya and A. Stratigea), 275–292. Cham, Switzerland: Springer.

Coopman, A., Osborn, D., Ullah, F. et al. (2016). *Seeing the Whole: Implementing SDGs in an Integrated and Coherent Way*. London: Stakeholder Forum.

Crawford, G. and Williams, C. (1985). A note on the analysis of subjective judgement matrices. *Journal of Mathematical Psychology* 29: 387–405.

Cumberbatch, J. and Moses, J. (2011). Social carrying capacity in beach management in Barbados. *Journal of Coastal Research*, Special Issue 61: 14–23.

Doczy, R. and AbdelRazig, Y. (2017). Green buildings case study analysis using AHP and MAUT in sustainability and costs. *Journal of Architectural Engineering* 23: 87–88.

Ferretti, V. and Pomarico, S. (2012). Integrated sustainability assessments: a spatial multicriteria evaluation for siting a waste incinerator plant in the Province of Torino (Italy). *Environment, Development and Sustainability* 14: 843–867.

Francis, J.M. (1996). Nature conservation and the precautionary principle. *Environmental Values* 5: 257–264.

Garrod, B. and Fyall, A. (1998). Beyond the rhetoric of sustainable tourism. *Tourism Management* 19: 199–212.

Henscher, D., Shore, N., and Train, K. (2005). Household's willingness to pay for service attributes. *Environmental and Resource Economics* 32: 509–531.

Henscher, D., Shore, N., and Train, K. (2006). Water supply security and willingness to pay to avoid drought restrictions. *The Economic Record* 82: 56–66.

Herath, G. (1999). Estimation of community values of lakes: a study of Lake Mokoan in Victoria, Australia. *Economic Analysis and Policy* 29: 30–44.

Herath, G. (2002). Research methodologies for planning ecotourism and nature conservation. *Tourism Economics* 8: 77–101.

Herath, G. (2004). Incorporating community objectives in improved wetland management: the use of the analytic hierarchy process. *Journal of Environmental Management* 70: 263–273.

Herath, G. (2012). *Institutional Change in Water Resources Management: The Past, Present and the Future*. New York: Nova Publishers.

Herath, G. (2017). Sustainable development: is this the panacea for Sri Lanka's ills. *The Daily Mirror* (June 16).

Houghton, K. (2014). *A Sustainble Development Goal for the Ocean: Moving from Goal Framing Towards Targets and Indicators for Implementation*. Potsdam, Germany: IASS Potsdam.

IPCC (2001). Intergovernmental Panel on Climate Change, climate change 2001. Impacts, adaptation and vulnerability. In: *Contribution of Working Group II to the Third Assessment Report of the Intergovernmental Panel on Climate Change* (eds. J.J. McCarthy, O.F. Canziani, N.A. Leary, et al.). Cambridge, UK: Cambridge University Press.

Jalil, A. (2010). Sustainable development in Malaysia:a case study on household waste management. *Journal of Sustainable Development* 3: 91–102.

Jungcurt, S. (2016). *Towards Integrated Implementation: Tools for Understanding Linkages and Developing Strategies for Policy Coherence, SDG Knowledge Hub*. Winnipeg, Canada: IISD.

Kangas, J., Karsikko, J., Laasonen, L., and Pukkala, T. (1993). A method for estimating the suitability function of wildlife habitat for forest planning on the basis of expertise. *Silva Fennica* 27: 259–268.

Kayat, K. and Radzi, R.M. (2012). Social tourism capacity in Kapung Kilin, World Geopark, Langkawi. In: *Island Sustainability* (eds. S. Pavro and C.A. Brebbin). Southampton, UK: WIT Press.

Keeney, R.L. and Raiffa, H. (1976). *Decisions with Multiple Objectives: Preferences and Value Tradeoffs*. New York: Wiley.

Le Blanc, D. (2015). Towards integration at Last? The Sustainable Development Goals as network of targets. DESA Working Paper, No. 141, ST/ESA/2015/DWP/141. New York: DESA. https://www.un.org/esa/desa/papers/2015/wp141_2015.pdf (accessed November 2019).

Lindberg, K., McCool, S.F., and Stankey, G.H. (1997). Rethinking carrying capacity. *Annals of Tourism Research* 24: 461–464.

Malczewski, J. (2010). Multiple criteria decision analysis and geographic information systems. In: *International Series in Operations Research & Management Science* (eds. M. Ehrgott, J. Figueira and S. Greco), 369–395. New York: Springer.

Meigh, J.R., McKenzie, A.A., Austin, B.N., Bradford, R.B. and Reynard, N.S. (1998). Assessment of global water resources – Phase II, Estimates of present and future water availability for Eastern and Southern Africa. DFID report 98/4. Wallingford, UK: Institute of Hydrology.

Nilsson, M., Griggs, D., Visbeck, M., and Riegler, C. (2016a). Map the interactions between the sustainable development goals. *Nature* 534: 320–322.

Noor, Z.Z. (2013). AHP application in Urban sustainability indicators prioritization in Johor, Malaysia. *Resources and Environment* 3: 1–5.

Peng, X., Xiuyun, G., and Xuehui, H. (2005). Future prospects for population development in Shanghai. A population carrying capacity analysis. *Asian Population Studies* 1: 261–281.

Prato, T. (2001). Modeling carrying capacity for national parks. *Ecological Economics* 39: 321–331.

Prato, T. (2006). Adaptive management of national parks. *George Wright Society* 23: 72–86.

Prato, T. and Herath, G. (2007). Multiple criteria decision analysis for integrated catchment management. *Ecological Economics* 63: 627–632.

Prato, T. and Herath, G. (2010). The precautionary principle and environmental protection: The Australia experience. In: *Handbook of Environmental Policy* (eds. J. Meijer and A. Berg). New York: Nova Science Publishers.

Prosser, G. (1986). The limits of acceptable change. *Australian Parks and Recreation* 22: 5–10.

Qureshi, M.E. and Harrison, S.R. (2001). A decision support process to compare Riparian revegetation options in Scheu Creek catchment in North Queensland. *Journal of Environmental Management* 62: 101–112.

Qureshi, M.E. and Harrison, S.R. (2003). Application of the analytic hierarchy process riparian revegetation policy options. *Small-Scale Forest Economics, Management and Policy* 2: 441–458.

Randall, A. (2011). *Risk and Precaution*. Cambridge, UK: Cambridge University Press.

Rasul, G. (2016). Managing the food, water and energy nexus for achieving the sustainable development goals in South Asia. *Environmental Development* 18: 14–25.

Ready, C.R. and Bishop, R.C. (1991). Endangered species and the safe minimum standard. *American Journal of Agricultural Economics* 73: 309–312.

Saaty, T.L. (1977). A scaling method for priorities in hierarchical structures. *Journal of Mathematical Psychology* 15: 234–281.

Saaty, T.L. (1980). *The Analytic Hierarchy Process*. New York: McGraw-Hill.

Sachs, J., Schmidt-Traub, G., Kroll, C., Durand-Delacre, D. and Teksoz, K. (2016). SDG Index and dashboards – Global Report. New York: Bertelsmann Stiftung and Sustainable

Development Solutions Network (SDSN). https://www.sdgindex.org/reports/sdg-index-and-dashboards-2016 (accessed November 2019).
Saveriades, A. (2000). Establishing the social carrying capacity for the tourists resorts of the east coast of the Republic of Cyprus. *Tourism Management* 21: 147–156.
Seidel, I. and Tisdell, C.S. (1999). Carrying capacity reconsidered: from Malthus's population theory to cultural carrying capacity. *Ecological Economics* 31: 395–408.
Seung, C. and Zhang, C. (2011). Developing socioeconomic indicators for fisheries off Alaska: a multi-attribute utility function approach. *Fisheries Research* 112: 117–127.
Shah, T., Bhatt, S., Shah, R.K., and Talat, J. (2008). Groundwater governance through electricity supply management: assessing an innovative intervention in Gujarat, Western India. *Agricultural Water Management* 95: 1233–1242.
Shelby, B. and Heberlain, T. (1984). *Carrying Capacity in Recreation Settings*. Covallis, OR: Oregan State University Press.
Singh, R.K., Murty, H.R., Gupta, S.K., and Dikshit, A.K. (2012). An overview of sustainability assessment methodologies. *Ecological Indicators* 15: 281–299.
Stankey, G.H., Cole, D.N., Lucas, R.C., Peterson, M.E. and Frissell, S.S. (1985). The limits to acceptable change (LAC) system for wilderness planning. General Technical Report INT-176 Forest Service. Washington, DC: US Department of Agriculture.
Talukder, B. and Hipel, K.W. (2016). Multi Attribute Utility Theory (MAUT) as a tool to develop index and dashboard for goal 2 of SDGs: a hypothetical case study. International Conference on Sustainable Development (ICSD), New York (September 21–22). https://www.researchgate.net/publication/313360951_Multi_Attribute_Utility_Theory_MAUT_as_a_Tool_to_Develop_Index_and_Dashboard_for_Goal_2_of_SDGs_A_Hypothetical_Case_Study (accessed November 2019).
Tian, W., Bai, J., Sun, H., and Zhao, Y. (2017). Application of the analytic hierarchy process to a sustainability assessment of coastal beach exploitation: a case study of the wind power projects on the coastal beaches of Yancheng, China. *Journal of Environmental Management* 115C: 251–256. https://doi.org/10.1016/j.jenvman.2012.11.015.
Wackernagel, M., Hanscom, L. and Lin, D. (2017). Sustainable Development Goals consistent with sustainability. Global Footprint Network: Advancing the Science of Sustainability (September 2017, blog).
Wagar, J.A. (1964). *The Carrying Capacity of Wildlands for Recreation*. Forest Service Monograph No 7. Bethesda, MD: Society of American Foresters.
Wattage, P. and Mardle, S. (2005). Stakeholder preferences towards conservation versus development for a wetland in Sri Lanka. *Journal of Environmental Management* 77: 122–132.
WCED (1987). *Our Common Future. World Commission on Environment and Development*. Oxford, UK: Oxford University Press.
Wills, I. (1997). The environment, information and the precautionary principle. *Agenda* 4: 51–62.
Wolf, W.J. (1984). *The Effects of Recreation on the Wadden Sea Ecosystem*. Amsterdam: Free University of Amsterdam.

7

Resilience Engineering and Quantification for Sustainable Systems

Anita Talan, Bhoomika Yadav, L.R. Kumar, and R.D. Tyagi

INRS, Université du Québec, Québec, Canada

7.1 Introduction

Global policymakers and strategic planners face a lot of difficulties in making choices such as sustainable energy sources that will power cities in the future. These policymakers focus on the pros and cons of every plan before implementation of policies affecting resiliency and sustainability. Global sustainability requires a systems approach to the development of policies and intervention strategies. If full understanding of systems implications is not there, then risk factors increase. Sustainability and resiliency are closely coupled and is essential to quantify the sustainable systems. Sustainability cannot be applied to a particular industry without analysis of effects on other sectors and hence resiliency plays an important role in defining global sustainability. Therefore, setting the boundaries for sustainability and resiliency together for meaningful analysis has become a tough challenge. And a more robust approach is to explore sustainability issues globally using integrated models of resiliency.

Based on the experiences and insights of intellectual global policymakers and leaders from academia, industry and government, this chapter seeks to provide better understanding of global sustainability based on resilience engineering and quantification approaches that are needed for achievement of global sustainability. This chapter focuses on sustainable development, the relationships between sustainable development and resilience, elements of resilience and sustainable development, ways to achieve sustainable development including assessment approaches, resilience indicators and associated risk factors, and different policies adopted in different countries to achieve global sustainability based on resilience engineering and quantification systems, and what needs to be done in order to achieve global sustainability.

Sustainability: Fundamentals and Applications, First Edition.
Edited by Rao Y. Surampalli, Tian C. Zhang, Manish Kumar Goyal, Satinder K. Brar, and R. D. Tyagi

7.2 Resilience and Sustainability: Definitions, Differences and Similarities

In recent years, there have been many uses of the terms resilience and sustainability, with some researchers treating both terms as the same, while others define them as unrelated terms. Many studies have been done to understand their similarities and differences in order to incorporate use of these terms into systems administration. Sustainability is defined through the triple bottom line (TBL) of environmental, social and economic system concerns, while resilience is defined as the capability of a system to get ready for threats, absorb impacts, mend and adapt to persistent stress or disruptive events (Hosseini et al. 2016). The three organizational frameworks are there for organizing sustainability and resilience:

- resilience as an element of sustainability;
- sustainability as an element of resilience;
- sustainability and resilience as diverse terms with dissimilar objectives.

The objectives and executions of these frameworks have the common aims of providing aid to society and environment under usual and life-threatening conditions. The executions and objectives work together to provide the examples that confer similarities to both the terms by reducing the conflicts between them. The meanings of both are confused owing to the different definitions that exist and that are used across different disciplines. The definitions, differences and similarities of sustainability and resilience are described below.

7.2.1 Definitions

Defining resilience and sustainable development is like speaking about two highly abstract and complex theories, each of which has a great variety of explanations and definitions.

Many meanings of sustainability are vague. Most of the researchers define sustainability as everything that ensures the wellbeing of the social order and environment (Marchese et al. 2018) or as an ethical concept with the emphasis on things to be better in future than they are at present (Seager 2008). Ohers define sustainability as durability, which means that more sustainability can be attained if a longer system can be completed (Lew et al. 2016). In 1987, the Brundtland Commission highlighted the concept of permanency in their report "Our Common Future" to define sustainability as meeting the needs of the present without compromising the needs of the future. However, this definition does not give a clear idea about the needs, as needs fluctuate in different conditions. Therefore, the TBL definition was given in 1997 as part of a sustainable corporate model that defines sustainability as a combination of wellbeing in terms of social, economic and environmental realms (Hosseini et al. 2016). The TBL definition originated as a way to attain responsible business achievements, but we use it usually to define environmental sustainability, with social and economic wellbeing of individuals interacting with the environment.

Similar to sustainability, there are multiple definitions of resilience from corner to corner such as in psychology, urban development and civil engineering (Linkov et al. 2014). This heterogeneity in definitions arises due to two deviating views of resilience, i.e. business resilience and environmental resilience. Ecological resilience is referred to as "bounce back" resilience, i.e. a resilient system is one that returns to its pre-disturbance state at the first opportunity (Manyena et al. 2008). Alternatively, the environmental definitions define resilient systems as having several stable states or commands in which the systems shift due to disturbance. Engineering resilience is relevant to systems operating within one constant state, while ecological resilience is applicable in structures with multiple stable states such as socio-ecological systems, or statistics and energy complexes. The National Research Council in 2012 (Linkov et al. 2014) defined resilience as the ability to prepare and plan for, absorb, recover from and more successfully adapt to adverse events.

7.2.2 Differences Between Sustainability and Resilience

While sustainability seeks to address major problems in society without sacrificing the needs of future generations, resilience, on the other hand, enables a system to respond to changing circumstances. Table 7.1 discusses the differences between the resilience theory approach and the sustainability science approach in detail (Redman 2014).

7.2.3 Similarities and Association Between Sustainability and Resilience

To understand the association between sustainability and resilience, one may consider the following four cases:

Resilience is essential and sufficient for sustainability: If the sustainability of the system is a subset of a basin of attraction containing a stable symmetry, and if the first state of the system is within the sustainability set of the system, the uninterrupted part of the system relics within the set and joins the locally stable equilibrium of the system. If the system is disturbed to such a degree that the system flips into a different field of attraction, it will unite with the external equilibrium of the sustainability set. In this case, the resilience of the environmental–economic system in the initial state, which is inside the sustainability set of the system, is essential and sufficient for sustainable development.

Table 7.1 Difference between resilience theory approach and sustainability science approach.

Resilience theory approach	Sustainability science approach
A system can respond to changing circumstances	Sustainability science seeks to address major problems in the society without sacrificing the needs of future generations
Focus is to maintain system dynamics	Focus is on attaining sustainability
Resilience approach does not evaluate results of any situation. Instead, it builds natural and social capital and cultivates adaptive capacity to cope with unknown futures	Evaluates future options, assigns values to sustainability indicators, and customizes its strategies to attain those options
Outcome of change is open-ended	Desired results are specified in change
Maintain ecosystem services	Transformation of society

Resilience is necessary but not sufficient for sustainability: Consider the situation that the sustainability set of the system is a subset of a basin of attraction, but does not comprise the stability of that basin of attraction. With an initial state within the sustainability set, the system will ultimately move out of the sustainability set on the track of stability. In this case, resilience is needed for sustainable development, because no other area of attraction is present within the sustainability set. On the other hand, resilience is not enough for sustainable development, because no stable symmetry is involved within the sustainability set. Here, although the primary state is in a resilient basin of magnetism, the system moves out of the sustainability set owing to its dynamics.

Resilience is sufficient but not necessary for sustainable development: In this case, the sustainability set includes more than one stable equilibrium in the vicinity. Assuming that initially the system is within the basin of magnetism meeting the stable equilibrium, if the system is not resilient to an exogenous distortion, it will flip from its primary field of attraction into another basin of attraction and unite with a new stable equilibrium. However, if the system leaves its domain of attraction it will not leave the sustainability set as stable equilibrium is still within the sustainability set of the system. Thus, resilience is needed but not necessary for sustainable development, since in alternative fields of attraction sustainable development is also possible.

Resilience is neither essential nor sufficient for sustainable development: The system is initially in the area of attraction that does not satisfy the sustainability requirements. If the system is resilient to an exogenous disturbance it will continue to be in this area of attraction and consequently outside the sustainability set of the system. As a result, the resilience of the system is neither essential nor sufficient for sustainability. Divergently, resilience stops sustainable development.

As can be seen from the different cases described above, the association between resilience and sustainable development depends on the position of the different equilibriums with respect to the sustainability set of the system. Since the important distinctions between all the cases discussed are not certain according to the rules of the model, the different connections between the resilience and sustainable development of ecological–economic systems can be useful in any ecological–economic model that structures more than binary equilibriums.

7.3 Technical Purposes of Resilience Engineering

Resilience engineering is "the internal ability of a system to adjust its functioning to the changes in the system, making it able to handle desired operations under any circumstance" (Hollnagel 2014, p. 5). Resilience engineering is concerned with efficient and safe functioning of the industrial systems. Here, "the system" means the infrastructures, buildings and cultural landscapes that make up the natural, physical, economic, social and cultural capital of a society. The idea of resilience is concerned with both the system and its surrounding environment; not just the system. The main focus of resilience engineering is on industrial systems, which have the purpose of producing or transforming something. The purpose of resilience engineering is to make the performance of the system e resilient; it can make people aware of and thus able to adjust to the conditions and situations (either individual or in a group), identify and solve the problems, acknowledge the demands and make appropriate adjustments. Here, the key is that resilience engineering allows the operators or managers to detect when things are not going in an expected way and improve the situation before it becomes worse (Hollnagel 2014). Resilience is not only related to management of disasters and sustainability but also a property of the system to handle desired operations under any circumstance (both expected and unexpected). The unexpected conditions are opportunities for any system. Put simply, the basic purposes of resilience engineering are:

- *Knowing what to do*: Preparing sets of responses or adjusting normal functioning when regular and irregular disruptions and disturbances occur. It is the ability to address the reality.
- *Knowing what to look for*: Monitoring the events in the environment and the performance of the system itself and realizing something has the potential to be a threat in the near future. It is the ability to address the critical.
- *Knowing what has happened*: Learn the right lessons from both positive and negative experiences. It is the ability to address the factual.
- *Knowing what to expect*: Anticipating threats and their consequences in the future. It is the ability to address the potential.

It should be noted that there is a difference between safety and resilience engineering. According to the traditional point of view, safety is defined as the condition where the numbers of unpleasant results are controlled to be as minimal as possible. According to the traditional safety view, when everything works as expected, the outcomes will be acceptable. But when something doesn't go as per expectations, this will lead to a failure. This is where change from normal to abnormal takes place. It is never possible in practical situations for the performance of a system or operation to comply with design assumptions because real conditions will never be completely identical to the expected conditions (Hollnagel 2014). On the other hand, resilience engineering makes sure that systems work because people in the system are able to adjust according to the situation at hand. They can identify and overcome flaws and functional faults; they can acknowledge the actual demands and take the steps accordingly; they can foresee the damage and make proper adjustments to match the conditions. They can also detect when something doesn't go as expected or is about to go wrong and improve the situation before it worsens (Hollnagel 2014).

7.4 Resilience Assessment Approaches

A resilience assessment consists of three main stages: (i) preparation; (ii) assessment; and (iii) follow-up. The preparation stage consists of planning and organization of a community-based resilience assessment workshop. This may include steps such as clarification of the purpose of the assessment, determination of the assessment area, as well as information acquisition about the system, resident communities and stakeholders. The workshop consists of introductory presentations, scoring based on the indicators and discussion, summary and next steps. The follow-up stage is used for the management and planning process for the areas to focus on based on the assessment results.

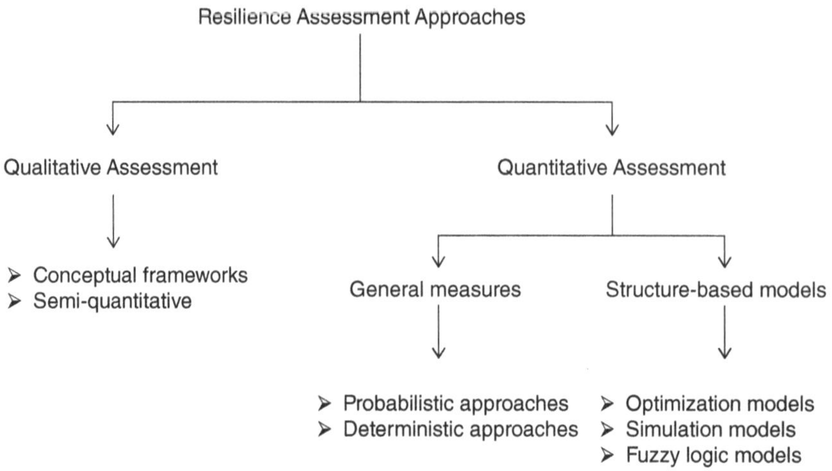

Figure 7.1 Classification of resilience assessment approaches.

The resilience assessment can be done qualitatively and quantitatively. The qualitative method does not use numerical values for analysis. As shown in Figure 7.1, both the approaches have their own subcategories. The two categories of qualitative approach are: (i) conceptual frameworks; and (ii) quantitative indices. The quantitative method also has two subcategories: (i) the general resilience approach; and (ii) the structure-based modeling approach. Details of each approach are described below.

7.4.1 Qualitative Assessment

As said above, the qualitative assessment approach is categorized into two subunits as intangible bases and measurable indices.

7.4.1.1 Intangible Bases

These constitute mainly the qualitative methods for measuring system resilience. The seven steps in the structure for assessing the resilience of a social–ecological system are:

1) understanding and defining the system to be studied;
2) recognizing the right scale to assess resilience;
3) identifying the system parameters responsible for external and internal disturbances;
4) identifying the lead roles in the system including people and governance;
5) developing theoretical prototypes for identifying essential retrieval activities;
6) implementation of the results at every step in policymaking;
7) integrating the outcomes of preceding steps.

The framework for assessing the resilience of livelihoods provides three aspects of resilience:

1) quantified change that a system can go through;
2) self-organization of the public based on inherent communal structure;
3) ability to learn and adapt.

On the other side, a conceptual framework was proposed with eight guiding principles for assessing system resilience:

1) risk and threat assessment;
2) strength;
3) significant mitigation;
4) flexibility;
5) risk knowledgeable planning;
6) risk-informed investment;
7) coordination of purposes;
8) comprehensive of scope.

Some studies have recommended a holistic resilience framework for assessing resilience systems consisting of two resilience categories: internal resilience and external resilience using resilience polices and subpolicies. For example, the structure for resilience and survivability of communication networks is reliant on six factors: defend, identify, analyze, remediate, improve and recover. This framework was designed in such a way that it can be used in other fields as well. However, this structure only provides theoretical insight and does not compute system resilience. Similarly, the properties of resilience in the field of telecommunication networks are common to most networks: consistency, convenience, welfare, privacy, maintainability, integrity and performance (Vlacheas et al. 2013). The assessment criteria for measuring the resilience of mass transportation systems in the case of explosive outbreaks have also been given (Bruyelle et al. 2014). To achieve resilience in medication delivery, three factors have been anticipated:

- innovative information-imagining techniques;
- scenario-based strategy and assessment treatments;
- cooperation during the elicitation of necessities.

Some studies bring aspects of resilience together as a function of absorptive capacity, adaptive capacity and restorative capacity (Vugrin et al. 2011). Absorptive capacity is defined as the degree to which an organization is able to adapt itself momentarily to new disturbed conditions, while restorative capacity is the degree to which a system is able to restore itself if adaptive capacity is not active. Adaptive and restorative capacities are referred to as recovery activities. This assessment approach is different from other studies as it doesn't come from the perspective of the resilience cost index, which is composed of two elements:

- loss prices posed by disruptive events;
- reclamation costs.

7.4.1.2 Barriers

There are many defined barriers to achieving resilience of a system, such as:

- lack of knowledge about resilience engineering;
- insubstantiality of resilience engineering level;
- selecting manufacture over safety;
- lack of reporting structure;

- religious beliefs;
- out-of-date techniques and manuals;
- poor response loop;
- economic complications.

To overcome these barriers, seven indicators have been established for system resilience from a safety culture viewpoint, including: program delays, safety teams, meeting efficiency, safety training, worker participation, safety drills and capability (Shirali et al. 2012). The resilience capacities are dependent on centralization or decentralization of controlling systems, management transformation, risk controlling, accident inquiry and a management board on safety and resilience. The community resilience framework of volcanic activity-prone areas comprises: recognizing the calamity or hazard, determining the susceptibility factor for an individual or the public, risk consciousness, as well as improving the social, economic and physical resources.

7.4.1.3 Semi-quantitative Indices

A semi-quantitative index is a method that is based on a set of questions to assess different resilience-based system features such as redundancy, resourcefulness on a 0–10-point scale or a 0–100 percentage scale. Assessments of the characteristics are done on the basis of expert's views and are combined in some way to produce an index of resilience. There are 36 well-recognized resilience variables of communities for natural tragedies, including originality, redundancy and strength (Cutter et al. 2008). These five variables are grouped into five subindices including economic, substructure, social, institutional and community capital. The overall score for each subindex is calculated using the unweighted average of each variable, and the total score is calculated by taking the unweighted average of all subindex scores into account. There are two key factors of resilience in an industrial supply chain: (i) level of the resource chain liability; and (ii) ability of the supply chain to hold out and recover from disturbance. Based on the resilience system a certain set of questions is designed to assess system resilience in areas such as susceptibility. For example, the capability of supply chains was assessed by a questionnaire with 152 questions in a study by Pettit et al. (2010). The significance of each factor is weighed by policymakers of the system and answers to the questions are calculated based on the weighted sum approach. The semi-quantitative approach is also used to assess the resilience engineering in the development industry based on six parameters: top administration; assurance; learning culture; responsiveness; alertness; and flexibility.

7.4.2 Quantitative Assessment

7.4.2.1 Structure-Based Models

This model for assessment is based on how the structure of a system impacts its resilience. The performance of the system is observed, and characteristics of a system are demonstrated or simulated. Based on this, structure-based models are further classified into three types:

Optimization Models Optimization models are used by various studies for evaluating system resilience by optimizing the system parameters. For example, a mathematical model was proposed for evaluating and optimizing airport resilience, aiming to maximize the resilience of an airport's runway and taxiway network. The take-off and landing capacities of an airport's runway are used for mathematical modeling by assessing the space, funds and capabilities before disruption; time, physical parameters, operational parameters, resources and budget restrictions are also taken into account (Faturechi et al. 2014).

Simulation Models To assess system resilience, simulation models have been designed and used in discrete events. Different studies use simulation models for different systems. For example, a discrete event simulation model was used to evaluate the preparedness of a fire and rescue service department in the presence of terrorist attacks (Albores and Shaw 2008). In this study, the preparedness was the pre-event disruption resilience. Two models were developed: the first model mimicked the mass decontamination of a population following a terrorist attack, while the second model dealt with the harmonization of resource allocation across regions. Similarly, the discrete event simulation model was also applied to assess the resilience of a supply chain (Carvalho et al. 2012). Flexibility and redundancy were taken as basic elements of resilience. Redundancy was modeled by keeping additional inventory to successfully withstand disruptions, and flexibility was modeled by restricting the extent of the disrupted transportation system. However, these results are not universally accepted. Similarly, for critical infrastructure networks such as railway transportation networks a time-dependent simulation model has been developed to measure resilience indicators. A set of disruptive events is modeled through a simulation model with consequences of increased travel time and reduction of train capacity (Adjetey-Bahun et al. 2014). An approach based on integrating analytical simulation, topology generation and experimental emulation has been used to improve the resilience and survivability of internet networks. The resilience of the internet network is defined as the ability of the network to provide a desired service level when it is challenged by large-scale disasters or intense failures.

Fuzzy Logic Models This model has been used for various systems to assess system resilience. It has been used for assessing organizational resilience, and fuzzy linguistic variables are used to define the relative importance of the organizational resilience factors. The study showed the use of fuzzy cognitive maps (FCMs) for assessing engineering resilience. Nine factors (teamwork, flexibility, redundancy, awareness, preparedness, learning, reporting, management and fault tolerance) are used to assess engineering resilience. The FCM is represented by a fuzzy graph structure, and the result is obtained as neural network and fuzzy logic approaches (Azadeh et al. 2014). Fuzzy architecture is used for the assessment of resilience of critical infrastructure. The primary components of critical infrastructure are redundancy and adaptability. These two components are the inputs of fuzzy architecture while the outputs are expressed using linguistic variables (Muller 2012). Tadić et al. (2014) integrated fuzzy forms of the Analytic Hierarchy Process (AHP) and the multicriteria discrete comparison technique TOPSIS (Technique for Order of Preference by Similarity to Ideal Solution) for evaluating and ranking organizational resilience based on qualitative assessments. This approach is used to assess several resilience factors such as planning strategies, capacity and capability, internal and external monitoring and reporting, human factors, quality, design, potential and emergency response.

7.4.2.2 General Measures

General measures provide a quantitative means to assess system resilience by measuring system performance, regardless of the system. These measures are comparable across different system contexts with similar underlying logic. General measures assess system resilience on the basis of system performance before and after disruption without concentrating on system-specific characteristics. These measures can be characterized into:

1) *Dynamic vs static*: A dynamic-based performance approach is time-dependent behavior while a static approach is not dependent on time for measuring the resilience of the system.
2) *Deterministic vs probabilistic*: A deterministic approach for assessing the performance of the system does not incorporate uncertainty into the metric, while a probabilistic approach is dependent on the stochasticity associated with system behavior.

7.5 Indicators for Quantifying Successful Resilience and Sustainability

An indicator is a quantitative or qualitative measure that reflects the reality of a complex situation. Indicators reveal the relative position of the situation being evaluated. When measured over a period of time, an indicator can illustrate the magnitude and direction of change. The resilience indicators can be helpful to stakeholders, local communities, policymakers and researchers. The indicators of resilience can enhance communication among stakeholders, identify and track social practices, empower communities in decision-making processes and also identify areas to focus on, such as food security, livelihood, governance, etc. In addition, the indicators can be helpful for local communities (increasing their understanding), researchers (identify research gaps and deepen their understanding of resilience), policymakers/project planners (identification of areas of improvement) and non-governmental organizations (NGOs) (Supply and Programme 2014).

Indicators for quantifying resilience depend on the systems to be evaluated. Here, we use resilience in socio-ecological production of landscapes and seascapes (SEPLS) as an example to illustrate the major principles involved in quantifying resiliency. Ecosystems, along with local communities, may experience disturbances and pressures of different types and magnitude, such as market decline, extreme weather conditions, demographic and socio-economic changes. Forests, farmlands, lakes and other habitats are affected by natural disasters like storms and droughts. Actually, all natural resources are affected to some degree by human-induced pressures such as deforestation, pollution and soil erosion. Human societies are impacted by events such as economic crises and political unrest, causing changes in the way ecosystem goods and services are used. These disturbances can affect the living of local communities, e.g. reduced production and increased prices of commodities and crops. In addition to these impacts, ecosystems are influenced by continuous climatic changes and political practices. Some of the indicators of SEPLS are described as follows:

1) *Ecosystem protection*: From the perspective of climate change, the restoration and protection of forests, watershed and coastal ecosystems help regulate water ecology and

climate, thereby providing a safeguard against extreme weather events, floods and droughts.

2) *Biodiversity*: Biodiversity provides material for adaptation, innovation and experimentation. Conditions of saline tolerance, drought, cold and persistence against pests and diseases help biodiversity in adapting to various soil and climatic conditions.
3) *Livelihood and wellbeing*: The resilience of a system/organization is related to the presence of functional and efficient infrastructure such as communication, health and education to meet various community needs and aspirations.
4) *Governance and social equity*: Youth, women and the elderly have skills and knowledge related to biodiversity that should not be ignored.
5) *Knowledge and innovation*: Innovation, experience and learning within and between different age groups, cultures and knowledge systems increase the resilience of communities.

To assess sustainability, one should use leading indicator sets that are based on: management commitment, learning culture, awareness, preparedness and flexibility (Wreathall 2006). The leading indicators like empowerment, problem identification, individual responsibility, individual feedback, anonymous reporting, vessel responsibility and, at the organizational level, like prioritizing for safety, organizational structure and effective communication have been identified (Herrera and Hovden 2008). Many countries have developed their own tools to assess the infrastructure to sustainability requirements (Table 7.2). The tools comprise a number of indicators that reflect the sustainability criteria of each country.

Table 7.2 International sustainability tools.

Tool	Criteria	Country
CASBEE®	• Indoor environment • Resources and materials	Japan
LEED	• Sustainable sites • Indoor environment quality	USA
AQUA™	• Acoustic and visual comfort • Water use and building operation	Brazil
HQE™	• Acoustic and visual comfort • Construction processes	France
Green globes™	• Resources, energy and indoor environment	Canada
DGNB	• Ecological quality • Health and wellbeing	Germany
BREEAM	• Health and wellbeing	UK
Green star	• Indoor environment quality	Australia
Green mark	• Energy efficiency	Singapore
BEAM	• Indoor environment quality	Hong Kong

Source: Adapted from Achour et al. (2015).

Table 7.3 Score guide for resilience assessment.

Scores	Trends
Very high (5)	Upward trend
High (4)	
Medium (3)	No change
Low (2)	Downward trend
Very low (1)	

Source: UNU-IAS, Bioversity International, IGES and UNDP (2014).

Table 7.3 gives guidelines for assessing resilience indicators based on a 5-point scale. If the indicator score is 5 or 4, then the system is considered to be following an upward resilience trend when compared with the previous assessment. If the indicator score is 3, it can be concluded that no change has occurred in the system in terms of resilience since the last assessment. If the indicator score is 1 or 2, it can be concluded that the system is following a downward resilience trend compared with previous the assessment. The score is assigned based on the questionnaire and survey. For example, *for recovery and regeneration of landscape*, the question in the survey could be: "Does the landscape have the ability to recover after extreme environmental conditions?"

7.6 Some Resilient Systems and Associated Codes/Standards

Resilient systems are systems that are designed in such a way that they are able to resist, recover and adapt effectively and quickly from disruptive events (natural disasters or man-made threats). Such a system includes interacting communities, operations, infrastructures, stakeholders, networks and processes that collectively work together to function as a whole. Efficient decision making in designing, preparing, responding to, recovering from and adapting toward disruptions helps in building resilience in systems. Resilience in an engineered system should cover the following:

- Identifying capabilities of the system required.
- Identifying the adverse situations in which the capabilities would be delivered.
- Identifying the system engineering to make sure that it can deliver the capabilities that are required.

As shown in Table 7.4, several agencies like the International Organization for Standardization (ISO) and the National Institute of Standards and Technology (NIST) are charged with developing standards, policies and metrics regarding resilience. They provide guidelines for applying measurements and standards to the resilience of physical, organizational systems or industry practice. NIST is developing new standards and guidelines to fill the remaining gaps along with forming and promoting standards that can be

Table 7.4 Agencies responsible for issuing standards and codes to build resilience in a system.

Agencies/committees governing standards	Category
• International Organization for Standardization (ISO) • National Institute of Standards and Technology (NIST)	Social security, risk management, systems and software engineering, telecommunications, networks, infrastructures
• ASIS International	Organizational resilience standards
• Department of Homeland Security (DHS) • International Association of Plumbing and Mechanical Officials (IAPMO) • International Code Council (ICC) • National Institute of Building Science (NIBS) • Building Owners and Managers Association (BOMA)	Buildings and infrastructures
• American Association of State Highway Transportation Officials (AASHTO)	Transportation
• American Lifelines Alliance (ALA) • American Water Works Association (AWWA) • Institute of Electrical and Electronic Engineers (IEEE) • North American Electric Reliability Corporation (NERC) • ASCE Technical Committee for Lifeline Earthquake Engineering (TCLEE)	Water, electrical and communication systems
• National Academics, Disaster Management and Homeland Security • United Nations International Strategy for Disaster Reduction (UNISDR)	Resilience of the built environment

Source: Young and Hanners (1973) and Marchese et al. (2018).

implemented to enhance resilience. The codes and standards used in the built environments can be categorized as:

- Codes and standards for buildings and infrastructures that ensure minimum performance criteria and structural stability to meet minimum life safety requirements during hazard events.
- Codes and standards for roads, railways, bridges and ports that provide minimum life safety performance criteria during disruptive events. They must also account for the degradation and aging of the facilities that will be exposed to such events.
- Standards for electricity, water and communication systems that ensure reliability.

Communities need certain guidelines to adopt codes and standards for achieving a minimum life safety and identifying situations where these standards can be applied. Adopting current codes and standards for a model is necessary. However, this may not be enough to achieve resilience. This gap between implementing the codes and achieving resilience has to be reduced. Some resilient systems and associated code/standards are introduced below.

7.6.1 Infrastructures and Building Systems

Building systems comprises architectural, structural, mechanical, electrical, safety, security, technological, communication, plumbing and information technology systems. To achieve resilience, all the components of a building should function at the same level. Gaps between the performance criteria for structural and non-structural systems should be identified and reviewed. Performance goals and requirements for resilient buildings need to be identified and, accordingly, codes and standards need to be achieved.

- *ASCE 7-10*: The American Society of Civil Engineers (ASCE) has issued standards for minimum load requirements for buildings and other structures that are set for strength design and allowable stress design. They include the structure's own weight, rain loads, loads associated with occupancy, and loads from flooding, earthquake and winds. There are several risk categories (I–IV), which relate to structural failure and life safety or welfare. ASCE 7–10 list the material standards (e.g. wood, steel) for obtaining structural resistance.
- *ASCE 41-06*: This was developed from FEMA 356 (Federal Emergency Management Agency), Pre-standard and Commentary for Seismic Rehabilitation of Buildings. ASCE 41-06 consists of the guidelines for seismic rehabilitation of existing buildings which uses the latest generation of performance-based techniques. It provides knowledge on earthquakes and serves as a valuable tool in improving the performance of buildings and infrastructures in the case of future earthquakes (Poland and Mitchell 2007).
- *ISO 22316*: The ISO is a worldwide federation of national standards bodies. It provides guidelines to enhance organizational resilience of any size or type. It includes principles for enhancing resilience in an organization, attributes that describe the characteristics of an organization, and activities to guide the utilization and enhancements of attributes. The standards don't promote uniformity across all organizations, as specific objectives are altered to suit an organization's needs (ISO 2017).

7.6.2 Transportation Systems

These include roads, railways, bridges, ports and tunnels. Various agencies like the Federal Highway Administration (FHWA), the Federal Transit Administration (FTA) and the Federal Aviation Administration (FAA) are responsible for providing financial and technical support for transportation. Design standards and guidance are provided by the American Association of State Highway and Transportation Officials (AASHTO) for rail, air, highways and public transportation and the Department of Defense (DOD) for piers, seawalls, ferry terminals, etc. Codes and standards developed by these agencies focus on addressing challenges, which include improvements in safety, infrastructure and work efficiency. Intermodal transportation such as ship to truck should also be considered.

- *AASHTO T 307*: Standard methods of testing to determine the resilient modulus of soils and aggregate materials are issued by AASHTO. These consist of certain procedures and guidelines for preparing and testing untreated soils and untreated base materials for determination of resilient modulus under stimulated conditions and stress states of materials underneath pavements subjected to load and stress (Groeger et al. 2003).

- *ALA*: The American Lifelines Alliance (ALA) is a partnership project funded by FEMA and managed by the National Institute of Building Sciences (NIBS), who govern certain guidelines with the goal of reducing risks from hazards and improvement and development in transportation systems. ALA is developing guidelines to improve transport systems in hazardous events, which can be manmade or natural.

7.6.3 Electrical Systems

These include generation, transmission and distribution systems of electricity, which are owned and managed by electrical cooperatives or municipalities. Standards and regulations are governed by the Federal Energy Regulatory Commission (FERC), the North American Electric Reliability Corporation (NERC) and the Nuclear Regulatory Commission (NRC) for reliability, cost of structures and energy utilities. They address maintenance and improvements in systems in the case of disruptive events and implement coordinated management and efforts to improve the system. Disruptive events can result in partial failures, which are addressed by redistribution of power, or complete failures, which result in loss of power. Plans for small events like lightning strikes or falling trees, as well as hazard events like tornadoes, earthquake and hurricanes have to be properly made, identified and enacted.

- *ANSI-HSSP*: The American National Standards Institute's (ANSI) Homeland Security Standards Panel (HSSP) Steering Committee has issued standards and guidelines based on performance for electric power systems. HSSP was established by ANSI in February 2003 and was originally developed for establishment of voluntary standards in relation to homeland security and emergency preparedness. It addresses both traditional security threats (e.g. nuclear detectors, decontamination standards, etc.) and modern security field approaches (e.g. cybersecurity etc.).
- *FERC*: FERC has issued mandatory standards to support resilience and reliability of the nation's bulk system. The proposed standards provide accurate reporting of events to examine the impact on reliability of the grid, clarify the procedures to execute system restoration and to refine the elements of the plan if the primary control is lost.

7.6.4 Water and Wastewater Systems

These include both water and wastewater systems, which are composed of treatment facilities like filtration, disinfection, sedimentation and transmission systems, and storage tanks. Water and wastewater systems are owned by government and municipalities. The American Water Works Association (AWWA), (ASCE) and the International Building Code (IBC) govern the design of water and wastewater systems. Certain standards evaluate the threats, consequences and vulnerabilities in the system such as the Risk Analysis and Management for Critical Asset Protection (RAMCAP) Standard for Risk and Resilience Management of Water and Wastewater Systems (ANSI/ASME-ITI/AWWA 2010). The performance goals for water systems are also considered based on the requirements of the Environmental Protection Agency (EPA). Seismic standards for pipelines and sewers are needed and should be developed and designed for resisting and recovery practices.

- *ANSI/ASME-ITI/AWWA J100*: The RAMCAP Standard for Risk and Resilience Management of Drinking Water and Wastewater Systems was issued by ANSI and was effective from July 1, 2010, and reaffirmed in December 2013. It was the first voluntary consensus standard developed by AWWA consisting of all hazards, risk and resilience management processes for water and wastewater utilities (Marchese et al. 2018). RAMCAP standards identify and evaluate threats and vulnerabilities by using qualitative or quantitative techniques.

7.6.5 Communication Systems

Communication systems include radio towers, cable systems and satellite systems, which consist of networks, data transmission systems based on optical, radio or electronic signals that operate phone, televisions, internet and radio systems. They can be privately held or owned and regulated by state and local agencies. Due to hazardous events, a long-term loss of communication systems can prevail. Thus, performance criteria based on risk assessments with the priority of resilience and consistency in the community must be developed.

- *TIA/EIA-568*: The Telecommunication Industry Association (TIA) issued a set of standards related to telecommunication that address commercial building cabling for networking services. It was first revised in 1991. These standards provide practices for design and installation of cable systems to support and improve existing and future services.
- *ISO/IEC 11801*: Information technology – generic cabling for customer premises specifies multi-vendor cabling systems for use within premises, which may comprise single or multiple buildings. The international standard specifies structure and confirmation for generic cabling, performance requirements for cabling links and channels, conformational requirements and verification procedures. The standard defines various channel classes and cable categories of copper interconnect.

In addition to telecommunications, gas, water, transportation or electricity networks, improving social networking also serves as a component of resilience. In order to act as a resilient society, communities must come together to adapt, recover and move forward. Resilience must be divided into different time zones, such as immediate (evacuation and healthcare services), intermediate (forming communities for social connections) and long-term (recovering, rebuilding) as followed by Japan during and after the Fukushima Daiichi nuclear disaster on March 11, 2011.

7.7 Communal Resilience and Built Environment Sustainability

Climate change and various demographic movements in cities are causing disasters to occur more frequently and with higher intensity. In many cases, government and supporting organizations and agencies cannot always help people immediately when there is a disaster. The official emergency preparedness guide of Canada asks families to take care of themselves and their families for a minimum period of 72 hours. Thus, the focus has been shifted to helping communities to help themselves with an emphasis on understanding the factors that make a community resilient.

Community resilience was first developed by Judith Kulig and other colleagues working in health promotion in Alberta, Canada (Kulig et al. 2009). Community resilience was derived from theoretical work on risk and was applied to individuals. Disaster resilience and recovery have been studied from a range of different perspectives like sociology, geography, policy implementation, decision making and urban planning. For this reason, it is challenging for researchers as there is a wide range of definitions. Still, there is no universally accepted definition of community resilience, and the way it is defined affects how it is measured and enhanced (Cohen et al. 2010). The Conjoint Community Resiliency Assessment Measure (CCRAM) defines community resilience as "the ability of a community to withstand disruptions and crises" with a focus on certain variables related to leadership, preparedness, collective efficacy and social trust. The Communities Advancing Resilience Toolkit (CART) describes it as "the community which has the ability to transform the environment through deliberate, collective actions and requires that the community as a whole must cope effectively with and learn from adversity" (Patel et al. 2017; Scheunpflug and Asbrand 2006). A frequently used definition of community resilience is given by the National Academies. It states community resilience is "the ability to prepare and plan for, absorb, recover from, and more successfully adapt to adverse events" (the four abilities) (National Research Council 2012). The remainder of this section describes several topics on community resilience.

7.7.1 Characteristics/Elements of Community Resilience

The role of communities in coping with disturbance or crises is dependent on certain dimensions. These dimensions have important impacts and influence on participating individuals' behavior and ability to overcome an emergency. Such dimensions are called as community characteristics or elements, which are proposed to be significant within the concept of community resilience (Cohen et al. 2010). These concepts are related to each other and together form a particular social character of the community. These concepts include:

1) *Collective efficacy and empowerment*: The effects of crises can be made less severe if a community understands its existing vulnerabilities. If these vulnerabilities are addressed properly, then this could build resilience within a community. Collective efficacy can be defined as a community's shared belief in its ability to overcome hardships caused by a disturbance or disaster. The community should know and understand their processes in order to respond better to crises, for example, through self-reliance.
2) *Social capital*: Social capital is a feature of social life and used to describe aspects of networks, relations, norms, trust and power that help individuals to come together and work on a shared objective. Its basic idea is that individuals access and use resources involved in social networks to gain returns (Seibert et al. 2001). Social capital can be defined as the degree to which the community's resources are invested in social relations. Originally, the concept was developed for urban and suburban communities, but it has been stretched to villages, rural areas, neighborhoods, etc. Social

capital is a useful dimension for community resilience because it offers a dynamic image to understand external and internal relationships of community, and also it captures social elements like sharing and reciprocating, which are important for a community perspective.

3) *Social networks and social support*: Social networks refer to the nature of networks and linkages among individuals. They help to increase the ability of the community to use materials, resources and information effectively to respond and recover during and after crises. This will help to improve an individual's decision-making ability and, thus, reduce damage from a catastrophic event. They can include families, friends, colleagues, work groups, recreational, religious and other organizations. Social networks serve as a theoretical framework for assessing the adversity of crises. Social support refers to the emotional and instrumental support that the individuals receive from others, especially during the time of crises. Received support helps in mobilization during and after crises by associating positivity with the difficulty of threats.
4) *Communication*: Effective communication is an important element in building resilient communities. Effective communication will occur if the community members use common meanings for everyone to understand with regard to their needs, attitude and views, and provide opportunities for open dialogue. Effective communication will be dependent on the establishment of appropriate communication infrastructures to coordinate in both pre and post-disaster events. A diversity of mode and content has to be used, for example, using a trusted and established source on social media to promote emergency messages. Accurate information about possible present and future threats should be provided both before and after crises. During a disaster, up-to-date information about impact and relief plans should be provided. During risk communication, norms and beliefs of the community should also be considered in order to ensure that the message addresses the expectations of the members and is placed in a way that helps people understand it.
5) *Governance/leadership*: Good governance or leadership has a significant role in every step of dealing with a crisis: pre-event, i.e. while constructing emergency plans, during the crisis itself, and after the crisis and during the recovery period. The authority should be able to guide the members in a better and direct way, and also adjust the procedures and policies for the aftermath. Governments have to develop plans that are comprehensive to establish continuity after the disturbance. Trust among the members and the leader can help to build better resilience processes as it leads individuals to develop more informal relations. With more public involvement, a sense of community empowerment is an added advantage with good leadership.
6) *Economic investment*: Direct and indirect economic costs due to threats can affect a community long after the disruptive event. Thus, it is very important to address the economic situation after the crisis, which involves:
   a) supply of financial funds;
   b) economic planning to make it cost effective
   c) economic development of infrastructures and building systems after the crisis.

   The community's economy has to be rebuilt by identifying a useful economic policy framework. Addressing the post-disaster economy of a community is significant not just for recovery, but also for reducing future crisis risks.

7) *Preparedness*: Management systems in a community should plan crisis-response processes before a disaster. Risk assessment, involving community stakeholders in planning and running practice drills based on the risk management plan will contribute to community resilience. Activities like reinforcing buildings for better resistance, conserving resources and using sustainable methods will boost resilience. Thus, planning measures and overall preparedness will allow increased response and recovery by the community.
8) *Mental outlook*: Mental outlook can be defined as feelings, attitudes and viewpoints while facing crises or uncertainty or contemplating future threats. The search for meaning after the crises can affect an individual's or community's attitude. It is more concerned on the long-term impact on the affected population. Thus, a fit mental outlook is important in influencing the willingness of individuals to continue after the crises. Hope is a vision of community which describes a better future (Urmson et al. 2016). In addition, adaptability also helps the community members to accept change, recover and adapt to the new environment.

7.7.2 Community Resilience as a Strategy for Disaster Readiness

Disasters like Hurricane Katrina on August 29, 2005 and the terrorist attacks of September 11, 2001 have increased the efforts for better preparedness against disasters over the century. The perspective of community resilience does not decrease the importance of individual agency. Community resilience provides different ways in which individuals can use their surrounding resources for healing and coping. It has always been an important topic in the discussions and debates on the challenges and complexities of disaster management strategies. Some authors theorize about the role of community resilience in local readiness and preparedness, while others have shown how community resilience can help build post-disaster psychosocial relations (Kirmayer et al. 2009). There are several adaptive ways to provide direction for enhancing community resilience against disasters (Norris and Elrod 2006):

1) To increase resilience, communities should work on developing economic resources, reduce resource inequities, give special importance to the areas of greatest vulnerabilities and thus reduce risks. To decrease vulnerability, Godschalk (2003) recommended that urban areas should set aside resources for making poor areas with high risks, safer. The needs and mitigation approaches have to be identified for each neighborhood, and mitigation activities should be combined with social and economic development in order to achieve the multiple objectives required for resilience. People who need support and have the least access to it can be targeted in this way. Political action and problem-solving methods could be used to reach such dynamics.
2) Being one of the primary capitals of community, local people should be involved in every step of the mitigation process. Community members should be able to address their own vulnerabilities to disaster. They should identify and invest in their networks and improve their own capacities to solve the difficulties (Longstaff 2005). Practices should focus on empowerment; i.e. communities should emphasize strengths, use the community's capabilities and make it self-sufficient.

3) Involvcments are needed that will improve and protect the social supports of community in trauma and in the aftermath of a disaster. Social support helps families and communities to maintain their capacity for emotional and mental support. Also, it will help people to build better networks and keep members informed about each other's needs; thus maintaining resource equity. If communities are able to build social skills and mutual support during social involvement, then these social supports are most effective as they help individuals to improve their knowledge, skills, attitude and emotional stability, which can be used as their own support systems (Hogan et al. 2002).
4) To rapidly mobilize support and aid services during an emergency for survivors, pre-existing relationships and organizational networks serve as a key point. Organizational networks that are cooperative provide flexibility and linkages. Various case studies of mental health conditions in response to major disasters, including the September 11, 2001 terrorist attacks and other disasters (Norris and Elrod 2006), have revealed that developing networks and organizational agreements is very important and how key stakeholders will be involved in the case of a disaster. After a disaster, in order to work effectively, it is essential to understand and trust each other, which can be challenging if the systems have not worked together before.
5) Communities planning for a disaster should also be ready for uncertainties. There will be times when there will be no plan available; in that case, communities should be flexible and build information and communication to face the unknown. Flexibility remains one of the important resources for disaster management (Norris and Elrod 2006). Using trusted sources of information for decision-making and problem-solving approaches will allow improvement.

Thus, for a resilient city, its organization should be planned ahead and acted on flexibly and spontaneously. Economic resources, organizational networks and problem-solving approaches and initiatives should be developed. Communities should set goals to prepare and adapt to risks in the process.

7.7.3 Integrating Communal Resilience and Environmental Sustainability

Environmental sustainability is important to increase the efficiency of buildings and overall system design, and to reduce the impacts of threats due to any disruptive event. A resilient community will be able to respond and thrive in stress conditions and shows an ability to plan and mitigate in contexts of change, thus making resilience primary to sustainability. Communities cannot control the events that affect them. They constantly adapt, recover and change with respect to the threats and shocks of a disruptive event. Thus, a community's resilience is said to be dynamic, and changes with the changing external forces. The more resilient a community is, the more it is able to absorb and recover from larger shocks and, thus, can be more sustainable. Managing resilience within communities increases sustaining development where sudden shocks and threats are most likely. It will not only help the overall system to cope with the surprise for the present generation, but also meet the needs of future generations. In recent years, resilience has been integrated into urban planning and design to make infrastructures and the built environment more adaptive to disruptive changes.

The idea of sustainability does not solely focus on immediate and future needs, but also on a central ground to improve short- and long-term resilience within communities. For better environmental sustainability, understanding various components of community resilience is necessary. Interrelations within the various subsystems of the community help to maintain the balance and flexibility required to improve resilience within the community, and contribute toward sustainability. The resources on which the stakeholders of the community depend and contribute to are called community capital. A comprehensive framework among all the community capitals (environmental, human, social, cultural, public structural and commercial capital) is required to build total community sustainability. A sustainable built environment aims at reducing the consumption of resources, decreases degradation in the environment, and generates a better environment and overall system for living, which can be done by planning and building resilience within a community.

Built environment sustainability will only be achieved if commitments are long-term; that is, planning efforts of the community must be ongoing. Proper planning, support from stakeholders, powerful networking among society, communication at different levels, problem-solving approaches and initiatives of the resilient community serve as the characteristics of sustainability. Resilient communities empower groups to come together and adapt to shocks or threats to survive any disruptive event, improve infrastructural systems and build efficient networks, improve wellbeing, form a long-term healthy environment and maximize sustainability.

7.8 Risk Analysis and Resilience

In the case of a disruptive event, the infrastructure system is damaged, and the overall community is affected, damaging its functionality and altering the recovery rate of the system. The resilience and the performance of a system largely depend on the codes and standards that are adopted and enforced. However, even if a community adopts all the current codes and standards, there still will be inconsistencies as they are independently developed through various different processes and with varying performance criteria. With the increase in frequency and intensity of climatic events, there is no method to precisely calculate the long-term effects, the distribution of disruptive events and the impacts on infrastructure. Building resilience for entire large complex systems serves as the best method in the wake of such uncertainties and unanticipated shocks.

Risk analysis identifies vulnerabilities that the components of a system suffer from and identifies the loss of functionality due to the occurrence of adverse events (Linkov et al. 2014). Risk management focuses on managing the components of a system such that the system withstands the identified shocks and system failure is prevented. However, the problem with such forms of protection is that the risk analysis of components of a large complex system is dependent on time and money, and our ability to understand and manage climatic extremes is still not clear due to the vulnerabilities and unpredictability of the events. To address these problems resilience should be built in along with risk analysis and management in a specific system. Risk analysis will help to prepare for and prevent the consequences of disruptive events; however, resilience in systems will help to recover and adapt from adverse events. Developing such a capability should include (Linkov et al. 2014):

- defining and measuring resilience using specific methods;
- developing new and better modeling techniques to be used for complex systems;
- developing resilience engineering covering all the individual components of a large system;
- approaches and methods for communicating with stakeholders.

Resilience cannot be said to be a substitute for risk management or system design and is differentiable from, but complementary to risk analysis. Resilience serves as an attribute that improves traditional risk management by using adaptation, mitigation and recovery techniques. Risk analysis is dependent on identification and characterization of three components: threat, vulnerability and consequences of adverse events to determine the lowest point of the system's critical functionality. Once risk analysis has been done, a proper resilience management framework should be built that includes risk analysis as a central component. Highly resilient systems can also improve the functionality of the system with respect to its initial performance and enhance the overall system resilience to future disruptive events.

Threats from disruptive events do not affect an individual domain alone, but multiple domains of a system. However, our current efforts focus on one domain rather than defining systems in multiple domains. Various domains (e.g. physical, social, information domain) should be combined with the definition of resilience to build a matrix where each stage, i.e. planning, absorbing, recovering and adapting, is assessed in each individual domain (Linkov et al. 2014). Clearly, resilience serves as a strategic tool. The more a system is resilient, the less vulnerable it is. Use of the matrix approach requires us to have a detailed understanding of how the system behaves and functions during disruptive events. Use of modeling and simulation techniques allows us to fix errors, plan measures, identify weak spots in the system and, thus, prepare for threats.

After defining the system and developing models and simulation techniques, the focus should be shifted to resilience engineering. It builds resistance, adaptability and the ability to recover from adverse events. Resilience engineering not only focuses on building resilient infrastructure but also identifies critical functionalities of a system. Adaptive materials, automated sensors and energy self-sufficient techniques are some examples of techniques for resilience engineering that decrease the risks associated with threats and improve the odds of efficient recovery. The functionality of one network often depends on another network. For example, during hurricanes or floods, the power remains off in affected areas, and the flooding prevents repair of the damaged facilities. Resilience can be improved by identifying and enhancing the interconnectivity between networks. In such cases, alternate power systems or functional capacities of a given network have to be identified where the original networking systems are impacted.

Disruptive events (natural or manmade) might result in shocks or threats that cannot be precisely analyzed or predicted using traditional risk analysis and management techniques. Resilience will be a complementary attribute to these techniques and can also serve in developing sustainability, as resilient systems work on minimizing the negative and adverse impacts of disruptive events and help to improve the functionality of the system after the events. Various standards, policies and regulations for risk analysis and management have been built by organizations and nations. However, such an approach is no longer sufficient to address the threats, adapt and recover from a disruptive event due to the increased complexity in society's critical networks. Therefore, there exists an urgent need for a framework that will act as a complementary system to risk analysis and management. Integration of resilience will help to identify the complexity and uncertainties associated with an event. Furthermore, developing models and framework for measuring resilience, implementing engineering and management systems and establishing transparency on resilience with stakeholders are essential for consistency and addressing the complexities and uncertainties of future threats.

7.9 Integration of Sustainability and Resilience

7.9.1 Resilience as Part of Sustainability

Although sustainability and resilience are different ideas, incorporating resilience as part of sustainability can be beneficial for the system and its surrounding environment. For example, the sustainability model was first introduced in the USA in 1969 and was later adopted in many countries; the model is about assessing the impact of a project or an activity on the environment, society and economics. Since 2000, several extreme weather events have taken place, which have affected hundreds of thousands of people across the UK. These events increased concern for improved critical infrastructure resilience. Disasters are now on the priority list of many countries, specifically after the earthquake experienced in Japan in March 2011 along with rising climate change events across the globe. The vulnerability and risks associated with critical infrastructure and the built environment are also accounted for in many countries (Achour et al. 2015). Incorporating resilience as part of sustainability will ensure the efficacy of disaster management and safety measures as well as the social and economic development of a country.

7.9.2 Sustainability as Part of Resilience

The idea of resilience takes into account both the system and its relation with the surrounding environment. Thus, sustainability is an integral part of resilience as the environment will be taken care of for the current and future generations (Achour et al. 2015). Incorporating sustainability as part of resilience will ensure that buildings and infrastructure are economically, environmentally and socially viable besides being resilient enough to cope with disruptions.

Sustainability and resilience are still dealt with as two different aspects, which is a major problem leading to improper budgeting and inappropriate allocation during project planning. As shown in Figure 7.2, there are possibilities for integrating sustainability and resilience. This combination will ensure that buildings and infrastructure are resilient enough to cope with disruptions and natural disasters along with being economically and socially viable (Achour et al. 2015). However, integration has not been achieved in many countries due to a lack of awareness and motivation among key decision makers (Achour et al. 2015).

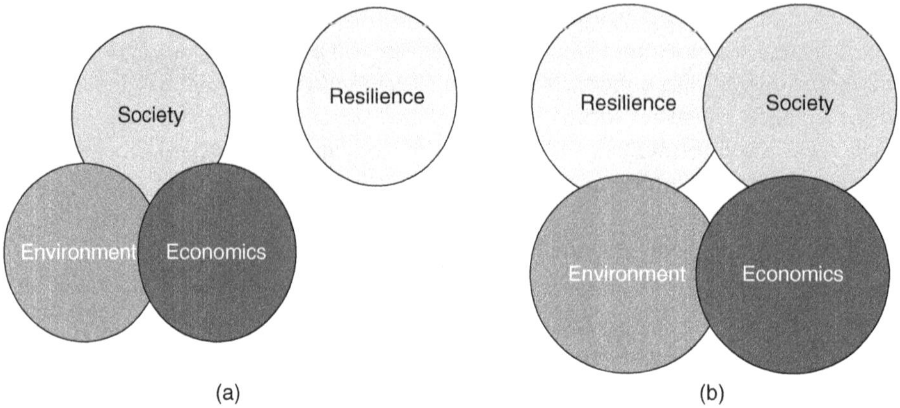

Figure 7.2 (a) Sustainability and resilience before integration and (b) integration of sustainability and resilience.

There have been several theories for the integration of sustainability and resilience. Nishida and Hua (2011) reported that stakeholders were involved to design the *Tokyo Cap-and-Trade Program* (TMG 2010), which aimed for carbon reduction in public, commercial and private buildings through the involvement of stakeholders in the design and planning (Nishida and Hua 2011). Du Plessis and Cole (2011) stated that buildings are complex and stakeholders have different interests. For the sake of increasing stakeholder participation, the following objectives should be emphasized:

- identification of appropriate combination of policies and incentives;
- cooperation and innovative decision-making processes;
- changing attitudes of people and providing standards for the behavioral judgment of individuals and society.

The Hyogo Framework for Action (HFA) 2005–2015 was developed for the integration of sustainability and resilience. In this framework, five priorities were identified for integration and were supported by guidelines to integrate sustainability and resilience, as shown in Table 7.5. The guidelines were further classified into four major elements: political (including legal), social, technical and strategic planning (Figure 7.3). The initiative can be taken from any side (technical, political or social) to inform strategic planning, which develops the necessary strategies, targets/gives feedback to technical and political teams for enforcement and execution of the strategy, as well as informing social elements about the emerging strategies (Achour et al. 2015).

7.10 Conclusions

Over the past few decades, the importance of the concepts of resilience and sustainability has been well documented among researchers and practitioners. Efforts have been made to measure the resilience of engineering systems, but challenges always exist. In this chapter, we have tried to provide better explanations of two diverse concepts, i.e. resilience and

Table 7.5 Integration of resilience and environmental preservation in the Hyogo Framework for Action (HFA).

Element	HFA priorities	Guidelines
Political	Assurance that disaster risk reduction (DRR) is a national and a local priority	• Community participation • Foster political commitment • Legislation to DRR
Technical	Identification, assessing and monitoring disaster risks	• Early warning • Data and information sharing • Climate modeling and forecasting • Scientific and technological development
Social	Awareness, cooperation, training and learning	• Information sharing and cooperation • Public awareness and media • Inclusion of DRR in school curriculum and education
Strategic planning	Reducing the underlying risk factors	• Food security • Protection of critical public facilities • Rural development plans and DRR • Land use planning and building roles • Public–private partnership • Recovery schemes of social safety
Technical	Strengthening disaster preparedness	• Voluntarism and participation • Emergency funds • Disaster management

Source: Adapted from Achour et al. (2015).

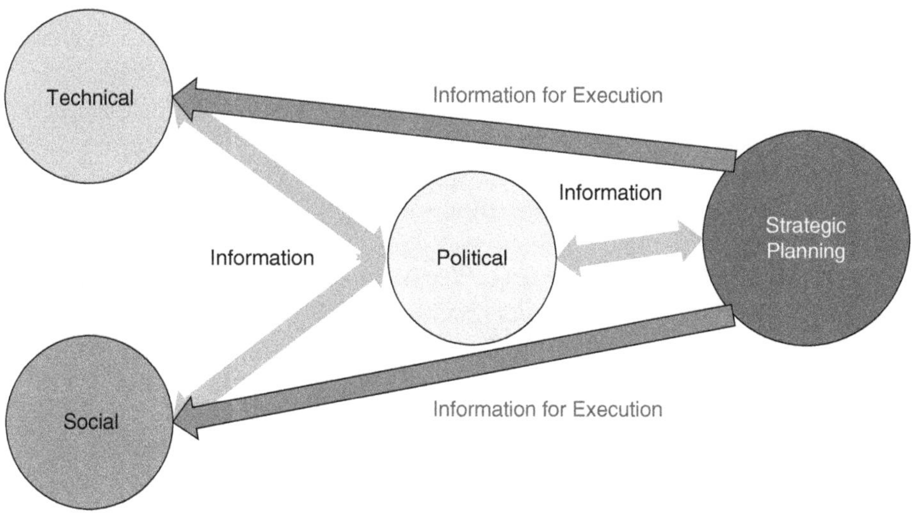

Figure 7.3 Resilience and sustainability integration process.

sustainability. Efforts have been made to explain resilience and quantification systems to measure system resilience and its linkage with sustainable development. The different terms and their quantification are well explained inside the chapter, using diverse examples taken from the studies done up till now. All the available assessment approaches are explained, such as conceptual frameworks, general measures, semi-quantitative indices and many more. Some approach measures are static and stable in nature, while other approaches measure system performance dependent on time.

Nowadays the term "resilience" is increasingly used in systems and emphasis has been given to using the facts and figures obtained after resilience assessment. Still, the terms are not fully explored and further research is needed, including intangible resource allocation models, trades among different domains and their relationship in the context of a sustainable environment to ensure resilience. Resilience engineering strongly supports the pre-requirements for sustainable and efficient systems to standby disruption or adverse conditions. Resilience engineering meets numerous of the most challenging technical objectives of modern system growth and overall risk control systems.

References

Achour, N., Pantzartzis, E., Pascale, F., and Price, A.D. (2015). Integration of resilience and sustainability: from theory to application. *International Journal of Disaster Resilience in the Built Environment* 6 (3): 347–362.

Adjetey-Bahun, K., Birregah, B., Châtelet, E., Planchet, J.-L. and Laurens-Fonseca, E. (2014). A simulation-based approach to quantifying resilience indicators in a mass transportation system. Proceedings of the 11th International ISCRAM Conference – University Park, PA (May). https://pdfs.semanticscholar.org/35f2/d6da64694182b59b9e2e1b1c013fa02ecfaf.pdf (accessed November 2019).

Albores, P. and Shaw, D. (2008). Government preparedness: using simulation to prepare for a terrorist attack. *Computers and Operations Research* 35 (6): 1924–1943.

ANSI/ASME-ITI/AWWA (2010). *AWWA*. New York: ANSI/ASME-ITI.

Azadeh, A., Salehi, V., Arvan, M., and Dolatkhah, M. (2014). Assessment of resilience engineering factors in high-risk environments by fuzzy cognitive maps: a petrochemical plant. *Safety Science* 68: 99–107.

Bruyelle, J.-L., O'Neill, C., El-Koursi, E.-M. et al. (2014). Improving the resilience of metro vehicle and passengers for an effective emergency response to terrorist attacks. *Safety Science* 62: 37–45.

Carvalho, H., Barroso, A.P., Machado, V.H. et al. (2012). Supply chain redesign for resilience using simulation. *Computers and Industrial Engineering* 62 (1): 329–341.

Cohen, M.J., Brown, H.S., and Vergragt, P.J. (2010). Individual consumption and systemic societal transformation: introduction to the special issue. *Sustainability: Science, Practice, and Policy* 6 (2).

Cutter, S.L., Barnes, L., Berry, M. et al. (2008). A place-based model for understanding community resilience to natural disasters. *Global Environmental Change* 18 (4): 598–606.

Du Plessis, C. and Cole, R.J. (2011). Motivating change: shifting the paradigm. *Building Research and Information* 39 (5): 436–449.

Faturechi, R., Levenberg, E., and Miller-Hooks, E. (2014). Evaluating and optimizing resilience of airport pavement networks. *Computers and Operations Research* 43: 335–348.

Godschalk, D.R. (2003). Urban hazard mitigation: creating resilient cities. *Natural Hazards Review* 4 (3): 136–143.

Groeger, J.L., Rada, G.R., and Lopez, A. (2003). *AASHTO T307 – Background and discussion. In Resilient Modulus Testing for Pavement Components*. West Conshohocken, PA: ASTM International.

Herrera, I. and Hovden, J. (2008). Leading indicators applied to maintenance in the framework of resilience engineering: a conceptual approach. Paper presented at the 3rd Resilience Engineering Symposium, 30. https://www.researchgate.net/publication/228651319_Leading_indicators_applied_to_maintenance_in_the_framework_of_resilience_engineering_A_conceptual_approach (accessed November 2019).

Hogan, B.E., Linden, W., and Najarian, B. (2002). Social support interventions: do they work? *Clinical Psychology Review* 22 (3): 381–440.

Hollnagel, E. (2014). Resilience engineering and the built environment. *Building Research and Information* 42 (2): 221–228.

Hosseini, S.A., de la Fuente, A., and Pons, O. (2016). Multi-criteria decision-making method for assessing the sustainability of post-disaster temporary housing units technologies: a case study in Bam, 2003. *Sustainable Cities and Society* 20: 38–51.

International Organization for Standardization (2017) International Organization for Standardization: Security and resilience – Organizational resilience – Principles and attributes, ISO 22316, 2017. https://www.iso.org/standard/50053.html (accessed November 2019).

Kirmayer, L.J., Sehdev, M., and Isaac, C. (2009). Community resilience: models, metaphors and measures. *International Journal of Indigenous Health* 5 (1): 62.

Kulig, J.C., Hegney, D., and Edge, D.S. (2009). Community resiliency and rural nursing: Canadian and Australian perspectives. In: *Rural Nursing: Concepts, Theory and Practice*, 3e. New York: Springer.

Lew, A.A., Ng, P.T., Ni, C.-C., and Wu, T.-C. (2016). Community sustainability and resilience: similarities, differences and indicators. *Tourism Geographies* 18 (1): 18–27.

Linkov, I., Bridges, T., Creutzig, F. et al. (2014). Changing the resilience paradigm. *Nature Climate Change* 4 (6): 407.

Longstaff, P.H. (2005). Security, resilience, and communication in unpredictable environments such as terrorism, natural disasters, and complex technology. Harvard, CT: Center for Information Policy Research, Harvard University.

Manyena, S.B., Mutale, S.B., and Collins, A. (2008). Sustainability of rural water supply and disaster resilience in Zimbabwe. *Water Policy* 10 (6): 563–575.

Marchese, D., Reynolds, E., Bates, M.E. et al. (2018). Resilience and sustainability: similarities and differences in environmental management applications. *Science of the Total Environment* 613: 1275–1283.

Muller, G. (2012). Fuzzy architecture assessment for critical infrastructure resilience. *Procedia Computer Science* 12: 367–372.

National Research Council (2012). *Disaster Resilience: a National Imperative*. Washington, DC: National Research Council.

Nishida, Y. and Hua, Y. (2011). Motivating stakeholders to deliver change: Tokyo's cap-and-trade program. *Building Research and Information* 39 (5): 518–533.

Norris, F.H. and Elrod, C.L. (2006). Psychosocial consequences of disaster: a review of past research. In: *Methods for Disaster Mental Health Research* (eds. F.H. Norris, S. Galea, M.J. Friedman and P.J. Watson), 20–42. New York: The Guilford Press.

Patel, S.S., Rogers, M.B., Amlôt, R. and Rubin, G.J. (2017) What do we mean by 'community resilience'? A systematic literature review of how it is defined in the literature. *PLoS Currents, 9.*

Pettit, T.J., Fiksel, J., and Croxton, K.L. (2010). Ensuring supply chain resilience: development of a conceptual framework. *Journal of Business Logistics* 31 (1): 1–21.

Poland, C.D. and Mitchell, A.D. (2007). A New Seismic rehabilitation standard – ASCE/SEI 41-06. In: *New Horizons and Better Practices* American Society of Civil Engineers (ed.), 1–2. Reston, VA: American Society of Civil Engineers.

Redman, C.L. (2014). Should sustainability and resilience be combined or remain distinct pursuits? *Ecology and Society* 19 (2): 37.

Scheunpflug, A. and Asbrand, B. (2006). Global education and education for sustainability. *Environmental Education Research* 12 (1): 33–46.

Seager, T.P. (2008). The sustainability spectrum and the sciences of sustainability. *Business Strategy and the Environment* 17 (7): 444–453.

Seibert, S.E., Kraimer, M.L., and Liden, R.C. (2001). A social capital theory of career success. *Academy of Management Journal* 44 (2): 219–237.

Shirali, G., Motamedzade, M., Mohammadfam, I. et al. (2012). Challenges in building resilience engineering (RE) and adaptive capacity: a field study in a chemical plant. *Process Safety and Environmental Protection* 90 (2): 83–90.

Supply, W.U.J.W. and Programme, S.M. (2014). *Progress on Drinking Water and Sanitation: 2014 Update.* Geneva: World Health Organization.

Tadić, D., Aleksić, A., Stefanović, M., and Arsovski, S. (2014). Evaluation and ranking of organizational resilience factors by using a two-step fuzzy AHP and fuzzy TOPSIS. *Mathematical Problems in Engineering* 2014.

UNU-IAS, Bioversity International, IGES and UNDP (2014). Toolkit for the Indicators of Resilience in Socio-ecological Production Landscapes and Seascapes (SEPLS). https://comdeksproject.files.wordpress.com/2014/11/toolkit-indicators-web.pdf (accessed November 2019).

Urmson, K.A., Johnston, D.M., and Kemp, S. (2016). Asking for help and receiving support after a disaster. *Australasian Journal of Disaster and Trauma Studies* 20 (1): 3.

Vlacheas, P., Stavroulaki, V., Demestichas, P. et al. (2013). Towards end-to-end network resilience. *International Journal of Critical Infrastructure Protection* 6 (3–4): 159–178.

Vugrin, E.D., Warren, D.E., and Ehlen, M.A. (2011). A resilience assessment framework for infrastructure and economic systems: quantitative and qualitative resilience analysis of petrochemical supply chains to a hurricane. *Process Safety Progress* 30 (3): 280–290.

Wreathall, J. (2006). Properties of resilient organizations: an initial view. *Resilience engineering: Concepts and precepts*: 275–285.

Young, W. and Hanners, R. (1973). Compound-two-stage resilient isolation mounting for use in attenuating mechanical vibrations. Google Patents.

Part II

Dimensions and Different Aspects

8

Economic Development and Sustainability

L.R. Kumar, Anita Talan, and R.D. Tyagi

INRS, Université du Québec, Québec, Canada

8.1 Introduction

One of the difficult issues in sustainable development is determining what needs to be done to attain sustainable development, assuming it as an overarching goal. The focus of sustainable development has always been on intergenerational equity, but the length of any sustainability time parameter is the subject of keen debate. Some argue that it must be few generations at least, but the number cannot be infinite. Some "coefficient of concern" can be set on how far into the future we should look to meaningfully assess the impact of such developmental activities. Sustainable development can be expanded into "sustainable economic development." Economic development itself encompasses several key indicators of development, such as education, health and quality of life. The human development index (HDI) is an important reflection of economic development. The HDI relatively measures education, life expectancy and gross domestic product (GDP) to indicate a country's achievement vis-à-vis other countries (Pearce et al. 1994).

This chapter focuses on economic development, the relationships between economic development and sustainability, elements and indicators of economic sustainable development, ways to achieve sustainable development including carbon mitigation and using a bio-based economy (BBE), different policies adopted in different countries to achieve economic sustainable development and what needs to be done in order to realize sustainable development in developing countries.

8.2 Different Perspectives of Economic Development

8.2.1 Micro-perspective

The micro-perspective of economic development is all related to microeconomics, which is a branch of economics that studies the behavior of people and organizations in decision

Sustainability: Fundamentals and Applications, First Edition.
Edited by Rao Y. Surampalli, Tian C. Zhang, Manish Kumar Goyal, Satinder K. Brar, and R. D. Tyagi

making, keeping scarcity of resources in mind and the interactions among these people and organizations. The main objective of this is to analyze the market chains that determine the relative price of a commodity and check the use of limited resources. It focuses on the conditions under which free markets gets desirable allocations and that determine the market failure and product efficiency. In in terms of economic development, this market chain needs checking from time to time to understand the market and to analyze the steps toward the economic development of the nation (Hodgson 2017). The elements of microeconomics are:

- supply and demand chain;
- change of economic variable into another variable;
- consumer demand and its theory;
- production theory, i.e. process of conversion of inputs into outputs;
- costs of production;
- time management for high production;
- market structure and its analysis;
- competition:
  - perfect competition;
  - imperfect competition.

8.2.2 Macro-Perspective

In economic development, macroeconomics also plays a crucial role, and is defined as a branch of economics dealing with the function, structure and performance of an economy as a whole. This includes economic parameters at every level: GDP, income ratio, price stability, price distribution, interdependence and interrelations, and unemployment rates are the indicators of macroeconomics. Researchers have developed various models to understand the relationship between different factors of the economy (La Porta and Shleifer 2014). There are many factors defining the macro-development of an economy. But primarily among those factors, three are basic macroeconomic elements, including producers, consumers and workers.

8.3 Indicators for Economic Development

8.3.1 National Output and Per Capita Income

The national output is defined as the total production of a country during a given timeframe. The cost of production and the amount of the product that has been sold is equal to the income of employees in a nation. Therefore, these terms are interchangeable. The output is measured as total cost of the product or sum of all values added to the economy. It is usually measured in terms of GDP or other national accounts. The economic output is increased over time due to technological advancement, machinery accumulation, better education, human resources and other capitals. However, output is not a consistent process, but faces financial declines called recessions. Thus, microeconomics finds solutions to prevent these financial drops, leading to a faster growing economy with long-term effects.

8.3.2 Unemployment

Many economies around the globe are facing many issues of unemployment. The scarcity of jobs is responsible for poverty, which is the main hindrance in sustainable development. There are four types of unemployment in an economy (Figure 8.1). Different causes create one or another type of unemployment. Lack of jobs in an economy increases the third group of society minimizing living needs and quality of life.

The main causes of unemployment within an economy vary from nation to nation. However, there are certain reasons that are the same at a global level and accepted as the main reasons for unemployment worldwide, such as:

- globalization;
- technological advancement;
- competition between international and domestic products;
- lack of financial services;
- increasing competitive advantages;
- lack of education.

8.3.3 Inflation and Deflation

Inflation and deflation are interlinked terms. When the price of a commodity is increased across the entire economy, it is called inflation, while a price decrease defines deflation. These changes are measured by price indexes. Inflation is mostly found in growing countries and during the scarcity of a product when the amount of available product fails to meet the demands of the people, while deflation is mostly seen in declining economies. The government officials of central banks use monetary policies to avoid price change and to balance the nation's money supply chain. Inflation is also controlled by these officials by increasing interest rates or by reducing money supply as inflation has negative consequences and increases the chances of uncertainty (Brink et al. 2016). Therefore, central banks always try to balance the money flow and stabilize the prices to protect the economies from negative consequences of hiked prices.

Price changes are the result of several factors. Price fluctuations in the short term are largely related to monetary factors. But demand–supply mismatch can also trigger price fluctuations. For example, shortfall in demand due to a recession often results in deflation, while a negative supply shock lowers supply and causes inflation. For long-run changes, economists believe the quantity theory of money, which states that the price of goods is directly proportional to the money supply.

The quantity theory of money works on Fisher's equation of exchange (Neely et al. 2014):

$$M \times V_T = \sum_i (p_i \times q_i) = p^T q \tag{8.1}$$

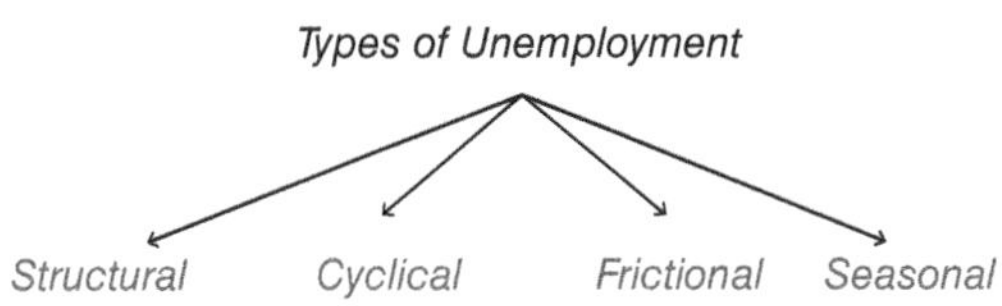

Figure 8.1 Types of unemployment in an economy.

where M is the total quantum of money in circulation; V is the velocity of money, i.e. average frequency of all transactions; p_i is the price of i-th transaction of money; q_i is the quantity of i-th transaction of money; p is the column vector of p_i; and q is the column vector of q_i. Thus, the equation for quantity theory of money is:

$$M \times V_T = P_T \times T \tag{8.2}$$

where P_T is the price value; and T is the real value index.

8.3.4 Gross Value Added (GVA) and Local Value Added (LVA)

GVA and LVA are the indicators providing information about how much the production chain adds to the value of raw materials when they are processed into final sales products. GVA is needed to calculate GDP, and both are overlapping indicators. While GVA describes the economic contribution in broader terms (e.g. as a sector in national accounts), LVA is further restricted to describing impacts on a local (community) level (Herder et al. 2012). These indicators are used in combination with different models to support country-tailored quantitative analysis that can effectively inform the policymaking process for national development planning, which has been addressed in Section 8.7.

8.4 Economic Development for Sustainability

Economic development or economic growth is the main theme defining sustainability and is also most problematic issue of sustainability. The major part of society is not economically balanced, which directly affects the economic growth of the whole nation by reducing employment and thus slowing down the corporate sector. This all also affects the normal pathway required for growth of the economy by sticking to sustainability guidelines. Sustainability guidelines emphasize supporting and encouraging incentives for employees, these guidelines have a cumulative effect on every employee, i.e. even providing incentives for less efficient employees. The high demand in the market and supply is exploiting the natural resource base. Therefore, to prevent damaging the environment and natural resource base it is necessary to control consumption (Buys et al. 2014). The guidelines of economic sustainability seek to meet the needs of society without compromising the environment, natural resources, needs and quality of life especially in developing nations (Ameen et al. 2015). Economic growth is very important for economic sustainability. Economic sustainability can be achieved by:

1) Making new advancements in national and international markets by creating an opportunity for vending things.
2) Reducing energy consumption and inputs, i.e. raw material usage.
3) Reducing cost by using new efficient techniques and improving production.

8.4.1 Elements of Economic Sustainable Development

8.4.1.1 Environmental Economics

When we talk about the environment, it does not just mean the biosphere of earth, air and water, but also includes human interactions with the natural components of the earth and with manmade surroundings. The rapid economic advancement around the globe has put an excessive strain on the ability of the natural environment to absorb and recycle pollutants that are produced as a result of the unprecedented economic growth. Therefore, sustainable solutions are needed in order to maintain global economic growth and to preserve the natural environment as well. In terms of economics, the parameter of environmental quality is considered as a scarce resource. Therefore, it is considered as a resource that needs to be conserved and protected (Steffen et al. 2015). In this scenario, a thorough cost–benefit analysis (CBA) is conducted to evaluate the possible outcomes of any policy decisions involving the scarce resources. This type of analysis primarily focuses on different methods of resource mobilization, a complete evaluation of the decisions taken and the resulting consequences of such decisions.

The CBA considers several methods of solving a problem and then considers the best possible solution paradigm based on the set of consequences that would result from taking a particular course of action. This kind of analysis facilitates us to choose the optimized method that would minimize the damage to environmental quality. Rigorous analysis is conducted to understand the interrelationships of the various components of the environment that are affected by various other outcomes of a particular action which are not accounted for in the CBA (Davoudi et al. 2012).

8.4.1.2 Energy

Energy utilization is as important as other elements of sustainability. Sustainable energy is sourced from renewable resources and can be utilized over a considerably longer duration of time. Non-renewable sources of energy such as fossil fuels and biofuels are responsible for more pollution as they provide a major source of energy as compared with renewable sources of energy like hydroelectric energy, solar energy and wind energy. Keeping sustainability in mind, the use of renewable energy sources is increasing day by day, e.g. most street lights, parking meters and roof buildings use solar energy. Also, wind power is expanding quickly to replace non-renewable sources of energy. In developing countries, the economically weaker section of society largely depends on fossil fuels for their day-to-day energy needs. Traditionally, they have been dependent on fossil fuels as they have limited knowledge about cleaner energy resources. These communities are also not involved during the decision-making process and are even left out of CBA. As sustainable energy becomes the major part of the global energy mixture, the fossil fuel-based energy ventures are replaced by renewables, as a result providing better social equity and inclusive development (Mancarella 2014). By reducing the overall ecological footprint of energy generation process, sustainable development also goes a long way in ensuring economic sustainability and achieving national developmental goals. Thus, such energy-efficient efforts should be encouraged using prudent investment policies and technological know-how.

8.4.1.3 Technology

In the modern world, technology has become the most important element of sustainable development as it is deployed to improve energy efficiency and fulfill the developmental imperatives of society. The technology meeting the needs of sustainable development is often termed as *appropriate technology*, and covers a wide range of technologies. Many studies have emphasized that such technology should be environmentally sustainable, economically viable and citizen-centric. Today, appropriate technology is often developed using open sources, which has given an impetus to Open Source Appropriate Technology (OSAT) and thus every plan to harness such technology and entire methodology can readily be found on the internet (Pearce 2012). Thus, technological development in energy and sustainability has become more transparent and equitable. Many scientists have supported and backed the idea of OSAT as a new model of innovation and a new driver of sustainable development.

8.4.1.4 Transportation

Transportation is the major hindrance to achieving environment sustainability as it is the biggest source of greenhouse gas emissions. Against this backdrop, it is imperative to understand the environmental impact of the entire transportation industry as a third of global greenhouse gas emissions are generated by the transportation system. The emissions produced also include particulate matter, which has a pernicious effect on human health and is directly responsible for climate change. Therefore, sustainable transport solutions will have countless social, economic and environmental merits that would directly enhance local and global sustainability. As per many industrial reports and studies, it has been proved that sustainable transport can enhance employability and make socio-economic opportunities more equitable and efficient. It saves time, increases the per capita income and lessens government expenditure, i.e. government budgets. Steps have been taken to implement sustainable transportation for the long term as well as in the short term. Many countries have implemented sustainable transport and embraced the idea of sustainable cities in a big way. For instance, the city of Freiburg in Germany has developed an extensive network of public transport along with tracks for walking and cycling to minimize the ecological footprint at city level (Moeinaddini et al. 2015). In order to reduce greenhouse gas emissions, the state government has been taking certain steps, such as:

- Encouraging walking and biking.
- Improving public transit by covering larger areas to provide mobility and accessibility.
- Increasing taxes on personal transportation and its services.
- Educating people about behavioral economics to transmit knowledge.
- Developing evidence-based sustainable development policies.

8.4.1.5 Corporate Field

The efficient use of capital is required and is accepted as a criterion for corporate sustainability. This efficiency is calculated in terms of the economic value aggregated to have ecological impact. This is defined as *eco-efficiency*, which is attained by supplying competitively priced goods and services in the market that cater to human basic needs and, at the same time, reducing ecological footprint and preserving natural resources. Another concept considered vital for corporate sustainability is *socio-efficiency*, which considers the relation between an organization's value addition and its impact on society at large. Corporate sustainability is affected by corporate impacts that may be positive or negative for the environment but always have positive social impacts or maximal social impacts. Both *eco-efficiency* and *socio-efficiency* are important for economic sustainability (Hahn and Lülfs 2014). The emphasis is placed on different aspects of sustainable development to ensure economic profitability of the corporates coupled with social effectiveness and ecological efficacy.

8.4.1.6 Income

Reducing poverty is essential for sustainable development. An increase in income is required to fulfill the basic living needs. Income is responsible for the satisfaction of human basic needs. Income sustainability needs proper distribution of services to people to prevent major economic consequences. To meet the basic necessities of an expanding world it is necessary to have income distributed in such a way that people can fulfill their basic requirements. In order to have sustainable income, there must be employment. Other income-linked basic necessities of housing, water, sanitation and health are also economically important for sustainable development (Mamabolo 2017). Therefore, people taking initiative for sustainable development must find ways to provide the basic necessities to every domain of the society using low-cost technologies.

8.4.1.7 Architecture

The paradigm of sustainable cities envisions resilient and environmentally friendly urban landscaping. To achieve the goals of sustainable architecture, recent movements such as *new urbanism* and *new classical architecture* focus on sustainable *modus operandi* toward urban planning and construction. This can be viewed as a responsible approach to the development of smart urban growth, architectural tradition and classical design (Elmualim and Gilder 2014). It focuses on the economic domain and at the same time on long-term architectural landscaping that is more responsible for sustainability in the ecological domain (Wu et al. 2016).

8.4.2 Economic Growth vs. Economic Development: Fundamental Difference

Economic growth is often defined as "the incremental increase in the monetary value of goods and services produced by the country which include contributions from all the sectors of the economy during a particular time period" (Banerjee et al. 2015). Strictly speaking in economic terms, it is a quantitative attribute that measures the increase in the number of business transactions taking place in an economy. Furthermore, this can be expressed in terms of GDP and gross national product (GNP), which determine the volume of such transactions in the economy. It indicates the overall change in the size of an economy. It enables economists to bring out meaningful comparisons in the absolute change and percentage change that capture the growth in an economy (Banerjee et al. 2015). Economic growth is basically a quantitative measure to track the outcome of the increase in resource mobilization and advancement of technology. As the population grows, the size of the economy also increases, putting a considerable strain on the ecological resources available in the country. Figure 8.2 shows the factors responsible for economic growth.

On the other hand, economic development is defined as the process of qualitative improvement in the living conditions of people at large. It is basically defined in terms of progress in the social sphere of a country, such as improvement in education and literacy, enhancement of quality of life, better access to healthcare, etc. Since the last decade of the last century, HDI has been the tool used by economists and policymakers to measure economic development. HDI is calculated as a composite score across three parameters to rank different countries to mark their progress in terms of economic development. It considers all the qualitative parameters of the population at large, such as life expectancy, education and per capita income (Kemp and Nurius 2015). A comparison between economic growth and economic development has been drawn in Table 8.1.

8.4.3 Growth and Sustainability, Mutual Existence and Exclusive

In present-day society, we are totally dependent on natural resources such as water, oxygen and other natural elements, yet still people are not able to understand the dependence of the economy on natural resources. People are unaware and not concerned about this interdependence as they search for happiness through materialism and economic growth or maybe they do not know the consequences of increased GDP or businesses that make a loss.

The significance to knowing about this interdependence is to prevent indiscriminate use of natural resources and this also means keeping a check on unbridled economic growth that is dependent on them. This is the major concern while achieving sustainability. In recent development to attain sustainability, governments and other private organizations have taken actions to prevent environmental degeneration and damage to natural resources. In recent years, many people and nations have started to understand this interdependence between the economy and the sustainability of society and the environment. This is of benefit to economic growth as more the corporate field flourishes, the more the nation will grow and individuals will be given more profits, employment, resources and better quality of life (Almutairi 2015). With advancements in technology the consumer demand and supply ratio increases, giving a high economic value to the nation.

Yet, the economy needs natural resources and other reinforcing materials to facilitate economic growth and technological advancement. But this advancement is also creating a sinkhole for the economy, environment and society as the use of newly developed technology-based products emits more harmful pollutants. Thus, the goals of economic growth are in disjunction with the sustainability of our planet in one way or another. In

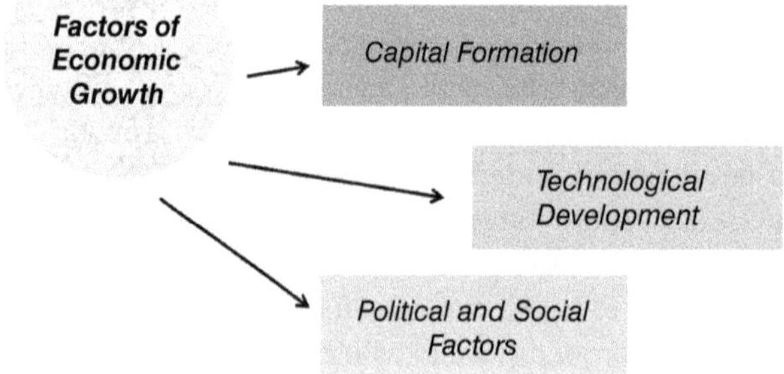

Figure 8.2 Factors of economic growth.

Table 8.1 Comparison between economic growth and economic development.

Basis	Economic growth	Economic development
Meaning	Growth is defined as the quantitative change in the economic output of a country in a particular period of time	Development involves a qualitative increase in production and technological advancement that focus on the quality of life
Area of application	Narrow	Broad
Effects	Increase in GDP and per capita income indicators	Improvement in life expectancy, literacy and overall living standards
Duration	Short term	Long term
Type of economies	Developed	Developing
Measured by	Periodic improvement in national income	Improvement in real national income
Changes	Quantitative indicators	Qualitative as well as quantitative indicators
Type of process	Automatic	Manual
Process time	Particular time period	Continuous process

spite of developments in alternative energy resources, the use of carbon-based energy resources has not been fully stopped. Therefore, the need of the hour is to rethink economic growth and environmental sustainability and to reevaluate priorities in order to conserve nature and sustain the health of the planet. In order to fully address the dichotomy between economic growth and ecological sustainability, we must find lasting solutions to preserve the environment and the progress of human civilization. This goal cannot be achieved by reprioritizing judicious use of different energy sources alone. But we also must take into consideration population stability, depletion of resources and many more factors affecting the sustainability of the environment (Jeyasushma 2017). The effects of these decisions and actions will definitely affect the quality of life, population growth and resource usage for the growth of many nations across the globe. The balancing must be accompanied by a deep understanding of the consequences of disturbed nature in short-term growth and long-term survival.

8.5 Economic Policies for Sustainable Development

When it comes to energy consumption, Canada is an energy-intensive country. Canada's energy sector is an important segment of the economy both in terms of investment and employment generation. In the past few decades, energy demand and energy production in Canada have witnessed a steady increase. Primary energy production in Canada has risen from 11 495 PJ to 16 594 PJ between 1990 and 2004, an increase of 44% (Hofman and Li 2009). In the same period, primary energy consumption increased from 9229 PJ to 11 617 PJ, registering a growth of almost 26%. Similarly, the global energy consumption has been increasing rapidly because energy is vital to all sectors of economy, and energy plays a crucial role in the economic and social development of a country (Hofman and Li 2009). Overreliance on energy derived from fossil fuels has serious environmental repercussions. The major issue regarding fossil resources is that they are finite and in the foreseeable future will no longer be available. Another major environmental problem caused by major energy consumption is the emission of toxic chemical pollutants, greenhouse gases like CO_2 and other air contaminants. These incessant toxic emissions cause climate change and environmental degradation in the form of air, land and water pollution. The polluted environment becomes unconducive to human beings and, in the longer run, adversely affects the health of species living in such contaminated conditions (Farrelly et al. 2013). To alter this dangerous trend, sustainable development should be accorded the highest priority. A switch to bio-based renewable energy sources can result in sustainability at various levels and bring about inclusive and balanced development in a geographically vast country like Canada.

The dramatic transition from a fossil fuel-based energy production to biofuel-based resource planning requires concomitant formulation of new strategies and policies. As due emphasis is being placed on enhanced research and innovations in bio-based energy forms, other promising fields such as new materials, bio-economy (BE) and BBE have also evolved. Taking into account various environmentally sustainable technologies and trends, Canada's pioneering role will be determined by three major indicators: the green economy as a percentage of GDP; the share of Canada's bioenergy production at the global level; and the acknowledgment and adoption of Canadian biotechnology by other countries (Staffas et al. 2013). When it comes to quantitative assessment of economic success in the domain of biofuel-driven resource mobilization, several indicators are considered, namely research expenditure on biotechnology, the number of people working in the clean energy sector, revenue generated by the sale of green products and other parameters such as commercial viability of innovative ideas, existing successful products, services and/or technologies in this domain, and the number of graduates in biotechnology and allied sectors.

To date, different countries have adopted separate strategies and policies related to biotechnology and bio-based energy production, but in contemporary times more and more countries are developing strategies under the conceptual paradigm of the BE. A paradigmatic shift toward a more advanced BE will have multitudinous implications for different aspects of the economy, environment and society at large. Conservation planning is embedded within overarching BE strategy, which involves the protection of biodiversity, ensuring food security and climate change mitigation. Now some of the world's largest economies have adopted nationally determined goals vis-à-vis the BE to enhance research efforts in this domain for a better tomorrow (Farrelly et al. 2013; House 2012; Staffas et al. 2013).

With the publication of the Organisation for Economic Co-operation and Development (OECD) document "The Bio-economy to 2030: Designing a Policy Agenda," a new phase of publications of national BE strategies and policies has commenced. This document states that advancement in the biological sciences can offer a plethora of solutions for numerous health, socio-economic and environmental challenges that the world is facing right now. These technologies can act as a catalyst to achieve increased sustainability in economic development without jeopardizing the environment. It is imperative to understand that to achieve such an ambitious goal we need to put in place a policy agenda directed toward implementing the research findings and innovations that form the core of the BE. The OECD strongly suggests that it is the responsibility of both the public sector and private sector to partake in designing and developing prospective agenda in order to realize the full potential of the BE (OECD 2009).

8.5.1 Bio-economy and Bio-based Economy

The OECD has pioneered the concept of the BE and defined it as "transforming life science knowledge into new, sustainable, eco-efficient and competitive products." The OECD has highlighted the potential of innovations in this transformation and more efficient use of bio-based resources. The US has defined the concept as: "A bio-economy is one based on the use of research and innovation in the biological sciences to create economic activity and public benefit" (OECD 2009). The conceptual paradigm of the BE is focused on the methods of conversion of raw materials into value-added products without according much significance to sustainability (OECD 2009).

The BE has been previously defined as being an economy based on the use of research and innovation in the biological sciences to create economic activity and public benefits. The key drivers behind the notion of BE are economic growth, societal benefits, health and environment, and currently the USA assumes a leadership position in this field. Three technologies that form the bedrock of BE are: genetic engineering, DNA sequencing and automated high-throughput manipulations of biomolecules. During the incipient stage the European Union (EU) came up with a working definition of the BBE. In 2005, the BBE was defined as "the sustainable, eco-efficient, transformation of renewable biological resources into food, energy and other industrial products" focusing on food security and energy sustainability (Staffas et al. 2013). The conceptual definition has undergone some changes since then and, in more recent parlance, the EU has changed the definition to "a bio-based economy integrates the full range of natural and renewable biological resources – land and sea resources, biodiversity and biological materials (plant, animal and microbial), through to the processing and the consumption of these bio-resources." This new definition of the concept lays greater emphasis on the raw material rather than the conversion processes and methods. Many European countries such as Germany, Finland and Sweden are using the same concept of the BBE (Staffas et al. 2013).

8.5.2 Case Studies in Canada

The year 2011 is considered as a watershed as an interesting development took place in Canada, where the province of British Columbia (BC) formed a BE committee as per the administrative rules given by the Ministry of Jobs, Tourism and Innovation. The central role of the committee was to objectively assess the opportunities for the province in the emerging BE and also the possibilities for BC to expedite the growth of the BBE in the province and eventually publish a BE strategy for BC. The committee primarily focuses on the economic value of biological systems. As the province has significant forest resources, BC envisions the increased use of naturally available biomass for energy generation. Earlier in

2008, the province has also taken in the work to reduce the dependence on fossil resources, and a bioenergy strategy for the province was published (Staffas et al. 2013).

In the BC strategy, the forest sector plays a crucial role along with agriculture, life sciences and clean technology. The document identifies five key areas in which collective action is needed: establishment of a clearly defined long-term BE vision; improvement of access to biomass and feedstock; establishment of a meticulous technology development strategy; development of markets for BC bio-products; and integration of the requirements of the BE with provincial-level initiatives. The committee also recommends the creation of a BE team to formulate the vision for the entire province, and proffer advisory inputs from time to time in order to provide directions to the market and other stakeholders (Staffas et al. 2013). Subsequently, in 2013, Alberta also published a BE policy document, in which clear recommendations for developing the province's BE were given. The main drivers for the development of a BE in the province include ensuring Alberta's economic future, advancing world-leading resource leadership, and investing in families and communities. Alberta is one of Canada's largest producers of agricultural products and it is also a leading producer of varied forest products, which constitute favorable pre-requisites for the development of BE. Biomaterials, biochemical and bioenergy are the areas that have marginal shares in Alberta's economy but are destined to expand in a big way. Ecosystem services are now prioritized and knowledge gaps are identified based on research, and the systemic capacity to develop BE is being strengthened (Staffas et al. 2013).

8.6 Economic Development and Environment

8.6.1 Carbon Mitigation, Transfer and Energy Solutions

Recently, growing concerns over global warming and carbon emissions have sparked interest in carbon sequestration. Anthropogenic activities are already directly and indirectly affecting almost half of the terrestrial biological carbon cycle. Through proper management of the carbon cycle, mitigation of CO_2 emissions can be achieved efficiently. A large proportion of CO_2 emissions results from the combustion of fossil fuels for energy generation. As energy needs are increasing day by day, especially in the emerging economies of the developing world, CO_2 emissions are expected to rise significantly in the foreseeable future. It has been well documented that a high concentration of CO_2 in the atmosphere has some detrimental effects on the environment and human health. Thus, in order to reduce carbon emission various techniques have been adopted. The following are the most commonly used mechanisms for CO_2 removal that have been previously reviewed:

- absorption (Alie et al. 2005);
- membrane separation (Thiruvenkatachari et al. 2009);
- desiccant adsorption (Stewart and Hessami 2005);
- cryogenic separation (Meratla 1997).

In chemical removal, amine solvents are used to absorb the CO_2 released from flue gases. It is the most commonly used technique for CO_2 removal, while the other methods are less energy-efficient and more expensive than chemical absorption (Stewart and Hessami 2005). Various methods for carbon mitigation have been enumerated in Table 8.2.

Table 8.2 Different methods reported in literature for carbon mitigation.

Carbon sink		Mitigation period	Pathway back to atmosphere
Soil	Biomass	Short	Plant respiration, microbial decay
	Soil	Long	Carbon sequestrated
Ocean	Biomass, ocean acidification	Short	Air-water exchange, respiration, decay
	Deep ocean	Long	Carbon sequestrated
CCS (carbon capture and storage)	Aquifers	Long	Carbon stored
	EOR (enhanced oil recovery)	Long	Carbon stored
	Biochar	Long	Carbon stored

8.6.1.1 Biological Carbon Mitigation (BCM)

BCM is the process where autotrophic organisms and plants convert CO_2 present in the atmosphere into organic carbon through the process of photosynthesis, thereby producing large amounts of biomass. Produced biomass in turn could be used in other areas such as: biofuel production, food, feeds and biochemical production. In biofuel production, the CO_2 is essentially being recycled. In this process, the use of biomass in the "biomass-to-energy" chain displaces the use of fossil fuels for energy production (Farrelly et al. 2013). Photosynthesis is the process by which certain organisms utilize anthropogenic (emanating from human activity) carbon for biomass production, and it was the original process that fixed carbon millions of years ago. Using this process, the flue gases containing high concentrations of CO_2 are captured and can be used to cultivate large amounts of biological media.

BCM utilizes biological media for carbon sequestration. Photosynthetic organisms and plants utilize this process as their food source and convert CO_2 into organic carbon. Photosynthesis is the process by which green plants make their own food using energy from sunlight to form glucose in their chlorophyll-containing tissues. There are two stages in photosynthesis, namely the light-dependent stage and the light-independent stage, in which complex reactions occur that can be simplified to the formation of glucose and O_2 from CO_2 and water in the presence of light (Farrelly et al. 2013). The produced glucose is then converted into starch and cellulose, which store carbon in the plant cells, thus reducing the inorganic carbon by creating organic carbon. BCM has been proposed as an environmentally friendly method of alleviating carbon emissions and simultaneously producing many useful by-products in the process. Nitrogen and phosphorus are utilized by algae from wastewater in the presence of sufficient light and the CO_2 is converted into biomass through photosynthesis (Figure 8.3). The produced biomass is harvested periodically and used in various applications such as co-firing, biofuels, biochemicals, animal feed and food supplements (Packer 2009). This method of co-processing creates an efficient, cost-effective system that reduces carbon emissions from the point sources. This system also minimizes nitrogen and phosphorus effluent from the wastewater treatment system. This leads to an enhanced yield of high-value biomass for the various applications.

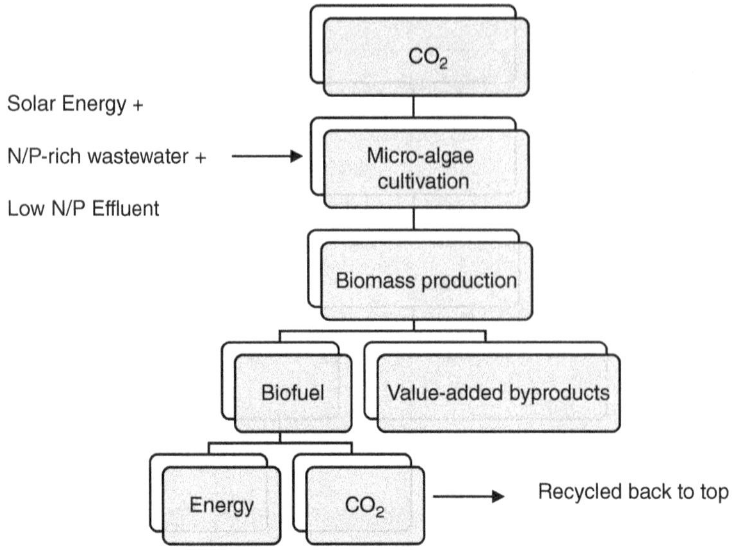

Figure 8.3 A microalgal system for simultaneous biofuel production and CO_2 mitigation.

8.6.1.2 Terrestrial Carbon Sink

The soils of the earth contain over 2000 gigatons of carbon (GtC) and up to 3 GtC per year is sequestrated by the vegetation cover, making it a significant natural carbon sink. Trees work as a good carbon sink in the medium term. Based on the timescale of sequestration, terrestrial carbon mitigation can be divided into three groups. Out of the total, 120 GtC is absorbed by photosynthetic organisms and 60 GtC returns to the atmosphere almost immediately via plant respiration. A further 10 GtC is required by plants for growth and maintenance, making up the total of 550 GtC sequestered in growing biomass (Farrelly et al. 2013). The rest is dropped into the soil, where it is decomposed by microorganisms releasing nutrients into the soil and CO_2 into the atmosphere. An increase in atmospheric CO_2 increases the rate of photosynthesis and, carbon sequestration (Lal 2005). However, plant maintenance and soil respiration both increase with temperature, an effect of elevated atmospheric CO_2 levels, thus reducing the carbon mitigation and storage of plants.

8.6.1.3 Biochar

The application of biochar to soil was proposed as a novel approach to establish a significant, long-term sink for atmospheric CO_2 in terrestrial ecosystems. Biochar is charcoal that is created under anaerobic conditions by low-temperature pyrolysis of biomass. Biochar application to soil has many advantages like long-term carbon sequestration, improvement of soil properties and biomass waste management. Moreover, bio-oil produced by pyrolysis can also be used as a renewable energy source (Lehmann et al. 2006). There is a strong link between climate change, soil degradation and food insecurity, but applying biochar for increasing soil organic carbon (SOC) levels improves soil quality and fertility. Agriculture residues, wood chips, organic wastes, microalgae and macroalgae are the feedstocks for pyrolysis. The presence of sufficient SOC is of great importance because low levels of carbon reduce crop yields, the soil's ability to retain water and applied nutrients as well as the ability for holding CO_2 (Lal 2005). Biochar derived from the pyrolysis of algae has high nitrogen and phosphorus content, which further enhances soil fertility and reduces the need for fertilizers on agricultural land (Bird et al. 2011). Despite the many benefits of biochar in carbon sequestration, there are issues with the increased requirement for nutrients and water in the soils, the possible effect of increased microbial activity in soils as well as possible saturation of the terrestrial carbon sink. The perception of these barriers is of great importance as once biochar has been applied to soil it cannot be removed (Lal 2011).

8.6.1.4 Oceanic Nourishment

Ocean nourishment is described as a process for developing the sequestration of atmospheric CO_2 in the ocean by providing the nutrients needed to enhance the production of phytoplankton. Photosynthetic organisms in the upper ocean have the ability to remove large quantities of CO_2 from the atmosphere (Farrelly et al. 2013). The availability of nutrients in ocean water is the only limitation to the rate of biological sequestration in the ocean. The main disadvantage of ocean nourishment methods is that they can change plankton structures having long-term effects on the ocean ecosystem. The production of methane gas triggered by large quantities of produced organic mass sinking to the bottom of the ocean is another disadvantage (Farrelly et al. 2013).

8.6.1.5 Natural Oceanic Carbon Sink

On a geological timeline, carbon is stored in the oceans in the form of magnesium and calcium salts. The geochemical reactions responsible were first described by Urey (Farrelly et al. 2013):

$$CO_2 + CaSiO_3 \rightarrow CaCO_3 + SiO_2 \tag{8.3}$$

$$CO_2 + MgSi_2 \rightarrow MgCO_3 + SiO_2 \tag{8.4}$$

Here, $CaSiO_3$ and $MgSiO_3$ represent any generic Ca or Mg silicate produced at high temperatures in the earth's crust. These are inorganic reactions driven by organic processes in microorganisms, such as cyanobacteria and microalgae, and, in the long term, they store large amounts of carbon. Table 8.3 highlights different microorganisms reported in the literature for carbon mitigation.

8.6.2 Climate Mitigation and Economics

BCM from point sources offers great potential in addressing the issue of climate change. BCM using microalgae offers many advantages over conventional carbon sequestration methods as the CO_2 is being utilized to produce high-value biomass that has a number of applications in energy production, biochemical generation as well as in food and feed applications (Farrelly et al. 2013). This is a more advantageous method of carbon sequestration when compared with geological storage where CO_2 is pumped into aquifers. This is because concerns over leaks are present in geological storage as a high cost is associated with the separation of CO_2 from flue gases without any economic benefit from the process. The cost of CO_2 mitigation depends on the productivity of algal biomass, its carbon content and the efficiency of the overall process. In a study carried out by Stewart and Hessami (2005), a carbon uptake rate of $1.5\,g\,l^{-1}\,day^{-1}$ was assumed for *Synechocystis aquatilis*, and under a natural light cycle, up to $2.2\,ktCO_2$ per pond per year could be mitigated in a pond of capacity $4000\,m^3$ (Stewart and Hessami 2005). Kadam concluded that a 1000 ha open pond system could mitigate $210\,000\,t\,year^{-1}$ CO_2 of the $414\,000\,t\,year^{-1}$ CO_2 generated by a 50 MW coal-fired power plant, which means that the system achieved a 50% reduction in CO_2 in flue gas emissions (Kadam 2001). Currently, carbon sequestration cost in the US is around US\$40/tCO_2 when buried in saline aquifers in contrast with US\$150/tCO_2 for processes such as amine scrubbers, which remove the CO_2 from the flue gases (Farrelly et al. 2013). The cost of mitigation using microalgae cultivated in open ponds has been calculated at US\$793/tCO_2. This is very cost intensive at almost 20 times the cost of CO_2 storage in saline aquifers. However, extra revenue may be created through the production of valuable by-products from the produced biomass thus reducing the cost of mitigation and leading to sustainability (Farrelly et al. 2013).

8.7 Economic Analysis of Sustainable Development

8.7.1 Economic Tools for Evaluating Sustainable Development

The aim of economic analysis is to provide quantified results for sustainable development. However, it is not possible to state everything in quantitative terms. Thus, qualitative approaches may also be needed. The impact of each economic indicator needs to be measured and quantified; based on it, evaluation of the results can be performed. Evaluation of results of an indicator can be conducted by using indicator weights and priorities according to the information provided by decision makers and stakeholders.

Table 8.3 Carbon mitigation using cyanobacteria, macroalgae and microalgae.

Strain	Type	CO_2 mitigation
Botryococcus braunii	*Microalgae*	$1\,g\,l^{-1}\,day^{-1}$
Chlorella kessleri		—
Chlorella sp. KR-1		—
Chlorella vulgaris		$0.075\,g\,l^{-1}\,day^{-1}$
Chlorococcum littorale		$4–16.7\,g\,l^{-1}\,day^{-1}$
Scenedesmus obliquus		13.56–28%
Gracilaria chilensis	*Macroalgae*	2.92–1280 ppm
Gracilaria sp.		320–1420 ppm
Porphyra yezoensis		—
Chlorogloeopsis	*Cyanobacteria*	$20.45\,mg\,l^{-1}\,day^{-1}$
Fischerella		—
Aphanothece microscopia		$109.2\,mg\,l^{-1}\,day^{-1}$

Source: Adapted from Farrelly et al. (2013).

Finally, weighted values can be aggregated to obtain an overall picture for economic analysis (Karvonen et al. 2017).

$$\text{Analysis results} = \sum_i \left(p_i \times e_i\right) \tag{8.5}$$

where p_i is priority assigned to the indicator, and e_i is impact/weight of economic indicator. The analysis is helpful in assessing sustainable development in a nation, and can be helpful in assessing the economic benefits of any policy/plan and project that is yet to be undertaken. There are several economic indicators for assessing sustainable development. Table 8.4 shows these indicators, the source of the data and their linkages.

Since economic analysis requires collection and computation of large amount of data, tools have been developed for economic analysis in sustainable development. Two tools are described as follows:

1) *CBA*: CBA estimates a benefit attained via monetary values, but can also incorporate non-monetary values. CBA works well in assessing monetary values, which is its strength since economic evaluation as such is objective-oriented and suitable for policymaking decisions. However, CBA has been criticized for challenges in monetizing all inputs and outputs. For assessing sustainability, terms like forestry, scenic beauty and recreation are difficult to evaluate in economic terms. Hence other tools are needed for incorporating values for intangible and non-monetary factors in CBA (Prokofieva et al. 2011).
2) *Input–output (IO) analysis*: IO analysis is a tool where an input (a resource) is converted into some output (a product), and the interdependence between the various sectors of the economy is assessed for its impacts (Leontief 1986). For example, input of wood to a sawmill may be a direct part of the forestry sector, although it utilizes inputs from other sectors such as energy and transportation. Naturally, other inputs (e.g. labor, water or chemicals) are also needed in these processes. To be more informative, IO has been extended to include also environmental and social aspects (Päivinen et al. 2012). Environmentally extended input–output analysis (EEIO) derives its environmental impacts by using economic tables used in traditional, material and economically focused IO. However, IO has a disadvantage that the market needs to be

Table 8.4 Economic indicators for sustainability and possible data sources.

Indicator	Indicator units	Strong indicator linkages	Other major connections	Main data sources
Gross domestic product (GDP)	€, % change in GDP	Gross value added (GVA)/ local value added (LVA)		National statistics
GVA/LVA	€	GDP, trade	Rural development	Industry, national statistics
Trade	Import/export change	National supply security and self-reliance	National self-reliance	National statistics

Source: Adapted from Karvonen et al. (2017).

well established and should have high certainty levels in order to get accurate results from the analysis. For new products, there are a lot of price fluctuations and market uncertainty (Mattila et al. 2011).

Besides economic tools, there are other tools like lifecycle assessment (LCA), material flow analysis (MFA) and multicriteria analysis (MCA) for assessing sustainability and sustainable development, which are discussed in other chapters.

8.7.2 Profitability Indicators for Investing in a Sustainable Project

Besides economic indicators, there are some profitability indicators that are taken into account before investing in any project (sustainable or non-sustainable) by the government or investor. The following are some of the indicators used to measure profitability of an investment. These indicators may vary for a sustainable and non-sustainable project.

Net present value (NPV) is estimated as the difference between cash inflow and cash outflow. It is used to assess a project's future returns at a particular period. NPV is 0 at the break-even point where the investment cost incurred by the company is completely recovered. NPV is calculated as:

$$\mathrm{NPV} = \sum_{t=1}^{T} \frac{C}{(1+r)t} - Co \tag{8.6}$$

where C is net cash inflow during the period t; C_o is total investment costs (cash outflow); r is discount rate; and t is number of years.

Payback time is defined as the period in which the company can recover all its investment costs through the profits earned by selling product. Mathematically, payback time is evaluated as:

$$\text{Payback time} = \frac{\text{Total investment}}{\text{Annual net profit}} \tag{8.7}$$

Since payback period is inversely proportional to net profit, annual net profits earned should be as high as possible so that the plant can recover its capital investment in the shortest time. Annual net profit is revenue earned by company (through sale of commodity) minus the unit production cost and taxes paid to the government. Revenue earned by the company is dependent on market prices and market demand.

Return on investment (ROI) is used to evaluate the efficiency of an investment. In purely economic terms, it is a way of relating profits to capital invested. Mathematically, ROI is calculated as:

$$\mathrm{ROI} = \frac{\text{Annual net profit}}{\text{Total investment}} \times 100 \tag{8.8}$$

ROI is dependent on the net profit generated by the company, which is dependent on the difference between annual revenues produced by the enterprise (through selling the commodity) and annual operating cost.

Let's consider two projects – one purely profitable but unsustainable project, and the second, a sustainable project having high benefits in the longer run but high initial investments; the government has to approve only one of them. The first example assumes that there are positive cash flows in the initial 5 years and the capital costs associated with the project only accrue in the future, after 5 years (unsustainable project). The second example assumes that costs occur at the beginning of the project and that positive cash flows only occur in the future, i.e. after 5 years (sustainable project, e.g. the creation of a national park or bird sanctuary or clean-up of a site that is used for a future project). The first example is a typical case of a project that leads to significant environmental or social demerits in the longer run. The second example is perhaps less typical as it is associated with significant costs that occur early. However, the second example emphasizes that projects require a significant investment at the beginning of the project's lifetime and see positive cash flows toward the end – even if these cash flows are both relatively larger and stable. Considering the payback period and ROI, the second project is more likely to be rejected (Baur and Lagoarde-Segot 2015). One solution to enhance sustainability may be to allow investors to hold shares for a longer period, e.g. 10 years. This would decrease the lack of commitment of investors and favor the realization of more sustainable projects (Baur and Lagoarde-Segot 2015).

8.8 Future Directions for Achieving Economic Sustainable Development

8.8.1 Balance Between Energy Production and Consumption

The first requirement for sustainability in the energy sector is to have a favorable energy balance (i.e. energy production – energy consumption = positive) for energy localization. For example, Canada has good possibilities to apply energy localization, but currently Canada is still exporting and importing major amounts of energy and is still dependent on fossil fuels. Analyzing the future energy projections, it can be concluded that in Canada the production of energy will still exceed the energy demand in the future due to which the export rate is higher than the import rate. This means that, from an energy localization point of view, there is still space for further development for Canada (Hofman and Li 2009). For other countries, energy localization and diversification might be a challenge, because they lack sufficiently reliable energy resources and are currently dependent on resources from other countries. These countries have to search for alternative energy resources like bioenergy and biofuels in order to achieve sustainable development (Hofman and Li 2009).

8.8.2 Role of Government

Government has an important role to play in policy formulation that involves both sustainability and economic growth. Government needs to take care of providing incentives for more renewable energy with higher efficiency. Instead of non-compulsory policies like subsidies and voluntarism, the government should have compulsory policies including tree planting, use of renewable resources of energy and promoting BBEs. More renewable energy sources and less energy demand will be the necessary step toward a more localized and more diverse energy system, which will be the basis for further sustainable development of a nation. Government should also take decision in granting funds to investors based on sustainability and long-term impacts of a project rather than based on a purely economic basis.

8.8.3 Role of Society

The philosophy of the separation of mind and body is related to the fundamental conception of the separation of human beings from the environment. Technology is often seen as separate from society and the environment. Sustainable development will require more than technical modifications leading to CBA. It needs a shift in how humans perceive the world. Humans are part of a web of connections within what is called the environment and society. We tend to separate the impacts of our actions into distinct compartments. Sustainable development, to have long-term meaning, will require changing the perception of human beings and how they relate technology and economic growth to environment. This awareness has been inculcated in the society of developed countries during the past decade but, for developing countries, people need to have awareness about sustainability, the environment and the interrelations among sustainability, the environment and the economic growth of a nation. Government campaigns, advertisements and rallies related to "technology and its relation to environment and sustainable development" can be helpful.

8.9 Conclusions

The following conclusions can be drawn from this chapter:

1) Economic development is an essential part of sustainable development. Energy, transportation, employment and technology are important elements that influence the economic development of a nation.
2) GDP, trade, GVA and LVA are important economic indicators used in economic analysis of sustainable development.
3) Several economic assessment tools have been used for performing economic analysis in sustainable development like IO analysis and CBA. But they have the disadvantages that they are unable to account for non-monetary things like forestry, scenic beauty and recreation. Therefore, for a complete picture, sustainable development, economic, social and environmental analysis needs to be performed. Tools like LCA, MFA and MCA for assessing sustainability and sustainable development can be used.
4) To achieve sustainable development, many developed countries have adopted BBE like employing renewable and bio-based energy, biochar, bio-oil, BCM using microalgae, etc.
5) In developing countries, people need to have awareness about sustainability, the environment and the interrelations among sustainability, the environment and the economic growth of a nation. Government campaigns, advertisements and rallies related to "technology and its relation to environment and sustainable development" can be helpful.
6) In developing countries, government should grant funds to investors based on sustainability and the long-term impacts of a project rather than based on a purely economic basis.

Acknowledgment

The authors would like to acknowledge Mr Tilak Raj (Alumni, Indian Institute of Technology Delhi, India) for his valuable suggestions in writing the chapter.

References

Alie, C., Backham, L., Croiset, E., and Douglas, P.L. (2005). Simulation of CO_2 capture using MEA scrubbing: a flowsheet decomposition method. *Energy Conversion and Management* 46 (3): 475–487.

Almutairi, D.O. (2015). The mediating effects of organizational commitment on the relationship between transformational leadership style and job performance. *International Journal of Business and Management* 11 (1): 231.

Ameen, R.F.M., Mourshed, M., and Li, H. (2015). A critical review of environmental assessment tools for sustainable urban design. *Environmental Impact Assessment Review* 55: 110–125.

Banerjee, A., Duflo, E., Glennerster, R., and Kinnan, C. (2015). The miracle of microfinance? Evidence from a randomized evaluation. *American Economic Journal: Applied Economics* 7 (1): 22–53.

Baur, D.G. and Lagoarde-Segot, T. (2015). The contradiction of the time value of money and sustainability. May 11. SSRI. https://papers.ssrn.com/sol3/papers.cfm?abstract_id=2605161 (accessed November 2019).

Bird, M.I., Wurster, C.M., de Paula Silva, P.H. et al. (2011). Algal biochar–production and properties. *Bioresource Technology* 102 (2): 1886–1891.

Brink, E., Aalders, T., Ádám, D. et al. (2016). Cascades of green: a review of ecosystem-based adaptation in urban areas. *Global Environmental Change* 36: 111–123.

Buys, L., Mengersen, K., Johnson, S. et al. (2014). Creating a sustainability scorecard as a predictive tool for measuring the complex social, economic and environmental impacts of industries, a case study: assessing the viability and sustainability of the dairy industry. *Journal of Environmental Management* 133: 184–192.

Davoudi, S., Shaw, K., Haider, L.J. et al. (2012). Resilience: a bridging concept or a dead end? "Reframing" resilience: challenges for planning theory and practice interacting traps: resilience assessment of a pasture management system in Northern Afghanistan urban resilience: what does it mean in planning practice? Resilience as a useful concept for climate change adaptation? The politics of resilience for planning: a cautionary note: edited by Simin Davoudi and Libby Porter. *Planning Theory and Practice* 13 (2): 299–333.

Elmualim, A. and Gilder, J. (2014). BIM: innovation in design management, influence and challenges of implementation. *Architectural Engineering and Design Management* 10 (3–4): 183–199.

Farrelly, D.J., Everard, C.D., Fagan, C.C., and McDonnell, K.P. (2013). Carbon sequestration and the role of biological carbon mitigation: a review. *Renewable and Sustainable Energy Reviews* 21: 712–727.

Hahn, R. and Lülfs, R. (2014). Legitimizing negative aspects in GRI-oriented sustainability reporting: a qualitative analysis of corporate disclosure strategies. *Journal of Business Ethics* 123 (3): 401–420.

Herder, M.D., Kolström, M., Lindner, M. et al. (2012). Sustainability impact assessment on the production and use of different wood and fossil fuels employed for energy production in North Karelia, Finland. *Energies* 5 (11): 1–22.

Hodgson, G.M. (2017). Karl Polanyi on economy and society: a critical analysis of core concepts. *Review of Social Economy* 75 (1): 1–25.

Hofman, K. and Li, X. (2009). Canada's energy perspectives and policies for sustainable development. *Applied Energy* 86 (4): 407–415.

House, W. (2012). *National Bioeconomy Blueprint*. Washington DC: The White House, April.

Jeyasushma, V. (2017). The relationship between principal transformational leadership practices, teacher organizational commitment, and school culture in primary cluster schools in Selangor/Jeyasushma Veeriah. DPhil Thesis. University of Malaya. http://studentsrepo.um.edu.my/7931/6/Thesis_-_sushma_veera.pdf (accessed November 2019).

Kadam, K.L. (2001). Microalgae production from power plant flue gas: environmental implications on a life cycle basis. Technical Report. Golden, CO: NREL. https://www.nrel.gov/docs/fy01osti/29417.pdf (accessed November 2019).

Karvonen, J., Halder, P., Kangas, J., and Leskinen, P. (2017). Indicators and tools for assessing sustainability impacts of the forest bioeconomy. *Forest Ecosystems* 4 (1): 2.

Kemp, S.P. and Nurius, P.S. (2015). Preparing emerging doctoral scholars for transdisciplinary research: a developmental approach. *Journal of Teaching in Social Work* 35 (1–2): 131–150.

La Porta, R. and Shleifer, A. (2014). Informality and development. *Journal of Economic Perspectives* 28 (3): 109–126.

Lal, R. (2005). Forest soils and carbon sequestration. *Forest Ecology and Management* 220 (1–3): 242–258.

Lal, R. (2011). Sequestering carbon in soils of agro-ecosystems. *Food Policy* 36: S33–S39.

Lehmann, J., Gaunt, J., and Rondon, M. (2006). Bio-char sequestration in terrestrial ecosystems–a review. *Mitigation and Adaptation Strategies for Global Change* 11 (2): 403–427.

Leontief, W. (1986). *Input–Output Economics*. Oxford: Oxford University Press.

Mamabolo, M.E. (2017). The impact of the implementation of government policy on the spatial distribution of emerging farmers in the Mopani district in the Limpopo province of South Africa. DPhil Thesis. University of South Africa. https://pdfs.semanticscholar.org/c169/62da385b160b4e027cb4d7d894c7bedbe8a3.pdf (accessed November 2019).

Mancarella, P. (2014). MES (multi-energy systems): an overview of concepts and evaluation models. *Energy* 65: 1–17.

Mattila, T., Leskinen, P., Mäenpää, I., and Seppälä, J. (2011). An environmentally extended input-output analysis to support sustainable use of forest resources. *Open Forest Science Journal* 4: 15–23.

Meratla, Z. (1997). Combining cryogenic flue gas emission remediation with a CO_2O_2 combustion cycle. *Energy Conversion and Management* 38: S147–S152.

Moeinaddini, M., Asadi-Shekari, Z., and Shah, M.Z. (2015). An urban mobility index for evaluating and reducing private motorized trips. *Measurement* 63: 30–40.

Neely, C.J., Rapach, D.E., Tu, J., and Zhou, G. (2014). Forecasting the equity risk premium: the role of technical indicators. *Management Science* 60 (7): 1772–1791.

OECD (2009). *The Bioeconomy to 2030: Designing a Policy Agenda*. Paris: OECD https://www.oecd-ilibrary.org/economics/the-bioeconomy-to-2030_9789264056886-en (accessed November 2019).

Packer, M. (2009). Algal capture of carbon dioxide; biomass generation as a tool for greenhouse gas mitigation with reference to New Zealand energy strategy and policy. *Energy Policy* 37 (9): 3428–3437.

Päivinen, R., Lindner, M., Rosén, K., and Lexer, M. (2012). A concept for assessing sustainability impacts of forestry-wood chains. *European Journal of Forest Research* 131 (1): 7–19.

Pearce, J.M. (2012). Building research equipment with free, open-source hardware. *Science* 337 (6100): 1303–1304.

Pearce, D.W., Atkinson, G.D., and Dubourg, W.R. (1994). The economics of sustainable development. *Annual Review of Energy and the Environment* 19 (1): 457–474.

Prokofieva, I., Lucas, B., Thorsen, B. J. and Carlsen, K. (2011). Monetary values of environmental and social externalities for the purpose of cost-benefit analysis in the EFORWOOD project. EFI Technical Report 50. Joensuu, Finland: European Forest Institute.

Staffas, L., Gustavsson, M., and McCormick, K. (2013). Strategies and policies for the bioeconomy and bio-based economy: an analysis of official national approaches. *Sustainability* 5 (6): 2751–2769.

Steffen, W., Richardson, K., Rockström, J. et al. (2015). Planetary boundaries: guiding human development on a changing planet. *Science* 347 (6223): 1259855.

Stewart, C. and Hessami, M.-A. (2005). A study of methods of carbon dioxide capture and sequestration––the sustainability of a photosynthetic bioreactor approach. *Energy Conversion and Management* 46 (3): 403–420.

Thiruvenkatachari, R., Su, S., An, H., and Yu, X.X. (2009). Post combustion CO_2 capture by carbon fibre monolithic adsorbents. *Progress in Energy and Combustion Science* 35 (5): 438–455.

Wu, P., Wang, J., and Wang, X. (2016). A critical review of the use of 3-D printing in the construction industry. *Automation in Construction* 68: 21–31.

9

Social Dimensions of Sustainability

Anita Talan[1], *R.D. Tyagi*[1], *and Rao Y. Surampalli*[2]

[1]*INRS, Université du Québec, Québec, Canada*
[2]*Global Institute for Energy, Environment and Sustainability, Lenexa, KS, USA*

9.1 Introduction

Social sustainability is a broad term that has been used for a long time but its meaning is still very far from developed. Social sustainability is one of the important dimensions of sustainability and is interlinked to other dimensions in various ways, which entails that change in any dimension of sustainability has an impact on social sustainability. From a social perspective, human wellbeing cannot be sustained without safe surroundings, a healthy environment and the presence of a vibrant economy that can fulfill all the basic needs of the communities residing nationally or globally. Government makes policies for the uplifting of social communities to achieve social sustainability but, alongside this, every nation needs informed citizens who participate actively. This is the essence of democracy and is essential for the protection of basic human rights. The three fundamental goals of social sustainable development are poverty reduction, social investment, and safe and caring communities. Nations, social communities and corporate communities are required to work together to achieve these goals.

Therefore, this chapter focuses on an in-depth analysis of the social dimension of sustainability with emphasis on factors affecting policies implemented in order to achieve the goal of social sustainability. The chapter also focuses on the gaps that are present in the dimension of social sustainability and measures to overcome these gaps. The government of any nation plays an important role in achieving social sustainability so policies have been implemented to move toward social sustainability. Social sustainability is based on an integrated approach of people, communities, the corporate world and other affecting factors, which is described in the chapter to give a whole-picture understanding of social sustainability globally, along with indicators and their characteristic features from a social perspective.

Sustainability: Fundamentals and Applications, First Edition.
Edited by Rao Y. Surampalli, Tian C. Zhang, Manish Kumar Goyal, Satinder K. Brar, and R. D. Tyagi

9.2 Concepts and Definitions of Social Sustainability

Recently, the social sustainability dimension of sustainability has expanded and gained increased acceptance as a most important component of sustainability with an emphasis on urban, corporate and community sustainability. The important environmental and economic issues have always dominated sustainable development discussion, but in the late 1990s social issues and their impacts were taken into consideration within the sustainability agenda. Being a new dimension, the concept of social sustainability is statically indistinct and is always restricted by various theoretical and methodological limitations arising from its framework and its dependent meanings and measurements.

Social sustainability is a social-historical ongoing process rather than a static one and, therefore, understanding of the social dimension of sustainability cannot be condensed to a binary state where "zero" is an unsustainable condition and "one" is a sustainable situation. The social dimension of sustainability being at a fundamental level cannot define whether it means the social prerequisites for sustainable development or the necessity to withstand specific factors and customs of societies. Various studies have been conducted on social sustainability to define the concept and to explore its evolutionary significance emphasizing the shift from outmoded themes to emerging softer research and policymaking themes. Some of the definitions of social sustainability are given below.

Social sustainability relies on the basic values of equality and democracy, which means the appropriation of human legal rights in terms of the political, civil, economic, social and cultural arenas by all people. Social sustainability is defined as the quality of societies or communities that signifies the relationship between nature and society mediated by work and relationships within the society. If work and related professional arrangements satisfy the essential needs of humans and are shaped to prevent and preserve generative abilities over a long run of time, and the normal rights of social justice, human self-respect and contribution are satisfied, this is said to be socially sustainable (Littig and Griessler 2005).

Social sustainability aims to achieve the minimal social requirements for effective and long-lasting sustainable development characterized by critical social capital and to recognize the challenges in the smooth functioning of society over the long term (Griessler and Littig 2005).

Social sustainability is also defined in terms of growth that is compatible with the smooth evolution of civil society to attain sustainability, nurturing a favorable environment for the compatible cohabitation of different groups that are culturally and socially diverse and at the same time focusing on social integration to improve the quality of life (QOL) for all segments of the population.

The definitions given do not provide any operational definition but show the interpretations of this concept in the past, present and possible future. For an operational definition, social sustainability should be looked at on the scale of planning new settlements and their global development. Practical understanding needs to be built up, and maturity and professional commitments are necessary for urbanization to create a sustainable environment and to enhance social sustainable development. Social scientists view this concept with respect to "sustainable communities," "social cohesion" and "corporate communities." Socially sustainable communities are a concept readily accepted as people find these societies safer to live in and work freely. In these sustainable communities, people meet their essential needs with regard to existing and future necessities and at same time reduce the load on the environment, which leads to high QOL. These places are considered to be very safe and inclusive, well planned and skillfully designed, built and operated, and offer equal prospects, opportunities and services for all members of society (Dempsey et al. 2011).

9.3 Social Sustainability

Due to the vast expansion of population, the role of cities and urbanization has become more prominent in social sustainability. The year of 2008 was the first time when half of the global population were urban residents, and between 2000 and 2030 this number is set to increase by 72% according to the United Nations Population Fund (UNPFA 2007). Since social sustainability arrived in the context of sustainability, urban social sustainability has gained importance and political momentum globally. Sustainable cities have broad social, economic and political strategies aiming toward the smooth operation of the city and these are related to poverty and urban decline as well due to the increase in urban population, pollution and competition for a better life. Social sustainability needs improvements in housing estates; policy modification is required to provide funded housing to the private sector for social regeneration and maintenance. The social policy focuses on empowerment and poverty eradication by involving governance and multiple agencies during the generation of urban communities (Power 2012). Social sustainability is the incorporation of various dependent concepts of QOL, social cohesion, social sustainability, sustainable communities, livability and social wellbeing, and depends on several physical and non-physical factors (Table 9.1). These factors illustrate the wide variety of inter-related concepts and dependency of conceptual juxtaposition among them.

Table 9.1 Factors affecting social sustainability.

Physical factors	Non-physical factors
• Urbanization • Communal relations • Accommodation • Personal relations • Local conservational and environmental quality and amenities • Local services and accessibility • Sustainable urban design • Neighborhood and social networks	• Learning and teaching • Communal integrity • Contribution and local social equality • Wellbeing and quality of life • Communal presence communal resources • Public • Protection, shelter and security • Equal division of income • Communal order • Social cohesion • Communal interaction • Occupation and work • Housing permanency and stability • Dynamic public societies • Traditional and cultural stability

Social sustainability can be achieved when communities prefer to work and live their life peacefully in the present as well as in the future in the places provided. These communities fulfill the different requirements of current and future residents and contribute a high QOL. The social sustainability of a community defines the physical context in which the communities live. These places are safe, well planned, perfectly designed and provide good services to the residents.

Studies on community stability have identified different main elements that are required to form new communities to achieve social sustainability and to be successful for a long time (Woodcraft et al. 2011). These important elements are:

Amenities and social structure: The social infrastructure is a basic necessity to maintain and support the establishment of new communities. Social structure includes both physical and social infrastructure. The interconnectivity between physical and social infrastructure leads to integrated participation of communities, establishing a sense of shared ancestors (Woodcraft 2012).

Social and cultural life: This concept is all about social capital of the community including QOL, and the quality and strength of the relationships between residents. These relationships are supportive for the community giving empowerment to its people. The diverse culture of people in the community gives a rich cultural life.

Voice and inspiration: This element focuses on the involvement of residents of the community in the development and designing of a new community that helps in empowering the community. The empowerment always contributes toward sustainability by enhancing QOL in three different ways (Woodcraft 2012): (i) creating opportunities for residents to influence their living standards; (ii) facilitating and strengthening the relationships between neighbors within the community; and (iii) building residents' confidence to control local happenings and to make them responsible for the development of community.

Space to grow: This element always focuses on planning and design that can be easily adapted.

Implementation of the above key elements is required to design new communities. People make decisions to attain a more sustainable world using three basic building blocks (Woodcraft 2012): (i) building environment and creating public space; (ii) supporting social architecture; and (iii) supporting social practices. The social dimension of sustainability is dependent on many factors, which are considered in the following sections.

9.3.1 Social Networks

Social interactions are the basis of social networks of social sustainability. Social interaction can be defined as the basic process involved in the foundation of human nature and social behavior. Hanging out together, in part, through social interactions forms social networks and has a direct impact on the social dimension of sustainability. Social networking is required for a community in order for them to be described as an individual group living separately with pride in and attachment to their surroundings. Social networks and interaction are fundamental features of social resources (Dempsey et al. 2011). Social capital is defined as the social systems and related standards of mutual benefit. Moreover, social capital refers to all the features of the social group such as trust, faith, solidity, education, maturity, obligations, expectations and knowledge of relationships within social networks. All the features of social organization simplify the processes of actions taken equally and voluntary support. Unlike physical resources, social capital is an insubstantial form of resources that exists in the relationships of the society. Social resources or capital is highly dependent on different factors of social organization such as communities, families, groups and individuals. These factors have an undeviating influence on social cohesion (Collins et al. 2011).

9.3.1.1 Contribution in Collective Groups

Social networks are required for the involvement of people in local and communal activities. This participation enhances the resources in terms of social capital and aspects of communal sustainability linked to community or social cohesion, networking and communal integration. This participation exhibits itself within the community and closely related to the understanding of community being associated with the concept of civic society. Civic society is a measure of communal sociability and culture that is regularly needed to participate in organized communal activities, influencing factors such as voting turnout in elections. A society is said to be socially unsustainable if there is no participation in organized activities. Humans have various types of communal networks around them, within and outside the community, social groups and neighborhood. These communal networks signify that people are active and take an interest in communal activities within as well as outside the community. To have a positive community sustainability people should desire to actively participate in the organized activities. People should mix with each other to have a positive outcome working in collective groups. Active participation is also linked to high density and the land available for mixing of residents. This in turn is directly linked to the ease of access to community facilities, which has a direct effect on participation. Empirical authentications from the US validate the inverse relationship between commuting time and participation, and have proved that communal individuals with stretched daily commuting are probably less likely to take part in communal activities (Dempsey et al. 2011).

9.3.2 The Corporate System

Corporate sustainability is defined as an approach to implementing various business strategies after considering every dimension of the running of the business to create long-lasting stakeholder values. It places an emphasis on the approaches and strategies to start an industry and promotes permanency by being transparent and by focusing on the development of employees. Corporate sustainability is also defined as development of business by modifying traditional business strategies so that they are ethically sound. Corporate social responsibility (CSR) and corporate citizenship are the two major terms that are frequently used in the corporate system. However, these terms have been succeeded by the wider term "corporate sustainability," which describes all the business methodologies and practices that are built after considering the social and environmental domains of sustainability. Describing sustainable development, the longing to rise without damaging the needs of future generations is becoming more centralized to business viewpoints. From the sustainability point of view, business goals are not separate from the environments and communities within which they function. For a while temporary commercial increases could be chased, but impacts on social and environmental sustainability make these business ideas unsustainable. Corporate sustainability can be measured by various indicators, such as corporate governance, social factors, and economic resources and measures (Amini and Bienstock 2014).

9.3.3 Community Stability

Community stability comes from inter-generational equity, social networks, personal and professional relationships and, most importantly, social capital. This aims at increasing good education and community facilities to enhance social sustainable development. In turn, it also minimizes corruption, criminality and anti-communal behavior, which is required in order for a community to establish permanency in terms of residents attaining social sustainability (Hightree et al. 2018). According to the literature, there is no evidence that residential turnover is responsible for the social sustainability of an urban setting but residential turnover improves the social status of the society and, hence, affects the community. Resident mobility is the main indication of community instability, worsened by very low communal unity or low emotional attachment brought on by people moving out of the community. Social organizations provide appropriate services for residents to settle down at specific life platforms characterized by increased resident movement. Low residential mobility increases the community stability by increasing attachment and social feelings. Directly, increasing attachment and feelings strengthens social networks and interaction (Winther 2017). It has frequently been the topic of debate whether community stability is required to have communal order but it is broadly accepted as an optimistic communal quality. Still, the links between the stability of the community and the environment are blemished and indirect. However, a resident's decision to stay in the community or to move out of a community may be directly related to the quality, maintenance and sustainability of the environment, the level of social amenities and conveniences provided.

9.4 Dimensions of Social Sustainability

Social life has given the basic framework for social sustainability and its development. Social sustainability is basically affected by amenities, communal and cultural life infrastructure, opinions, expressions and its influence, as well as the empty space for urbanization to develop (Dempsey et al. 2011). Nobel prize-winner Amartya Sen gives the following dimensions (Figure 9.1) for social sustainability, which helped develop the theory of "social choice" in the early 1970s. Social sustainability is affected by many factors and these factors are described in depth below to understand the diversity and importance of social sustainability.

9.4.1 Social Equity

The expanding horizon of sustainability needs development in every area with respect to all three pillars of sustainable development. Social equity is an important indicator that

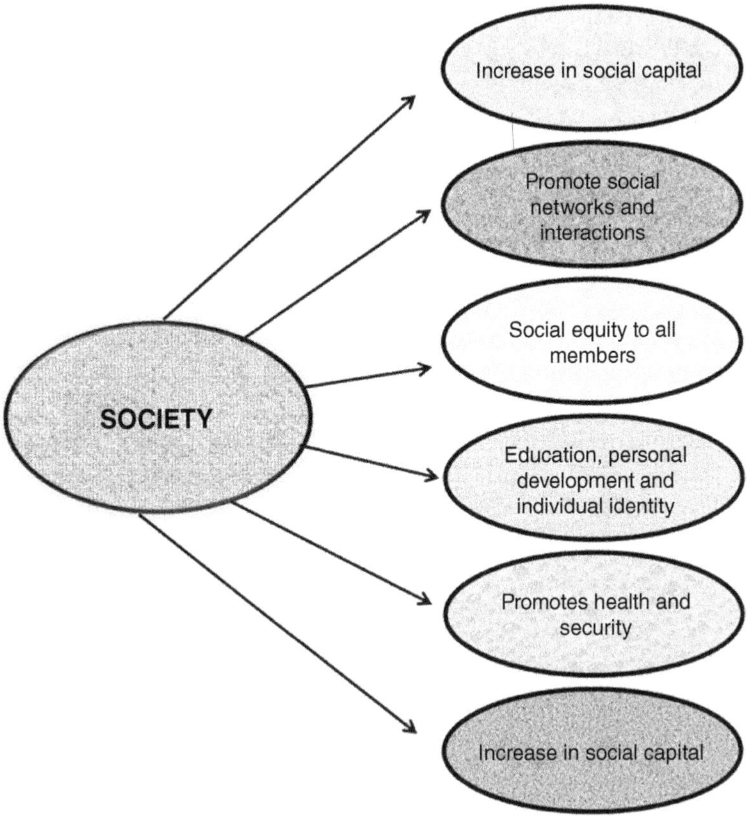

Figure 9.1 Dimensions of social sustainability.

determines social sustainability. It is an indispensable requirement for sustainable development. With social equity rises in living standards, economic growth and the balance of use of natural resources are maintained. From the global sustainability point of view the goals of equity are advanced. Thus, growth cannot be only about economic progress, nor can sustainability only be about environment protection and preservation. Global development must be people-centered and should promote human rights, empowerment, prospects, dignity, pride and variability (Dempsey et al. 2011). Green growth is not only about protecting the environment and switching to green energy but also about social progress contributing to eradicating poverty and achieving high QOL.

Hence, social equity is a concept that focuses on equal justice and objectivity for all members of society. In the early years of sustainability social equity was used in various institutional contexts, such as education and public organizations. With advancement in the goals and objectives of sustainability the new concept of social equity was developed, defining the fair and equal delivery of public facilities and policies.

Later, social equity was considered as the third pillar of public management. Similar economic and social conditions should be provided to every individual in society. The three main areas of social equity are considered equally important for social sustainability (Dempsey et al. 2011):

1) *Equality in terms of gender and sexuality:* As per social equity objectives, there should be equal rights for all individuals of a society irrespective of gender and sex.
2) *Racial equality:* For social sustainability racial equality is an important factor. It focuses on giving biological equality and social equality to all human races of the world.
3) *Religion equality:* Religion equity is as important as gender and race equality. Religion is defined as all aspects of religious observance and practice as well as religious beliefs. Religion equity is important to protect human rights in the corporate sector and in other religion-dominated societies.

9.4.2 Diversity/Multiplicity

Diversity is attained after a long period of time influenced by environment and social factors; it is a process of recognizing differences and similarities and creating an environment that capitalizes on them. Diversity has environmental, social and economic impacts. Development in social sustainability is greatly affected by diversity. Diversity enhances QOL by promoting different ways of accepted living. The multiplicity of people and agencies located at different levels of community entails an attention to social integration, housing provisions, social infrastructure and services for resource generation. It is relevant for various public policies and decision-making sectors such as education, health, social and community services, communication, transport services and resource management. Diversity is required for better understanding of the concept and its application to an urban context for better QOL.

9.4.3 Quality of Life

QOL is the state of wellbeing of persons, societies and communities, focusing on every aspect of life. It is all about attaining satisfaction in life in terms of education, sacred principles, family, health, employment, economy, wealth and natural resources (Li et al. 2018). From the global development point of view, QOL has a wide range of contexts including areas of global development such as international relations, health services, employment and legislation (Dempsey et al. 2011). Some important dimensions of QOL are described as follows.

Economic and physical wellbeing: Safety and security are crucial aspects of the life of any individual. The main concern of security is to provide safety to the life of citizens, which is measured in terms of physical wellbeing and economic safety. Economic and physical wellbeing are measured on the basis of three sub-dimensions: earnings; depletion of resources; and housing environments.

Health and maintenance: Health is a key element in achieving a good QOL of any individual. Poor wellbeing affects the growth of an individual and hence of society. The wellbeing of an individual is defined in terms of complete physical and mental state. Healthiness is measured in terms of health indicators such as healthy life span, infant impermanence, life expectancy and more subjective indicators such as self-evaluation of own fitness and access to healthcare facilities (Verdugo et al. 2012).

Social interactions: Social networks and connections are equally important for measuring the health of an individual affecting life gratification. They are measured in terms of leisure activities, cultural events and other group activities. These are necessary to receive social support and increase social contact.

Education and personal development: Education plays a key role in achieving life satisfaction in knowledge-based economies and is an important factor on which the QOL of an individual relies. The nature of jobs and salary levels are greatly dependent on the level of education one has. Educated communal groups have a wider range of job opportunities and growth parameters.

Basic rights: To have the basic right to take part in public debates influencing the outlines and decisions of public policies is an important dimension of QOL. A democratic society guarantees the right legislation for citizens as a decent governing body is highly dependent on the involvement of citizens in communal and governmental life, such as participation of citizens in administration, political events and parties, trade and labor unions, etc. (Lafont 2010). The correct implementation of rights and policies can be seen from the measurable levels of trust and faith of citizens in the governing system, satisfactory public facilities and services. The absence of discrimination such as gender discrimination is measured in terms of the salary gap of men and women (Andorno 2007).

9.4.4 Interdependent Social Cohesion

Social cohesion is the capacity of a society to ensure the wellbeing of all its members, reduce inequalities and avoid partitions of the community. A cohesive society is an equally supportive and caring community of free individuals achieving common objectives by self-governing means. Societies have to live with the pressures and tensions caused by partitions within the community. Cohesion is at risk when disparities in social indicators are excessive or widen with time. Societies often have ethnic and cultural diversity, and the key to cohesion is to manage diversity so that it becomes a source of mutual enrichment rather than a factor for partition or clash. A cohesive society is one that has developed satisfactory ways of coping with the strains and tensions of the society in an open, democratic manner. It takes action to enhance equity to manage and stabilize itself in order to grow. Social cohesion needs to be encouraged, enhanced and modified for the betterment of society. Each generation has to find a fresh, controllable, stable force as this is a constantly shifting equilibrium, which has to adapt the changes of social and economic environment with advances in technology and political systems (Dempsey et al. 2011).

Social cohesion is not only about fighting social exclusion and poverty but also about creating unity in society so that exclusion can be reduced. If poverty and prohibition remain in society then there is a need to take specific action to help those affected. Hence, social cohesion is a strategy to tackle exclusion by means of both prevention and treatment. The dimensions of social cohesion are identified as: belonging; inclusion; participation; recognition; and legitimacy. Other methods are also there for mapping social cohesion that cover the issues of the society based on the concept of values, identity and culture differences. The issues of society required for mapping social cohesion include inequalities and inequities, cultural diversity and geographical diversity. The three ways distinguished to define social cohesion are: the absence of social exclusion; interactions and connections based on social capital; and shared values and communities of interpretation based on group identity.

9.4.5 Integrated Governance

For development of the social dimension of sustainability, governance issues have a place alongside increasing social issues. Integrated governance (Figure 9.2) is required to ensure effective accountability toward stakeholders, national and international regulations. National and international relations have largely proliferated due to increasing attention given to compliance issues within the governance debate, while some studies have also emphasized that in order to implement effective governance compliance is not enough. In this context, a broader view of governance is suggested that combines compliance with the achievement of financial and non-financial objectives, ethical issues, environmental concerns and risk awareness (Parkes et al. 2010). This broadened perspective suggests going beyond formal structures of governance to consider the actual processes and mechanisms through which governance principles are operationalized. Good governance is a key element for sustainable development, and this view is also embraced by Sustainable Development Solutions Network of the United Nations launched in 2012.

With an integrated approach to the governance system, integrated sustainability requires implementation of ad hoc strategies and business models that should include social, environmental, relationships and financial dimensions of sustainability. Based on founding principles of sustainable development new strategies have been emerging resulting in business models for charity and social programs. In this context, sustainability and corporate performance are managed separately representing objects of two distinct strategies carried out by an organization. This separation between sustainability and corporate models is a hindrance in actual implementation of sustainability, relegating it to ineffective solutions or to mere "sustainability rhetoric" in the search for legitimization. This situation asks for implementation beyond mere rhetoric. Generally, business models comprise internal and external patterns of interactions that shape a company's value chain, value proposition and value system. Given the tensions between the business needs of corporations and the social and environmental dimensions of sustainability, the implementation of sustainable business models is a challenging process. Despite these, new business models have emerged as "hybrids" or "hybrid organizations" or "hybrid business models," in

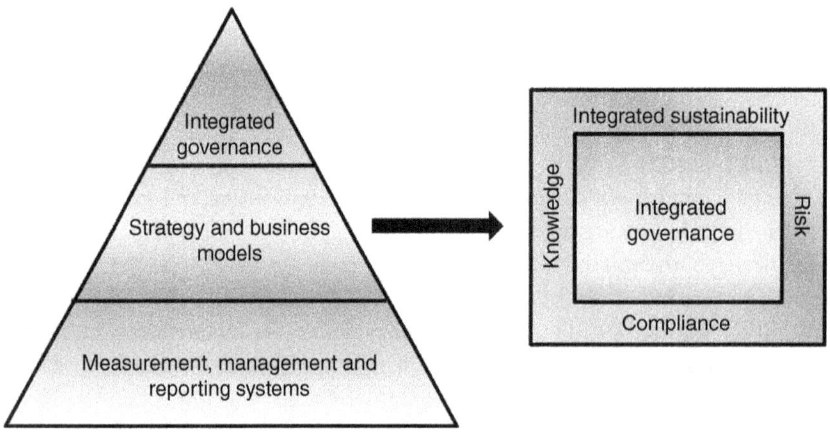

Figure 9.2 Integrated governance. Source: Adapted from Busco et al. (2005).

which social and commercial performance are sought after simultaneously through a single, unified strategy (Smith et al. 2013). Hybrid business models are gaining acceptance and experiencing rapid growth in every sector.

Social entrepreneurship gives rise to many hybrid organizations by searching for autonomy from donations and subsidies as well as by attempting to scale up to stand in a larger market. Similarly to hybridization, hybrid business models also face many challenges and opportunities that come from the integration of diverse elements within the same strategy. This diversity is considered as a unique source of innovation totally different from pure social and pure commercial models. Hybrid organizations should have a balance between social and economic objectives, to avoid any mission drift. Therefore, in this case focus is on the profits to the detriment of social good. For this reason, hybrid business models are characterized by a ubiquitous and unstable nature (Miller et al. 2008), which does not allow one-off solutions to take place but rather requires the continuous search for the "hybrid ideal." When organizations combine social missions with commercial agendas, they create unfamiliar combinations of activities for which supportive systems may not exist (Smith et al. 2013).

9.4.6 Maturity

Maturity is defined as an ability of an individual to respond to environment in an appropriate manner. That person accepts the responsibility for consistent growth and improvement through social attributes such as communication, behavioral patterns, indirect education and philosophical explorations. Maturity is also being aware of the correct time and place to behave in a particular way and knowing when to act as per the circumstances and the culture of the society one lives in. It has different definitions in different contexts such as legal, political, social, religious, sexual, emotional and intellectual. In social sustainability, the maturity of an individual plays an important role in the development of community sustainability by enhancing social networking. It is responsible for social integration and interactions. It is required to work in collective groups of an organization or in societal activities.

9.5 Gaps in Dimensions of Social Sustainability

The main purpose here is to trace the evolution of social sustainability by studying the gaps in the process of development that connects the bio-physical environment and the economic challenges and issues of society. These gaps are defined in terms of three schemes of social sustainability comprising: (i) development sustainability – which addresses the basic needs of society, the creation of social capital, justice, equity and resource distribution; (ii) bridge sustainability – its main focus is on changes in behavior to achieve the bio-physical environmental goals; and (iii) maintenance sustainability – its main objective is to preserve or to sustain socio-cultural characteristics to lead to persistent changes and to challenge how society resists these changes.

Therefore, development sustainability is defined as the needs of the society that must be fulfilled to remove the gap between rich and poor based on poverty and inequity; Bridge sustainability is concerned with changes in behavior to reach bio-physical environmental goals; and maintenance sustainability is all about the preservation of social–cultural practices with respect to social and economic change.

9.5.1 Development Versus Bridge Sustainability

These two forms of sustainability share the common objective to raise and maintain the integrity of the bio-physical environment, on which the species survival and "survival of the fittest" theory depend. Bridge sustainability takes advantage of nature and a raft of techno-scientific estimation that contributes to the bio-physical environment, social health and thoughts of social consequences and its importance. Development sustainability places an emphasis on the poverty and inequity that are responsible for social environmental degradation. The alleviation of poverty and inequality is important for environmental social wellbeing. These two sustainabilities are aligned, such as when a house is constructed it should be affordable and green, which stimulates interest in bio-physical environmental issues. Several recent studies have highlighted the need to be much more aware of the social implications of the solutions to bio-physical problems. For example, the use of public transport depends on efficient, clean and safe services but these services are likely to be more expensive and limited to high-demand routes. Such situations exacerbate the exclusion of some marginalized groups and therefore act against the principles of sustainability (Vallance et al. 2011). These relationships point to the need to address bio-physical environmental and social concerns together and to expand sustainable development beyond the "brown agenda" in order to think about the consequences of poverty and inequity in developed countries as well.

9.5.2 Development Versus Maintenance Sustainability

These two forms of sustainability conflict with each other as what is good for the individual and groups may not always be good for the wider groups or community. In the context of housing infrastructure, for example, the establishment of gated communities is designed to exclude certain groups, and thus may promote the inequitable location of public goods and services. The gap in requirements in non-gated subdivisions allows affordable housing to be provided. While in some cities, the desire for a large number of houses puts an increased pressure on land supply and drives up the prices, denying poorer people the opportunity to own their own house and leading to overcrowding. Thus, many recent studies in the field of political ecology focus on the inequitable distribution of environmental goods, which has become the impetus driving the environmental justice movement (Vallance et al. 2011).

Maintenance social sustainability is all about traditions, practices, culture preferences and places people would like to see sustained or improved, such as the preservation of natural landscapes or historical places. These practices of maintenance underpin QOL, social networking, corporate relations, pleasant work, finest living spaces, leisure opportunities and so on. It is, therefore, concerned with the ways in which social, cultural, and environment characteristics and preferences are maintained over time. This maintenance occurs through habit, movement and usage of both local and global connections and the influence they exert via technological innovations, resource shortage, immigration, employment and other forces of change.

9.5.3 Maintenance Versus Bridge Sustainability

Maintenance sustainability is important as it builds on work contributing toward maintenance and understanding of the context within which the environment is required with a focus on the ways people adopt and resist changes developed by the bio-physical environment. Sometimes changes of the environment are good for people personally, which makes it easy to adopt the changes. However, in many cases the changes that suit the environment are not suitable for small-scale communities due to clashes with important habits and preferences of people's everyday lives. These clashes are due to practical consciousness, which allows people to live their daily life without making decisions constantly, and discursive consciousness, which is related to the actions of which people are more aware. Small changes in our practical consciousness can yield good results and higher decision making to accept changes when they are difficult to accept (Vallance et al. 2011). The major point is that particular practices are maintained in different ways and bridge sustainability measures needs to be better if they are to be successful. Rather than hoping for positive outcomes for the bio-physical environment will follow development, another approach of social sustainability actively and explicitly explores ways of promoting eco-friendly behavior or stronger environmental ethics. The social conditions range from transformative to non-transformative approaches that challenge fundamental ways in which the environment is socially constructed. This distinction is important as transformative approaches re-imagine an individual's relationships with the environment, other individuals and animals, while non-transformative methods are all conventional methods, which are limited in scope and aspire only to small, incremental changes. Although the list of supposed environmental benefits of compaction is long, the advantages may be lost on those urban residents who prefer low density and suburban living. Therefore, it is important to overcome the distance between these different versions of the concept, which might require not just better science, but also better understanding of how to frame sustainability goals so that they seem to be more consistent with what people value and would like to preserve.

9.6 Research-based Policies and Federal Perspective Toward Social Sustainability

With large gaps in the evolution of social sustainability and challenges in its achievement, government bodies have made certain policies based on research to achieve social sustainability by removing social exclusion and marginalization of poor people. The effect of these policies is observed in terms of large-scale settlement growth, employment and livelihood activities. Therefore, to examine the effectiveness of these policies some of the examples and policies are studied in more detail in the following text.

As defined above, government plays an important role in achieving social sustainability so it is necessary to know the federal perspective toward sustainability. This federal perspective decides the effectiveness of policies and steps taken to reach forward in terms of social sustainability. The implementation of developed research-based policies ensures social inclusion in society. Some major policies affecting the evolution of social sustainability are defined below based on global analysis of social sustainability.

9.6.1 Tanzania's Policy and Its Impact

Tanzania is an East African country known for its diversity and wilderness. Since independence in 1961 Tanzania has gone through three distinct phases of macroeconomic changes. These changes had varying impacts on its people. The first phase was between 1961 and the mid-1980s; during this phase socialism policy was supported by social welfare principles that emphasized equity to access all the basic services, livelihood opportunities and a communal way of living. People of the society were assured of free health services, education and water supply. Several programs were also implemented toward achieving the goals including Universal Primary Education (UPE) for all in 1974. This policy orientation could not be sustained because of poor economic performance amid rapidly urbanizing cities (Embuldeniya 2014).

The second phase was between 1985 and 1995, and a shift was observed from a controlled to a liberal economy. During this period the Structural Adjustment Programme (SAP) was implemented, which was characterized by changes of retrenchment and freezing of public employment and participation of the private sector in the economy (Mutahaba 2014). The private services provision in education, health, water supply, electric supply and transport was given more impetus. However, for poor households the affordability of privately run services had become an issue. The third phase is from 1995 to the present time, and focuses on sustaining liberal macroeconomic policies, restoring fiscal disciplines and creating an institutional base for supporting greater participation of the private sector. Today all these macroeconomic changes have contributed to shaping inclusion observable in many cities around the globe.

9.6.2 Binary Theory

McElroy et al. (2008) give a combined approach to measuring social sustainability in their essay about "Sustainability Quotients and the Social Footprint." They called this approach the "binary theory" in combination with specified quotients to create a social footprint. Binary theory gives the scale of sustainability performance depending on which side of a demarcation point an operation or entity falls (Figure 9.3).

This theory of sustainability is a variant of the absolute goal orientation that is completely anti-relativistic. According to the binary orientation the entity is either sustainable

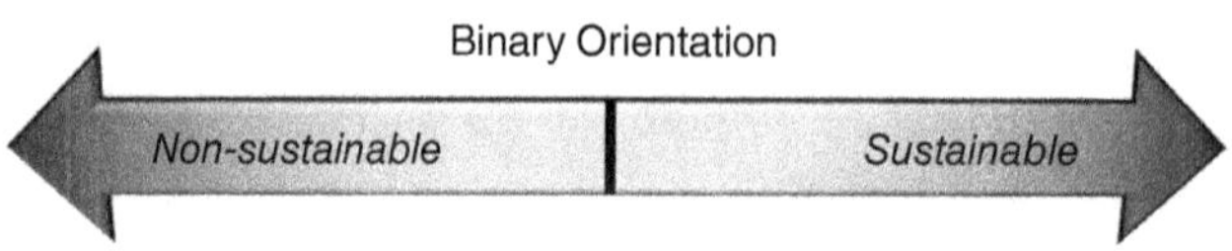

Figure 9.3 Binary orientation.

or not as there is no option in between. There are only higher and lower scores for sustainability performance: some fall on the sustainable side while others fall on the non-sustainable side. Using binary orientation, the social researchers derive quotients for the impacts of actions on sustainability. These quotients are then used to create their social footprint. This is the most applicable scientific approach to measuring the sustainability performance of an entity or artifact. The Social Sustainability Footprint (SSF) consists of a quotient-based method for measuring sustainability performance and for reporting its impact. The quotient context can be social, ecological and can be much more diversified. The SSF is:

- Based on activity – Sustainability is measured in terms of impacts that arise from organized activities and actions.
- Practically, confines the measurement and reporting to organizational boundaries the same as for financial reports.
- Conceptually, the quantitative quotients approach to sustainability where performance is measured relative to standards of performance.
- Based on the principle of personal and collective responsibility, people are responsible for their mutual and collective decisions and actions, including professional tie ups with an organization.

9.6.3 Macroeconomic Social Policies in Place

There are many long-lasting and evolving challenges in economic management during the process of promoting economic growth, accurate collection of information/data and structural information in the world, developing appropriate fiscal management strategies, and managing natural resources and conflicts. The overall functioning of the economy is studied under the branch of economics known as macroeconomics. Macroeconomic policies are required for developing the outline within which factor markets such as labor and capital and product markets operate. These policies have a critical effect on decision making by companies for production such as hiring and firing of workers, or export and import of goods. The policies also determine household decisions to consume, save and borrow as well as government decisions to capitalize infrastructure, education and many other aspects of sustainable development.

Macroeconomic policies include excises, government expenditure and borrowing, exchange rate factors, and economic and credit rules. The prime goal of an effective macroeconomic environment is to develop growth prospects and improved living principles. Stability is not the only concern of these policies as they have an important impact on income distribution across economic classes and generations. The primary goals of World Bank Groups are to eliminate poverty and to ensure mutual prosperity and its macroeconomists are no exception (Blanchard et al. 2010). World Bank Groups work to support decision makers to better recognize and manage macroeconomic policies so that policies work to reduce poverty and to increase incomes. Governments and macroeconomists face complicated situations in choosing macroeconomic and fiscal policies, especially in countries facing fragility and conflict. Over the past few decades economists have observed a renaissance in conflict across the countries having both low and middle incomes. Macroeconomists consider macroeconomics as a driver or enabler of conflict in some circumstances. This conflict is mainly present in the countries where the distributions of natural resources are challenged and is the root cause for income differences within the country. Macroeconomic policymakers are working to achieve the goals of development, growth, poverty elimination and shared wealth in an environment constrained by instability and conflict. The efficiency of various approaches depends on the challenges that a country is facing at the diverse phases of conflict (Blanchard et al. 2010).

Macroeconomic policies are based on a country's environment, such as level of incomes, human resources, physical capital and natural resource base. Macroeconomics and Fiscal Management (MFM) plays a major role for macroeconomic analysis for the World Bank Group (Blanchard et al. 2010). Macroeconomists engage clients to study the development process and economic management efforts at every level and also take local context into consideration.

The World Bank's MFM experts who are employed in insubstantial conflict and post-conflict states are assisting governments in making policies to address the macroeconomic drivers of conflict, to design and develop macroeconomic policies in contexts forced by fragility and conflict, and to develop a path of reconstruction after the end of conflict for the growth of the community. The MFM decision makers use their knowledge and experience working in fragile and conflict environments around the globe to help national and state governments to learn new approaches that have succeeded and failed in other parts of the world, and to support them in designing and developing new approaches as per a nation's own environment for growth and stability in their own countries.

The major challenge for a government of a post-conflict nation is in achieving a social peace. Macroeconomic policies are not the primary mechanism to achieve social peace, but the fundamental goal of macroeconomic policies is to contribute to achieve the desired result and not to make it more difficult (Blanchard et al. 2010). A summary of policies adopted after conflict in various countries is given in Tables 9.2 and 9.3.

For resource-rich countries, at a superficial level the design of macroeconomic policies is a simple task. The goals of the policies are strong growth, poverty reduction and economic diversification. Resource-rich countries are not constrained by the supply of foreign exchange; neither are the governments budget constrained. The combination of goals and lack of constraints is responsible for growth and success in resource-rich countries.

Table 9.2 The list of short-term macroeconomic policies for post-conflict nations.

Policy	Objectives and actions
Fiscal policy	To pay compensation for provisional lack of private investment and trades; focus on work creation by organizing public works activities; and to maintain excises on imports and charges on price of inflexible exports
Monetary policy	Tie sales to enhance fiscal development, or by monetization that also ropes exchange rate competitiveness
Exchange rate policy	Slow down real fall or depreciation through modification in the nominal rate compared to export opponents

Table 9.3 The list of long-term macroeconomic policies for post-conflict nations.

Policy	Objectives and actions
Fiscal policy	Public investment required to repair and reconstruct the infrastructure and to diversify tax base and tax settlements
Monetary policy	To reduce the use of foreign currencies or capital; to support the development of financial mediators; creating stores of financial wealth
Exchange rate policy	To consider "crawling peg" to get a bit-by-bit depreciation of real exchange rate

Table 9.4 Short-term macroeconomic policy for resource-rich countries.

Policies	Objectives and actions
Fiscal policy	To create reserve funds for intervention and investment by maximizing the income from resource flows by setting suitable fees and tax rates
Monetary policy	To accept the central bank rates to regulate the commercial bank reserves to prevent unnecessary inflationary burdens
Exchange rate policy	To employ fixed exchange rate to prevent nominal indebtedness

Table 9.5 Long-term macroeconomic policy for resource-rich countries.

Policy	Objectives and actions
Fiscal policy	To encourage public investment and to facilitate diversification into non-resource trading that help to broaden direct tax base preventing inequality
Monetary policy	To facilitate private investment to expand non-resource trading and to support exchange rate management
Exchange rate policy	To maintain constant real exchange rate to expand trading

The fundamental elements of the macroeconomic policies framework for resource-rich nations are given in Tables 9.4 and 9.5.

9.6.4 Universal Declaration of Human Rights

The Universal Declaration of Human Rights (UDHR) is a document that was adopted by the United Nations General Assembly in 1948 as "Resolution 217" in Paris, France. The document consists of 30 articles encouraging an individual's rights, which were elaborated and explained in subsequent universal treaties, national policies and other laws. This declaration was the first step in framing the International Bill of Human Rights that was completed in 1966 and implemented in 1976 (Elkins et al. 2013). This declaration was concerned with the duty and responsibilities of the individual toward society. Since the declaration was adopted on December 10, this date is accepted as "Human Rights Day" or "International Human Rights Day" (Waltz 2002). The UDHR declaration consists of 30 articles with a preamble:

- *Preamble:* It sets out the chronological and social causes that were required for drafting of the document.
- *Articles 1–2:* Focused on the concepts of self-esteem, freedom, equality and union.
- *Articles 3–6:* Established right to life and ban on slavery.
- *Article 6–11:* All about the fundamental freedom of an individual.
- *Article 12–17:* Established the individual's rights in terms of society and community and a right freedom of movement.
- *Article 18–21:* Focused on the concepts of *constitutional liberties*, such as spiritual, public and political choices.
- *Article 22–27:* These articles focus on an individual's economic, social and cultural rights. Health and healthcare services are also included in these articles. As per these articles everyone has the right to a good QOL, and to basic amenities such as food, clothing, housing, social services and health services. It is also about orphan rights, and care for the very poor people of society.
- *Article 28 to 30:* These articles define the ways of using these rights, where the rights can be used without exploitation of an individual.According to the UDHR governments must commit themselves to people by taking progressive measures to secure, observe and recognize human rights. The UDHR was adopted as a declaration rather than a treaty as it was believed that it would have the same kind of influence on global society as the United States Declaration of Independence (USDI) had within the US. The UDHR has served as the basis for a growing number of national laws, international laws and agreements, as well as for a growing number of regional and local organizations protecting and encouraging human rights. The UDHR was never accepted as a treaty but was adopted to help understand the meanings of "human rights" and "fundamental freedoms." The UDHR is an important constitutive document; many international lawyers believe that the UDHR forms a part of regular international law and is a controlling tool in applying diplomatic and ethical pressure to governments that violate any of its articles.

Still today the political account of the UDHR is not well acknowledged, and its insignificance has raised a number of assumptions that require assessment. Recent challenges have accepted that the UDHR was supported and encouraged by the Western countries, and certainly raised questions about unlimited power backing for efforts to craft international human rights standards. There are four political myths about UDHR, each of which is true but also misleading. Its legality ranges from states to whole nations that share a major part of development in the history of development (Waltz 2002).

9.7 Social Sustainability – an Integrated Approach

The social dimension of sustainability cannot be achieved by one authority or profession. Rather, it is delivered by an integrated, multidisciplinary approach supported by a range of policies to achieve its desired impacts and implementation. The integrative approach is the interrelationship between both the designed outcome and the social infrastructure, as well as the relationship between the process and the community of the society (Goworek 2011). Thus, there is no single policy or project or responsible authority that can deliver social sustainability. All the national governments, communities of the society, decision makers in all areas and fields and creators play a role in social sustainability. The levels of integrated participation for social sustainability according to Berkes and Ross (2013) are as follows.

9.7.1 National Government

The national government plays a major role in achieving social sustainability. It allows the best national approach at every level to facilitate a focus on the issues present within the country. To prevent hindrance in achieving social sustainability the federal government is required to establish important planning principles with appropriate support of the state and local governments for the delivery of established government policies. For example, the Australian government is well placed to centralize a national approach to social sustainability. A fundamental role is to assist on the issues across a range of government strategies such as population goals, social enclosure, heritage and cultural policy. The national government establishes important planning policies with the support of state and local government for implementation. The Council of Australian Governments (COAG) Capital City Strategic Planning Criteria, the National Urban Policy and national Urban Design Protocol are the models that can be duplicated and in relation to social sustainability. The National Urban Policy and Urban Design Protocols quantify indicators of social sustainability and wellbeing including QOL, communication and improved urbanization. The social part of development is well-defined by the national government in the designed framework of "Our Cities, Our Future": Australia's National Urban Policy for a fruitful and sustainable future to achieve the objective of enhanced livability of cities by promoting sustainable design and planning, and services that can be easily accessed by community. This project also emphasized the need to support social wellbeing. Better urbanization is responsible for better employment and good QOL (Berkes and Ross 2013). Hence, the implementation of national policy with the support of state and local bodies is necessary to achieve social sustainability.

9.7.2 State and Regional Government

The state and area governments are responsible for implementation of national polices or projects across the country. They are responsible for effective planning and management of cities and communities. State levels are the best place to enable different levels of government to work together to achieve better cities, communities, QOL and employment. State government acts as a bridge between national and local governments and so state governments are able to work directly with local government in allowing the community and appropriate corporate sectors and stakeholders to focus on equity when developing policies or planning strategies. The state government works with the national government to support the integrated designs, approaches to achieve sustainable society and high QOL.

9.7.3 Local Government

The local government is the mediator between the state government and lives of people. The local government is responsible for the needs and services of its community. It plays an important role in supporting social sustainability by providing physical and easy infrastructure within the community. State governments recognize and facilitate community participation in strategic planning concerns and coordinate service suppliers within the community to work effectively to address local issues present in society. Local governments with state governments play a role in accountability and non-governmental developments to ensure proper implementation of policies and delivery of all projects that have a focus on social sustainability (Berkes and Ross 2013).

A good example of integration between state and local governments is seen in the Community Indicators Project of Australia. Let us have a look at Australian state and local government integration where Community Indicators Projects are expanding across the country to achieve social sustainability. The Australian Community Indicators Network provides a forum of the groups to collaborate and share the same resource base. The Community Indicators Victoria was the first project started in Australia, which provides a framework to quantify QOL and wellbeing. The community wellbeing framework has almost 80 community wellbeing indicators that are used to engage with communities and identify issues, and provides a baseline for local government decision making. Similarly, Community Indicators Queensland has developed a wellbeing framework for measuring wellbeing in Queensland. Its aim is to provide a platform for investigating and measuring patterns to define the effectiveness of a policy, community participation in policymaking, the development of national frameworks and implementation of these frameworks at local levels (Berkes and Ross 2013). Further, these frameworks are used for the development of indicators useful in the designing of community resilience frameworks.

9.7.4 Communities and Non-Governmental Organizations

The citizens always play a vital role in creating social sustainability. The people are the activators of the society or community and the major participants creating networks, social integration and relationships, and they cooperate in maintaining these indicators of social sustainability. To promote social sustainability, it is very important for citizens to be aware of their rights and responsibilities toward society in the process of creating sustainable societies. This all acknowledges them as the community creators of their local areas. Help and assistance is provided to communities from government and non-government organizations to enhance community capacity and services provided. During urbanization, communities have the right to be heard and to be represented by community union groups. Communities should always be encouraged to actively contribute in the design process rather than simply be present.

9.8 Safety and Security

The perceived safety of a community and its residents is a fundamental part of social sustainability. In its definition of social cohesion, the British House of Commons Committee placed perceived safety within Maslow's "hierarchy of needs" to fulfill the basic necessities required before achieving social cohesion. Therefore, it is an antecedent of any positive social activity within the community. Giving security to residents of a community and, with it, feelings of safety in a society, is closely interlinked to other dimensions of social sustainability. The community should be free from crime and illegal activity so that residents can feel secure and safe in their social networks and interactions during their participation in organized social activities. Feelings of safety within residents enhance trust and reciprocity between people and a sense of being in a good place with an understanding neighborhood (Winther 2017). The linkage between safety and the built environment is not clear, but examples such as parks nearby housing areas that provide an easy way to mix with others have cited benefits of social survival and interactions. Poor environments with poor maintenance of services are said to have psychological effects on the residents' sense of safety and security (Stacchezzini et al. 2016). The thought of nobody caring is closely linked to the "broken window syndrome" where even cosmetic damage can give rise to more serious anti-social or criminal behavior.

9.9 Future Perspective and Conclusions

The long-term consequences are very significant in defining the processes and policies for social sustainability. The formation of new urban society needs a good quality of infrastructure, opportunities, shared spaces and activities, support for community development and job opportunities. If these supports are not provided by the government, the new communities struggle to become cohesive and sustainable. A wide range of resources is required for the development of sustainable societies and for communities to flourish. Social sustainability requires development in four areas: amenities and social infrastructure, social integration and cultural life, voice and influence, and the space to grow for new community establishments (van der Krieke et al. 2017).

If the new communities and places do not become successful communities, then a risk for the future develops. Social designing and development are issues of public value as well as community satisfaction. Implementation of new corporate policies and new models are needed to support investment in social infrastructure during the formation of new communities if they are to survive long term.

Apart from social infrastructure, the maintenance of QOL is equally important while aiming for social sustainability. The issues of the society should be addressed during the designing of frameworks and indicators in ensuring QOL. The following issues need to be addressed in the making of a sustainable society:

- Accessibility or availability of public services and consequences of particular social, cultural and political economic contexts.
- Limits of urban superiority and carrying capacity.
- Issues of supply, poverty, equality and eccentricity.
- Lack of significant components of analysis.
- QOL indicators and social diversity.
- QOL and incorporation of cities into the global system.

Cooperation and good relationships between residents of the society and cultural events or local activities are the keys for the growth of communities. QOL is also defined by integration, shared identities, participation and good relationships. This gives a feeling of being safe and secure in the community. Good neighbors always support each other throughout good or bad phases of life, which strengthens the local social networks of the community. The strength of these relationships directly reduces the crime percentage. Social capital and quality of relationships between people make a community more supportive, empowered and with a rich cultural life. This is important to build the roots of a community based on responsibilities and integration to feel safe and secure (Porio 2015).

Therefore, keeping social sustainability in mind and getting the balance right at the planning stage is very crucial, which is a very challenging and tedious task. As discussed above the new developments need to be well planned and organized to ensure the proper distribution of basic amenities and public services. However, it very difficult and almost impossible to anticipate all the future needs of society. Many features of social development cannot be planned in advance, such as community projects or policies, government arrangements or administration. Other local administrative institutions need to develop and grow, building on relations within the community and outside the community. It is important to recognize mutual benefits, have a sense of sympathy and support, and trust people and other stakeholders, and to realize that all of these need time to develop and flourish. In the process of social sustainability, urban politics and development should be of very high priority.

Acknowledgment

We are appreciative to Lalit R. Kumar for his support, guidance and modification of the earlier drafts to improve the chapter.

References

Amini, M. and Bienstock, C.C. (2014). Corporate sustainability: an integrative definition and framework to evaluate corporate practice and guide academic research. *Journal of Cleaner Production* 76: 12–19.

Andorno, R. (2007). Global bioethics at UNESCO: in defence of the Universal Declaration on Bioethics and Human Rights. *Journal of Medical Ethics* 33 (3): 150–154.

Berkes, F. and Ross, H. (2013). Community resilience: toward an integrated approach. *Society & Natural Resources* 26 (1): 5–20.

Blanchard, O., Dell'Ariccia, G., and Mauro, P. (2010). Rethinking macroeconomic policy. *Journal of Money, Credit and Banking* 42 (s1): 199–215.

Busco, C., Frigo, M.L., Giovannoni, E. et al. (2005). Beyond compliance: why integrated governance matters today: the idea is for financial professionals to help their companies address performance and knowledge management, too. *Strategic Finance* August: 35–44.

Collins, S.L., Carpenter, S.R., Swinton, S.M. et al. (2011). An integrated conceptual framework for long-term social–ecological research. *Frontiers in Ecology and the Environment* 9 (6): 351–357.

Dempsey, N., Bramley, G., Power, S., and Brown, C. (2011). The social dimension of sustainable development: defining urban social sustainability. *Sustainable Development* 19 (5): 289–300.

Elkins, Z., Ginsburg, T., and Simmons, B.A. (2013). Getting to rights: treaty ratification, constitutional convergence, and human rights practice. *Harvard International Law Journal* 54: 61.

Embuldeniya, D. (2014). Ethnic conflict, horizontal inequalities and development policy: the case of Sri Lanka. Doctoral Thesis, University of South Africa. http://uir.unisa.ac.za/bitstream/handle/10500/13540/thesis_embuldeniya_d.pdf?sequence=1&isAllowed=y (accessed November 2019).

Goworek, H. (2011). Social and environmental sustainability in the clothing industry: a case study of a fair trade retailer. *Social Responsibility Journal* 7 (1): 74–86.

Griessler, E. and Littig, B. (2005). Social sustainability: a catchword between political pragmatism and social theory. *International Journal for Sustainable Development* 8 (1/2): 65–79.

Hightree, J., Kliskey, A., Higgins, L. et al. (2018). Themes in community resilience: a meta-synthesis of 16 years of Idaho community reviews. *Community Development* 49 (1): 65–82.

van der Krieke, L., Blaauw, F.J., Emerencia, A.C. et al. (2017). Temporal dynamics of health and well-being: a crowdsourcing approach to momentary assessments and automated generation of personalized feedback. *Psychosomatic Medicine* 79 (2): 213–223.

Lafont, C. (2010). Accountability and global governance: challenging the state-centric conception of human rights. *Ethics and Global Politics* 3 (3): 193–215.

Li, N., Chan, D., Mao, Q. et al. (2018). Urban sustainability education: challenges and pedagogical experiments. *Habitat International* 71: 70–80.

Littig, B. and Griessler, E. (2005). Social sustainability: a catchword between political pragmatism and social theory. *International Journal of Sustainable Development* 8 (1-2): 65–79.

McElroy, M.W., Jorna, R.J., and van Engelen, J. (2008). Sustainability quotients and the social footprint. *Corporate Social Responsibility and Environmental Management* 15 (4): 223–234.

Miller, P., Kurunmäki, L., and O'Leary, T. (2008). Accounting, hybrids and the management of risk. *Accounting, Organizations and Society* 33 (7): 942–967.

Mutahaba, G. (2014). Achieving good governance and development in Tanzania: is leadership the missing link? *Journal of Public Administration* 49 (Special issue 1): 568–590.

Parkes, M.W., Morrison, K.E., Bunch, M.J. et al. (2010). Towards integrated governance for water, health and social–ecological systems: the watershed governance prism. *Global Environmental Change* 20 (4): 693–704.

Porio, E. (2015). Sustainable development goals and quality of life targets: insights from Metro Manila. *Current Sociology* 63 (2): 244–260.

Power, A. (2012). Social inequality, disadvantaged neighbourhoods and transport deprivation: an assessment of the historical influence of housing policies. *Journal of Transport Geography* 21: 39–48.

Smith, W.K., Gonin, M., and Besharov, M.L. (2013). Managing social-business tensions: a review and research agenda for social enterprise. *Business Ethics Quarterly* 23 (3): 407–442.

Stacchezzini, R., Melloni, G., and Lai, A. (2016). Sustainability management and reporting: the role of integrated reporting for communicating corporate sustainability management. *Journal of Cleaner Production* 136: 102–110.

UNFPA (2007). *State of World Population 2007: Unleashing the Potential of Urban Growth*. New York: United Nations Population Fund.
Vallance, S., Perkins, H.C., and Dixon, J.E. (2011). What is social sustainability? A clarification of concepts. *Geoforum* 42 (3): 342–348.
Verdugo, M., Navas, P., Gómez, L., and Schalock, R.L. (2012). The concept of quality of life and its role in enhancing human rights in the field of intellectual disability. *Journal of Intellectual Disability Research* 56 (11): 1036–1045.
Waltz, S. (2002). Reclaiming and rebuilding the history of the Universal Declaration of Human Rights. *Third World Quarterly* 23 (3): 437–448.
Winther, A.M. (2017). Community sustainability: a holistic approach to measuring the sustainability of rural communities in Scotland. *International Journal of Sustainable Development and World Ecology* 24 (4): 338–351.
Woodcraft, S. (2012). Social sustainability and new communities: moving from concept to practice in the UK. *Procedia-Social and Behavioral Sciences* 68: 29–42.
Woodcraft, S., Hackett, T., and Caistor-Arendar, L. (2011). *Design for Social Sustainability: A Framework for Creating Thriving New Communities*. London: The Young Foundation.

10

Social Engineering and Sustainability: Revisiting Popper's "Piecemeal Approach"

Neeraj Mishra

Center for Wage Employment, National Institute of Rural Development and Panchayati Raj, Hyderabad, India

10.1 Introduction

With rapid industrialization in developed and developing countries alike, combined with the consumerist lifestyle of the modern society, complex environmental problems have surfaced that threaten the very existence of human society, for example, global warming, climate change, depleting ozone layer, desertification and environmental degradation, among others. In today's world of rapidly depleting natural resources, nothing is more desirable than to build a society that has sustainability embedded into its everyday practices, thought and lifestyle.

It can be observed from the reports published by both national and international organizations that earlier efforts to promote sustainable development of resources have achieved sporadic and limited success. It has been widely claimed that such efforts need to be complemented with other innovative solutions to widen the reach of sustainable developmental practices. This chapter proposes that the concept of piecemeal social engineering can be usefully employed to bring about a paradigm shift in the way processes of policymaking and institutional reforms are being debated to achieve Sustainable Development Goals (SDGs).

Following the works of Karl Popper, this chapter introduces the concept of "piecemeal social engineering," which acknowledges the context, plurality of actors, institutions and objectives of socio-economic development, as an effective point of departure toward building a sustainable planet and society. This chapter first defines the concept of social engineering, outlines its historical trajectory of development and critically examines its different variants applied in the processes of social transformation over the past century. It then provides the contexts and circumstances in which social engineering has been applied with successful outcomes. The next section explains how the idea of social engineering overlaps with some of the other social science concepts such as "bio-politics" and governmentality. Finally, the chapter suggests some of the ways in which this concept can be usefully applied in the creation of a sustainable society and recommends the tools that can be used to embed sustainability and sustainable practices into our everyday lives.

Sustainability: Fundamentals and Applications, First Edition.
Edited by Rao Y. Surampalli, Tian C. Zhang, Manish Kumar Goyal, Satinder K. Brar, and R. D. Tyagi

In brief, this chapter explains the concept of piecemeal social engineering and explores how this concept can be applied in the creation of a sustainable society. It helps to understand how a man's social environment can be altered in such a way that his natural environment remains unaltered.

10.2 Emergence of Sustainability on the Global Agenda

The Brundtland Report of 1987 brought the term "sustainability" into popular usage in recognition of the fact that earth is a fragile planet and its survival depends on the sustainable management of its resources by humans in alignment with the laws of nature. The physical boundaries demarcating the nations were irrelevant in the goal of building a sustainable planet. This new reality "from which there is no escape" was to be recognized and "managed" (United Nations 1987).

Neil Armstrong's pictures of earth from the Apollo spaceship in 1969 had already confirmed, beyond all doubts, the fragility of the planet, and that all inhabitants of the planet were equal stakeholders in this ecosystem. For the first time, it has become possible to speak of *our* planet (Sachs 1992). In a way, this laid the foundations for a "world society" or a "global community" and it became possible to imagine that the earth was just one big living space for humans, without much relevance for geographical boundaries, when it came to global problems such as climate change or global warming. Environmental degradation was already identified as a problem in the making at the United Nations Conference on Human Environment in Stockholm (1972), followed by the World Conservation Strategy (1980) and the World Charter of Nature (1982). This was also recognized by the World Commission on Environment and Development, and published as the Brundtland Report in 1987, which firmly established the issue of sustainability of the planet on the global agenda as one of the most urgent priorities for mankind. Other declarations and protocols such as Global Biodiversity Strategy (1991), Rio Declaration (1992), Kyoto Protocol (1998) and Cartagena Biosafety Protocol (2000) also strengthened the environmental discourse in the global public sphere, which has remained in the forefront ever since, reiterated again in the formulation of SDGs in the year 2015. A series of international conferences on sustainable development have also provided increased visibility to the issue of the environment and the need to focus on sustainable practices.

At the same time, technological advancements in the past three decades and the growth of knowledge in social and natural sciences have also been unprecedentedly higher than at any other period in human history so far. This has been accompanied by an increased awareness of the environmental consequences of human action spearheading large-scale changes in the social and political behavior of the people and governments in some parts of the world. While developing countries still struggle to provide their people with basic needs such as water and food, some advanced nations such as Germany and the Netherlands have invoked intelligent policymaking to induce sustainable everyday lifestyles among the people. These countries provide a good template for developing nations to contextually emulate this complex blend of piecemeal social engineering toward developing sustainable lifestyles, embedded in the everyday activities and behavior of the people.

Population growth and increasing economic activities in many of the developing countries along with rapid urbanization in the past three decades have unleashed exponential increase in the consumption of natural resources. The World Bank has calculated the global GDP growth to be around 3–4% annually since 1965, mostly in the developing countries of Asia and the Pacific. The cities have progressively acquired a disproportionate share of economic powers as only 600 urban centers are generating 60% of the global GDP (Dobbs et al. 2011). This also implies that urbanization is experiencing a faster rate of growth, especially in developing countries and is projected to be 1.6% annually between 2025 and 2050, which would mean a doubling of the population in these countries from 2.5 billion in 2010 to over 5 billion in 2050. Managing such pressures effectively requires a profound change in how cities operate and, while cities do indeed set targets to reduce such pressures, the systematic means to assess the attainability of these targets are currently lacking (Walker et al. 2017). The Intergovernmental Panel on Climate Change (IPCC), among others, has proposed that if the very worst impacts of global warming are to be avoided, rich and rapidly industrializing societies will need to reduce their aggregate greenhouse gas emissions by as much as 80% by the middle of the twenty-first century (IPCC 2007). While a compromise on the modern lifestyle is an option few people are willing to take, there could be alternative routes through which new patterns of consumption and rationality can be embedded into the everyday lives of people using new forms of rationality. In this chapter, the concept of "piecemeal social engineering" has been suggested as one such alternative route that can be useful in developing contextually sensitive sustainable practices among the residents of a community, village, city or a nation.

10.3 Concept of Social Engineering

In his 1944 published book *United World* (Scott 1944, p. 221), the author proclaims "social engineering" to be a "new science," which would help in the transformation of a haphazard world society into a "deliberative and experimental" one. This deliberative society would be planned to its last detail, oriented toward achieving specific objectives with the help of available knowledge in the contemporary world concerning the principles of social organization.

The *Oxford English Dictionary* defines social engineering as the "use of centralized planning in an attempt to manage social change and regulate the future development and behavior of a society." In other words, social engineering is the deliberate modification of human behavior, opinions and habits with the help of well-defined stimuli to produce desirable outcomes.

The terms "social technology" and "social engineering" are perceived to have a negative connotation and hence fail to inspire immediate trust in those whom it reminds of the "social blueprints of the collectivist planners, or perhaps even of the technocrats" (Popper 1957, p. 58). Hence to offset the undesirable negative associations with this term, Popper uses the prefix "piecemeal" to distinguish it from other forms of social engineering and argues that, in combination with critical analysis, piecemeal social engineering has immense potential to produce practical results in everyday social life as well as knowledge systems.

The earliest mention of the term "social engineering" can be found in the writings of William Tolman. He defined it as an ideal, reciprocal relationship of capital and labor, whereby the reciprocity boils down in the end to a conditioning of workers to serve the interests of entrepreneurs (Tolman 1909). The concept of social engineering can also be traced back to the works of Henry Ford, Charles Taylor and, especially, Ferdinand Tönnies and his "applied" sociology, since social engineers pursued the rationalization of life in order to replace the "atomized" society through an "organic," harmonious community (Brückweh et al. 2012).

The Ford Motor Company established the Sociological Department under the leadership of Samuel Marquis in 1913, which kept a close eye on the personal lives and habits of the employees to ascertain their eligibility for full wage, and transformed their living conditions in the process. As argued by behaviorists such as B.F. Skinner, if human action can be studied as being primarily determined by various triggers or stimuli in the environment, it can then also be transformed and reformed by changing the environment to achieve a sustainable society.

The word "social engineering" also finds mention in the essay by Dutch industrialist J.C. Van Marken published in 1894, which emphasized the development of technical expertise in the management of specifically human problems. With the objective of aligning the interests of employer and the workers in his yeast and oil factories, Van Marken created institutions, policies and organizations that worked specifically for the welfare of his employees. These included a profit-sharing system, insurance funds, institutions such as schools, libraries, sports, athletic and social clubs, etc. aimed at the improvement of the condition of the people working in his company.

Van Marken firmly believed that men cannot be trusted to take necessary action to safeguard their future of their own accord, therefore the employer should find indirect ways to ensure that they invest in securing their future (Östlund 2007). The term "social engineering" in this context meant engineering of the social organizations with the aim of solving social problems, especially of the poor and working classes. The employers used social engineering methods in the early stages of its development in the late nineteenth century to secure the loyalty of the workers.

The twentieth century began with grand schemes for improving the human condition in accord with the scientific laws. In the early parts of this century, the term "social engineering" was used by the followers of the enlightenment school of thought to understand the faith in, and application of, modern scientific knowledge to resolve all social problems. James Scott refers to this faith in modern science as the "high modernist ideology," which he describes as a powerful faith in the adequacy of the scientific progress to expand production and meet the growing needs of the human race, as well as acquire mastery of both human as well as external, non-human nature. Scott argues that state-initiated social engineering projects culminated with unintended tragic consequences due to such an ideological aim. The goal of a perfect society based on Utopian ideals thus disregarded all criticism and treated society as an object for the transformative policies of the state (Scott 1998).

The term later shifted in meaning from "engineering for social purposes" to "social policy in the style of engineering." Due to the positive associations of engineering, the term was appropriated for various purposes (Hansson 2006). The phrase social engineering is also employed to mean designing and erecting social structures in which human beings serve as the raw material, without being asked what their opinion about all that is. The past century saw endeavors to change the social behavior of people on a large scale in accordance with the national priorities and vision of the leaders, who sometimes resorted to violent measures and exercises of power against the will of the people. Whenever mega social engineering programs were initiated, it met with resistance, leading to the growth of dictatorships and repressive regimes in different countries of Europe. The history thus points us in the direction of "piecemeal social engineering" instead of mega social engineering, as radical reforms in terms of property ownership, means of production, patterns of consumption, application of penalties and rewards, etc. to transform humans, as if they were mindless materials, have always been unsuccessful.

Friedrich von Hayek, who won the Nobel Prize in Economics in 1974, published his critique of centralized planning in his book *The Road to Serfdom* in 1944. In his Nobel Prize lecture called "The Pretense of Knowledge" Hayek criticized the agency of government and its belief that it could successfully plan the economic affairs of society. Hayek argues that as humans learned to organize the forces of nature with the growth of science and technology in the past hundred years or so, this has led to the widespread belief that the human condition can also be improved by a similar control of the forces of society:

> That, with the application of engineering techniques, the direction of all forms of human activity according to a single coherent plan should prove as successful in society as it has been in innumerable engineering tasks, is too plausible a conclusion not to seduce most of those who are elated by the achievement of the natural sciences. However, it must be acknowledged that a belief that has proved so eminently successful in producing advances in so many fields has limits to its usefulness and can become positively harmful if extended beyond these limits.
>
> *(Hayek 1949, p. 426)*

While Hayek's arguments have some merit in the area of large-scale social engineering projects, the problems posed by him can be overcome with the help of Popper's piecemeal engineering approach, which works within well-defined contexts and limits.

10.4 Karl Popper's Description of Social Engineering

This section elaborates upon the concept of social engineering as proposed by Karl Popper, and argues that his idea of "piecemeal social engineering" can be usefully employed to bring about favorable behavioral changes in modern societies, especially in the developing countries, leading to the development of a sustainable planet. This approach can be understood as a stepwise process of introducing incremental changes in the everyday lives and habits of the people, which have a far-reaching impact on environmental protection and sustainability in the long run.

Popper's ideas on social engineering are premised in his conviction that the course of human history cannot be predicted by scientific or rational methods as we cannot predict the future of human knowledge and therefore the historical path of mankind. Taking his cue from Plato's theory of Forms/Ideas, Popper introduces the idea of "social engineering" as the means of democratic social reconstruction, through which the historical destiny of mankind can be altered using critical and rational empirical knowledge.

For Popper, there exists a "social engineering attitude" that does not believe in the destiny of man as predetermined, or imposed by historical trends. The social engineer only believes in the power of man as the creator and master of his own destiny, and that in accordance with our aims, we can influence or change the history just as we have changed the face of the earth (Popper 1945).

As the natural sciences provide us today with factual information about our planet and its deteriorating environment, it is proposed that policymakers and planners use the socio-political knowledge available today to pursue predetermined ends such that it leads to the reduction of environmental threats to the planet. Without taking much interest in the origin of the institutions, or in the original intentions of their founders, the social engineer asks if the given institution is well designed and organized to serve our specific and chosen aims, such as reduction of pressure on natural resource consumption. For a social engineer, institutions are rationally organized means to serve certain ends. Their merit lies only in their appropriateness to serve these ends in the most efficient and simplified manner, argues Popper.

For Plato, the history of society was actually a history of social decay from the original and primitive form of the "ideal state." This theory formed the basis of rational social engineering, which Popper calls "Utopian engineering." Plato's theory of Forms or Ideas argues that, since purely rational and universal knowledge of sensible things is unattainable, it was more useful to obtain at least such pure knowledge that was applicable to sensible things to a certain extent.

Popper makes a distinction between "Utopian social engineering," which seeks to create an ideal society and begins with a blueprint of the society that it aims to create before considering the best rational ways and means of its realization, and "piecemeal social engineering," which does not begin with the idea of creating a perfect society but searches for removable injustices and evils in the society and fights them one by one, rather than fighting for some grand Utopian plan for greatest ultimate good. He advocates the usefulness of the latter approach and rejects the former, citing the example of democratic states, which are not perfect, but do uphold certain basic freedoms and democratic rights that were achieved in a piecemeal manner, which makes it a far more desirable form of social engineering.

10.4.1 Utopian Social Engineering

The Utopian approach to social engineering is premised on the fact that every rational action is aimed toward a certain goal that it seeks to achieve. Rational action thus begins with the choice of the end as the first step, distinguishing ultimate ends from the intermediate ends, which are only the means toward achieving that envisioned ultimate end. The same principle of determining the ends before taking the first step when applied to the realm of sustainability engineering would require an all-encompassing definition and concept of a truly sustainable society. A working concept of the "Ideal Scenario Society" is required before any action is initiated. Once the blueprint of the desired society is ready, the best ways and means of its realization can be planned and practical action initiated (Popper 1945).

Popper describes these as the essential preliminaries and the methodological approach of Utopian engineering. He is critical of this approach for its attempts at large-scale reconstruction of society and grand changes whose environmental and sociological impacts are difficult to fathom, due to our limited sociological knowledge, which is a prerequisite for such large-scale social engineering projects. Based on the above argument, Popper criticizes the sociologists of the nineteenth century, especially Comté and Marx and his successors, who sought to establish an universal explanation and science of human societies, which according to him was necessarily doomed to failure. Such grand narratives that would be valid for all societies and contexts, and their usage in bringing about social change, have very limited applicability, argues Popper.

To overcome this problem of instituting grand changes in the social behavior of people, Popper introduces a methodology of *technological social science* that would seek to establish the general laws of social life and discover facts that would be useful in bringing about social and institutional reform. Technological methodology would be aimed at furnishing means of avoiding such unrealistic constructions, and instead of finding laws of social development, it would look for the various laws that impose limitations upon the construction of social institutions, or for other uniformities. From a broader perspective, the Utopian approach may be seen as against the Marxist tradition of historical materialism. Marx condemns all social engineering and considers rational planning of social institutions unrealistic because, in his view, society does not grow according to rational plans but only follows the laws of history. Though he believes in sweeping societal changes such as a proletarian revolution, he does not believe that these can be triggered by human interventions, but can only be brought about by historical conditions of the society.

10.4.2 Piecemeal Social Engineering

"Piecemeal social engineering" differs a great deal from the "Utopian engineering" of grand reforms or revolution. Given his emphasis on the fallibility of universal ideas, Popper was a natural opponent of grand schemes for social reform. He felt they often misfired and that their impacts could never be accurately predicted (Popper 1945). Instead, he proposed a piecemeal approach to social engineering as rationally justifiable, deriving his analogy from the success of piecemeal approach in natural sciences, as it was small-scale, incremental and offered the possibility of amendment with increased experience and experimentation. This method implied that the institutions of the society must only be changed bit by bit and monitored carefully for the effects of such changes to achieve positive results.

A piecemeal social engineer does not require a blueprint of the society before him to begin his work, nor does he aim for the achievement of a perfect society and an ideal state of existence. He acknowledges the fact that perfection, if at all achievable, is a time-taking process, spread over many generations. He is satisfied with searching for, and fighting the removable injustices from the society in a systematic and stepwise manner, rather than aiming for some ultimate justice. Popper argues that reforming society by piecemeal social engineering would improve institutions and maintain social stability more effectively than Utopian strategies (Bunning and You 2004). Moreover, piecemeal social experiments can be undertaken in our everyday lives without involving the entire social structure or institutions. It also permits repeated experiments and continuous readjustments, a possibility to learn from the mistakes on a real-time basis.

Piecemeal social engineering also derives strength from Amartya Sen's idea of justice, which reiterates the same point in different words: that removal of visible injustices are far more practicable and cognizable than the imagining and building of a perfectly just society. "What moves us, reasonably enough, is not the realization that the world falls short of being completely just, which few of us expect, but that there are clearly remediable injustices around us which we want to eliminate" (Sen 2009, p. vii).

Popper's idea of piecemeal social engineering has several applications in building a sustainable society. For example, the specific needs of modern society can be prioritized and each such need can be then taken as an area of small-scale social engineering intervention. Various examples – from the products of daily usage to larger issues of transportation, water supply, electricity generation, etc. – which are now working based on the model of centralized planning, and consume large amounts of fossil fuels, would benefit immensely from the introduction of sustainability-based models of renewability and eco-design introduced in a phase-wise manner.

The scope of the piecemeal approach extends from legal reforms and constitutional amendments to minor changes in the everyday life of people. In societies such as India, which has an accumulated burden of inequality and discrimination, providing the deprived communities with socio-political justice in one go with the help of constitutional amendments has been found to be unfruitful. A piecemeal approach to meeting their demands one by one has a better chance of success, based on the identification of needs on a priority basis. It can be extended to the factories and production units of consumable goods, provision of eco-friendly alternatives for transportation, rational social reforms, among other things. The chance of success for piecemeal social engineering is much higher as problems can be approached with an open mind and willingness to make changes along the way. It has been rightly pointed out by Popper that a systematic fight against concrete forms of injustice and social problems is a much more plausible goal than the attempt to realize a distant ideal blueprint of a Utopian society (Popper 1957).

10.5 Social Engineering as a Tool for Sustainable Development

In his 1953 book *Childhood's End* Clarke (1953) proclaims that the world and its people can be changed beyond recognition in about 50 years, if we have a sound knowledge of social engineering, a clearly defined goal and power. Given the fact that sustainable development of the planet has emerged as the most urgent issue facing mankind, the equations can be reversed in the favor of human society in the next 50 years subject to the fulfillment of one more condition. As we do have a clearly defined goal, i.e. the need for a sustainable planet, all that is required is a sound knowledge of social engineering to achieve this goal.

Popper proposes piecemeal engineering as a method of bringing about incremental changes in human society, which can be applied at any moment and has already proved to be successful. Similar to Sen's idea of remediable injustices, the piecemeal social engineer claims that it is more sensible to initiate a fight against injustice, which has the potential to win more support than the search for a completely ideal and just society.

In the twenty-first century, the existence of genuine threat to the environment due to unsustainable human practices can be fairly well established and some extreme threats can be easily identified. It would be useful to build a consensus about addressing these environmental threats in a piecemeal manner, based on their socio-economic context, as it is infinitely more difficult to build consensus about the contours of a fully sustainable and environmentally sensitive society. Compared with the grand blueprints for the creation of a sustainable world in one go, it is more advisable to develop an agreement about the current environmental problems that have emerged in human interaction with nature, and the means by which they can be countered in a piecemeal manner.

While the Utopian engineering approach would demand strong and centralized concentration of power for the purpose of large-scale reconstruction of the social fabric, the piecemeal approach proposes a fairly concrete and realizable end that provides simultaneous justification for the measures undertaken, in a way that a more distant ideal never could. Hence we must approach the institutional and social reforms for a sustainable planet in a stepwise fashion, scaling up our efforts only when we have more knowledge and experience in social engineering.

Deriving from the piecemeal engineering approach, in developing countries such as India it would be useful to focus on the sustainability of rural areas first, rather than aiming to transform the whole country in one go. Within the cities also, the initial efforts must be targeted on specific problems such as transportation, which contribute directly to the emission of greenhouse gases. A piecemeal approach of social engineering in the cities of the developing countries could begin by learning from the examples followed in the developed world such as Germany and the Netherlands and initiate some of the measures followed there, the such as the construction of cycling tracks in every nook and corner of the country, which provides people with the option of using a greener source of transportation. As this is also economically advantageous compared with the soaring prices of fuel, it is highly likely that a large number of people in the developing countries would exercise this option.

Similarly, providing incentives for the usage of electric vehicles, or making rainwater harvesting mandatory in every building, or providing incentives for using solar panels for electricity, etc. would encourage people to incorporate such measures into their everyday living. Some of these measures have already been imposed by the government in favor of the greater common good.

This would be one of the many ways to bring sustainable practices into the lives of the people using piecemeal social engineering. It is also important to avoid the wholesale rejection of all things traditional, and it would be useful to salvage some of the wisdom acquired by our ancestors. Some good examples for such revival of traditional and environmentally sensitive knowledge can be seen in the area of water management and water harvesting techniques used in different geographical contexts by different civilizations, or the usage of windmills in Europe, before the Industrial Revolution made them redundant. Such ancient techniques of water and power generation can play a leading role in building a sustainable society, if their usage is revived and embedded in the daily lives of the people, either voluntarily or by legal, even coercive measures.

10.5.1 Eugenics, Social Engineering and Sustainability

Along with the increasing amount of greenhouse gas emissions, the increasing number of people on the planet in the past 150 years has also been a cause of environmental degradation due to the increased pressure on natural resources such as water, forests, agriculture, land, etc. Currently the world population is estimated to be around 7 billion and it is likely to grow to over 9 billion by 2043. However, population growth has not been uniform across all regions and countries. While the poorer countries are observed to be having a faster rate of population growth, developed countries are experiencing population stagnation and even decline. As noted by the United Nations Population fund's Deputy Director of Management, the challenge of this century is to solve the problem of meeting the increasing needs and expectations of a growing population while at the same time modifying the present patterns of consumption to achieve a more sustainable model of development (Albrectsen 2013).

There exist obvious connections between the population growth and virtually all sustainable development objectives as it increases the pressure on the planet's finite resources, contributing to global warming and reducing environmental assets. It also complicates the efforts of governments and international organizations to ensure universal education, health, housing, sanitation, water, food and other basic developmental requirements.

It has been widely realized by the planners of the past as well as those of the present that slowing down of population growth is possible by mass sterilization measures. This slowdown results in an increased concentration of the people of working age, which enables countries to reap a demographic bonus and jumpstart economic development. However, whether population dynamics pose challenges or bring opportunities depends largely on the policies that are in force (Albrectsen 2013).

It is at this juncture that social engineering tools come into the picture to tackle different aspects of population dynamics. The Scandinavian countries, especially Sweden, embarked upon an extensive program in the 1930s and 1940s to transform its society and the people, and experienced positive growth. Swedish scientists, Alva and Guunar Myrdal published a book called *Crisis in the Population Question* in 1943, which made them famous as pioneers in the field of social engineering. This book dealt with the organization of society in Sweden with population as the central question. The Myrdals investigated the declining number of children in Sweden and concluded that many women preferred to remain single and childless because it was easier to work and most of them could not afford the expenses of rearing a child. This conclusion led to the development of wide-ranging programs of social reform including rationalized housing construction, the collectivization of child-rearing, and the socialization of consumption, among others.

The Myrdals' goal was not "controlling people" but "emancipation through self-normalization." With a comparative glance at the Nazi regime, Etzemüller describes Sweden from the 1930s to the 1960s as a "normalizing society." Sweden was a "normalizing society" focused on "the systematic production of knowledge in order to differentiate between normality and problem areas" and to develop rational techniques of steering society (Etzemüller 2012, p. 102).

The Swedish Population Commission of 1935 and 1941 presented 17 reports that thoroughly examined the population-related issues of Sweden from a micro- as well as macroperspective (Etzemüller 2012). This report was to serve as the crystallization point for efforts of various actors and institutions to reform their country. Some of the measures taken by the state included regulation of everyday life of the people through unwarranted housing inspections and rationing of alcohol. The state also activated a series of interventions to oversee and correct the social behavior of individuals. The program was also designed to exclude anti-social individuals from accessing social security funds as they were considered to be a burden on the society contaminating the overall genetic quality of the national body as they passed on their genetic and social defects. Sweden therefore passed two sterilization laws, one in 1935 and the other in 1941, which regulated the compulsory sterilization of "feebleminded" persons. More than 60000 individuals were sterilized in Sweden between 1935 and 1975, about 20000 of them coercively. No one questioned the right to life, "merely" the right to *give* life. Some of its democratic neighbors – Denmark, Norway and Finland – had also implemented sterilization laws earlier than Sweden. Swedish politicians had drafted only general framework laws, which were supposedly oriented simply toward practical needs, not ideologies (Etzemüller 2012).

Eugenics-based compulsory sterilization has a long history in USA as well as many European countries. Even India had its share of sterilization programs, which did not become popular and were later discontinued. Indian society and its laws of caste-based marriages can also be seen as a form of selective gene pool associations.

The importance of eugenics-based piecemeal social engineering program in controlling asocial populations to ascertain sustainability cannot be overlooked. Family planning programs and one-child laws in countries such as China have proved to be effective in limiting the population boom. Family planning programs in India have also been successful in bringing about an attitudinal change among the people, which has helped in reducing the average family size in the country in the past three decades. Though sustainability of environmental resources is not addressed directly in this approach, the indirect effects of controlling overpopulation are reflected in the reduced pressure on natural resources and decreased production of carbon emission, thereby leading to a more sustainable habitat in the long run.

10.6 Promoting Sustainable Consumption and Bio-Politics of Sustainability

Since the beginning of the industrial age, human development has been altering the earth at an unprecedented and unsustainable rate (Hoekstra and Wiedmann 2014). Thus individual consumption is seen as a relevant factor for a turn toward a more sustainable development since societal change requires also individual change (Pollex 2017). This has led to a "consumerist turn" in environmental as well as political research (Spaargaren and Mol 2008). With the consumerist lifestyle of contemporary society, the depletion of natural resources and non-renewable fossil fuels is taking place at an exponential rate, which has already initiated some unforeseen problems for mankind, namely climate change, global warming, depletion of the ozone layer, among others. At this pace of life, the planet earth would soon run into a shortage of the basic ingredients of human life, such as water, subsequently leading to the disappearance of humans and other forms of life. In a nutshell, the modern way of life is unsustainable in the long run. The main issue that emerges from this situation is to explore how sustainability and sustainable practices can be embedded and inbuilt into our everyday lives. This has also been referred to as "sustainable citizenship," the idea that individuals should increase social justice and safeguard nature through individual and collective practices, and become caretakers of the planet in their daily lives (Micheletti and Dietlind 2012).

The global discourse on sustainable development is increasingly emphasizing the crucial role of sustainable consumption and thus the contribution of individual behavior and consumption choices to preserve the environment and guarantee sustainability. This is amply reflected in the current trends such as abstinence from public smoking, eating organic food, usage of electrically powered vehicles, etc., which exemplify the individual actor's efforts toward building a sustainable planet. However, some thinkers such as Žižek are critical of such sustainable citizenship lifestyle changes:

"Who really believes that half-rotten and overpriced 'organic' apples are really healthier than the non-organic varieties? The point is that, in buying them, we are not merely buying and consuming, we are simultaneously doing something meaningful, showing our capacity for care and our global awareness, participating in a collective project" (Žižek 2009, p. 54).

He contends that such examples of sustainable consumption are actually a product of "cultural capitalism," where we are egged on to buy commodities to get the experience provided by them; we consume them to render our lives pleasurable and meaningful. To what extent such individual efforts contribute toward mitigating the environmental problems of the earth remains an open question.

Many countries, especially in the European Union and Organisation for Economic Co-operation and Development (OECD) have identified consumers as key to driving sustainable production and have taken initiatives to promote sustainable consumption. An important component of sustainable consumption is to ascertain the social and ethical dimensions of production processes as well as their environmental footprints. While governmental policies have focused their energies on reducing the environmental burden created by industrial production units, mainly by legal and financial penalties, sustainable habits of consumption are equally important to reduce the negative environmental and social externalities (OECD 2008). For example, it has now become mandatory in many countries to provide labels showing energy consumption levels of different household appliances. Such labels help the individual consumer in choosing an energy-efficient appliance for his/her use. Labels also provide information to consumers on their disposal methods and recycling obligations (OECD 2008). The labels have thus brought about a significant change in consumer behavior, and can be seen as an example of piecemeal social engineering with a positive outcome. It has helped in generating awareness among the common people about their individual energy consumption and efforts that they could make in reducing their carbon footprint.

10.6.1 (Bio)politics and Law as Tools of Social Engineering

Any attempts toward creating a sustainable world with the help of social engineering presupposes the idea of the state as a supreme power ruling over its population or citizenry. The state then enacts policies or laws to elicit the desired (sustainability-oriented) behavior from the population. Promoting the usage of green vehicles or banning those that do not pass the pollution check, promoting usage of non-plastic bags and even passing laws to prohibit the use of plastic are instances where the state devises policies to elicit a certain type of behavior.

The state as the absolute authority is also vested with the responsibility to provide amenities and infrastructure that inculcate habits and behavior toward sustainable consumption and lifestyles. For example, if a policy were to be made to connect all the villages of India with cycling tracks of very high quality, with proper signposts, speed limits, warning boards, roadsigns, etc. strictly meant only for cyclists, there would be a considerable rise in the usage of bicycles and cycling culture in India, which was once a popular form of travel. In this sense, social engineering may also serve the purpose of reviving some of the old traditions, which are much more environmentally friendly that the modern ways of doing the same thing.

Michael Foucault's engagement with the theories of governmentality and power has given rise to works that show how the birth of modern forms of education and welfare provisions correspond to the emergence of bio-politics, defined as a "form of politics entailing the administration of the processes of life of populations" (Dean 1999). Bio-politics refers to those interventions of the government that seek to improve the quality of population as a whole and their living conditions. These can be positive and related to questions of public health and standards of living, or even to incarceration for the purposes of reform or improvement. But they can also be negative, as when they are concerned with the purity of the group or class (Corbridge et al. 2005). This chapter engages with the powers of positive bio-politics in the form of social engineering for building a sustainable future for the planet and its people. Following the works of Foucault, Dean identifies the expanse of bio-politics as concerned with the social, cultural, environmental, economic and geographic conditions under which humans live, procreate, become ill, maintain health or become healthy and die. From this perspective, bio-politics is concerned with the family, with housing, living and working conditions, with what we call "lifestyle," with public health issues, patterns of migration, levels of economic growth and the standards of living. It is concerned with the biosphere in which humans dwell (Dean 1999).

Following Dean's conceptualization, the sustainability of the environment falls within the purview of bio-politics. In one sense, therefore, a bio-politics of carbon involves moving the world's high carbon consuming populations toward modes of such lifestyles and practices that consume less and generate lower levels of carbon dioxide. Whether this can succeed will only become clear if it is put into usage in a piecemeal manner. Effective management of environmentally unsustainable human habits will be extraordinarily complex, but the case of reduction in public smoking suggests that such a bio-politics is not impossible. Perhaps, some day, driving a car will become as socially distasteful a practice as cigarette smoking in public is today (Lipschutz 2009). Bio-politics thus involves a reshaping of desire and morality of every citizen on the globe based on the basic principles of promoting sustainability. All such practices and habits that have contributed toward global warming, carbon production, ozone depletion, etc. need to be denormalized, no matter how deeply embedded or widely accepted they may be today. It is possible to do so as the case of denormalization of social smoking confirms.

Given that planet earth has only a limited stock of resources that are essential for human life, and the fact that there are already more people in the world than these limited resources could sustain, many governments have enacted laws to bring about a reduction in the speed of natural resource appropriation. Roscoe Pound defines law as a task of social engineering designed to eliminate friction and waste in the satisfaction of unlimited human interests and demands out of a limited store of goods in existence (Pound 1942). Laws are instrumental in achieving predetermined ends in the interest of the common good and maintenance of order in human societies. Modern society cannot be imagined without an ever-increasing set of rules, regulations and controls governing everything from property, marriage, noise levels, recycling to online transactions. Legal instruments can be used in a variety of ways to control polluting industries and unsustainable human behavior. Some examples of the law-based piecemeal social engineering include the ban on public smoking, the odd–even rule for driving cars, the ban on plastic bags, the ban on styrofoam, among others, which can directly reduce the environmental damage effected by human societies.

There is also a need for transparency at the highest level, where the neoliberal agenda attached to the discourse of sustainability promotes a type of development that appears sympathetic but continues to instigate environmental degradation through exploitation, profiting and overconsumption. While the responsibility for making the planet green again is shifted to the individual, with the promotion of fashionable organic products and lifestyle, there is no perceptible change in the way of the working of the corporations and large industries in the developing countries. This is problematic as the focus on "green" lifestyle choices is largely predicated on privilege and incorrectly assigns blame to individuals, who become redefined and recognized as consumers only (Berg 2016). From this point of view, "sustainable development" has also been read as an oxymoron given the fact that modern understanding of development itself, premised on industrial production and increased consumption, is the biggest threat to the sustainability of the planet.

10.6.2 Reorienting Education and Engineering Studies

Education of the citizens and creating awareness toward the environment has been one of the popular strategies to foster sustainable development. National governments as well as international organizations have stressed the need for embedding environmental education in the school and college curriculum, so that it becomes an integral part of the thought process of all citizens irrespective of their profession and lifestyle. Governments and corporations employ various discursive strategies to stimulate sustainable behavior among the people, at both individual and societal levels. It involves the invocation of preference shaping for sustainable and environment friendly lifestyles through recourse to rhetorical claims. In practice, preference shaping for an organic and sustainable living involves the construction of a new environment-centric reality in which the global environmental problems are placed center stage.

This preference for sustainable development is also reshaping the engineering education and its traditional methods. The usage of non-biodegradable materials and chemical compounds, which would have passed unnoticed only a couple of decades ago, is now under scrutiny by environmentalists. The disposability of the products and their long-term effects on people and the planet have become important considerations for the engineering community. Products designed to solve a problem must not end up creating another one, when seen from an environmental point of view. The example of CFLs (compact fluorescent lamps) is aptly quoted by many authors in this connection, as they were able to reduce the overall energy consumption but came under attack due to their increased mercury content. The challenge, then, is to design products that can be successful in the marketplace without damaging the environment in the long run. With the pervasive usage of technology in our everyday lives, engineers also have the ability to change the way people think and act. The way a product is engineered and designed changes the way it is used by the consumers, and has a direct impact on the everyday habits and lifestyle of the people. For example, if an electric vehicle with comparable features to a conventional car is made available and affordable, not only will it reduce the carbon footprint of the planet, it will also bring about a major social reorganization in terms of replacement of gas stations by charging posts, redrafting of emission norms, redesigning of service stations, etc. gradually leading to improved quality of the environment.

The environmental discourse is based upon scientifically established phenomena of climate change and global warming, driven largely by increased production of greenhouse gases by industries and individuals to meet their daily requirements of income, food and leisure. The reduction of these gases without compromising the living standards of people is an engineering challenge, which has been addressed by innovations such as capturing solar and wind energy, reduction in the usage of plastic and petroleum products, automation of home appliances, replacement of lithium-ion batteries with silver–zinc, among others. As many more such innovations are expected to come from the engineering community, the broader paradigm within which the discipline emerged is now undergoing a complete paradigm shift with the arrival of ecological discourse and market preference for eco-sensitive products. The engineers have an important contribution to make in building a sustainable society, which can be accomplished by treating one problem at a time with all its interconnections, and designing products that not only reduce carbon emissions and save energy, but also bring about fundamental changes in fostering a sustainable lifestyle among the people. This entails determining the carbon and freshwater footprint of the product, its impact on energy usage, the usage of environment friendly materials and chemicals, maximizing the recyclability of the product, its impact on social behavior of the people, among others.

10.7 Social Engineering, Sustainability and Industrial Production

The discourse of sustainable development has been focusing on corporate producers of goods to evolve more environmentally sensitive production processes. With the changes in the economic landscape due to globalization, industries now face international competition, which has lured them into adopting cost-cutting measures that involve increased economies of scale, shedding of labor, and often ignoring health, safety and environmental hazards (OECD 2008). Producers differ in their motivation and capacity to make the transition toward sustainable means of production, which could be supported by policy-based social engineering methods. Corporations have also taken measures to reduce their carbon footprints and find innovative solutions to reduce their environmental impacts, mostly under pressure from their local governments. Some of the popular innovations in corporate management and production philosophy are as follows:

10.7.1 Cradle-to-Cradle (C2C) Efforts for Sustainability and Lifecycle Analysis

There is a common belief among people that every product that is manufactured will inevitably end up creating some amount of waste, and the best approach in these circumstances would be to devise methods to reduce the quantity of waste created by the product in its entire lifecycle. the United States Environmental Protection Agency has provided the concept of "lifecycle assessment" (LCA) as an approach for assessing different industrial systems for their environmental effects from the collection of raw materials for making the product, to its final disposal: cradle to grave. LCA evaluates all stages of a product's life from the perspective that they are interdependent, and enables the estimation of cumulative environmental impacts resulting from all stages of a product's lifecycle (USEPA 2006).

At each stage of the lifecycle, three primary aspects of environmental impact are considered: energy consumption and resultant emissions, chemicals/materials and waste, and other natural resources used in the product. Other considerations include supply chains, consumables, hidden impacts, services, and the effects of design and prototypes. LCA can be undertaken with the help of tools that are now available, such as the Economic Input–Output Lifecycle Assessment (EIO-LCA) method, the process-sum approach, etc. (Douglas et al. 2010). In other words, one calculates the energy expenditure for each discrete activity involved in making, using and renewing the product, and adds them all up. Both methods discussed above may also be used in combination to achieve a better understanding of the product's lifecycle.

In 2002, William McDonough and Michael Braungart published their book that provided a reconceptualization of the "cradle to grave" concept to a "cradle-to-cradle" (C2C) concept ensuring the full recyclability of products (McDonough and Braungart 2002). It demands that manufacturers of products plan in advance the final disposability of the product keeping in view the natural assimilation and reusability of the product in the natural ecosystem so that the "waste" generated is usable as a raw material for other products or industries. The idea of C2C is indeed compelling and inspiring in theory, however the actual implementation of this idea in practice and product manufacturing remain a challenge for the engineers and industrialists. At best, it serves as an ideal toward which all engineering education must strive, and hope to bridge the gap that exists between this ideal and the real constraints of the production process.

10.7.2 Sustainable and Closed-Loop Supply Chain Management

With the increasing environmental degradation due to the carbon dioxide emissions, the issue of sustainability of the planet has gained unprecedented prominence. This has given birth to numerous approaches where scientists have proposed different ways to counter this problem and build a sustainable society that aims to achieve an equilibrium between socio-economic development and ecological protection. Firms and corporations producing different types of industrial goods have also come under pressure to redesign their supply chains so that they are sustainable and environment friendly. Sustainable supply chain management (SSCM) is one such approach. It refers to the management of logistics and information flow as well as the upstream and downstream firms in a supply chain from three dimensions (i.e. economic, environmental and social) of sustainable development in accordance with the needs of stakeholders (El-Haggar 2007). SSCM is defined as the set of supply chain management policies held, actions taken and relationships formed in response to concerns related to the natural environment and social issues with regard to the design, acquisition, production, distribution, use, reuse and disposal of the firm's goods and services (Carter and Easton 2011). Firms are increasingly aware of the importance of their responsibility toward their stakeholders to address sustainability in their strategic development, and this environmental sustainability is impossible without incorporating SSCM (Haake and Seuring 2009). From a theoretical perspective, the concept of SSCM integrates the environmental, social and economic aspects that allow an organization to achieve long-term economic viability in supply chain management (Preuss 2005). In particular, SSCM includes the movement and storage of sustainable raw materials and green products from the point of origin to a point of sustainable consumption (Tseng and Chiu 2013). SSCM leads to a reduction in resources, material and waste by enabling better resource utilization and plays a significant role in achieving social, environmental and economic performance; thus, it contributes to the field of sustainable development (Brandenburg et al. 2014). Currently, operations, purchasing and supply chain managers are seeing the integration of environmental and social issues, including those embedded in related standards (e.g. ISO 9001 and ISO 14001, etc.), into their daily tasks. In addition, some authors emphasize that there is a need for increasing cooperation along the supply chain, if sustainability goals are to be reached (MingLang et al. 2015).

Owing to the increasing concerns related to the integration of social, economic and environmental dimensions of sustainability into LCA, the traditional LCA approach has been transformed into a new concept, which is called "life cycle sustainability assessment" (Seuring and Müller 2008). John Elkington proposed the triple bottom line as an accounting framework to evaluate the performance of firms in their social, economic and environmental impacts in his book *Cannibals with Forks: The Triple Bottom Line of 21st Century Business*, published in 1997. This book argued that firms should measure and report not only their financial performance, but also their social wealth creation and environmentally responsible behavior. The triple bottom line approach has now become a benchmark to assess the overall sustainability and environmental sensitivity of a firm's production process.

Other possible ways of reorienting industrial production processes to address the environmental question include: eco-design, i.e. a modern approach toward designing products in which the environmental impact of the product in its entire lifecycle is the central concern; social innovation, training and integration of engineers on social aspects of product development; integration of triple bottom line LCA; popularization of corporate social responsibility; and the development of ISO standards such as the ISO 14000 series and ISO 26000, among other measures.

Such initiatives can be seen as examples of "piecemeal social engineering" that push the developers and producers of industrial goods to use innovative methods so that the issue of sustainability is explicitly addressed.

10.8 Conclusions

Sociologists as well as behaviorists argue that human beings are shaped by the social and cultural contexts in which they are brought up and hence are a product of society. This implies that changing the contexts and conditions of the people can also change their behavior and attitudes toward a desired goal. By using specific inputs, one can hope to achieve predetermined outputs in human behavior. Views about the "nature of human nature" form the backbone of public policy formulation, which is implicit in the policy measure and may not be readily recognized. A certain conceptualization of human nature is embedded in the measures that are proposed as public policy to deal with social, political and economic problems, and a corresponding set of assumptions about the likely consequences of governmental efforts to encourage, redirect, limit or even prohibit a given type of social, political or economic behavior (Somit and Peterson 2003). The so-called Standard Social Science Model (SSSM) insists that, for all practical purposes, human nature – and thus human behavior – is shaped by culture (Harrison and Huntington 2000).

Hence, building a sustainable lifestyle culture is our only hope of saving the earth from the threats of global warming, which can be achieved with the help of piecemeal social engineering. Given the fact that human nature is a product of learning and socialization, therefore malleable, the task for planners and policymakers is to provide the right socio-political and economic structures that would automatically direct mankind toward a sustainable future. Social engineering can be usefully applied in creating some of these spaces and structures, not only to (re)create veritable institutions for fostering sustainability, but also to bring about paradigmatic changes in the thinking and behavior of the people.

This chapter discusses the type of "institutional technologies" and piecemeal social engineering that could foster the development of a sustainable society. The piecemeal engineering in this context refers to the removal of obvious obstacles to the development of a sustainable society rather than seeking ideal conditions and institutions that could foster sustainability on a grand scale. While ends and means could be easily formulated for single issues such as conservation of natural resources, educational reform, transportation, urban expansion, etc., a grand plan to transform the contemporary world into a sustainable society in a short space of time would appear to be unrealistic. As history would tell us, such grand plans to transform society overnight have often led to violence and establishment of fascist dictatorships. In fact, social engineering acquired a negative connotation in the postwar period due to the cruelty unleashed by such dictatorships in the name of creating a perfect society by using social engineering approaches. Such associations and allusions can be overcome with the usage of piecemeal social engineering. While piecemeal social engineering could be a useful tool in creating a sustainable society, an all-embracing centralized planning for social engineering runs the risk of turning into a totalitarian regime that could even bypass democratic ethics and cause distortions to privacy and human rights.

In this sense, piecemeal social engineering involves effecting such appropriate measures in society that serve as means toward achieving desired ends. In the democratic nations, for example, the state has been able to reduce inequalities among its people, reduce discrimination based on caste or color and foster women's empowerment, among other things, by enforcing the right kind of social engineering through legal and socio-cultural means. Given the positive outcome of such measures, it can be hoped that engineering problems of sustainable living would not be impossible, once the consensus and awareness about the fragility of the planet is developed among the inhabitants of all nations. If the problem of sustainability is presented clearly, it should be possible to convince people that protective measures are necessary, both at the national level and on a world-wide scale. With the development of the internet and mobile phones, a new era of information sharing and communication has begun, which was unknown during the period of the printing press, radio, telephone and telegraph. Now ideas and messages can travel instantaneously over the entire planet to an unlimited number of people. The need of the hour is to exploit this technology toward building a consensus for developing a sustainable world among the citizens of the global community.

Sustainability of the planet is a tough target to achieve, given the patterns of consumption in our modern lifestyles. However, once the problem is approached rationally and the issue is presented clearly to the citizens of all nations in a language that they understand, it should be possible to arrive at a consensus that institutions for the protection of environment are required, both on a national and on an international scale. With the global consensus for such measures, the institutional technologists can begin to improve the internal institutional set up, as well as build organizations for the prevention of environmental damage at the international scale. Not only science, technology and engineering have important contributions to make in this effort: the role of politics, bio-politics, laws, education and policymaking also cannot be overemphasized.

However, one must be aware that interventions activated by the state and a world society should not be based on oversimplified understandings of social processes and simplistic conceptions of what an intervention could achieve. The very complexity of social phenomena in human societies means that unintended effects could also emerge from social engineering projects, which requires that a feedback system be incorporated. This would help in the organic growth of sustainable living in a socially engineered society of the future.

References

Albrectsen, A.B. (2013). Sustainable development and population dynamics: placing people at the centre. Speech by UNFPA's Deputy Director of Management, at the Open Working Group on Sustainable Development Goals, June 19, 2013.

Berg, K.v.d. (2016). Neoliberal sustainability? The biopolitical dynamics of "green" capitalism. Colloquium Paper No. 31. Global governance/politics, climate justice and agrarian/social justice: linkages and challenges. ISS, Hague, February 4–5, 2016.

Brandenburg, M., Govindan, K., Sarkis, J., and Seuring, S. (2014). Quantitative models for sustainable supply chain management: developments and directions. *European Journal of Operational Research* 233 (2): 299–312.

Brückweh, K., Schumann, D., Wetzwell, R.F., and Ziemann, B. (eds.) (2012). *Engineering Society: The Role of the Human and Social Sciences in Modern Societies, 1880–1980*. Basingstoke, UK: Palgrave Macmillan.

Bunnin, N. and You, J. (2004). *The Blackwell Dictionary of Western Philosophy*. Oxford, UK: Blackwell Publishing.

Carter, C.R. and Easton, P.L. (2011). Sustainable supply chain management: evolution and future directions. *International Journal of Physical Distribution & Logistics Management* 41 (1): 46–62.

Clarke, A.C. (1953). *Childhood's End*. New York: Random House.

Corbridge, S., Williams, G., Srivastava, M., and Véron, R. (2005). *Seeing the State: Governance and Governmentality in India*. New York: Cambridge University Press.

Dean, M. (1999). *Governmentality: Power and Rule in Modern Society*. New Delhi: Sage Publications.

Dobbs, R., Smit, S., Remes, J. et al. (2011). *Urban World: Mapping the Economic Power of Cities*. New York: McKinsey Global Institute.

Douglas, D., Papadopoulos, G., and Boutelle, J. (2010). *Citizen Engineer*. Upper Saddle River, NJ: Prentice Hall Press.

El-Haggar, S.M. (2007). *Sustainable Industrial Design and Waste Management: Cradle-to-Cradle for Sustainable Development*. Burlington, VA: Elsevier Academic Press.

Etzemüller, T. (2012). Rationalizing the individual: engineering society: the case of Sweden. In: *Engineering Society: The Role of the Human and Social Sciences in Modern Societies* (eds. K. Brückweh, D. Schumann, R.F. Wetzwell and B. Ziemann), 1880–1980. Basingstoke, UK: Palgrave-Macmillan.

Haake, H. and Seuring, S. (2009). Sustainable procurement of minor items exploring limits to sustainability. *Sustainable Development* 17 (5): 284–294.

Hansson, S.O. (2006). A note on social engineering and the public perception of technology. *Technology in Society* 28: 389–392.

Harrison, L.E. and Huntington, S.P. (eds.) (2000). *Culture Matters: How Values Shape Human Progress*. New York: Basic Books.

Hayek, F.A. (1949). The Intellectuals and Socialism. *University of Chicago Law Review* 16 (3): 417–433.

Hoekstra, A.Y. and Wiedmann, T.O. (2014). Humanity's unsustainable environmental footprint. *Science* 344 (6188): 1114–1117.

Intergovernmental Panel on Climate Change (IPCC) (2007). Climate change 2007 – synthesis report – summary for policymakers. Geneva: IPCC Secretariat, World Meteorological Organization.

Lipschutz, R.D. (2009). The governmentalization of "lifestyle" and the biopolitics of carbon. Annual Conference of the British International Studies Association, University of Leicester, December 14–16, 2009.

McDonough, W. and Braungart, M. (2002). *Remaking the Way We Make Things: Cradle to Cradle*. New York: North Point Press.

Micheletti, M. and Dietlind, S. (2012). Sustainable citizenship and the new politics of consumption. *The Annals of the American Academy of Political and Social Science* 644 (1): 88–120.

OECD (2008). *Promoting Sustainable Consumption: Good Practices in OECD Countries*. Paris: OECD Publications.

Östlund, D. (2007). A knower and friend of human beings, not machines: the business career of the terminology of social engineering, 1894–1910. *Ideas in History* 2 (2): 43–82.

Pollex, J. (2017). Regulating consumption for sustainability? Why the European Union chooses information instruments to Foster sustainable consumption. *European Policy Analysis* 3 (1): 185–204.

Popper, K.R. (1945). *The Open Society and Its Enemies, Vol 1: The Spell of Plato*. London: Routledge.

Popper, K.R. (1957). *The Poverty of Historicism*. New York: Harper and Row Publishers.

Pound, R. (1942). *Social Control Through Law*. London: Transaction Publishers.

Preuss, L. (2005). Rhetoric and reality of corporate greening: a view from the supply chain management function. *Business Strategy and the Environment* 14 (2): 114–139.

Sachs, W. (1992). *The Development Dictionary: A Guide to Knowledge as Power*. London: Zed Books.

Scientific Applications International Corporation, USEPA (2006). *Life Cycle Assessment: Principles and Practice*. Reston, VA: USEPA.

Scott, N. (1944). *United World*. New York: Island Press.

Scott, J.C. (1998). *Seeing Like a State: How Certain Schemes to Improve the Human Condition Have Failed*. London: Yale University Press.

Sen, A. (2009). *The Idea of Justice*. Cambridge, MA: Harvard University Press.

Seuring, S. and Müller, M. (2008). From a literature review to a conceptual framework for sustainable supply chain management. *Journal of Cleaner Production* 16 (15): 1699–1710.

Somit, A. and Peterson, S. (2003). *Human Nature and Public Policy: An Evolutionary Approach*. New York: Palgrave Macmillan.

Spaargaren, G. and Mol, A.P.J. (2008). Greening global consumption: redefining politics and authority. *Global Environmental Change* 18: 350–359.

Tolman, W. (1909). *Social Engineering: A Record of Things Done by American Industrialists Employing Upwards of One and One-Half Million People*. New York: McGraw Publishing Company.

Tseng, M.L. and Chiu, A.S.F. (2013). Evaluating firm's green supply chain management in linguistic preferences. *Journal of Cleaner Production* 40: 22–31.

Tseng, M.L., Lim, M., and Wong, W.P. (2015). Sustainable supply chain management: a closed-loop network hierarchical approach. *Industrial Management and Data Systems* 115 (3): 436–461.

United Nations (1987). *Report of the World Commission on Environment and Development: Our Common Future*. New York: Oxford University Press.

Walker, R.V., Beck, M.B., Hall, J.W. et al. (2017). Identifying key technology and policy strategies for sustainable cities: a case study of London. *Environmental Development* 21 (2017): 1–18.

Žižek, S. (2009). *First as Tragedy, Then as Farce*. London: Verso.

11

Environment Modeling for Sustainable Development

Lalit Borana[1], *Bhaskar Jyoti Deka*[2], *Jiaxin Guo*[2], *and Alicia Kyoungjin An*[2]

[1] *Discipline of Civil Engineering, Indian Institute of Technology Indore, Indore, India*
[2] *School of Energy and Environment, City University of Hong Kong, Kowloon, Hong Kong, China*

11.1 Introduction

There has been a dramatic revolution in environmental development over the past few years. Many potential questions are being asked about the domain of development and environmental concerns and how sustainable development can be possible. Sustainable development has become the key slogan for many leading conferences, worldwide government agencies, standard regularity bodies, etc. It is essential to be aware of the application of the sustainability concept in various day-to-day activities, the construction industry, transportation and the waste management sectors. In this chapter, the authors provide a brief explanation of the basic concepts of sustainable development together with modeling aspects.

11.1.1 Theoretical Background: Sustainability and Sustainable Development

Sustainability is the practice of indefinite productivity without degrading or endangering natural biotic systems. Sustainability can be understood as the capacity to preserve, endure and nurture (Gallopin 2003). It means identifying, developing and promoting sustainable mindsets, practices and policies in order to maintain a healthy natural environment along with economically sound as well as socially viable models. The environmental aspect is the ability of a natural system to sustain itself and life on earth (Goodland and Daly 1996).

The three pillars of sustainability, i.e. economic, social and environmental, are the dominant tools for complete sustainable development; non-responding or inefficiency of any of these pillars will lead to unsustainable development (Goldman and Gorham 2006; Harris

Sustainability: Fundamentals and Applications, First Edition.
Edited by Rao Y. Surampalli, Tian C. Zhang, Manish Kumar Goyal, Satinder K. Brar, and R. D. Tyagi

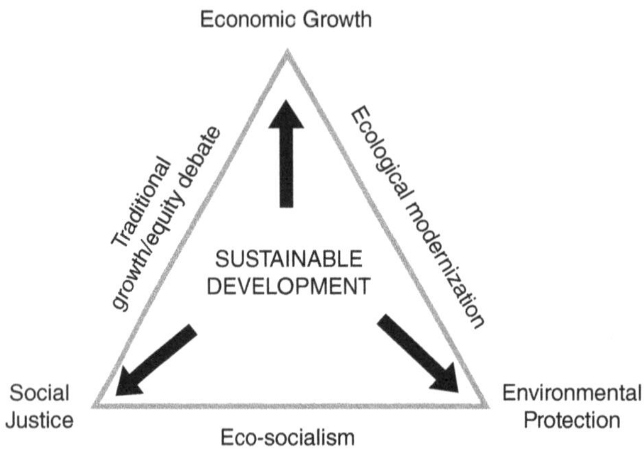

Figure 11.1 Relationship of sustainable development with social, environmental and economic aspects (Peeters 2012).

2003; Pero et al. 2017; Goodland and Daly 1996). Figure 11.1 shows a popular graphical representation of the three pillars, together with a brief description as follows:

- *Economic perspective*: An industry or organization can produce goods economically in a sustainable manner.
- *Environmental perspective*: Not only avoid the overuse of depleting natural resources, but also try to find and use possible alternative ecofriendly resources.
- *Social perspective*: Socially sustainable organizations confirm reasonable spreading of prospects, acceptable facility of social services and gender fairness.

Sustainable development places an emphasis on the capability of natural systems to deal with the political, social and economic challenges encountered by mankind. Thus, the science of sustainability is the combined study of the environmental and sustainable aspects of development. Alarming concerns has been raised on regenerating, preserving and recovering planetary resources for the future. In recent years, there has been a significant transformation in the debates involving the environment and development. The question raised is no longer whether developmental activities and environmental concerns mutually contradict each other. Rather, it is "By what means can sustainable development be achieved?" In a flash the term sustainable development has become pervasive, particularly owing to the need to resolve the present day requirements of various social and environmental concerns. In fact, the term "sustainable development" has become the catchphrase for international aid agencies, development planners, theme of conferences and a key mantra for activists related to environment and development. Overall, the term sustainable development has certainly gained popular support when compared with earlier concepts of development, for instance, "eco-development" (Lele 1991).

11.1.2 Historical Evolution of Sustainability

Sustainable development is achieved through human development as well as sustaining natural resources and the ecological system. Millions of people have adopted the consumption

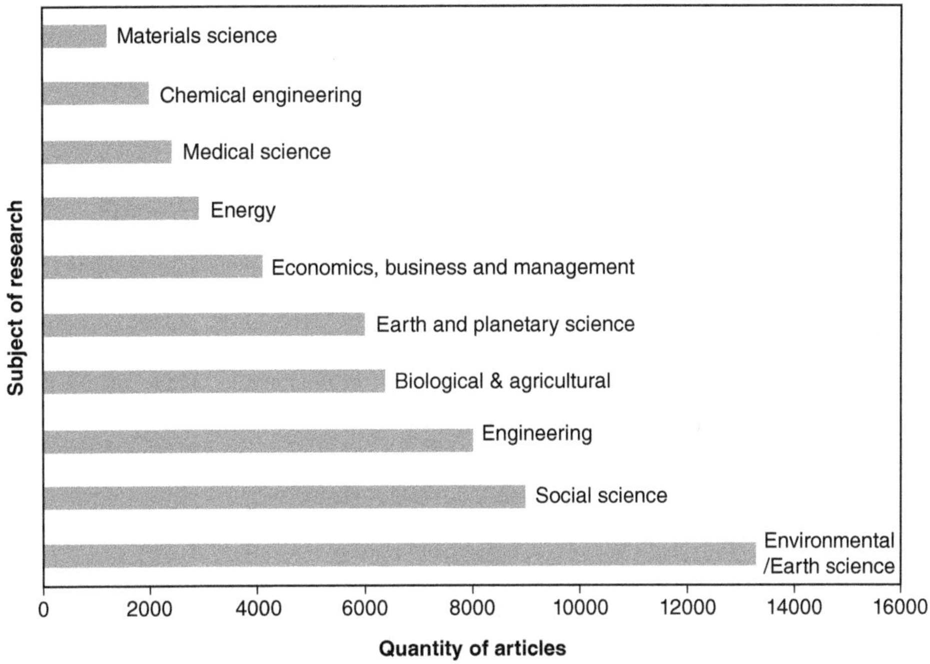

Figure 11.2 The number of articles on sustainability in various fields (Linton et al. 2007).

pattern of Western countries, with the rapid globalization of arcades (Prahalad and Hart 2002). Due to the increase in consumption rate, these arcades are facing pressure such as damage to the biodiversity, inadequate energy resources and natural resource contamination (Nunes et al. 2016). Subsequently these factors exert additional pressures on countries with large populations, such as China, India, Indonesia and Brazil. For instance, rapid growth in terms of energy consumption by countries like China and India is alarming in the context of global sustainability practice for policymakers, business leaders and scientists. Sustainability or sustainable development has been widely discussed by many policymakers and popular journals in various technical fields as summarized in Figures 11.2 and 11.3.

In recent years, operations and management science researchers has started to place an emphasis on sustainable operations management (SOM), such as green supply chains (Linton et al. 2007), green procurement (Walker et al. 2009) and reverse logistics (RL) (Lambert et al. 2011). SOM plays a vital role in complex sustainability challenges confronted by many organizations (Rajapakshe et al. 2005)

11.2 Environmental Indicators in the Context of Sustainable Development

With limited non-renewable resources and interactions between civilization and the biosphere, it is evident that economic development alone is not suitable to propagate the concept of environmental sustainability. Further, based on economic development patterns, it is necessary to implement a flexible framework of sustainability with depletion of

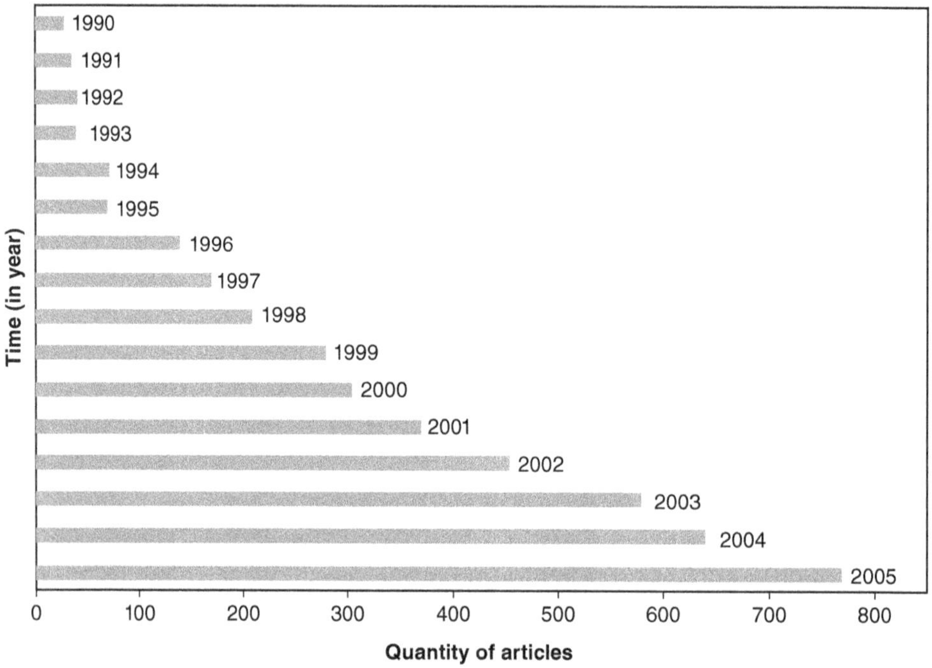

Figure 11.3 An increase in publishing frequency on the concept of sustainability (Linton et al. 2007).

non-renewable resources. More attention is needed to be paid to environmentally friendly technologies.

To mitigate further damage to the environment, the Stockholm Resilience Centre suggested the practicing of planetary boundaries (PBs) to enhance the planet's renewability (Rockström et al. 2009). However, lately the planetary concept has been criticized mainly due to the involvement of arguable thresholds and fundamental principles of irreversible damage. The Breakthrough Institute claimed that there was no evidence for lowering the impact from human activities and that implementation of boundaries was arbitrary, and may lead to increased poverty (Nordhaus et al. 2012). Figure 11.4 shows the components of a sustainability model. In this section, three environmental sustainability assessment methods and their indicators have been discussed.

11.2.1 Planetary Boundaries

PBs describes an innocuous functional space for civilization to regulate the stability of the earth system. The main aim of PBs is to protect the earth's functioning system and ensure future sustainability. Keeping in view the prime focus of the PB method, it is primarily related to various impacts on nature and the environment. It does not reflect the impacts of the earth system process on human health. Figure 11.5 shows nine PBs; considering the whole earth as system boundary, the PB method offers an approach to assess environmental impacts.

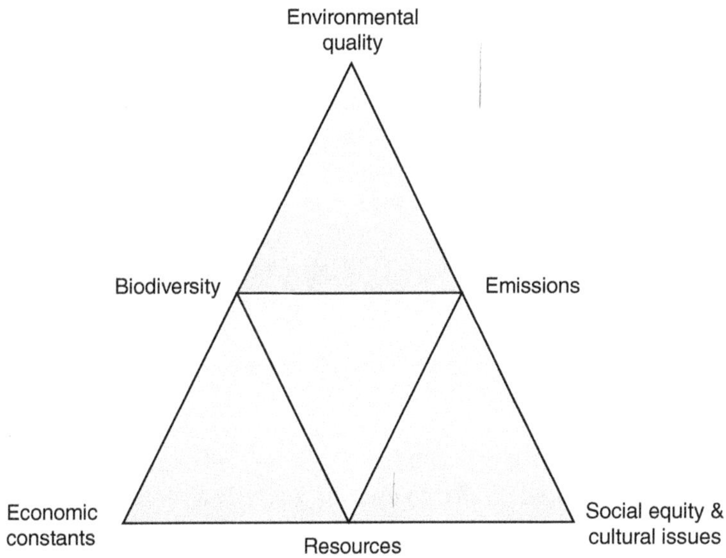

Figure 11.4 Sustainability paradigm (Basiago 1999; CIB 1999).

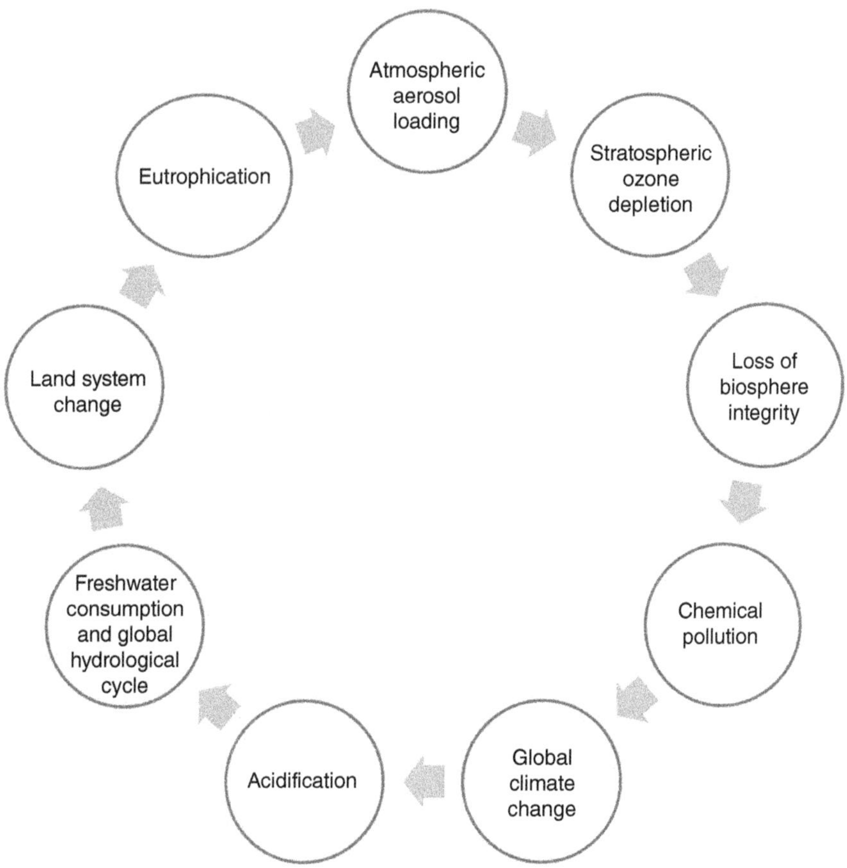

Figure 11.5 The nine planetary boundaries.

11.2.2 Lifecycle Assessment (LCA)

With regard to any goods or services, LCA computes all the emissions during resource consumption, the impact on the environment and health, and depletion of resources as per ISO standards (ISO 14040, 14041, 14042, 14043, TS14048). LCA facilitates the choice of environmental friendly options on the basis of environmental impact assessment (EIA). Like PB, LCA is inadequate to judge the solutions for absolute sustainability. LCA primarily focus on global concerns but some relevant regional scale impacts are also modeled. The four basic steps of LCA are shown in Figure 11.6. A set of impact categories with multiple indicators is used to described the quality of environment (Posada n.d.).

$$\text{Category indicator} = \Sigma(\text{Characterization factor}(S) \times \text{Emission inventory}(S)) \qquad (11.1)$$

where S denotes the components.

For example, during production of 1,3-propanediol (PDO), the overall impact assessment can be calculated as illustrated in Table 11.1 after multiplying lifecycle inventory results by respective characterization factors (emissions values) for non-renewable energy use (NREU) (MJ kg^{-1}) and greenhouse gas (GHG) emission (kg CO_2eq/kg) (Eq. (11.1)).

To address the impacts, selection of indicators and modeling, various methods are available, such as ReCiPe, IMPACT 2002+, etc. Environmental sustainability of an activity is evaluated by endpoint or midpoint indicators. The time scale of each impact is different depending on the categories. It ranges from (i) a few years, for instance acidification, eutrophication and ecotoxicity, to (ii) very long time scales, such as climate change, fossil, ozone and mineral depletion. Table 11.1 illustrates in detail the various processes commonly practiced in the sustainable construction industry.

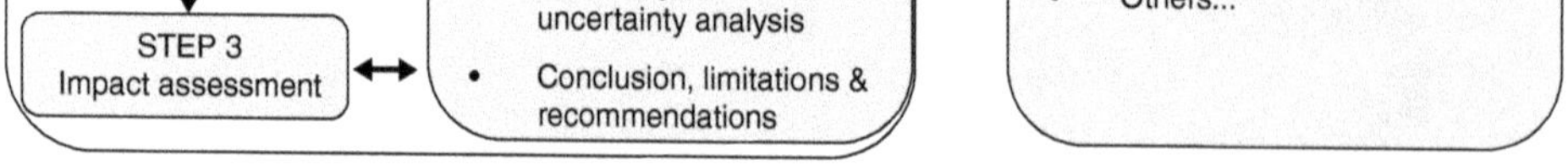

Figure 11.6 Stages for an LCA (International Organization for Standardization 2006; Posada n.d.).

11.2.3 Sustainable Development Goals (SDGs)

Several regulatory frameworks discuss environmental targets and indicators for a livable and sustainable earth, a perspective for better human society. SDGs are an essential component of the United Nations' action plan to encourage all countries to "heal and safeguard our planet" and "shift the world to a sustainable and resilient path" (Mkumbo 2016). In a sustainably developing world, SDGs play the driving role in achieving the shared goals

Table 11.1 Characterization of category indicator results (International Organization for Standardization 2006; Posada n.d.)

Characterization of factors			Overall	Category indicator results	
	NREU	GHG	Flow	NREU	GHG
	($MJ\ kg^{-1}$)	(kg CO_2eq/kg)	(kg/kg PDO)	(MJ/kg PDO)	(kg CO_2eq/kg PDO)
Components	1	2	3	4 (1 × 3)	5 (2 × 3)
Glucose	0.721	0.234	3.48	2.51	0.81
Ammonia	41.672	2.0962	0.05	2.23	0.11
Glycerol	24.246	2.8416	0.07	1.76	0.21
Wastewater treatment	0.321	0.181	3.26	1.05	0.59
Electricity (MJ/MJ)	0.421	0.0638	2.13	0.9	0.14
Steam	3.952	0.2343	5.94	23.47	1.39
Cooling water	0.006	0.0003	44.83	0.25	0.01

NREU, non-renewable energy use; GHG, greenhouse gas; PDO, 1,3-propanediol.

and understandings among various stakeholders, including the policymakers, local residents and other partners. The system boundaries of SDGs are considered within a particular place or country but, depending on the application context, some of them are global. The specified indicators require a definite level to be achieved as per the scheduled time. In order to achieve environmental sustainability, it is essential to strengthen technology transfer, capacity building and public awareness in various countries.

The Driver-Pressure-State-Impact-Response (DPSIR) system is presented in Figure 11.7. It starts with "driver," which stands for the needs of individuals and industries. Further, the key drivers in the system lead to human activities that essentially increase "pressures" on the environment (Table 11.2).

11.3 Overview of Sustainable Development

Control and management of environmental impacts is the main focus for sustainable environmental development. Reduction of energy use, emissions and waste generation is the prime objective. With the key concern being environmental suitability, sustainable development should also attempt to deliver value for clients, shareholders and other investors. The rapid increase in the global population along with depletion in natural resources create pressures on our planet's ecosystems. Climate changes due to manmade activities are being felt by communities around the world.

Technological innovation and industry have been increasing worldwide for the past two centuries. In the past three decades, continuous development in the industry sector has led to severe impacts on the global environment, causing serious natural disasters, which

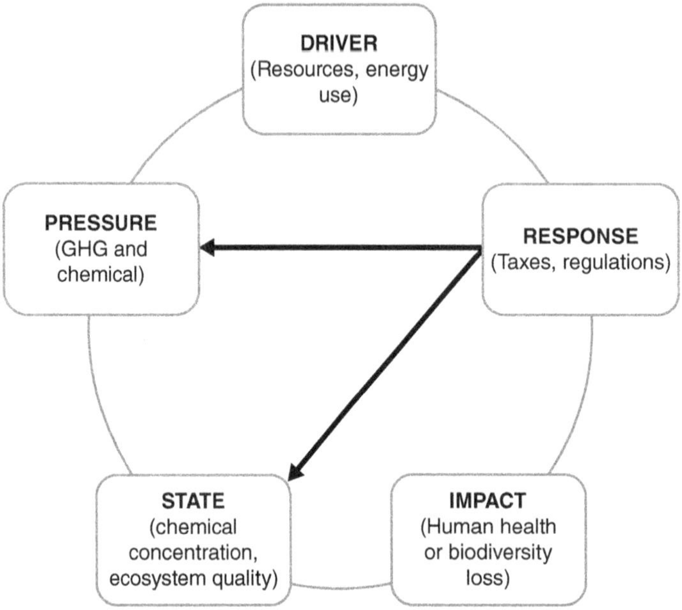

Figure 11.7 DPSIR system (Dong and Hauschild 2017).

receive international attention and priority; severe academic concerns has been raised on sustainable approaches over material developments (Haque 2000). In the general concepts and models for development, the environmental aspects were ignored, which has caused massive deterioration in the environment and increased in natural disasters. Subsequently, awareness of the "sustainable development" model has been rising. Pioneers, policymakers and researchers into sustainable development tend to explore and prescribe ecofriendly policies, environmental protection laws, and publish the sustainability concepts via global media and magazines. Irrespective of recognition of the need for environmental sustainable development, models for sustainable development have severe deficiencies which needs to be addressed. This section will briefly discuss a few key environmental impacts and emphasize the severe challenges for sustainable human development.

Excessive emissions of carbon dioxide mainly due to global warming or the greenhouse effect and deforestation have become a major concern in context of environmental sustainability (Yan et al. 2016). Around 57% and 17% CO_2 is emitted due to fossil fuel use and deforestation/decay of biomass, respectively, as shown in Figure 11.8a. GHG emissions by buildings contribute 8%, green forest 17%, the industrial sector 19%, transportation 13%, agricultural farming 14%, the energy sector 26%, and waste and wastewater 3% (IPCC 2007; Sovacool and Brown 2010) as shown in Figure 11.8b.

Global warming causes an increased in the earth's surface water due to thermal expansion and glaciers and ice caps melting, lead to rises in sea water level, loss of land by flooding or erosion intrusion of saline seawater to existing freshwater sources and overall adverse impacts on the environment. Sustainability or sustainable development can control and reduce the impact of global warming. Already the coastal regions of many countries like Thailand, India, Pakistan, Bangladesh, Vietnam and China have faced adverse impacts due

Table 11.2 Sustainability practice in the construction industry (Pero et al. 2017).

Process		Sustainability practices
	Lifecycle assessment (LCA)	
Design	Ecodesign	Decreasing energy consumption
		Decrease in raw materials consumption
		Limiting the consumption of dangerous materials
		Anatomization, reuse and return
		Design guidelines for suppliers
	Training on ecodesign	
Procurement	Supplier selection	Green criteria
		Social criteria
		Certification
		Local suppliers
	Sustainable supplier rating	Audit at the supplier
		Audit in the yard
	Green material purchasing	Recycled materials
		Materials without any dangerous components
		Smart materials
	Contractors/subcontractor awareness as well as obligation to code of conduct	
	Providing technical support and training to "green suppliers"	
Authority	Corporation oriented toward sustainability	
	Organizational responsibilities dedicated to sustainability	
	Exercise on sustainability	
	Ethical code	

to global warming such as hotter summers triggering severe drought, extreme flooding leading to serious erosion of riverbanks and coastal regions, and a rise in the occurrence of catastrophic storms (Jensen 1990). Therefore, global climate change and the greenhouse effect pose an alarming risk to sustainability (Galdos et al. 2013).

11.4 Role of Sustainability in Environmental Development

In addition to national and international proceedings, institutional concern for environmental sustainable development is also mandatory to investigate applied platforms and theoretical keystones. The relationship between development and sustainability has been discussed on many occasions such as at world conferences, in legal frameworks, government policies, global organizations, etc. Sustainable environment development establishes a balance among society, the economy and the environment to maintain a quality of life for present and future generations.

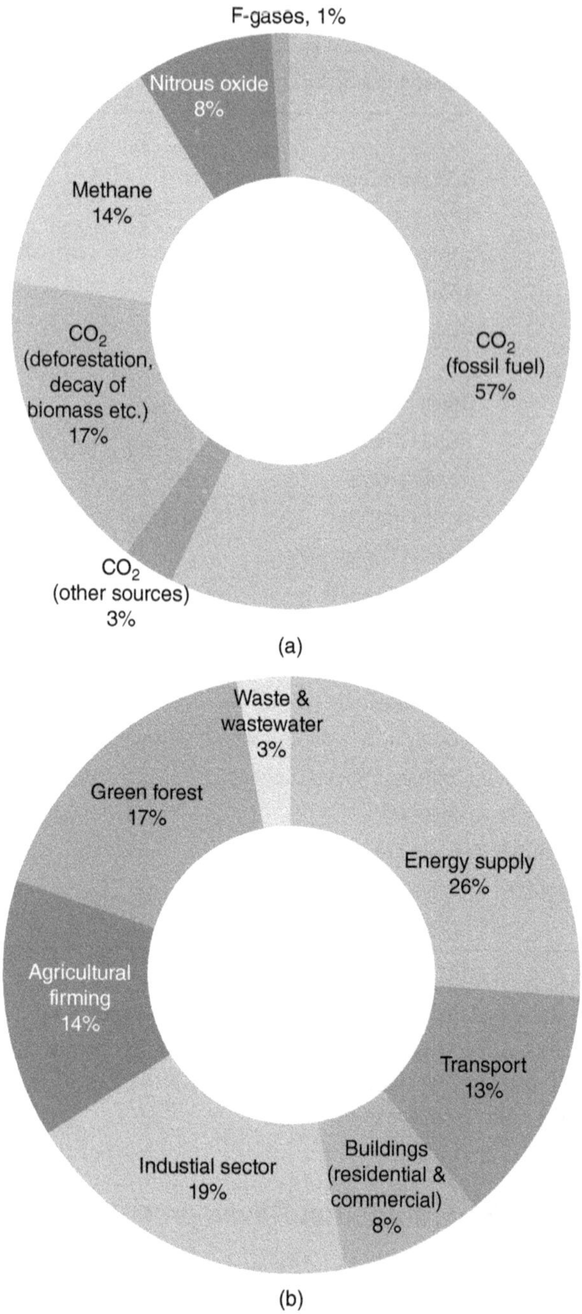

Figure 11.8 (a) Global greenhouse gas emissions by gas (IPCC 2007). (b) Global greenhouse gas emissions by source (IPCC 2007).

The sustainability of planet has several unexplored issues that needs to be addressed urgently to achieve the common goal. Real-life application of the sustainable development concept is difficult for many nations, cities, organizations and human beings due to complex and uncertain design, operation and assessment of environmental sustainability policies (Barber 2007; Boyko et al. 2012).

Foundations for sustainability (Shriberg 2000) are as follows:

1) *Morality and intergenerational equity*: Living with available resources is ethically desirable instead of borrowing from future generations of humankind or biota.
2) *Survival*: The present levels of stresses on ecosystems are too large for comfort of any creature. For long-run future survival and prosperity, sustainability of ecological aspects will play a great role.
3) *Organizational benefits and risks*: Practicing strategic action plans to adopt sustainability as an important goal for organizations to avoid penalties due to a rise in economic and social liability.

11.4.1 Sustainable Buildings

Much has been articulated about sustainable development and various social, economic and environmental implications associated with it. However, with numerous definitions available for defining and characterizing different aspects of sustainable development, it becomes difficult to actually define what is sustainable, primarily because it is a matter of personal opinion. American Society for Testing and Materials (ASTM) International (BEAM 2012, p. 3) defines a "green building" as: "a building that provides the quantified building performance requirements while reducing disturbance to and improving the functioning of local, regional, and global ecosystems at all time intervals of its life span, which involves the construction phase and design service life." Likewise, "a green building enhances proficiencies involved in operational performance and resource management; it reduces the risks related to health and the environment." To this can be added social equity and economic viability. According to Hong Kong's building environmental assessment method (BEAM), a sustainable building is one that is, in priority order, safe, healthy, comfortable, functional and overall efficient. Around 40% of energy is consumed by building, out of which 30% is related to GHG emissions. The infrastructure industry has a great role to play in mitigating emissions of GHGs at low or no cost or with surplus savings to the economy. The infrastructure sector is consuming over one-third of the world's natural resources, out of which 12% is fresh water and 40% solid waste production. On average this sector is responsible for more than 10% of employment. Due to the continuous increase in the world population, building sustainability plays an important role in overall sustainable development. A socio-economic aspect of building has been shown in Figure 11.9.

Table 11.1 shows the relationships among performance level and approaches to building under various ranges of goals. It ranges from regulatory levels (or "business as usual"), to "good practice," "green level," and "sustainable practices." One of the messages rooted in this classification of performance is that a broader view needs be considered, with respect to an increase in performance levels. Thus, sustainable building approaches must consider social and economic issues, as well as those relating to scale. Scales mentioned in Table 11.3 are associated with "buildings" and/or "community"-related matters.

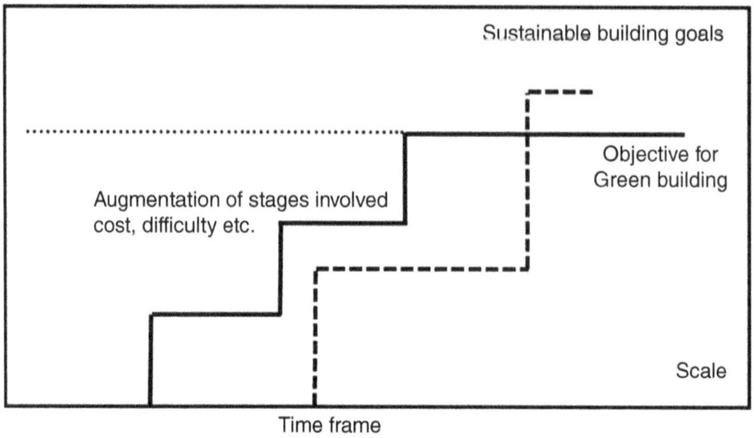

Figure 11.9 Social and economic aspect of building (BEAM 2012).

Table 11.3 Relationships between levels of performance and approaches to building.

Goals	Community scale	Building scale
Sustainable		
No negative impact on ecosystems	Minimal use of vehicles, pedestrian oriented, green space	Very high performance, feed electricity to grid
Green		
Substantial improvement in environmental performance	Focus on higher densities, diverse uses, control of car, setting up of public transport	
Good practice		
Substantial improvement in environmental performance	Community planning follows conventional suburban path	Performance levels achieved by top 25%
Regulation		
Minimum performance according to regulation and/or industry practice	Transport, water, sewer, etc. seen as fairly separate	Emphasis on energy performance

11.4.2 Principles of Sustainable Buildings and Construction

The model of "green building" is presently evolving as sincere efforts are being made globally to assist the basic understanding and facilitate further development of broader concepts related to "sustainable building" (Mokhlesian and Holmén 2010) (Figure 11.10).

Such an evolution requires that all the associated key aspects (namely, environmental, social and economic aspects) should be equally considered, and this must be alongside technical, aesthetic, functional and urban development issues that are within the scope of a variety of activities and processes. This includes activities ranging from design, planning, construction and management, valuation, risk assessment, investment analysis, etc.

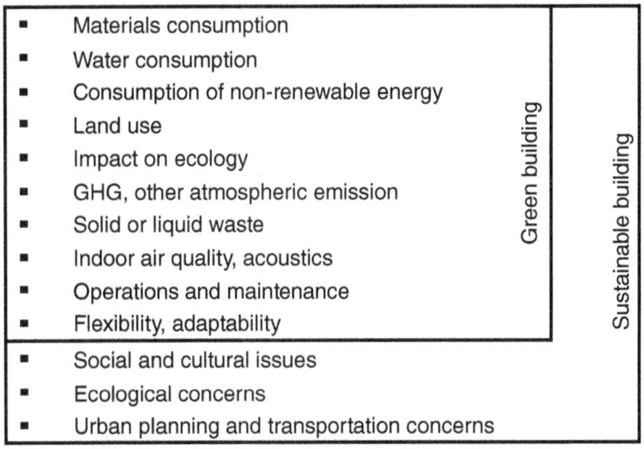

Figure 11.10 Variances in scope between green and sustainable building schemes (CIB 1999).

To achieve sustainable development, the various strategies and action plans have been shown in Figure 11.11.

11.5 Case Studies: Sustainability Management

Environmental sustainability or sustainable development is the key factor to maintain ecosystem balance. The concept of sustainability needs to be adopted in every sector of practice so that future generations are also able to use resources fairly. In this chapter, the authors have tried to present the broad concept of the applicability of the sustainability concept across different areas, from the construction industry to water treatment as well as waste disposal processes. To illustrate, three case studies are considered.

11.5.1 Maintenance Planning Project for Residence and Commercial Buildings, Northern Sweden

In this case study (Persson 2009), a commercial office building (Riksbanken 1) of approximately 3700 m^2 and residential apartments of around 9700 m^2 are considered. These building were constructed between 1950 and 1972. The main objective was to frame strategy for future maintenance works of all buildings under the purview of a sustainable approach. With the assistant of the Swedish tool "Miljöbedömning av fastigheter," (environmental assessment) established by a Swedish consultancy company, environmental assessment was carried out. The assessment was performed by a certified reviewer for various environmental features of a building and its surroundings. The results were evaluated as per definite environmental norms founded by authorized national standards and recommendations of Swedish national legislation.

The assessments were conducted for the indoor and outdoor environment for 12 different groups as follows:

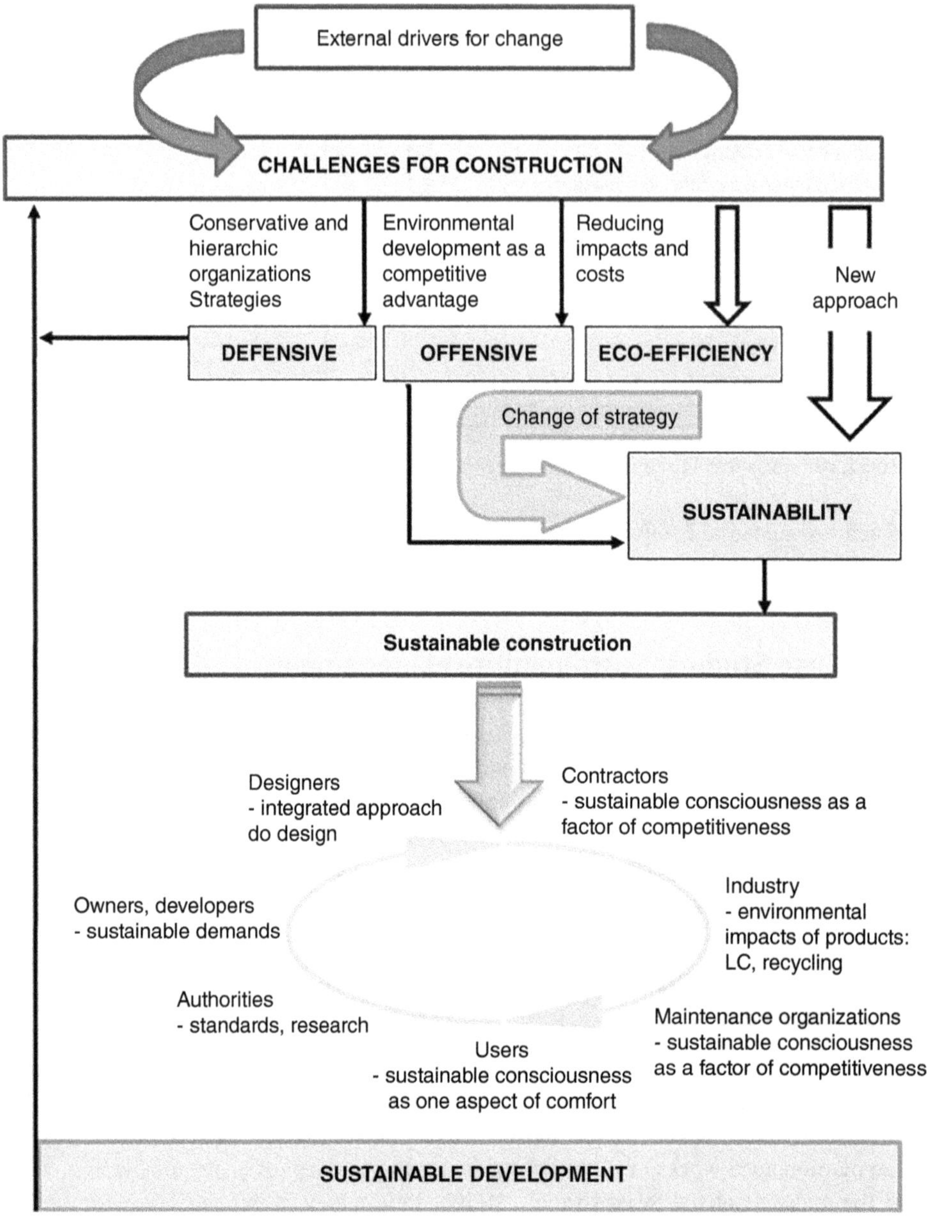

Figure 11.11 Strategies and actions for sustainable development (CIB 1999).

The outdoor environment comprises the following:

- biodiversity;
- consumption of water;
- consumption of energy;
- hazardous waste regulated by legislation;

- dangerous waste which is regulated by legislation;
- component and material recirculation.

The indoor environment comprises the following:

- questionnaire to tenants regarding actual indoor environmental conditions of the building;
- electromagnetic radiation;
- indoor noise;
- indoor lighting;
- indoor radon radiation;
- actual indoor climate.

Depending on the assessment, each single group collects individual points. On the basis of individual weight, all the groups are rated between 1 and 10, where 6 and 4 points are from the outdoor and indoor environment groups, respectively. Higher points indicate better environmental condition of the building, whereas lower points indicate the opposite. Performance assessment of the outcomes was prepared by considering the lifecycle standpoint. The assessment is shown in Figure 11.12.

The Riksbanken-1 commercial building was constructed in 1972 and the remaining buildings between 1950 and 1963; therefore, it is probable that different materials were used for commercial building than the rest. As compared with the old residential buildings, the commercial building consists of new and environmentally hazardous materials. Indoor climate, consumption of energy and hazardous waste were found to be substantial environmental features to be evaluated in the overall building performance.

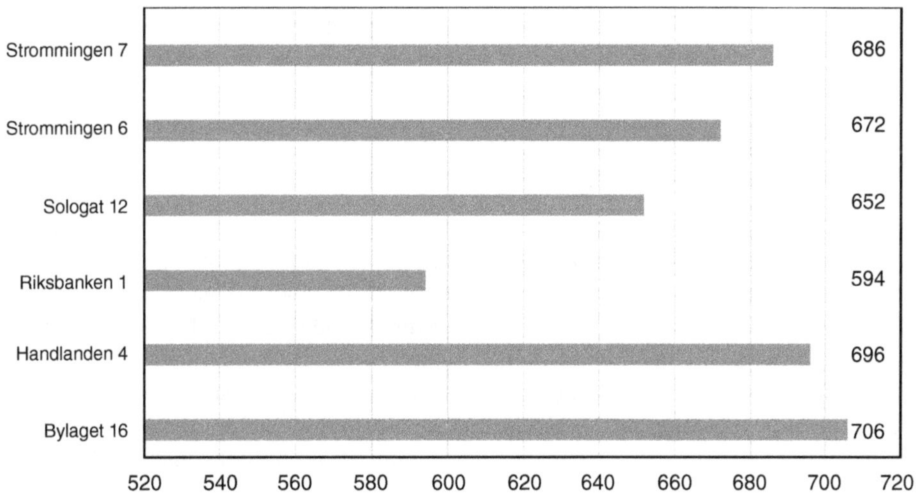

Figure 11.12 Evaluation of the performance of the buildings.

11.5.2 Renewable and Sustainable Approaches for Desalination

Energy and potable water are the basic commodities for living species on earth. Due to the rapid increases in population and industrialization, presently the earth is facing serious stress as well as the need to provide for future generations. Innovative sustainable desalination technologies are one of the solutions to deal with this alarming global issue. Existing seawater reverse osmosis (SWRO) technology has great water production capacity but needs significant energy input. Also membrane fouling, especially during the occurrence of harmful algal bloom (HAB), has limited its further application (Deka et al. 2018). A combination of renewable energy sources with desalination technologies could be a sustainable and ecofriendly methodology to recover, reclaim and reprocess water as well as energy. Energy requirements and emissions of GHGs for different desalination processes are provided in Table 11.4.

The present technologies for desalination are as follows:

1) *Phase change processes*: Here feed (seawater) is boiled and the generated steam is further condensed to collect the fresh water. For example, solar distillation (SD), mechanical vapor compression (MVC), multi-effect distillation (MED), multi-stage flash distillation (MSF) and thermal vapor compression (TVC).
2) *Non-phase change processes*: Potable water is produced by using a membrane-based separation system between the feed (seawater) and permeate (potable water) by means of mechanical or chemical/electrical processes. For example, electrodialysis (ED) and reverse osmosis (RO) (Deka et al. 2018).
3) *Hybrid processes*: The combined application of phase change and separation is used in this process. Examples include: membrane distillation (MD) (Drioli et al. 2015; Lee et al. 2018, 2017; Wang and Chung 2013); RO combined with MED or MSF processes.

Plant installation and operating costs could be minimized along with a significant reduction in GHGs by coupling desalination with renewable resources as shown in Figure 11.13. In this way, the scarcity of fresh water and energy consumption can be reduced. The energy and fuel savings of the hybridization of various desalination processes are shown in Table 11.5.

11.5.3 Reduction in Food Waste in Korea by Smart Bin

Smart bin is an advanced technology used for the sustainable treatment and disposal of municipal solid waste (MSW) through community collaboration. Smart bin has become very popular in many developed countries. The scope of the modern smart bin covers automatic systems for food waste, fermentation procedures, and offers a full database of food waste management.

The basic main objectives of any smart bin are as follows: sustainability modeling for good practice regarding food waste management (food waste monitoring, reduction and disposal practice); and sustainable generation of ecofriendly end products for agricultural use.

Table 11.4 Energy requirements and greenhouse gas emissions in desalination processes (Gude et al. 2010).

Process		Multi-effect solar still (MESS)	Multi-stage distillation (MSF)	Multi-effect distillation (MED)	Mechanical vapor compression (MVC)	MED- thermal vapor compression (TVC)	Reverse osmosis	Electrodialysis
Requirement of energy	Thermal energy ($J\,kg^{-1}$)	1 500 000	250 000–300 000	150 000–220 000	—	220 000–240 000	—	—
	Electrical energy ($Wh\,m^{-3}$)	0	3500–5000	1500–2500	11 000–12 000	1500–2000	5000–9000	2600–5500
Greehouse gas (GHG) emissions (kg CO_2 m^{-3} H_2O)	Total electric equivalent ($Wh\,m^{-3}$)	0	15 000–25 000	8000–201 000	11 000–12 000	215 000–22 000	5000–9000	2600–5500
	Maximum value	0	24 000	19 200	11 500	21 000	8600	5300

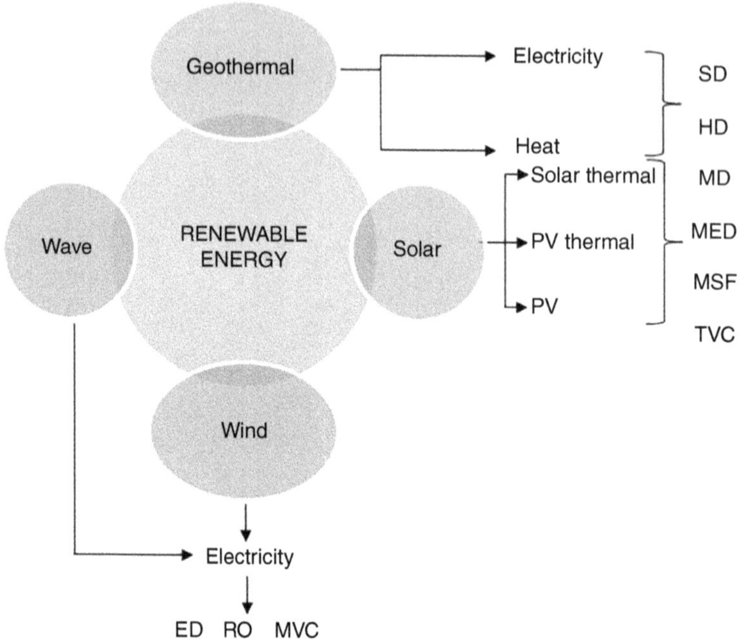

Figure 11.13 Possible combinations of desalination technologies with renewable energy sources (Gude et al. 2010).

Table 11.5 Energy and fuel savings in different hybridization of desalination processes (Gude et al. 2010).

Process hybridization	SWRO/MED (LT-TVC)	SWRO/BPT/MED (LT-TVC)	SWRO/BPT/ SWRO	MVC/MED (LT-TVC)	SWRO
Capacity ($m^3\,day^{-1}$)	14 300	14 800	14 700	19 000	11 000
Energy ($kWh\,m^{-3}$)	9.58	9.23	9.34	7.27	12.5
Fuel savings ($tons\,year^{-1}$)	4937	5319	4162	11 094	

BPT, backpressure turbine; LT-TVC, low temperature thermal vapor compression; SWRO, seawater reverse osmosis; MED, multi-effect distillation; MVC, mechanical vapor compression.

11.6 Modeling of Sustainability

In the mission to attain sustainable development, countries need to implement policies in the economic, social and environmental dimensions. It is tough to recognize the interlinkages that are responsible for trade-offs and collaborations among these dimensions. The modeling of sustainability or sustainable development modeling supports the research analysis by statistical validation with empirical data.

11.6.1 Basic Definitions

Numerical modeling, simulation and data management tools are some of the most important tools to facilitate the process to gather the important information needed for decision making in different tasks and projects. These tools are based on qualitative judgments and are quantitative in nature: they provide the details of various parameters and their importance.

Based on its significance, it can be decided whether to use the parameter or not. The most "rational" or "objective" of decisions rest on choices that are subjective. These choices are influenced by what decision makers think is vital, the objectives or standards that are to be achieved and for whom. The selection of parameters for environmental modeling is very important as it can facilitate to understand the natural systems with respect to different conditions. Also, the accuracy and reliability of certain information can be evaluated. Generally, the simulation models can either be statistical- or process-oriented; however, in some cases they can be a mixture of both. Statistical-oriented models are solely based on the data obtained from field measurements, whereas process-oriented models are based on information gathered from the ongoing fundamental processes. The simulation model subsequently discussed is a process-oriented model, which incorporates and simulates various physical processes that take place in the system.

The objective of sustainability modeling for a system is to stimulate scenarios for evaluation by considering the system boundary conditions (Nunes et al. 2016). As per the demonstration of Liu et al. (2008), the various guidelines have been shown in Table 11.6.

11.6.2 Description of Environmental System

The establishment of compassionate physical, institutional or psychological environments comprises living and working, amenities, technologies, strategies and platforms, and artistic support. For instance:

- As an example, the impact of transport usage by an organization on social, economic as well as environmental aspects could be reduced with adaptation of habits, e.g. an office shuttle bus facility for transport, a yearly package for transport tickets with incentives, a secure bicycle route and parking facilities. Also shifting the organizational offices nearer to public transport facilities. Assimilative capacity: extreme contamination of the environment.
- *Overutilization of resources*: Deforestation of forest, agricultural land.
- *Global issues*: Transfer of pollutants from one nation to other nation by means of atmosphere and rivers, hazardous waste.
- *Incapability to integrate*: Misinterpretation of environmental processes during industrial and economic activities with natural processes.

The goals and respective indicators for sustainability need to be well defined after consultation with a diverse range of community members and the executing organizations.

11.6.3 Systems Modeling and Simulation for Sustainability Assessment

Typical modeling and simulation mainly include the following points (Parker et al. 2002):

- Data models for measurements and experiments.

Table 11.6 Guidelines in implementing the modeling framework (Liu et al. 2008).

Steps	Guidelines
Problem design	• Understanding the conclusion framework in the contexts of natural and human features • Identifying the vital strategy related concerns which are of interest to relevant stakeholders • Developing multidisciplinary science and integrating focus
Scenario definition	• Extensive involvement of all the stakeholders in this action • Identifying the crucial factors of the system; for instance, global climate and environmental change, demography or increase in population, etc. • Determining spatial ranges and time horizons of different conditions • Adopting suitable situation categories in agreement to participants' concerns
Hypothetical modeling	• Establishment of system mechanisms/component/processes with detailed modeling • Postulating suitable temporal as well as spatial resolutions • If required, using different hypothetical models to explain the ambiguity of our actual real knowledge • Clear limits and assumptions of hypothetical model • All stakeholders must be communicated with via commonly used language and same should be the case for recorded documents
Development of model	• Developing or choosing appropriate computer models to execute various scenario conditions • Abilities or limitation of modeling codes must be transparent • User-friendly model interfaces must be designed and developed with various stakeholders
Authentication, standardization and validation	• Defining performance principles appurtenant to the explicit decision context • Acceptable design of system boundaries with great performance • Authentication of code, standardization of model and its validation must be established with independent data • The model precision and applicability must be recorded in documents
Development of model scenarios and its simulation	• Sufficient essential datasets for model input to execute diverse scenarios • Output of model scenarios must be collected that meets the requirement criteria of various stakeholders
Evaluation and assessment	• Ambiguity assessment of scenario outputs, considering all potential probability bases • Establishing the most prominent and essential indicators after discussion with all stakeholders for decision making • Performing indicators analysis to evaluate the possible threat related to a respective model or scenario

Table 11.6 (Continued)

Steps	Guidelines
Application/decision making	• Share the simulated scenario results with uncertainty evaluations with all the stakeholders • Provide dynamic and continuous support and communicate with all stakeholders
Observance with post-auditing	• Constant observance of important decisive variables and evaluating indicators for imminent unfolds, in collaboration with the stakeholders • Determining whether the present strategies should be altered or if new scenarios are required

- Qualitative conceptual models used as vocal or graphic descriptions of systems and processes.
- Quantitative numerical models.
- Mathematical modeling to analyze and interpret the numerical results.
- Governing models for transforming information and data into action.

11.6.4 System Dynamics for Sustainability Assessment: An Overview

The dynamics of a system are typically considered as a "scheme and program laboratory" and a "socio-economic laboratory" they are responsible for several strategic effects and system guidelines, particularly concerning socio-economic aspects.

System dynamic models involve various mathematical (differential) equations with solutions to validate dynamic system performance and for future policy options. Various interconnected processes are responsible for technology development, as presented below (Aizstrauta and Ginters 2015):

- System's socio-technical characteristics.
- Establishment in a designated time period.
- Participation of several decision-making bodies, for example, industries, institutes, clients and consumers.
- The development of individual revolution courses due to internal/external socio-technical aspects.
- Opportunity to add or remove constraints.

11.7 Modeling Tools for Sustainable Development Policies: Vision 2030 – United Nations Department of Economic and Social Affairs

The complex challenges interconnected across the three dimensions of sustainable development can be addressed by the launch of the 2030 Agenda, in particular the challenges whose dynamics are driven by the accidental consequences of human action (United

Nations General Assembly 2015). The abilities of quantitative modeling tools have been developed by the United Nations Department of Economic and Social Affairs (UNDESA) over the past decade. This has been done with special emphasis on program possibilities to accomplish growth objectives for human beings, comprising the Millennium Development Goals (MDGs). Thereby, models have been furnished to Member States from other developed regions. The implementation of the 2030 Agenda possesses some challenges, and to meet these challenges UNDESA is increasing its capacity to support Member States. This can be done by expanding the scope of the modeling tools and the number of beneficiary countries. The challenge of examining various policy scenarios across the three dimensions of sustainable development needs better technical capacities to understand the trade-offs and potential synergies. Modeling of policy opportunities through scenario scrutiny is a valuable method to inform the design of integrated and comprehensible policy frameworks to provide growth prospects for all. Responding to these challenges, UNDESA has encouraged reinforcing the capacities of nations to tackle economic, social and environmental aspects of sustainability via diverse modeling tools available on a public web platform.

These modeling tools assist in raising awareness about the approaches available. They also provide a valuable access point for policy analysis. Modeling tools are simply extended and tailored to specific country requirements, which helps to build in-depth capacity at the country level. By using the open-source coding and software to the maximum extent that is possible, the modeling tools are completely transferred to Member States or countries. Additionally, the web-based platform vows to inspire the community of scientists, academicians and development practitioners with the aim of improving the modeling tools available and ensuring their coherence/significance to modify the development policy plan. A set of modeling tools and visualizations are showcased on the website, supported by the UNITE Concepts platform, Office of Information Communication Technology (OICT), UN. It illustrates and assists in the investigation of planning schemes, along with their interconnectivity, which characterizes sustainable development. The result emphasizes four of the models: (i) the Fuel Tax and Development tool, which provides information on the impact of a fuel tax on the economy as a whole and on several socio-economic indicators, with the help of an economic-wide model; (ii) the Growth and Social Inclusion tool, which uses an economy-wide model and socio-economic microsimulations to evaluate the influence of policies and of external shocks on poverty and income distribution; (iii) the updates and extensions of the Universal Access Electricity tool, which helps the computation of the lowest investment cost and combination of technologies to provide general contact to power; and (iv) weather, use of ground, policies on energy and water, by which an integrated valuation can be organized to demonstrate the relationships between strategic resources at the global, national, or subnational level. The event aims at demonstrating the true potential of modeling-based scenarios to support the analysis of diverse economic, social and environmental strategies. Also, it aids in evaluating the interlinkages among the three dimensions of sustainable development, and thereby ensuring that nobody is left behind in meeting sustainable development priorities (https://sustainabledevelopment.un.org).

11.8 Summary

Economic, environmentally feasible and socially acceptable are the three important pillars for any sustainable business. To control the use of natural resources, the concept of sustainability is applied to various projects irrespective of specialization, nature, etc. The goals, targets and strategies for any sustainable development are controlled by the SDGs. Practical applications of SDGs in real-world scenarios are still needed to have better understanding and to develop innovative approaches to fulfill the SDG targets in different geographical, cultural and organizational situations.

However, in the case of industry, it is more fruitful to promote sustainability-impeding projects – there are great examples why various sectors like the solid waste management, water and power industries can follow the path of sustainability. UNDESA has developed tools for sustainable environmental modeling. Environmental modeling presents wide scope for various sectors to apply and fulfill the aims of sustainability by integrating social, ecological and economic aspects. Sustainability has four stages: ethics, extensive objectives, computable indicators and extensive scheme for execution. The objectives and guidelines facilitate in stimulating scenarios for evaluation by considering the system boundary conditions for implementation of sustainability modeling.

Sustainability also provides a comprehensive theoretical framework and strategy for practical implementation, comprising government at maximum levels, professional and civic establishments. Additionally, sustainability can involve and provide empowerment to people during the implementation of environmental sustainable development by changing the overall system.

References

Aizstrauta, D. and Ginters, E. (2015). Integrated acceptance and sustainability assessment model transformations into executable system dynamics model. *Procedia Computer Science*, Elsevier Masson SAS 77: 92–97.

Barber, J. (2007). Mapping the movement to achieve sustainable production and consumption in North America. *Journal of Cleaner Production* 15 (6): 499–512.

Basiago, A.D. (1999). Economic, social, and environmental sustainability in development theory and urban planning practice. *The Environmentalist* 19: 145–161.

BEAM (2012). BEAM Plus New Buildings Version 1.2. 1.2, 234pp. https://www.beamsociety.org.hk/files/download/download-20130724174504.pdf (accessed November 2019).

Boyko, C.T., Gaterell, M.R., Barber, A.R.G. et al. (2012). Benchmarking sustainability in cities: the role of indicators and future scenarios. *Global Environmental Change* 22 (1): 245–254.

CIB (1999). Agenda 21 on sustainable construction. CIB Report Publication 237. Rotterdam, the Netherlands: CIB General Secretariat.

Deka, B.J., Jeong, S., Alizadeh Tabatabai, S.A., and An, A.K. (2018). Mitigation of algal organic matter released from *Chaetoceros affinis* and *Hymenomonas* by in situ generated ferrate. *Chemosphere* 206: 718–727.

Dong, Y. and Hauschild, M.Z. (2017). Indicators for environmental sustainability. *Procedia CIRP* 61: 697–702.

Drioli, E., Ali, A., and Macedonio, F. (2015). Membrane distillation: recent developments and perspectives. *Desalination* 356: 56–84.

Galdos, M., Cavalett, O., Seabra, J.E.A. et al. (2013). Trends in global warming and human health impacts related to Brazilian sugarcane ethanol production considering black carbon emissions. *Applied Energy* 104: 576–582.

Gallopin, G. (2003). *A Systems Approach to Sustainability and Sustainable Development*. Gobierno, Chile: Sustainable Development and Human Settlements Division, CEPAL.

Goldman, T. and Gorham, R. (2006). Sustainable urban transport: four innovative directions. *Technology in Society* 28 (1–2): 261–273.

Goodland, R. and Daly, H. (1996). Environmental sustainability: universal and non-negotiable. *Ecological Applications* 6 (4): 1002–1017.

Gude, V.G., Nirmalakhandan, N., and Deng, S. (2010). Renewable and sustainable approaches for desalination. *Renewable and Sustainable Energy Reviews* 14 (9): 2641–2654.

Haque, M.S. (2000). Environmental discourse and sustainable development: linkages and limitations. *Ethics and the Environment* 5 (1): 3–21.

Jonathan M. Harris. (2003). Sustainability and sustainable development. International Society for Ecological Economics. http://isecoeco.org/pdf/susdev.pdf (accessed November 2019).

International Organization for Standardization (2006). *ISO 14040-Environmental Management – Life Cycle Assessment – Principles and Framework*. Geneva: International Organization for Standardization.

IPCC (2007). *Climate Change 2007: Impacts, Adaptation and Vulnerability*. Cambridge, UK: Cambridge University Press.

Jensen, A.J. (1990). Growth of young migratory brown trout *Salmo trutta* correlated with water temperature in Norwegian rivers. *Journal of Animal Ecology* 59 (2): 603–614.

Lambert, S., Riopel, D., and Abdul-Kader, W. (2011). A reverse logistics decisions conceptual framework. *Computers and Industrial Engineering* 61 (3): 561–581.

Lee, E.J., Deka, B.J., Guo, J. et al. (2017). Engineering the re-entrant hierarchy and surface energy of PDMS-PVDF membrane for membrane distillation using a facile and benign microsphere coating. *Environmental Science and Technology* 51 (17): 10117–10126.

Lee, E.-J., Deka, B.J., and An, A.K. (2018). Reinforced superhydrophobic membrane coated with aerogel-assisted polymeric microspheres for membrane distillation. *Journal of Membrane Science* 573: 570–578.

Lele, S.M. (1991). Sustainable development: a critical review. *World Development* 19 (6): 607–621.

Linton, J.D., Klassen, R., and Jayaraman, V. (2007). Sustainable supply chains: an introduction. *Journal of Operations Management* 25 (6): 1075–1082.

Liu, Y., Gupta, H., Springer, E., and Wagener, T. (2008). Linking science with environmental decision making: experiences from an integrated modeling approach to supporting sustainable water resources management. *Environmental Modelling and Software* 23 (7): 846–858.

Mkumbo, W. (2016). The role of school libraries in realizing the achievement of inclusive and equitable quality education in Tanzania: SDGs by 2030. *Journal of Library and Information Science* 6 (2): 184–190.

Mokhlesian, S. and Holmén, M. (2010). Business model changes in green construction: a literature review. In: *ARCOM 2010 – Proceedings of the 26th Annual Conference* (eds. C.O. Egbu and E.C.W. Lou), 997–1006. Leeds, UK: Association of Researchers in Construction Management.

Nordhaus, T., Shellenberger, M. and Blomqvist, L. (2012). The planetary boundaries hypothesis. A review of the evidence. Technical report, June, pp. 1–43.

Nunes, B., Alamino, R.C., Shaw, D., and Bennett, D. (2016). Modelling sustainability performance to achieve absolute reductions in socio-ecological systems. *Journal of Cleaner Production* 132: 32–44.

Parker, P., Letcher, R., Jakeman, A. et al. (2002). Progress in integrated assessment and modelling. *Environmental Modelling & Software* 17: 209–217.

Peeters, J. (2012). Social work and sustainable development: towards a social-ecological practice model. *Journal of Social Intervention: Theory and Practice* 21 (3): 5–26.

Pero, M., Moretto, A., Bottani, E., and Bigliardi, B. (2017). Environmental collaboration for sustainability in the construction industry: an exploratory study in Italy. *Sustainability* 9 (1): 125.

Persson, U. (2009). *Management of Sustainability in Construction Works*. Lund, Sweden: Lund Institute of Technology, Lund University.

Posada, J.A. (n.d.). *Environmental Assessment of a Process Design*. Delft, the Netherlands: Delft University of Technology.

Prahalad, C.K. and Hart, S.L. (2002). The Fortune at the bottom of the pyramid. *Strategy+Business Magazine* 26: 1–16.

Rajapakshe, T.K., Vakharia, A.J., Wang, L., and Yenipazarli, A. (2005). Sustainable operations management. *Production and Operations Management* 14 (4): 482–492.

Rockström, J., Steffen, W., Noone, K. et al. (2009). Planetary boundaries: exploring the safe operating space for humanity. *Ecology and Society* 14 (2): 32.

Shriberg, M. (2000). Sustainability management in campus housing: a case study at the University of Michigan. *International Journal of Sustainability in Higher Education* 1 (2): 1467–6370.

Sovacool, B.K. and Brown, M.A. (2010). Twelve metropolitan carbon footprints: a preliminary comparative global assessment. *Energy Policy*, , Elsevier 38 (9): 4856–4869.

United Nations General Assembly (2015). Transforming our world: the 2030 Agenda for Sustainable Development. New York: United Nations. https://www.un.org/sustainabledevelopment/development-agenda (accessed November 2019).

Walker, H.L., Gough, S., Bakker, E.F. et al. (2009). Greening operations management: an online sustainable procurement course for practitioners. *Journal of Management Education* 33 (3): 348–371.

Wang, P. and Chung, T.S. (2013). A new-generation asymmetric multi-bore hollow fiber membrane for sustainable water production via vacuum membrane distillation. *Environmental Science and Technology* 47 (12): 6272–6278.

Yan, M.R., Chien, K.M., and Yang, T.N. (2016). Green component procurement collaboration for improving supply chain management in the high technology industries: a case study from the systems perspective. *Sustainability* 8 (2): 105.

12

Biodiversity and Sustainability

Akshaya K. Verma[1], Prangya R. Rout[2], Eunseok Lee[2], Puspendu Bhunia[3], Jaeho Bae[2], Rao Y. Surampalli[4], Tian C. Zhang[4], Rajeshwar D. Tyagi[5], Pengzhi Lin[6], and Yu Chen[6]

[1]*Department of Civil Engineering, Siksha 'O' Anusandhan, Bhubaneswar, India*
[2]*INHA University, Incheon, Republic of Korea*
[3]*School of Infrastructure, Indian Institute of Technology, Bhubaneswar, India*
[4]*Department of Civil & Environmental Engineering, University of Nebraska-Lincoln, Omaha, NE, USA*
[5]*INRS, Université du Québec, Québec, Canada*
[6]*Key State Laboratory of Hydraulics and Mountain Engineering, Sichuan University, Chengdu, China*

12.1 Introduction

Biodiversity is defined as the variability in life forms and their interactions on the planet. The Convention on Biodiversity (CBD) at the United Nation Conference on the Environment and Development (UNCED), "The Earth Summit" at Rio de Janeiro, Brazil in 1992 articulated biological diversity as "the variety and variability among living organisms from all sources and the ecological complexes in which they occur including diversity within species, between species and of ecosystem" (UNEP 1992). It encompasses all the variation of life forms expressed in genes, individuals, populations, species, communities and ecosystems. These dimensions are closely interlinked, influencing the existence, services and productivity of the ecosystem (Schneiders et al. 2012), making biodiversity not only an ecological, but also a socio-economic entity (Laurila-Pant et al. 2015). Its quantitative measurement focuses mostly on the species within the taxa; however, ecological diversity can also be quantified, and is an onerous task. Biodiversity is a dynamic entity, which is changing continuously and very slowly throughout the life of the planet, e.g. the end of the dominance of dinosaurs and the origin of mammals (Lacher and Roach 2018).

On the other hand, sustainability is defined as "development that meets the needs of the present without compromising the ability of future generations to meet their own needs" (WCED 1987). In a broader sense, "the objective of integrating economic activity with environmental integrity, social concerns and effective governance systems, while maximizing the contribution to the well-being of the existing generation, fairly sharing the cost and benefits, without compromising the potential for the upcoming generations

Sustainability: Fundamentals and Applications, First Edition.
Edited by Rao Y. Surampalli, Tian C. Zhang, Manish Kumar Goyal, Satinder K. Brar, and R. D. Tyagi

to meet their needs" (IIED 2002). In other words, sustainability is an advancement to safeguard the forthcoming and existing potential to satisfy vital human requirements within the ecological and resource limits of planet earth (Blum et al. 2017).

The mechanisms accountable for change in biodiversity are evolutionary processes of extinction and speciation including ecological processes (Sechrest et al. 2002). Species interactions, climate change and even cosmic disturbances have also played key roles in influencing past and present biodiversity (Pimm et al. 1995). As per the International Union for the Conservation of Nature (IUCN), 22.4% of all evaluated animals are threatened by extinction (IUCN 2010). During the past decade, several researchers have well documented that the numerous threats to biodiversity are mostly associated with humans. The major threats pertaining to the human-associated factors are population growth, habitat degradation, overexploitation of resources, pollution, climate change, incidental mortality, invasive alien species and disease outbreaks. Foin et al. (1998) have categorized nine distinct threats (i.e. habitat reduction, habitat modification, introduction of exotic species, species extinction due to human, specialized habitats, succession and disturbance, hybridization, biotic interaction and co-evolution) for their quantification in their study. Researchers have concluded that habitat reduction and modification are among the major threats to their recovery plans. According to Chu et al. (2003), habitat conversion is identified as one of the major indicators of stress for the depletion of biodiversity in the watersheds. Yiming and Wilcove (2005) have investigated the threats (e.g. overexploitation, habitat destruction, pollution, introduced species and disease) to vertebrates present in the United States and China, and reported the significant differences being observed due to the different threats, contributing to the loss of biodiversity.

Infectious diseases are also reported as a major biotic factor that can threaten biodiversity and may lead to extinctions (Brito et al. 2012). Smith et al. (2006) stated that infectious diseases may transform the ecosystem, leading to species extinction and endangerment (based on the subjective data). However, the IUCN Red List reports that infectious disease is not the only cause of species extinction. For example, Dextrase and Mandrak (2006) reported that the habitat loss and introduction of alien invasive species could be regarded as the primary and second threats to imperiled fish, respectively. Habitat modification in terms of flow adjustment, urbanization and fragmentation of agricultural land further damages the inherent species (Light and Marchetti 2007). Numerous researchers support the link between the conversion of habitat for agricultural and urban use and threatened biodiversity, and further recognize the increase in pollution associated with conversions as one of the major threats (Kerr and Cihlar 2004; Kerr and Deguise 2004). Kerr and Deguise (2004) and Brown and Laband (2006) found that change in land use is more closely associated with species extinction than the conversion of area (land use change is a process of transformation of the natural landscape by anthropogenic activities (emphasizing the functional role of land for economic activities) while conversion of area means changing the status of land from agriculture to non-agriculture).

This chapter focuses mainly on the aspect of the threat to biodiversity as sustainability is covered in detail in other specialized chapters. The chapter highlights the major threats to biodiversity like deforestation and loss of habitat, climate change, overexploitation, invasive species and environmental pollution, with an emphasis on how to enhance sustainability by combating these threats through implementation of conservation programs, policies and regulations including awareness. The chapter also discusses the role of biodiversity in attaining and maintaining sustainable development. Also included is a section addressing the trends of biodiversity, particularly emphasizing current as well as future trends and pointing out the role of the CBD in unifying nearly all the countries worldwide in a commitment to reduce the loss of biodiversity in the coming decades.

12.2 Threats to Biodiversity

12.2.1 Deforestation and Habitat Loss

One of the most crucial threats to biodiversity is the loss of habitat associated with human activity. Habitat loss mainly encompasses conversion of land for other purposes including urbanization and agriculture. In tropical regions, vegetation loss over a large area can lead to permanent land transformation, where soils exhibit significant nutrient deficiency. Tropical forests are extremely diverse in nature as they support over 66% of the world's biodiversity though they cover less than 10% of earth's land area (Raven 1988; Bradshaw et al. 2009). Natural habitat is generally lost if the transformation is forced. However restoration of some habitats is possible over a period by adopting some of the reclamation methods (natural or artificial). Terrestrial ecosystems suffer habitat degradation in a number of ways, including deforestation.

12.2.1.1 Deforestation

Deforestation occurs as a result of growing human population and consumption of resources, and, therefore, it may be considered as one of the root threats to biodiversity (Figure 12.1). Deforestation reduces biodiversity by destroying habitat, interfering with reproduction in plants and by creating ecotones. A large number of different types of herbivores are forced to move away in the search for food once their food chain becomes affected. Some species can make their way to alternative food sources, while others vanish. Consequently, the complete food chain is disrupted, including herbivores and carnivores, at the majorly affected trophic levels. For example, when a small portion of the rainforest is affected, it can result in extinction of a significant number of animal species. Regardless of the undesirable effects of deforestation, which results in diminishing of biodiversity, forests are constantly being removed as a result of human greed (for economic and consumer benefits). On account of the above, species lose their natural habitat and fight against different hostile factors for their survival. However, some species are unable to survive the change and may perish due to the hostile changes. As a result, the forests are losing biological diversity.

Increase in population and resource consumption (to satisfy their needs) endanger the existence of many species and ecosystems due to land and soil degradation, pollution, habitat fragmentation, overexploitation and climate change. Overexploitation of resources in terms of overharvesting and extraction of excessive mineral resources leads to significant destruction of biodiversity. Implementation of awareness programs and preservation measures are the major approaches to curbing overharvesting. Forest ecosystems of the planet survive through a continuous global climate change over a

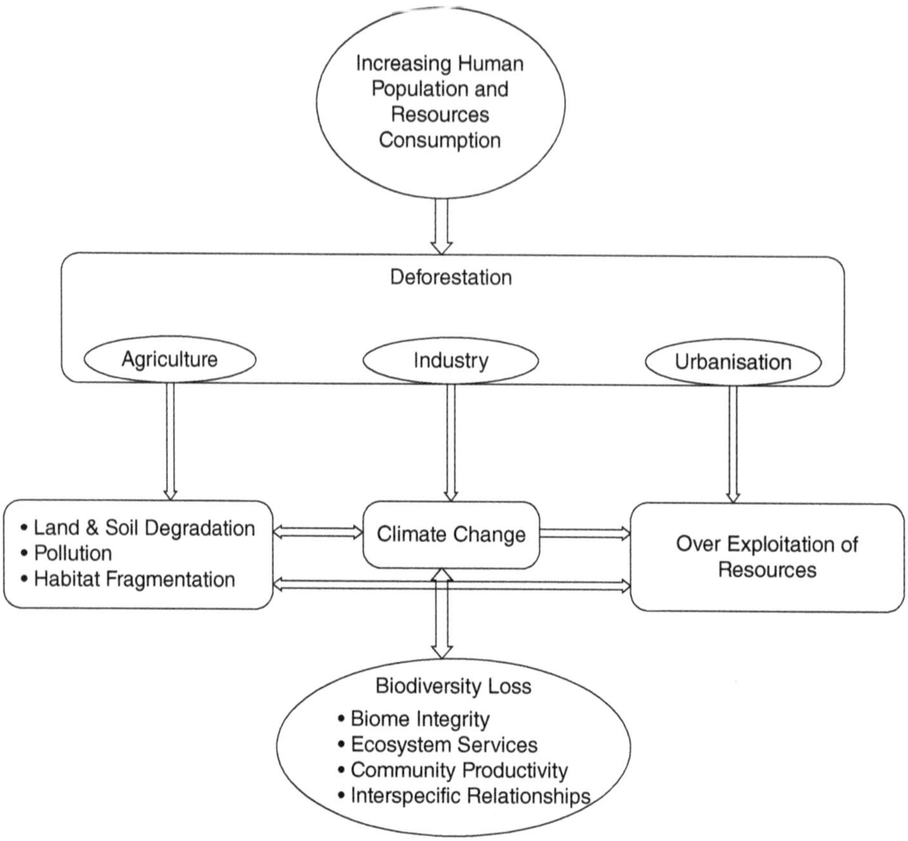

Figure 12.1 Deforestation as a root threat to biodiversity.

long period. As a result, a number of species have gone extinct while others are also on the verge of extinction. Anthropogenic activities are the major cause of the global temperature rise; if this trend continues, the threats to biodiversity will increase, leading to the destruction of species and ecosystem. For example, the increase in ocean temperature due to the decrease in Arctic ice is among the major causes of change in the vegetation zones and decline of marine wildlife.

Furthermore, human activities, such as deforestation (destruction of a major carbon sink) and increased dependency on fossil fuels, have aggravated the problem of climate change. In order to diminish the carbon footprint and also to assure a better tomorrow for future generations, strict norms and legislations should be implemented by policy-makers. Environmental pollution owing to industrialization is continuously harming the biosphere by releasing and introducing many toxic chemicals. These toxicants are potentially affecting the habitats of various animals and plant species and their ecosystems. According to a recent report, deforestation has significantly affected the Amazon rainforest, Sumatran jungles and forests of British Columbia (Greentumble 2018). These forests have been specifically targeted for deforestation for a few reasons, but mainly for human economic gain. Deforestation is being carried out in many different ways, such as:

- The Amazon rainforest is being removed at an alarming rate to transform the forest land for agricultural purposes and grazing livestock.
- Sumatran rainforests are being destroyed and replaced by massive palm plantations to help with the human obsession with palm oil.
- Canada's British Columbia is a hotspot for old-growth forests that contain some of the most sought-after lumber in the world.

Deforestation may create higher risk of wildfires and droughts due to the alteration in the moisture retention of trees and other vegetation in the forest. This consequently might affect the hydrological cycle, and further creates dry and warm climatic conditions, leading to accelerated temperature rises.

12.2.1.2 Oceans and Fisheries

Ocean biodiversity, which is the combination of living resources and ecological processes that sustain it, is also affected by human activity (Hourigan 1999). The ocean itself delivers many vital socio-economic, socio-cultural, environmental and aesthetic benefits to society. It is known as an endless reservoir of marine productivity with the ability to assimilate different kinds of waste. It is a specialized ecosystem, which is diverse in species, thereby ensuring the stability of the marine ecosystem by providing several pathways for energy flow. Marine biodiversity is one of the major primary sources of protein food, and a reduction in it will have a great impact on human society.

Ocean resources are continuously being overharvested for economic gain, which threatens marine biodiversity. Marine ecosystem services are quantified as being worth $8400 billion per annum for open ocean ecosystems, and $12 600 billion per annum for coastal ecosystems, globally (Costanza et al. 1997). Fishery services are estimated as being worth $80 billion annually, which provides employment to 0.2 billion people and fulfills the requirement of 16% of the protein across the globe (Global Opportunity Network 2018). The coastal ecosystem delivers a wide range of important goods and services, i.e. food, medicines, coastal protection, entertainment and ecological services. Oceans generate a significant amount of oxygen and act as a major sink for CO_2. However, marine biodiversity is under constant threat, presumably because of the habitat destruction, overfishing (overharvesting), hydrocarbon and mineral exploration, toxic chemicals, eutrophication, rise in sea water temperature and acidification due to anthropological negligence. The 2018 Global Opportunity Network report states that "the marine Living Planet Index (LPI), an indicator of the state of global biological biodiversity, shows a decline of 39% between 1970 and 2010" (Global Opportunity Network 2018, p. 38). The enormous amount of CO_2 released due to anthropogenic activities does not limit itself to causing global warming as it also ends up increasing the acidity of oceans. Furthermore, coral reefs are expected to vanish by 2050 if these unsustainable practices continue. The primary threats to ocean biodiversity are impact from natural disturbances, anthropogenic disturbances (e.g. rise in sea level, temperature change and acidification), fisheries operations (including habitat alteration and overfishing), naval operations (e.g. maritime transport and tourism), nutrient and toxic chemical pollution, as well as alien species invasions and damage to plankton by ultraviolet radiation (Figure 12.2). Most of these factors have also been identified by the CBD (UNEP/CBD 1995).

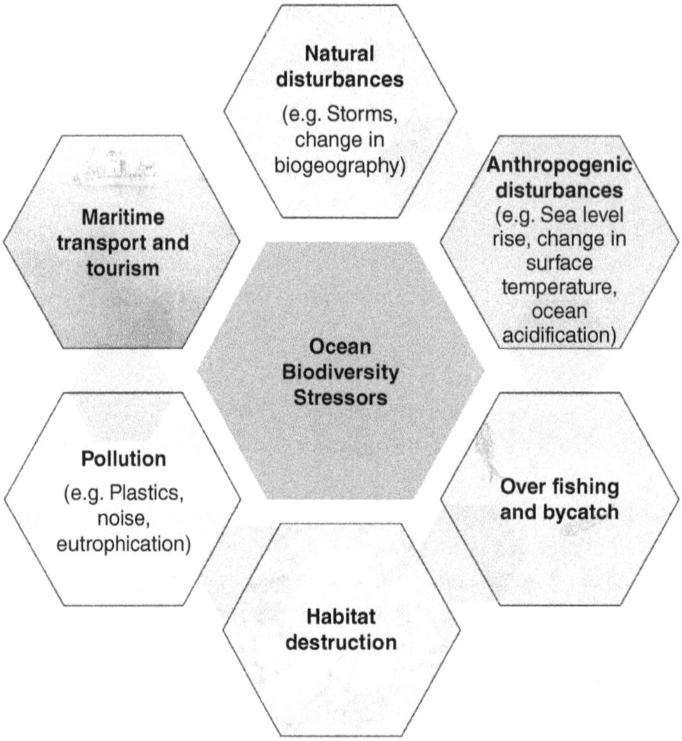

Figure 12.2 Major indicators of ocean biodiversity stressors.

The Ocean Atlas detailed that several oxygen-deficient zones present in the world's oceans lie in tropical regions, on the coasts of Peru, Namibia and the Arabian Peninsula (*The Ocean Atlas* 2017). Some specifically acclimatized microbes (anaerobic or facultative) are present in these zones. The formation of anaerobic zones occurs due to the introduction of huge quantities of agricultural and animal husbandry wastes containing nitrate and phosphate, which overfertilize the sea and result in eutrophication. Eutrophication may generally be found at several locations around the world including the South China Sea and the Bay of Bengal. Due to lack of proper waste management facilities, huge quantities of plastic wastes flow along the sewers and further to rivers, ending up in the sea. Removal of plastics cannot be done very easily if plastics wastes have started to accumulate in the sea. A maximum proportion of these plastics is transformed into microplastics, and therefore, it becomes impossible to remove it even after filtration. Consequently, it poses severe threats to human and aquatic animals through biomagnification. Furthermore, during the early 1950s, several countries around the world legally dumped huge amounts of radioactive waste, including chemical and common weapons, into the sea; some of those were generated by the nuclear power plants (*The Ocean Atlas* 2017).

Oil pollution also appears to be one of the most detrimental factors for aquatic life. The major sources are the leakage of oil during different operations of on-shore exploration, spilling of oil due to exploration accidents, as well as illegitimate cleaning of vessels and shipping. Moreover, the magnitude of noise in the ocean is increasing continuously because of increasing traffic over the oceans. Marine mammals are severely affected during the interaction and navigation with sound. The coral reef ecosystem is known to have a very high degree of biodiversity, being home to a third of the population of marine fish species and over 10 000 other species. According to Bryant et al. (1998), 58% of coral reefs are under moderate to severe risk from coastal pollution and overexploitation. Corals are very sensitive to the rise or fall of temperature and intense lighting, since they possess a low degree of tolerance to these alterations. An increase in the temperature by only 1.1°C can start the bleaching effect and result in the death of corals (Veron et al. 2009). It has been predicted that the ocean temperatures will increase by 1.0°C and up to 4°C, respectively, by 2029 and 2099, while warming will persist for decades thereafter (IPCC 2007). As a result, marine ecosystems will be changing due to the migration of temperature-sensitive species. Some of them may vanish owing to the alteration in their habitat and food chain (Planque and Frédou 1999; Cheung et al. 2009). The erosion of ocean biodiversity affects various ecosystem services such as shoreline stabilization, water quality, hydrologic functions, tourism and recreation activities (Burke et al. 2001; Worm et al. 2006).

12.2.1.3 Land and Soil Degradation

Like a fragile skin, land is the solid covering on the earth that sustains all forms of life (Young and Crawford 2004). Soil is made up of complex biomaterials and is a major constituent of the terrestrial ecosystem functioning alongside the segments of the environment in an integrative manner (Szabolcs 1994). As compared with plant species, far fewer soil micro-organisms have been recognized (Breure et al. 2012). Global biodiversity has suffered an extraordinary threat due to the limited knowledge of soil biodiversity. Out of the 63.8 million investigated species, 32% are either extinct or threatened (Li et al. 2017). Numerous evidence also exists related to the decrease in soil biodiversity in several countries (Gardi et al. 2013; Abdul Rida and Bouché 1995).

The Food and Agricultural Organization (FAO) have stated that "Soils are fundamental to life on Earth but human pressures on soil resources are reaching critical limits. Further loss of productive soils will amplify food-price volatility and send millions of people into poverty. This loss is avoidable. Careful soil management not only secures sustainable agriculture, it also provides a valuable lever for climate regulation and a pathway for safeguarding ecosystem services" (FAO 2015, p. 4). Soil plays a crucial role in the functioning of the ecosystem by providing numerous ecosystem services, such as contributing to food security, climate stabilization, nutrient cycling, water holding capacity, filtration of water and hosting biodiversity (Adhikari and Hartemink 2016). Soil characteristics regulate these ecosystem services, which are significantly influenced by its use and management. Healthy soils ensure food security, climate regulation, environmental quality, water retention, provide habitat to biodiversity and prevent landslides (Smith et al. 2015). Degradation of land and soil is a serious global challenge not only for long-term food security but also for sustainability of the ecosystem owing to the serious risks it poses for wildlife diversity (Godfray et al. 2010).

Researchers have reported that there is a considerable decline in the biota of many areas, which contributes to soil biodiversity and ecosystem services (Jones et al. 2003; Piotr 2013). Reduction of these services could affect mankind drastically since ecosystems provide at least $33 trillion worth of services annually, and a major chunk of this comes from the terrestrial ecosystem, which is an important part of the soil biota (Piotr 2013). Land degradation can be induced as a result of unsustainable land management practices (e.g. deforestation, overgrazing, soil nutrient mining) and some of the biophysical features, such as natural topography of an area including climatic factors. Several factors are responsible for degradation of land and soil and, thus, are considered as the major threats affecting biodiversity of an ecosystem (Figure 12.3).

Additionally, due to soil erosion, substantial loss of soil nutrients has been observed and the reclamation of these lands requires significant economic and environmental expense. The impact of erosion is very intensive on sloping land, where the major portion of soil can be lost with the water flowing downhill into valleys. Soil structure can determine the erosion potential: soils of fine texture containing fewer organisms with poor structure can be eroded quickly (Bajracharya and Lal 1992). Soils are extremely diverse in nature, and are a home to millions of fauna within a single ecosystem including soil bacteria, fungi, protozoa, nematodes and earthworms. These organisms form a food web and maintain the proper functioning of the soil ecosystem by driving different ecological processes such as nutrient recycling, carbon sequestration, etc.

Desertification and the associated loss of vegetation is another important stressor, which reduces the ecosystem biodiversity significantly and accelerates climate change by hampering the carbon cycle. It is a slow process that diminishes productivity and resilience due to change in soil properties, vegetation and climate. As a result, dryland formation occurs and ecosystem services that are fundamental to sustaining life are severely affected. Dryland covers about 41% of the available land surface, and most of this is in developing nations (UNDDD 2010). Anthropogenic activities related to the transformation of natural land to human-modified landscapes are the major driver behind land degradation and biodiversity loss (Halada et al. 2008; Theobald 2014).

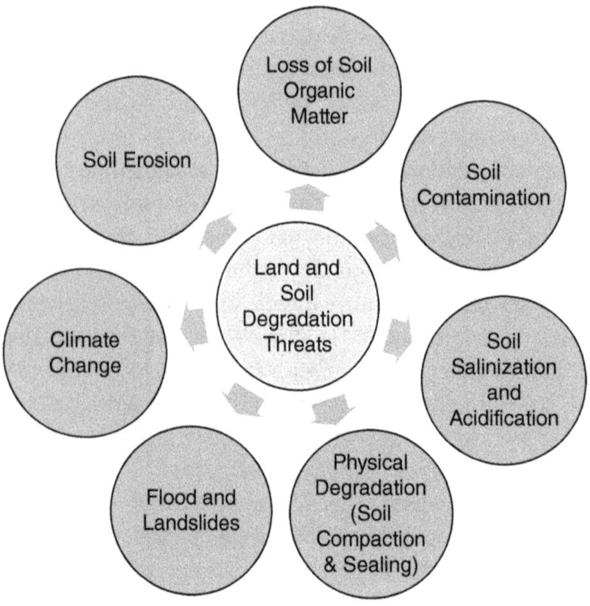

Figure 12.3 Major threats related to land and soil degradation.

Forests and tree plantation help in stabilizing soils, reducing erosion and maintaining nutrient cycling in soils. As a result, desertification and land degradation can be minimized. The Inter-governmental Panel on Climate Change states that "the conservation and restoration of forests can considerably reduce emissions at a low cost. Further co-benefits can also be achieved when steps to combat land degradation and conserve biodiversity are included in forest conservation and restoration" (UNFCCC 2018, p. 4).

12.2.2 Climate Change

Climate refers to the average pattern of weather phenomena, including temperature, precipitation (rainfall, snowfall), winds, glaciations, storms, etc. Long-term variation in this pattern is called climate change. The overall climate of the earth is changing slowly and gradually. The most remarkable change is its increasing temperature, which is often referred to as global warming. Global warming is caused by trapping of outgoing long-wave solar radiation by greenhouse gases (such as water vapor, carbon dioxide, ozone, methane, etc.), which in turn increases the temperature of the planet. The natural concentrations of greenhouse gases are being enhanced by emissions resulting from numerous anthropogenic activities. As a result, the temperature of the earth's surface and troposphere is increasing along with many other changes. According to the IPCC Fourth Assessment Report, an average global temperature increase of 0.74°C was recorded over the past 100 years. Consequently, an increase of 12–22 cm of global mean sea level was assessed. Rainfall patterns have changes and frequency of extreme weather events has increased drastically. Further, an increase in temperatures of 1.4–5.8°C by 2100 is predicted, which may enhance the associated impacts, for example, further rises in global mean sea level, alteration in precipitation patterns and mass risk from "vector-borne diseases"(IPCC 2007).

Climate change has produced a noticeable impact on biodiversity at all levels (species, genetic and ecosystem) and has affected distribution, composition and functions significantly. Populations of many species have declined, with some species either thriving or adapting. A recent study conducted at James Cook University in Australia and Independent University Bangladesh states that the habitat of the iconic Bengal tiger and the world's largest mangrove forest "Sunderban" could be destroyed by climate change and rising sea levels over the next 50 years and will vanish entirely by 2070 (Laurance 2019). At the species level, the observed changes may be related to the timing of events referred as phenology; for example, many birds and insects are showing changes in migration, egg-laying and breeding. Desert ecosystems have expanded, vegetation in mountains has altered and lianas (climbing vines) have increased in tropical rainforests. Such noticeable changes may affect the functions and services of ecosystems. These observed changes may be modest, but the future predicted changes due to an increase by 1°C in global temperature may produce detrimental impacts resulting in the extinction of at least 10% of species. Aquatic and terrestrial ecosystems are mostly vulnerable because of their associated limitations such as narrow geographic areas and limited dispersal opportunities. Ecosystem alteration as a result of climate change often produces negative impacts for socio-economic and socio-cultural values. However, the extent and impact rate is still uncertain. Different methods and tools may be used to assess the impacts of climate change on biodiversity (Guisan and Zimmermann 2000). Vulnerability assessments are applicable to monitor natural and socio-economic hazards, while modeling tools such as climate envelope modeling are commonly used for the assessment of potential impacts on a particular species. Use of climate envelope modeling is limited since it provides the instant magnitude of climate change impact. Assessment of future impacts and vulnerability on the ecosystem can be monitored using dynamic models, population models and mechanistic models (Phillips et al. 2008). Dissolution of excess atmospheric CO_2 in oceans alters the oceanic water chemistry, which result in decreased pH leading to the acidification of the oceans. Oceanic acidification is one of the major impacts of climate change, which has resulted in a significant decrease in marine biodiversity. Global projection models show that many species of lizards are on the verge of extinction due to climate change (Shah 2014). Biodiversity can combat the adverse impacts of climate change. Habitat restoration can remove CO_2 from the atmosphere by sequestration. In addition, devastating impacts of climate change such as flooding and storm surges can be reduced by conserving intact ecosystems (UNFCCC 2018).

12.2.3 Overexploitation

Harvesting of plants and animals is crucial to fulfill our daily requirements and to sustain our lives, but in the process we have crossed the limits and practiced harvesting to a new extreme. Overexploitation is the harvesting of existing resources with a rate that exceeds its natural replacement rate, and that cannot sustain viable populations. As a result, species become endangered and threatened with detrimental impacts on earth's biodiversity. In general, the magnitude of overexploitation by humans in the marine ecosystem is much greater than that of the land ecosystem. During the past few decades, the human population has grown exponentially, which has resulted in the overexploitation of food resources to meet the demand. The use of advanced techniques to enable harvesting, transport and storage has further enhanced this trend. According to a study, 29% of aquatic food resources (fish and seafood species) have vanished and are predicted to disappear by 2048, unless immediate measures are taken (Worm et al. 2006). The trend of overexploitation of marine fish stocks has alarmingly increased over the past few decades. Consequently, 52% of marine stocks were completely harvested at their extreme sustainable level and 24% were overexploited or exhausted by 2005 (FAO 2007). Overexploitation also threatened ecologically significant species such as sharks, tunas and sea turtles. Globally, the populations of sharks have declined by 90% in the past 15 years (Baum et al. 2003; Myers and Worm 2003; Myers et al. 2007). Indonesia is the main exporter of frog's legs and up to 1 billion wild frogs are estimated to be harvested every year for human consumption (Warkentin et al. 2009). As a result, overexploitation also threatens such amphibia. In a specific habitat, the species are interdependent to maintain their food chain. Extinction of a single species can alter the structure of that habitat by reducing the possibility of survival of the remaining species. The highly overexploited species are marine fish, invertebrates, plants and mammals harvested for the purpose of medicinal and commercial trading (UNEP 2006). Species traded around the globe include endangered creatures such as apes, rhinoceros and elephants, etc. including popular flora and pet animals. Commercial trading can be in the form of meat, skins, fur, organs or other body parts of the animals because of their additional values.

Unfortunately, sustainable management of species utilization is very difficult and poorly understood at regional, national and global levels. For example, the rarer the creature, the greater the demand and the higher the price, leading to increased chances of extinction. Therefore, both of these areas should be addressed effectively to combat the overexploitation of different resources and species and conserve biodiversity. Extraction of timber and other wood products eliminates millions of hectares of forests each year, including the habitat of uniquely adopted species. Unsustainable forest management and intensive agriculture produce negative impacts on biodiversity leading to a decrease in soil fertility and an increase in soil erosion. Drug discovery and availability of medicinal products are the significant accomplishments derived from biodiversity. Many medicinal products are of natural origin and derived from marine ecosystems (Jain et al. 2008). Nevertheless unregulated bioprospecting could lead to overexploitation of resources and loss of biodiversity. Hunting of higher vertebrates also produces an inconsistent effect on the structure of an ecosystem, since they function as ecosystem engineers and their elimination may impact their role in landscaping (Jones et al. 1994). According to an assessment performed by Sheil and Salim (2004) in Uganda, two similar forests with and without elephants demonstrated different succession and regeneration patterns. Different kinds of measures may be adopted to curb the overexploitation of natural resources and to maintain biodiversity. Some such measures are as follows:

- Implementation of laws and policies at regional, national and global levels.
- Awareness raising campaigns and programs.
- Responsible consumption of natural resources.

Strict regulations may be implemented for restoration of priority habitat and degraded ecosystems and to expand the protected area network significantly. Awareness campaigns should be focused to encourage the use of public transport, to foster the culture of recycling and, thereby, reducing the carbon footprint. Moreover, companies may be rewarded for promoting conservation, sustainable resource management and invention in their activities.

12.2.4 Invasive Species

Invasive species refer to biota that spread quickly due to their preadaptation ability and successfully establish themselves in areas where they have no prior evolutionary history and sometimes they may have negative impacts on local, regional and global biota through competitive and/or predatory influences. The introductions of invasive/exotic species are primarily due to natural phenomena and also driven by intentional or unintentional anthropogenic spread. In most cases the exotic species probably fail to proliferate because of poor adaptation to the new environmental condition and introduction of low number of individuals. However, some exotic species often exhibit a dramatic population boom in the new ecosystem they enter and alter the environmental conditions in the new habitat through predation, competition for resources, etc., thereby threatening the other species that exist there. The non-presence of co-evolved parasites and predators of the invasive species in the new ecosystem enables the invaders to outcompete the native species. For instance, in British Columbia, the knapweed plant species introduced to arid grasslands and the purple loosestrife introduced to riparian areas progressively became dominant plant species in the respective ecosystems. The non-native species alter the native food web resulting in a changed nutrient cycle of the ecosystem. Subsequently, the interactions between native species are altered leading to the loss of native biodiversity. Some researchers even argue that invasive species are the second greatest threat to biodiversity after habitat destruction (Wilcove et al. 1998). For example, the intentional introduction of the fresh water fish Nile perch in Lake Victoria was the main reason behind the extinction of about 200 species of cichlids. Likewise, the unintentional introduction of the brown tree snake from the Solomon Islands to Guam in 1950 via aircraft has led to the extinction of five species of reptiles and three species of birds endemic to the island. Introduction of a fungal pathogen, *Geomyces destructans* from Europe is responsible for white-nose syndrome, the disease that decimated cave-hibernating bats in eastern North America. Another fungus of African origin, *Batrachochytrium dendrobatidis*, the causative organism of chytridiomycosis disease, is the main reason behind the global decline in amphibian species. The fungus is believed to have been spread throughout the world by the African clawed frog, a commonly used laboratory species. The green crab, *Carcinus maenas*, has significantly reduced the richness of native clams and crabs along the coast to Washington after its introduction from San Francisco Bay, likely through ship ballast water. However, the introduction of zebra mussel to the Great Lakes from Europe though threatened native mollusk populations radically improved water quality due to its filter feeding habit. Its introduction as an invasive species not only saved millions of dollars of water clean-up costs but also benefited some species by facilitating enhanced water quality-mediated growth of aquatic plants that ultimately provided shelter for young fish such as smallmouth bass that did not exist before in that habitat. As invasion biology is a relatively new field, there are ongoing debates about the assessment of the consequences of invasive species. Some ecologist have argued that biological invasions are a natural process, whereas others have counterargued that anthropogenic-mediated invasions pose a threat because of their rapid and widespread nature as compared with the natural invasions (Young and Larson 2011). Therefore, more in-depthstudies of species invasions are essential to fully understand the mechanisms of how invasive species appear, thrive in changed environmental conditions and the impacts of the invaders on community composition, biodiversity and native environments.

12.2.5 Environmental Pollution

Environmental pollution is a major global concern and the consequences of environmental pollution pose serious threats to biodiversity. Biodiversity faces an extinction crisis from all forms of pollution, such as air pollution, water pollution, soil pollution, etc. However, air and water pollution have further-reaching negative effects on biodiversity than the other pollution forms. This section deals primarily with air and water pollution as soil degradation has already been discussed in earlier sections.

Ecosystems are impacted by different types of human-caused air pollution through the emission of greenhouse gases, volatile organic compounds, compounds of nitrogen, and sulfur, etc. primarily contributed by fossil fuel burning, industrial emissions and vehicular emissions. Greenhouse gases heat up the earth by trapping heat in the atmosphere resulting in a rise in sea level by faster melting of glaciers. Rising sea level affects the biodiversity in coastal areas. Likewise, organic compounds mainly in the form of halocarbons are the causative agents of the hole in the ozone layer. Depletion of ozone in the upper atmosphere lets ultraviolet radiation reach the earth's surface, which can harm biodiversity by impairing the DNA and proteins of land-based and aquatic organisms, causing skin cancer in human beings and damaging cell membranes of plants and inhibiting key processes required for their growth. Air pollutants, compounds of nitrogen and sulfur in particular, cause acidification of the environment. Sulfur dioxide and nitrogen oxides emitted into the environment recycle back in water, on vegetation and on soils as acid rain, thereby adversely affecting flora and fauna. Aquatic organisms in streams face difficulty in surviving in acidic conditions. Fish cannot hatch their eggs in acidic pH (≤ 5); fish, snails and insects even die in acidic conditions, leaving fewer food options for predators and ultimately disturbing the food chain and food web of that particular aquatic ecosystem. Acid rain also severely affects trees by damaging their leaves and making soil more toxic to plants. Conversely, the trees help in absorbing pollutants like nitrogen dioxide and particulate matter, thus help to improve air quality. The loss of trees means less filtering capacity to clean the air, ultimately causing harm to biodiversity. Therefore, acidification severely impairs biodiversity by disturbing related ecosystem services, like nutrient cycling, carbon cycling and water provision, on which the planet and biotic community is dependent.

Several attempts have been made to date to analyze the impacts of air pollution on biological diversity. Research for the World Wildlife Fund (WWF) assessing air pollution impacts on wildlife conducted through a literature survey identifying effects on 1300 species is the most detailed report to date (Tickle et al. 1995). The trends of the effects of air pollutants on biological diversity can be summarized as: (i) lower life forms like lichens, bryophytes, fungi, etc. are usually more affected than higher forms; (ii) plants are more affected than animals by air pollution in the terrestrial ecosystem, but in the aquatic ecosystem the reverse trend is observed; (iii) though the majority of species are adversely affected by air pollution and their numbers start to decline, there are a few species like aphids that benefit from air pollution and start to proliferate.

Water pollution with chemical, physical and biological contaminants originating from diversified sources like industrial effluents, domestic sewage, agricultural runoff, etc. is a major threat to aquatic biodiversity. Though the effects of water pollution on biodiversity differ based on the pollutant types, nutrient loading in particular is a major and increasing cause of biodiversity loss. Most water environments have very low range of nutrients and a minor variation in the nutrient burden can result in noticeable changes in function and structure of the biotic community of the ecosystem. Nitrogen is often the nutrient that regulates aquatic ecosystem productivity, whereas phosphorus is the limiting nutrient that determines growth. So unceasing inputs of phosphorus ($PO_4^{3-} > 0.5$–1 ppm) and/or nitrogen compounds ($NO_3^- > 10$ ppm) to aquatic ecosystems lead to enhanced eutrophication rates, the nutrient fortification of water environments triggering disproportionate growth of undesirable aquatic life like algae, rooted aquatic vegetation duckweeds and cyanobacteria (Rout et al. 2016a,b; Rout et al. 2017a,b). Other indications of eutrophication include algal bloom-mediated obstruction of sunlight, which deters the photosynthesis process of submerged aquatic vegetation and disrupts the reproductive ability of fish. The oxygen-consuming decomposition process of the algal biomass also decreases the dissolved oxygen content in water, due to which most of the oxygen-sensitive aquatic organisms either move to oxygen-rich places or they suffocate. Furthermore, the presence of nutrients in wastewater accelerates the production of an environmental toxin, microsystin, through the proliferation and metabolic activity of *Pfisteria*, and the toxin is primarily responsible for the hepatocellular carcinoma in human beings (Yuan et al. 2006). Also, excess nitrate nitrogen in drinking water may contribute to methemoglobinemia or "blue baby syndrome," inhibit iodine uptake and, thereby, harmfully affect the thyroid gland, and cause miscarriages in pregnant women and critical poisoning in livestock. Nitrite can react with amines enzymatically or chemically to form nitrosoamines and nitrosoamides, which are similarly carcinogenic in nature. Nitrogen in the form of ammonia is very toxic for fish, amphibians and invertebrates such as Unionidae mussels, and its concentration from 0.1 to 10 ppm can hamper the mortality, health and reproduction of the abovementioned aquatic fauna (Rout et al. 2018). The alarming decline in the population of Unionidae mussels has recently been related to the ammonia content of inland waters. Succinctly, the detrimental effects of nutrient overburdening are becoming more and more observable in recent years in the form of noticeable alterations in food web structure with likely extirpations of certain fish species and other organisms, changes in the natural composition of the aquatic biota, reduced biodiversity of aquatic species through depletion of desirable flora and fauna, impairment of human health, etc. Hernández et al. (2016) in their survey on nutrient-induced biodiversity imperilment considering 1400 endangered species, reported that 78 species face known hazards from water pollution caused by nutrient overloading.

12.3 Role of Biodiversity in Sustainable Development

Biodiversity is a vital asset for sustainable development. Through a constant supply of numerous ecological services that reinforce sustainable development in many significant ways, biodiversity plays an imperative role in the economic wellbeing of societies and maintaining a healthy human civilization. Biodiversity's contribution to sustainability is associated either with the actual genetic and species variability in a given place or with the ecological functional aspects. Species diversity provides the goods for livelihoods, sustenance, medicines and for other diverse sectors like agriculture, forestry, fisheries, pharmaceuticals, biotechnology, etc. Therefore, biodiversity is a central player in the global economy, directly supporting jobs and key economic activities in the abovementioned industrial sectors. Likewise, genetic diversity provides the basis for better crops, enriched agricultural production and improved food security through new breeding programs. Food production is dependent on numerous diverse animal breeds and crop varieties that are originated as a result of a rich genetic pool of species. For example, over the course of time around 7000 plant species have been used for food and edible parts of additional 70 000 plants are used as food by human beings (Dale 2009). Other important aspects of food production like pest control, soil fertility, pollination, etc. are also reliant on biodiversity. For instance, one-third of the world's food is derived from plants pollinated by wild pollinators like birds, bats, bees, butterflies, etc. Over 100 000 different species take part in the pollination process to ensure the continuation of reproduction in flowering plants (Dale 2009). In other words, most of our foods, and about one-fourth of drugs, result from wild organisms. More species diversity leads to greater varieties of genetic materials and chemical compounds, which subsequently results in a greater resource of information from which new medicines and crops with advantageous nutritional properties can be developed. So far as functional aspects of biodiversity are concerned, natural ecosystems provide a variety of services like draught and flood control, water and air purification, carbon sequestration, nutrient and hydrological cycling, soil preservation and fertility improvisation, climate stabilization, functional diversity maintenance, etc. For example, the global carbon cycle is controlled primarily by terrestrial and oceanic ecosystems, climate change mitigation is mediated via long-term carbon sequestration by a number of biomes, forest ecosystems protect water supplies and, together with wetlands and mangroves, forests play a vital role in reducing disaster risks like floods, droughts and tsunamis. In this context, it is essential to look forward and realize the importance of biodiversity for the attainment of sustainable development.

It is discussed in the previous paragraph that sustainable development, an integral component for survivability of human civilization, is indispensably dependent on the services of biodiversity that are delivered through the populations of existing species in living communities. Regardless of these facts, biodiversity is being lost at a faster pace, which has been assessed to be 100–1000 times faster than the prehistoric rate, fundamentally due to human activities (Rockström et al. 2009). Many of the world's biodiversity hotspots are located in the most densely populated developing countries where underprivileged communities depend on them for their survival and livelihood. Globally, approximately 2.6 billion people make their livings from agricultural activities, while over 3 billion people depend on marine biodiversity and more than 1.6 billion rely on forest ecosystems. In the case of developed countries, urbanization-mediated increased consumption of natural resources and demand-based land conversions are other major factors of biodiversity loss globally (Jeannine et al. 2013). In brief, biodiversity loss is basically attributed to disproportionate consumption in developed countries and the resource manipulation in developing countries essential to meet that consumption. Such biodiversity loss can reinforce the loss of ecosystem services and other nature-offered benefits to human society, which consequently can derail efforts for sustainable development. Therefore, taking a serious note of the human impact-mediated current rate of extinction, conservation of existing biodiversity, sustainable use of natural resources and decreasing threat to endangered species should be prioritized as necessary steps toward sustainable biodiversity management. In addition, inducing new "diversities" must be attempted not only to reverse the present-day extinction rates, but also to ensure maintenance of sustainable development (Lockie and Ransan-Cooper 2015).

12.4 Trends of Biodiversity

Conservation and sustainable management of biodiversity is one of the most intimidating global challenges of recent times. The Anthropocene, the period of human dominance, is ranked as the "Sixth Extinction" due to the alarming biodiversity extinction (100–1000 times more than the prehistoric rates) trends in the current epoch of humanity (Leakey and Lewin 1995). For instance, during the 1990s, global forest losses were at a rate of 16 million hectares per year. Though the deforestation rate has decreased over the past decade (2000–2010), still it continues at an alarmingly high rate of around 13 million hectares per year in many countries worldwide (Jeannine et al. 2013). Likewise, in coastal ecosystems, 40% of fish population have dropped below 10% of their maximum population in last 200 years, whereas in ocean ecosystems 80% of invertebrates and fishes have collapsed in the span of the past 50 years and more than 50% of the coral reefs have faced coral bleaching, obliteration, changes in species composition, etc. subsequently affecting the world's coral abundance (Worm et al. 2006). Summarily, terrestrial and marine species have declined by 39% each, whereas freshwater species extinction has been as high as 76% since 1970 (Lacher and Roach 2018). Taxonomic group-wise assessments have revealed that 58% of cycads, 34% of conifers, 32% of amphibians and 21% of mammals are facing extremely high levels of extinction risks (Butchart et al. 2007). Low reproduction rates, low population density and long life span are the characteristics of the species that tend to face the greatest extinction risk. Moreover, significant loss of species diversity is witnessed in the islands and tropical areas that accommodate large number of species per acre. For instance, all the great apes in Asia and Africa are highly endangered and are confined primarily to small areas. Furthermore, homogeneousness of domestic crops and animals at a global scale is resulting in declined genetic diversity. As a matter of fact, the lack of genetic diversity pushes species toward greater risk of disease and other population stresses, leading to diversity loss. It is clear from these facts that, despite national and international efforts, the world's biodiversity continues to be threatened. Based on the current status of endangered species, it is predicted that extinction rates will continue to rise up to 10 times their current rates in the near future (MEA 2005). Despite these alarming forecasts, many researchers believe that in the past, earth's biodiversity has recuperated following mass extinctions and, given time, it will do so again. This trend may also be changed in the succeeding several decades as human misappropriation of ecosystem services approaches maximum limits and with the declining growth rate of the human population. However, the preservation of biodiversity may be possible by equal prioritization of the conservation of biodiversity and human wellbeing. With this ethical aspect, the CBD has started unifying nearly all the countries worldwide in a commitment to reduce the loss of biodiversity in the coming decades.

12.5 Conclusions

Looking at the causal link among major stressors, ecosystem function and biodiversity, it is revealed that a large number of parameters such as climate change, soil erosion, flood and landslides, pollution, deforestation, natural and anthropogenic disturbances, etc. are involved in the degradation of ecosystems, which not only influence biodiversity but also the Sustainable Development Goals. Most of these parameters are associated with the exploitation of the available natural resources and, consequently, habitat degradation. In several countries, including China and India, policymakers and the government have taken many necessary steps to combat land degradation in the dryland ecosystems, which have improved ecosystem productivity and revenue. Thus, opportunities often exist to mitigate the worst effects of ecosystem degradation in several ways, ultimately contributing to sustainability. These mitigation measures sometimes include the conservation of structural and functional attributes of local or regional landscapes, which are disrupted by habitat loss and fragmentation due to human development. Ecosystem-based adaptation strategies may be cost-effective, generate socio-economic and cultural benefits, and contribute to the conservation of biodiversity and maintenance of sustainability.

Direct pressures on biodiversity imposed by stressors (e.g. population growth, climate change, unsustainable economic growth and unsustainable consumerism) require immediate attention. These pressures on natural ecosystems are not insignificant and may be aggravated in the coming years. Therefore, the impacts of these stressors (e.g. climate change) on the ecosystem, specifically on vulnerable species, should be taken into immediate consideration while planning for conservation of biodiversity in a long run. The knowledge gaps between the status of biodiversity and the associated stressors' impacts could be a major hurdle in implementing comprehensive conservation planning within the context of rapid environmental degradation and unsustainability. There is a pressing need to balance regulatory systems that protect biodiversity for sustainable development. Implementation of policies and laws is one of the most crucial steps to curb the problems associated with ecosystem degradation-mediated biodiversity loss (e.g. deforestation and habitat loss). The industrial fraternity can adopt sensible practice by not manufacturing the products that are associated with deforestation. Similarly, awareness programs should be conducted, aiming to encourage consumers not to use products extracted from unsustainable natural resources. Governments and regulatory agencies should ensure the effective enactment of environmental regulation through coordinated efforts. Individuals and organizations should be encouraged to participate in these efforts through charities and awareness programs to support environmental conservation. Providing incentives is by far the best way to facilitate cooperation and to enhance the compliance of environmental regulation. Simultaneously, these incentives should be granted selectively and violators should be faced with negative incentives (penalties). Therefore, it is concluded in this chapter that biodiversity and sustainable development are interdependent, and attaining and maintaining sustainability requires the participation of a complex mix of stakeholders in policy formulation, decision making and implementations for sustainable biodiversity management.

References

Abdul Rida, A.M. and Bouché, M.B. (1995). Earthworm contribution to ecotoxicological assessments. *Acta Zoologica Fennica* 196: 307–310.

Adhikari, K. and Hartemink, A.E. (2016). Linking soils to ecosystem services – a global review. *Geoderma* 262: 101–111.

Bajracharya, R.M. and Lal, R. (1992). Seasonal soil loss and erodibility variation on Miamiam silt loam soil. *Soil Science Society of America Journal* 56 (5): 1560–1565.

Baum, J.K., Myers, R.A., Kehler, D.G. et al. (2003). Collapse and conservation of shark populations in the Northwest Atlantic. *Science* 299: 389–392.

Blum, C., Bunke, D., Hungsberg, M. et al. (2017). The concept of sustainable chemistry: key drivers for the transition towards sustainable development. *Sustainable Chemistry and Pharmacy* 5: 94–104.

Bradshaw, C.J.A., Sodhi, N.S., and Brook, B.W. (2009). Tropical turmoil: a biodiversity tragedy in progress. *Frontiers in Ecology and the Environment* 7: 79–87.

Breure, A., Deyn, G.D., Dominati, E. et al. (2012). Ecosystem services: a useful concept for soil policy making! *Current Opinion in Environmental Sustainability* 4: 578–585.
Brito, D., Moreira, D.O., Coutinho, B.R., and Oprea, M. (2012). Ill nature: disease hotspots as threats to biodiversity. *Journal for Nature Conservation* 20: 72–75.
Brown, R.M. and Laband, D.N. (2006). Species imperilment and spatial patterns of development in the United States. *Conservation Biology* 20: 239–244.
Bryant, D., Burke, L., McManus, J., and Spaulding, M. (1998). *Reefs at Risk: A Map-Based Indicator of Threats to the World's Coral Reefs*. Washington, DC: World Resources Institute.
Burke, L., Kura, Y., Kassem, K. et al. (2001). *Pilot Analysis of Global Ecosystems: Coastal Ecosystems*. Washington, DC: World Resources Institute.
Butchart, S.H.M., Akçakaya, H.R., Chanson, J. et al. (2007). Improvements to the Red List Index. *PLoS One* 2 (1): e140. https://doi.org/10.1371/journal.pone.0000140.
Cheung, W.W.L., Lam, V.W.Y., Sarmiento, J.L. et al. (2009). Projecting global marine biodiversity impacts under climate change scenarios. *Fish and Fisheries* 10: 235–251.
Chu, C., Minns, C.K., and Mandrak, N.E. (2003). Comparative regional assessment of factors impacting freshwater fish biodiversity in Canada. *Canadian Journal of Fisheries and Aquatic Sciences* 60: 624–623.
Costanza, R., d'Arge, R., de Groot, R. et al. (1997). The value of the world's ecosystem services and natural capital. *Nature* 387: 253–260.
Dale, A. (2009). Biodiversity and sustainable development. In: *Area Studies (Regional Sustainable Development Review): Canada and USA*, vol. I, 254–279. Oxford, UK: EOLSS Publishers.
Dextrase, A.J. and Mandrak, N.E. (2006). Impacts of alien invasive species on freshwater fauna at risk in Canada. *Biological Invasions* 8: 13–24.
FAO (2015). Revised world soil charter. Rome: FAO. www.fao.org/fileadmin/user_upload/GSP/docs/ITPS_Pillars/annexVII_WSC.pdf (accessed August 22, 2018).
FAO (2007). *State of the World's Forests*. Italy: Food and Agriculture Organization of the United Nations.
Foin, T.C., Riley, S.P.D., Pawley, A.L. et al. (1998). Improving recovery planning for threatened and endangered species. *BioScience* 48: 177–184.
Gardi, C., Jeffery, S., and Saltelli, A. (2013). An estimate of potential threats levels to soil biodiversity in EU. *Global Change Biology* 19: 1538–1548.
Global Opportunity Network (2016). Global Opportunity Report 2016. www.globalopportunitynetwork.org/report-2016/loss-of-ocean-biodiversity (accessed August 16, 2018).
Godfray, H.C.J., Beddington, J.R., Crute, I.R. et al. (2010). Food security: the challenge of feeding 9 billion people. *Science* 327 (5967): 812–818.
Greentumble (2018). How does deforestation affect animals? https://greentumble.com/how-does-deforestation-affect-animals (accessed August 16, 2018).
Guisan, A. and Zimmermann, N.E. (2000). Predictive habitat distribution models in ecology. *Ecological Modelling* 135: 147–186.
Halada, L., Ruzickova, H., David, S., and Halabuk, A. (2008). Semi-natural grasslands under impact of changing land use during last 30 years: Trollio-Cirsietum community in the Liptov region. *Community Ecology* 9: 115–123.

Hernández, D.L., Vallano, D.M., Zavaleta, E.S. et al. (2016). Nitrogen pollution is linked to US listed species declines. *BioScience* 66 (3): 213–222.

Hourigan, T.F. (1999). Conserving ocean biodiversity: trends and challenges. In: *Trends and Future Challenges for U.S. National and Coastal Policy*, 45–50. Washington, DC: NOAA.

IIED (2002). Breaking new ground: mining, minerals and sustainable development. London: International Institute of Environment and Development (IIED) with support from the World Business Council on Sustainable Development. http://pubs.iied.org/pdfs/9084IIED.pdf (accessed November 2019).

IPCC (2007). *Climate Change 2007: The Physical Basis. Summary for Policy-Makers.* Contribution of Working Group I to the Fourth Assessment Report of the Intergovernmental Panel on Climate Change. Cambridge, UK: Cambridge University Press.

IUCN (2010). Red List of Threatened Species. https://www.iucnredlist.org (accessed August 16, 2018).

Jain, R., Sonawane, S., and Mandrekar, N. (2008). Marine organisms: potential source for drug discovery. *Current Science* 94 (3): 292.

Jeannine, C.B., James, H., Elizabeth, K. et al. (2013). Sustainability and biodiversity. In: *Encyclopedia of Biodiversity*, 2e, vol. 7 (ed. S. Levin), 71–84. New York: Academic Press.

Jones, C.G., Lawton, J.H., and Shachak, M. (1994). Organisms as ecosystem engineers. *Oikos* 69: 373–386.

Jones, D.T., Susilo, F.X., Bignell, D.E. et al. (2003). Termite assemblage collapse along a land-use intensifi cation gradient in lowland central Sumatra, Indonesia. *Journal of Applied Ecology* 40: 380–391.

Kerr, J.T. and Cihlar, J. (2004). Patterns and causes of species endangerment in Canada. *Ecological Applications* 14: 743–753.

Kerr, J.T. and Deguise, I. (2004). Habitat loss and the limits to endangered species recovery. *Ecology Letters* 2004: 1163–1169.

Lacher, T.E. Jr. and Roach, N.S. (2018). The status of biodiversity in the Anthropocene: trends, threats, and actions. In: *Encyclopedia of the Anthropocene*, vol. 3 (eds. D.A. DellaSala and M.I. Goldstein), 1–8. Amsterdam: Elsevier.

Laurance, B. (2019, February 14). Climate change could wipe out Bengal tigers in 50 years. *The Times of India*: 25.

Laurila-Pant, M., Lehikoinen, A., Uusitalo, L., and Venesjärvi, R. (2015). How to value biodiversity in environmental management? *Ecological Indicators* 55: 1–11.

Leakey, R. and Lewin, R. (1995). *The Sixth Extinction: Biodiversity and Its Survival.* New York: Anchor Books.

Li, B., Wu, S., Zhou, S. et al. (2017). Quantifying and mapping threats to soil biodiversity in Nanjing, China. *European Journal of Soil Biology* 82: 72–80.

Light, T. and Marchetti, M.P. (2007). Distinguishing between invasions and habitat changes as drivers of diversity loss among California's freshwater fishes. *Conservation Biology* 21: 434–446.

Lockie, S. and Ransan-Cooper, H. (2015). Biodiversity and Sustainable Development. In: *The Routledge International Handbook of Sustainable Development* (eds. M. Redclift and D. Springett), 1–15. London: Routledge.

Millennium Ecosystem Assessment (MEA) (2005). *Ecosystems and Human Well-Being: Biodiversity Synthesis.* Washington, DC: World Resources Institute.

Myers, R.A. and Worm, B. (2003). Rapid worldwide depletion of predatory fish communities. *Nature* 423: 280–283.

Myers, R.A., Baum, J.K., Shepherd, T.D. et al. (2007). Cascading effects of the loss of apex predatory sharks from a coastal ocean. *Science* 315: 1846–1850.

Phillips, O.L., Lewis, S.L., Baker, T.R. et al. (2008). The changing Amazon forest. *Philosophical Transactions of the Royal Society B* 363: 1819–1827.

Pimm, S.L., Russell, G.J., Gittleman, J.L., and Brooks, T.M. (1995). The future of biodiversity. *Science* 269: 347–350.

Piotr, S. (2013). Biodiversity and ecosystem services in soils under threats. *Journal of Pollution Effects and Control* 1: 1. http://dx.doi.org/10.4172/jpe.1000e101.

Planque, B. and Frédou, T. (1999). Temperature and recruitment of Atlantic cod (*Gadus morhua*). *Canadian Journal of Fisheries and Aquatic Sciences* 56: 2069–2077.

Raven, P.H. (1988). Our diminishing tropical forests. In: *Biodiversity* (eds. E.O. Wilson and F.M. Peter), 119–122. Washington, DC: National Academy Press.

Rockström, J., Steffen, W., Noone, K. et al. (2009). A safe operating space for humanity. *Nature* 461: 472–474.

Rout, P.R., Dash, R.R., and Bhunia, P. (2016a). Development of an integrated system for the treatment of rural domestic wastewater: emphasis on nutrient removal. *RSC Advances* 6 (54): 49236–49249.

Rout, P.R., Bhunia, P., and Dash, R.R. (2016b). Assessing possible applications of waste organic solid substances as carbon sources and biofilm substrates for elimination of nitrate toxicity from wastewater. *Journal of Hazardous, Toxic, and Radioactive Waste* 21 (3): 04016027.

Rout, P.R., Bhunia, P., and Dash, R.R. (2017a). Response surface optimization of phosphate removal from aqueous solution using a natural adsorbent. In: *Trends in Asian Water Environmental Science and Technology* (eds. F. Kurisu, A.L. Ramanathan, A.A. Kazami and M. Kumar), 93–104. Cham, Switzerland: Springer.

Rout, P.R., Bhunia, P., and Dash, R.R. (2017b). Evaluation of kinetic and statistical models for predicting breakthrough curves of phosphate removal using dolochar-packed columns. *Journal of Water Process Engineering* 17: 168–180.

Rout, P.R., Dash, R.R., Bhunia, P., and Rao, S. (2018). Role of *Bacillus cereus* GS-5 strain on simultaneous nitrogen and phosphorous removal from domestic wastewater in an inventive single unit multi-layer packed bed bioreactor. *Bioresource Technology* 262: 251–260.

Schneiders, A., Van Daele, T., Van Landuyt, W., and Van Reeth, W. (2012). Biodiversity and ecosystem services: complementary approaches for ecosystem management? *Ecological Indicators* 21: 123–133.

Sechrest, W., Brooks, T.M., da Fonseca, G.A.B. et al. (2002). Hotspots and the conservation of evolutionary history. *Proceedings of the National Academy of Sciences of the United States of America* 99: 2067–2071.

Shah, A. (2014, January 19). Climate change affects biodiversity. *Global Issues*. www.globalissues.org/article/172/climate-change-affects-biodiversity (accessed February 12, 2019).

Sheil, D. and Salim, A. (2004). Forest trees, elephants, stem scars and persistence. *Biotropica* 36: 505–521.

Smith, K.F., Sax, D.F., and Lafferty, K.D. (2006). Evidence for the role of infectious disease in species extinction and endangerment. *Conservation Biology* 20: 1349–1357.

Smith, P., Cotrufo, M.F., Rumpel, C. et al. (2015). Biogeochemical cycles and biodiversity as key drivers of ecosystem services provided by soils. *Soil* 2 (1): 537–586.

Szabolcs, I. (1994). *The Concept of Soil Resilience, Soil Resilience and Sustainable Land Use*. Wallingford, UK: CAB International.

The Ocean Atlas (2017). *Facts and Figures on the Threats to Our Marine Ecosystems*. Germany: Heinrich Böll Foundation Schleswig-Holstein, the Heinrich Böll Foundation (National Foundation), and the University of Kiel's Future Ocean Cluster of Excellence.

Theobald, D.M. (2014). Development and applications of a comprehensive land use classification and map for the US. *PLoS One* 9: e94628.

Tickle, A., Malcolm, F., and Graham, D. (1995). *Acid Rain and Nature Conservation in Europe: A Preliminary Study of Protected Areas at Risk from Acidification*. Gland, Switzerland: WWF International.

UNDDD (2010). The United Nations Decade for Deserts and the Fight Against Desertification (2010–2020), www.un.org/en/events/desertification_decade/background.shtml (accessed August 16, 2018).

UNEP (1992). The United Nations Convention on Biological Diversity. Article 2. Nairobi: UNEP.

UNEP (2006). *Millennium Ecosystem Assessment, Ecosystems and Human Well-Being: Biodiversity Synthesis*. Washington, DC: World Resources Institute.

UNEP/CBD (1995). The Jakarta Mandate on Marine and Coastal Biological Diversity: Decisions of the Second Meeting of the Conference of the Parties to the Convention on Biological Diversity. Jakarta, Indonesia.

UNFCCC (2018). Forests – climate change, biodiversity and land degradation. http://unfccc.int/resource/docs/publications/forest_eng.pdf (accessed August 24, 2018).

Veron, J.E.N., Hoegh-Guldberg, O., Lenton, T.M. et al. (2009). The coral reef crisis: the critical importance of <350 ppm CO_2. *Marine Pollution Bulletin* 58 (10): 1428–1436.

Warkentin, I.G., Bickford, D., Sodhi, N.S., and Bradshaw, C.J.A. (2009). Eating frogs to extinction. *Conservation Biology* 23: 1056–1059.

Wilcove, D.S., Rothstein, D., and Dubow, J. (1998). Quantifying threats to imperiled species in the United States. *BioScience* 48: 607–615.

Worm, B., Barbier, E.B., Beaumont, N. et al. (2006). Impacts of biodiversity loss on ocean ecosystem services. *Science* 314: 787–790.

Yiming, L. and Wilcove, D.S. (2005). Threats to vertebrate species in China and the United States. *BioScience* 55 (2): 147–153.

Young, A.M. and Larson, B.M.H. (2011). Clarifying debates in invasion biology: a survey of invasion biologists. *Environmental Research* 111: 893–898.

Young, I. and Crawford, J. (2004). Interactions and self-organisation in the soil-microbe complex. *Science* 304: 1634–1637.

Yuan, M., Carmichael, W.W., and Hilborn, E.D. (2006). Microcystin analysis in human sera and liver from human fatalities in Caruaru, Brazil 1996. *Toxicon* 48 (6): 627–640.

World Commission on Environment and Development (WCED) (1987). *Our Common Future*. New York: Oxford University Press.

13

Sustainability of Ecosystem Services (ESs)

Carlos S. Osorio-González[1], Niranjan Suralikerimath[1], Krishnamoorthy Hegde[1], and Satinder K. Brar[1,2]

[1] *INRS, Université du Québec, Québec, Canada*
[2] *Department of Civil Engineering, Lassonde School of Engineering, York University, Toronto, Canada*

13.1 Introduction

Ecosystem services (ESs) provide a range of benefits to human societies that grow from natural resources, and ESs explain the human dependency on and trust in nature (Daily 1997; Guerry et al. 2015). The main categories that make up ESs are shown in Figure 13.1. Meanwhile, human societies have been exploiting the use of ESs such as water, food, fodder, medicines, timber, non-timber forest products and a range of indispensable goods derived from nature for basic needs, with the unique purpose of improving human wellbeing (HWB). Nevertheless, it must be borne in mind that the foregoing can be confused with other processes that are immersed within the ecosystems themselves (biomass productivity, nutrient cycles, energy flow and hydrological regulations). These components are important but do not fulfill the function of providing or satisfying any basic need for the survival of any part of the ecosystem. The relationship between human societies and nature has been studied from end to end through the lens of a subsistence economy. Nevertheless, there is a need for critical analysis and a change in thinking patterns to mitigate growing challenges like food security and the need for energy. Collective efforts from policymakers, academicians, practitioners and development planners have been made to accept and renew their approach to sustainable solutions for both the survival of future generations and natural resources (Alcamo and Bennett 2003).

The origin and popularity of the ESs concept are the collective results of modern-day thinking, which provides the definitive framework for the decisions we make. ESs provide the range of services that enable us to address the growing number of challenges that societies are facing today (Boyd and Banzhaf 2007). The ESs concept potentially brings prime issues together, and ESs enable values to be more visible on the international platform for development of global initiatives with the purpose of strengthening the studies and knowledge of ESs. ESs also help in individual preferences and are an ideal concept to improve conservation, while taking development into consideration. Some of the relevant report details of the Millennium Ecosystem

Sustainability: Fundamentals and Applications, First Edition.
Edited by Rao Y. Surampalli, Tian C. Zhang, Manish Kumar Goyal, Satinder K. Brar, and R. D. Tyagi

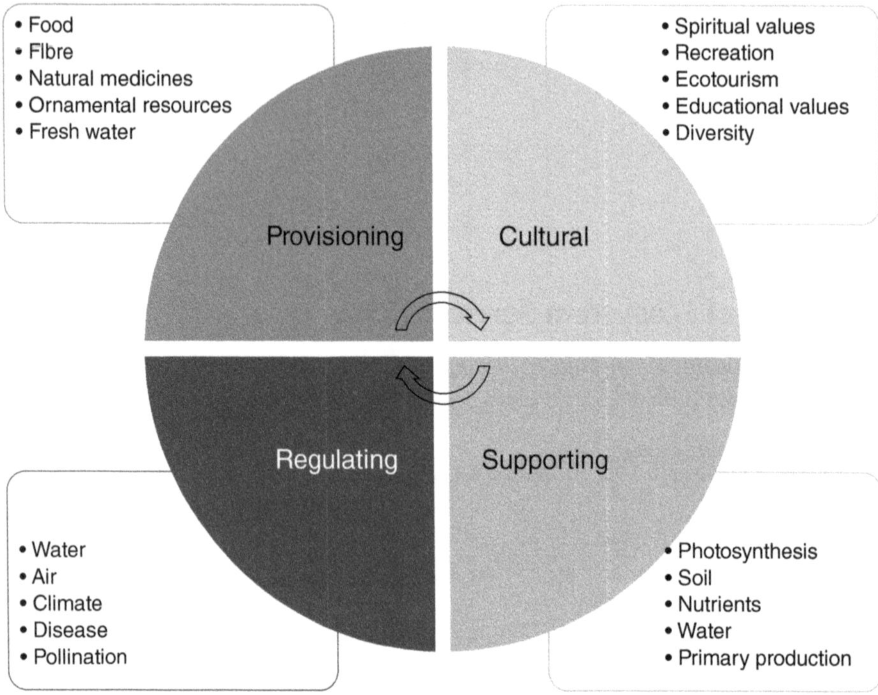

Figure 13.1 Main categories of ecosystem services.

Assessment (MA) in 2005, and The Economics of Ecosystems and Biodiversity (TEEB) in 2010 are explained in various sections of this chapter. ESs provide different types of functions throughout the world, contributing billions of dollars per year. For instance, Costanza et al. (1997) carried out elaborative analyses and listed 14 ESs from forest ecosystems that were estimated to be contributing at least US$33 trillion per year. From these analyses, it is evident that ESs play a major role in socio-economic development. Nevertheless, many articles and literature criticize the practicality and use of the ESs concept (Norgaard 2010; Carpenter et al. 2009; Bennett et al. 2015) despite visible advancement in this field compared with previous decades.

This chapter offers an overview of the history, classification and sustainability of ESs. Likewise, it provides a perspective on different tools and methods used for their evaluation, as well as an analysis of their potential of their contribution to sustainable development in general. Finally, it provides a perspective of different challenges that ESs face and the importance of addressing them to improve the quality of the services that ESs offer to the general population.

13.2 Historical Evolution of Ecosystem Services Definition

The very first use of the term ESs was in the 1960s. In general, ESs are the assistance or benefits that a human society gets/obtains from the ecosystem (King 1966; Helliwell 1969). Daily et al. (2000) defined ESs as: "The conditions and processes through which ecosystems and their component contribute to life-sustaining, without damage to biodiversity, and the production of goods such as food, biomass, wood, fuels, and different products of industrial interest (chemical commodities, pharmaceutical compounds, or precursors for both of them)" (Daily 1997). Costanza et al. (1997) explained that ecosystem goods and services benefit the human population, directly or indirectly from ecosystem functions (Costanza et al. 1997). ESs can be considered in different ways, such as:

1) *Organizational groupings*: Services related to certain species and regulating exogenous responses related to biotic entities come under organizational groupings (Norberg 1999).
2) *Functional grouping*: Regulation of carrier, habitat, production and information services (De Groot et al. 2002; Lobo 2001).
3) *Descriptive groupings*: Renewable and non-renewable resources, biogeochemical services, physical structure services, biotic services, social and cultural services (Moberg and Folke 1999).

The dependence of human society on nature for prosperity and basic survival has been well studied. Initially, the history of civilization defined that people seek protection from the elements and access resources. For example, managing natural services like husbandry and agriculture and increasing productivity was how we used ESs. Welfare and ecology come under the concept of ESs. For the past few decades, researchers have tried to link these two components together. Westman (1977) said that "society should make more self-evaluated and informed policies at the individual and organizational level for the benefit we receive from the ES," and, furthermore, he explained that it is the enumeration that we can offer in this time. In this context, the concept was called Westman's ESs, a term that was used for the first time by Ehrlich and Ehrlich (1981).

In 2007, Boyd and Banzhaf gave an alternative definition for the above-mentioned statement. With respect to their definition, rather than describing ESs' benefits to humans, ESs are the ecological components enjoyed or consumed for human wellbeing. The defined services of Boyd and Banzhaf have directly consumed components. Therefore, functions and indirect processes are not ESs. More importantly, Boyd and Banzhaf elucidated that benefit and service are not the same. ESs are actually the benefit of multiple entries being used as recreation instruments generating an economic gain for the community or the institution entrusted to administer it. Figure 13.2 depicts and suggests the various issues and their connections in ESs by clearly defining how ESs are important so that we know how synergies can be made across projects. To get a meaningful classification of ESs requires a clear definition of ESs and understanding the behaviors of these ecosystems.

Figure 13.2 shows four significant components for a better understanding of ecosystem services, where the definition of an ES is that set of processes (direct or indirect) which contributes greatly to human wellbeing. Although this is a very general definition, it has been evolving over the past 30 years, changing from being a heuristic model to becoming a powerful tool in decision making and politics. Nowadays, the concept of ESs can be described in two main categories: first, in a descriptive concept that focuses on the interdependence of nature and human beings, and second in a normative sense, used for the attribution of value judgments of the above-mentioned interaction (Abson et al. 2014). Another key factor is a better understanding of the characteristics that make up the ESs, which, once understood, can be managed, implemented, evaluated, maintained and corrected if

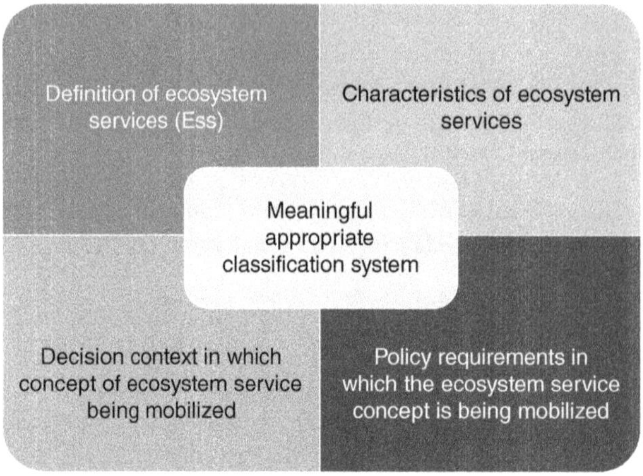

Figure 13.2 Classification of ecosystem services.

necessary. This will allow better understanding and management of ESs and their interaction with society for better decision making. Some of the main features are spatial and temporal dynamism, product and service obtention, a resilient dependence on the benefits obtained, the cost–benefit analysis, among others (Fisher et al. 2009). Regarding policy requirements, the perspective of the ESs is very extensive and depends on the context in which it is used. Nevertheless, current policy is interdisciplinary collaboration in three main contexts that allow it to interrelate with multiple actors. These contexts are: interpretive flexibility of different situations inside and outside the system, compilation of information from different processes that the system contemplates, and the limiting capacity that the system has in terms of its application and development (Steger et al. 2018). Finally, the context in decision making is a big challenge, since it is determined mainly by the criteria of the person who applies it and sometimes these differ from those of the person who receives it. Likewise, in ESs, there are other factors that largely influence this parameter since they are determined by the impacts that humans, nature, climate and other factors have on the system (Daily et al. 2009).

Research related to the different relationships (synergistic, negative, positive and neutral), as well as their interactions at different levels of the ESs, is crucial in decision making for environmental, economic and sustainability policies (Wu and Li 2019). Some of these studies have focused on improving the understanding of relationships, such as provision, service, compensation, management, environmental and social impact, as well as the interaction between them (Turkelboom et al. 2018).

13.3 Framework for Assessing Ecosystem Services

In the past, researchers recognized the need for global valuation of ESs, mainly to establish a body that is something like the Inter-governmental Panel on Climate Change but is exclusively for the ESs. Since the 1990s, advances in knowledge about ESs have increased in relation to their application in a general manner in various fields. This has resulted in the evolution of the concept and its application to finally give way to programs such as the MA (UNEP 2018). The MA was the efforts of 1400 researchers over 5 years from around the world that evaluated the ESs.

The MA established an ES framework to channel the relationship between biodiversity and ESs (Alcamo and Bennett 2003). Briefly, the ESs framework is associated with the MA regarding the interaction between human beings and changes that they generate to the ecosystems to obtain and maintain their wellbeing. On the other hand, various factors that are not related to the environment have effects on the human condition itself, and on well-being, for which MA also focuses on the links that are formed between ecosystems, and the wellbeing that is generated. In conclusion, it is a multiscale approach that reflects the interactions of ecosystems and the impact they have on people. In this sense, the general framework of MAs reminds us of the main concern, the objectives we are trying to achieve and the impact of ESs by addressing the main phenomena (ecological and social) happening at different levels and at different times. Further looking into the literature, it becomes clear that harvest and resource consumption can be greatly influenced directly by landowners or managers and, similarly, technological adaptation (Wilson et al. 1999; Folke et al. 2005). Likewise, local organizations should be used for better results of individual ES projects.

On the other hand, different researchers explain other types of system using different models (Norgaard and Baer 2005a,b) that help to understand and evaluate ESs. Nevertheless, some of these models do not fit the MA framework underlying the ESs concept, due to the fact that these systems do not have a pattern that adheres to MA or ESs guidelines. This directly affects the application and performance of the different processes involved in ESs to generate welfare for society.

13.4 Utilization of Ecosystem Services for Sustainable Development

There are many conflicts among environmentally Sustainable Development Goals. Generally, in most cases, developmental goals are driven based on economic growth and in the direction of controlling poverty, which clearly puts environmental sustainability on the backseat. Nevertheless, millions of dollars have been spent on recent studies from global institutions with the emergence of the ESs concept and valuation of these services. There has been a paradigm shift toward knowledge-based evidence, suggesting goals can only be achieved by integrating the goals of environmental sustainability. It is also a very easy and economical way to maintain the ESs rather than to increase the expansion of different alternatives as they are often less effective. So far, the ESs have formed the core-value-based building blocks for poor economic systems, which simply tends to collect stocks of facilities and services (including capital investments). Thus, ESs have been recognized as a cost-effective way to meet and support development issues and processes dedicated to people with less ability to acquire wealth (Emerton 2008).

In the framework of sustainable development, Fisher et al. (2014) explained how we can use the range of ESs wisely not only to uplift poverty but also provide opportunities for each zone's output with improved business performance. For example, they explain the importance of ecological factors from the points of view of energy and mass expenditure, as well as carbon and nutrient expenditure, within and through ecosystems and how this analysis influences processes, such as support, regulation and the different transformation cycles of ESs. On the other hand, they explain the difference between biological and geochemical processes and social or cultural processes, which are independent, but at the same time are intrinsically linked to the global system and are perceived by the people who are in contact with ESs. Likewise, they show diverse components to contribute to the reduction of poverty and increase the wellbeing of people, which, however, is generic and depends on the interests of each individual. On the other hand, the range of ESs and rich biodiversity are of great importance for the support of the most vulnerable. Several studies have shown that degradation impact of biodiversity directly affects poor people as they have far fewer options to explore. Therefore, it is evident that it is essential to maintain sustainable biodiversity and ESs, otherwise it will have a targeted effect on poverty alleviation. Studies suggest that sectoral outputs depend directly or indirectly on the ESs, by reducing expenditure. In general, these relationships are more representative in natural resource exploitation (agriculture, forestry and fisheries), and can be similar to other service-based industries such as the health sector, energy and urban development, as well as approaches to water and its treatment. For example, in the case of Mongolia's natural rainwater harvesting system (called catchment ecosystem conservation), they spend around US$1 million annually and eventually the system generates a minimum of US$15 million per year for downstream communities (Emerton et al. 2009). In another case in Vientiane, Laos, the use of wetlands can offer alternative services in the treatment of wastewater to different urban areas, which saves public funds and energy for the treatment of wastewater to the tune of around US$2 million per year (Gerrard 2004). As another example, Jaipur city in India faces a serious problem in replenishing groundwater during monsoon; the state pollution control board has developed an enlarged urban green space that is very cost-effective to reduce surface water runoff (Rodell et al. 2009; Singh et al. 2010).

When it comes to the influence of ecosystems on business performance, ESs continuously produce a growing number of opportunities to expand, build and create strength in business without provoking damage to biodiversity, which can be a real risk during business operations development. In 2011, the World Business Council of Sustainable Development (WBCSD) elaborated guidelines for collective evaluation of different ecosystems. More than 50% of the corporations considered that ESs must be considered within the top 10 priorities, while the remaining percentage believed that the ESs are only an opportunity instead of a risk for their corporate agenda. The current world ES offset is worth around US$3 billion and is expected to grow rapidly. Similarly, along parallel lines, the global carbon trade grew by US$140 billion from 2004 to 2009 (WBCSD 2011). Natural resources business opportunities will be in the order of US$2–6 trillion by 2050 (Wallace 2007). Meanwhile, the forestry ecosystem in poor counties represents more than 10% of their GDP but helps in providing employment for around 50 million people. On the other hand, depending on the type of ESs, some of the sustainable frameworks can be focused on three main relationships: (i) ESs, (ii) sustainability strategies and (iii) sustainability results. Within these three "features" aspects of appropriation of services within the ecosystem are considered as well as the impact they have within the framework of sustainability, particular strategies that are adopted for this appropriation inside and outside of ecosystems, and regulations that make up the regulatory strategies to achieve a sustainable environment within the ecosystem and the services provided in it (Schröter et al. 2017).

13.5 Recent Advances in Mapping and Measuring Multiple Services

In the past few decades, research has been focused on the framework as well as services of biodiversity. However, to make a decision requires advanced measurement tools in order to ensure HWB and both direct and indirect impacts of ESs on HWB (Fisher et al. 2009; TEEB 2010; Lele et al. 2013). The success of dependable measurement of ESs is based on how we approach systematically the interrelationships among multiple layers of ESs and how we estimate them with respect to time. In this context, Wolff et al. (2015) proposed five main models for mapping and measuring ESs. These models are shown in Figure 13.3, together with their main characteristics.

Even early literature tried to explain the multilayered relationships among biodiversity services and their connections with society by analyzing HWB. However, if one analyzes ESs with comprehensive approaches such as risks, demand, consumption or use, one would find there is an inability of the literature to explain the link between ESs and HWB. In other words, lack of consideration of all the aspects would eventually affect the results of the estimation directly or indirectly (Chan et al. 2006; Naidoo et al. 2008; Nelson et al. 2009). Many approaches can only explain the single-layered relationship to measure biodiversity services; for example, spatial correlations, management comparisons and functional experiments (Ricketts et al. 2016). Distinctly, decision makers as individuals or at the organizational level require information about multiple services and their impact on human society in order to make decisions efficiently to mitigate the declines in production of other services (Bennett et al. 2009; De Groot et al. 2012). In recent years, very few articles have explained multiple services linking with HWB. Bateman et al. (2013) explained a UK ecosystem assessment. This evaluation has contributed to enriching the conceived basic

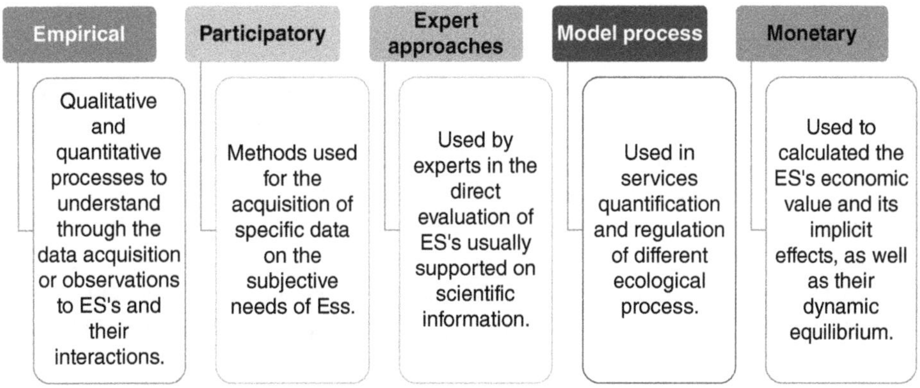

Figure 13.3 Five models for mapping and measuring ecosystem services.

idea, in which ecological models are linked with the ESs due to their economic value as well as their contribution to the internal economy. Scientists have keenly pointed out the critical points of different services and analyzed the different aspects that made them successful (Chan et al. 2006; Naidoo et al. 2008; Nelson et al. 2009). Qiu and Turner (2013) found that most linkages among ESs were synergies. At least six service hotspots were identified based on where they occurred (geographical location); high levels were rarely found, and the six hotspots corresponded to around 3.3% of the landscape. This study provided specific data about the ESs and services in a particular space. The study allowed the people in charge to make decisions in each of the different regions and to acquire general knowledge.

As a final point, mapping and evaluating ESs is indispensable and will continue in search of tools that critically understand and deliver information associated with case-specific issues. These tools can be used to analyze past trends and future projections that will eventually affect policy and decision makers. However, decision and policymaking authorities often require more information than the predicted values, and these people want to increase the sustainable use of all biodiversity. To meet our needs and maintain services in important locations, there is a demand for investment in ecological and landscape restoration and explanations about the repercussions so that different types of development can affect all the stakeholders involved. The mitigating steps toward sustainable use of ESs are detailed below (Mitchell et al. 2015; Oteros-Rozas et al. 2015).

13.6 Possible Approaches for Sustainable Use of Ecosystem Services

As we discussed in the above sections, we must be acquainted with the relationships among ESs, HWB and developmental goals, simply because ESs have tremendous potential in contributing to the development goals. Despite the fact that there are a number of bottlenecks in the process to finding solutions, we are striving to incorporate ESs into the reality of sustainable development and action-oriented planning. A few approaches to sustainable use of ESs are explained below.

13.6.1 Tools and Frameworks for Decision Support

At present, many frameworks and tools are available to support the decision-making process in sustainable management. Some of these assessment frameworks or models for decision making are shown in Table 13.1. Multiple-approach methods and integration of diverse methods in developing diverse frameworks (Wegner and Pascual 2011) provide the path for future research opportunities. For example, regarding the balance sheets approach, Turner et al. (2015) advocated that different results obtained using different methods can be implemented in order to carry out ES assessment, which demonstrates the importance of supporting the development of the ES, including

Table 13.1 Tools for decision support in ecosystem services.

Tool	Description	Who is involved	References
Ecosystem Valuation Toolkit®	Values ecosystem services in order to provide sustainable decisions to people and companies	All the components of ecosystem services	Earth Economics (2019)
EcoMetrix®	A tool used for valuation of services within the ecosystem services	All the components of ecosystem services	Ecometrix (2019)
Multiscale integrated models of ecosystem services (MIMES)	It is a multiscale dynamic framework or model for the evaluation of ecosystem services	Specific components of the ecosystem to be evaluated, and people who interact directly within the ecosystem	Boumans et al. (2015)
Ecosystem portfolio model (EPM)	Used for modeling of quality life, use of natural resources and their economic impact on society	Government and federal agencies	Labiosa et al. (2013)
Ecosystem services review (ESR)	Production units, marketing, suppliers and customer service	Managers, analysts, consultants, workforce	World Resources Institute (WRI) (2012)
Social values for ecosystem services (SolVES)	Application about geographical information services that allow the evaluation, mapping, as well as the quantification of social values of ecosystem services	Government and non-governmental organizations, the private sector and scientists	Sherrouse et al. (2011)

feasibility studies (cost–benefit) that allow the use of numerous evaluation measures such as effectiveness, social effects, equity and ecological sustainability.

13.6.2 Methods for Data Collection

Biodiversity is a dynamic system in which raising concerns regarding environmental change at the global level necessitates ecological research across whole communities and landscapes (Mace 2013). Experimental studies may helpful for understanding the relationships between biodiversity and global change as they enable the manipulation of environmental conditions and producer diversity. Past responses to global and regional changes can be studied with *ex post facto* analysis, such as mathematical analysis and validation of a phenomenon after it occurred, which may not have been impacted directly by human society (Beier et al. 2004; Tobner et al. 2014). Dendrochronological techniques are a scientific method of dating tree rings or growth rings to obtain weather data and atmospheric conditions during different periods in the history of the wood, and may be very advantageous in extreme environmental conditions such as in polar areas and tropical countries (Myers-Smith et al. 2015; Wils et al. 2011). It can be very useful to study ecological responses at microscale levels with respect to different environmental changes. However, this needs global level collaborative efforts in terms of funding and creation of quality databases. Development opportunities in remote sensing will also support the collection of high-quality data considering different scales (spatial and temporal); this will help in extrapolation and data comparison between scales.

On the other hand, exploiting the application of mobile and smartphone technology to develop ESs apps will allow researchers to find and understand the wide-scale audiences in sensitive areas. For instance, leaf-watch is the mobile application used in the UK for documenting the leaf-mining moth across the country (Pocock and Evans 2014). Recently, South Africans developed a mini Stream Assessment Scoring System (mini SASS) for the local river catchment quality. Mini SASS allows the uploading of interactive data results obtained with Google Earth that provide essential information about maps and databases and contribute to a better distribution of the drinking water supply (Water Research Commission 2015).

13.6.3 Statistical Approaches

Statistics is an indispensable tool, providing convincing links for researchers to understand the environmental problems that are linked to societal consequences. Marcot et al. (2006) explained the usefulness of Bayesian Belief Networks in creating a model that intersects different disciplines. It also helps us to link many other parameters like ecological factors along with causal relationships among the habitat and environmental variables and to perform the analysis of non-linear models in ESs using generalized additive models (GAMs) (Hossain et al. 2017). The vector autoregressive, structural equation models and different econometric methods (e.g. the Granger causality test) can be used to analyze HWB and ecosystems and to gain insights into the dynamic interrelationships between ESs and HWB (Santos-Martin et al. 2013; Chen et al. 2013).

13.6.4 Linking Science to Policy

The relationships between science and policy include diverse relationships among decision makers and scientists, as well as communicating knowledge to the public (Ishii 2014). Establishing these relationships requires working proactively with policymakers and researchers to integrate formal or informal exchange of knowledge, collaboratively with decision-making bodies to determine the content of research programs (Mitlin 2008). The urban planning of Birmingham in alliance with Birmingham City Council and Birmingham City University is one such example. In order to understand the communication between various levels of the system, they included different working groups on science and society issues while conducting qualitative and quantitative studies. Figurative language and the terminology used usually play a crucial role in the exchange of knowledge, taking into account that preferences often vary according to the public type. This allows the establishment of an exchange of knowledge with a certain sensitivity owing to the different audiences involved (Raymond et al. 2013; Mitton et al. 2007). With a detailed analysis of traditional decision making and public consultations, it is possible to evaluate different communication methods used for ESs research and biodiversity in general.

13.6.5 Enhancing Interdisciplinary Research

In order to support the new framework of approaches, the quality of data is very important for the researchers to correlate ESs and HWB. In a dynamic world, to strengthen interdisciplinary links and to secure healthy biodiversity and sustainable human societies, training for social-ecologists is essential, with the aim of providing support for long-term monitoring and research projects.

13.7 Challenge and Interlinkage Themes for Researchers

Broadly, challenges in interdisciplinary research for ESs can be addressed under four headings as follows:

1) Understanding the fact that this field of research inherently demands a comprehensive approach for researchers and continuous updating.
2) A global network must be created in order to facilitate researchers from different areas to work on a common platform without any gaps in understanding (Hirsch-Hadorn et al. 2006).
3) Better integration of various components of ESs with the help of appropriate methods and the adoption of a scalable model for research projects.
4) Making research data or reports available on a common platform for multiple scientific communities, ensuring that they can be questioned or challenged for many reasons: fundamental, factual and/or application-based.

Other than the challenges mentioned above, there may be many other bottlenecks to overcome to achieve better interlinkages between themes and researchers. An additional key bottleneck is in collaboration: not only collaboration with researchers from different fields but also with the local communities and policy- and decision makers. Most importantly, as mentioned above, we should work for real-world situations. Figure 13.4 and Table 13.2 explain the connection between the different themes for establishing sustainable management, the future directions of ESs and their importance, respectively (Raworth 2012).

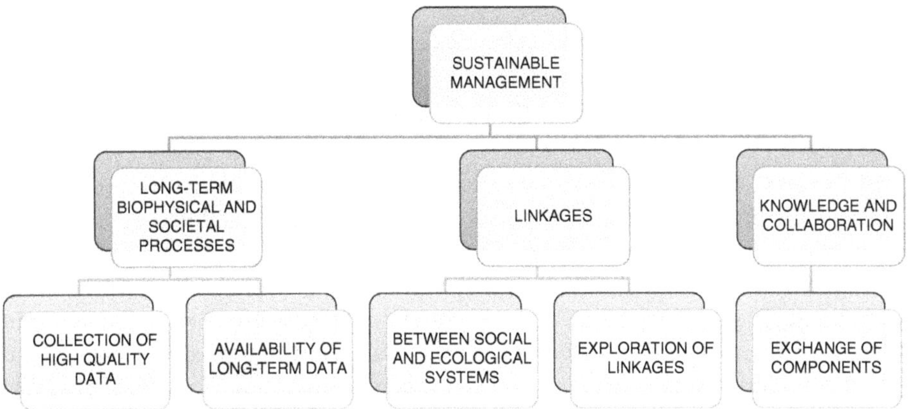

Figure 13.4 Connection among different themes for establishing sustainable management.

Table 13.2 Future directions of ecosystem services and their importance.

Future direction	Key problems for decision maker	Significance	Possible solutions	References
Role of multiple forms of capital in the provision of ecosystem services and human wellbeing	When can various forms of human capital substitute for natural capital in the provision of ecosystem services?	It is difficult to know how important natural capital is	Build models that estimate the role of multiple forms of ecosystem services	Rieb et al. (2017)
Focus on ecosystem services provision over longer time frames	Plan for the decisions made to play out in terms of ecosystem services and human wellbeing	Sustainability is about the long term; snapshot events of ecosystem services	Integration of models of climate and other drivers of long-term change	Dearing et al. (2012) and Tomscha et al. (2016)
Measure and monitor ecosystem services governance and outcomes	To be tracked over time so we can continue to learn together and adapt our future questions and research	We must proceed with what we do know	Funding for monitoring of both ecological and social outcomes as an integral part	Cundill et al. (2015)
Plan for sustainable use of ecosystem services	How will climate change affect the ecosystem service outcomes of land use decisions?	Climate, biogeochemical cycles, a prevalence of plastics – we know the world is changing	Uncertainty analysis of different scenarios of decision making	Liu et al. (2016)
Role of teleconnections and trade in the provision of ecosystem services and human wellbeing	When can ecosystem services be purchased from elsewhere and when must they be locally produced	Trade is an increasingly important part of ecosystem function in a globalized world	Expand system boundaries of ecosystem services studies to include traded goods	Meyfroidt et al. (2013) and Liu et al. (2013)

13.8 Future Directions and Conclusions

For ESs science to expand beyond the boundaries advanced research development in multiple directions is required. Much of the recent literature and many conferences or workshops on ESs suggest including the ongoing effect on global change, in linearity and long time frames. ESs increase the opportunities to include multiple areas of knowledge about capital, to benefit social systems and also to support involvements in designing and implementing different research plans. In the next step, expansion of physical boundaries to include various types of services around the globe is critical to understanding the planetary ESs and their consequences for sustainability and

Sustainable Development Goals. Ultimately ES science should develop in such a way to respond to all the significant inquiries raised by decision makers.

The available scientific literature and evidence make generalized statements about ESs, which mainly express that there is a great need for ESs on a large scale of operation, and none of the human intervention technologies will be able to provide a substitute for ESs. The very nature of human civilization is damaging the flow of ESs directly and indirectly. Fortunately, constant development of understanding ESs and their significance in human societies has taken place over the decades. There is a huge gap between understanding and implementing those understandings in reality for integrated development and planning at any level. Evaluation of ESs is one of the most important and primary challenges despite its popularity. Monitoring and implementing sustainable services at the local to global level or in developing or developed countries and incorporating them into the decision-making process are all in need of further understanding. The outcomes of the MA in 2005 are very important milestones that clearly elucidate the targets and challenges still to be achieved in all the themes involved of ecological, cultural and economic ESs. The general perception is that, as time passes, there are considerable developments to mitigate the shrinking natural resources on the one hand and the ever-growing human population on the other. With the common understanding of all key players (decision makers, academicians, human society) in the business of ESs, improved development processes will be realized from grassroots up to the global level. A specific important aspect is to improve the integration of ESs into spatial and sectoral planning. Working with all the key players for a common agenda will definitely catalyze more integrated and user-inspired development processes. Finally, the success of any process depends on (i) trans-disciplinary research to improve understanding of complex socio-ecological challenges; (ii) simplifying research outcomes to raise social awareness and concerns from the individual level to all the hierarchy levels about the significance of ESs; and (iii) facilitating an informed multi-stakeholder process into development planning.

Acknowledgments

Financial support from the Natural Sciences and Engineering Research Council of Canada (No. 284111, Discovery; No. 476649-14, Strategic Research Grant) and Institut National de la Recherche Scientifique-Eau Terre Environnement have been acknowledged.

References

Abson, D.J., von Wehrden, H., Baumgärtner, S. et al. (2014). Ecosystem services as a boundary object for sustainability. *Ecological Economics* 103: 29–37.

Alcamo, J., Bennett, E.M., and Millennium Ecosystem Assessment (Program) (eds.) (2003). *Ecosystems and Human Well-Being: A Framework for Assessment*. Washington, DC: Island Press.

Bateman, I.J., Harwood, A.R., Mace, G.M. et al. (2013). Bringing ecosystem services into economic decision-making: land use in the United Kingdom. *Science* 341: 45–50.

Beier, C., Emmett, B., Gundersen, P. et al. (2004). Novel approaches to study climate change effects on terrestrial ecosystems in the field: drought and passive nighttime warming. *Ecosystems* 7: 583–597.

Bennett, E.M., Peterson, G.D., and Gordon, L.J. (2009). Understanding relationships among multiple ecosystem services: relationships among multiple ecosystem services. *Ecology Letters* 12: 1394–1404.

Bennett, E.M., Cramer, W., Begossi, A. et al. (2015). Linking biodiversity, ecosystem services, and human well-being: three challenges for designing research for sustainability. *Current Opinion in Environmental Sustainability* 14: 76–85.

Boumans, R., Roman, J., Altman, I., and Kaufman, L. (2015). The multiscale integrated model of ecosystem services (MIMES): simulating the interactions of coupled human and natural systems. *Ecosystem Services* 12: 30–41.

Boyd, J. and Banzhaf, S. (2007). What are ecosystem services? The need for standardized environmental accounting units. *Ecological Economics* 63: 616–626.

Carpenter, S.R., Mooney, H.A., Agard, J. et al. (2009). Science for managing ecosystem services: beyond the Millennium Ecosystem Assessment. *Proceedings of the National Academy of Sciences of the United States of America* 106: 1305–1312.

Chan, K.M.A., Shaw, M.R., Cameron, D.R. et al. (2006). Conservation planning for ecosystem services. *PLoS Biology* 4: e379.

Chen, H., Cummins, J.D., Viswanathan, K.S., and Weiss, M.A. (2013). Systemic risk and the interconnectedness between banks and insurers: an econometric analysis. *Journal of Risk and Insurance* 81: 623–652.

Costanza, R., d'Arge, R., de Groot, R. et al. (1997). The value of the world's ecosystem services and natural capital. *Nature* 387: 253–260.

Cundill, G., Leitch, A.M., Schultz, L. et al. (2015). Principle 5 – encourage learning. Principles for building resilience: sustaining ecosystem services. In: *Social-Ecological Systems* (eds. R. Biggs, M. Schlüter and M. Schoon), 174–200. Cambridge, UK: Cambridge University Press.

Daily, G.C. (1997). Introduction: what are ecosystem services? In: *Nature's Services: Societal Dependence on Natural Ecosystems* (ed. G.C. Daily), 1–10. Washington, DC: Island Press.

Daily, G.C., Söderqvist, T., Aniyar, S. et al. (2000). The value of nature and the nature of value. *Science* 289: 395.

Daily, G.C., Polasky, S., Goldstein, J. et al. (2009). Ecosystem services in decision making: time to deliver. *Frontiers in Ecology and the Environment* 7 (1): 21–28.

Dearing, J.A., Yang, X., Dong, X. et al. (2012). Extending the timescale and range of ecosystem services through paleoenvironmental analyses, exemplified in the lower Yangtze basin. *Proceedings of the National Academy of Sciences of the United States of America* 109: E1111–E1120.

Earth Economics. (2019). Mission and vision. www.eartheconomics.org/vision (accessed October 25, 2019).

Ecometrix (2019). Homepage. www.ecometrix.ca (accessed November 2019).

Ehrlich, P.R. and Ehrlich, A.H. (1981). *Extinction: The Causes and Consequences of the Disappearance of Species*. New York: Random House.

Emerton, L. (2008). *Investing in Ecosystems and Poverty Reduction Infrastructure*. Gland, Switzerland: IUCN.

Emerton, L., Erdenesaikhan, N., de Veen, B., Tsogoo, D., Janchivdorj, L., Suvd, P., Gandolgor, G., Dorjsuren, C., Sainbayar, D. and Enkhbaatar, A. (2009). The economic value of the Upper Tuul ecosystem, Mongolia. Discussion paper, East Asia and Pacific Sustainable Development Department. Washington, DC: World Bank.

Fisher, B., Turner, R.K., and Morling, P. (2009). Defining and classifying ecosystem services for decision making. *Ecological Economics* 68: 643–653.

Fisher, J.A., Patenaude, G., Giri, K. et al. (2014). Understanding the relationships between ecosystem services and poverty alleviation: a conceptual framework. *Ecosystem Services* 7: 34–45.

Folke, C., Hahn, T., Olsson, P., and Norberg, J. (2005). Adaptive governance of social-ecological systems. *Annual Review of Environment and Resources* 30: 441–473.

Gerrard, P. (2004). Integrating wetland ecosystem values into urban planning: the case of That Luang Marsh, Vientiane, Lao PDR. Report. https://iwlearn.net/files/pdfs/Gerrard%202004_Integrating%20wetland%20ecosystem%20into%20urban%20planning.pdf (accessed October 25, 2019).

de Groot, R.S., Wilson, M.A., and Boumans, R.M. (2002). A typology for the classification, description and valuation of ecosystem functions, goods and services. *Ecological Economics* 41: 393–408.

de Groot, R., Brander, L., van der Ploeg, S. et al. (2012). Global estimates of the value of ecosystems and their services in monetary units. *Ecosystem Services* 1: 50–61.

Guerry, A.D., Polasky, S., Lubchenco, J. et al. (2015). Natural capital and ecosystem services informing decisions: from promise to practice. *Proceedings of the National Academy of Sciences of the United States of America* 112: 7348–7355.

Helliwell, D.R. (1969). Valuation of wildlife resources. *Regional Studies* 3: 41–47.

Hirsch-Hadorn, G., Bradley, D., Pohl, C. et al. (2006). Implications of transdisciplinarity for sustainability research. *Ecological Economics* 60: 119–128.

Hossain, M.S., Eigenbrod, F., Amoako Johnson, F., and Dearing, J.A. (2017). Unravelling the interrelationships between ecosystem services and human wellbeing in the Bangladesh delta. *International Journal of Sustainable Development and World Ecology* 24: 120–134.

Ishii, T. (2014). Potential impact of human mitochondrial replacement on global policy regarding germline gene modification. *Reproductive Biomedicine Online* 29: 150–155.

King, R.T. (1966). Wildlife and man. *New York Conservationist* 20 (6): 8–11.

Labiosa, W.B., Forney, W.M., Esnard, A.-M. et al. (2013). An integrated multi-criteria scenario evaluation web tool for participatory land-use planning in urbanized areas: the ecosystem portfolio model. *Environmental Modelling and Software* 41: 210–222.

Lele, S., Springate-Baginski, O., Lakerveld, R. et al. (2013). Ecosystem services: origins, contributions, pitfalls, and alternatives. *Conservation and Society* 11: 343.

Liu, J., Hull, V., Batistella, M. et al. (2013). Framing sustainability in a telecoupled world. *Ecology and Society* 18: 1–15.

Liu, J., Yang, W., and Li, S. (2016). Framing ecosystem services in the telecoupled Anthropocene. *Frontiers in Ecology and the Environment* 14: 27–36.

Lobo, G. (2001). Ecosystem functions classification. http://gasa3.dcea.fct.unl.pt/ecoman/delphi (accessed November 2019).

Mace, G. (2013). Global change: ecology must evolve. *Nature* 503: 191–192.

Marcot, B.G., Steventon, J.D., Sutherland, G.D., and McCann, R.K. (2006). Guidelines for developing and updating Bayesian belief networks applied to ecological modeling and conservation. *Canadian Journal of Forest Research* 36: 3063–3074.

Meyfroidt, P., Lambin, E.F., Erb, K.-H., and Hertel, T.W. (2013). Globalization of land use: distant drivers of land change and geographic displacement of land use. *Current Opinion in Environmental Sustainability* 5: 438–444.

Millennium Ecosystem Assessment (2005). *Ecosystems and Human Well-being: Synthesis for Decision-Makers*. Washington, DC: Island Press.

Mitchell, M.G.E., Bennett, E.M., Gonzalez, A. et al. (2015). The Montérégie connection: linking landscapes, biodiversity, and ecosystem services to improve decision making. *Ecology and Society* 20: article 15.

Mitlin, D. (2008). With and beyond the state – co-production as a route to political influence, power and transformation for grassroots organizations. *Environment and Urbanization* 20: 339–360.

Mitton, C., Adair, C.E., McKenzie, E. et al. (2007). Knowledge transfer and exchange: review and synthesis of the literature. *The Milbank Quarterly* 85: 729–768.

Moberg, F. and Folke, C. (1999). Ecological goods and services of coral reef ecosystems. *Ecological Economics* 29: 215–233.

Myers-Smith, I.H., Elmendorf, S.C., Beck, P.S.A. et al. (2015). Climate sensitivity of shrub growth across the tundra biome. *Nature Climate Change* 5: 887–891.

Naidoo, R., Balmford, A., Costanza, R. et al. (2008). Global mapping of ecosystem services and conservation priorities. *Proceedings of the National Academy of Sciences of the United States of America* 105: 9495–9500.

Nelson, E., Mendoza, G., Regetz, J. et al. (2009). Modeling multiple ecosystem services, biodiversity conservation, commodity production, and tradeoffs at landscape scales. *Frontiers in Ecology and the Environment* 7: 4–11.

Norberg, J. (1999). Linking nature's services to ecosystems: some general ecological concepts. *Ecological Economics* 29: 183–202.

Norgaard, R.B. (2010). Ecosystem services: from eye-opening metaphor to complexity blinder. *Ecological Economics* 69: 1219–1227.

Norgaard, R.B. and Baer, P. (2005a). Collectively seeing climate change: the limits of formal models. *BioScience* 55: 961–966.

Norgaard, R.B. and Baer, P. (2005b). Collectively seeing complex systems: the nature of the problem. *BioScience* 55: 953–960.

Oteros-Rozas, E., Martín-López, B., Daw, T.M. et al. (2015). Participatory scenario planning in place-based social-ecological research: insights and experiences from 23 case studies. *Ecology and Society* 20: 32.

Pocock, M.J.O. and Evans, D.M. (2014). The success of the horse-chestnut leaf-miner, *Cameraria ohridella*, in the UK revealed with hypothesis-led citizen science. *PLoS One* 9: e86226.

Qiu, J. and Turner, M.G. (2013). Spatial interactions among ecosystem services in an urbanizing agricultural watershed. *Proceedings of the National Academy of Sciences of the United States of America* 110: 12149–12154.

Raworth, K. (2012). A safe and just space for humanity: can we live within the doughnut. *Oxfam Policy and Practice: Climate Change and Resilience* 8: 1–26.

Raymond, C.M., Singh, G.G., Benessaiah, K. et al. (2013). Ecosystem services and beyond: using multiple metaphors to understand human–environment relationships. *BioScience* 63: 536–546.

Ricketts, T.H., Watson, K.B., Koh, I. et al. (2016). Disaggregating the evidence linking biodiversity and ecosystem services. *Nature Communications* 7: 13106.

Rieb, J.T., Chaplin-Kramer, R., Daily, G.C. et al. (2017). When, where, and how nature matters for ecosystem services: challenges for the next generation of ecosystem service models. *BioScience* 67: 820–833.

Rodell, M., Velicogna, I., and Famiglietti, J.S. (2009). Satellite-based estimates of groundwater depletion in India. *Nature* 460: 999–1002.

Santos-Martin, F., Martín-López, B., García-Llorente, M. et al. (2013). Unravelling the relationships between ecosystems and human wellbeing in Spain. *PLoS One* 8: 1–12.

Schröter, M., Stumpf, K.H., Loos, J. et al. (2017). Refocusing ecosystem services towards sustainability. *Ecosystem Services* 25: 35–43.

Sherrouse, B.C., Clement, J.M., and Semmens, D.J. (2011). A GIS application for assessing, mapping, and quantifying the social values of ecosystem services. *Applied Geography* 31 (2): 748–760.

Singh, V.S., Pandey, D.N. and Chaudhry, P. (2010). Urban forests and open green spaces: lessons for Jaipur, Rajasthan, India. Rajasthan State Pollution Control Board, Occasional Paper No. 1/2010.

Steger, C., Hirsch, S., Evers, C. et al. (2018). Ecosystem services as boundary objects for transdisciplinary collaboration. *Ecological Economics* 143: 153–160.

The Economics of Ecosystems and Biodiversity (TEEB) 2010. TEEB for local and regional policy makers. Geneva: TEEB.

Tobner, C.M., Paquette, A., Reich, P.B. et al. (2014). Advancing biodiversity–ecosystem functioning science using high-density tree-based experiments over functional diversity gradients. *Oecologia* 174: 609–621.

Tomscha, S.A., Sutherland, I.J., Renard, D. et al. (2016). A guide to historical data sets for reconstructing ecosystem service change over time. *BioScience* 66: 747–762.

Turkelboom, F., Leone, M., Jacobs, S. et al. (2018). When we cannot have it all: ecosystem services trade-offs in the context of spatial planning. *Ecosystem Services* 29: 566–578.

Turner, R.K., Schaafsma, M., Mee, L. et al. (2015). Conceptual framework. In: *Coastal Zones Ecosystem Services: From Science to Values and Decision Making* (eds. R.K. Turner and M. Schaafsma), 11–40. Cham, Switzerland: Springer International Publishing.

UNEP. (2018). https://www.unenvironment.org/explore-topics/sustainable-development-goals.

Wallace, K.J. (2007). Classification of ecosystem services: problems and solutions. *Biological Conservation* 139: 235–246.

Water Research Commission (2015). miniSASS: acommunity river health monitoring tool. www.minisass.org/en (accessed March 21, 2018).

WBCSD (2011). *Guide to Corporate Ecosystem Valuation – A framework for Improving Corporate Decision Making*. Geneva: World Business Council for Sustainable Development (WBCSD).

Wegner, G. and Pascual, U. (2011). Cost-benefit analysis in the context of ecosystem services for human well-being: a multidisciplinary critique. *Global Environmental Change* 21: 492–504.

Westman, W.E. (1977). How much are nature's services worth? *Science* 197: 960–964.
Wils, T.H.G., Sass-Klaassen, U.G.W., Eshetu, Z. et al. (2011). Dendrochronology in the dry tropics: the Ethiopian case. *Trees* 25: 345–354.
Wilson, J., Low, B., Costanza, R., and Ostrom, E. (1999). Scale misperceptions and the spatial dynamics of a social–ecological system. *Ecological Economics* 31: 243–257.
Wolff, S., Schulp, C.J.E., and Verburg, P.H. (2015). Mapping ecosystem services demand: a review of current research and future perspectives. *Ecological Indicators* 55: 159–171.
World Resources Institute (WRI) (2012). Homepage. https://www.wri.org (accessed October 25, 2019).
Wu, S. and Li, S. (2019). Ecosystem service relationships: formation and recommended approaches from a systematic review. *Ecological Indicators* 99: 1–11.

14

Sustainable Infrastructure

Gilbert Hinge[1], *Rao Y. Surampalli*[2], *and Manish Kumar Goyal*[1, 3]

[1]*Department of Civil Engineering, Indian Institute of Technology, Guwahati, India*
[2]*Global Institute for Energy, Environment, and Sustainability, Lenexa, KS, USA*
[3]*Discipline of Civil Engineering, Indian Institute of Technology, Indore, India*

14.1 Infrastructure – An Introduction

Infrastructure consists of basic systems and facilities, which include transportation, buildings, communication systems, water supply, power supply, etc. Infrastructure development is essential to facilitate a country's economic development and reduce poverty. In recent years, globalization has led to an increase in domestic and international infrastructure development and connectivity at an unprecedented rate. Globalization can be defined simply as: "the processes of collaboration and coordination among the people, organization, and governments of various countries, a method determined by universal trade and investment and supported by information technology" (Léautier and Lemer 2006).

Globalization has made the development of transport and communication feasible. This means that the world is becoming more and more interconnected, and people and countries can exchange information, goods, knowledge and services with ease. Nowadays, data and information are promptly conveyed and universally available, making geographical vicinity less important in the success of any infrastructure project. With globalization, the way people think about infrastructure and decision making related to infrastructure has also undergone various changes (Léautier and Lemer 2006). Strong evidence of this can be seen in the case of China. China has invested a huge amount in infrastructure since 1990, particularly in road infrastructure. Li (2010), in his report, cited the example of Lanzhou-Xinjiang railroad. He reported that, within 3 years of completion of this road, trade volume was observed to increase by 40%, and trade cost was reduced by 30%, indicating a net social return to society of roughly 30% per year. Thus, it can be seen that well-designed infrastructure can facilitate a country's economic development and can reduce the cost of trade: it's the key to raising the living standard of the people and reducing poverty (Ali and Pernia 2003).

Sustainability: Fundamentals and Applications, First Edition.
Edited by Rao Y. Surampalli, Tian C. Zhang, Manish Kumar Goyal, Satinder K. Brar, and R. D. Tyagi

However, infrastructure development often comes at the cost of the local environment. These are well documented in energy construction work, transportation, sanitation, irrigation works, etc. as explained in the following sections.

14.1.1 Energy Sector

The local environment generally is subjected to risks from construction and operation of infrastructure that would result in environmental damage if not mitigated or remunerated effectively. This is well acknowledged especially in operations that involve power plants. The crop yield was reported to decline by 5–30% for around 70% of all crops in China due to acid rains and other consequences resulting from coal combustion (Chameides et al. 1999). A similar observation was reported in India, where in areas located near large power plants, there was a ~50% reduction in crop yield (Shalizi 2007). Particulates and smog coming out of large power plants and from traffic are thought to cause around 427 000 and 107 000 annual deaths in China and India, respectively (Shilling 2007).

14.1.2 Transportation Work

Similarly, with road construction, despite the widely accepted fact that road construction will promote both economic and social benefits (Amadi 1988), it can likewise add to environmental damage. For instance, it can cause an increase in immigration and increase in harvesting of environmental resources needed for urban market supply. These can lead to deforestation, increases in landslides, extraction of fuelwood and extinction of wildlife (Forman and Alexander 1998). Road construction can also directly impact the environment, e.g. erosion and sedimentation caused by the unpaved roads (Amos 2008).

14.1.3 Irrigation Work

Irrigation infrastructure, such as in the case of road construction or the energy sector, can be both advantageous and harmful. Irrigation infrastructure helps to store and transport water, thus helping farmers to reduce their dependencies on rain, allowing a greater intensity of cultivation on their existing land. However, it can also lead to excessive water usage, salinization of groundwater, land degradation and can even lead to downstream pollution (due to use of pesticides etc.). In its report the Millennium Ecosystem Assessment (MEA 2005a) claimed that irrigation has already consumed around 20–30% of the world's available fresh water. A study carried out by the United Nations Environment Programme (Morris et al. 2003) reported that almost 50% of the world's irrigated land is affected by salinity, alkalinity and waterlogging.

14.1.4 Sanitation Work

In contrast, construction of sanitation structures presents an example of infrastructure that is good for public health and the environment. The United Nations Task Force in its report (Lenton 2005) stresses the importance of improved sanitation infrastructure, points out its benefits related to health, poverty and the environment, and recommends steps to promote this infrastructure.

The impact of infrastructure on the environment is more pronounced in developing countries. Development of infrastructure without proper planning leads to environmental and infrastructure problems (Shilling 2007). Improper planning and construction of roads leads to erosion, preventing access to waste collection and firefighting. Lack of electricity will lead to the cutting down of more trees to be used for cooking and heating purposes; lack of sanitation and good drinking water supply will lead to diseases (Shilling 2007). The United Nations Environment Programme in its reports claims that environmental damage because of lack of sustainable infrastructure has increased and will pose a great threat p to developing countries in particular, and may undermine their potential for sustainable growth (Lenton 2005).

14.1.5 Global Warming and Climate Change in a Globalized World

Climate change due to global warming is the biggest environmental problem faced by mankind living in the age of globalization. With the increase in the globalization of goods and services, we are also observing more fossil fuels being exchanged to sustain this production, leading to more carbon emissions, which is the main cause of global warming. Greenhouse gases, which were identified as the major cause of global warming, have now reached the highest levels ever recorded in history. Climate change is now influencing every nation on the planet. It already has a considerable impact on the ecosystem, life of infrastructure, national economies and lives of communities (Pachauri and Reisinger 2007). Climate change will affect all kinds of infrastructure, including the water, energy and transport sectors. If no actions are taken, these greenhouse gases will further increase in the atmosphere and will threaten the efficiency and reliability of all kinds of infrastructure (IPCC 2014), with the consequences of social and economic impacts.

Sustainable development is the need of the hour in our effort to slow down climate change. Sustainable development is defined as "development that addresses the requirements of the present generations without compromising the ability of the future generations to meet their own needs" (Brundtland 1985). Sustainable development does not necessarily mean that a nation should stop economic growth for the sake of protecting the environment; however, it aims at promoting economic development that is ecofriendly. Dealing with climate change alone could be very expensive; therefore, integrating development policies as part of the solution can be one of the ways to slow down climate change.

14.1.6 Adapting Infrastructure to Climate Change

Over the past few decades, efforts and investment have been made to lessen the amount of CO_2 emissions from the infrastructure sector. Yet, in spite of these endeavors, scientific reports show that the current CO_2 emissions generated from the existing infrastructure are still responsible for significant climate change (Pachauri and Reisinger 2007). Most of the existing infrastructure, comprising buildings, water, roads and energy, is not designed to endure changes in environmental conditions (IPCC 2014).

Over these past few years, unpredictable changes in weather conditions have been observed to cause some challenges to the current infrastructure. The literature has reported that extreme weather conditions such as snowfall have resulted in travel disturbances, slower speeds, travel delays and accidents (Knapp and Smithson 2000; Datla and Sharma 2010). Changes in precipitation and runoff have influenced the water supply and water quality (Frederick and Major 1997); various infrastructure assets and services have been affected by extreme climate, making them vulnerable to floods and drought (Adger 1999).

In the coming 30–50 years, more climatic changes are expected to occur (Pachauri and Reisinger 2007). We might further experience drastic changes in temperature, rainfall patterns and more frequent extreme events like drought and freezing winters. Hence, we need to act today. We need to plan and decide what necessary steps are to be taken to ensure that our infrastructure is sustainable and resilient enough to meet any unpredictable challenges in the future. Planners, designers and engineers must plan and design sustainable infrastructure that integrates decarbonization as well as being adaptive against various extreme climates. Moreover, most infrastructure will have a considerable lifespan; therefore it will be subjected not only to adverse climate condition at the time of construction but also to unpredictable climate variations in the future. Therefore, in order to enhance the sustainability and resilience of the structures, they should be designed in such a way they can withstand any extreme climate changes.

14.1.7 Building Climate-Resilient Infrastructure

Newly built infrastructure can become climate resilient by guaranteeing that the structure is situated, designed, constructed and used keeping both current and future climates in mind. Existing infrastructure can be made climate resilient by making sure that maintenance factors in integrated resilience to the effects of climate change over the structure's lifespan. To accomplish this, promising adaptation actions consist of:

- Checking and verifying that the infrastructures are resilient enough for the likely increase in extreme weather events such as heatwaves, floods, drought, storms, etc.
- Building flexibility into infrastructure so that, in the future, it can be modified without excessive cost.
- Ensuring that companies/administration and the professionals who built and implemented the infrastructure have the proper skills and capacity to carry out the work.

If these actions are implemented, we will have a more resilient and robust infrastructure that can cope with the projected changing climate. Case example 14.1 demonstrates the importance of implementing these actions.

Thus, with the likely frequent occurrence of various extreme weather events, decision makers in both the public and private sectors should carry out a test for climate resilience at each stage of their decision making to evaluate whether they consider new or retrofitted infrastructure projects. Introducing climate resilience to the life cycle of any infrastructure project should become the mainstay of the future.

Case Example 14.1 How One Japanese Village Was Able to Resist the 2011 Tsunami by Building Climate-Resilient Infrastructure

Background. In 1933, an earthquake known by the name "Sanriku" hit the coast of Japan. In spite of the fact that the earthquake did little damage, a gigantic tsunami followed this earthquake with waves recorded up to a height of 28.7 m. This tsunami destroyed over 7000 homes and many human lives were taken (1522 citizens were killed and 12 053 were injured).

Solutions. In 1967, the mayor (the highest-ranking official) of Fudai planned to construct a concrete seawall of about 15.5 m so as to prevent a similar disaster in the future. The mayor also planned to construct concrete floodgates as tall as the seawall in locations where the majority of the people resided. In the beginning, the village people were against his ideas; however, eventually he persuaded them. So, in 1967 the seawall was constructed and in 1972, 52 concrete floodgates were also constructed.

Results. On March 11, 2011, a huge earthquake followed by a tsunami hit Japan. The national police agency of Japan reported that this natural disaster took lots of lives (15 884 lost their lives, 6147 were critically wounded and 2636 people went missing), and many buildings and structures collapsed. However, in Fudai, despite the waves reaching up to a height of 20.1 m, the village remained safe and untouched, thanks to the concrete seawall built in 1967.

Source: www.nbcnews.com/id/43018489/ns/world_news-asia_pacific/t/how-one-japanese-village-defied-tsunami/#.Uxj2yuNdUw1

14.2 Mitigation Policies

Infrastructure development is a double-edged sword: it provides income benefits, but an environmental cost is also associated with it. The environmental cost can be reduced to a large extent if the infrastructure is properly planned. There are various mitigation policies that, if properly followed, can help to reduce the costs and increase environmental benefits of infrastructure projects. These policies are available at both the project- as well as national level.

14.2.1 Project Level

Proper choices made while designing various projects can help to reduce the environmental effects. A few examples include the provision and maintenance of side drains along the road, which will enhance the longevity of the road. Erosion rates during road construction are directly proportional to the amount of land exposed, resulting from the cuts and fills needed for road construction. The extent of the cuts and fills depends upon the set-up of the road alignment (horizontal, vertical). Thus, proper analysis will reduce the extent of cuts and fills needed for construction of roads. Proper water management can help to reduce floods and a shortage of water supply, while still providing access to good-quality water supply. A properly oriented building can provide a pleasant and suitable environment to live in, thus saving money that would otherwise be needed for providing heating and cooling. These few examples demonstrate that a wise decision made at the project level can help to reduce both environmental and economic costs to a sizable extent. Operation and maintenance are also equally important to enhance social and economic benefits at the project level.

14.2.2 National and Sector Level

There are more possibilities for improving the social, economic and environmental benefits of infrastructure investments at the national level compared with the project level because some of the available options might not be apparent if an environmental assessment is carried out only at the project level. These options require planning at national and sector levels. Some of these options include:

1) *Moving toward sustainable and ecofriendly infrastructure*: This involves changes in direction of consumption and production patterns in a country. For example, the electric generation sector of a nation could shift away from its dependencies on fossil fuels and move toward clean and sustainable sources. Even though some of these changes might cost more initially than their counterpart options, , one should consider their positive environmental impacts and how they can generate benefits in the long run.
2) *Environmental assets superseding physical assets*: The need for investment in various sectors can be reduced by adopting suitable environmental management practices. For example, the need for major investments in the water sector for the removal of sediments and pollutants can be reduced by construction of wetland. Wetlands are a low-cost and environmentally friendly measure that plays a key role in removing point- and non-point-source pollutants. Healthy wetland consists of rich, diverse plants and animal species that can act as filters and help in removing pollutants from water (Bassi et al. 2014). Thus, a wetland can prevent the need for huge investment and can serve as natural infrastructure. Another example is the managing of floodplains to increase their water retention capacity so that they can act as a buffer against flooding and help in reducing flood hazard.
3) *Spatial planning and siting of infrastructure*: Proper planning and siting of infrastructure can have a profound effect on the environment. For example, the decision on road siting is often a trade-off between encouragement for development and induced deforestation. Siting a new road in an already populated and degraded environment and avoiding forested areas can help to protect the environment. Similarly, the ecological and social effect of a dam varies with the size and population of the basin in which it is flooded.
4) *Investing in water and energy efficiency*: Enhancing water and energy efficiency can help to lessen the requirement for investment in environmentally degraded and costly infrastructures. The efficiency of water application in the irrigation sector is still reported to be very low. In its reports MEA (2005b) mentioned that only 40–50% of diverted water is being used efficiently for irrigation. Also, often a large quantity of water is consumed on lower-value crops rather than investing it for high-value crops or in industry sectors. Likewise, there is greater work needed to improve energy efficiency.

14.2.3 For Long-term and Forward-looking Projects

Governments form the future of communities by defining and implementing development strategies for the nation or for regions. These plans and policies defined by the government will form the basis of various decisions related to various projects in diverse fields such as energy, agriculture, land use, transport, regional development, water management and waste treatment. For instance, in the case of a national energy plan, the government sets up a plan on how a country should generate and distribute energy. These plans and policies then define what kinds of energy are to be produced for years to come. Based on this, various investment decisions for construction, expansion and for demolition are established. To develop such long-term and forward-looking programs and policies, the decisions makers have to consider multiple factors such as resources available, available infrastructure, energy requirement, etc. Also, because such plans and policies can have a big influence on people's lives, both at present and in the future, it is crucial that environmental and health impacts are among the decision-making factors. To ensure this, many countries have introduced and implemented strategic environmental assessments (SEAs) to protect the environment and to ensure sustainability for various proposed plans and policies. The next section includes a brief introduction to SEA, the steps involved in it, it's advantages and disadvantages, and some case examples that demonstrate the effectiveness of SEA.

14.3 Strategic Environmental Assessment

SEA can be seen as an: "instrument for evaluating environmental and health-related consequences of various proposed plans, policies or programs so as to make sure that they are included and properly addressed, right from the initial sustainable stage of the decision-making process" (Von Seht 1999).

SEA came as an improvement over environmental impact assessment (EIA) for making sound environmental decisions. It integrates socio-economic values with environmental values that are viable for better decision making. It deals with concepts and goals rather than with specific activities. It aims to prevent unacceptable environmental damages. It gets involved early enough in the decision making to avoid any irreversible effect and costly errors due to poor planning. SEA additionally enhances transparency and public trust in its decision-making process. It aids countries to foster economic development and also to pursue green economy targets and environmental goals, such as in protecting biodiversity and increasing energy efficiency. The scope of SEA is vast: it can be applied to various disciplines and subsectors. With infrastructure development being the main activities of a nation, this section explains the concept of SEA, steps involved in forming SEA, and its benefits and drawbacks, which can be adopted in various infrastructure projects for sustainable development.

14.3.1 Steps Involved in SEA

Figure 14.1 shows the steps involved in decision making and SEA processes and their relationships. The steps involved in an SEA are:

Figure 14.1 Steps involved in strategic environmental assessment and how it can be integrated into decision making.

- screening (determining whether or not SEA is needed);
- scoping (deciding the scope of the environmental problem to be deal by the SEA);
- preparation of an Environmental Report;
- consultations;
- implementation of environmental considerations into the plan or program;
- publication of decision details (SEA Statement).

The following two case examples (Case Examples 14.2 and 14.3) demonstrate a few examples of effective SEA practice in Africa.

14.3.2 Advantages of Strategic Environmental Assessment

Application of SEA can assist the decision maker:

1) *To attain environmental sustainability and sustainable development*: SEA supports the idea of ecological and social aspects being good as they support economic development. Moreover, the utilization of SEA facilitates the reconciliation of various goals sought

Case Example 14.2 Strategic Environmental Assessment for Power Sector in Nile River Basin

Uganda, located in East Africa, and the entire regions of the Nile equatorial lake suffer from a lack of a proper supply of electricity. To solve the problem, the World Bank agreed to the arrangement for preparation of an SEA. The SEA helped in the development of hydropower in the Nile equatorial regions. It did so by first analyzing the likely impact of drought over the lifespan of the proposed project through a detailed analysis of climate change and its likely effect. The results of the analysis indicated that rising temperatures due to climate change will result in increases inboth precipitation and runoff and, ultimately, will provide a better potential for development of power generation. Based on these results, along with social and EIA, Bujagali Energy Limited (BEL) developed the 250-MW Bujagali Hydropower Project for this region.
Source: www.ifc.org/disclosure

Case Example 14.3 Strategic Environmental Assessment for Transport Planning

In Mozambique, a country located in southern Africa, construction of a road to the coast was suggested in order to convey some large bulk resources from an inland titanium smelter. The government of this country decided to opt for SEA analysis, which might provide better alternatives to serve their country not only for transportation of the mining resources, but also to promote ecotourism and, ultimately, improved life conditions for the local people. Based on the SEA result, the government decided that the concession for the mining firm would be allowed only for 20 years. Throughout this period, an alternate infrastructure plan with the north–south connection was to be established. Also, meanwhile the east–west road was to be developed.
Source: www.eia.nl

after by various administrative levels and various sectors. SEA can be thought of as an expert dynamic instrument that addresses the causes of environmental issues rather than just treating its side effects. It encourages the early evaluation of various environmental and sustainable goals right from the stages of planning of policy, plans and programs. It encourages the assessment of entire sets of options and choices; it assists in choosing and designing environmental sustainable plans for favored strategic alternatives. Also, by promoting the protection of the environment, SEA not only helps to deal with the prosperity of the current generation, but also considers the benefits that future generations may obtain.

2) *SEA bolsters strategies for the development of plans, policy and programs*: SEA encourages the distinguishing proof and examination of the probable significant environmental and health effects of the suggested approaches, plans and programs. This pushes the organizers and decision makers to carefully consider both likely local environmental problems and worldwide environmental issues, such as climate change, etc.
3) *SEA can help to spare time and money*: SEA cautions policymakers right from the early stages against making unsustainable development choices. This helps to spare time and

money as their poor choices are recognized and discarded at a point in time when less time and money have been spent on their development. SEA helps to diminish the requirement for exorbitant remediation of environmental issues that happen while executing poor development decisions. Thus, SEA helps to provide social, environmental and economic benefits both for the present and future generations.
4) *SEA improves transparency and public trust in decision making*: SEA improves the overall transparency of all strategic decision making. This assists the organizers and decision makers to create and obtain public trust in their decision making and planning process. SEA enables the policy- and decision makers to obtain the views of important stakeholders ahead of their planning schedule. This helps to diminish the danger of deadlock in their decision making on various projects.

14.3.3 Limitations of Strategic Environmental Assessment

Some of the limitations of SEA are:

1) *SEA requires both time and resources*: The resources and time required depend upon the nature of the strategic action carried out and how effectively SEA is implemented.
2) *SEA is still in its early stages*: Hence, baseline data may not exist. A public consultation center may not have been established. Therefore, the policymakers and planners may have to undergo t a learning stage for every new ideas and decision implemented.
3) *SEA covers broad areas accompanied by huge potential projects*: It needs to deal with a large range of decision-making situations, right from global issues down to almost project-level situations. Often these projects last for longer periods of time; hence many uncertainties will be encountered during the course of their development. For example, rises in petrol prices, drought and many other extreme events may occur. In such cases, decisions need to be taken very quickly. Hence SEA needs to be adaptable, responsive and fast. This means that SEA may not be robust and scientific.

In addition to the above-mentioned mitigation assessment and policies, other mitigation practices include recycling, reuse, reclamation, adaptive reuse and deconstruction of infrastructure materials at the end of their lifespan. These mitigation policies are explained in the next section.

14.4 Recycling Reuse and Reclamation

Another mitigation practice includes recycling, reuse and reclamation of infrastructure materials at the end of their lifespan. These practices offer an immense opportunity to improve resource efficiency, helping in reducing energy consumption along with their associated carbon emissions, reducing the quantity of waste generated and again creating value out of these materials (Wahlström et al. 2013). Reuse is a process of using the same discarded components again (or sometimes the component is upgraded) for the same purpose or for different functions without destroying the component (Cooper and Allwood 2012). Recycling is a process during which the disposed of materials are reprocessed to form a completely new product (Thormark 2000). Reclamation can be defined as the process of returning the components to the original state or to a better state.

With respect to the recycling of materials, the environmental benefits obtained by recycling a product vary from one infrastructure component/material to another. The reason for this is that there are differences in the amount of energy essential for transportation as well as the mechanical energy required to reprocess the components. All the factors, such as the distance required to transport the materials, the weight of the materials, the type of methods needed to reprocess it and availability of raw materials, can make a difference concerning the environmental properties of the original vs recycled materials (Thormark 2000). For example, in term of energy consumption required to recycle a material, a huge amount of energy is required for remelting of steel materials as compared with the energy required to recycle concrete to be used as an aggregate for road construction materials (Hradil et al. 2014).

In the case of reuse of materials, though it requires energy for reprocessing or for upgrading, the energy consumption is comparatively much less compared with the energy required for the recycling of materials (Hosseini et al. 2015). For example, twice the amount of energy consumption per ton for fabrication of a particular steel section is required for recycling compared with reusing the same steel sections (Geyer and Jackson 2004). In addition, with respect to carbon emissions, reuse of materials such as glass and steel can save up to 60% of the carbon that would otherwise be needed to recycle them (Gorgolewski 2008). In addition to these environmental benefits, reuse of materials also provides social and economic benefits because of reductions in cost (instead of using new materials, which will require additional costs for extraction, processing, operation and storage) and because of the creation of jobs and benefits (Horvath 2004). Regardless of all these benefits, one must consider the cost of transportation of the materials from the location where they were discarded to the processing sites, the storage needs, and the cost and materials required for repairing their secondary components before considering the reuse of any materials (Anderson 2002).

14.5 Reverse Logistics

Reverse logistics, an operation associated with the reuse of components and products, is one of the developing concepts in the construction sector. It aims at preventing the movement of construction materials into waste products, instead recirculating them in the construction of a new project. Some of the interventions that support this concept (in order to prolong the life of construction materials) include those related to construction and demolition of the structure. A few of these are described in the next sections. All these methods aim at providing adaptable design, reusing materials in an optimized way that can enhance the lifespan of materials.

14.5.1 Adaptive Reuse

Adaptive reuse refers to the method of reusing part or the whole of a structure for a purpose different from that which it was originally designed for. When old buildings and structures become unsuitable for the purpose for which they were initially constructed, adaptive reuse is the perfect way to reclaim these structures and breathe new life into them while preserving their heritage and resources. Adaptive reuse offers a sustainable solution along with environmental, social and economic benefits that can be obtained by extending the useful lifespan of a structure (Bullen 2007). Some of the key advantages include the reuse of structural components without the need to reprocess them, thus saving energy and carbon. In some cases, materials like marble floors and stone walls that are used in architectural buildings can even contain more value than some of their new counterpart components (Langston et al. 2008). Despite the effectiveness of this method, however, the cost involved in the process may sometimes be even more than constructing a new structure. Therefore, there are always some criteria that determine whether a structure can be reused or is to be demolished.

14.5.2 Criteria for Adaptive Reuse

Some of the determining criteria for adaptive reuse include:

1) The physical damage sustained to the structure/site and its suitability in terms of proposed reuse.
2) The historical and social values of a structure/site as well as its role in society's understanding of the past.
3) The economic cost involved.
4) The environmental condition of the area. Whether the proposed modified structure will prevail in the existing environmental conditions.
5) The presence of hazardous materials that require special attention during construction.

14.5.3 Examples of Adaptive Reuse

A few examples where adaptive reuse has been successful applied include reuse of industrial and commercial buildings (Anderson 2002; Bullen 2007), reuse of power plant and water plants (Loures 2015), reuse of car packs (Gorgolewski et al. 2006) and reuse of residential buildings (Haidar and Talib 2015). Case Example 14.4 demonstrates the effectiveness of adaptive reuse of Kolkata town hall.

Case Example 14.4 Adaptive Reuse of Kolkata Town Hall

Kolkata town hall was built by the British in the year 1893 in Kolkata, India. The building was made up of various ornaments in its floor and is regarded as one of the most significant heritage buildings, reflecting the various important historical events that took place inside this building.

Over time, the building started showing sign of deterioration; however, the Indian government, through adaptive reuse, was able to preserve this important building.

Through various stages of adaptive reuse, this building at present contains the first high tech storytelling museum, known by the name "Kolkata Panorama." It also consists of a socio-cultural event area, an administrative area and an academic seminar area. (*Source:* Bose 2012)

14.6 Deconstruction

Deconstruction is another mitigation strategy. It is a process of carefully breaking down a structure with the aim of retaining maximum values (Guy et al. 2006). It has several advantages as compared with normal demolition process. The materials retained from deconstruction work are mostly reused or recycled, while the remaining components that are unfit for recycling or reuse are disposed of. This permits the consumers to maintain the embedded energy and carbon contents in the materials by prolonging the useful life of the materials with the addition of just a small energy input. Thus, deconstruction is capable of providing social, economic and environmental benefits (Geyer and Jackson 2004). However, deconstruction takes much longer than the conventional demolition process; thus, it becomes a problem in terms of economic and time perspectives. High costs for labor, the cost of machines and time constraints often make deconstruction practically questionable (Couto and Couto 2010). In addition, the presence of hazardous materials may add additional cost, as trained labor will be required for their removal.

Deconstruction has been carried out successfully in many parts of the world. In a study carried out in Canada (Gorgolewski 2008), it was reported that the cost of deconstruction of a residential building was 21% more than that of demolition. However, when taking into account the reuse and revenues of the materials obtained from deconstruction, the net cost of deconstruction was 37% less than that of conventional demolition. Geyer and Jackson (2004) compared the cost of recovery of steel sections for both deconstruction and demolition processes. They reported that the cost of deconstruction needed for reclamation of steel per ton was £50 more than that of demolition. However, when taking into account the net revenues, they reported that deconstruction cost £350 less than demolition. This shows that deconstruction can offer higher environmental as well as economic benefits.

14.6.1 Design for Deconstruction

Design for deconstruction works on the principle of green design. Its main objective is to design a new structure in a manner such that the economic value of the structural components along with its associated values can be easily recovered (Crowther 2014). Here, the lifecycle stages of components right from the planning stages to design, operation, even up to maintenance are designed in an optimized way with an aim to maximize the recovery of the components. Therefore, this method is sometimes referred to as design for adaptability and deconstruction (Guy and Shell 2002). Some of the benefits of design for deconstruction include:

- Allowing a structure to function adequately and expanding its lifetime by designing structures that can easily adapt.
- Minimizing the cost of construction.
- Providing environmental benefits by minimizing the amount of construction waste.
- Minimizing the amount of carbon impact.
- Reducing the amount of energy needed for transportation.
- Making it easier to change their uses by designing such a structure.

As with other methods, this method is obstructed by many challenges in term of economy, technical and logistical challenges. For instance, the economic viability of this method differs from one project to another, because the monetary values assigned to different projects are different. This method also cannot be standardized as the cultural, geographical and climatic features of each location will command different materials and components for construction of a structural component. For example, the design for construction of a building where wood is generally the main material will be different in regions where steel frames are commonly used. As this method is still in its beginning stages, there is still very literature related to this method, and the available literature is also all related to building the structure. The Delft University of Technology in the Netherlands has designed and constructed many buildings based on the concept of design for adaptability and deconstruction (Gorgolewski 2008). The concept of open building, as described by Ness et al. (2015), is another example of design for adaptability and deconstruction approach.

14.6.2 Design for Manufacture and Assembly

This is a newly developing method intended to efficiently deconstruct materials of a structure, particularly components of a building at the end of its useful life. The use of this method was inspired by its effectiveness in the automotive sector. Different components of automobiles are designed and manufactured with high technical abilities, proper surface finish, proper shape and tolerance, and are finally put into a complete structure. A similar concept was applied in the case of a building where concrete slabs, structural columns and beams were properly designed and manufactured and finally assembled to form a complete building (Kalyun and Wodajo 2012). A major advantage of this approach is in the way that it allows easy assembly and disassembly in a time- and cost-saving way, encouraging deconstruction and recovery of structural components at the end of its life time. Design for manufacture and recovery is, therefore, one of the ways to promote sustainability of structures or products (Pasquire and Connolly 2003). The method ensures that the structural components after destruction are still in a good physical and technical condition as compared with those materials recovered by traditional methods; thus encouraging the reuse of materials. As with any other methods, this method also has major limitations. The components of a structure after deconstruction will have a specific size and dimension, thus creating a barrier to their suitability for reuse, unless they are applied and reused for a similar purpose (Lahtinen 2011).

14.7 Summary

Infrastructure development can facilitate a country's economic development and reduce poverty. Building good infrastructure and the selection of appropriate policies can help to solve the world's environmental problems. However, infrastructure often imposes environmental problems that need to be dealt with or mitigated. This chapter presents a brief background of sustainable infrastructure, with some case studies spanning sustainable infrastructure underlining the importance of the nexus. It also presents some of the mitigation approaches and opportunities that can alleviate environmental issues while meeting the world's demand. Building right and resilient infrastructure is thus necessary, especially with the changing climate. By doing so, it can even promote environmental values, especially if infrastructure projects are legitimately incorporated into a national environmental strategy that is well designed and prudently managed.

References

Adger, W.N. (1999). Social vulnerability to climate change and extremes in coastal Vietnam. *World Development* 27 (2): 249–269.

Ali, I. and Pernia, E.M. (2003). *Infrastructure and Poverty Reduction: What is the Connection?* Mandaluyong, Metro Manila: Asian Development Bank.

Amadi, B.C. (1988). The impact of rural road construction on agricultural development: an empirical study of Anambra state in Nigeria. *Agricultural Systems* 27 (1): 1–9.

Amos, P. (2008). Safe, clean, and affordable... transport for development: the World Bank Group's transport business strategy for 2008–2012. Washington, DC: The World Bank.

Anderson, J. (2002). *The Green Guide to Specification: An Environmental Profiling System for Building Materials and their Components.* Oxford, UK: Blackwell Science Publishing Inc.

Bassi, N., Kumar, M.D., Sharma, A., and Pardha-Saradhi, P. (2014). Status of wetlands in India: a review of extent, ecosystem benefits, threats and management strategies. *Journal of Hydrology: Regional Studies* 2: 1–19.

Bose, S. (2012). Restoration of town hall in Kolkata for adaptive reuse: a case study. *Structural Survey* 30 (3): 280–291.

Brundtland, G.H. (1985). World Commission on Environment and Development. *Environmental Policy and Law* 14 (1): 26–30.

Bullen, P.A. (2007). Adaptive reuse and sustainability of commercial buildings. *Facilities* 25 (1/2): 20–31.

Chameides, W.L., Xingsheng, L., Xiaoyan, T. et al. (1999). Is ozone pollution affecting crop yields in China? *Geophysical Research Letters* 26 (7): 867–870.

Cooper, D.R. and Allwood, J.M. (2012). Reusing steel and aluminum components at end of product life. *Environmental Science and Technology* 46 (18): 10334–10340.

Couto, A. and Couto, J.P. (2010). Guidelines to improve construction and demolition waste management in Portugal. In: *Process Management* (ed. M. Pomffyová). Rijeka, Croatia: InTech.

Crowther, P. (2014). "Investigating design for disassembly through creative practice."

Datla, S. and Sharma, S. (2010). Variation of impact of cold temperature and snowfall and their interaction on traffic volume. *Transportation Research Record: Journal of the Transportation Research Board*, Transportation Research Board of the National Academies 2169: 107–115.

Forman, R.T.T. and Alexander, L.E. (1998). Roads and their major ecological effects. *Annual Review of Ecology and Systematics* 29 (1): 207–231.

Frederick, K.D. and Major, D.C. (1997). Climate change and water resources. In: *Climate Change and Water Resources Planning Criteria* (eds. K.D. Frederivk, D.C. Major and E.Z. Stakhiv), 7–23. New York: Springer.

Geyer, R. and Jackson, T. (2004). Supply loops and their constraints: the industrial ecology of recycling and reuse. *California Management Review* 46 (2): 55–73.

Gorgolewski, M. (2008). Designing with reused building components: some challenges. *Building Research and Information* 36 (2): 175–188.

Gorgolewski, M., Straka, V., Edmonds, J. and Sergio, C. (2006). Facilitating greater reuse and recycling of structural steel in the construction and demolition process. Ryerson University, Canadian Institute of Steel Construction.

Guy, B. and Shell, S. (2002). Design for deconstruction and materials reuse, design for deconstruction and materials reuse. Proceedings of the CIB Task Group 39 – Deconstruction Meeting, CIB World Building Congress, New Zealand (Vol. 272). Gainesville, FL: CIB/ University of Florida.

Guy, B., Shell, S., and Esherick, H. (2006). Design for deconstruction and materials reuse. *Proceedings of the CIB Task Group* 39 (4): 189–209.

Haidar, L. and Talib, A. (2015). Adaptive reuse practice in tower houses of Old City Sana'a Yemen. *Procedia – Social and Behavioral Sciences* 202: 351–360.

Horvath, A. (2004). Construction materials and the environment. *Annual Review of Environment and Resources* 29: 181–204.

Hosseini, M.R., Rameezdeen, R., Chileshe, N., and Lehmann, S. (2015). Reverse logistics in the construction industry. *Waste Management and Research* 33 (6): 499–514.

Hradil, P., Talja, A., Wahlström, M., Huuhka, S., Lahdensivu, J. and Pikkuvirta, J. (2014). Re-use of structural elements. Environmentally efficient recovery of building components. VTT Technology.

IPCC (2014). Climate Change 2014 Synthesis Report Summary Chapter for Policymakers. Geneva: IPCC.

Kalyun, M. and Wodajo, T. (2012). Application of a design method for manufacture and assembly, flexible assembly methods and their evaluation for the construction of bridges. Chalmers University of Technology, Gothenburg, Sweden.

Knapp, K. and Smithson, L. (2000). Winter storm event volume impact analysis using multiple-source archived monitoring data. *Transportation Research Record: Journal of the Transportation Research Board* 1700: 10–16.

Lahtinen, T. (2011). Design for manufacturing and assembly rules and guidelines for engineering. https://trepo.tuni.fi//handle/123456789/20737 (accessed October 27, 2019).

Langston, C., Wong, F.K.W., Hui, E.C.M., and Shen, L.-Y. (2008). Strategic assessment of building adaptive reuse opportunities in Hong Kong. *Building and Environment* 43 (10): 1709–1718.

Léautier, F. and Lemer, A.C. (2006). Globalization and urban infrastructure. In: *Cities Globalizing World*, vol. 9 (ed. F. Leautier). Washington, DC: World Bank.

Lenton, R.L. (2005). *Health, Dignity and Development: What Will it Take?* London: Earthscan.

Li, Z. (2010). Some evidence on the performance of transport infrastructure investment in China. Presentation given at the conference "The Economics of Infrastructure in a Globalised World: Issues, Lessons and Future Challenges," Sydney, March 18–19.

Loures, L. (2015). Post-industrial landscapes as drivers for urban redevelopment: public versus expert perspectives towards the benefits and barriers of the reuse of post-industrial sites in urban areas. *Habitat International* 45: 72–81.

Millennium Ecosystem Assessment (MEA) (2005a). *Ecosystems and Human Well-Being: Wetlands and Water*, vol. 5. Washington, DC: World Resources Institute.

Millennium Ecosystem Assessment (MEA) (2005b). *Ecosystems and Human Well-being: Synthesis*. Washington, DC: Island Press.

Morris, B.L., Lawrence, A.R.L., Chilton, P.J. C., Adams, B., Calow, R.C. and Klinck, B.A. (2003). Groundwater and its susceptibility to degradation: a global assessment of the problem and options for management. Nairobi: UNEP.

Ness, D., Swift, J., Ranasinghe, D.C. et al. (2015). Smart steel: new paradigms for the reuse of steel enabled by digital tracking and modelling. *Journal of Cleaner Production* 98: 292–303.

Pachauri, R.K. and Reisinger, A. (2007). IPCC fourth assessment report. Geneva: IPCC.

Pasquire, C.L. and Connolly, G.E. (2003). Design for manufacture and assembly. Nottingham Trent University, Nottingham.

Shalizi, Z. (2007). Energy and emissions: local and global effects of the rise of China and India. World Bank Policy Research Working Paper No. 4209. Washington, DC: World Bank.

Shilling, J.D. (2007). The nexus between infrastructure and environment. Working paper. Evaluation Cooperation Group. Washington, DC: World Bank.

Thormark, C. (2000). Environmental analysis of a building with reused building materials. http://citeseerx.ist.psu.edu/viewdoc/download?doi=10.1.1.831.3339&rep=rep1&type=pdf (accessed October 27, 2019).

Von Seht, H. (1999). Requirements of a comprehensive strategic environmental assessment system. *Landscape and Urban Planning* 45 (1): 1–14.

Wahlström, M., Laine-Ylijoki, J., Järnström, H., Kaartinen, T., Erlandsson, M., Palm Cousins, A., Wik, O., Suer, P., Oberender, A. and Hjelmar, O. (2013). Environmentally sustainable construction products and materials: assessment of release and emissions. Nordic Innovation. https://www.nordicinnovation.org/2014/environmentally-sustainable-construction-products-and-materials-assessment-release-and (accessed October 27, 2019).

15

Industrial Practices in Sustainability

K.K. Brar[1], Dalila Larios Martinez[2], Mitra Naghdi[2], Satinder K. Brar[3], Bhupinder Singh Chadha[1], Preetinder Singh[4], and Rao Y. Surampalli[5]

[1] *Department of Microbiology, Guru Nanak Dev University, Amritsar, India*
[2] *INRS, Université du Québec, Québec, Canada*
[3] *Department of Civil Engineering, Lassonde School of Engineering, York University, Toronto, Canada*
[4] *Ludhiana Beverages, Pvt Ltd, Doraha (Ludhiana), India*
[5] *Department of Civil Engineering, University of Nebraska-Lincoln, Lincoln, NE, USA*

15.1 Introduction

The performance and throughput of business operations depend upon economic structures, policies, availability of resources, distribution of goods and services being provided. The technologies, quality of raw materials and the processing activities being carried out for the manufacturing of goods have a substantial impact on the quality of the environment (www.sustainablescale.org). In the late nineteenth and early twentieth centuries, manufacturing processes were basically dependent on non-renewable sources of energy (www.sustainablescale.org). With the increase in world population, the need for resources has also been increased and, in order to meet the increasing demand, the scale of industrial activities has also grown, which leads to a gradual decrease in the quality and quantity of environmental resources. Over the years, the main focus has been on how to take the maximum output from the industrial sector without considering its impact on the environment. With this outlook, economic activities continue to grow at the cost of environmental degradation, which causes society to suffer. The impact of these increased industrial activities has been so devastating that many ecosystems have been destroyed completely and several others are adversely affected to the point that they can no longer recover (Attah 2010). A recent example of such an impact of environmental change is the bleaching of the coral reefs at the Great Barrier Reef (Kennedy et al. 2018).

This chapter aims to assess the impact of industrialization on the environment and society, and to evaluate the efforts being made by the corporate world, industrial units and governments to control environmental degradation. The chapter examines a number of strategic initiatives being adopted by industries to achieve a balance between the environment, economic growth and society.

Sustainability: Fundamentals and Applications, First Edition.
Edited by Rao Y. Surampalli, Tian C. Zhang, Manish Kumar Goyal, Satinder K. Brar, and R. D. Tyagi

15.2 Causes of Environmental Deterioration

Environmental deterioration is an output of the dynamic interplay of institutional, technological and socio-economical activities. There are many factors that have contributed toward this degradation (Johnson et al. 1997). Table 15.1 lists these factors and their main characteristics.

Table 15.1 Main factors contributing to environmental degradation.

Factors	Characteristics	Reference
Population growth	• It is a source of development, but it exceeds the threshold limits of available natural resources and becomes the major source of environmental degradation • It puts pressure on the environment to fulfill the needs of the population by using natural resources • It has led to increased demand for agricultural land, building human colonies, dumping solid wastes and thereby resulting in deforestation, air and water pollution, and loss of biodiversity	(Harte 2007; Chopra 2016)
Urbanization	• It results from the sudden, rapid and unplanned expansion of cities resulting in the movement of poor families from rural areas toward urban areas • It results in widening of the gap between demand and supply of basic necessities such as living spaces, logistics, transport, communication, energy, medical facilities, education, clean water supplies, sewerage and other amenities, which results in the depletion of environmental resources • It leads to the environmental pollution and overexploitation of land and other natural resources • Increased use of automobiles, which are contributors to air pollution worldwide	(Chopra 2016)
Industrialization	• Produces more to meet the increased demand • Environmental degradation • Risk of public health	(www.sviva.gov)
Discarding of civic norms	• Lack of strict civic norms • No strict environmental policies to regulate industrial activities • The weakness lies in the enforcement of environmental protection and control • Current policies are divided among various government agencies having different policy mandates • Non-availability of comprehensive database and lack of trained personnel resulting in delay of many projects	(Chopra 2016)

Table 15.1 (Continued)

Factors	Characteristics	Reference
Decline in quality and quantity of most of our natural environments	• Intensive agricultural practices due to the increase of population and food supply; lead to deforestation, which contributes to climate change, global warming and loss of habitat for millions of species • Land degradation due to overloading the land with fertilizers, overgrazing and shifting agriculture. Hence leads to accumulation of toxic substances like pesticides and heavy metals • Soil erosion leading to silting in major rivers and water reservoirs, desertification, destruction of fertile lands and nearby vegetation cover • Runoff' of waste into the freshwater and marine environment impacting aquatic life • Emissions of air pollutants such as greenhouse gases, particulate matter and other atmospheric contaminants are the main causes of air pollution and climate change	(www.eartheclipse.com) (www.nationalgeographic.com) (Van der Werf and Petit 2002; Rahman 2015)

15.3 Effects of Environmental Degradation

Since the beginning of their existence, humankind has always been benefited from nature. Initially, the relationship between humans and the environment was very good and was never negative. However, with the advancement of science and technological activities, there have been disruptive results for the environment. For a certain period of time, human beings were not aware of the detrimental effects of their activities toward nature. In that time, the increased human activities resulted in increased harm to the environment in each and every aspect, which was far beyond the renewability capacity of the environment. As a result, the environment started to deteriorate rapidly with significant resultant effects on human health, biodiversity and ecosystems as illustrated in Figure 15.1.

The environment plays an important role in human life and impacts human health in various ways. The interactions between human health and the environment have been studied, and there is evidence that environmental risks have significantly impacted human health by either exposing humans directly to harmful agents or indirectly by disrupting the ecosystem (Remoundou and Koundouri 2009). For example, deterioration of water quality results in billions of cases of illness annually across the world and around 2 million deaths (www.unicef.org). Air quality reduction has resulted in millions of chronic diseases and more than 300 000 deaths annually United Nations Research Institute for Social Development (UNRISD). The most common method of solid waste disposal is landfill, which increases the risk of hazardous materials getting incorporated into the food chain, resulting in biomagnification and more risk of chronic diseases. Altogether, toxic materials and other by-products generated by automobiles, industrial units and agricultural activities have resulted in severe illness and death in children and adults (Donohoe 2003).

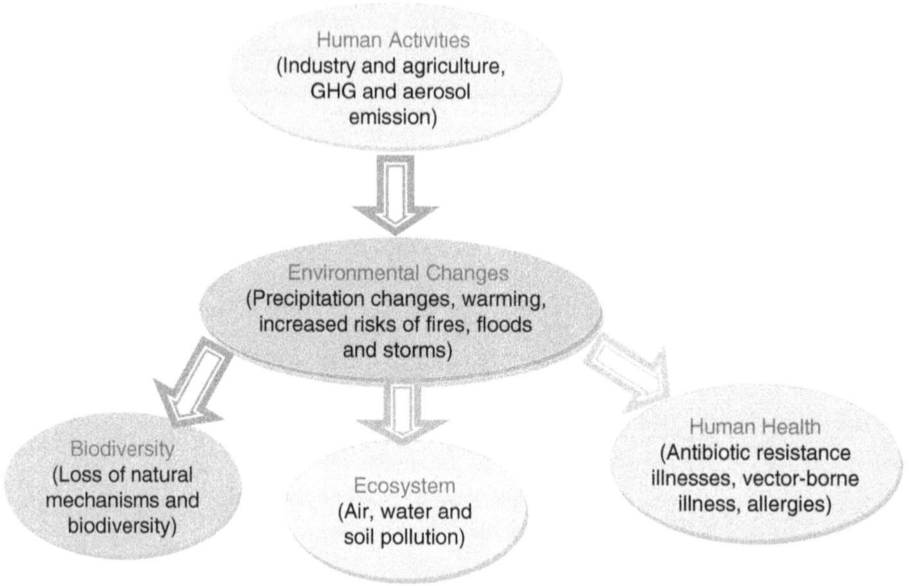

Figure 15.1 Impact of human activities on environment, human health and biodiversity.

The physical and psychological wellbeing of people is affected by air quality to which they are exposed. Air pollution is one of the major factors contributing toward issues related to human health. It is estimated that annually around 2 million premature deaths are caused by poor air quality worldwide. The reduction of air pollution and improvement of air quality would reduce the global burden of various diseases such as respiratory infections, lung diseases and heart failure (Remoundou and Koundouri 2009).

Biodiversity plays an important role in maintaining equilibrium between living beings and the environment by cleaning the environment, combating pollution, maintaining the water cycle of the area, restoring essential nutrients and stabilizing the climate. Human beings have altered the environment and overexploited natural resources to such an extent that they have led to the destruction of the natural habitats of several species. As a result, the species have been forced to migrate to new habitats, modifying the new territory and changing the biogeochemical cycle of that area; thereby bringing more exploitation to that area. Some of the major causes for loss of biodiversity are deforestation, overexploitation of natural resources, population growth at an alarming rate, pollution and climate change (Bellard et al. 2012).

The depletion of the ozone layer is also an indicator of the environmental deterioration. The ozone layer acts as a shield that protects the earth from harmful ultraviolet radiation by preventing them from entering the earth's atmosphere. The production and emission of chlorofluorocarbons (CFCs), hydrochlorofluorocarbons (HCFCs) and volatile organic compounds (VOCs) along with other ozone-depleting substances (ODSs) is the most important reason for the depletion of ozone layer. The major sources for the generation of these ODSs are vehicular emissions, air conditioning units, by-products of chemical processes, refrigerants and aerosols. These ODSs are stable in the lower region of the atmosphere but they are not ideal when they reach the upper atmosphere. In the upper atmosphere these

CFCs reacts with UV radiation and undergoes degradation resulting in the generation of free chlorine atoms, which further react with ozone resulting in depletion of the ozone layer (Chopra 2016).

Global warming is an increase in the average temperature of the earth's surface due to increased levels of greenhouse gases. Since 1880, it has been found that the average temperature of the earth's atmosphere has been increased by 0.8°C (1.4°F) around the world (https://earthobservatory.nasa.gov). With the rise in industries, industrial pollution has also been increased tremendously, resulting in release and build-up of greenhouse gases, which primarily include carbon dioxide (CO_2), methane (CH_4), nitrous oxide (N_2O) and CFCs. These greenhouse gases block some of the heat being radiated from the earth's surface to cause an effect called the greenhouse effect, leading to the warming of the earth. Furthermore, the burning of fossil fuels deforestation also contribute toward global warming.

Global warming has led to global climate change, which has already had observable negative effects on the environment. These include the melting of glaciers at an alarming rate resulting in an increase in the sea level, which could lead to the flooding of major coastlines of the world along with the submerging of islands (Shukla et al. 2017).

Secondly, increased temperature will lead to changes in the evaporation and precipitation pattern, which will lead to severe rainfall in certain areas causing frequent flooding, while other areas will receive less rainfall leading to severe drought. It is one of the most crucial factors that contributes significantly to environmental deterioration. There is evidence to show that the earth's atmosphere is warming at a rate that has no substantial precedent in the history of human civilization. If the earth's temperature keeps on rising at such a pace it might break the balance of the human ecosystem, which was established a long time ago at a lower temperature. Due to the melting of glaciers and polar ice caps, the sea level has been rising; rainfall patterns have been changing drastically leading to severe droughts in some areas and heavier rainfall in others; and the number and frequency of cyclonic storms have been increased.

Evidence has shown that in the past few decades, around 90% of natural disasters have been caused by climate-related natural hazards. The effects of droughts in some regions and floods in others will impact the agriculture potential and reduce the effectiveness of "ecosystem services" such as availability of clean water and fertile soil. Thus, global warming has a detrimental effect on the environment, which ultimately results in mass migration (Isaksson et al. 2010).

With marginal changes in the climate it remains fairly stable over the timescale of decades or centuries, but at the present time it fluctuates continuously, which is influenced by a plethora of variables (Cunningham et al. 1998). There have been noticeable changes in the climate in the past couple of decades, which have adversely impacted the environment, human health and the economy. Hence, finding achievable solutions to climate change has been taken as a top priority for discussion at various national and international forums.

Poverty and the environment are interlinked with each other. If we don't conserve the environment and natural resources on which people rely, we cannot lift them out of poverty. The people who are primarily dependent on natural resources for their livelihood are the ones who suffers the most from the impacts of climate change. Poverty puts stress on the environment and, in turn, environmental issues cause severe suffering to poor people. The non-availability of basic necessities (water and food) is directly responsible for inducing poverty (Barbier 2010; Nwagbara et al. 2012).

Environmental issues cause more suffering among poor people as environmental damage increases the impact of floods, drought and other environmental catastrophes. These catastrophes lead to declining food production and cause shortages of fuelwood, resulting in higher inflation (Duraiappah 1998).

Water is one of the essential elements required by the human beings for their survival. Non-availability of water has resulted in the displacement of millions of people all over the world. Increased seawater level and extended areas of saline water have resulted in the contamination of groundwater and estuaries; thereby reducing the availability of freshwater for humans and ecosystems. Furthermore, the changed precipitation patterns have created pressure on the availability of clean water supplies (Le Houérou 1996).

15.4 Human-Induced Eye-Opening Environmental Incidents

Since the latter part of the twentieth century, with the increasing evidence of environmental pollution, concerns have been aroused regarding the impact of industrial activities on the environment. A variety of industrial accidents are impacting humans as well as different species of plants and animals. The most important human-induced disasters range from wars to nuclear explosions, oil spills, chemical spills and leakage of toxic gases. Example of the worst human-induced environmental disasters include Minamata disease in 1950, the great London smog in 1952, the Love Canal tragedy in 1953, the Great Pacific Garbage Patch, the Amoco Cadiz oil spillage in 1978, the Bhopal gas tragedy in 1984, the Chernobyl nuclear tragedy in 1986, the Exxon Valdez oil spillage in 1989, the E-waste dump in Guiyu, China (Pidgeon and O'Leary 2000) and the BP oil spillage in 2010. In many cases, the loss of property and lives remains beyond repair or full compensation (Pidgeon and O'Leary 2000).

Thus, environmental concerns have led individual nations to establish monitoring and regulating agencies. Some of these include United Nations Educational, Scientific and Cultural Organization's (UNESCO) Intergovernmental Conference for Rational Use and Conservation of the Biosphere, established in 1968; the Environmental Protection Agency, established by the USA in 1969; and the European Environment Agency, established by the European Union in 1993. Along with these governmental agencies, there are many other non-governmental organizations (NGOs) such as the Environmental Defense Fund, established in 1967, and Greenpeace, established in 1972. They have also played a substantial role in striving to take the utmost care of the environment with conclusive steps. In the late twentieth century, efforts by these external forces have pushed businesses to take issues related to the environment more seriously (www.sustainablescale.org).

Human-induced activities have overexploited environmental resources resulting in more extreme weather changes, global warming and challenges beyond human control. If we don't change, the day is not far away when the fossil fuels run out, and when a number of plant and animal species will become extinct.

It is becoming increasingly hard for businesses to focus on achieving economic objectives while neglecting social and ecological issues (Jackson 2011). Moreover, while striving to achieve economic goals, businesses have to incorporate environmental goals into their missions, and strategies have to be defined to achieve them. Corporations have to be mindful about their business strategies and their actions should not have negative impacts on future generations (Isaksson et al. 2010).

15.5 Sustainable Development

The issues pertaining to sustainability arise whenever there is a threat to a valued system or process or object or any attribute that could result in a serious decline of quality or quantity (Sutton 2004). Sustainability can be defined as "to meet the needs of the current generation to fulfill their goals without compromising the ability of future generations to fulfill their own goals" (United Nations 1987). Since human beings are dependent in countless ways on the environment for their needs such as water, food, shelter, leisure activities, entertainment and quality of life, we cannot afford to cause damage to our environment or deplete resources that we can't replenish (DeSimone and Popoff 2000; Dyllick and Hockerts 2002; Rainey 2010).

Sustainable development is the basic principle of meeting the human development goals by providing the required resources. The desired aim is to develop a society where utilization of resources will not compromise the integrity and stability of natural systems (European Commission 1999). Sustainable development requires stern mutual actions by business corporations to reflect a balance between social, economic and environmental considerations both in the present and the distant future. This can be achieved through the integration of policies that connect the environment, economy and society (Dyllick and Hockerts 2002). This concept has given rise to the "triple bottom line," which is used by organizations to evaluate their performance in a broader perspective with respect to people, planet and profit.

Both environmental and industrial policies play an important role in achieving sustainable development. These policies will boost environment protection and provide better opportunities for competition, innovation and employment. Increased environmental and economic performance will further provide better resources to improve social and economic standards. Better profits from economic growth and development will provide better options to companies to upgrade with new technologies to increase efficiency, quality and quantity, which in turn will provide better environmental performances and create more employment opportunities (Bonini et al. 2006).

Organizations are actively working to integrate the needs and expectations of society into their business objectives. This is done not only to meet the expectations of the consumers, customers, employees and other stakeholders, but also to find better opportunities to gain competitive advantage and achieve long-term goals (Rogers 2016). A sustainable business means that an organization is actively participating in environmentally friendly initiatives and introducing better practices and technologies to improve its manufacturing processes, so that current environmental concerns are adequately addressed and have minimal impact on the community, society and economy.

A business is called sustainable if it is using its resources to assess the feasibility of designing and manufacturing products that are environmentally friendly in nature. Secondly, sustainability is assessed by how well a company is utilizing and integrating renewable resources into product development processes. A business is said to be sustainable only if it follows the below-mentioned criteria:

1) Incorporation of environment sustainability principles into business decisions.
2) Providing products or services that are environment friendly by replacing non-green products and services.
3) Greener in comparison with traditional competition.
4) Commitment to principles of environmental protection in its business operations.

Companies are leading in business sustainability if they incorporate environment protection strategies into their core values, administrative policies, goals/objectives and corporate culture. The risks associated with society and the environment are categorized under business-related risks, and strategies are drafted accordingly to control the impact on business operations and objectives (McDonach and Yaneske 2002). As a result decisions related to sustainability become an essential part of the processes related to decision making, commercialization, business planning, customer satisfaction and supplier management.

The current business environment requires corporations to act swiftly in finding efficacious solutions to problems (Dyllick and Hockerts 2002). These challenges include the impact of climate change on business practices, consumption of non-renewable resources, environmental pollution and many more. Furthermore, globalization has increased the concerns related to the deteriorating quality of the environment. Also, with increased stakeholder share in business, their influence on the business practices has also increased (Rainey 2010). Instead of focusing on business performance indicators such as profitability and return on investment, companies are concentrating on the entire value chain and developing more business partnerships that are beyond the traditional financial motives. As a result, socio-environmental issues are not treated as regulatory requirements; instead, they are treated as vital for business. Accordingly, businesses are carrying out their activities with the awareness that socio-environmental concerns are as significant as economic objectives (DeSimone and Popoff 2000).

15.6 Benefits of Being a Sustainable Business

Sustainable businesses are likely to do well by creating and attracting business, revenue and empathy toward the environment and communities at the same time. Some of the major benefits of sustainable business include:

- *Improved brand image and competitive advantage*: A survey has found that 58% of consumers consider a company's behavior toward the environment and are more likely to purchase goods and services from companies that practice sustainable habits. Also, it has been reported that consumers favor those companies that are actively supporting communities and show a positive track record toward personal, social and environmental values (Jasch 2000).

- *Increased productivity and cost reduction*: Development and incorporation of sustainable business practices lead to an efficient operation that streamlines efforts and conserves resources, resulting in enhancement of employee productivity and cost reduction. Energy conservation strategies such as switching off unnecessary lights or incorporation of sensor-operated lights, insulating walls/roofs, and installation of geothermal heating and cooling systems are some of the basic techniques that contribute toward cost reduction. The techniques with a greater overall impact are more costly to implement, but they have more long-term benefits, which justify the investment (Jasch 2000).
- *Increased business ability to comply with regulation*: With the current scenario toward climate change, dwindling energy resources and impact of industrial activities on the environment, state and central governments are enacting regulations to protect the environment. Integrating sustainability into business practices will allow companies to meet changing regulations in a timely manner (Jasch 2000).
- *Attract employees and investors*: It is natural that people like to associate with companies that have a positive attitude toward the environment and do not want to be linked with companies that are implicated in ecological disasters and social welfare scandals. By incorporating sustainability into business, the company is showing a positive response toward the environment and will definitely attract the caliber of people whom they want to employ and the funds required to expand (Jasch 2000).
- *Waste reduction*: This is the simplest way of engaging in sustainable practices. Some of the basic examples include collecting empty cans and polyethylene terephthalate (PET) bottles for recycling, the effort to go paperless by eliminating or greatly reducing paper consumption by going digital and developing processes that use fewer raw materials. Many of these practices are incorporated by industries to minimize waste generation (Jasch 2000).

15.7 How to Achieve Sustainable Development

To achieve sustainable development, companies are developing indicators to evaluate an organization's environmental performance. However, some challenges to face are related to the approaches used to measure and improve their environmental sustainability performance (European Commission 1999).

Some employed approaches are responsible entrepreneurship, eco-efficiency, lifecycle thinking, and sustainable product designing. Also, environment management systems have being implemented by some companies in order to "get a leading edge in environmental issues through effective implementation of responsible entrepreneurship and eco-efficiency (European Commission 1999, pp. 6–7). Table 15.2 presents the main proposed approaches used for companies to achieve sustainable development. These approaches are used by companies to develop environmental impact indicators by using tools such as lifecycle analysis, cost–benefit analysis, benchmarking techniques and environmental friendly responsible image corporation.

15.8 Commitment Toward the Environment

Businesses from various sectors and different associations are developing and incorporating a set of approaches to achieving sustainable business practices. Many of these strategies include improvements in business practices and manufacturing processes in order to meet customer expectations and redesigning the manufacturing process (www.sustainablescale.org). Industries are working on good environmental performance and environmental leadership as well as on making their operations more efficient and cost-effective, leading to a high-quality product. Environmental policies are being created where commitment from the business operations has been made to carry out their operations in a way that protects, preserves and enhances the environment in which their employees are working. In addition, they are committed to continually improving their environmental performance by minimizing or preventing environmental impacts of the activities interacting with the environment, optimizing the use of resources and minimizing generation of waste by employing responsible pollution prevention and control practices.

Business units are incorporating technologies and practices into their business operations to make them more energy efficient, minimizing the usage of non-renewable resources, increasing the dependency on renewable sources, reducing water consumption, minimizing solid/hazardous waste generation, initiating awareness generating programs and many more examples. These companies are defining sustainability goals with relevant action plans and timelines to achieve them. Some of these programs or strategies are described in the following sections.

15.8.1 Sustainability Development Goals

On September 25, 2015, a resolution was adopted by the UN General Assembly on Sustainable Development, in which 17 Sustainable Development Goals and 169 targets were defined for different stakeholders around the world. Industries are incorporating particular Sustainable Development Goals into their own business planning and organizational goals; objectives and efforts are being made to engage in the development of leading sustainability practices and standards to ensure they have significant environmental and social impact.

15.8.2 Climate Protection

With the increase in global energy demand, companies, politicians, society and science are joining forces to defeat climate change and use energy resources more efficiently. Energy efficiency is the key for improving economic development, conservation of resources and climate protection (www.basf.com). Under climate protection agenda, industries are majorly focusing on the areas described in Figure 15.2.

Energy reduction and recovery is a win–win opportunity for sustainable companies due to its positive impacts, decreasing energy cost and the return on investment. Energy reduction programs can be started through "energy audits," which are performed by expert third-party consultants. These audits are prepared to enable the identification of key issues to potentiate saving opportunities, incentives and corrective action to decrease cost

Table 15.2 Approaches to achieve a sustainable development.

Approaches	Characteristics	Reference
Responsible entrepreneurship	• Improves positive influence toward society • Decreases negative impacts on people and the environment • Treating honestly and caring about the wellbeing of employees and consumers • Preserves natural resources and environment protection	(İyigün 2015)
Eco-efficiency	• Produces more products and at the same time decrease resource utilization (energy and materials) and environmental impacts • Promotes the improvement of product quality and the incorporation of energy-efficient processes	(DeSimone and Popoff 2000)
Environmental management systems (EMS)	• Implementation of a structured framework in order to control environmental impacts such as waste, emissions, energy usage, transport and material consumption • This includes the supply chain and environment impact analysis in order to improve resource efficiency, performance, prestige and sustainability	(McDonach and Yaneske 2002)

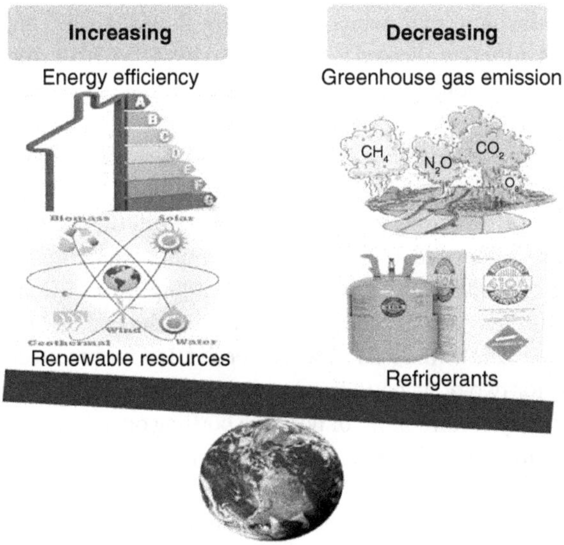

Figure 15.2 Focused areas to achieve climate protection.

and environmental impacts such as greenhouse gases, because a reduction in the amount of energy being consumed in manufacturing operations has a direct effect on greenhouse gas emissions (Lynn and Kjelleren 2015).

The use of renewable energy resources allows the fossil fuel dependency of companies to be decreased. Solar and wind turbine technologies are being increasingly adopted by companies in order to promote climate protection programs across their value chain (agriculture, packaging and transportation) to reduce greenhouse gases emissions, carbon footprints and generation of solid and hazardous waste. At the same time, these companies are targeting the adoption of carbon-offset projects, renewable resources and phasing out hydrofluorocarbon (HFC) refrigerants by replacing them with non-HFC refrigerants. For example, the Coca-Cola Company is working to reduce 25% of CO_2 emissions in 2020 throughout its entire value chain. In the year 2015 and 2016, respectively, the company decreased greenhouse gas emissions by 12% and 14% by changing ingredient sourcing, manufacturing and packaging processes, delivery fleet, refrigeration equipment and the use of renewable resources (www.coca-colaindia.com). This company had replaced harmful refrigeration systems that contain HFCs and installed 1.4 million HFC-free coolers in the global marketplace. Also, Coca-Cola has invested in energy management systems to decrease CO_2 emissions and improved energy efficiency (www.cocacolacompany.com). The eKOcool Solar Cooler program, which is based on the use of solar coolers, is an example of the programs that Coca-Cola Company has promoted in order to maintain energy efficiency, while helping to boost revenue and enhancing lives (www.coca-colaindia.com). The consumption of energy per liter of the beverage being produced is another initiative promoted by Coca-Cola branch offices such as Coca-Cola India. The purpose of the initiative is to decrease the energy usage ratio (EUR) by replacing conventional compact fluorescent lamps (CFLs) with more efficient LED lights, condensate recovery, modernization of steam and heating systems, incorporating energy-efficient technologies into manufacturing processes, installation of daylight roof panels and natural ventilators, and installation of solar power plants. This initiative allowed a reduction in its global EUR from 0.47 in 2007 to 0.40 in 2016 and in its emissions ratio from 40.60 in 2007 to 35.29 in 2016 (www.coca-colacompany.com).

In addition, to generate steam from boilers, instead of conventional furnace oil (FO), briquette and pellets from agricultural waste are being used as biofuels in the industrial boilers to generate steam, which is more eco-friendly and provides a better solution in managing agricultural wastes. Biomass use comprised 48% of the total primary energy use of the Coca-Cola Company (www.coca-colaindia.com). Recently, the bottling firm of Coca-Cola India planned to supply 40% of its energy from renewable sources by the end of 2018. To meet the target, they signed an agreement with Atria Solar Power Pvt. Ltd for procuring 30 million solar power units per year for two of its bottling units in Karnataka, India. This agreement will fulfill 85% of the electricity requirement for both the bottling operations, off the grid. Also, this company has inaugurated its first compressed natural gas (CNG) boiler to decrease carbon emission (www.livemint.com).

In May 2017, Coca-Cola European Partners (CCEP) unveiled its major renewable energy project, with all electricity being generated by a new 8-ha solar farm at Wakefield. The electricity produced is used to support production at its facility. The project delivers 15% of the total electricity being used at the site and resulting in a reduction of the site's operational carbon footprint by 8.6%. The amount of renewable electricity being produced is sufficient to produce approximately 900 cans and 330 PET bottles in every single minute (https://economictimes.indiatimes.com). Thus, by continuing to invest in energy-efficient technologies to reduce energy consumption and reduce carbon footprint, the Coca-Cola Company is an example of a globally sustainable business.

Other companies like Apple are working to decrease its carbon footprint through the use of lower carbon materials to make mobile devices and by partnering with utility suppliers to increase the share of clean energy. In 2016, the carbon footprint of this company was around 10 million metric tons, less than the year before. This was achieved by the implementation of programs to decrease emissions from aluminum manufacturing and also by replacing outdated or inefficient heating, cooling and lighting systems, identifying and repairing compressed air leakages, recovering waste heat, building technical capabilities, conducting energy training programs and increasing awareness toward environmental sustainability (https://images.apple.com). Energy audits were carried out at the supplier facilities to identify the gaps, which resulted in identifying saving opportunities worth US$55 million. Reengineering the product manufacturing process and reincorporating the scrap aluminum into the manufacturing process has resulted in a reduction of greenhouse gas emissions by 60% in the manufacturing of iPhone 7 as compared to iPhone 6. By applying the same approach to the manufacturing of the 13-in. MacBook Pro, there is 48% less emission of greenhouse gases compared to the previous generation MacBook Pro. Another contribution of Apple company is the installation of 485 MW of solar and wind systems across six provinces of China (https://images.apple.com).

Taking a holistic approach toward energy, BMW group is working continuously to reduce CO_2 emissions by a schematic reduction in energy consumption and by focusing on renewable energy generation systems. To power its manufacturing operations at Rosslyn, South Africa, the group has started its first biogas-based power plant, and the energy generated by the power plant will cover around 30% of the manufacturing plant's energy needs. Similar to the Rosslyn plant, the BMW plant in Spartanburg in the USA also uses biogas to generate heat and energy to power the manufacturing operations in the plant. In this way, 50% of the plant's energy needs are fulfilled. Tapping the wind energy, the BMW group has commissioned four wind turbines at its manufacturing location in Leipzig, Germany, with an estimated output of 10 MW, which is sufficient to produce 26 000 MWh of electricity, fulfilling one-fifth of the plant's total power demand (www.climateactionprogramme.org).

P&G has also promoted sustainable projects, such as the installation of the Tyler Bluff wind farm in Texas, and Albany Biomass combined heat and power facility in Georgia. This company is also partnered with EDF Renewable Energy and Southern Company to install 52 wind turbines to produce more than 125 MW of electricity to the grid. In collaboration with Constellation, P&G has also commissioned 50-MW produced from biomass fuel to produce 100% of the steam energy used to fabricate its products (https://us.pg.com).

Microsoft is another company promoting renewable energy projects by using eolic energy in order to support local communities to remove the load from the local grid as much as possible and to supply energy demands for an upcoming data center in Ireland (https://blogs.microsoft.com).

Google is also working on an initiative for the purchase of renewable energy and buying carbon offsets. For example, it is working on a carbon-offset project in collaboration with Oneida-Herkimer Solid Waste Management Authority in order to collect the methane from organic waste decomposition to produce electricity. The electricity being produced is used to power more than 3300 local households. This project enabled the removal of 500 000 metric tons of carbon dioxide (https://environment.google/projects/landfill-NewYork). Furthermore, Google became the first non-utility company to get permission from the Federal Energy Regulatory Commission (FERC) to buy eolic and solar energy. By 2016, the company had signed more than 20 projects to purchase renewable energy (Google Environmental Progress Report 2017).

15.8.3 Water Stewardship

Water scarcity, water pollution and its conservation are also important environmental concerns and a priority for sustainable companies. Some water conservation initiatives include the improvement of water treatment and its reuse, in order to profit and keep water availability for their business in the near future (www.un.org). Water stewardship is one strategy to manage the complexities of the balancing of water needs and water availability. This strategy includes the identification of water-related risks, integrating water strategy into operational plans, engaging stakeholders and suppliers in water-saving initiatives, and achieving compliance with all relevant policies (http://awsassets.panda.org). Business units are working toward water stewardship by incorporating the following practices into their business activities:

- Performing water footprint assessment.
- Water resource management.
- Improving overall water usage efficiency: "reduce" and "reuse."
- Managing wastewater and storm water discharge.
- Local relationships.
- Water replenishment projects.
- Source vulnerability assessment/source water protection plan.

Thus, to support the sustainability of water resources, local communities and companies are engaging with local stakeholders to do source vulnerability assessment (SVA) to identify the risks and prepare source water protection plans (SWPPs) to avoid impacts on the sustainability of local water sources and ensure the industrial facility's water usage and wastewater treatment (http://awsassets.panda.org). To reduce water requirements and operating costs, some companies like Coca-Cola are using the best technologies to optimize manufacturing processes by implanting water recovery strategies like backwash water recovery, reject water recovery, "cleaning in place" rinse water recovery, secondary use of water by stream segregation, water reusability in processing and other water reduction interventions. The recycled or reused water is then used in secondary operations like boiler water make-up, cooling towers, floor cleaning, gardening and toilet flushing (www.coca-colacompany.com). The rest of the water is returned to nature at a level that supports aquatic life. The Coca-Cola water efficiency system has enabled the company to decrease its water use ratio from 2.7 to 1.96 l of water to manufacture 1.0 l of product, and its target is to reduce this to 1.7 l of water per liter of the product by 2020 (www.sustainablebrands.com).

Groundwater replenishment is another strategy used by companies to save water. Companies are working in collaboration with local communities and governments on projects related to the construction of check dams, restoration of ponds and other natural water bodies (www.sustainablebrands.com). For example, in August 2016, Coca-Cola Company returned around 200 billion liters of water to nature and communities, which is equivalent to 115% of the water being used by the company (www.coca-colaindia.com).

Nestle India is also a company developing sustainable action plans for water use. It has decreased by 53% its water consumption per ton of production. This company has saved 65 841 m^3 of water by implementing water recovery strategies and the optimization of water consumption. The company is also promoting with farmers manage and save water strategies, such as the decreasing of water usage by improving irrigation methodology and reducing flood irrigation (www.nestle.in). This company has influenced different water management initiatives in agriculture to reduce overall water footprints throughout the supply chain. For example, through the NESCAFÉ program, Nestle India has started an awareness program under which technical support is being provided to coffee growers on various water conservation techniques in irrigation. Technical assistance is also being provided for rainwater harvesting and recharging practices (www.nestle.in). In partnership with the International Water Management Institute, Nestle India has carried out a study on water footprints of agricultural and dairy production in areas of Punjab, India. Results from this study indicated that changing the cropping pattern, laser leveling of agriculture land, delayed transplanting of paddy and the use of intermittent irrigation could significantly reduce water usage. The company has also developed an awareness program for farmers to raise their awareness of the importance of water, challenges likely to be faced in the upcoming years and mitigation actions that can be taken to avoid any severe consequences (www.nestle.in).

15.8.4 Sustainable Packaging and Recycling

All processes require resource input, which is then converted into the final product. These processes are not always 100% efficient and hence result in waste generation. Some of the waste is reused in the manufacturing cycle; some is sold profitably to other businesses; some is sold or given away at a loss; and some is disposed of (Lynn and Kjelleren 2015). Working toward sustainability, companies are looking to improve their resource utilization to product conversion ratio by incorporating reuse and recycling of materials (www.basf.com). A packaging material or packaging is defined as sustainable only if it meets the following four principles:

1) *Package effectiveness*: Whether the packaging material or the package is fit for goal and achieves its functional requirements with minimal social and environmental impact.
2) *Package efficiency*: Whether the packaging material is designed to use a minimal amount of material with energy efficiency throughout the product lifecycle.
3) *Package recyclability*: Whether the packaging material is designed to reduce reliability on non-renewable resources and is made up of material that is recyclable or reusable.
4) *Package reliability*: Whether the packaging material is designed in such a way that it is non-toxic and non-polluting and does not cause any risks to humans or the environment (www.nestle.in).

Sustainable packaging involves increased use of product lifecycle assessment (LCA) and lifecycle inventory (LCI), which provide effective guidance for effective utilization of the packaging and, in turn, result in reduced environmental impact and ecological footprint (Lee and Xu 2005). The basic concept of LCA is to assess all types of environmental burdens connected with a product or service, starting with the raw materials and moving down to waste removal (Klöpffer 1997).

Business units are working on the concept of the three R's, i.e. "Reduce," "Reuse" and "Recycle." Efforts are being made to reduce the overall weight of packaging material without compromising the quality of the product. This can be achieved by optimizing the packaging efficiency, implementing efficient packaging techniques, by installing better packaging equipment and by investing in preventive maintenance. Coca-Cola has taken several steps for "light-weighting" of its products, by making some fundamental changes in the designing of the bottles, resulting in an effective reduction in the weight of PET and returnable glass bottles. Innovations like replacing the conventional round bottle base with the conical bottom in the Kinley packaged drinking water PET bottle and introducing nitro filling in hot-fill juice-based drinks/Kinley packaged drinking water have significantly resulted in grammage reduction while keeping the shape and pressure intact. As compared with the earlier bottle designs, these changes have led to 15% weight reduction in sparkling PET bottles, 14% in juice PET bottles, 30% in packaged drinking water PET bottles and 33% in returnable glass bottles. All of these have a significant impact on the logistics of operations as well as on the environment. When the weight of the resin used to make the preform is reduced, this reduces the overall logistics cost to manufacture the "drink in your hand," which in turn helps to further cut down the carbon emissions (www.coca-colacompany.com).

Nowadays, companies are moving toward extended producer responsibility (EPR), in which the ownership of the product is superseded by the service being proposed by the product, keeping the environmental impacts of the services in mind (Lee and Xu 2005). Businesses have now become much more interested in the designing of their products and goods as, once the useful life of the product is over, the goods are no more in use and are thrown away as waste, which has the potential to create environmental issues. This has increased the producers' interests in designing goods that are biodegradable or recyclable, more durable, can be easily disassembled, can be reused or composted and leave no toxic or chemical residues (Klöpffer 1997). What was previously a waste and a liability has now become an asset and a source of new product material. This is known as closed-loop production, where efforts are made to reach the highest productivity from materials in repeated loops via the production cycle (www.sustainablescale.org).

Recycling is the process of collecting used items, products, material, disassembling and separating them into different categories and finally processing them into recycled products. During the recycling process, the identity, originality and functionality of the original material are lost hence impacting the overall quality of the recycled material being produced (https://en.wikipedia.org). Nowadays, the packaging materials are being manufactured by using recyclable materials. The recycled materials can be reused in a number of other products by using various advanced technologies without much impacting the overall quality of the product. With regard to sustainable packaging, the Coca-Cola Company has set an industry-first aim to collect and recycle the equivalent of every can or bottle it sells globally by 2030. To achieve this, the company has initiated the holistic plan "World Without Waste" based on the entire packaging lifecycle, starting from how cans and bottles are designed and made, to how they are recycled and repurposed (www.coca-colacompany.com).

In 2002, Coca-Cola in collaboration with Mexican Plastics Industry and others created a non-profit organization, Ecology and Corporate Commitment (ECOCE), which is dedicated to encouraging a culture of recycling. The company has also invested to create the country's two food-grade PET plastic recycling facilities, IMER and PetStar. With the help of these recycling facilities, Mexico has increased its recycling capability drastically. In 2002, Mexico recycled only 9% of PET plastic, while in the year 2016 it recycled 57% of the PET plastic it produced; making it a leading country globally for recycling PET (www.coca-colacompany.com).

To help tackle PET packaging waste and to help achieve a debris-free environment, the Coca-Cola Company has made two strategic partnerships: first with Ocean Conservancy under "Trash Free Seas Alliance" and second with the World Wildlife Fund (WWF) and the Ellen MacArthur Foundation under the "New Plastics Economy initiative." Under its continuous efforts toward environmental sustainability, Coca-Cola has been the leading sponsor of the world's largest volunteer effort for the marine environment under "Ocean Conservancy's International Coastal Cleanup." Over the course of this partnership, the company has helped to mobilize around 20 million people to clean up around 100 000 metric tons of trash from 603 000 km of coastline (www.coca-colacompany.com). In 2005, Coca-Cola initiated a buyback and recycling scheme for used PET bottles irrespective of their source or brand in cooperation with Alkem Nigeria Limited, a fiber processing manufacturer, and the two together launched Nigeria's first bottles-to-fiber recycling operation. Through this program, the volume of bottles that are recycled has risen from 135 metric tons in 2005 to 6200 metric tons in 2012 and the initiative has also helped to create jobs for an estimated 1500 people (www.coca-colaindia.com).

PepsiCo is working on developing new packaging materials with a lower environmental impact and, for this, it is working with Danimer Scientific to develop compostable packaging material for PepsiCo's snack brands. Furthermore, the company is working on a reducing and removing concept wherein the use of conventional packaging material is being reduced and use of recycled material is being increased. In 2016, the company used 143 million lb (6.49 million kg) of recycled PET (rPET), an increase of approximately 3% as compared to 2015. Also, the company is converting shrink sleeves (a frequently used label on beverage containers and other packages) to recyclable material on products like Gatorade and Lipton Pure Leaf products (www.pepsico.com).

The renowned automotive company BMW is continuously working on the concept of intelligent application of raw materials by incorporating lifecycle engineering during the design and development of their vehicles. The company is using up to 20% of thermoplastic materials made from recyclates. The company has implemented an end-of-life vehicle recovery and recycling system, under which, instead of disposing of the vehicle, the company is recovering end-of-life vehicles, components and materials, and then reintegrating them into the raw material cycle. Furthermore, the group has introduced innovative lightweight material carbon fiber reinforced plastic (CFRP) to manufacture their vehicles, which contribute toward efficient dynamics, including reduced emissions and fuel consumption (www.bmwgroup.com).

P&G, in partnership with recycling experts TerraCycle and SUEZ, has produced the world's first recyclable bottle of shampoo, which is made up of 25% recycled beach plastic. The initiative has received support from thousands of volunteers and NGOs for collecting waste plastic from the beaches. To further expand its positive packaging impact, the company has developed technology to improve the quality of recycled polypropylene (PP). P&G has licensed the technology to Pure Cycle Technologies (PCT) to produce recycled PP plastic of "virgin-like" quality (www.coca-colacompany.com).

Companies are continuously working on improving the quality of the material being used in packaging. Efforts are made to make packaging from more biodegradable materials. Renewable polymers for new biodegradable plastic packaging have been developed. The material for this packaging comes from maize and sugarbeet. Later on, these renewable polymers will be replaced by more advanced biodegradable cellulosic packaging material. Furthermore, synthetic biodegradable polymers, edible materials based on natural polymers and partially degradable starch-synthetic polymer blends, have also been developed (Klöpffer 1997).

With the increased demand for plastic packaging and containers, the concept of bio-plastics has been introduced. Bio-plastics behave in a similar way to normal plastics but have superior technical performance as compared with regular plastics, are completely biodegradable and have a reduced impact on the environment. These starch-based bio-plastics are used in a vast number of applications such as in the fabrication of fast food serviceware, industrial packaging's, laminated papers, film wrapping and food containers (Klöpffer 1997). Coca-Cola has introduced the concept of "PlantBottle packaging," which is the first ever recyclable PET beverage bottle that is partially made from plant parts. The material of these bottles functions and looks just like traditional PET plastic, but has a lighter footprint on the planet and its scarce resources. The bottle is made through a process that turns sugarcane into a key component of PET plastic (www.bioplasticsnews.com). Since its launch, the Coca-Cola Company has distributed more than 35 billion bottles of its current version of PlantBottle packaging in nearly 40 countries. These bottles are made from up to 30% plant-based materials and it is estimated that the use of these bottles has helped to save more than 315 000 metric tons of potential CO_2 emissions, which is equivalent to the emissions being produced annually by burning 41 million gallons of gas or by burning around 845 000 barrels of oil (www.coco-colacompany.com). In 2013, Coke expanded the use of PlantBottle technology beyond packaging by partnering with the Ford Motor Company. Together, Ford and Coke unveiled the first ever fabric covering made with Coke's PlantBottle Technology used in the seat cushions, head restraints and more for the Ford Fusion Energi plug-in hybrid vehicle (www.coca-colacompany.com).

15.8.5 Sustainable Transportation

With population expansion, more and more vehicles are coming onto the road and hence increasing their contribution to carbon emissions; thus polluting the environment. In order to help positively the environment we live in, sustainable transportation is a must and the introduction of electric cars is a great step forward. Electric vehicles have the potential for significant contributions in achieving environmental protection aims in transport and hence they are considered as the most sustainable form of transportation.

There are basically four types of factors that are considered to evaluate the impact of electric cars on the environment:

1) *Tailpipe emissions*: Tailpipe emissions are also called direct emissions and result from the burning of fuel. Since, electric vehicles run on electricity, there are no tailpipe emissions. In the case of plug-in hybrid electric vehicles, there are no tailpipe emissions when the vehicle is running in electric mode but tailpipe emissions are produced when they convert to internal combustion engine mode. When compared with conventional gasoline-powered vehicles, the electric vehicles are much more eco-friendly (https://us.pg.com).
2) *Well-to-wheel emissions*: This includes the overall emissions of greenhouse gases and air pollutants, including those that are emitted during the production and distribution of electric energy, required to power these electric cars. Considering well-to-wheel emissions, all-electric cars used to emit around 4500 lb (2041 kg) of CO_2 equivalent, while conventional gasoline cars used to emit twice that amount annually. The emissions produced by electric vehicles are mainly associated with the source of electricity used for recharging these cars. The emissions will be higher if the electricity is supplied from coal-based thermal power plants as compared with electricity being supplied from power stations that are using renewable sources of energy such as wind, solar, nuclear or hydropower (https://us.pg.com).
3) *Source of energy*: This includes the source of electricity being used for recharging electric cars. The electricity is being produced and distributed through local grids and also includes well-to-wheel emissions. If the electricity is produced by using renewable sources of energy, then there will be zero emissions. Companies and governments are providing community solar charging stations, where solar energy is being utilized to produce electricity and is stored in the batteries. Even the owners of electric cars are installing rooftop solar panels on their homes to recharge their cars (https://us.pg.com).

Car's efficiency: Another reason why electric cars are considered more sustainable is that they are more efficient compared with conventional cars. When gasoline is used in traditional vehicles, about 17–21% of energy is converted to provide power to the car, whereas electric vehicles are capable of converting 59–62% of energy to power the vehicle. Furthermore on average the miles per gallon equivalent (of gasoline) for conventional cars is around 24.8 (10.5 km l^{-1}), while on the other hand, electric vehicles have a fuel economy of 100 miles gal^{-1} (42.5 km l^{-1}), which is four times the efficiency of conventional cars (https://us.pg.com).

In addition, an increase in electric vehicles will lead to a decrease in the demand for and consumption of gasoline and hence decrease the burden on non-renewable sources of energy. Furthermore, with the introduction of solar panel-based electric vehicle recharging stations, the emissions corresponding to vehicle recharging by using electricity generated from thermal power plants will be eliminated or reduced to desirable limits (www.energysage.com).

Many leading automobile manufacturers are investing in the electric platform. BMW is working on hybrid cars instead of purely electric cars and introduced two models i3 and i8 in 2013 and 2014, respectively (www.afdc.energy.gov). Tesla has launched its Model S, Model X and Model 3 electric car models (www.bmwusa.com).

The transportation sector is also actively working to enhance and reduce the adverse environmental impact of transport through cooperation with transport and distribution providers, to improve efficiencies in product sourcing, distribution networks, modal shifts, vehicle choice and route planning.

15.8.6 Sustainable Sourcing

The food and beverage industries are majorly dependent on agricultural produce; thus, for the long-term growth and prosperity of the industry, it is crucial that farming systems should be sustainable. Significant initiatives have been taken to improve the sustainable supply of key commodities with the development of many new technologies, schemes and policies in recent years. Many companies are integrating sustainable sourcing into their business strategy, corporate policy and product design.

The ultimate goal of sustainable sourcing is to develop strong and long-term relationships with suppliers. The major part of the overall process is to improve performance in environmental, social and ethical issues. The approach toward sustainable agriculture seeks utilization of natural resources in such a way that they can regenerate their productive capacity by minimizing harmful impacts on the ecosystem. Working toward sustainable sourcing has become a part of the overall business structure and model. The companies concerned are working on the following factors, which are having an impact on sustainable sourcing:

- *Supplier guiding principles*: Working toward sustainable sourcing, the Coca-Cola Company has developed "Sustainable Agriculture Guiding Principles (SAGP)" consisting of 15 principles that provides a framework for the company's commitment to sustainable sourcing. The company is engaging the suppliers of all key ingredients through a "Supplier Engagement Program" to ensure that they meet all the requirements mentioned in the SAGP. Globally in 2016, the company sourced more than 1 million tons of sustainable sugar and procured nearly 100% of tea and coffee from sustainable sources (www.tesla.com).

  In India, through project "Unnati," Coca-Cola is partnering with experts from Jain Irrigation to provide training to around 25 000 mango growers with respect to the new farming techniques like Ultra-High Density Plantation (UHDP), which helps them to increase the quality and quantity of the fruit they grow (www.coca-colacompany.com). The company has also initiated another project named "Meetha Sona Unnati," which is focused on sustainable sugarcane development. The company has partnered with DCM Shriram, IFC and Solidaridad to help address the challenges faced by the local farmers with respect to soil health, poor yield and increased cultivation costs (www.coca-colacompany.com). Through this project, the company is investing in training and capacity building of sugarcane growers along with strengthening the extension services and sugar mill management systems to speed up adoption of sustainable production standards. The project is aimed at providing interventions with respect to techniques to improve water efficiency, plantation methods, mechanization and educating about the negative impacts of crop residue burning (www.coca-colacompany.com).

  Working toward sustainable agriculture programs, food and beverage companies are adopting Voluntary Sustainability Standards so as to promote sustainability across the supply chain. These standards are legally non-binding and have a positive impact on the general development of responsible corporate conduct. These standards act as a principal instrument to promote sustainability by limiting environment damaging practices, improving livelihood of farmers and providing new opportunities for product marketing. Fairtrade, Rainforest Alliance, UTZ Certified and Organic/Bio are some of the major sustainability standards that are working toward sustainable supply chains, meeting the needs of farmers, companies and consumers (www.coca-colacompany.com). Unilever has developed the "Unilever Sustainable Agriculture Code (SAC)," which provides a set of rigorous standards to farmers and suppliers to drive improvements across the supply chain. The suppliers have to comply with the SAC requirements through self-assessment or through any third-party certification (Rainforest Alliance Certification) that has more stringent requirements than Unilever (Von Hagen et al. 2010).

- *Improved irrigation techniques*: Unilever has initiated a project in the Gastouni region of Greece to save water by incorporating simple tools in the tomato fields and the technique has resulted in reducing water consumption by 28%. The tool is called a Tensio-meter and is used to measure the amount of water present in the soil. By reading the data shown by the device, the farmers get to know instantly when their fields need irrigation and they water them accordingly. The technique has resulted in a reduction in water consumption from 86 $l\,kg^{-1}$ of tomatoes in 2010 to 56 $l\,kg^{-1}$ in 2017. Furthermore, with the introduction of this new approach in irrigation, the crop yield has increased from 76 to 81 tons ha^{-1} (Von Hagen et al. 2010).

  PepsiCo has initiated an agriculture process in India in which direct seeding of rice is being done in the region that faces a severe shortage of water. The initiative has resulted in a reduction of water usage by avoiding three crucial water-consuming steps, i.e. puddling, transplanting and standing water. After extensive research and development, field trials were successfully done in 2010, and the company has expanded the direct seeding to 10 000 acres (4047 ha) of farmland, thereby saving around 7 billion liters of water (www.unilever.com).

  Companies are also working with farmers in water stress regions on techniques such as drip irrigation and sprinkler irrigation. PepsiCo is helping farmers in water-scarce areas by promoting drip irrigation in over 3000 acres (1214 ha). The initiative includes providing incentives to the farmers for adopting drip irrigation and providing help in designing agriculture equipment to make drip irrigation more viable for farmers (www.unilever.com). In India, PepsiCo has increased the area under cultivation of potato crop through drip irrigation from 200 acres (80.9 ha) in 2009 to 1600 acres (647 ha) in 2013. This is around an eightfold increase and constitutes more than 50% of the total area under drip irrigation supported by PepsiCo. Technology has helped the company to save more than 1.5 billion liters of water since 2009 (www.pepsicoindia.co.in).
- *Better pest management*: The companies concerned are working on optimizing agrochemical application and trying to minimize the usage of insecticides and pesticides in the fields. PepsiCo has initiated a project on the same in several countries under which an online data application advises the grower on the type of pesticide, the correct amount required and the best time to apply it to the fields. This tool has resulted in improved usage efficiency and also lowering the amount of active ingredient applied (www.unilever.com).

Implementation of pest management programs encourages the use of tools such as weather monitoring, which helps the farmers to predict the arrival of pests and hence results in the effective identification and application of target pesticide. This type of targeted approach results in lowering the risk to other non-target insects that are beneficial to the crops (www.unilever.com).

- *Optimization of fertilizers and agrochemicals*: The companies are working with the aim of optimizing the use of fertilizers and other agrochemicals by substituting them with natural controls. The companies in collaboration with farmers are developing and implementing policies and procedures to calculate the number of pesticides being used in their supply chains and efforts are being made to minimize their impact (www.unilever.com).

  PepsiCo used to encourage suppliers to use safer insecticides and alternative types of fertilizers that are more natural. With the aim of reducing synthetic fertilizers for potato production by 40%, the company is taking part in trials on potato farms in Turkey in cooperation with local farmers by using a natural fertilizer, Naturalis, prepared from the waste being generated from the chip factory (www.pepsicoindia.co.in).

  Working on sustainable agriculture practices, in association with its mushroom supplier in Shanghai (China), Unilever has resulted in increased yields of mushroom and better quality by eliminating pesticides from the growth chain. This is achieved by converting waste organic matter into a compost soil mix and then sterilizing it with steam to control weeds and pests. The program won the Green Development Excellence Award in the 2016 Invest in Green Future Corporate Social Responsibility awards (www.pepsicoindia.co.in).

15.8.7 Sustainable Building

Sustainable building, known as green building, refers to both an application and the structure of processes that are environmentally responsible and resource efficient throughout a building's lifecycle: from planning to design, maintenance, renovation, construction, operation and demolition (www.pepsico.com).

An innovative total systems approach is being used to develop sustainable buildings under which special attention is being given not only to the types of materials being used in the construction but the overall building designing process. Creating a design team that combines various technical groups and professionals with the people who will use the building and making a commitment to make it as environmentally benign as possible, can lead to remarkable outcomes in terms of energy needs and reducing materials (www.sustainablescale.org). The ING Bank's headquarters in Amsterdam is a good example of this approach: it has 12 times more energy efficiency than its predecessor despite having more agreeable space, fresher air and comparable cost to a traditional building of comparable size (www.sustainablescale.org).

The common objective of the green building concept is to develop buildings in a way that uses the minimal amount of natural resources at the time of construction and during operation using the following methods (Khasreen et al. 2009):

- Effective and efficient utilization of water, energy and other resources.
- Improving occupant health and productivity.
- Reducing wastage, increasing reusability and recyclability of material.

These green buildings include measures to reduce energy consumption for infrastructure set-up such as the energy required to extract, transport, process and install building materials and to provide services such as heating and power for equipment. This is done by installing high-performance windows by minimizing air leakages between conditioned and unconditioned spaces and by providing extra insulation in walls, ceilings and floors (Khasreen et al. 2009). In addition, effective window placement and installing natural light sources can provide more natural daylight and thereby eliminate the need for electricity-based lighting during the daytime. Installation of solar panels for solar water heating further reduces energy costs. Furthermore, utilization of technologies such as solar power or biomass for onsite generation of renewable energy significantly reduces the environmental impact of the building.

The Leadership in Energy and Environmental Design standards (LEED) were developed by the US Green Building Council in an effort to propel green building design in the United States (Chen and Huang 2012). LEED certification is a very prestigious title and can be achieved through compliance with all environmental regulations and laws, occupancy scenarios, building permanence and site boundaries, pre-rating completion and area-to-site ratios, and obligatory 5-year sharing of whole-building energy and water use data from the start of occupancy (for new construction) or date of certification (for existing buildings) (Chen and Huang 2012). Some of the world's most sustainable buildings include the following:

- *Manchester's One Angel Square*: This is powered by rapeseed oil, which is fed into a combined heat and power system. LED lighting is used to light the building and there is a system in place to recycle the waste and rainwater. The building was awarded "Outstanding Building Research Establishment Environmental Assessment Methodology (BREEAM) (Dobiáš, and Macek 2014).
- *East London's The Crystal*: This building runs entirely on electricity majorly generated by photovoltaic solar panels. A combination of LED and fluorescent lighting is used to light the building, which automatically gets switched on or off depending upon the amount of natural daylight present. The building has its own rainwater harvesting structure and the sewage generated is treated, recycled and reused onsite. The building is LEED and BREEAM certified (Dobiáš and Macek 2014).
- *Manhattan's Bank of America Tower*: This is one of the world's greenest skyscrapers and is the world's first high-rise building that has received LEED certification. The best features of the building include CO_2 monitors, LED lightning and waterless urinals. The building has its own power generation plant that produces 4.6 MW of clean sustainable energy (Dobiáš and Macek 2014).

15.9 Future Action Needed

The sustainable business development movement is an important step toward developing a sustainable economy. Widespread adoption of various approaches and incorporation of advanced technologies in industrial and business operations have resulted in significant reduction in the use of non-renewable resources and elimination of waste/toxins (www.sustainablescale.org). Although much progress has been achieved by industrial units and business with respect to the improvement of environmental performance, it is necessary to keep up the pace by further encouraging a more proactive approach on the part of the business. In order to develop such proactive behavior, it is essential that policymaking is designed in such a way that the achievement of sustainable development becomes a concern for the whole of society. This implies that the policy is developed and implemented in close partnership and cooperation and the responsibility of all stakeholders is well identified and acknowledged (www.sustainablescale.org).

Moreover, environmental externalities should be internalized by way of price signals, as the industry best knows how to appropriately react to changing prices. To make the industry more responsible for environmental sustainability, along with responsible entrepreneurship, eco-efficiency, innovation and preferences should also be given to market-based measures and cost effectiveness (Harte 2007). Moreover, it is essential to apply new and integrated appraisal methodologies along with existing tools such as cost effectiveness and cost–benefit analysis for assessing the impacts of policy measures on the environment, competitiveness and innovation, as well as implementing policy and developing performance indicators to measure progress toward sustainable industrial development. Finally, the effective implementation and promotion of an integrated product lifecycle approach will certainly contribute to increasing the synergy between environmental protection and industrial development, especially as far as measures and activities in the field of industrial policy are concerned (Harte 2007).

References

Attah, N.V. (2010). Environmental sustainability and sustainable growth: a global outlook. MSc thesis. https://repository.upenn.edu/cgi/viewcontent.cgi?article=1033&context=od_theses_msod (accessed October 27, 2019).

Barbier, E.B. (2010). Poverty, development, and environment. *Environment and Development Economics* 15 (6): 635–660.

Bellard, C., Bertelsmeier, C., Leadley, P. et al. (2012). Impacts of climate change on the future of biodiversity. *Ecology Letters* 15 (4): 365–377.

Bonini, S.M., Mendonca, L.T., and Oppenheim, J.M. (2006). When social issues become strategic. *McKinsey Quarterly* 2: 20–32.

Chen, S.-Y. and Huang, J.-T. (2012). A smart green building: an environmental health control design. *Energies* 5 (5): 1648.

Chopra, R. (2016). Environmental degradation in India: causes and consequences. *International Journal of Applied Environmental Sciences* 11 (6): 1593–1601.

Cunningham, W.P., Cooper, T.H., Goreham, E. et al. (1998). *Environmental Encyclopedia.* Farmington Hills, MI: Gale Research.

DeSimone, L.D. and Popoff, F. (2000). *Eco-Efficiency: The Business Link to Sustainable Development.* Cambridge, MA: MIT press.

Dobiáš, J. and Macek, D. (2014). Leadership in energy and environmental design (LEED) and its impact on building operational expenditures. *Procedia Engineering* 85: 132–139.

Donohoe, M. (2003). Causes and health consequences of environmental degradation and social injustice. *Social Science and Medicine* 56 (3): 573–587.

Duraiappah, A.K. (1998). Poverty and environmental degradation: a review and analysis of the nexus. *World Development* 26 (12): 2169–2179.

Dyllick, T. and Hockerts, K. (2002). Beyond the business case for corporate sustainability. *Business Strategy and the Environment* 11 (2): 130–141.

European Commission. (1999). Commission staff working paper on sustainable industrial development. http://ec.europa.eu/environment/archives/action-programme/pdf/sec991729_en.pdf (accessed October 27, 2019).

Google Environmental Progress Report (2017). Google environmental progress report: 2017 progress update. https://storage.googleapis.com/gweb-environment.appspot.com/pdf/google-2017-environmental-report.pdf (accessed October 27, 2019).

Harte, J. (2007). Human population as a dynamic factor in environmental degradation. *Population and Environment* 28 (4): 223–236.

Isaksson, R., Johansson, P., and Fischer, K. (2010). Detecting supply chain innovation potential for sustainable development. *Journal of Business Ethics* 97 (3): 425–442.

İyigün, N.Ö. (2015). What could entrepreneurship do for sustainable development? A corporate social responsibility-based approach. *Procedia – Social and Behavioral Sciences* 195: 1226–1231.

Jackson, T. (2011). Societal transformations for a sustainable economy. *Natural Resources Forum* 35: 155–164.

Jasch, C. (2000). Environmental performance evaluation and indicators. *Journal of Cleaner Production* 8 (1): 79–88.

Johnson, D.L., Ambrose, S.H., Bassett, J.L. et al. (1997). Meanings of environmental terms. *Journal of Environmental Quality* 26: 581–589.

Kennedy, E.V., Alvarez, A.O., and Diaz-Pulido, G. (2018). Coral bleaching in the southern inshore great barrier reef: a case study from the Keppel Islands. *Marine and Freshwater Research* 69 (1): 191–197.

Khasreen, M.M., Banfill, P.F., and Menzies, G.F. (2009). Life-cycle assessment and the environmental impact of buildings: a review. *Sustainability* 1 (3): 674–701.

Klöpffer, W. (1997). Life cycle assessment. *Environmental Science and Pollution Research* 4 (4): 223–228.

Le Houérou, H.N. (1996). Climate change, drought and desertification. *Journal of Arid Environments* 34 (2): 133–185.

Lee, S. and Xu, X. (2005). Design for the environment: life cycle assessment and sustainable packaging issues. *International Journal of Environmental Technology and Management* 5 (1): 14–41.

Lynn, G.E. and Kjelleren, G.G. (2015). The business case for environmental sustainability. *Business Law Today* January https://environblog.jenner.com/files/the-business-case-for-environmental-sustainability.pdf (accessed November 2019).

McDonach, K. and Yaneske, P.P. (2002). Environmental management systems and sustainable development. *Environmentalist* 22 (3): 217–226.

Nwagbara, E.N., Eucharia, N., Raphael, A.P. et al. (2012). Poverty, environmental degradation and sustainable development: a discourse. *Global Journal of Human-Social Science Research* 12 (11-C).

Pidgeon, N. and O'Leary, M. (2000). Man-made disasters: why technology and organizations (sometimes) fail. *Safety Science* 34 (1): 15–30.

Rahman, S. (2015). Green revolution in India: environmental degradation and impact on livestock. *Asian Journal of Water, Environment and Pollution* 12 (1): 75–80.

Rainey, D.L. (2010). *Sustainable Business Development: Inventing the Future through Strategy, Innovation, and Leadership*. Cambridge, UK: Cambridge University Press.

Remoundou, K. and Koundouri, P. (2009). Environmental effects on public health: an economic perspective. *International Journal of Environmental Research and Public Health* 6 (8): 2160.

Rogers, M. (2016). 6 benefits of becoming a sustainable business. Environmental Leader. https://www.environmentalleader.com/2016/03/6-benefits-of-becoming-a-sustainable-business (accessed October 27, 2019).

Shukla, J.B., Verma, M., and Misra, A.K. (2017). Effect of global warming on sea level rise: a modeling study. *Ecological Complexity* 32: 99–110.

Sutton, P. (2004). A perspective on environmental sustainability. Paper on the Victorian commissioner for environmental. *Sustainability*: 1–32.

United Nations (1987). *Report of the World Commission on Environment and Development: Our Common Future*. Oslo, Norway: United Nations General Assembly, Development and International Co-operation: Environment.

Van der Werf, H.M.G. and Petit, J. (2002). Evaluation of the environmental impact of agriculture at the farm level: a comparison and analysis of 12 indicator-based methods. *Agriculture, Ecosystems and Environment* 93 (1): 131–145.

Von Hagen, O., Manning, S., and Reinecke, J. (2010). Sustainable sourcing in the food industry: global challenges and practices. *Moderne Ernährung Heute* 4: 1–9.

16

Challenges of Sustainability in Agricultural Management

Jew Das[1], *Srinidhi Jha*[1], *Manish Kumar Goyal*[1], *and Rao Y. Surampalli*[2]

[1]*Discipline of Civil Engineering, Indian Institute of Technology, Indore, India*
[2]*Global Institute for Energy, Environment, and Sustainability, Lenexa, KS, USA*

16.1 Introduction

Agriculture is derived from the Latin word "ager," which means land and "culture," which means cultivation. Hence, agriculture is the art or science of producing crops and livestock for the benefit of mankind. It occupies more than one-third of earth's terrestrial surface and also utilizes a huge amount of water resources, providing for the basic needs of mankind. Agriculture affects all aspects of human life directly or indirectly. In the present scenario above and beyond farming, agriculture incorporates forestry, poultry, horticulture, beekeeping, etc. Nowadays processing, marketing, distribution of agricultural products, etc. are acknowledged as intrinsic parts of agriculture. Hence, it is worth mentioning that present agricultural practices not only include production but also processing, promoting and distribution of agricultural products. Moreover, agriculture plays a vital role in economic development and can be considered as the backbone of a country's economy. Agriculture creates employment opportunities and is the source of livelihood for a large number of the population. It enhances livelihoods through better medication, education and improvements in infrastructure. The role of agriculture in developing countries like India has been observed through its contribution to the GDP and employment. The sustainability of agriculture in any country depends on the available natural resources, and the ongoing pace of climate change has a predominant impact on natural resources. In India, two-thirds of the total population live in rural areas with agriculture as their livelihood. Due to urbanization, industrialization and limited arable land, increased productivity is more important than the area of cultivation and cropping intensity in India.

Agriculture, an essential for survival and societal development, is causing ecological degradation and environmental pollution as a result of using various chemicals and fertilizers to accomplish the demands of the increasing population and economic growth (Zhao et al. 2008). Reducing the agricultural footprint while ensuring food security will be a serious problem (Foley et al. 2011) as population growth rates will continue to rise in developing regions.

Sustainability: Fundamentals and Applications, First Edition.
Edited by Rao Y. Surampalli, Tian C. Zhang, Manish Kumar Goyal, Satinder K. Brar, and R. D. Tyagi

According to the Food and Agriculture Organization (FAO 2011), 70% of grassland, 50% of savannahs, 45% of temperate forests and 27% of tropical forests have been converted to agricultural land; thus agriculture has become the major driver of deforestation and contributes a major share of greenhouse gas emissions (Kissinger et al. 2012). In this sense, sustainability in agriculture has become a paramount concern and is essential to the sustainable development of social, environmental and economic systems.

Sustainable agriculture is a subject of great concern, and the varying viewpoints regarding the understanding of sustainable agriculture has led to lively debate across the globe (Crews et al. 1991; Keeney 1989; Lowrance et al. 1986). Therefore, due to the lack of a proper definition of sustainable agriculture, it is viewed as management practices, rather than the method of operation (Macrae et al. 1993). Lowrance et al. (1986) segregated the definition of sustainable agriculture into three different parts, namely sustainability as food security, sustainability as community and sustainability as stewardship. Moreover, they stated that the increasing population growth in the present and in the foreseeable future will cause a significant increase in the demand for food, and hence the proper definition of agricultural sustainability needs to be addressed. Abubakar and Attanda (2013) defined sustainable agriculture as the commitment to fulfill the demand for food and enhance society and the quality of life of farmers as a whole, now and in future. The Board for International Food and Agricultural Development (BIFAD 1988) proposed various definitions for sustainable agriculture based on different resources, which include: (i) effective management of agriculture to satisfy human needs without affecting natural resources and environmental degradation; (ii) the ability of the system to maintain production under the situation of social and economic pressures; and (iii) preserving and protecting natural resources and maintaining long-term economic growth by managing all the resources that are exploited for sustainable yields. Therefore, most agriculturalists accept that for the sustainability of the biosphere with increasing population growth the concept of sustainable agriculture plays a significant role and is of paramount importance. The combined focus on the environment, economic value and environmental impacts on society brings us to the concept of sustainability.

Agriculture changed drastically after the end of World War II. Agricultural productivity has increased significantly due to new methodologies, technologies, use of chemicals, application of fertilizers and favorable government policies to maximize productivity. However, agriculture is highly vulnerable to climate change and its effects. In addition to climate change, the rapid population growth has imparted additional threats to the available agricultural resources. For instance, with an increase in global population, the demand for food will increase significantly. To fulfill the increasing demand by 2020 agricultural production should increase by 90–140% for oilseeds, cereal and meat, and 60% for cotton (Farrell et al. 1984). A report by the Intergovernmental Panel on Climate Change (IPCC) stated that climate change has significant impact on the global maize production and has more significant negative impacts on food production than positive impacts. Hence, for most countries, climate-resilient agriculture has become a top priority for suitable adaptation to climate change (OECD 2012).

Moreover, the world's population will reach 9 billion by 2050. Climate change impacts such as increase in drought events, floods, extreme precipitation and temperatures make the situation even more challenging. In order to fulfill the inevitable demand, agriculture will need to produce as much in the next 25 years as it has produced in the past 10 000 years. Therefore, sustainable agriculture has become a major challenge for researchers and policymakers. The prime motive of agriculture is to provide food, and it is quite reasonable to ensure food production through sustainable agricultural management for present and future scenarios. The sharp increase in the demand for food has been augmented due to the change in the standard of living, economic growth, as well as the variability in the available water resources, limited arable land and soil degradation. In addition, ineffective policy planning and management systems constrain the increase in agricultural productivity. Therefore, the idea is to address various issues that degrade the agricultural sustainability and various approaches that efficiently increase the agricultural sustainability, quality of life and economic growth. It is of paramount importance to ensure agricultural sustainability incorporating different conservation techniques, improved irrigation practices, genetically improved varieties and improved soil quality (Dover and Talbot 1987). In this chapter, we discuss the various challenges to agricultural sustainability and the appropriate measures to overcome those challenges from the perspective of Indian agriculture by developing new technologies to sustain and increase yield without degrading the environment. The reasons for choosing Indian agriculture include: (i) three-quarters of India's families rely on rural income mostly from agriculture; (ii) 70% of the total population in India live in rural areas; and (iii) food security in India depends on the production of cereal crops, fruit and vegetables to satisfy the demands of the growing population.

16.2 An Overview of Indian Agriculture

The agriculture sector in India continues to be the mainstay of its economy. The reasons that agriculture is described as the backbone of the country include: first, it contributes a large share of the country's GDP; second, it provides employment to a large proportion of the populations; and third, overall development in other sectors considerably depends on the performance of agriculture. More than 70% of the members of the rural population adopt agriculture as the prime source of their livelihood, and agriculture provides the highest employer among all the workforce i.e. 49%. Figure 16.1 presents the variation in agricultural workers (including cultivators and agricultural laborers).

The agricultural statistics depend on the utilization of the land use potential. It is important to quantify the land use utilization in India to understand whether the land use is at its full potential or far away from the available potential. Initially, the land use classification of India were categorized into five types, namely, area under forest, area under current fallows, uncultivated lands, area not available for cultivation and net area shown. However, later it was realized that the five classifications of land use were not adequate for agricultural planning and hence the land types were reclassified. Figure 16.2 depicts the percentage of change in different land use classifications during 1950–2013 according to the updated classifications.

It should be noted that percentage of change for area under utilization is computed based on the total geographical area, i.e. 328.73 million ha, and percentage of change for other land use classifications is computed based on the area under utilization. It can be noted from

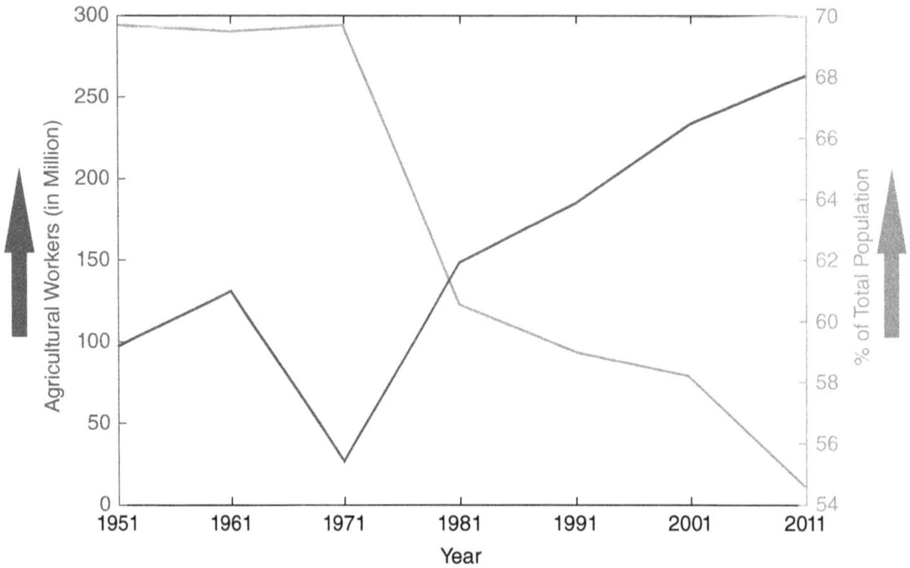

Figure 16.1 Agricultural workers from 1951–2011. *Source:* Government of India 2016.

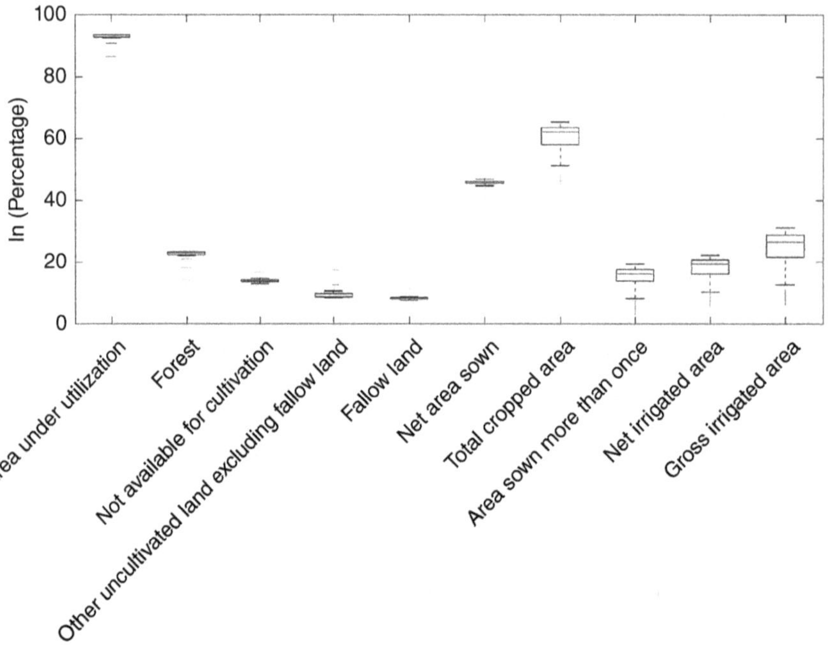

Figure 16.2 Percentage of change in different land use classifications. *Source:* Government of India 2016.

Figure 16.2 that the variation in the change in percentage for the land use classifications such as total cropped area, area shown more than once, net irrigated area and gross irrigated area

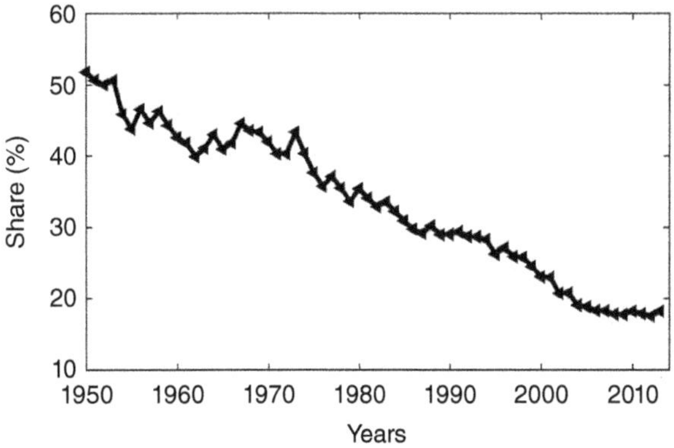

Figure 16.3 Contribution of agriculture to GDP. *Source:* Planning Commission 2017.

is significant. In addition, it can be noted that the percentage change increased during 1950–2013 for these classifications, suggesting an increase in the agricultural demand.

Agriculture contributes 18% of the total GDP of the country (Department of Agriculture and Cooperation and Statistics 2014). The average GDP from agriculture for the 2011–2018 period is reported as INR 4052.74 billion. The sector-wise contribution to GDP reveals a sharp decline in the agricultural sector during 1950–2014 (Planning Commission 2017). During the period of independence, the contribution from the agricultural sector was more than 50% and has decreased to 18% in recent years (Figure 16.3).

In addition to employment, agriculture in India also provides food security: an average Indian spends more than half of their income to ensure ood security (NSSO 2013). Nevertheless, the growth rate in the agricultural sector highly fluctuates and is highly sensitive to meteorological variability. India has made significant improvements in agricultural growth during the post-independence period. In addition, the introduction of the green revolution during the late 1960s significantly improved the agricultural growth rate. Figure 16.4 presents the agricultural growth rate from 1975 to 2017.

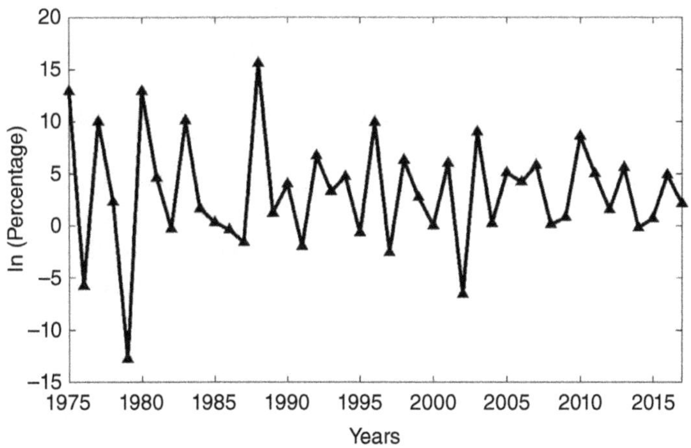

Figure 16.4 Agricultural growth rate. *Source:* www.pincodeindia.net/agriculture-growth-rate.php.

Since independence, India has made significant progress in terms of yields, growth in output and area utilized by different crops. In addition, to bring about significant growth in different agricultural products, India has gone through the green revolution, blue revolution, yellow revolution and white revolution. Currently, India is the largest producer of jute, milk and pulses in the world. Moreover, India has secured second position in rice, sugarcane, cotton, wheat, fruit and vegetable production. Figure 16.5 represents the area and yield of different major crops in India during the period from 1950–2015. It can be noted from the figure that, with the exception of jowar, the yield of major crops has increased over time. More interestingly, despite a decrease in the area, the yield of coarse cereals has increased significantly. The area under cultivation for both coarse cereals and jowar has been decreased, but for rice, wheat, maize and pulses it has been increased since independence.

Figure 16.6 shows the per capita availability of food grains during the period from 1950–2015. The average availability of annual food grains per capita from post-independence to the end of the twentieth century was 167.78 kg and decreased to 165.87 kg during the initial period of the twenty-first century (2000–2015). It should be noted that food grains include rice, wheat, other cereals and all pulses. The availability of cereals, wheat and rice has been increased; that of gram and pulses has shown no significant change; and that of other cereals has been decreased post-independence.

The growth rate of the agricultural sector significantly depends on investment from different sectors. The government plays a vital role in boosting the agricultural sector through its increased investment and encouraging the private sector to do the same. An economic survey report by the government suggests the necessity for mutual involvement of government and private sectors toward the growth of agriculture. The overall growth of the Indian economy primarily depends on the performance of the agricultural sector. Public investment has not contributed significantly to its development: a steady decrease in public investment over the years has been reported in capital formation in Indian agriculture. On the other hand, private investment has been increased over the years. According to Economic Survey (1999) the total investment in gross capital formation in agriculture during 1960–1961 was 1668 crores, of which the public sector share was 35.6%, while the private sector share was 64.7% (at 1980–1981 prices). However, during 1996–1997 the total investment was 6999 crores, of which 16.2% was from the public and 83.8% was from the private sector. Moreover, from Figure 16.7 it can be noted that the investment from the private sector increased and from the public sector, the contribution decreased during 1960–1997.

Foreign trade in agricultural commodities plays a vital role in the country's economic development. With subsequent economic reforms, removing the restrictions, and with free trade agreements, India's share in the global agricultural trade has increased over the years. Figure 16.8 represents agricultural exports and imports as a percentage of the total national exports and imports. It can be noted that the export of agricultural commodities is significantly higher than the import of commodities, making India's contribution significant in overall world imports and exports.

From this overview of Indian agriculture, it can be concluded that agriculture continues to be the mainstay of Indian economy, providing a livelihood to more than half of its population. The agriculture sector plays a significant role as a major contributor to the country's

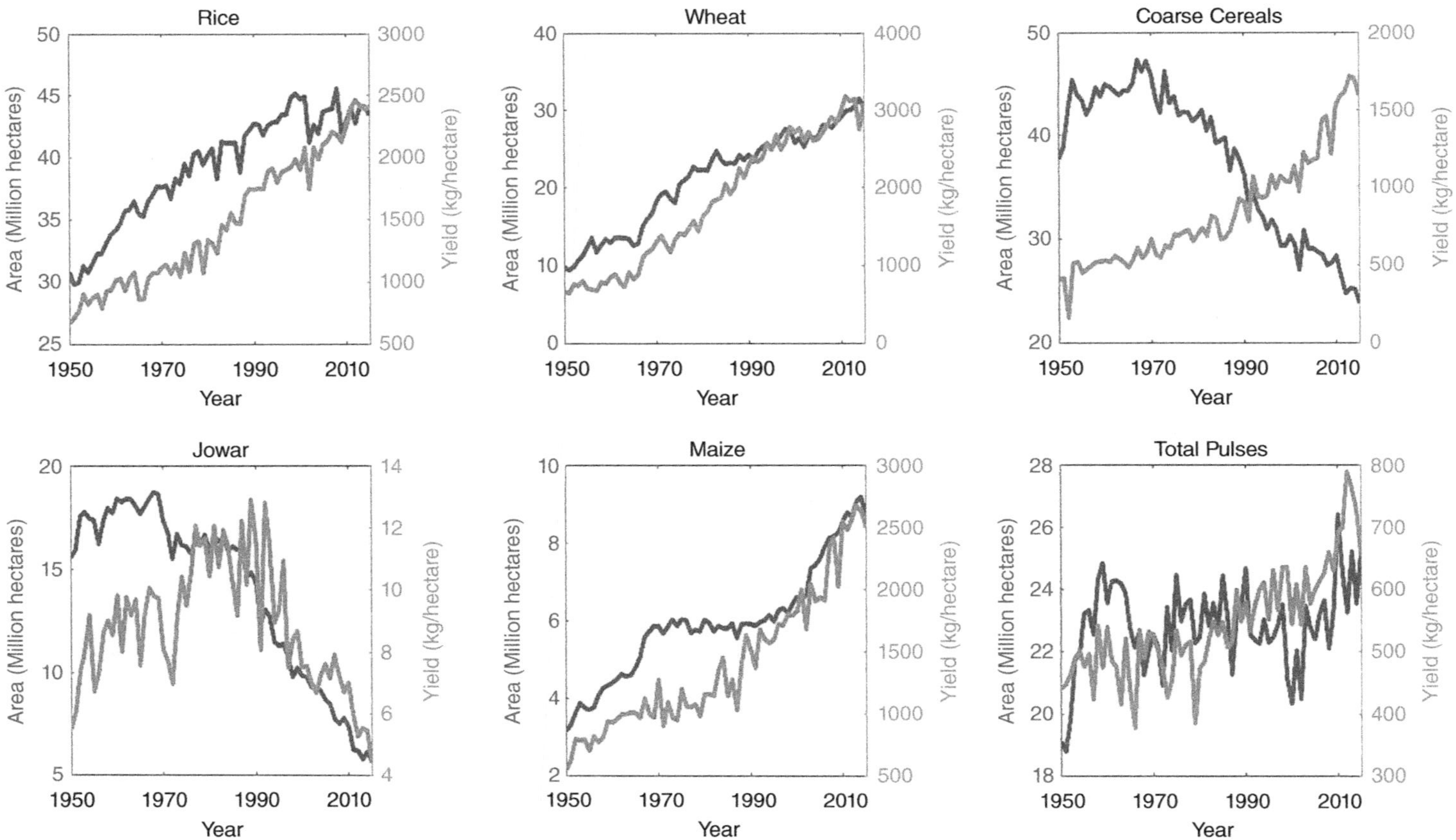

Figure 16.5 Area and crop yields of different major crops in India during 1950–2015. *Source:* Government of India 2016.

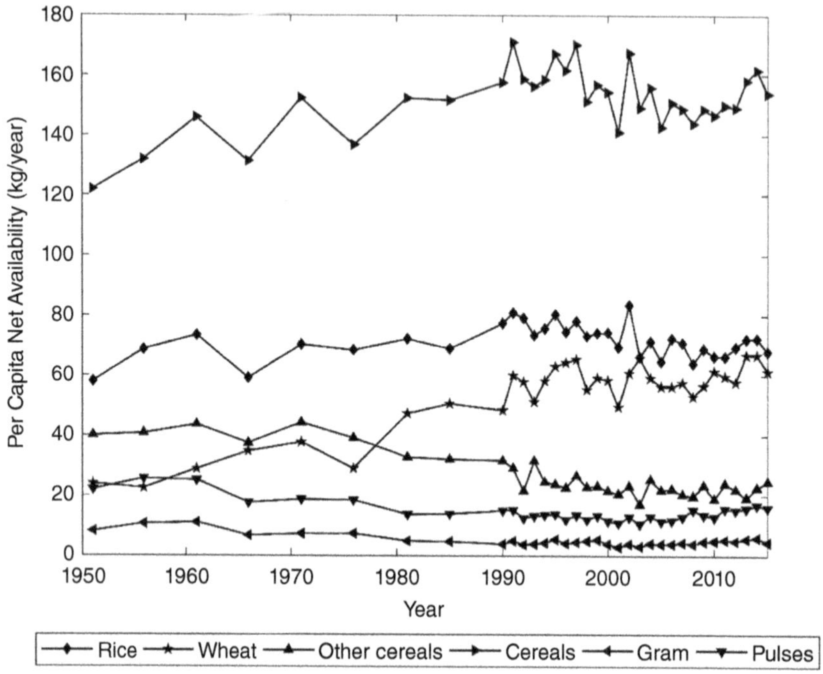

Figure 16.6 Per capita net availability of food grains (kg/year) during 1950–2015. *Source:* Government of India 2016.

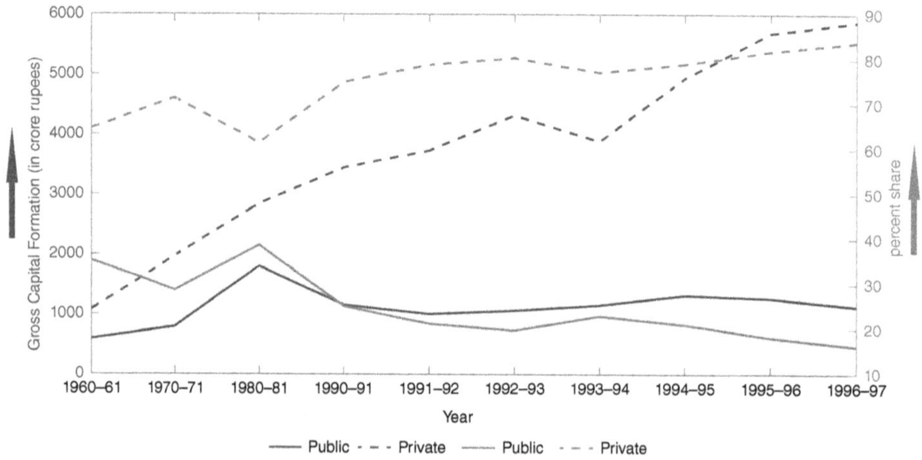

Figure 16.7 Gross capital formation in agriculture. *Source:* Economic Survey (1999).

GDP and foreign exchange. Globalization of agriculture brings profound changes in Indian agriculture, and the economy of the country depends upon the overall development of the agriculture sector. For sustainable development of agriculture, identification of various constraints or challenges is of paramount importance.

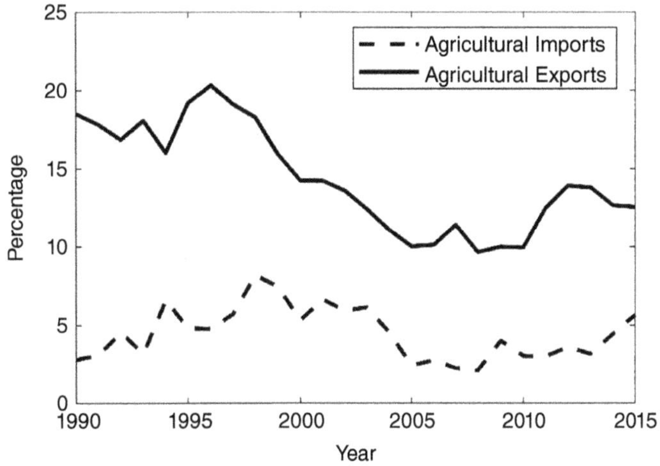

Figure 16.8 Percentage of agricultural exports and import during 1990–2015.
Source: Government of India 2016.

16.3 Agricultural Sustainability: Environmental, Economic and Social Perspectives

One of the major goals in agricultural sustainability is to maintain environmental quality both on and off the farm. Modern agricultural systems, intensive tillage and cropping practices, use of chemicals and fertilizers have resulted in excessive soil and nutrition loss impairing the productive capacity of soil (Dover and Talbot 1987). Such agricultural practices aiming at sustainability are regarded as unsustainable. Therefore, sustainability from an environmental perspective refers to the preservation of producibility of soil resources. One more important concern of agricultural sustainability in the context of the environment comes from beyond the farm gate. The transport of different chemicals and fertilizers used in agricultural practices pollutes the surface and groundwater resources, and the resultant pollution causes the loss of habitat for different biological species and negatively affects biological diversity. Therefore, agricultural sustainability should take place in such a way that natural resources should be protected, and practices aiming toward the sustainability should be environmental friendly and ecologically oriented (Neher 1992).

In addition to confirming long-term food productivity and environmental quality, sustainability also refers to the rural community and ensuring the economic return of sustainable farming (Brklacich et al. 1991). Therefore, sustainable agriculture with economic return is adequate to sustain the farming business and the maintenance of the rural community. From environmental and economic perspectives, sustainable agriculture can be viewed as environmentally friendly when there is sufficient productive capacity to build a healthy rural community with improved economic growth. Similarly, from the context of the social perspective, sustainable agriculture refers to creating more opportunities, providing basic food and education, and improving the standard of living of less advantaged groups in society. Consequently, environmental, economic and social equitability can be considered as a triangle of agricultural sustainability, which equitably distributes the agricultural benefits among and between social groups, regions and countries.

16.4 Challenges to Agricultural Sustainability

The main purpose of agriculture is to provide food and livestock for human survival and the main aim behind developing sustainable agriculture is to secure enough food for the present as well as future generations. In recent years, there has been sharp change in the demand and supply balance in food all over the world due to increasing population, a change in people's living standards, economic growth and changing climatic conditions. A region's agricultural potential is mainly constrained by heavy shortages of water and cultivated land, low productivity, inefficient irrigation or agricultural planning and management. In the Indian context, Indian agriculture is a vast industry and has a significant impact on every citizen, directly or indirectly. The importance of the agricultural contribution in Indian is evident in its contribution to employment generation and GDP. This sector also plays a very important role in sustainable economic development of the country. The sustainable development of agriculture of any country rests upon the sensible balanced use of the available natural resources. Agriculture governs the fate of a nation like India where more than 60% of the population still resides in rural areas with agriculture as its major livelihood source, although industrialization has been gaining good pace in the past few decades. The major objectives for the betterment of the agriculture sector can be realized through fast progress in agriculture, which depends on increasing the area of cultivation, cropping intensity and productivity. However, for a country like India, increasing productivity is more important than the rest. India is still fighting with the issue of food security. According to the FAO, more than 190 million Indians sleep hungry every day. This is because of the misbalance caused by factors like climate change, the growing population, industrialization, the exponential growth rate of urbanization and the limited infrastructure for sustainable agriculture. Other major challenges are the inadequate rural transport system, the lack of awareness related to crop treatment, land degradation, inefficient irrigation facilities and limited access to technology. Here, we discuss some major challenges to the development of sustainable agriculture in India.

16.4.1 Climate Change

Studies about Indian climate suggest an increasing temperature trend in surface temperature over the past century (Hingane et al. 1985; Kumar et al. 2006). Similarly, increased chances of changes in extreme rainfall patterns have also been reported (Rajeevan et al. 2008; Vittal et al. 2013). It is well known that agriculture productivity is directly influenced by climate change and weather. Crop development, growth, water use efficiency and crop yield under normal circumstances are determined by climate conditions during the growing season. Even a small deviation in normal conditions can seriously impair crop productivity.

Indian summer monsoon and the crop production of the country have a strong relationship. Global warming may alter the hydrological cycle, affecting the magnitude and frequency of rainfall in India. India is experiencing increased frequency of extreme events and has already experienced 23 large-scale droughts in the period from 1891–2009; the frequency of droughts is increasing (Goswami et al. 2006; Kumar and Gautam 2014). Availability of water becomes the most crucial factor for agriculture as most of the cultivable land lacks irrigation facilities. Furthermore, warmer air accelerates the evaporation process and holds more moisture, which in turn influences the crop evapotranspiration (ET). This affects the arid regions of the country, where soil moisture scarcity limits agricultural production. For example, Goyal (2004) predicted a rise of 14.8% in total ET demand with increasing global warming for Rajasthan state. In addition, even a slight increase in ET demand due to global temperature rise would have a large impact on the scarce water resources of the arid region ecosystem of Rajasthan.

Climate variability and increased occurrence of extreme events are creating new challenges for agriculture on the Indian subcontinent. It is necessary to understand the response of yield and growth and propose a proper land use pattern for sustaining agricultural productivity under the influence of existing climatic variations. Research shows that the direct impacts of changing climate would make crops more vulnerable to increasing extreme weather events, changes in the number of wet days, rainfall magnitude and intensity, frequency of floods and droughts, temperature anomaly and change in humidity. The changes induced due to climate variability on the agriculture of the country will be small in the recent future, but from the sustainability point of view, it should be kept in mind that long-term productivity will depend upon the plans and measures that the decision-making bodies take to mitigate the forthcoming challenges of climate change.

16.4.2 Population Dynamics

Growing population is a major concern for India, as availability of water is declining with population growth. A study by Mall et al. (2006) suggests a sharp decline in annual mean availability of freshwater and heavy growth of the population during 1951–2050. The study predicts negative impacts on water resources as the increase in population will cause a rise in water demand further causing faster withdrawal of water, which in turn will reduce the recharge time window of water tables. This forthcoming misbalance of water availability will put stress on agricultural water, causing a setback to agricultural production. A recent study of 482 of the world's largest cities by Flörke et al. (2018) explains that 27% of cities, comprising a population of 233 million, will put forth demands beyond the available surface water volumes. It should be noted that Jaipur in India has been predicted to become the second most water-deficient city in the 2050s. Moreover, 19% of the cities, which highly depend on water transfer to fulfill the needs of their population, will have a high probability of experiencing a struggle between the agricultural and urban and agricultural sectors, because neither sector will be able to keep up with the pace of their estimated demands. As a possible solution, it was suggested that in 80% of the investigated catchments, improving water use efficiency could lead to surplus water availability, enough to fulfill the urban requirement.

Increase in population not only leads to encroachment of water availability for the agriculture sector, but also the increasing demand can be a threat to sustainable agricultural practices. The major social and economic factors that augment rising demand for food are growing population, rampant urbanization, better standards of living and availability of more capital to the widening middle class of developing countries. It is believed that food demand will increase by between 60 and 98% by the end of 2050. This will have a huge impact on the agriculture markets, and farmers across the globe will need to increase crop production either by increasing the area of cultivable land or alternatively by improving crop productivity on currently available land holdings by means of adopting better techniques and infrastructures.

16.4.3 Poor Technology and Lack of Knowledge

It is known that the agriculture pattern has evolved over the years. It has of course reduced the price of food across the globe, sometimes supported by government policies, fed a large number of populations, reduced manual laboring and offered a large number of food options to consumers irrespective of season. The role of technological advancement is indispensable in the progress of agricultural evolution. Even in the current scenario, technology has a crucial role to play in order to achieve sustainable agriculture goals. The selection of technology by farmers is majorly influenced by the need to increase productivity, production and profit. However, these choices are restricted by the lack of sufficient knowledge about the technical and marketing aspects and availability of required capital. In India, existing government agencies simply focus on increasing productivity by introducing new technologies and methodology, which is not feasible or accessible for a large number of farmer population. This transfer of technology is not focused around improving the efficiency of small-scale farmers, which proves to be a huge setback. In order to achieve these objectives, technology and knowledge transfer has to be up to international standards, which in turn would produce a high-quality agricultural harvest at the same time keeping sustainability targets in mind. Though the availability of new technology provides new opportunities, it should be kept in mind that these are also accompanied by several limitations. It is often observed that the environmental and sustainable regulations are ignored either directly by individual farmers or indirectly by both consumers and private sector players. This unseen pressure also forces them to select technology and methodology without proper investigation, and the limited reach of private and government sector agencies fails to further help them.

16.4.4 Policy and Management Issues

In a dynamic agriculture system, the unevenness of agriculture productivity is often observed depending upon the economic, climatic, social and other such conditions. Improving productivity proves to be a difficult task in terms of collecting and managing resources for investment in backward areas. In addition to this, adjusting the laws and policies and implementation of regulation also becomes clumsy. Oftentimes, the institutional structure that is supposed to play a major role in implementing government policies at the ground level is also observed to be stagnant. These limitations constrict the availability of much-needed resource, knowledge, technologies and facilities to farmers. To overcome these limitations, needless to say, a very strong government commitment and desire is required, which seem to be caught in the web of the complex socio-political scenario in developing countries like India. This being the case, organizations of various types and levels, from micro to multilateral, need to understand their role in realizing the much-needed shift in policy.

Since the Indian economy is exclusive and is semi-planned, price control of agricultural products has been one of the very crucial policy interventions for the agriculture community. The famous public distribution system (PDS) is the most important tool for providing food to a large number of populations in the country. PDSs are supposed to act as a full-time available medium for ensuring food security. This bridge between total production and its distribution is achieved through collecting major food commodities like rice and wheat and setting up various levels of distribution networks. It is well understood that such organizations can be the most important tool in achieving food security targets. However, issues such as poor storage, management of supplies and rampant corruption in this sector have come into playe, more often in crisis. A better PDS could help achieve sustainable food security targets in an efficient way.

In India, the issues of pricing, policy, subsidies and management issues have been topics of great debate. Although recent discussions seek to relieve both consumers and producers by planning various measures, this is not easy to achieve. The institutions often seem to be ignoring the fact that if they are successful in achieving the target of increased productivity, the increase in productivity will be accompanied by better buying power of the rapidly widening middle class of India, which will again normalize the situation. Hence, the question of how best to formulate sustainable agriculture management policies remains open for consumers, producers as well as institutions.

16.4.5 Fragmented Land Holding

According to Bentley (1987), land fragmentation is known as scattering, or pulverization. The literature definition of land fragmentation is a single farm containing numerous spatially segregated parcels (Binns 1950; Van Dijk 2003; King and Burton 1982). In the literature, different types of land fragmentation are illustrated, including (i) fragmentation due to land use; (ii) fragmentation within the farm itself; (iii) fragmentation of land ownership; and (iv) fragmentation due to separation of ownership. Fragmentation due to land use suggests a greater number of tenants use the same piece of land. Similarly, fragmentation of land ownership mentions a greater number of landowners use the same land. Fragmentation within the farm refers to internal fragmentation, i.e. the land is spatially divided into a number of parcels of varying shape and size. The discrepancy among the landowners results in fragmentation due to separation of ownership. A study by Manjunatha et al. (2013) reported that land fragmentation in India is significantly associated with agricultural efficiency. Increase in land fragmentation decreases efficiency. They also stated that land fragmentation has a negative impact on agricultural profit. Moreover, the average profit efficiency is higher for unfragmented and large farms, and large farms are more susceptible to land fragmentation than small farms. Land consolidation can be an effective solution to the problem of land fragmentation; however, due to the high population, lack of other income-generating sources and inheritance of parental property, land consolidation is a very difficult task in India (Jha et al. 2005; Niroula and Thapa 2005, 2007). Due to land fragmentation, it is not viable to use efficient agricultural instruments effectively. Irrigation, mechanized farming and soil conservation practices are not possible on such fragmented farms. In addition, fragmentation leads to an increase in the travel time between fields, which in turn will cause lower labor productivity and increased transport costs.

16.5 Road to Sustainable Agriculture: Possible Solutions

Along with the provision of improving the current situation of stakeholders, long-term planning focused around better environmental, economic and social policies is required. This can be achieved by arranging good education, R&D, marketing and risk, technology, infrastructure, mitigation, etc. and will give us the agricultural empowerment we need. We have already discussed that the biggest challenges to the agricultural sector are faced by small-scale farmers. Most of these farmers have fragmented land systems where implementing any improvement becomes difficult. The first step in easing the burden of poor cropping systems and improving agricultural efficiency is the consolidation of small areas of land. Furthermore, cooperative farming can be employed for better outputs, especially for economically poor farmers. The concept of cooperative farming or other such organized approaches can further be helpful in paving the way for more consolidated and simplified connections between institutions and individuals. Banks or other such organizations would be more willing to help organizations or cooperative societies than individuals. Any economic or technical help from the government or organization can be utilized to implement large-scale reforms to improve crop productivity, adopting sustainable agriculture techniques, limiting the use of chemical fertilizers and helping each other during any extreme weather event crisis. With this approach, the impact of crop failure is less severe in adverse conditions and achieves better incomes in favorable seasons.

A good national-level policy to manage micro agriculture can prove to be of great help and will tackle a whole lot of problems related to sustainable agriculture management. Credit relief for marginal and small farmers is necessary and should be the first priority before moving ahead. Other problems such as irrigation and electricity can be tackled by the government at various levels under a single dedicated framework. Although forcing any individual to growing a particular crop is not advisable, educating them about the various pros and cons of using different crops could be a good idea. A good water management plan in combination with a good crop management policy at the regional to national level will surely be a key to the betterment of both farmers and consumers. The problem of improper irrigation and tradition crop dependence can also be addressed in more detail by a centralized policy for agricultural management. Suggesting a well-researched conventional crop will not only help in increasing productivity but will also motivate farmers to shift their agriculture pattern from traditional crops to cash crops, which will further help the nation to ease the stress on importing such materials.

Sometimes simple innovation at the community level can solve a large-scale complex problem in a given region. Government agencies need to overcome their lethargic state and must ensure that such innovations do not fade unnoticed. Officers at the regional levels must encourage such innovation and provide the knowledge and necessary technologies to improve the efficiency of such innovations. Another innovative solution to sustainable agriculture management is proper crop management in combination with organic farming. As we discussed, a cooperative approach is more efficient: the fully sustainable organic farming approach is very productive in village areas where there is plentiful availability of organic fertilizers. Nevertheless, it has been observed that people are reluctant to adopt it on a large scale. Hence, for efficient adoption of large-scale organic farming, farmers need to be made aware, educated and motivated. Organic farming will limit the overutilization of chemical fertilizers, increase water use efficiency and prove to be better from the environmental perspective. This method will also help farmers to retain the fertility of the land for a longer period of time, which, in the long run, will result in less economic pressure.

The aforementioned PDS is a crucial element of food security plans in the India. However PDSs encounter a lack of management and storage facilities, resulting in a large-scale loss of food grains every year. The storage facilities need to be improved, which can be done by providing granaries or cold storage at the village or panchayat scale. This can be jointly managed by the government agencies and village bodies or simply a group of farmers. If providing such facilities is not possible for some reason, quick transit of agriculture products should be a priority. Nowadays, internet penetration is quite significant in the rural areas of the country. This can be utilized to improve the marketing of products by using the simple concept of internet marketing. This will enable farmers to limit the role of middlemen and earn more profit. Large-scale planning and implementation of such an idea seems difficult; however, the successful example of Walmart encourages us to see its possibilities.

To summarize the role and responsibilities of every person involved in the food system with the proper integration of the environmental, economic and social perspectives, Table 16.1 highlights different practices for sustainable food and agriculture systems.

With the importance of sustainable agriculture, any workable strategy must take the following steps:

- Cultivators need to adopt advanced agricultural practices, and awareness regarding sustainability should be developed among farmers through regular campaigns.
- Steps to be taken to improve the sustainability of water resources significantly through proper management and conservation practices, and awareness should be created to control water pollution.

Table 16.1 Role and responsibilities of different practitioners in promoting sustainable agriculture.

Practitioners	Role and responsibilities
Growers/ cultivators	Promote soil health, minimal and optimal utilization of water resources, reduce pollution levels on farm, diversification of crops
Consumers and retailers	Promoting farmworker wellbeing with environmentally friendly methods, strengthening the local economy, industrialization of agricultural markets
Researchers	Understanding the impact of climate change on crop yield, developing a sustainable framework with the integration of engineering, economics, biology and many others
Policymakers	Support for water harvesting, regional integration of agricultural and rural policies, incentives for biofertilizers, soil management support, pesticide reduction policies, participatory irrigation management, promoting organic farming
Government	Regulatory and approval systems for new products in the market, policies to stimulate technological innovation and hence international competitiveness, restriction of some products regarded as damaging to the environment

- The improved agricultural practices need to be backed and supported by agricultural extension in order to create a link between research and agriculture.
- Facilities need to be created to promote and provide facilities for agro-processing to enhance the economic value of agricultural products.
- Farmers should be provided with technical and financial assistance, to improve the quality and quantity of their production levels. This will certainly enable opportunities for export.
- It is imperative to create vibrant environmental, economic, social and agricultural policies to meet the criteria for sustainable agriculture.
- Last, but not least, better coordination, cooperation and accommodation policies need to be framed for overall sustainable development.

16.6 Conclusions

The agriculture sector is undoubtedly one of the most vital elements of the rapidly growing Indian economy. This sector is the most important field, where improvement will lead to uplifted social and economic conditions. Any negative change in the development of this sector may cause further deterioration in the already bleak condition of the poor community of the country. Indian sustainable agriculture management plans are hugely hindered by factors such as climate change, resource availability, lack of advanced technical knowledge, management issues and the growing demand of the large population. Since sustainable development is the conservation and management of natural resources for present and future generations, it requires planning at several individual, organizational and institutional levels. Furthermore, to attain the goal of sustainable agriculture development, the optimum utilization of human resources, natural resources, capital, technical advancement and proper management is required. In India, agricultural productivity is hugely reliant on monsoon rainfall, uncertainty in which is one of the main reasons for causing the reduced growth of the agriculture sector. This reduction or impairing of the agriculture sector most severely damages the interests of small-scale farmers and landless laborers, sometimes costing them their lives. Hence, strong steps are required to be taken in order to prevent these farmers from ongoing catastrophe and to reduce the chances of such conditions in future. Until adequate facilities are provided to free farmers from the present state of insecurity owing to climate variability, the future dream of sustainable agriculture seems to be unrealistic. India is the second largest country of the world in terms of population; of this large population, a large proportion of people are dependent on agriculture, which is currently very disorganized. The farmers are neither well equipped with advanced technologies nor do they have proper knowledge. Indian agriculture practices have been carried out for a long period of time with the same inefficient technology and knowledge. In addition to this, a lack of positive interference from public and private organizations is also a major blockage in the road toward sustainable agricultural development goals. For a developing country like India, the issue of sustainable agriculture management is important from the environmental, economic and social perspectives. With forthcoming challenges already in play, it is necessary that we take up a large-scale planning effort to ensure the needs of present as well as future generations.

References

Abubakar, M.S. and Attanda, M.L. (2013). The concept of sustainable agriculture: challenges and prospects. *IOP Conference Series: Materials Science and Engineering* 53 (1): 012001.

Bentley, J.W. (1987). Economic and ecological approaches to land fragmentation: in defense of a much-maligned phenomenon. *Annual Review of Anthropology* 16 (1): 31–67.

BIFAD (1988). *Environment and Natural Resources: Strategies for Sustainable Agriculture*. Washington, DC: US Agency for International Development.

Binns, O. (1950). *The Consolidation of Fragmented Agricultural Holdings*. Washington, DC: FAO Agricultural Studies.

Brklacich, M., Bryant, C.R., and Smit, B. (1991). Review and appraisal of concept of sustainable food production systems. *Environmental Management* 15 (1): 1–14.

Crews, T.E., Mohler, C.L., and Power, A.G. (1991). Energetics and ecosystem integrity: the defining principles of sustainable agriculture. *American Journal of Alternative Agriculture* 6 (3): 146–149.

Department of Agriculture and Cooperation and Statistics (2014). *Agricultural Statistics At a Glance 2014*. New Delhi: Government of India.

Dover, M. and Talbot, L. (1987). *To Feed the Earth: Agro-Ecology for Sustainable Development*. Washington, DC: World Resources Institute.

Economic Survey (1999). *Economic Survey 98–99*. New Delhi: Government of India.

FAO (2011). *State of the World's Land and Water Resources for Food and Agriculture*. Rome: FAO.

Farrell, K.R., Sanderson, F.H. and Vo, T.T. (1984). Feeding a hungry world. Report no. 67. Washington, DC: Resources for the Future.

Flörke, M., Schneider, C., and McDonald, R.I. (2018). Water competition between cities and agriculture driven by climate change and urban growth. *Nature Sustainability* 1 (1): 51–58.

Foley, J.A., Ramankutty, N., Brauman, K.A. et al. (2011). Solutions for a cultivated planet. *Nature* 478 (7369): 337–342.

Goswami, B.N., Venugopal, V., Sengupta, D. et al. (2006). Increasing trend of extreme rain events over India in a warming environment. *Science* 314 (5804): 1442–1445.

Government of India (2016). *Agricultural Statistics at a Glance*. New Delhi: Government of India.

Goyal, R.K. (2004). Sensitivity of evapotranspiration to global warming: a case study of arid zone of Rajasthan (India). *Agricultural Water Management* 69 (1): 1–11.

Hingane, L.S., Kumar, K.R., and Murty, V.R. (1985). Long term trends of surface air temperature in India. *Journal of Climatology* 5: 521–528.

Jha, A., Nagarajan, H.K. and Prasanna, S. (2005). Land fragmentation and its implications for productivity: evidence from southern India. ASARC Working Paper. Canberra: ASARC.

Keeney, D.R. (1989). Toward a sustainable agriculture: need for clarification of concepts and terminology. *American Journal of Alternative Agriculture* 4 (3–4): 101–105.

King, R. and Burton, S. (1982). Land fragmentation: notes on a fundamental rural spatial problem. *Progress in Geography* 6 (4): 475–494.

Kissinger, G., Herold, M., and De Sy, V. (2012). *Drivers of Deforestation and Forest Degradation: A Synthesis Report for REBD + Policymakers*. Vancouver, Canada: Norwegian Ministry of the Environment, DFID and Department for Energy and Climate Change.

Kumar, R. and Gautam, H.R. (2014). Climate change and its impact on agricultural productivity in India. *Journal of Climatology & Weather Forecasting* 2 https://doi.org/10.4172/2332-2594.1000109.

Kumar, K.R., Sahai, a.K., Kumar, K.K. et al. (2006). High-resolution climate change scenarios for 21st century. *Current Science* 9: 334–345.

Lowrance, R., Hendrix, P.F., and Odum, E.P. (1986). A hierarchical approach to sustainable agriculture. *American Journal of Alternative Agriculture* 1 (4): 169–173.

Macrae, R.J., Henning, J., and Hill, S.B. (1993). Strategies to overcome barriers to the development of sustainable agriculture in Canada: the role of agribusiness. *Journal of Agricultural and Environmental Ethics* 6 (1): 21–51.

Mall, R.K., Gupta, A., Singh, R. et al. (2006). Water resources and climate change: an Indian perspective. *Current Science* 90 (12): 1610–1626.

Manjunatha, A.V., Anik, A.R., Speelman, S., and Nuppenau, E.A. (2013). Impact of land fragmentation, farm size, land ownership and crop diversity on profit and efficiency of irrigated farms in India. *Land Use Policy* 31: 397–405.

Neher, D. (1992). Ecological sustainability in agricultural systems. *Journal of Sustainable Agriculture* 2 (3): 51–61.

Niroula, G.S. and Thapa, G.B. (2005). Impacts and causes of land fragmentation, and lessons learned from land consolidation in South Asia. *Land Use Policy* 22 (4): 358–372.

Niroula, G.S. and Thapa, G.B. (2007). Impacts of land fragmentation on input use, crop yield and production efficiency in the mountains of Nepal. *Land Degradation and Development* 18 (3): 237–248.

NSSO (2013). *Income, Expenditure, Productive Assets and Indebtedness of Agricultural Households in India*. New Delhi: NSSO.

OECD (2012). Building resilience for adaptation to climate change in the agriculture sector. Proceedings of a Joint FAO/OECD Workshop, Rome, April 23–24. www.fao.org/3/i3084e/i3084e.pdf (accessed November 2019).

Planning Commission (2017). *Sector-wise Contribution of GDP of India*. New Delhi: Planning Commission.

Rajeevan, M., Bhate, J., and Jaswal, A.K. (2008). Analysis of variability and trends of extreme rainfall events over India using 104 years of gridded daily rainfall data. *Geophysical Research Letters* 35 (18): L18707.

Van Dijk, O. (2003). *Dealing with Central European Land Fragmentation*. Delft, the Netherlands: Eburon.

Vittal, H., Karmakar, S., and Ghosh, S. (2013). Diametric changes in trends and patterns of extreme rainfall over India from pre-1950 to post-1950. *Geophysical Research Letters* 40 (12): 3253–3258.

Zhao, J., Luo, Q., Deng, H., and Yan, Y. (2008). Opportunities and challenges of sustainable agricultural development in China. *Philosophical Transactions of the Royal Society B: Biological Sciences* 363 (1492): 893–904.

17

Food Security and Sustainability

Preetika Kuknur Pachapur[1], *Vinayak Laxman Pachapur*[2,3], *Satinder K. Brar*[1,3], *Rosa Galvez*[4], *Yann Le Bihan*[3], *and Rao Y. Surampalli*[5]

[1] *INRS, Université du Québec, Québec, Canada*
[2] *Department of Civil Engineering and Water Engineering, Université Laval, Québec, Canada*
[3] *Department of Civil Engineering, Lassonde School of Engineering, York University, Toronto, Canada*
[4] *Centre de recherche industrielle du Québec (CRIQ), Québec, Canada*
[5] *Global Institute for Energy, Environment and Sustainability, Lenexa, KS, USA*

17.1 Introduction

Around 800 million people around the world are food insecure, lacking the basic intake of food necessary to lead a healthy life. In the developing countries around 20% of the world's population lacks the financial and physical access to provide basic food to its citizens (Ferranti 2016; Pinstrup-Andersen and Pandya-Lorch 1998; Songlin and Ruihong 2010). Due to food insecurity, children in the developing countries are facing the risk of vitamin and nutrient deficiency and are underweight with risk for many diseases. The rising world population is causing major food insecurity, along with afforestation, rises in food prices, a lower supply of food, agriculture biomass conversion into bioenergy production, and decreased land and water resources for food production (Mohr et al. 2016; Toledo and Burlingame 2006; Waldron et al. 2017). Food insecurity across the world will cause human suffering, create global/national imbalances and will be increasingly constraining to human existence.

In order to create food security, the International Food Policy Research Institute has developed a 2020 Vision of "a world where every person has economic and physical access to sufficient food to sustain a healthy and productive life, where malnutrition is absent, and where food originates from efficient, effective, and low-cost food and agricultural systems that are compatible with sustainable use and management of natural resources" (Pinstrup-Andersen and Pandya-Lorch 1998). FAO defined food security as "a complex situation requiring a worldwide treatment towards malnutrition, revenues of food producers, food products yield/demand, ability of food production, proper utilization of biodiversity resources" (Chowdhury et al. 2017; Fan and Brzeska 2016; FAO 2018; Ferranti 2016). The 2020 vision can be achieved and sustained in the future with equal contributions from

Sustainability: Fundamentals and Applications, First Edition.
Edited by Rao Y. Surampalli, Tian C. Zhang, Manish Kumar Goyal, Satinder K. Brar, and R. D. Tyagi

national and international communities, households, farmers, and the local and private sector in developing sustainable food resources for everyone in the present and beyond 2020 (FAO 2018). To attain the food security vision 2020, several policies, priorities, technologies and sustainable solutions need to be implemented and followed across the countries (Fan and Brzeska 2016; Waldron et al. 2017).

In this chapter, we discuss the conditions, necessary steps, appropriate policies, technological improvements and sustainable use affecting food security. The chapter provides an overview of global challenges along with Canada's plan of action for sustainable food security.

17.2 Food Needs

The FAO has recently estimated that 815 million people go hungry every day; the number of people at risk of food needs, hunger and malnutrition is set to rise by 10 billion by 2050 (FAO 2018). Recent research suggested that the experience of hunger leaves an ineradicable mental depression in children, and stunting affects 155 million children as represented in Table 17.1 and Figure 17.1. The conditions of child malnutrition in Africa (31%) for the year 2016 are higher in comparison to Asia (24%) and Latin America (11%). The child malnutrition rate decline over the period 2015–2016 was higher in Asia (10%) than Africa (5%) and Latin America and the Caribbean (5%). According to the numbers presented in Table 17.1, around 155 million children around the world suffer from malnutrition; 640 million suffer from adult obesity; and around 613 million women are affected by anemia during pregnancy. Food insecurity in adults causes chronic diseases and women during pregnancy can be affected by anemia. This is evident from the fact that the amount of money spent on health-related issues in each country is directly related to food insecurity (Schut and Florin 2015; Tarasuk et al. 2016). Access to the basic amount of food necessary for a healthy life must be achieved by every person in the world. The poor community accounts for 30% of the world's population. These people have limited income and are unable to grow or to buy the basic food necessary for a healthy life (Chowdhury et al. 2017). During the early 1990s, areas of sub-Saharan Africa, the Latin American countries, Gulf countries and South Asian countries were below the poverty line (currently US$1.90/day/person) with an income of around US$1/day/person (Conceição et al. 2016; Grindle et al. 2015; Schut and Florin 2015).

Table 17.1 Conditions of malnutrition in the World for the year 2016, data derived from FAO.

Conditions of malnutrition in the world	Number in millions	Percentage of the reference population
Children <5 years who suffer from stunted growth	155	23
Adult obesity	641	13
Women of reproductive age affected by anemia	613	33

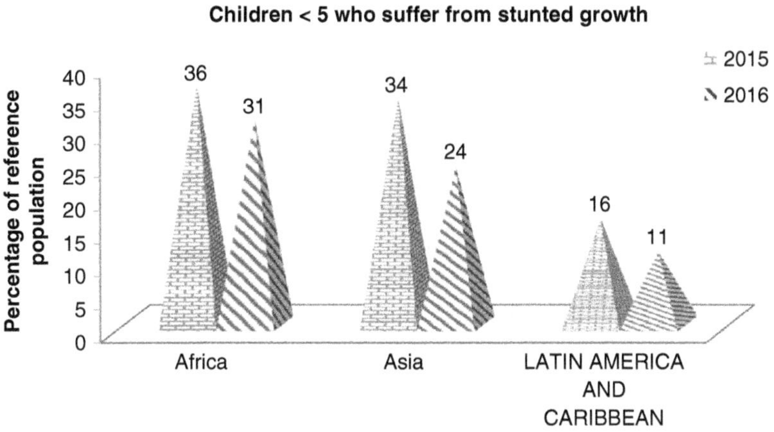

Figure 17.1 Percentage details of children <5 years who suffered from stunted growth during 2015 and 2016 in different regions of the world. *Source:* Data derived from (FAO 2018).

Food needs not only define a country's poverty but also, with increasing food needs, there is an effective market demand for growth and purchase of food products (Schut and Florin 2015). The per capita demand for food products has shown a sharp increase of 11% between 1990 and 2020 (Fanzo 2015; Pinstrup-Andersen and Pandya-Lorch 1998). Along with the increase in demand, an alternate source of income for poor people through food production plays an important role in food security. Sustainable action is necessary to build the food industry, packaging industry and agriculture in regions easily accessible to the poor, leading to the benefit of employment for the poor through food growth (Fan and Brzeska 2016; Garnett 2014; Schut and Florin 2015).

In addition, the poor community needs to have adequate information on education, healthcare, clean water, sanitation, community developments and the nutritional requirement for a productive lifestyle in society (Schut and Florin 2015). Different policies directed toward better lives for the poor need to be facilitated, with increased small-scale industry set-ups, empowerment for women and various poverty relief programs to boost the food security of the poor community.

17.3 Population Growth

The world population is set to increase to 9 billion in 2020 with 80 million people added every year to the growing population. Ninety-five percent of the increase in the population is expected to be contributed by the developing countries due to urbanization. Population rise results in a rise in demand for feedstock and food products. The balance between the increasing population and the increasing food demand will challenge food security across countries (Fan and Brzeska 2016; Maxwell et al. 2008). The increase in the population will directly affect insecure households and children across the countries. Statistical surveys by the Canadian Community Health Survey (2012) provide a comprehensive review of the state of food insecurity across the provinces and territories in Canada as represented in

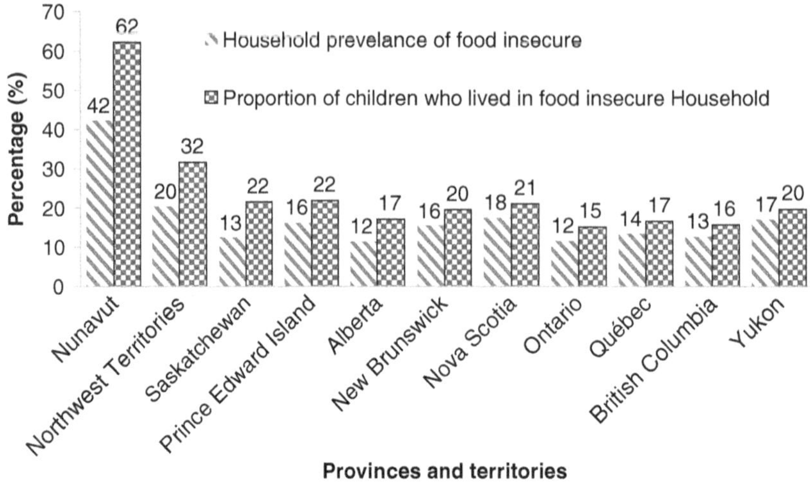

Figure 17.2 Statistic surveys by the Canadian Community Health Survey (2012) provide a comprehensive state of food insecurity across the provinces and territories in Canada.

Figure 17.2. The household food insecurity for the Nunavut was the highest at 42%, which resulted in 62% children suffering from food insecurity. The least food insecure provinces were Alberta, Ontario and Québec. The extreme living conditions, much less industry and fewer work opportunities are among the reasons for the food insecurity around the Nunavut and Northwest Territories. Strategies must thus be made to control population growth by educating people on reproductive health to balance family planning with food needs.

17.4 Impact of Climate Change on Food Security

The two main important factors in food supply are natural resources and agriculture industrialization. In the case of natural resources, degradation of water, soil, forest and fisheries resulting in declined production capacity directly affects food security (Maxwell et al. 2008) (Fanzo 2015). Similarly, improvements in fertilizer, pesticide, technology and agriculture research will boost the production capacity of agriculture industrialization to benefit food security.

Deforestation, forest conversion, overgrazing, productivity losses, unsuitable agricultural practices, overharvesting, restrained access and disputes over fish stocks are the reasons for challenged natural resources (Fanzo 2015). Climate-related events, especially droughts, floods and tsunami, can damage food resources, resulting in risks to food security. Increased food production must come from land already under cultivation using economical and environmentally feasible options. In the case of fisheries improvement, proper management, maintaining the harvest level and legitimate access to improved sustainable aquaculture production will meet the food security vision. Climate change moves millions of people into a cycle of food insecurity and poverty. However, we can fight food insecurity and climate change by adopting climate-improved agriculture practices (FAO 2018; Ferranti 2016).

Depletion of soil nutrients, crop losses owing to pests, fertilizer effects on water quality, environmental concerns regarding fertilizers, pesticide effects on human health and existing technology all have an impact on food production (Ferranti 2016). Fertilizer input needs to be efficient for plant nutrient uptake, available pest control needs to be improved with no negative environmental effects and biotechnology research is necessary to support the technology gap in sustainable agriculture (Schuenemann et al. 2018). Biotechnology research will bring about promising developments in agriculture industrialization by collaborating with traditional methods with improved breeding, higher resistance to pests, better nutritional value, increased shelf-life of products, and increased tolerance to disease and weather conditions, leading to increased biodiversity for food security (Garibaldi et al. 2017; Maxwell et al. 2008).

Agriculture accounts for a large proportion of employment and income in developing countries, so investment in agriculture research will lead to increased food yield, increased agriculture productivity per unit of land and will strengthen agriculture research for sustained food security (Fan and Brzeska 2016; Schuenemann et al. 2018). Strategic investments from private, regional and national stakeholders for agriculture research with enhancements to the national budget should also have a potential benefit for international research.

17.5 International Case Studies

Food insecurity is not only determined by natural resources and agricultural inputs, but also by different international and national situations. The conflict in southern Sudan in 2015–2016 resulted in serious food insecurity for more than 4.9 billion people, leading to a humanitarian emergency and catastrophe. During the crisis, violence led to depletion of food sources, and food transference was used as a weapon of war with the blocking of emergency food, the hijacking of food aid trucks (Blanchard 2013; FAO 2018; Ferranti 2016).

The crisis in Yemen affected 17 million people (60% of the population) with food insecurity, child undernutrition and acute undernutrition, altogether requiring continuous human support for food and health. The situation coincided with a breakdown of the healthcare system, an outbreak of cholera and epidemic diseases, affecting infrastructure, with the downfall of the government from 2016 to 2017 (FAO 2018; Maxwell et al. 2008).

Six years of civil war across the borders of Syria plunged 85% of the population into poverty (Callahan et al. 2017). The crisis stretched food insecurity, with acute malnutrition, anemia-affected women and children, and also affected agricultural infrastructure and food production, leaving Syrian refugees to migrate to border countries (Callahan et al. 2017; Taleb et al. 2015).

The traditional agriculture practices in Kenya have led to overgrazing and ecological degradation, resulting in environmental degradation. The traditional system is unable to cope with drought and climate changes, leading to soil degradation, soil erosion and soil cover loss (FAO 2018; Lal and Stewart 2010).

The clash between the government and the army in Uganda led to population movements, poverty and food insecurity. After the crisis, the government identified and supported agricultural growth and was supported by multiple nations with agriculture support and policies for enhanced food security. The agricultural support for the people of Uganda improved the sustainability of food security and decreased the poverty line by 10% (Conceição et al. 2016; FAO 2015).

Libyan economic redevelopment in response to the food crisis in 2007 and 2009 was slow. Libyans decided to take a "better safe than sorry" approach to the reconstruction with research initiatives. A research organization in Libya initiated steps to overcome food security to reduce, reuse and recycle resources into rebuilding in order to mitigate extreme events and climate change. Libyans launched the water–energy–food initiative: the framework on water and climate security to support sustainability of water and food security (Bindra et al. 2014).

Agricultural development to enhance food security includes increased agricultural production in addition to increased income and improved access to the basic lifestyle needs of farmers. Positive impacts also lead to risks such as heavy workloads, greater maintenance costs, competition among farmers, loss of conservation awareness and social clashes (Schindler et al. 2016). To solve the farmers' issues, technological improvement and agricultural extension from the researchers need to be transferred. The agriculture extension "reduction for gains" worked at minimizing the input budget by decreasing plant growth supplements to produce higher-quality food products. Similarly, agricultural extension practices play a very important role in the development of sustainable production and strategic planning to ensure food security (Garnett 2014; Sattaka et al. 2017).

To make agriculture more nutrition sensitive, European Union (EU) policymakers developed sustainable food and nutrition security (SUSFANS) aimed at obtaining food production from agriculture/marine sources and ensuring nutrient availability in the long run (Rutten et al. 2016). The SUSFANS project bridged researchers from various streams, working on nutrition impacts, health benefits and increasing food system outcomes. SUSFANS promotes a balanced diet to EU citizens via competitive food and agricultural businesses with minimum environmental impact and contributing toward global nutrition security (Rutten et al. 2016). The framework of SUSFANS includes different policies as represented in Figure 17.3. The framework of SUSFANS connects consumers, producers and food chain actors with different policies, such as food, environment, agriculture and nutrition policies, in order to establish the conditions to prevent food insecurity. The objectives of SUSFANS are to provide competitive food and agricultural businesses, balance the diet of EU citizens and also maintain the global food security with minimum impact on the environment, as represented in Figure 17.3.

Canada's plan for food security constitutes federal governments, provinces, different non-governmental organizations (NGOs), private groups, civil society and government-level groups. The plan focuses on strengthening international relationships, and increasing domestic economic programs to address the reduction of food insecurity in areas such as nutritious food supply, sustainable ways to increase food products, and promoting education at home and abroad (North 2001; Rideout et al. 2007; Tarasuk et al. 2016). One in eight Canadian households is food insecure, which is a serious problem and impacts the social healthcare system throughout Canada. The rise of food insecurity from 11% to 12% during 2007–2008 to 2011–2012 is a major concern since few provinces and territories have opted to monitor food insecurity. According to national statistics, of food-insecure households,

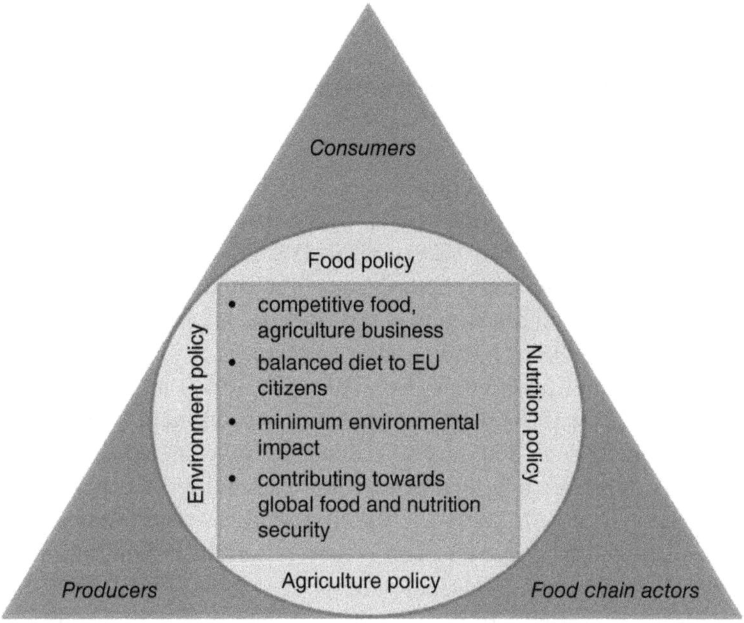

Figure 17.3 The framework for sustainable EU food and nutrition security (SUSFANS).

43% are single adults, living alone or with others, 25% are couples with children, 16% are female lone parents and 13% are couples with no children. Monthly wages, salaries or self-employment contribute around 63% of the main sources of income for food-insecure households in Canada. Family income is directly linked to household food insecurity: a decline in or very low household income increases food insecurity (North 2001; Rideout et al. 2007; Tarasuk et al. 2016). Household incomes between \$10 000 and \$19 000 are at a higher risk of 15% food insecurity, in comparison to 0–1% of the risk for household incomes between \$70 000 and \$79 000. Modest changes to income, public policies and different policy actions can improve the condition of low-income households in Canada (Rideout et al. 2007). Senior pension policies, such as Old Age Security, the Canada/Quebec Pension Plan and Income supplement, etc. provide a source of income to help lower the rates of food insecurity. The amount received through pension policies is sufficient for senior citizens to fulfill the right to food and nutrition in comparison to the amount received through social assistance (North 2001; Rideout et al. 2007; Tarasuk et al. 2016). Likewise, Canada Child Benefit (CCB) provides \$100–200 per month to households with a child under the age of 6 years. The introduction of CCB in 2006 reduced food insecurity, benefited households led by single parents and supported low-income families. Different social assistance programs have been undertaken by provinces and territories to decrease the risk of food insecurity via poverty reduction strategies. The programs include increased income support rates, increased health benefits, decreased tax for those on a low income, increased affordable housing and increased liquid asset limits (Tarasuk et al. 2016). Federal poverty reduction strategies, national food policies, pension schemes, child benefit programs and interactions of income-based organizations such as Food Banks Canada are all working simultaneously for the reduction of household food insecurity in Canada.

17.6 Biofuels and Food Security

Bioethanol, biodiesel and biohydrogen are classified as biofuels, which are renewable sources of energy and an alternative to fossil fuels (Ayadi et al. 2016; Sarma et al. 2016). Biofuel production increased from 45×10^3 million l to 117×10^3 million l in 2016, bioethanol represented 75% of total biofuel production, with the remaining 25% made up by biodiesel (Finco and Doppler 2010; Koizumi 2015). The production of biofuel and biodiesel around the world for the year 2016 (Koizumi 2015) is illustrated in Figure 17.4. The US (56%) and Brazil (27%) are the major contributors to bioethanol production and the EU (44%) contributes to biodiesel production at the world scale (Finco and Doppler 2010; Pachapur et al. 2016). Generation (I, II, III and IV) feedstocks are used and treated via different methods for biofuel production (Renzaho et al. 2017) as presented in Table 17.2. The significant factor in biodiesel production is represented by the wide range of available feedstock with their potential oil yields (Atabani et al. 2012). Much of this biodiesel feedstock is presented as a first-generation biodiesel feedstock, such as oil-based crop seeds (Pachapur 2017). These are the first crops used to produce biodiesel, which accounted for 95% of edible oils. It has created serious ecological imbalances, as countries began a process of deforestation in order to grow feedstock leading to extensive damage to the ecosystem. To address this problem, non-food feedstocks have been developed (Ahmad et al. 2011; Tan et al. 2013). Second-generation feedstock energy crops are accounted for by non-edible oils, restaurant cooked oils and animal-derived waste oil (Pachapur 2017). These feedstocks are not appropriate for household use and eliminate the competition for food. These second-generation feedstocks are not abundant enough to meet the demand levels, show poor performance in cold conditions and also working with animal waste causes process difficulties along with biosafety issues (Ahmad et al. 2011; Tan et al. 2013). Third-generation biodiesel feedstocks were created to improve efficiency and sustainability in large-scale commercial biodiesel production. Microalgae are the next-generation feedstock and the most promising alternative source for use in biodiesel production (Pachapur 2017). Microalgae are highly

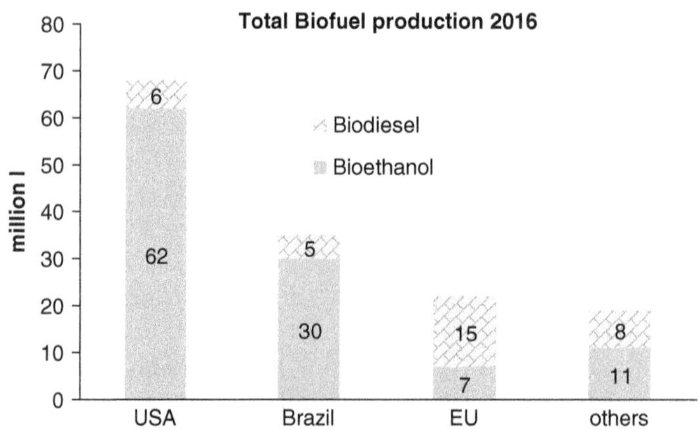

Figure 17.4 The production of biofuel and biodiesel around the world for the year 2016.

Table 17.2 The generation (I, II, III and IV) feedstocks and production methods for biofuel.

Generation biofuels	Feedstocks used	Production method
First	Animal fats, vegetable oils, oily seeds such as soybean, rapeseed, mustard or sunflower, well as starch crops such as maize, sugarcane, sorghum or cassava	Transesterification or chemical conversion of animal/vegetable oil into biodiesel and fermentation of starch/sugar into bioethanol
Second	Cereal straw, sugar cane bagasse, forest residues, organic wastes, vegetative grasses, short rotation forests and other energy crops	Biochemical or thermochemical processes to convert lignocellulosic biomass to biofuels
Third	Species such as yeast, fungi, algae biomass and other aquatic biomass	Microbiological processes, genetic engineering, microalgae with chemical treatment of bio-oil extract to produce biofuels
Fourth	Genetically optimized feedstocks	Hydroprocessing, oxy-fuel combustion or thermochemical processes to produce biofuels

efficient in producing biomass with increased growth rates and productivity in comparison to earlier generation feedstock fuels (Atabani et al. 2012; Tan et al. 2013). Microalgae have an oil content up to 25 times higher than other feedstocks but require minimal land to produce biodiesel (Atabani et al. 2012). There has been increasing interest in microalgae farming and the substantial processing steps required for petroleum fuels (Pachapur et al. 2017a,b).

Future biofuels must be environmentally friendly with minimal release of greenhouse gases (Pachapur et al. 2017a,b); however, the production of biofuels requires caution with regard to food security (Finco and Doppler 2010). Biofuel production has led to a change of cultivated crops to oilseed crops, deforestation to grow oil seed crops and concerns regarding local food production (Finco and Doppler 2010). Biofuel production affects the environment with a shift of forest land to biofuel production, which causes a negative carbon balance shift (Finco and Doppler 2010). In addition, it causes climate change, biodiversity shifts, and imbalance of land and water resources, affecting agricultural production and leading to a decrease in food feedstock, thereby raising food insecurity.

Countries such as China and Japan are promoting biofuel programs for rural development. Several major companies in China and Japan produce biofuels using corn, wheat and cassava; in the case of biodiesel production cooking oil and vegetable oil are used (Koizumi 2015). With the excessive use of generation feedstocks for biodiesel production, the world faced a "food versus fuel" crisis due to the food imbalance created by directing these food crops to producing biodiesel. In China 40% of corn was used in bioethanol production in 2011 (Koizumi 2013); for the US it was 46%; in the case of Brazil it was around 55% of sugarcane production (Zhang et al. 2017). The rise in biofuel demand indirectly increases the requirements for feedstock resulting in competition with food security. China and Japan have subsequently proposed to use non-food, cellulosic-based bioproducts, unused agriculture products and soft cellulose feedstocks for biofuel production (Koizumi 2013, 2015; Zhang et al. 2017). Brazil has expanded and cultivated sustainable land for sugarcane production, promoting an increase in the food market and in bioethanol production, creating numerous jobs and also using bagasse for electricity production. The bioethanol and sugar industries are growing together, successfully contributing to Brazilian rural development to maintain food security (Koizumi 2015).

The current global agriculture system is for monoculture grain crops, such as cereal grains, legumes and oilseeds, covering 80–90% of food consumption, along with 10–20% as feedstock for biofuels. In order to minimize competition and increase security between food and biofuel, sustainable agriculture and biorefinery play a very important role. Perennial cultivation is the new trend: the crops store more carbon, require less water and low nutrient additions, tolerate adverse climate conditions and have resistance to pests to deliver higher biomass yield (Chen and Zhang 2015). Perennial crops such as grass, bamboo and fertilizer trees can provide biofuel feedstock to 75% along with 25% of food feedstock to minimize competition with food crops. Improvements in sustainable agriculture can come alongside breakthroughs in biofuel production based on the new directions in biotechnology (Chen and Zhang 2015). The advancement of biotechnology includes a breakdown of non-food lignocellulosic biomass into multiple components, similar non-food cellulose into starch and production of oil from microbes, along with new biorefinery that can produce biofuels (biobutanol, biohydrogen and bioethanol) (Chen and Zhang 2015). The new biorefinery concepts of using non-food biomass, such as lignocellulosic wastes, agriculture wastes and renewable feedstock, to produce multiple food products for industrial applications has helped to minimize food insecurity. The concept of the biorefinery is similar to a modern oil refinery. The feedstock biomass containing a targeted substance is pre-processed and the method is optimized to acquire the targeted substance at higher concentration (Schuenemann et al. 2018). In the case of cellulosic ethanol refineries, simultaneous saccharification, fermentation and enzymatic transformation convert cellulosic feedstock into glucose, ethanol, amylose, butanol, etc. In the case of a biorefinery, the process conditions and methods can vary for different industrial applications (Chen and Zhang 2015).

There are numerous problems with biofuel production, such as the interaction of climate change, land requirements for biofuel crops, government rules, fragmented policies, socio-economic impacts, technological impacts and deforestation. To eliminate these problems, the biofuel production system needs strong policies in partnership with food security for sustainable development and achievements (Renzaho et al. 2017). Biofuel policy works on developing agro-industrial biofuels, creating more jobs, rural development, increasing biofuel crop production and improving agro-industrial producers for sustainable biofuel production (Schuenemann et al. 2018).. Before and during the operation of biofuel industry, a risk assessment of food security needs to be carried out in the local region, investigating any negative impacts for the present and future caused by the operation of biofuel production (Schut and Florin 2015). The framework for sustainable biofuel operations should not compete with food security; there should be continuous accessibility to basic food both before the operation and after the establishment of biofuel industry (Schut and Florin 2015). The Human Right to Adequate Food (RTaF) across the world ensures the biofuel policies on food security are fulfilled and certified before operation in food-insecure regions.

17.7 Water–Energy–Food Security Nexus

The global community and policymakers are increasingly focused on new approaches and solutions to the complex relationships among water, energy and food – the so-called water–energy–food (WEF) nexus. Human existence is dependent on food and water utilization for day-to-day activity, and energy is necessary for human development. Access to these resources, recognizing efficient use, identifying declining resources and managing the resources is the basis for the sustainable development of the nexus (Bizikova et al. 2013). International sustainable development recognizes the WEF linkages for poor people to gain access to water, energy and food both presently and in future at sufficient quantity. The WEF approach is crucial to recognize the dependency of water, energy and food production and aims to assess all the resources to manage the synergies locally and regionally (Bizikova et al. 2013; Rasul 2014).

The concept of the WEF nexus includes elements of food security such as food availability, access to food, utilization, self-production, food aid and food stability. The elements of water security include water access, safety, water aid, water distribution and affordability. The elements of energy security include continuity, energy aid, availability, distribution and supply to demand (Rasul and Sharma 2016). A number of international, regional conferences, dialogues, workshops and meetings have been held to focus on the elements of the WEF nexus for guidance and policy developments. The WEF policies improve our knowledge on the nature of interdependencies among the elements, the consequences of any diverse changes and enable the development of responses to ensure the security of these elements (Rasul 2014; Rasul and Sharma 2016).

The elements of the WEF nexus are interlinked and dependent on one another in numerous ways. Water for food, energy for food: both water and energy policy influence food demand. Water for energy and energy for water: water and energy policy influence each other. The nexus framework is divided into the energy–water nexus, the water–energy nexus, the water–food nexus and the energy–food nexus. The interface among water, energy and food is adapted for enhanced water, energy and food security (Mannan et al. 2018).

The challenges for the WEF nexus depend on several factors, such as increasing population, declining agricultural land, low levels of food production, water-/energy-intensive food production, water/energy scarcity and the use of biomass for energy (Rasul 2014). Additional risks include economics (deflation, financial failure, inflation), geopolitical (interstate conflict, terrorist attacks, state crisis), societal (spread of diseases, urban failure) and technological (cyberattacks, information breakdowns) factors (de Amorim et al. 2018). The challenges of the WEF nexus can be overcome by integrating sustainable livelihoods as noted by Environmental Livelihood Security (ELS). The framework also supports decision making (assessing transparency with trade-offs), policy assessments (cost-effective policies), policy harmonization (integration of conflicting policies), technology assessments (inclusive options) and scenario developments (elaborate socio-economic development) (Bazilian et al. 2011; Biggs et al. 2015). The framework of sustainable livelihood identifies the interactions of WEF as a single system in assessment with human population, governance and natural environment (climate change).

17.8 Genetically Modified Foods for Food Security

The application of biotechnology using genetically modified (GM) organisms in the field of agriculture has resulted in sustainable productivity and developed strategies for improved farming (Dibden et al. 2013). The rapid global spread of GM crops increased from 2 million ha (in 1996) to 120 million hectares (in 2006), and presently around 100 million farmers grow GM crops across 25 developed and developing countries (Qaim 2010). The choice of GM crops differs among developing countries: insect-resistant cotton is grown in Asia and Africa, with herbicide- and insect-resistant soybean grown across Latin America (Qaim 2010). The impact of GM crops in India has been to deliver 89% of profit, with a 42% decrease in insecticide usage, contributing 90% of total cotton production, providing employment opportunities, supporting agricultural laborers' income and boosting regional transport with inter-state trading business with the USA (Qaim 2010). GM tomatoes gain the nutritional quality with ripening, can be harvested simultaneously and tolerate increased storage time. GM rice, also called "golden rice," is rich in vitamin A, provides 80% of daily caloric uptake and is a staple food for half of the world's population (Mathur 2018).

GM seeds/food crops possess improved quality, extended shelf-life, and are farmer, consumer and environmentally friendly. GM food crops deliver a potential solution to food security now and even for the future. GM seed/food technology companies promote food products claiming to be healthier, cheaper, stable, tastier and high quality with essential nutrients. The introduction of GM technology in developing countries will decrease the growth duration with higher yields, decrease the amount of fertilizers required, and allow plant breeding for new crop varieties, antibiotic resistance to drought conditions and elevated vitamin/nutrient quality (Azadi and Ho 2010). GM crops in developing countries will ensure a stable, sustainable food supply, increase production, improve livelihoods and enhance food quality (Dibden et al. 2013).

The introduction of GM seeds/food in several countries still needs public acceptance for several reasons, such as public risk perceptions, varied responses, socio-political attitudes, the methodologies adopted, risk involved, trust in institutions, public worry, acceptability and concerns over GM foods (Frewer et al. 2004). There are some disadvantages of GM foods, such as potential toxicity and allergenicity, unintentional gene transfer to non-GM crops, synthesis of new toxins/viruses, limited access to patented GM seeds, the possibility of new as-yet-unrealized concerns and religious/cultural/ethical concerns (Azadi and Ho 2010).

The food safety system has introduced different regulations, principles and guidelines for more transparency on the safety of GM crops for public acceptance. The food safety system has introduced regulatory frameworks, defined principles of risk analysis, elaborated safety assessment, and introduced guidelines for toxicity measurement, investigating potential adverse effects, as well as developing an animal testing and monitoring system for post-marketing of food products from GM crops. The introduction of safety assessments on GM crops and food products in comparison with traditional counterparts has determined a high level of safety assurance for human food use (König et al. 2004; Qaim and Kouser 2013).

The use of GM technology has reduced food insecurity by 25–40% among farmers, will help to solve the food security problem and is also an important element for sustainable food security across the world.

17.9 Horizontal Gene Transfer for Food Security

With the introduction of GM crops, there is a movement from species-rich natural ecosystems of plant and animal diversity to species-poor agro-ecosystems. Previously, the species-rich ecosystem was composed of diverse cultures that carried major genetic diversity. However, GM crops have adapted high-yielding dwarf cultures, with a reduction in heterogeneity, and carry very little genetic diversity (Farre et al. 2011). Due to homogeneous genetics agro-ecosystems favor crop pathogens that are new, virulent, transmitting and wild in comparison to the species-rich ecosystems (Pritchard et al. 2016). These pathogens start mutating with higher host density, decreasing genetic drift, increasing the likelihood of plant multi-infections and co-infections of horizontal gene transfer (HGT), leading to rapid global spread of highly damaging pathogens. HGT plays an important role in the emergence of new pathogens, with adaptation composed of local pathogenic activity emerging and evolving rapidly into the agro-ecosystems. Pathogens such as *Phytophthora infestans* causing potato blight, the fungal pathogen causing cereal crop triticale and *Pyricularia graminis-tritici* causing wheat blast originated in Brazil and are well-known global pathogens causing crop production to drop by 40–100% (McDonald and Stukenbrock 2016).

It is necessary to re-engineer GM agro-ecosystems to minimize HGT and control the spread of infectious diseases across the globe (Farre et al. 2011). The fundamental steps are to disable the mechanism of genetic uniformity in the environment and minimize the evolution of the pathogen to prevent disease epidemics. The evolution of pathogens can be controlled by changing the culturing patterns of agro-ecosystems to include more environmental and genetic heterogeneity in crops in a dynamic way. Introduction of dynamic agro-ecosystem diversity, which changes over time and space, will impose and force plant resistance to pathogens (McDonald and Stukenbrock 2016). Dynamic diversity includes rotations of crops, improving the fertility of soil, strategies for gene resistance, antibiotic resistance, fungicide deployment and disruptive selection, and leads to a decrease in pathogen virulence (Fisher et al. 2018). It is also necessary to carry out biosafety investigation of re-engineered HGT of GM crops in order to understand the unexpected harmful effects (Arber 2017). It is necessary to minimize and control the pathogen infections resulting from HGT to avoid the disruption of crops and include new dynamic technologies to secure global food security.

17.10 Future Plans for Food Security

Food security principles and rights ensure the availability of food and protect the right to food at a basic level. Each country needs to analyze the conditions and incorporate strategies with improved policies on agricultural development for better food security. Countries with food-secure conditions need to transfer any technology developments in agriculture improvement to food-insecure countries. Countries with food-secure conditions must provide the necessary help with food and nutrition aids during periods of international food insecurity. Food-insecure countries must adapt, set up the improved agriculture technologies, leave behind the traditional agriculture practices, include citizen-friendly policies and increase cultivated forest land with improved food

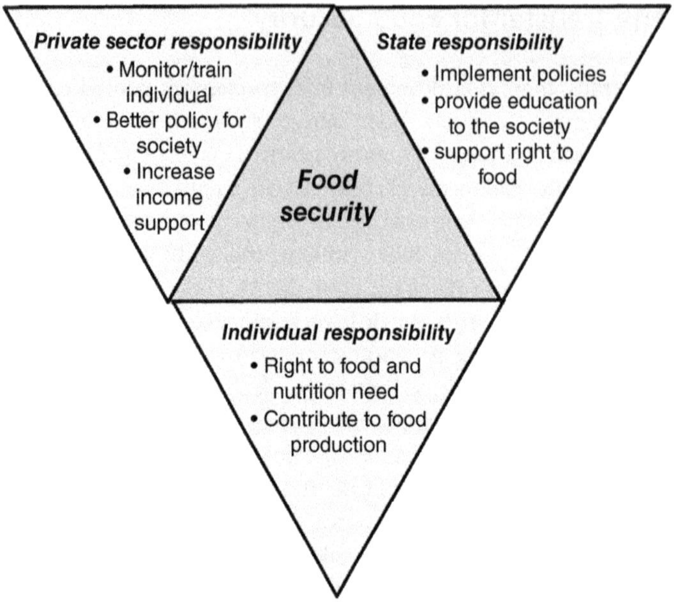

Figure 17.5 The role of the individual, the private sector and the state concerning food security.

crop cultivation. It is the responsibility of the individual, the state and the private sector, working together in synchronization, to help to minimize food insecurity across the world. The role of different commodities for food security is represented in Figure 17.5. The individual must fulfill their basic food and nutrition needs, must be active, capable of work and contribute to food production. The individual must be aware of their family responsibilities, should contribute and secure a healthy lifestyle for their family. The role of the private sector must be to monitor individuals to secure their right to food, train individuals in food production and amend the laws for the betterment of society. The role of the state is to provide the framework of food security, implement food policies, support the individual's right to food and provide education to society (Dou et al. 2017; Mohr et al. 2016).

The framework and principles demonstrate the scope of food security and biomass sustainability and provide guidance for regional and national certification to improve the food security system. The framework of food security needs improved national and international strategies concerning food security, market system investments to provide jobs, and to control or minimize natural and human-made disasters, adapt to ecological sustainability, adopt economic development policies and enhance the nutritional value of local food.

Acknowledgments

Financial support from INRS-ETE and LAVAL has been acknowledged. Vinayak Laxman Pachapur is greatly thankful to FQRNT for financial assistance under foreign scholarship for post-doc. research.

References

Ahmad, A., Yasin, N.M., Derek, C., and Lim, J. (2011). Microalgae as a sustainable energy source for biodiesel production: a review. *Renewable and Sustainable Energy Reviews* 15 (1): 584–593.

de Amorim, W.S., Valduga, I.B., Ribeiro, J.M.P. et al. (2018). The nexus between water, energy, and food in the context of the global risks: an analysis of the interactions between food, water, and energy security. *Environmental Impact Assessment Review* 72: 1–11.

Arber, W. (2017). Genetic engineering represents a safe approach for innovations improving nutritional contents of major food crops. *Journal of Innovation & Knowledge* 2 (2): 87–89.

Atabani, A.E., Silitonga, A.S., Badruddin, I.A. et al. (2012). A comprehensive review on biodiesel as an alternative energy resource and its characteristics. *Renewable and Sustainable Energy Reviews* 16 (4): 2070–2093.

Ayadi, M., Sarma, S.J., Pachapur, V.L. et al. (2016). History and global policy of biofuels. In: *Green Fuels Technology*, 1–14. New York: Springer.

Azadi, H. and Ho, P. (2010). Genetically modified and organic crops in developing countries: a review of options for food security. *Biotechnology Advances* 28 (1): 160–168.

Bazilian, M., Rogner, H., Howells, M. et al. (2011). Considering the energy, water and food nexus: towards an integrated modelling approach. *Energy Policy* 39 (12): 7896–7906.

Biggs, E.M., Bruce, E., Boruff, B. et al. (2015). Sustainable development and the water–energy–food nexus: a perspective on livelihoods. *Environmental Science & Policy* 54: 389–397.

Bindra, S.P., Hamid, A., Salem, H. et al. (2014). Sustainable integrated water resources management for energy production and food security in Libya. *Procedia Technology* 12: 747–752.

Bizikova, L., Roy, D., Swanson, D. et al. (2013). *The Water-Energy-Food Security Nexus: Towards a Practical Planning and Decision-Support Framework for Landscape Investment and Risk Management*. Winnipeg, Canada: International Institute for Sustainable Development.

Blanchard, L.P. (2013). *The Crisis in South Sudan*. Washington, DC: Library of Congress, Congressional Research Service.

Callahan, J., Chung, S., O'Connor, D., Demeke, M.S., Eberhardt, C., Lu, C., Miller, R., Qian, Y., Quinn, C. and Rangel, A. (2017). Syrian women and children: identifying gaps and goals for reconstruction. Task Force Report. Washington, DC: University of Washington.

Canadian Community Health Survey (2012). Household food insecurity in Canada statistics and graphics (2011 to 2012). https://www.canada.ca/en/health-canada/services/nutrition-science-research/food-security/household-food-security-statistics-2011-2012.html (accessed November 2019).

Chen, H.-G. and Zhang, Y.-H.P. (2015). New biorefineries and sustainable agriculture: increased food, biofuels, and ecosystem security. *Renewable and Sustainable Energy Reviews* 47: 117–132.

Chowdhury, R.B., Moore, G.A., Weatherley, A.J., and Arora, M. (2017). Key sustainability challenges for the global phosphorus resource, their implications for global food security, and options for mitigation. *Journal of Cleaner Production* 140: 945–963.

Conceição, P., Levine, S., Lipton, M., and Warren-Rodríguez, A. (2016). Toward a food secure future: ensuring food security for sustainable human development in Sub-Saharan Africa. *Food Policy* 60: 1–9.

Dibden, J., Gibbs, D., and Cocklin, C. (2013). Framing GM crops as a food security solution. *Journal of Rural Studies* 29: 59–70.

Dou, Z., Toth, J.D., and Westendorf, M.L. (2017). Food waste for livestock feeding: feasibility, safety, and sustainability implications. *Global Food Security* 17: 154–161.

Fan, S. and Brzeska, J. (2016). Sustainable food security and nutrition: demystifying conventional beliefs. *Global Food Security* 11: 11–16.

Fanzo, J. (2015). Ethical issues for human nutrition in the context of global food security and sustainable development. *Global Food Security* 7: 15–23.

FAO (2015). *WFP (2014) the State of Food Insecurity in the World 2014: Strengthening the Enabling Environment for Food Security and Nutrition*. Rome: FAO.

FAO (2018). Food security and nutrition around the world: hunger is on the rise. Rome: FAO. www.fao.org/state-of-food-security-nutrition/en (accessed November 2019).

Farre, G., Twyman, R.M., Zhu, C. et al. (2011). Nutritionally enhanced crops and food security: scientific achievements versus political expediency. *Current Opinion in Biotechnology* 22 (2): 245–251.

Ferranti, P. (2016). Food Sustainability, Security, and Effects of Global Change. Reference module in food science. Elsevier. https://doi.org/10.1016/B978-0-08-100596-5.03445-4.

Finco, M.V.A. and Doppler, W. (2010). Bioenergy and sustainable development: the dilemma of food security and climate change in the Brazilian savannah. *Energy for Sustainable Development* 14 (3): 194–199.

Fisher, M.C., Hawkins, N.J., Sanglard, D., and Gurr, S.J. (2018). Worldwide emergence of resistance to antifungal drugs challenges human health and food security. *Science* 360 (6390): 739–742.

Frewer, L., Lassen, J., Kettlitz, B. et al. (2004). Societal aspects of genetically modified foods. *Food and Chemical Toxicology* 42 (7): 1181–1193.

Garibaldi, L.A., Gemmill-Herren, B., D'Annolfo, R. et al. (2017). Farming approaches for greater biodiversity, livelihoods, and food security. *Trends in Ecology & Evolution* 32 (1): 68–80.

Garnett, T. (2014). Three perspectives on sustainable food security: efficiency, demand restraint, food system transformation. What role for life cycle assessment? *Journal of Cleaner Production* 73: 10–18.

Grindle, A.K., Siddiqi, A., and Anadon, L.D. (2015). Food security amidst water scarcity: insights on sustainable food production from Saudi Arabia. *Sustainable Production and Consumption* 2: 67–78.

Koizumi, T. (2013). Biofuel and food security in China and Japan. *Renewable and Sustainable Energy Reviews* 21: 102–109.

Koizumi, T. (2015). Biofuels and food security. *Renewable and Sustainable Energy Reviews* 52: 829–841.

König, A., Cockburn, A., Crevel, R. et al. (2004). Assessment of the safety of foods derived from genetically modified (GM) crops. *Food and Chemical Toxicology* 42 (7): 1047–1088.

Lal, R. and Stewart, B.A. (2010). *Food Security and Soil Quality*. Boca Raton, FL: CRC Press.

Mannan, M., Al-Ansari, T., Mackey, H.R., and Al-Ghamdi, S.G. (2018). Quantifying the energy, water and food nexus: a review of the latest developments based on life-cycle assessment. *Journal of Cleaner Production* 193: 300–314.

Mathur, R. (2018). Genetic engineering and biosafety in the use of genetically modified foods. *International Journal of Advanced Scientific Research and Management*, Special Issue (1): 76–82.

Maxwell, D., Webb, P., Coates, J. and Wirth, J. (2008). Rethinking food security in humanitarian response. Proceedings of the Food Security Forum, Rome, pp. 16–18.

McDonald, B.A. and Stukenbrock, E.H. (2016). Rapid emergence of pathogens in agro-ecosystems: global threats to agricultural sustainability and food security. *Philosophical Transactions of the Royal Society B: Biological Sciences* 371 (1709) https://doi.org/10.1098/rstb.2016.0026.

Mohr, A., Beuchelt, T., Schneider, R., and Virchow, D. (2016). Food security criteria for voluntary biomass sustainability standards and certifications. *Biomass and Bioenergy* 89: 133–145.

North, I. (2001). Food insecurity in Canadian households. *Health Reports* 12 (4): 11.

Pachapur, V.L. (2017). *Fermentation du glycérol brut en co-culture pour la production d'hydrogène*. Québec, Canada: Université du Québec, Institut national de la recherche scientifique.

Pachapur, V.L., Sarma, S.J., Brar, S.K. et al. (2016). Energy balance of hydrogen production from wastes of biodiesel production. *Biofuels* 9: 129–138.

Pachapur, V.L., Brar, S.K., Le Bihan, Y., and Buelna, G. (2017a). Alternate green approach of spent media utilization for hydrogen and for lipid production. *International Journal of Hydrogen Energy* 42 (9): 5832–5839.

Pachapur, V.L., Das, R.K., Brar, S.K. et al. (2017b). Valorization of crude glycerol and eggshell biowaste as media components for hydrogen production: a scale-up study using co-culture system. *Bioresource Technology* 225: 386–394.

Pinstrup-Andersen, P. and Pandya-Lorch, R. (1998). Food security and sustainable use of natural resources: a 2020 vision. *Ecological Economics* 26 (1): 1–10.

Pritchard, L., Glover, R.H., Humphris, S. et al. (2016). Genomics and taxonomy in diagnostics for food security: soft-rotting enterobacterial plant pathogens. *Analytical Methods* 8 (1): 12–24.

Qaim, M. (2010). Benefits of genetically modified crops for the poor: household income, nutrition, and health. *New Biotechnology* 27 (5): 552–557.

Qaim, M. and Kouser, S. (2013). Genetically modified crops and food security. *PLoS One* 8 (6): e64879.

Rasul, G. (2014). Food, water, and energy security in South Asia: a nexus perspective from the Hindu Kush Himalayan region. *Environmental Science & Policy* 39: 35–48.

Rasul, G. and Sharma, B. (2016). The nexus approach to water–energy–food security: an option for adaptation to climate change. *Climate Policy* 16 (6): 682–702.

Renzaho, A.M., Kamara, J.K., and Toole, M. (2017). Biofuel production and its impact on food security in low and middle income countries: implications for the post-2015 sustainable development goals. *Renewable and Sustainable Energy Reviews* 78: 503–516.

Rideout, K., Riches, G., Ostry, A. et al. (2007). Bringing home the right to food in Canada: challenges and possibilities for achieving food security. *Public Health Nutrition* 10 (6): 566–573.

Rutten, M., Achterbosch, T.J., De Boer, I.J. et al. (2016). Metrics, models and foresight for European sustainable food and nutrition security: the vision of the SUSFANS project. *Agricultural Systems* 163: 45–57.

Sarma, S.J., Pachapur, V.L., Brar, S.K. et al. (2016). Biohydrogen. In: *Green Fuels Technology* (eds. C.R. Soccol, S.K. Brar, C. Faulds and L. Pereira Ramos), 407–429. New York: Springer.

Sattaka, P., Pattaratuma, S., and Attawipakpaisan, G. (2017). Agricultural extension services to foster production sustainability for food and cultural security of glutinous rice farmers in Vietnam. *Kasetsart Journal of Social Sciences* 38 (1): 74–80.

Schindler, J., Graef, F., König, H.J. et al. (2016). Sustainability impact assessment to improve food security of smallholders in Tanzania. *Environmental Impact Assessment Review* 60: 52–63.

Schuenemann, F., Msangi, S., and Zeller, M. (2018). Policies for a sustainable biomass energy sector in Malawi: enhancing energy and food security simultaneously. *World Development* 103: 14–26.

Schut, M. and Florin, M.J. (2015). The policy and practice of sustainable biofuels: between global frameworks and local heterogeneity. The case of food security in Mozambique. *Biomass and Bioenergy* 72: 123–135.

Songlin, M. and Ruihong, L. (2010). Evaluation on sustainable food security in Henan against the background of low-carbon economy. *Agriculture and Agricultural Science Procedia* 1: 412–416.

Taleb, Z.B., Bahelah, R., Fouad, F.M. et al. (2015). Syria: health in a country undergoing tragic transition. *International Journal of Public Health* 60 (1): 63–72.

Tan, H., Aziz, A.A., and Aroua, M. (2013). Glycerol production and its applications as a raw material: a review. *Renewable and Sustainable Energy Reviews* 27: 118–127.

Tarasuk, V., Mitchell, A. and Dachner, N. (2016). Household food insecurity in Canada, 2012. Research to identify policy options to reduce food insecurity. Report. University of Toronto. https://proof.utoronto.ca/resources/proof-annual-reports/annual-report-2012 (accessed November 2019).

Toledo, Á. and Burlingame, B. (2006). Biodiversity and nutrition: a common path toward global food security and sustainable development. *Journal of Food Composition and Analysis* 19 (6–7): 477–483.

Waldron, A., Garrity, D., Malhi, Y. et al. (2017). Agroforestry can enhance food security while meeting other Sustainable Development Goals. *Tropical Conservation Science* 10 https://doi.org/10.1177/1940082917720667.

Zhang, J., He, C., Chen, L., and Cao, S. (2017). Improving food security in China by taking advantage of marginal and degraded lands. *Journal of Cleaner Production* 171: 1020–1030.

18

Sustainable Healthcare Systems

Carlos S. Osorio-González[1], *Krishnamoorthy Hegde*[1], *Satinder K. Brar*[1 2], *Antonio Avalos-Ramírez*[1 3], *and Rao Y. Surampalli*[4]

[1]*INRS, Université du Québec, Québec, Canada*
[2]*Department of Civil Engineering, Lassonde School of Engineering, York University, Toronto, Canada*
[3]*Centre National en Électrochimie et en Technologie Environnementales (CNETE), Shawinigan, Canada*
[4]*Global Institute for Energy, Environment, and Sustainability, Lenexa, KS, USA*

18.1 Introduction

In 1948, the World Health Organization (WHO) defined health as "a state of total well-being (mental, physical and social) and not just as the absence of disease." Subsequently, healthcare systems were defined, within which all the participants were included, as well as the institutions and resources that are part of the delivery of health, where the main intention is to improve the health of the population in general (Mohrman et al. 2012; Grin and Broerse 2017).

Currently, most of the countries around the world have a priority to provide quality health services with a sustainable approach, with the aim of generating a productive society that contributes to the growth of the economy. However, even though during the past few years different initiatives have been implemented to improve these services, they remain an unsustainable social and economic problem for most countries. This is mainly due to the constant increase in the population that requires quality medical services.

A complex task within healthcare systems is decision making. This complexity is mainly based on the nature of the different interactions between the components of these systems (financial resources, technology and human resources) (Fleiszer et al. 2015). Currently, the demand for reliable, timely and affordable health services for all sectors of the population has increased. This has led governments and different health organizations to search for and implement innovative and creative healthcare systems at different levels (local, regional, national and international) (Qudrat-Ullah 2017).

Today, recent advances in organization, economics and technology applied to public and private healthcare systems have allowed larger access to the general population. This is reflected in the reduction of the death rate worldwide for every 1000 people (17.7 in 1960 and 7.6 in 2015) (World Bank 2015). Behind this reduction, there are several key factors,

Sustainability: Fundamentals and Applications, First Edition.
Edited by Rao Y. Surampalli, Tian C. Zhang, Manish Kumar Goyal, Satinder K. Brar, and R. D. Tyagi

such as better availability of medicines and better equipped hospital facilities, which affect the improvement of the treatments applied and are essential for the best recovery of patients (Imran et al. 2018).

The concept of sustainability, as well as its practices, has expanded within the health industry (Schulz et al. 2017). Principles of sustainability in healthcare systems have focused mainly on two components, capital management and the efficient use of resources (Goh and Marimuthu 2016). However, some authors have studied the role of sustainability exclusively within the context of healthcare services, looking at how healthcare can be improved as a dynamic process with a balanced approach to social, economic and environmental development (Ament et al. 2012; Buffoli et al. 2013; Ramirez et al. 2013; Mendel et al. 2014). The above perspective is reflected in the evolution of healthcare systems (patient safety, quality and access to the service) (Astolfi et al. 2012; Greenhalgh et al. 2012). The main discussion in this chapter centers around a general perspective of health models that are offered around the world, as well as an examination of the role played by sustainability within these healthcare systems.

18.2 Classification of Healthcare Systems

To classify healthcare systems, it is convenient, first, to define the healthcare system in a general way. According to Quadagno (2010), they are "institutions, doctors' offices, clinics, and hospitals that offer some type of medical service and that in turn administer the economic resource granted by the government, private agencies, provinces, communities, and insurance companies."

Nowadays, classification of healthcare systems by the academy has diversified. Currently, these systems can be classified according to the funding source, institution type or population to which the service is focused (Beckfield et al. 2013). Recently, some health researchers have tried to classify healthcare systems in all countries (Wendt et al. 2009). The main models in the healthcare systems used in industrialized countries and their main characteristics are shown in Table 18.1.

An example of these systems is reflected in the provincial insurance plans that exist in Canada, where taxpayers make up 70% of health spending (medical care, hospital care, medications and articles needed in long-term treatments), compared to France and Germany with 76% and 78%, respectively (Himmelstein et al. 2018).

However, all healthcare systems, no matter if they are taxable or not, have had difficulties during the past decade, due to their lack of sustainability (maintaining quality, service coverage and low costs). The foregoing situation is mainly caused by the increase in the population (particularly the elderly population $\geq$65 years old) and the development of new technologies to satisfy the existing demand. In addition, the recent economic crises have aggravated the problem of access and equity in healthcare systems (Lega et al. 2013). These inefficiencies are very common, even in countries with great technological and economic advances, such as the United States. Although in this country the healthcare system is competitive, it presents a high degree of inefficiency, due to its high administration costs, poor attention to equality (quality of care varies depending on race, income and place of residence) and limited improvement in the quality of the treatment of patients (Ryan et al. 2017).

Table 18.1 Main health models used in the world (International Health Care System 2018).

Health model	Features	Countries
Beveridge	This model offers medical care to all citizens of the country where it is used. This health model is financed by the government through the payment of taxes	UK, Spain, and New Zealand
Bismarck	The model uses an insurance system that is financed by employers and employees, through payroll deduction. Nevertheless, the insurance plans in this model are not profitable, since they do not include all citizens. One of the characteristics that harm this model is that hospitals and doctors who work in them are usually private	Germany, France, Belgium, the Netherlands, Japan, the US and Switzerland
National health insurance or Tommy Douglas	This model is mixed because it has characteristics of the two previous models (Beveridge and Bismarck). In this system, the providers are from the private sector. However, the payment of these comes from an insurance payment that is administered by the government, which all citizens finance through the payment of a premium or tax. These types of programs are usually less expensive	Canada
Out-of-pocket	The model is used in countries with a precarious economy, which does not allow the government to provide any kind of medical attention. In these countries, members of the population who can afford it are able to pay for their medical care and those who can't do so remain sick or die	Africa, India, China, South America

18.3 Sustainability and Healthcare Systems

The terms sustainability and sustainable development are used interchangeably; however, we could define sustainability or sustainable development *as a capacity to meet different needs such as food, energy, social and economic welfare, without compromising the ability to meet the same needs of the following generations*. Figure 18.1 shows the three components of sustainable healthcare systems. Currently, these three components are widely accepted, and the concept of sustainability is applied in different areas and disciplines. Of course, healthcare systems are not the exception as stated in Table 18.2. In Figure 18.1 the conventional terms that make up a sustainable system (economic, social and environmental) are not mentioned literally since, in healthcare systems, these terms are broader and can be approached from different perspectives. From our point of view, "equitable" corresponds to social inclusion as well as economic factors. In addition, this concept must include all environmental factors acquired by the healthcare systems.

However, there is no doubt that healthcare systems need to be sustainable, because everyone should have access to quality medical care, both now and in the future. In 2006, the HNA (Alliance for Natural Health) in the USA presented the definition of a sustainable healthcare system, which in summary did not mention an interactive approach between the management, optimization and restoration of human health based on the three pillars of the sustainability (Faezipour and Ferreira 2011).

On the other hand, it is very important to consider several factors when trying to create sustainable healthcare systems. The above implies considering and defining to a certain extent what the value means. Generally, valuable things are those with a high monetary

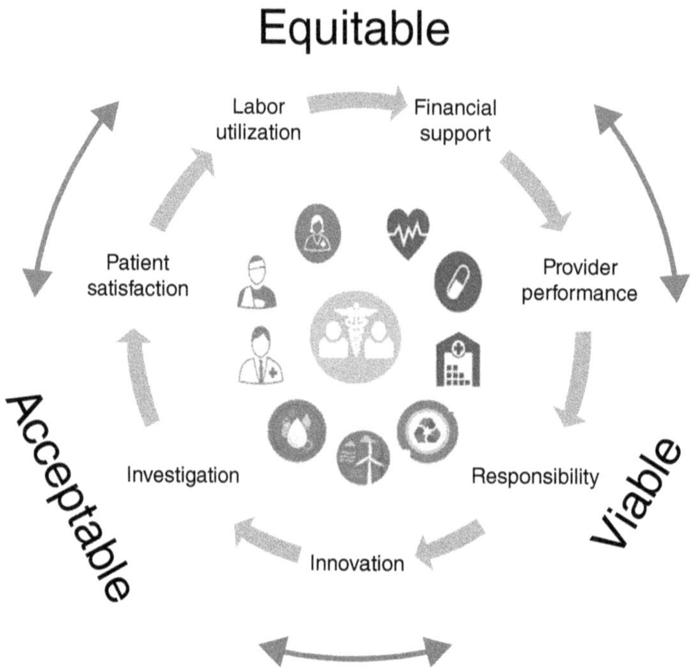

Figure 18.1 Sustainability healthcare components.

Table 18.2 Features from the three main components of sustainability in healthcare systems.

Component	Features	References
Social	The main objective of this component is the inclusion and equity of all people in the different systems. One of the great challenges to achieving social sustainability in health systems involves satisfying the basic needs of and for each individual. In particular, the concept of social sustainability implies promoting actions that allow access to healthcare systems without cultural distinction, including race, color, nationality and economic stratum. However, social sustainability in healthcare systems also implies certain commitments on the part of society such as supporting, social labor, volunteering, disseminating and participating in the resilience activities of the place where it is located	Capolongo et al. (2016)

Table 18.2 (Continued)

Component	Features	References
Economic	The forces that drive healthcare expenses in an unsustainable direction work regardless of how medical care is financed. Government resources are being stretched, which points to a probable redistribution of financial responsibility for medical care, which could have a significant negative social impact on the people who can least afford to pay. This type of policy discourages regular preventive care since the policy is probably only used for serious or acute injuries or illnesses. The social costs of chronic diseases are significant. The rapid growth in insurance and/or care costs affect consumption patterns and the strength of local economies. Medical care that does not meet the needs of the population generates greater sickness and greater demand, which further decreases the ability of society to pay	Tangcharoensathien et al. (2015) and Buse and Hawkes (2015)
Environmental	It is known that the health industry practices environmental sustainability and this approach to reducing environmental pollution (e.g. recycling water and minimizing the use of hazardous chemicals) in operations, which also helps to save operating costs. Some of the practices toward environmental sustainability in medical care are eliminating mercury, reducing the toxicity of sanitary waste and minimizing the use of hazardous chemicals. Meanwhile, recycling is also being introduced. In later years, environmental sustainability has extended to sustainable design and construction techniques	Marimuthu and Paulose (2016)

cost (clothes, cars, jewelry, trips), which incur high energy expenditure, waste production and contribute to the depletion of resources. Sustainable medical attention, on the contrary, focuses on different values that are unlimited in a certain way (compassion, care, fun, kindness, affection, attention), which have been shown to have beneficial effects on the health of patients (Bhatia and Orsini 2016).

18.4 Sustainability Challenges in Healthcare Systems

Nowadays, healthcare systems face endless challenges (e.g. diminishing resources, the development of medical technology, more expensive pharmaceutical products as well as a greater expectation on the part of patients). It is necessary to improve or find a balance between the demands of medical care, economy, resources and infrastructure available with respect to the needs of the population, to ensure a quality healthcare system for the current population and for future generations (Stirman et al. 2012). The main changes facing healthcare systems can be summarized in five specific challenges:

18.4.1 Technology Demand

Consumers demand new diagnoses and medications (mostly expensive once they are available on the market), for the treatment of their diseases. This directly impacts the profitability of the healthcare system, since two-thirds of the costs of these diagnoses and medicines are due to the development of technology to meet the growing demand of patients (Wallace 2013). However, there are often no economic arguments available to develop and adopt this type of innovation. In fact, the estimates suggest that only 21% of technological development in the health sector has proven scientific support (Buchan 2004).

18.4.2 Demographic Demand

Another factor that increases the cost of health services today is the elderly population (+65 years), which has been increasing over the past two decades and will continue to increase during the next two decades. For example, in Australia, per capita, health spending in 2003 was $1807 for people under 65, $5506 for people 65–75,and $15 690 for people over 85 years old (Coiera and Hovenga 2007). In general, this trend is present in all healthcare systems around the world, because this stratum of the population has a high rate of chronic diseases, requiring periodic medical examinations.

18.4.3 Workforce Capacity

For healthcare systems to work the availability of skilled labor and financial resources to meet the growing demand for medical care for the entire population are both important factors. These two components are severely affected by the increasing number of people, given that as the population ages, there will be fewer active people, which directly affects the contribution of taxes and fees to support the different public or private institutions that offer health services. Another critical point is the limited number of registered personnel in the labor field, especially in the United States and Canada. This is mainly down to two reasons: (i) the situation faced by universities around the world in terms of scarce resources for the training of new personnel; and (ii) the particular problems that do not allow them to perform better in their profession (job satisfaction, salary, expectations) (Girot and Albarran 2012).

18.4.4 Quality

It is currently recognized that one of the main causes of illness and death is mistakes made by the personnel working in different areas of the healthcare system (Carayon et al. 2014). Most of these errors are due to a lack of well-standardized processes, deficiency in communication channels, work overload and a lack of personnel within the healthcare systems (Kim et al. 2016). In most cases there is no assessment of patient satisfaction and/or no feedback from staff and therefore there is no improvement in the quality of healthcare systems (Kisliakovskii et al. 2017).

18.4.5 Supply Chain

Good management and understanding of the supply chain in healthcare systems is crucial because if this supply chain is deficient, many different areas that make up the system are affected (administrative, financial, pharmaceutical), which directly affects the users of the healthcare system. In situations where the supply chain works on a regular basis, it is possible to obtain previous results of the processes in which it directly intervenes. However, correct management of the supply chain used as a strategic, operational or tactical tool must have certain characteristics, such as sharing information, sharing risks and sharing the benefits, since the absence of any of these characteristics may result in a suboptimal performance of the process in which the chain participates (Kwon et al. 2016).

18.5 Categories of Sustainability in Healthcare Systems

The implementation of sustainability in healthcare systems has focused on identifying the main categories (as shown in Figure 18.2) that are necessary to make the system sustainable in a real-world context. Recently, some researchers have stated that sustainability in healthcare systems is a persistent challenge in the sectors that have implemented a sustainable service (Shelton et al. 2018). In this sense, it is necessary to maximize the efficiency of healthcare systems and to invest in the application of sustainability in these systems (Scheirer and Dearing 2011). However, there are sufficient reasons for the implementation

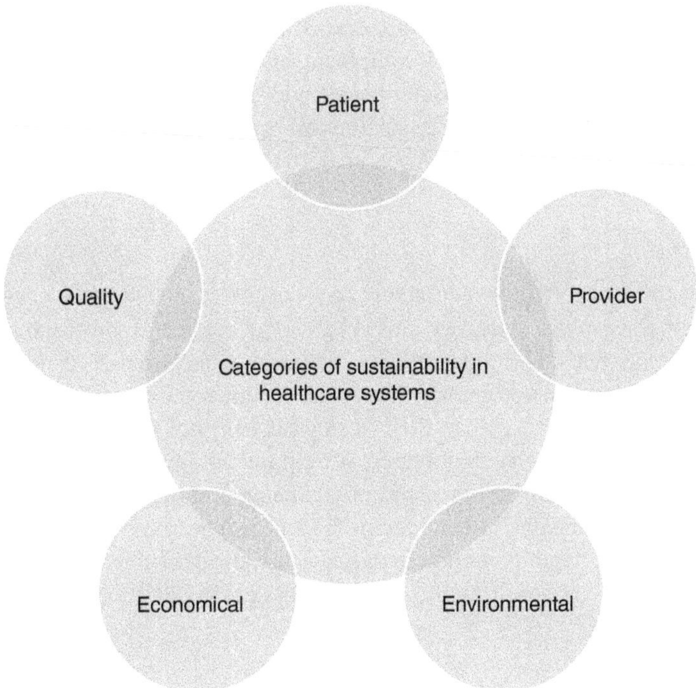

Figure 18.2 Main categories for sustainable development in the healthcare system.

of sustainability in healthcare systems, for the development of new public policies in the field of health and for long-term economic development (Proctor et al. 2015).

18.5.1 Patient

This section includes the entire population that requires or needs medical attention. The patient and the different categories that participate in the development of sustainable healthcare systems are interrelated according to certain criteria. One of the main factors with which the patient interacts is the measurement of quality inside and outside the healthcare system through evaluations of the satisfaction and wellbeing of the patient before, during and after using any healthcare system (Chen and Li 2010; Jordan et al. 2010).

18.5.2 Provider

This category includes all healthcare providers, medical personnel, nurses, laboratories, insurance companies, government institutions, hospitals, clinics, as well as all intermediaries that provide some type of health service to patients. The different types of suppliers play a key role in the development of sustainable healthcare systems because they largely influence all the processes that take place throughout each of the stages in the healthcare systems (Hawk et al. 2012; Sudhinaraset et al. 2013).

18.5.3 Resources

This section includes all material and physical resources, such as facilities for medical care, equipment for analysis and diagnosis, surgical equipment and medical supplies. This section is very important to include in the concept of sustainability in healthcare systems since the price and quality of resources directly impact the cost that patients must cover for access to health services and all their resources (Aktaş et al. 2007; Guindo et al. 2012).

18.5.4 Environment

It is necessary to address this category in two different ways: the first is related to the environment, which considers the ecological impact caused by energy usage and the elimination of waste by healthcare systems. This category also includes the factors that relate to research and innovation in terms of the management of environmental protection that healthcare systems have to develop to reduce their ecological footprint (Pinzone et al. 2012). The second is specifically responsible for measuring the environment within the healthcare system itself in terms of an environment of wellbeing and quality of care of patients within the healthcare system facilities and with the different personnel that interact with them. Here, it is prudent to point out that in this category, the personnel trained to carry out this measurement may not be within the healthcare system itself (Dunphy 2014).

18.5.5 Economic

This category includes the costs and revenues of all the resources and services offered by the healthcare system. This point is key to developing sustainability within a healthcare system since the development of all other factors will have an impact on this and the effect that will be generated will define if the healthcare system is economically sustainable and, in some cases, if it is profitable (Borgonovi and Compagni 2013; Minoglou et al. 2017).

18.5.6 Quality

In this category, there are different approaches to the criteria with which quality should be evaluated within healthcare systems. However, most of the research conducted on this topic focuses mainly on patient safety, the efficiency of medical and support staff, the efficiency of the service and equitable access to the healthcare system. One of the parameters with which quality is most evaluated in healthcare systems that are or aspire to be sustainable is patient satisfaction. On the other hand, quality can also be evaluated in terms of the supply chain within healthcare systems, which is nothing more than ensuring the supply of equipment, medical and surgical equipment, and the different consumable materials (Alrubaiee and Alkaa'ida 2011; Foley and Vale 2017).

On the other hand, there are some other general factors that contribute to the five main categories mentioned above. Some of these factors, as well as the role that they play in the sustainability concept of healthcare systems, are shown in Table 18.3.

Table 18.3 Complementary factors that contribute to sustainability in healthcare systems.

Complementary factor	Sustainable approach	References
Organizational	An increase in the demand for healthcare services has caused multiple participants to design, improve and innovate organizational strategies within their processes in healthcare systems. These actions have become indispensable to satisfy to a certain extent the health needs of all populations around the world. This action allows institutions to remain competitive within the system and the development of aspects such as counseling, leadership, motivational and informative culture, gender inclusion and sense of resilience, which improve the system and concept of sustainability within the healthcare system	Goh and Marimuthu (2016)
Innovation	For several decades, the concept of innovation has been changing and is currently considered as a new trend to regenerate and improve quality management processes for any system or process. This factor plays a critical role in sustainability process, since through it, elements, processes and technologies of healthcare systems are modified and improved. This is key for business, social and environmental development, as well as for the economic growth of the healthcare system in which it is implemented. The above generate a constant innovative process of all its elements (products, processes and organization)	Anttila and Jussila (2019)

(Continued)

Table 18.3 (Continued)

Complementary factor	Sustainable approach	References
Workforce	In general, healthcare systems occupy a large number of "workers," whether specialized, qualified or unskilled. Regardless of the position or activity that they perform within the health systems, there is a great difference among personnel in terms of training, ability, aptitude and attitude levels. These differences are more evident in countries with "deficient" healthcare systems due to poor distribution of resources or work overload. It is necessary to improve the organizational culture of personnel in the different departments with the purpose of improving the processes of healthcare services. An alternative to this could be the implementation of self-care workshops on health and integral health, with the aim of including the population and promoting autonomy and prevention of diseases	McKee (2018)
Marketing	One of the most important factors in the sustainability system is marketing. However, this factor has not been studied in depth in healthcare systems, as well as the impacts and relationships that this can have on different actors (patients, medical personnel and providers) that participate in all process of delivery and coverage of health services. The main impact that contributes to the sustainability system of healthcare systems is an economic increase of the system itself as well as the social interactions and diffusion that can be generated from proper use of a marketing program	Smyczek (2015)

18.6 Overview of Sustainable Healthcare Systems

In an international context, continuous improvement in terms of healthcare systems is of utmost importance and even more so the implementation of making these systems sustainable in all aspects. Many of these initiatives, although promising, are not sustainable themselves; the long-term benefits are not as expected and, in some cases, cause their rejection (Stirman et al. 2012; Lennox et al. 2018). Thus, it is imperative for governments and health institutions around the world to increase the inclusion of sustainable methods, products and technology in their healthcare systems. Below are some examples of these applications in different countries.

18.6.1 United States

This country does not have a consistent healthcare system since it does not have universal healthcare coverage for its entire population. The best way to describe the healthcare system used in the US is a hybrid system, in which the money is provided by private funds, beneficiaries and different types of governments (national, state and local). Another typical characteristic of this healthcare system is that, regardless of the economic source, private sector companies provide health services (Sustainable Development Unit 2018).

Currently, the United States is carrying out sustainable development in the value chain of healthcare systems, with the aim of improving the logistics in the different areas affected by it (patient organizations, regulatory agencies and health service providers). Likewise, it promotes a close collaboration with researchers to find better solutions to different diseases, which is very important for the innovation of the healthcare system. With these types of initiatives, interdisciplinary and inclusive work is developed within the healthcare system (dissemination activities and intersectoral projects) using communication technologies (Horgan et al. 2017).

Another sustainable initiative to improve healthcare systems is the implementation of the health park concept. This concept consists mainly of persuading the population, in general, to use parks and green areas as a platform to prevent and improve their health with respect to chronic diseases (cardiovascular, pulmonary diseases and metabolic syndromes). The two main benefits that this initiative generates for the healthcare system are: (i) a decrease in the demand for medical attention caused by these types of illnesses; and (ii) self-management of society in general with respect to their health (Arena et al. 2017).

18.6.2 Canada

In general, the healthcare system of this country is complete and efficient in many ways, since the healthcare systems are financed by the federal government and, in some cases, by the provincial governments. The system supported by the government covers the entire population under the same conditions (universal system); from the first dollar it covers the different hospital and medical services (portability and accessibility), as well as the health services provided by hospitals and doctors (completeness). However, the coverage of pharmaceutical products and non-medical providers (optometrists or physiotherapists) varies depending on the province (Hutchison et al. 2011).

The Canadian healthcare system applied a sustainable initiative (DREAM3), whose main objective is to provide health education to patients with hypertension problems through self-help and participatory care through text messages. This initiative greatly reduced the demand for regular home visits from nurses. In addition, the efficiency of all the medical team participants was improved. This program included different sustainable components for the treatment of chronic diseases in the northern territory of Canada. Some of these components are the inclusion of the whole community, an organization of health services, and control of the progress of the disease, planning and coordination of activities, among others (Yeates et al. 2017).

18.6.3 United Kingdom

The healthcare system in this country is run by three actors (parliament, the Secretary of Health and the Department of Health). Together, these three organizations have the obligation to provide a free health service, as well as the right to obtain care without discrimination. However, the responsibility for managing the healthcare system falls to an independent public body called the National Health Service (NHS). This organization has the task of managing the budget for health services and monitoring the efficient use of this budget to reduce the problems in healthcare systems (International Health Care System 2018).

The NHS has taken part in an initiative to reduce carbon emissions because this action generates fewer cases of respiratory diseases and consequently promotes an improvement in quality of life. The inclusion of the NHS in this initiative allows close observation of the impacts (diseases of abundance) that different lifestyles generate due to the excess in carbon consumption through the different activities that people perform daily. At the same time, resilience is promoted for other types of organizations; for example, by finding a solution to reducing the carbon footprint caused by healthcare systems and contributing to their sustainable development (Sustainable Development Unit 2009).

Studies have been conducted to evaluate the sustainability of the supply chain in the production system of technological developments applied to medicine and the ecological footprint that these processes leave in the environment. This initiative evaluates the different actors that participate in this supply chain (government institutions, suppliers, among others). This type of analysis allows us to evaluate the areas of an opportunity presented by the productive and technological development sectors that try to provide quality materials and services to the different healthcare systems and it can be seen that in some cases the damage caused by the process is greater than the benefits that it guarantees (Kumar et al. 2017).

18.6.4 Japan

The healthcare system in this country is a combination of different characteristics with high demand, excessive use of medications and high costs for the service. However, the impact that economic growth generated at the end of the Second World War on the Japanese healthcare system was highly beneficial since, currently, this country is one of the most advanced in the world in terms of health (Barata et al. 2012).

Currently, the country is going through a crisis in the healthcare system, since it has a population of more than 33.4 million people over 65 years of age and that will continue to increase, reaching 39 million by 2042. Now, with such a high proportion of aging population, there is a health problem since a high percentage of this population is not integrated into a healthcare system. Thus, it is necessary to look for a sustainable alternative solution for community care. For this initiative, five elements are necessary, of which three are specialized services (prevention, treatment and nursing) and the other two are non-specialized services (social assistance and housing). With these components, a medical–social–community prescription model was developed as a tool that links the specialized health sector with self-management of health by the community (Kumakawa et al. 2017).

18.6.5 India

This country is going through a process of critical challenges in its healthcare system. These challenges are mainly aimed at improving healthcare systems to protect the population from concurrent diseases, to make the healthcare system more efficient, as well as to improve user expectations, access to medical care, quality of services and sustainability of healthcare systems (Tavecchi and Rebecchi 2018).

One of the alternatives that the healthcare system has implemented is to improve the access to medical care in places where it is not easy to obtain. This project consists of providing mobile medical care, which has been developed with two main characteristics: mobility and modularity. This type of service is provided through semi-fixed modules that provide basic healthcare and, in a certain way, satisfy even more people. In general, the system has high adaptability to health needs because it can provide for the population living in areas where it is not possible to maintain adequate health services (Fugazzola et al. 2018).

In addition, as a result of the lack of equity in the delivery of health services, the government of this country implemented a program of social inclusion. This program consisted mainly of forming committees that promote, plan, manage and supervise to ensure that the healthcare system provides medical care to all regardless of their social or economic status. A key aspect of this process is the way in which citizen participation is perceived, although the approach is a conceptualization of the healthcare system, this also implicitly includes a new distribution of power within the healthcare system itself (Gaitonde et al. 2017).

18.6.6 Latin America

The situation in this part of the world regarding healthcare systems began to change in the late 1980s with social reform to reduce the poverty and inequality that prevailed at that time (Freire and Pan American Health Organization 2006). This change in their healthcare systems was due to the implementation of universal health coverage for the entire population. The principles of equity and collective action supported the trajectory of this change. Countries such as Mexico, Costa Rica and Colombia were a crucial part of this revolution (Titelman et al. 2015).

The main characteristics of these healthcare systems classify the population into three sectors: (i) low-income people, the unemployed and employees without access to a healthcare system; (ii) workers who receive a salary with access to a healthcare system; and (iii) people with the ability to acquire private health insurance (Atun et al. 2015). The reforms in terms of health policies are mainly focused on providing health coverage to the most disadvantaged segments of the population (farmers and the unemployed) (de Andrade et al. 2014).

Currently, in many of these countries, the main objective of healthcare systems is to better meet the demand for health services that the population needs. Because of this, the inclusion of sustainability to a certain extent is limited in the healthcare systems of most of these countries. However, there are initiatives like the one that was developed in Mexico, which is focused on patients who have overcome breast cancer. The program called "Young and Strong" aimed to optimize the different clinical, psychological and social needs of patients and their families, to make patients and their families aware of the special needs that arise during treatment and replicate the program in other health centers through the experiences that the patients themselves have lived, in order to raise the self-esteem of patients who have been newly diagnosed with this disease (Villarreal-Garza et al. 2017).

18.6.7 Nigeria

Like the previous countries discussed, the Nigerian government believes that maintaining a population in good health is essential for rapid economic development, so healthcare systems are a priority in the policies of this country with the same importance as education.

The total population in Nigeria amounts to 135 million inhabitants, of whom 55% live in rural areas and the remaining 45% in urban areas; due to this the health services face great challenges (Omoruan et al. 2009). The healthcare system in Nigeria is divided into three levels: primary, secondary and tertiary (administered by the local, state and federal government, respectively), as well as contributions and interactions from the private sector and traditional medicine (Abdulraheem 2012).

Currently, the country has implemented a system for healthcare called the National Health Insurance Scheme (NHIS), which aims to provide "easy access to medical care for all." This program is designed to carry out the easy financing of the costs that are generated from medical care, through a prepayment system that will distribute the costs to all persons (Okafor and Anigbogu 2017). The government has also intensified the development of the telecommunications sector and applied it to healthcare systems to strengthen said systems. Consequently, a unifying ICT system has been proposed and implemented to increase demand and access and improve the quality of health services. The main objective of this strategy is to coordinate all the information of the healthcare system in a timely manner, with the aim of improving the wellbeing of the entire population through technological advances and innovations (Gambo and Soriyan 2017). Additionally, a telemedicine system has been implemented, which allows the consultation of articles and information that have been used previously. These systems are for improving diagnosis and efficiency in terms of space (beds, operating rooms and healing material) since patients can be treated and monitored remotely (Ekanoye 2017).

18.6.8 Other Countries

During the past two decades, China has suffered a crisis in terms of health services. This is mainly due to the lack of qualified human resources, high demand for services and poorly designed payment systems (Wang et al. 2011). However, in the new coverage system the cost is higher for self-employed workers, dependents of insured persons, retired persons and employees of private companies with respect to the system maintained by the country until the 1990s, offered mainly by the Labor Insurance System (LIS) and the Scheme of Government Insurance (GIS) (Atella et al. 2014; Zheng et al. 2015). One of the options that is being implemented in some regions of the country in order to serve the healthcare needs of a greater number of people is a program called Long-term Care Medical Insurance (LTCMI). LTCMI was analyzed using demographic data to determine the costs of expansion and the social stratum in which this type of service has the greatest impact. One of its advantages is the decrease in hospital inflow (care can be institutional or residential) and a considerable decrease in investment costs within the healthcare system itself (Lu et al. 2017).

Another country that has suffered various transformations in the healthcare systems over the past 20 years is Russia. This is due to the fact that at the beginning of the 1990s (dissolution of the USSR), the country was left with a precarious and deteriorated healthcare system (facilities in poor condition, deficient equipment and a low budget) (Vlassov 2016). Notwithstanding the current healthcare system transition that the country is undergoing, studies are currently being carried out regarding the detection and improvement of diseases focused on certain groups of the population. For example, currently research is being undertaken in prisons for the detection and treatment of tuberculosis. This includes a social, economic and scientific research approach, with the intention of improving the healthcare systems inside and outside the prisons. The above strategy was implemented with the aim of preventing the contagion from spreading to the general population when the prisoners rejoin society (Winetsky et al. 2012).

South Africa as a country has a host of problems in its healthcare system. Among the main ones are the lack of qualified personnel (doctors and nurses), a high rate of racial discrimination in access to healthcare systems, the inequity of care between geographical regions, the difference in costs for medical care and that between the public sector and the private sector. In addition, the facilities where public health services are provided are generally deficient and in very bad condition (Coovadia et al. 2009; Mayosi and Benatar 2014). In response to the current situation and with the main objective of providing a health service of better quality that is available to the entire population, sustainable programs based on the use of informatic and communication technologies (mHealth) through mobile telephony are being carried out for the development of the existing eHealth strategy. However, these types of initiatives have various challenges (financial, organizational, cultural and technological) for its large-scale implementation (Leon et al. 2012).

Another important case is Australia. The healthcare system of this country (Medicare) was designed to meet the needs of the entire population with comprehensive coverage; with certain exceptions (dental care, physiotherapy and prescription drugs), care is free and available throughout the whole country (Van-Doorslaer et al. 2008). However, despite the efforts of the government and the private sector to provide health services to the entire population, there are endemic areas where, due to their geographical location, it is difficult to get access to a healthcare system like the one found in urban areas. Strategic plans have been proposed to combat this type of problem and to offer a primary health service (PHC) that is inclusive in these areas (Wakerman and Humphreys 2011). One of the alternatives that is currently being developed to combat these types of problems is sustainable systems that include the use of communication technologies, qualified medical personnel and the society to which it is addressed (mainly low-income people living in marginalized areas). Specifically, in this country, various models have been implemented on the use of communication technology for the monitoring of mental illnesses through telepsychology and tele-education in the rural areas of the north of the country (Bryant et al. 2018).

The public health services in France offer some advantages over other countries, such as shared costs (population and government) and exclusion from payment for services to people suffering from one of the 30 specific diseases that the government considers a special or high priority, which also covers certain types of medicines (Schoen et al. 2010). Nevertheless, there is a possibility that in the medium term the system will present problems, such as the lack of qualified personnel, increased costs associated with the increase in life expectancy and increase in the cost of prescribed medications, among others (Gille and Houy 2013). Derived from this context, a series of sustainable initiatives have been undertaken with the aim of contributing to the reduction of investment in the healthcare system that currently prevails in the country. An example is the use of electronic health records (EHR), which is a program intended to modernize health information and the technology that is used within health services. However, constant evaluations have to be carried out on the expenses and the capacity for improvement in the performance of this type of program within the healthcare system (Plantier et al. 2017).

Germany has a social healthcare system and is covered by social insurance, which is divided into five main categories (medical, pensions, long-term, unemployment and accidents), with contributions guaranteeing the support of this healthcare system (Sunyaev 2011). However, over the past few years, this healthcare system has faced a considerable increase in the costs of medical attention due to the increase in population size and the presence of two or more chronic diseases in some strata of society (mainly older adults) (Lang et al. 2017). In response to this situation, the government of the country has invested in telematic infrastructure, in order to interconnect the healthcare systems with the different actors involved in the supply chain and in turn provide a telemedicine service to integrate the entire population within the healthcare system, which should lead to a reduction in costs derived from a better control of the different departments (pharmaceutical, insurance, electronic records and others) within the healthcare system (Duennebeil et al. 2012).

In general, the need to provide a quality health service is imperative for each and every country around the world. Nowadays, the inclusion of sustainable systems within these healthcare systems is growing, with the purpose of providing health services with equality, quality and that are economically accessible to the majority of the population. However, some countries are in transition stages and the development and application of sustainable systems is still limited due to the fact that most of their current efforts and actions are directed toward the improvement of services and facilities.

18.7 Conclusion

Over the past few decades, great interest has been generated in the quality of medical care around the world and the different areas of opportunity that some countries present regarding their healthcare systems have been examined. In conclusion, this chapter provides a perspective on the incorporation of sustainable principles that can lead to improvement in the efficient performance of healthcare systems regardless of the model or philosophy on which the system is based. The implementation of sustainability in healthcare systems is one of the most important challenges in the world. Most of the challenges of sustainability in terms of healthcare are to be found in the difficulty in providing total and complete social inclusion without economically impacting the existing system, and with strict care for the environment.

Acknowledgments

Financial support from the Natural Sciences and Engineering Research Council of Canada (No. 284111, Discovery; No. 476649-14, Strategic Research Grant) and Institute National de la Recherche Scientifique-Eau Terre Environnement is acknowledged.

References

Abdulraheem, I.S. (2012). Primary health care services in Nigeria: critical issues and strategies for enhancing the use by the rural communities. *Journal of Public Health and Epidemiology* 4 (1): 5–13.

Aktaş, E., Ülengin, F., and Önsel Şahin, Ş. (2007). A decision support system to improve the efficiency of resource allocation in healthcare management. *Socio-Economic Planning Sciences* 41 (2): 130–146.

Albers Mohrman, S., Shani, A.B., and McCracken, A. (2012). Organizing for sustainable health care: the emerging global challenge. In: *Organizing for Sustainable Effectiveness*, vol. 2 (eds. S.A. Mohrman and A.B. Shani), 1–39. Bingley, UK: Emerald Group Publishing Limited.

Alrubaiee, L. and Alkaa'ida, F. (2011). The mediating effect of patient satisfaction in the patients' perceptions of healthcare quality–patient trust relationship. *International Journal of Marketing Studies* 3 (1): 103.

Ament, S.M., Gillissen, F., Maessen, J.M. et al. (2012). Sustainability of healthcare innovations (SUSHI): long term effects of two implemented surgical care programmes (protocol). *BMC Health Services Research* 12 (1): 423.

de Andrade, L.O.M., Pellegrini Filho, A., Solar, O. et al. (2014). Universal health coverage in Latin America. 3 Social determinants of health, universal health coverage, and sustainable development: case studies from Latin American countries. *Lancet* 385 (9975): 1343–1351.

Anttila, J. and Jussila, K. (2019). Striving for benefits of sustainability from the interactivity of quality and innovation. *Journal of Cleaner Production* 212: 409–419.

Arena, R., Bond, S., O'Neill, R. et al. (2017). Public park spaces as a platform to promote healthy living: introducing a HealthPark concept. *Progress in Cardiovascular Diseases* 60 (1): 152–158.

Astolfi, R., Lorenzoni, L., and Oderkirk, J. (2012). Informing policy makers about future health spending: a comparative analysis of forecasting methods in OECD countries. *Health Policy* 107 (1): 1–10.

Atella, V., Brugiavini, A., Chen, H., and Pace, N. (2014). Chinese healthcare system reforms and household saving patterns: some stylised facts. In: *The Globalization of Chinese Business* (ed. R. Taylor), 189–204. Amsterdam: Elsevier.

Atun, R., De Andrade, L.O.M., Almeida, G. et al. (2015). Health-system reform and universal health coverage in Latin America. *The Lancet* 385 (9974): 1230–1247.

Barata, A.N., Tokuda, Y., and Martins, H.M.G. (2012). Comparing healthcare systems: considering Japan and Portugal. *Revista Portuguesa de Saúde Pública* 30 (2): 180–192.

Beckfield, J., Olafsdottir, S., and Sosnaud, B. (2013). Healthcare systems in comparative perspective: classification, convergence, institutions, inequalities, and five missed turns. *Annual Review of Sociology* 39 (1): 127–146.

Bhatia, V. and Orsini, M. (2016). Narrating sustainability in Canadian health care reform discourse. *Social Policy & Administration* 50 (3): 297–315.

Borgonovi, E. and Compagni, A. (2013). Sustaining universal health coverage: the interaction of social, political, and economic sustainability. *Value in Health* 16 (1): S34–S38.

Bryant, L., Garnham, B., Tedmanson, D., and Diamandi, S. (2018). Tele-social work and mental health in rural and remote communities in Australia. *International Social Work* 61 (1): 143–155.

Buchan, H. (2004). Gaps between best evidence and practice: causes for concern. *Medical Journal of Australia* 180 (6 Suppl): S48.

Buffoli, M., Capolongo, S., Bottero, M. et al. (2013). Sustainable healthcare: how to assess and improve healthcare structures' sustainability. *Annali di Igiene* 25 (5): 411–418.

Buse, K. and Hawkes, S. (2015). Health in the sustainable development goals: ready for a paradigm shift? *Globalization and Health* 11 (1).

Capolongo, S., Gola, M., Di Noia, M. et al. (2016). Social sustainability in healthcare facilities: a rating tool for analysing and improving social aspects in environments of care. *Annali dell'Istituto Superiore di Sanita* 52 (1): 15–23.

Carayon, P., Wetterneck, T.B., Rivera-Rodriguez, A.J. et al. (2014). Human factors systems approach to healthcare quality and patient safety. *Applied Ergonomics* 45 (1): 14–25.

Chen, I.-C. and Li, H.-H. (2010). Measuring patient safety culture in Taiwan using the hospital survey on patient safety culture (HSOPSC). *BMC Health Services Research* 10 (1): 152.

Coiera, E. and Hovenga, E.J.S. (2007). Building a sustainable health system. *IMIA Yearbook of Medical Informatics* 2 (1): 11–18.

Coovadia, H., Jewkes, R., Barron, P. et al. (2009). The health and health system of South Africa: historical roots of current public health challenges. *The Lancet* 374 (9692): 817–834.

Duennebeil, S., Leimeister, J., and Krcmar, H. (2012). Business models for electronic healthcare services in Germany. In: *Critical Issues for the Development of Sustainable E-health Solutions* (eds. N. Wickramasinghe, R. Bali, R. Suomi and S. Kirn), 271–291. Boston, MA: Springer.

Dunphy, J.L. (2014). Healthcare professionals' perspectives on environmental sustainability. *Nursing Ethics* 21 (4): 414–425.

Ekanoye, F. (2017). Telemedicine diffusion in a developing country: a case of Nigeria. *Science Journal of Public Health* 5 (4): 341.

Faezipour, M. and Ferreira, S. (2011). Applying systems thinking to assess sustainability in healthcare system of systems. *International Journal of System of Systems Engineering* 2 (4): 290–308.

Fleiszer, A.R., Semenic, S.E., Ritchie, J.A. et al. (2015). The sustainability of healthcare innovations: a concept analysis. *Journal of Advanced Nursing* 71 (7): 1484–1498.

Foley, T.J. and Vale, L. (2017). What role for learning health systems in quality improvement within healthcare providers? *Learning Health Systems* 1 (4): e10025.

Freire, W. and Pan American Health Organization (2006). *Nutrición y vida activa: del conocimiento a la acción*. Washington, DC: Organización Panamericana de la Salud.

Fugazzola, E., Tognolo, C., and Rebecchi, A. (2018). Planning and designing impartial health systems through mobile healthcare. In: *Healthcare Facilities in Emerging Countries* (eds. S. Capolongo, M. Gola and A. Rebecchi), 113–123. Cham, Switzerland: Springer International Publishing.

Gaitonde, R., San Sebastian, M., Muraleedharan, V.R., and Hurtig, A.-K. (2017). Community action for health in India's National Rural Health Mission: one policy, many paths. *Social Science & Medicine* 188: 82–90.

Gambo, I.P. and Soriyan, A.H. (2017). ICT implementation in the Nigerian healthcare system. *IT Professional* 19 (2): 12–15.

Gille, L. and Houy, T. (2013). Forecast of the supply of healthcare in France. *Futures* 51: 50–58.

Girot, E.A. and Albarran, J.W. (2012). Sustaining the education workforce in healthcare: challenges for the future. *Nurse Education Today* 32 (1): 32–38.

Goh, C.Y. and Marimuthu, M. (2016). The path towards healthcare sustainability: the role of organisational commitment. *Procedia-Social and Behavioral Sciences* 224: 587–592.

Greenhalgh, T., Macfarlane, F., Barton-Sweeney, C., and Woodard, F. (2012). "If we build it, will it stay?" A case study of the sustainability of whole-system change in London. *The Milbank Quarterly* 90 (3): 516–547.

Grin, J. and Broerse, J.E.W. (2017) Chapter 1). Introduction. In: *Toward Sustainable Transitions in Healthcare Systems* (eds. J.E.W. Broerse and J. Grin), 1–20. Abingdon, UK: Taylor & Francis Group.

Guindo, L.A., Wagner, M., Baltussen, R. et al. (2012). From efficacy to equity: literature review of decision criteria for resource allocation and healthcare decisionmaking. *Cost Effectiveness and Resource Allocation* 10 (1): 9.

Hawk, C., Ndetan, H., and Evans, M.W. (2012). Potential role of complementary and alternative health care providers in chronic disease prevention and health promotion: an analysis of National Health Interview Survey data. *Preventive Medicine* 54 (1): 18–22.

Himmelstein, D.U., Woolhandler, S., Almberg, M., and Fauke, C. (2018). The U.S. health care crisis continues: a data snapshot. *International Journal of Health Services* 48 (1): 28–41.

Horgan, D., Baker, M., Riegman, P., and Bernini, C. (2017). Personalised medicine-bringing innovation to the healthcare system. *Biomedicine Hub* 2 (Suppl. 1): 5–5.

Hutchison, B., Levesque, J.-F., Strumpf, E., and Coyle, N. (2011). Primary health care in Canada: systems in motion. *The Milbank Quarterly* 89 (2): 256–288.

Imran, M., Kang, C., and Babar Ramzan, M. (2018). Medicine supply chain model for an integrated healthcare system with uncertain product complaints. *Journal of Manufacturing Systems* 46: 13–28.

International Health Care System Profile. (2018). The English health care system. http://international.commonwealthfund.org/countries/england (accessed November 2019).

Jordan, J.E., Buchbinder, R., and Osborne, R.H. (2010). Conceptualising health literacy from the patient perspective. *Patient Education and Counseling* 79 (1): 36–42.

Kim, R.H., Gaukler, G.M., and Lee, C.W. (2016). Improving healthcare quality: a technological and managerial innovation perspective. *Technological Forecasting and Social Change* 113: 373–378.

Kisliakovskii, I., Balakhontceva, M., Kovalchuk, S. et al. (2017). Towards a simulation-based framework for decision support in healthcare quality assessment. *Procedia Computer Science* 119: 207–214.

Kumakawa, T., Matsushige, T., and Tamaki, Y. (2017). Developing of social and community prescription model as the third system for sustainable community-based integrated health care system. *International Journal of Integrated Care* 17 (3): 106.

Kumar, M., Harrington, T.S., Srai, J.S., and Minakata, Y. (2017). Industrial system dynamics for environmental sustainability: a case study on the UK medical technology sector. *International Journal of Manufacturing Technology and Management* 31 (1–3): 100–132.

Kwon, I.-W.G., Kim, S.-H., and Martin, D.G. (2016). Healthcare supply chain management; strategic areas for quality and financial improvement. *Technological Forecasting and Social Change* 113: 422–428.

Lang, C., Kern, E.A.M., Schulte, T., and Hildebrandt, H. (2017). Integrated diabetes care in Germany: triple aim in Gesundes Kinzigtal. In: *Integrated Diabetes Care* (eds. D. Simmons, H. Wenzel and J.C. Zgibor), 169–184. Cham, Switzerland: Springer International Publishing.

Lega, F., Prenestini, A., and Spurgeon, P. (2013). Is management essential to improving the performance and sustainability of health care systems and organizations? A systematic review and a roadmap for future studies. *Value in Health* 16 (1): S46–S51.

Lennox, L., Maher, L., and Reed, J. (2018). Navigating the sustainability landscape: a systematic review of sustainability approaches in healthcare. *Implementation Science* 13 (1).

Leon, N., Schneider, H., and Daviaud, E. (2012). Applying a framework for assessing the health system challenges to scaling up mHealth in South Africa. *BMC Medical Informatics and Decision Making* 12 (1).

Lu, B., Mi, H., Zhu, Y., and Piggott, J. (2017). A sustainable long-term health care system for aging China: a case study of regional practice. *Health Systems & Reform* 3 (3): 182–190.

Marimuthu, M. and Paulose, H. (2016). Emergence of sustainability based approaches in healthcare: expanding research and practice. *Procedia – Social and Behavioral Sciences* 224: 554–561.

Mayosi, B.M. and Benatar, S.R. (2014). Health and health care in South Africa — 20 years after Mandela. *New England Journal of Medicine* 371 (14): 1344–1353.

McKee, M. (2018). Global sustainable healthcare. *Medicine* 46 (7): 383–387.

Mendel, P., Weinberg, D.A., Gall, E.M. et al. (2014). The national response for preventing healthcare-associated infections: system capacity and sustainability for improvement. *Medical Care* 52: S83–S90.

Minoglou, M., Gerassimidou, S., and Komilis, D. (2017). Healthcare waste generation worldwide and its dependence on socio-economic and environmental factors. *Sustainability* 9 (12): 220.

Mohrman, A.S., Shani, R.A., and McCracken, A. (2012). Organizing for sustainable health care: the emerging global challenge. In: Organizing for Sustainable Health Care (ed. A. Mohrman and A. Shani), 1–39. United Kingdom: Emerald Group Publishing Limited.

Okafor, N.I. and Anigbogu, K.C. (2017). National health insurance scheme and sustainable development in Nigeria. *International Journal of Health and Social Inquiry* 2 (1).

Omoruan, A.I., Bamidele, A.P., and Phillips, O.F. (2009). Social health insurance and sustainable healthcare reform in Nigeria. *Studies on Ethno-Medicine* 3 (2): 105–110.

Pinzone, M., Lettieri, E., and Masella, C. (2012). Sustainability in healthcare: combining organizational and architectural levers. *International Journal of Engineering Business Management* 4: 38.

Plantier, M., Havet, N., Durand, T. et al. (2017). Does adoption of electronic health records improve the quality of care management in France? Results from the French e-SI (PREPS-SIPS) study. *International Journal of Medical Informatics* 102: 156–165.

Proctor, E., Luke, D., Calhoun, A. et al. (2015). Sustainability of evidence-based healthcare: research agenda, methodological advances, and infrastructure support. *Implementation Science* 10 (1).

Quadagno, J. (2010). Institutions, interest groups, and ideology: an agenda for the sociology of health care reform. *Journal of Health and Social Behavior* 51 (2): 125–136.

Qudrat-Ullah, H. (2017). Innovative healthcare systems: an introduction. In: *Innovative Healthcare Systems for the 21st Century* (eds. H. Qudrat-Ullah and P. Tsasis), 3–12. Cham, Switzerland: Springer International Publishing.

Ramirez, B., West, J.D., and Costell, M.M. (2013). Development of a culture of sustainability in health care organizations. *Journal of Health Organization and Management* 27 (5): 665–672.

Ryan, A.M., Tompkins, C.P., Markovitz, A.A., and Burstin, H.R. (2017). Linking spending and quality indicators to measure value and efficiency in health care. *Medical Care Research and Review* 74 (4): 452–485.

Scheirer, M.A. and Dearing, J.W. (2011). An agenda for research on the sustainability of public health programs. *American Journal of Public Health* 101: 2059–2067.

Schoen, C., Osborn, R., Squires, D. et al. (2010). How health insurance design affects access to care and costs, by income, in eleven countries. *Health Affairs* 29 (12): 2323–2334.

Schulz, K.-P., Kawamura, T., and Geithner, S. (2017). Enabling sustainable development in healthcare through art-based mediation. *Journal of Cleaner Production* 140: 1914–1925.

Shelton, R.C., Cooper, B.R., and Stirman, S.W. (2018). The sustainability of evidence-based interventions and practices in public health and health care. *Annual Review of Public Health* 39: 55–76.

Smyczek, S. (2015). Relationship marketing performance of medical facilities. In: *Healthcare* (eds. M. Glowik and S. Smyczek), 89–104. Berlin, München: de Gruyter.

Stirman, S.W., Kimberly, J., Cook, N. et al. (2012). The sustainability of new programs and innovations: a review of the empirical literature and recommendations for future research. *Implementation Science* 7 (1): 17.

Sudhinaraset, M., Ingram, M., Lofthouse, H.K., and Montagu, D. (2013). What is the role of informal healthcare providers in developing countries? A systematic review. *PLoS One* 8 (2): e54978.

Sunyaev, A. (2011). Healthcare telematics in Germany with respect to security issues. In: *Health-Care Telematics in Germany*, 17–52. Wiesbaden, Germany: Gabler.

Sustainable Development Unit (2009). Fit for the Future Scenarios for low-carbon healthcare 2030. https://www.forumforthefuture.org/project/fit-future/overview (accessed February 2018).

Sustainable Development Unit (2018). What is sustainable health? www.sduhealth.org.uk/policy-strategy/what-is-sustainable-health.aspx (accessed February 2018).

Tangcharoensathien, V., Mills, A., and Palu, T. (2015). Accelerating health equity: the key role of universal health coverage in the sustainable development goals. *BMC Medicine* 13 (1).

Tavecchi, G. and Rebecchi, A. (2018). The current Indian healthcare system and West Bengal's health status. In: *Healthcare Facilities in Emerging Countries* (eds. S. Capolongo, M. Gola and A. Rebecchi), 13–31. Cham, Switzerland: Springer International Publishing.

Titelman, D., Cetrángolo, O., and Acosta, O.L. (2015). Universal health coverage in Latin American countries: how to improve solidarity-based schemes. *The Lancet* 385 (9975): 1359–1363.

Van Doorslaer, E., Clarke, P., Savage, E., and Hall, J. (2008). Horizontal inequities in Australia's mixed public/private health care system. *Health Policy* 86 (1): 97–108.

Villarreal-Garza, C., Castro-Sanchez, A., Platas, A. et al. (2017). "Joven & Fuerte": program for young women with breast cancer in Mexico – initial results. *Revista de investigacion Clinica* 69 (4): 223–228.

Vlassov, V.V. (2016). Russian experience and perspectives of quality assurance in healthcare through standards of care. *Health Policy and Technology* 5 (3): 307–312.

Wakerman, J. and Humphreys, J.S. (2011). Sustainable primary health care services in rural and remote areas: innovation and evidence: sustainable rural and remote phc services. *Australian Journal of Rural Health* 19 (3): 118–124.

Wallace, L.S. (2013). A view of health care around the world. *The Annals of Family Medicine* 11 (1): 84–84.

Wang, H., Gusmano, M.K., and Cao, Q. (2011). An evaluation of the policy on community health organizations in China: will the priority of new healthcare reform in China be a success? *Health Policy* 99 (1): 37–43.

Wendt, C., Frisina, L., and Rothgang, H. (2009). Healthcare system types: a conceptual framework for comparison. *Social Policy & Administration* 43 (1): 70–90.

Winetsky, D.E., Negoescu, D.M., DeMarchis, E.H. et al. (2012). Screening and rapid molecular diagnosis of tuberculosis in prisons in Russia and Eastern Europe: a cost-effectiveness analysis. *PLoS Medicine* 9 (11): e1001348.

World Bank.(2015). Death rate, crude (per 1,000 people). https://data.worldbank.org/indicator/SP.DYN.CDRT.IN?end=2015&start=1960 (accessed November 2019).

Yeates, K., Campbell, N., Maar, M.A. et al. (2017). The effectiveness of text messaging for detection and management of hypertension in indigenous people in Canada: protocol for a randomized controlled trial. *JMIR Research Protocols* 6 (12): e244.

Zheng, X., Wang, J., Li, X., and Yu, Y. (2015). On the supply of China's healthcare resources in a decentralized healthcare system. *The Social Science Journal* 52 (4): 449–458.

19

Ethical Aspects of Sustainability

Anita Talan[1], Rajwinder Kaur[1], R.D. Tyagi[1], and Tian C. Zhang[2]

[1]*INRS, Université du Québec, Québec, Canada*
[2]*Department of Civil Engineering, University of Nebraska-Lincoln, Omaha, NE, USA*

The most important human endeavor is the striving for morality in our actions. Our inner balance and even our very existence depend on it. Only morality in our actions can give beauty and dignity to life

–Albert Einstein

19.1 Introduction

A sustainable society always lives within the accepted carrying capacity of its natural and social domain. A sustainable society forms upon the pledge of its members to adapt and follow the rules willingly (Jennings 2013). Sustainable society has a system of rules, encouragements and incentives that support restocking of resources and limit the depletion and pollution present in the sustainable society. For sustainable development, the society needs to integrate its industrial, societal and environmental systems, and establish decision support systems that can not only evaluate environmental, societal and economic impacts, but also accept input from society and address the associated ethical implications.

In order to meet the global grant challenges, sustainable development has gained significant attention worldwide. However, many policies (e.g. about climate change and mitigation) related to sustainable development rely on various economic arguments, but ignore completely the ethical and justice limitations of these arguments. Sustainable development is a sensible choice with respect to ethical perceptions and thinking. However, most citizens and policymakers have not been trained in spotting ethically dubious claims that are often hidden in what appear at first glance to be "value-neutral" economic arguments. Thus, a systematical introduction to the ethical dimensions of sustainable development is important.

This chapter aims to introduce the concepts related to ethics and the ethical limitations of sustainable development. The chapter discusses the common grounds between choosing

Sustainability: Fundamentals and Applications, First Edition.
Edited by Rao Y. Surampalli, Tian C. Zhang, Manish Kumar Goyal, Satinder K. Brar, and R. D. Tyagi

ethics-based approaches and their perspective on sustainable development, followed by descriptions of the utilitarian and deontological approaches to ethics to attain sustainability. In addition, the chapter introduces the concepts of the neoliberal era and adaptation into the sustainable framework, discusses the importance of the conservation and sustainable use of global flora, and describes the role of government and individuals in global sustainability through ethical approaches and their impact assessments. The chapter also uses adaptation and mitigation of climate change to explain how to comply with the ethical requirements during the course of sustainable development.

19.2 Juxtaposing Ethics, Morality and Sustainable Development

The term ethics has several definitions, but according to Türkeri (2013) ethics can be defined as the system of principles according to which people organize their behaviors. Ethical thinking and concerned behavior are all interrelated to the welfare of society; rational behavior requires us to consider our own interests and those of others as well (Koçyiğit and Karadağ 2016).

Ethics or moralities are related to principles, standards, rules, justice, possible freedom and norms of conduct to have cooperation. Ethics are always inseparable from questions of cultural meaning and social power as they provide a philosophically based touchstone for an ideal justice, right relationships, and the proper use of authority and their power. Ethical analysis has four main components:

- evaluation of the character;
- evaluation of the intrinsic properties of an action;
- evaluation of the consequences;
- evaluation of the context within which actions take place.

Ethical analysis is always affected by the initial ontological starting point or coordination one accepts. Generally, there are three orientations:

1) the theocentric;
2) the anthropocentric (human-centered);
3) the biocentric.

The human-centered orientation accepts their own values and morals, and denies non-human inheritance and intrinsic moral values. The biocentric orientation sticks to the facts that resource value depends on not only people and communities but also natural and biotic components, of which the humans are a part. Thus, ethical rights and duties need to be implicated in terms of interdependency, relations, sustainability and resiliency.

The concept of sustainability is all about meeting the needs of the present generation without compromising the ability of future generations to meet their needs (WCED 1987). Sustainability has three pillars: environmental protection, economic development and social equity. The social equity pillar has the clearest ethical dimension (Munasinghe and McNeely 1995). Historically, sustainability ethics has been evolving from environmental (or ecological) ethics. Ecological ethics takes the idea of moral significance (someone or something should be the object of ethical concern) and extends it to the environment. Sustainability ethics refers not only to people (by focusing on the ethical dimension of socio-economic fairness or social justice) but also the human use of the earth. Sustainability science focuses on the dynamic interactions between nature and society (Nelson and Vucetich 2012). Sustainability is a moral response to an incredible gift given by the earth. Therefore, we need a more ethical way to carry out our relationship with the earth, for the planet's sake and for our own (Sibole 2013). Here, we need to discuss four issues related to ethics of sustainability.

First, some people believe that the ethical dimensions of sustainability are too general and vague for useful application to any specific problem, and thus, have become an obsolete of sustainability science (Jamieson 1998; Davison 2001; Thompson 2007). It is important to recognize that sustainability, much like "efficiency," does not have any real meaning in and of itself unless one clearly defines "the end goals of sustainability," e.g. what is to be sustained, for whom and for how long. This is why the social equity dimension of sustainability is so critical (Warner and DeCosse 2009). For example, from the perspective of sustainability ethics, we should evaluate the project by highlighting a wide range and complexity of ethical concerns that might emerge from such a project, rather than view it as a "technical" or "management" problem that technology, experts or the government can solve for us. After all, different cultures, value systems, life experiences and worldviews would always influence one's perception of sustainable development. From the sustainability ethical perspective, one can accept that developing countries would necessarily focus on economic development without attaching as much importance to the ecological aspect, while developed countries have no excuse, morally speaking, to develop without protecting the wellbeing of the planet (WCED 1987; Warner OFM and DeCosse 2009).

Second, ethics determines how we use technologies. Historically, the ethics of sustainability has been evolving with time (Jennings 2013). Nelson and Vucetich (2012) stated that the ethical dimensions of sustainability are closely related to the following five critical dimensions of sustainability:

1) Development of efficient technologies and markets for meeting human needs – the purview of engineering, physical science, biotechnology, economics and business.
2) Understanding the state and nature of ecosystems – the purview of ecology and environmental science.
3) Understanding how exploitation affects ecosystems – the purview of applied ecology and environmental science.
4) Understanding how exploitation affects human cultures – the purview of sociology, political science, policy, law, anthropology, and the arts and humanities.
5) Understanding the meaning of normative concepts such as human needs, socially just, depriving, ecosystem health – the purview of ethics and philosophy.

Dimensions 1 – 4 are inadequate for achieving sustainability as ethics determines how we use technologies. Different motivations would result in different worlds. Thus, the moral precepts should be closely linked with sustainable development.

Currently, people are more concerned with the scientific or engineering aspects of sustainability than with its social and ethical sides, but the opposite should be the case. In other words, sustainability is primarily, if not most importantly, about our moral relationships with the world, i.e. about ethics (Nelson and Vucetich 2012). Considering an example of a phenomenon known as Jevon's paradox, technological improvement in the 1970s providing more efficient home heating and insulation in the United States did not lead to less energy use, but to the building of larger houses. Therefore, without a moral obligation, technology can only increase our ability to exploit the environment and the efficiency of exploitation, but it does not determine how we ought to exercise that ability and efficiency (Nelson and Vucetich 2012). Actually, sustainability is more effective when people are engaged in moral cases rather than economic cases. During recent years, simple ideas to promote sustainability have been flourished. Moral values are leading to great changes in the public mindset and are used as a key element to get the public involved in factors affecting sustainable development. These moral values encourage the public to adopt sustainable behaviors to save natural resources.

Third, certain types of values cluster together and are always associated with pro-environmental attitudes and behavior. Studies show that factors encouraging environmental actions differ by country and by type of actions (Aoyagi-Usui et al. 2003). The work published by Bolderdijk et al. (2013) and others is the latest experimental evidence that focuses on associated pro-environmental attitudes and behavior of people. This study also raised very significant questions regarding several movements that started in spite of the assumption that individuals are more concerned about finances than about the environment. After this study, it is impossible to ignore the evidence for the benefits of being green. The negative consequences of people's attitudes and behavior toward the environment have made them more concerned about the environment than their economic worth. Therefore, to promote people's engagement with climate change the study broached the idea of "save money, save the planet" and targeted either economic or biospheric values to reach the people with slogans and messages comparing the effectiveness of both for the environment. Different messages and slogans presenting information are changing prime values that will increase the chances of pro-environmental behavior in both the long- and short term. This should also lead to a stable and morally good self-concept of people in relation to the environment. People react positively to environment safety or biospheric appeals as accepted the go-green policy makes them feel better than being greedy. However, as yet, these messages and empirical evidence are not included in government policy.

Fourth, making every person a moral citizen is the main challenge that any organization faces to promote sustainability. Many people are more comfortable about the economic case for sustainability than the moral one. However, if people can achieve moral purpose, it is easy to attain positive attitudes and behavior toward sustainable development. The green deal is the first step to achieving long-term sustainability and can be achieved when people become more responsible and moral citizens of the environment.

19.3 Approaches to Ethics

All of morality aims at the same thing, but there are several basic ways to get there. Generally, ethical theories include virtue ethics, consequentialist ethics and deontological or duty-based ethics. In most cases, different approaches need to be considered in order to reach the best ethical decision. These approaches to achieving sustainability ethics are briefly discussed below.

19.3.1 Utilitarian Approaches

Sustainable development is defended by government officials on the basis of a utilitarian approach to ethics. This approach supports the principle of greatest happiness with respect to the foundation of morals and considers actions appropriate if they create happiness or inappropriate if they tend to cause unhappiness. This theory involves the assessment of good and bad consequences by every individual, depending on individual situations and the consequences of their actions (Eid 2012). The theory of utilitarianism involves four ideas:

1) *Consequentialism*: This defines that the relevance of actions is only determined by their consequences.
2) *Hedonism*: This involves desire, happiness and pain.
3) *Maximalism*: This defines that consideration of bad consequences is also required in addition to that of good consequences.
4) *Universalism*: The consequences of any actions belong to all the stakeholders.

Act utilitarianism and rule utilitarianism are two major forms of utilitarian ethics. Environmental policy mostly relies on act utilitarianism ethical theories as it is mostly justified on the basis of cost versus benefits. However, both forms of utilitarianism are subject to criticism because (i) they both are related to human happiness, and non-human entities such as plants and animals are not considered; (ii) the utilitarian approach is not able to accommodate all the rights that individual must require for protection from some death-threating situations; (iii) many ethical issues are not considered, such as consequences of any action, including harm and benefits as well as time scale to assess these consequences; (iv) the division of cost or benefits among winners and losers is not considered and, therefore, the concept of utilitarianism should be expanded by involving the concepts of distributive justice and integrity; and (v) the influence of environmental problems on future generations is not considered (Hartman and DesJardins 2011).

Environment, social and economic welfare are looked at from a sustainable development prospective. The world environment is not affected by geographical and political boundaries; human development activities cannot be handled by the ecosystem of the earth because natural resources cannot be renewed further. Natural resources are depleting; however, the demand for them is continuously increasing. Therefore, environmental progression should be considered for everyone. Ecological sustainability should be supported because of environmental degradation and the depletion of natural resources. Preserving the ecological footprint could help human happiness (Winners 2009). The good consequences of environmental sustainability involve controlled consumption of natural resources according to utilitarian theory.

19.3.2 Deontological Approach

The deontological approach of ethics opposes the utilitarian approach concerning the claim that whether an actions is right or not depends upon the consequences of the action.

This approach believes that an action could be morally right or wrong independent of its consequences. The best-known deontological theory is that of German philosopher Immanuel Kant, who believed that morality should be assessed based upon reasoning alone (Betey and Essel 2013). Determining whether an action is correct or not should be based upon the rule authorizing the action and whether it could be universalized or not. Kant also believed that if one person chooses to follow a rule that is followed by every individual, the person is considered to be acting morally. Three formulations are built by Bowie (2017) based upon the principles of Kant:

1) *Universalizability*: This provides the theory of moral acceptability for social interactions.
2) *Respect for humanity*: This defines that human beings should be treated as autonomous individuals and not as means for the interests of other human beings.
3) *Moral community formulation*: This is a call for greater democracy and social intrusions.

Economic advancement, environmental enhancement and social welfare (together with inclusion of vulnerable groups in society) are three principles that are represented based upon the sustainability point of view. In terms of the economic aspect of sustainability, the careful use of natural resources provides many good practices, which are followed by developed economies and should be passed on to developing economies. Environmental sustainability follows the principle that damage to the environment and the associated biological imbalance should be controlled, while eco-efficiency and biomimicry should be idealized for support of the ethical principle. From the social perspective of sustainability, community inclusion and issues related to diversity and human rights are identified as the main focuses (Yüksel 2010).

The importance of the deontological approach lies in the basis that some actions should be considered wrong without knowing their exact consequences. However, people criticize the Kant approach to sustainable development for the following reasons: (i) the deontological approach is not applicable for decisions involving conflicts between two competing goods, such as conflict between food and shelter versus habitat protection. This approach instructs that the individual should act according to the rules that can be universalized, but advice for formation of these rules is not given. (ii) This approach only refers to humans; however, no value has been given to plants and animals.

19.4 The Neoliberal Era and Adaptation into the Sustainability Framework

Sustainability has become an important aspect in policymaking since the adoption of Agenda 21 at the United Nations Conference on Environment and Development (UNCED) in 1992 (Rio de Janeiro, 1992). Sustainability addresses problems such as threats posed to the environment due to changes in lifestyle, economic system and long-term effects of human intervention in the natural environment, which has further affected the survival and wellbeing of humans. Research and debates are continuous on the implications of sustainability and approaches to interlinking the theoretical concepts of sustainability with the practical demands of policy. For more than two decades, governments have recognized the importance of sustainable development, i.e. to meet present needs without compromising the ability of future generations to meet their own needs. However, unsustainable development is still continuous in some regions. Governments could help by imposing constraints, and individuals could help by creating awareness about sustainable development, which could further influence decision-making.

19.4.1 The Role of Government

Natural resources such as minerals, air, water and soil are depleting, and some legal constraints such as tight emission standards, reducing the use of fossil fuels, etc. could help to preserve them (Gorica et al. 2012). Governments can employ restrictions or regulations to affect sustainable development, such as land use controls and substance bans, etc. Moreover, governments can offer opportunities to make life better such as by providing convenient public transport so that the public can shift from the use of private cars toward public transport.

19.4.2 The Role of Individuals

Awareness about sustainable development spread in the form of education, conferences, seminars and advertisements could help in providing more targeted information. Markets can also help in creating awareness about the use of green products. Certification and labeling of products such as the carbon emissions created by using that particular product can help in creating sustainability or awareness of the sources of raw materials used. Campaigns can help by influencing the choice of some products over others, e.g. rejecting cars that consume more fuel (Chatterjee and Sharma 2013).

19.4.3 Impact Assessment

Impact assessment helps in raising awareness and could help to encourage stakeholders to implement policies and plans in favor of sustainable development. Impact assessment also identifies already employed constraints and imposes additional procedural constraints to information distribution and participation of the public as well as design constraints to prevent adverse impacts. Regulatory risk assessment provides information about the impacts of regulatory options to achieve a particular policy. Environmental impact assessment (EIA) is continuously evolving and has become a major part of project requirements since its first introduction in the United States in 1970 with the approval of the National Environmental Policy Act. The UNCED considers EIA an important tool for environment protection and sustainable development (Betey and Essel 2013). EIA explores all the possibilities for enhancing positive effects on the environment. It takes into account the impact of introduced projects, providing opportunities for comment by the authorities as well as the public in order to reduce negative impacts on the environment. In addition, a report is prepared and provided to the decision makers.

Impact assessment also considers the important role of the public in decision making. The tools that may influence plans and policies are put together in the form of "strategic environmental assessment (SEA)" (Calvey et al. 2016). To perform effectively, SEA should influence the development of any policy from its early stages to the final decision and should consider the impact on the environment throughout the development. Therefore, the emphasis is not on influencing the final decision of the policy alone, rather it is concentrated on influencing the whole cycle of policymaking, meaning that it is possible to change the policy to make it more favorable to sustainable development (Bonvoisin 2009).

Recently, the European Union (EU) has included sustainable development in a broad range of its policies. The sustainable development strategy will continue to provide an effective and long-term vision, and will establish the policy framework for all EU policies. The EU has taken the lead in the fight against climate change and environmental protection. A commission communication has considered the Europe 2020 strategy very effective to bring about sustainable development within EU countries. The strategy was adopted in 2010 and has helped to move Europe out of crisis and built the foundation for future with sustainable, smart and inclusive growth.

The world is adapting to the increased importance of sustainability in policymaking due to the fact that 17 Sustainable Development Goals (SDGs) of the 2030 Agenda for Sustainable Development were adopted by the world leaders at the United Nations (UN) summit held on September 2015. According to these goals, over the next 15 years, countries will apply efforts to eradicate all forms of poverty, handle climate change as well as fight inequalities. The newly formed goals are unique because these have been accepted by all countries to promote prosperity while keeping in mind the protection of the planet also. The goals state that poverty eradication must be considered while building strategies, which should help with economic growth as well as protecting education, health, job opportunities and dealing with climate change (Colglazier 2015).

Although impact assessment, SEA and other related tools are effective in identification of more sustainable development options, they are not sufficient to attain sustainable development. Many policies are still made without considering sustainable development. Throughout the policymaking cycle (ethical) sustainability needs to be considered as a continuous platform, starting from agenda setting, policy/decision making, implementation, evaluation and then moving toward agenda setting for the next policy (Figure 19.1).

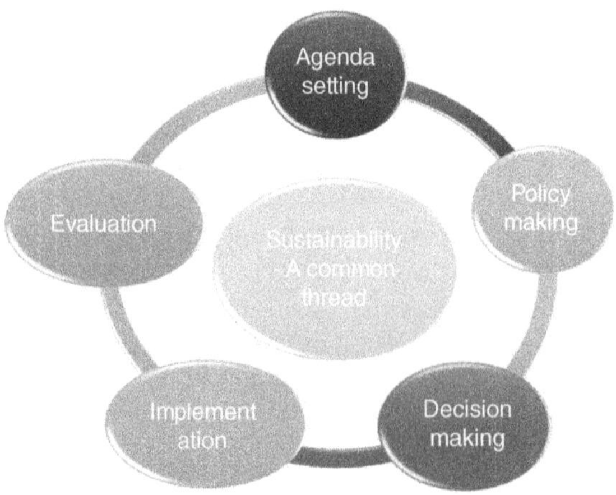

Figure 19.1 Sustainability as a continuous platform in policymaking.

The United Nations Environment Program (UNEP) is now engaged in initializing the integrative policymaking concept for sustainable development. The initiative is based on the UNEP experience with Integrative Assessment and Planning and upon the work carried out by the World Bank, European Commission and International Institute for Environment and Development. This initiative is helpful in bringing approaches, along with practical experience, for each step in the policymaking cycle sustainable development is considered, such as in the case of policymaking for climate change (Bonvoisin 2009). However, to consider sustainable development at every step of policymaking requires better understanding and more awareness about the importance of the environment, from trade and jobs to reduction in poverty as well as human wellbeing.

19.5 Global Flora and Sustainability

The global flora provides natural clean energy and water exchanges that remediate natural landscapes and improve land use. New, innovative ideas based on sustainability lead to increased flora value that directly impacts the environment in various ways such as in climate change. The global flora balances the net flow of gases in the environment by humid air transfer between biomes. Conservation and sustainable use of flora is required to meet the objectives of the Convention on Biological Diversity (CBD). Sustainability means granting the earth the ability to endure, and sustainable practices are designed to help ourselves and the world while keeping the future in mind.

Humans are linked to global flora in every aspect of their lives. The sustainable use of soils in various parts of the world connects global flora plant specimens to local topography and ecosystems, enhancing availability of floral public resources. Human beings completely depend on nature for all essential and life sustaining needs, but human activities are detrimental to the environment. Our activities directly or indirectly have adverse effects on global flora, and our own survival is at risk. The bitter truth is that we are meeting our present demands at the expense of the planet. To achieve sustainability the global challenge lies in ensuring growth and development without degrading the environment and natural resources. This has raised questions on how to value success and how to define growth.

19.6 Adaptation and Mitigation of Climate Change: Ethical Perspective

The concept of sustainability was first adopted by the World Commission on Environment and Development, which defined sustainability as a comprehensive and integrated approach to economic, social and environmental processes. Above all, growth was concentrated on the environmental and economic dimensions. Only in recent years, the social, political and cultural factors have been involved in development and growth. Integration is very important to follow the development trajectories to attain sustainability. Here, we use adaptation and mitigation of climate change to explain how to comply with the concept of ethical sustainability.

Strategic responses to climate change involve the consideration of climate-related risks along with the risks and co-benefits of adaptation and mitigation actions. Adaptation and mitigation are two complementary strategies for responding to climate change. Adaptation is defined as the process of adjustment to actual or expected climate and also its effects in order to either lessen or avoid harm or exploit beneficial opportunities. Mitigation is the process of reducing emissions or enhancing sinks of greenhouse gases (GHGs) to limit future climate change. Both adaptation and mitigation can reduce and manage the risks of climate change impacts. Yet adaptation and mitigation can create other risks as well as benefits. Adaptation, mitigation and climate impacts all results in transformations to and changes in systems. Based on the rate and magnitude of change and the vulnerability and exposure of human beings and natural systems, limited changes lead to change in ecosystems, food chains and systems, infrastructure, coastal, urban and rural areas, and human health and livelihoods. The adaptive responses to changing climate require actions that vary from incremental changes to more fundamental and transformational changes. Mitigation involves fundamental changes in the way that human societies produce and use energy services and land.

Climate science has shown that unabated climate change will lead to rising sea levels, frequent natural disasters and reduced agricultural productivity. For these reasons, it is necessary to call for measures to mitigate these impacts. Yet climate change mitigation has associated costs and risks. The mitigation costs include additional costs for incorporating low-carbon technology compared with fossil fuel-based energy, which is a main point of contention in the debate between dangerous climate change and dangerous mitigation. Data from integrated climate–energy–economy models suggest that in optimistic scenarios, coordinated efforts will reduce GHG emissions by incorporating new technologies. In addition to the above-discussed costs, there are certain risks involved when comparing climate change with those mitigations such as the increased use of nuclear power, which presumably would increase the probability of accidents, proliferation and problems with regard to storage and disposal of waste. If we use biomass for energy it raises pressure on land and adversely affects food security as well as biodiversity. Biomass use is related to higher water use and can increase water scarcity as well.

On the other hand, it has been argued that climate change mitigation achieves benefits in areas that are not directly related to climate change impacts; these benefits are known as "*co-benefits*." For example, while dealing with climate change problems human health is improved due to the improved air quality resulting from reducing fossil fuel usage. Similarly, to achieve climate mitigation low-carbon sources are used for energy production. Therefore, climate policies need to be assessed in a multi-objective framework, in which all climate impacts, adverse effects and co-benefits of mitigation and adaptation are taken into account.

19.6.1 Fundamentals of Decision Making About Climate Change

Effective decision making is important to limit climate change, and its effect can be analyzed using a wide range of analytical approaches that evaluate expected risks and benefits, recognizing the importance of governance, equity, value judgments, ethical dimensions, economic assessments, and diverse perceptions and responses to risk and uncertainty.

Sustainable development and equity are essential for decision making and assessing climate policies. To achieve ethical sustainability, it is necessary to limit the effects of climate change as it also impacts equity and poverty reduction. Every nation contributes differently to the accumulation of GHGs in the atmosphere, due to which countries face varying challenges and circumstances and have varied capacity to address mitigation and adaptation. The burden increases from the present to the future if mitigation shifts are delayed, and also insufficient adaptation responses to emerging impacts are eroding the fundamental of sustainable development.

Both adaptation and mitigation have varied distributional effects locally, nationally and internationally depending on who pays and who benefits. The process of decision making for climate change, and the degree to which it respects the rights and views of affected people, is also a concern in terms of justice for these people. Effective mitigation cannot be achieved if individuals advance their own interests independently as climate change is a collective action problem globally because most GHGs accumulate and mix globally over time and emissions have interdependent effects. Therefore, the effectiveness of adaptation can be enhanced through complementary actions across different levels that also include international cooperation (Team et al. 2014).

Decision making about climate change includes valuation and mediation among diverse values. Ethics analyzes the different values involved and relations between them. Economics and decision analysis provide quantitative methods to evaluate the estimated social cost of carbon, cost–benefit and cost effectiveness analyses for integrated model optimization. Economic methods reflect the ethical principles that take non-marketed goods, equity, ancillary benefits, behavioral biases and costs; and the varied wages for people doing the same work into account for sustainable economic growth. Analytical methods of valuation cannot identify a single best balance between adaptation, mitigation and climate impacts because climate change involves extremely complex natural and social processes. However, information on the consequences of emission pathways for alternative climate goals and risks are useful for decision-making processes.

Therefore, effective decision making and risk management are adjusted as new information and understandings are generated during implementation. Adaptation and mitigation affect the risks of climate change as the prospects of climate-resilient pathways for sustainable development depend on what is achieved through mitigation. The decision making about climate change is influenced by how individuals and organizations perceive risks and uncertainties. Thus, formalized analytical methods for decision making under uncertainties can accurately assess the risks with focused attention on short- and long-term consequences.

19.6.2 Climate Change Risks Abridged by Adaptation and Mitigation

Additional mitigation efforts with adaptation are required to prevent high risks and irreversible impacts of global warming globally. The risks of climate change, mitigation and adaptation differ in terms of their nature, magnitude, time scale and persistence. Maladaptation and negative ancillary impacts are the risks from adaptation; risks from mitigation include side effects of deployment of low-carbon technology options and economic costs. Mitigation and adaptation are complementary approaches for reducing risk of global warming or climate change impacts. They interact with each other and reduce the risks of climate change globally. Adaptation can analyze the present risks and emerging future risks. Adaptation has the potential to reduce climate change impacts in the future while mitigation has a relatively low influence on climate impacts long term. The potential for adaptation varies across different sectors and is limited by institutional and capacity constraints, increasing the long-term benefits of mitigation. The scale of mitigation influences the rate and magnitude of climate change due to GHG emissions exceeding adaptation limits.

Economic and climate systems and the possibility of irreversible impacts from climate change increase the benefits of mitigation efforts. The actions taken today affect the options available in the future to reduce GHG emissions, climate change and adaptation to climate change.

19.6.3 Adaptation Pathways Characteristics

Adaptation reduces the risks of climate change impacts, and its effectiveness is limited especially with greater magnitudes and rates of climate change. Adaptation contributes to the wellbeing of present and future generations. Adaptation and effective risk reduction strategies consider exposure, vulnerability and their linkages with processes of socio-economic development. Adaptation research has evolved from a dominant consideration of engineering and technological adaptations to considering more ecosystem-based, institutional and social measures. The implementation and planning of adaptation at all levels of governance are contingent on societal values, objectives and risk perceptions. Central government coordinates with local governments to scale up adaptation and its progress. Reducing the vulnerability and exposure to the present climate change is the first step toward adaptation to future climate change. Integration of adaptation into planning, policy design and decision making promotes synergistic effects with development and risk reduction (Team et al. 2014).

There are numerous constraints originated from limited finances, lack of human resources, coordination of governments, different perceptions of risks, and insufficient research, monitoring and evaluation. These constraints hinder the planning and implementation of adaptation. Therefore, underestimating the complexity of adaptation as a social process creates unrealistic expectations about achieving positive outcomes of adaptation. In addition, the increased rates and magnitude of climate change increase the chances of high adaptation limits and, thus, value-based judgments of intolerable risk may vary. Advantageous opportunities also decrease if limits to adaptation increase and synergies between adaptation and mitigation decrease. However, transformations in social, economic, technological and political decisions and actions enhance adaptation and promote sustainability.

19.6.4 Mitigation Pathway Characteristics

There are many mitigation pathways that will limit global warming to below 2 °C relative to pre-industrial levels. These mitigation pathways require substantial reductions in emissions of GHGs in the coming decades and almost zero emissions of carbon dioxide and other long-existing GHGs. The implementations of these pathways have substantial challenges in the technological, economic, social and institutional sectors that delay additional mitigation steps. Therefore, additional efforts are required to reduce the emission of GHGs, which are expected to increase with growth in global population and economic activities worldwide. Different combinations of technology, behavioral changes and policy can be used to reduce emissions and limit the temperature change. To limit the temperature change to less than 2 °C relative to pre-industrial levels would require substantial cuts in anthropogenic GHG emissions. Limiting temperature change to lower levels requires changes more quickly. To reduce the emissions of non-CO_2, climate change agents are an important element of mitigation strategies. The present GHG emissions and other climate-changing agents will affect the rate and magnitude of climate change over the coming decades. Therefore, reducing the emissions of certain short-life climate change agents would reduce the rate of global warming in the short term but would have limited effects on long-term global warming, which is mostly driven by GHG emissions.

19.6.5 Interactions Among Adaptation, Mitigation and Sustainability

Climate change poses an increasing threat to equitable and sustainable development. Mitigation, adaptation and sustainable development are interrelated with potential for synergies and trade-offs. Climate change is a threat to social and natural systems, affecting developmental pathways globally and thus eroding the basis for sustainable development. Aligning climate policy with sustainable development requires both adaptation and mitigation (Team et al. 2014). This interaction between adaptation, mitigation and sustainable development occurs at every region and scale and can bring substantial co-benefits. Co-benefits are seen in the following:

- improved air quality;
- reduced energy and water consumption in urban societies by emphasizing water recycling and greener cities;
- sustainable agriculture and forestry;
- enhanced energy security;
- protection of ecosystems for carbon storage.

Strategies and actions can be taken that move steps toward climate-resilient pathways for sustainable development and help toward improvements in livelihoods, social and economic wellbeing, and effective environmental management. Prospects for climate-resilient pathways are related fundamentally to what the world accomplishes with climate change mitigation. Since mitigation reduces the rate as well as the magnitude of climate change, it also increases the time available for adaptation to a particular level of climate change and delaying mitigation actions may reduce options for climate-resilient pathways in the future.

19.7 Conclusions

Up to now, sustainability science has focused on solving "problems;" however, it should focus more on its social and ethical sides because technology only determines our ability to, and efficiency at, exploitation of natural resources. Science alone cannot help us become sustainable. Sustainability without ethics is meaningless. Therefore, ethical sustainability needs to be considered as a continuous platform, starting from agenda setting, policymaking, decision making, implementation, evaluation, and then moving toward agenda setting for the next policy. To consider sustainable development at every step throughout the policymaking cycle requires better understanding and more awareness about the importance of the environment. In most cases, different approaches (e.g. utilitarianism and/or deontological ones) need to be considered in order to reach the best ethical decision. Sustainability and ethical principles are intertwined because sustainability concepts cannot be applied without strong ethical principles.

References

Aoyagi-Usui, M., Vinken, K., and Kuribayashi, A. (2003). Pro-environmental attitudes and behaviors: an international comparison. *Human Ecology Review* 10 (1): 23–31.

Betey, C.B. and Essel, G. (2013). Environmental impact assessment and sustainable development in Africa: a critical review. *Environment and Natural Resources Research* 3: 37–51. https://doi.org/10.5539/enrr.v3n2p37.

Bolderdijk, J.W., Steg, L., Geller, E.S. et al. (2013). Comparing the effectiveness of monetary versus moral motives in environmental campaigning. *Nature Climate Change* 3 (4): 413.

Bonvoisin, N. (2009). *Decision Making for Sustainable Development: How Assessment Can Help*. Geneva: UNECE.

Bowie, N.E. (2017). *Business Ethics: A Kantian Perspective*. Cambridge, UK: Cambridge University Press.

Calvey, C.H., Su, Y.-K., Willis, L.B. et al. (2016). Nitrogen limitation, oxygen limitation, and lipid accumulation in Lipomyces starkeyi. *Bioresource Technology* 200: 780–788.

Chatterjee, A.S. and Sharma, A. (2013). Role of individuals as change agent for sustainable development: integrating spirituality and ecology. *International Journal of Science, Engineering and Technology Research* 2 (5): 1199–1203.

Colglazier, W. (2015). Sustainable Development Agenda: 2030. *Science* 349 (6252): 1048–1050.

Davison, A. (2001). *Technology and the Contested Meanings of Sustainability*. Albany, NY: State University of New York Press.

De, J. R. (1992). United Nations Conference on Environment & Development Rio de Janerio, Brazil, 3 to 14 June 1992. Reproduction, 351(10.1007).

Eid, M.E.M. (2012). The ethical reasoning behind sustainable development: a paradoxical opportunity for the reform of developing countries. *International Journal of Urban Sustainable Development* 4 (2): 236–245.

Gorica, K., Kripa, D., and Zenelaj, E. (2012). The role of local government in sustainable development. *Acta Universitatis Danubius. Œconomica* 8 (2).

Hartman, L.P. and DesJardins, J. (2011). *Decision Making for Personal Integrity and Social Responsibility*. New York: McGraw Hill International.

Jamieson, D. (1998). Sustainability and beyond. *Ecological Economics* 24: 183–192.

Jennings, B. (2013). Ethical aspects of sustainability. Centre for Humans and Nature. www.humansandnature.org (accessed March 2019).

Koçyiğit, M. and Karadağ, E. (2016). Developing an ethical tendencies scale based on the theories of ethics. *Turkish Journal Of Business Ethics* 9 (2): 297–307.

Munasinghe, M. and McNeely, J. (1995). *Protected Area Economics and Policy*. Geneva and Washington, DC.: World Conservation Union (IUCM) and World Bank.

Nelson, M.P. and Vucetich, J.A. (2012). Sustainability science: ethical foundations and emerging challenges. *Nature Education Knowledge* 3 (10): 12–15.

Sibole, A. (2013). Why should we care? https://www.scu.edu/environmental-ethics/resources/the-ethics-of-sustainability (accessed March 2019).

Team, C.W., Pachauri, R.K., and Meyer, L. (2014). *IPCC, 2014: Climate Change 2014: Synthesis Report*. Contribution of Working Groups I, II and III to the Fifth Assessment Report of the intergovernmental panel on Climate Change. Geneva, Switzerland: IPCC.

Thompson, P.B. (2007). Agricultural sustainability: what it is and what it is not? *International Journal Agricultural Sustainability* 5: 5–16.

Türkeri, M. (2013). *Etik kuramları (2. basım)*. Ankara: Lotus Yayınları.

Warner, K.D. and DeCosse, D. (2009). The ethical dimension of sustainability. https://www.scu.edu/environmental-ethics/short-course-in-environmental-ethics/lesson-four (accessed March 2019).

WCED (World Commission on Environment and Development) (1987). *Our Common Future*. New York: Oxford University Press.

Winners, M.D.G. (2009). *The Performance of Sustainability-Focused Companies in the Financial Crisis*. Chicago, IL: AT Kearney.

Yüksel, C. (2010). *Siyasette Etik: Dünya'da ve Türkiye'de Siyasette Etik*. Istanbul: Boğaziçi University Publishing.

20

Education and Human Resource Development for Sustainability

Anita Talan, and R. D Tyagi

INRS, Université du Québec, Québec, Canada

20.1 Introduction

The ideology of sustainable development and its implications for human resource development (HRD) and the education system are of great importance as they respond to the dilemmas that confront sustainable projects worldwide. Sustainable development is a broad concept that integrates education and HRD to analyze global society and political economy. The education and HRD systems are implemented together at national and international levels toward global sustainability. The role of HRD and education systems in every sector can be seen from two points of view: on one side it is based on the influence on the social dimension, and, on the other, the indicators affecting sustainability. The focus is on the education system to produce world class students and human resources (HR) to attain sustainability defining the quality of learning outcomes. The outcomes can be examined by having in-depth information on policy levels and the contextual factors that shape these outcomes so that they have a high gain in investments from private and social organizations for improvements in education and HR. Therefore, increased levels of education empower the general public to change their own views and take responsibility for a sustainable future.

This chapter focuses on sustainable development, the relationships between education, HRD and sustainability, elements and indicators of HRD, integrated approaches for sustainable development, ways to achieve sustainable development including sustainability as HRD, and potential improvements via education for global sustainability based on global education and global actions of education for sustainable development (ESD).

20.2 Education for Sustainability

Education for sustainability (EfS) or sustainability education (SE) or ESD are all interchangeable terms for defining the practice of education to achieve sustainability. ESD is the term generally used by the United Nations (UN) and other government bodies. After

Sustainability: Fundamentals and Applications, First Edition.
Edited by Rao Y. Surampalli, Tian C. Zhang, Manish Kumar Goyal, Satinder K. Brar, and R. D. Tyagi

becoming aware of the significance of education to achieve sustainability, Agenda 21 (Michalos 2017) was the first international document that acknowledged education as a necessary tool to achieve sustainability. Agenda 21 highlighted all zones of action for education to be taken globally.

With advancement in scientific study and impact analysis, considerable development in ESD has been accomplished in recent years. The Organisation for Economic Co-operation and Development (OECD) and member nations are continuously developing new policies and projects based on the rising demand for education globally. Policy development and its implementation is a tedious process presenting several difficulties and trade-offs. First, the indicators are required to answer all educational concerns that top the list of national policy agendas and where global comparison of education systems can add value to achieve sustainability by national inquiry and assessment. Second, the indicators should be as good as possible and should specifically differentiate between ancient, universal and social dissimilarities between nations. Third, the indicators are required to be presented in a direct way as much as possible to prevent confusion arising from the complexities that reflect multi-faceted realities. Fourth, the indicators should be kept as a small set to have better impact assessment (IA) results but the set should be large enough to be considered useful to policymakers (OECD 2017).

The United Nations Educational, Scientific and Cultural Organization (UNESCO) defines "Education Sustainability as a tool that authorizes individuals to change the way they think and work towards a sustainable future" (Marope et al. 2015, p. 24). There are a number of different interpretations of "sustainability" and "sustainable future." Many of these versions relate to discovering a balance between economic development, protection of natural resource base and social wellbeing. The Quality Assurance Authority (QAA) and the Higher Education Authority (HEA) have tried to improve the definition of education sustainability as: "Education for Sustainability is the process of equipping student with knowledge and understanding skills and attributes needed to work and live in a way that safeguards ecological, social, and economic well-being both in the present and for future generations."

UNESCO defined EfS as an approach that empowers learners to take informed decisions and responsible actions for environmental integrity, economic viability and a just society, for present and future generations, while respecting cultural diversity.. Education changes thinking and promotes critical thinking, imagination and decisions (Latchem 2017). The objective to achieve sustainability requires government to resolve persistent societal problems and to identify new prospects. Large investments and creation of capital reduce global poverty rates. The rise in the economic conditions of the world is also responsible for increased inequality, violence, segregation and helplessness of poor people of the community throughout the world (Marope et al. 2015).

In 1989, the definition of sustainability was refined as follows: "Sustainable development involves devising a social and economic system, which ensures that these goals are sustained, i.e. that real incomes rise, that educational standards increase that the health of the nation improves, and that the general quality of life is advanced" (Pearce et al. 1989, p. 1).

Therefore, an increase in education levels empowers the general public to change their own views and take responsibility for a sustainable future. Educated people take action sensibly based on the information and understanding that present activities can have impacts on the lives of people and the planet as well. To increase access to valuable education at all stages in all social perspectives is to transform the world by reorienting teaching and helping people to develop knowledge, standards and behavior for education sustainability. The new policies developed for education systems encourage people to be responsible for identifying problems and creating a sustainable world.

Globally, ESD has been recognized as an integral component of superior education and an important concept for sustainability. The global community included ESD in its next 15year plan of Sustainable Development Goals (SDGs). Various Global Citizenship Education (GCED) programs have been initiated around the world. UNESCO has been given responsibility for the coordination of the Global Action Programmes (GAPs) on ESD (Latchem 2017).

The global human rights agenda has been reinforced over the past few years, but there are still unsustainable arrangements of economic production and consumption. Therefore, the implementation and protection of these policies and frameworks is very challenging. For example, today an emphasis has been placed on empowerment of women through access to education, but still women continue to face gender discrimination in public as well as in employment. Violence against women continues to weaken their rights and respect. With the increase in technology there is also a growth in traditional and spiritual intolerance, technology-based crimes and conflicts (Peña-López 2015). Sustainability is directly affected by education in three main areas (Paul and Mehera 2016):

1) *Decision making*: Educated people create a good community based on proper decisions. These decisions affect the social, economic and environmental wellbeing of individuals. Global development is directly proportional to education and improvements in the education system. For example, a nation with a high education rate would have large numbers of skilled laborers and technical workforce to work toward new gains in science and technology. Educated citizens of the country look after their communities by analyzing issues that create difficulties in the development of the country as well as affecting individual health. For example, today educated people are switching to organic farming rather than using herbicides and pesticides for high grain yield as the use of chemicals in agriculture is responsible for many health issues and adversely affects social sustainability.
2) *Implementation*: An educated citizen is very dynamic in implementation of developed decisions. Sustainability is always enhanced by education and it is not only useful nationally but enhances development globally. Nations with low literacy rates and unskilled employees have fewer growth opportunities. This slows down economic development as these nations have to buy energy and other manufactured goods from other countries with the currency that should be utilized for their own development. To acquire this capital, these countries are giving access to other countries to exploit their natural resources and land for cash-based activities, while an educated workforce is an elemental requirement for the development and growth of an economy.
3) *Quality of life*: Education is also responsible for giving a better quality of life and continuously working on its improvement. Education raises the economic position of people, improves life conditions, reduces mortality levels by enabling access to good health services and improves the education system for coming generations. Education is the key to both individual and national development.

Hence, education must find methods to respond to such challenges on a global scale. To achieve education sustainability worldviews should be changed; alternative education systems should be applied with advances in science and technology.

20.3 Objectives of Education for Sustainable Development

Quality education systems are needed to achieve a more sustainable world. ESD promotes the growth of the knowledge, skills, understanding, values and actions needed for development, ensuring conservation and protection of the environment, promoting social equity and boosting economic sustainability. The key goal of ESD is to empower citizens to take decisions and carry out actions to increase the quality of life without compromising the environment. It also aims to incorporate the intrinsic values of sustainability into all aspects and levels of learning. A number of themes are considered in ESD but the principal focus is on ecological resources; it also speaks about other themes, such as mitigation of poverty, morals, peace, responsibility, democracy and supremacy, justice, integrity, human rights, gender equivalence, corporate concerns, management of natural resources and biological diversity. Certain features are very important for the effective implementation of ESD, reflecting the importance of both the learning process and the consequences of the education process (Latchem 2017).

As per Education 2030, in 2017 UNESCO published SDGs specifying that the ESD should:

- *Be implanted in the core curriculum in an interdisciplinary and universal method*, agreeing on a whole-institution methodology for policy- or decision making.
- *Share the morals and values* that strengthen sustainable development.
- *Encourage critical thoughts, resolving a problem and action*, all of which develop self-reliance in addressing the challenges to sustainability.
- *Employ a range of educational systems and approaches*, such as literature, art, drama and debate to elucidate the courses and procedures.
- *Agree for participation of learners in decision making* on the proposal and content of educational programs for sustainability.
- *Report* local as well as international problems, and escape jargon-ridden language and expressions.
- *Look to the prospects*, confirming that the content has a long-term outlook and uses average and long-term design.

To encourage ESD, the UN Decade of Education for Sustainable Development (DESD) was adopted by the UN General Assembly with UNESCO designated as the principal organization for upgrading education throughout the decade. To promote ESD globally, the DESD of 2010 has a global vision "of a world where everyone should have the chance to get advantage from good education system and learn the morals, standards, conduct and way of life needed for a sustainable future and positive transformation of society in terms of education." The aim of the period as framed by UNESCO is to incorporate the principles, values and practices of sustainable development into all aspects of education and learning. The aim is to make changes in society and to make responsible citizens to create a more sustainable future. One of the most significant aspects of the DESD is the recognition that

ESD needs to participate in a wide range of investors from government, private segment, civil society, non-governmental associations and the general community. UNESCO, in its International Implementation Scheme (IIS) for DESD, states that ESD is basically about values, mainly respect for other people, including those of present and future groups, for transformation and multiplicity, for the surroundings and for the planet's natural resources. Education empowers us to recognize ourselves and others and our relationship with the extensive natural and social atmosphere; this understanding helps as a long-lasting basis for building respect. Along with the intelligence of impartiality, responsibility, assessment and negotiation, ESD aims to move us on the way to accept actions and practices that will enable us all to live a quality being without being deprived of basic human necessities (Combes 2005). The next parts of this section describe the objectives of ESD with respect to five important aspects.

20.3.1 Disaster Risk Reduction

Disaster risk is defined as the potential loss in terms of lives, health, resources, assets and belongings, amenities and livelihoods, which could take place in a particular area, community or society due to the impact of a natural calamity. Disaster risk reduction (DRR) is an organized approach to recognizing, evaluating and decreasing the risk. The main aim of DRR is to reduce susceptibilities and risks in order to prevent or mitigate the negative impacts of natural hazards throughout society to maintain/enable sustainability. DRR is also acknowledged as an important element in climate change adaptation policy. DRR focuses on the correlations among society, ecology, economy and culture, and their influences. In addition, it encourages rational thinking and problem solving as well as communal and emotional life aids that are necessary for the empowerment of individuals exposed to or affected by disasters (Apronti et al. 2015). The present understanding of DRR states that the right education and education systems have the potential to decrease disaster vulnerability. Various research has been conducted to establish the need for disaster education to learn about the causes and effects, prevention and reaction, and management of and recovery from disasters. The education provided must have diverse and practical techniques that strengthen disaster knowledge and build a culture of safety among the people of society. To evaluate DRR lessons innovative techniques for teaching and learning are required. Innovative and suitable efforts are required from all stakeholders to train instructors and to deliver the appropriate learning resources to improve the current DRR teaching. Being interdisciplinary and holistic in nature, ESD provides better learning to create responsible societies. It focuses on decision-making processes, critical thinking and innovative approaches to solve problems. Therefore, ESD directly contributes to DRR, and DRR increases its education and teachings in disaster-prone areas (Apronti et al. 2015).

UNESCO gives aid, advice and technical assistance to reactivate the education system in affected areas. It also plays an important role in advocacy, networking and participation in activities to ensure that educational needs are met in achieving post-disaster management. It is always actively involved in post-disaster programs across the world, such as the Myanmar Education Recovery Programme (MERP) in the Asia-Pacific. It plays a crucial role within the UN International Strategy for Disaster Reduction (ISDR) Thematic Platform on education and learning. It has employed the subject of DRR in national educational plans, school curricula, and national policies and approaches. UNESCO has promoted DRR education at international events and workshops on ESD, such as that on "building disaster-resilient societies" organized during the 2009 Bonn World Conference on ESD.

20.3.2 Climate Change Control

Researchers and experts across the world study the sources and effects of climate change. Various studies have been conducted on the greenhouse gas emissions that are responsible for global warming. Measures are being taken to protect the planet from the negative consequences of human activities. Rises in air pollution, climate change, increasing water levels and marine current modifications are all creating imbalance in the natural climate. To have a sustainable world to protect the planet is the foremost step. The signs of climate change and its influences are cumulative, so it it is clear that most of the causes of climate change are social activities, consumption of natural resources and activities that increase pollution and exploit the ecological resource base in an unsustainable manner. The health impacts of climate variation are well known and effects are also seen on agriculture and fisheries; this all may result in the collapse of ecosystems in the near future. The consequences of climate variation are primarily on health. Thus, health services are required to protect and promote the health of the public.

With this aim, the UK Public Health Association gave strategies for promoting health and sustainability. The Climate and Health Council has been established as an international organization with an aim to mobilize health experts to every corner of the world to take action to minimize climate change and its effects on human health (Nichols et al. 2009).

ESD is supported by UNESCO to provide maintenance and support to countries to integrate climate change into their education systems to facilitate critical thinking and exchange of experiences on climate change by organizing worldwide professional meetings and assemblies. It emphasizes the necessity for schools and educational organizations to implement climate change education (CCE), whereby sustainability policies are also integrated into the management and governance of the education system. This incorporation develops guidance material and training resources, such as a 6-day online course, "Climate change in the classroom: UNESCO course for secondary teachers on climate change education for development." UNESCO's clearing house on CCE provides free education and access to all of their CCE resources. It is also assisting other countries and member states to enhance public awareness and understanding of the cause and effects of climate change, and the steps countries and communities should take to adapt to these changes. The projects and policies implemented also define the do's and don'ts to respond to these threats; for example, in the publication *Climate Change in Africa: a Guidebook for Journalists*. Concerned with the negative impacts, today UNESCO, together with 13 UN agencies, promotes CCE and teachings with an emphasis on general awareness at high-level proceedings such as the annual United Nations Framework Convention on Climate Change – Conference of the Parties (UNFCCC-COPs) through the UN Alliance on Climate Change Education, Training and Public Awareness (Tanner and Allouche 2011).

20.3.3 Increase in Biodiversity

To promote students' commitment to protecting local biodiversity with the overall aim of protecting the planet's biodiversity is an important goal of ESD around the world. The main focus of biodiversity education is to spread knowledge, interest and skills to solve the problems of biodiversity with reference to local biodiversity. Emphasis is placed on action-oriented education to give new methods and approaches to prevent biodiversity, such as active classroom sessions, hands-on practice, practical knowledge and, most importantly, motivating students and people to protect and preserve their own biodiversity. ESD, with UNESCO's approval, developed a comprehensive framework to assess the efficacy of biodiversity education to enhance teaching and training in biodiversity conservation at every level of education. The real pre-experiment and post-experiment data is published in education resources to encourage all stakeholders to take appropriate efforts for sustainability. Biodiversity conservation through environmental education protects species, ecosystems and the natural resource base, and integrates policies to fight against poverty and gender inequality, and promote human rights, education for all, health, security and intercultural rights. The main aim of education is to provide knowledge to enable people to make decisions both for their own benefit and that of others, and to act upon these decisions (Combes 2005).

Education for climate change also focuses on interlinked issues such as livelihoods, agriculture, livestock, forestry, fisheries and many more. The DESD offers an opportunity to understand the impacts of consumption of natural resources on biodiversity at the local as well as global level. This makes people responsible for encouraging sustainability by preventing habitat loss, degradation, pollution and species loss. Everyone in the world depends on natural ecosystems. The increasing population has made unprecedented changes in ecosystems to meet needs. Human activities are responsible for taking the planet to the edge of extinction, threatening our own survival. Biodiversity can be preserved and protected by changing human activities and attitudes (Fredericks 2012).

20.3.4 Poverty Reduction

The complete removal of poverty is one of the largest global challenges. As the Millennium Development Goals (MDGs) will expire soon, a summit held in 2012 put forward the post-2015 SDGs, which replace the MDGs to support global development until at least 2030 (UN General Assembly 2015). The SDGs cover all the content within the framework of the Post-2015 Development Agenda and consider global sustainability as the main agenda (Rakodi 2014). As per the 2015 summit, the elimination of extreme poverty in the coming 15 years is the main aim behind the formulation of the SDGs. Therefore, poverty eradication and sustainability are inseparable terms. The new global partnership focuses on eradicating poverty and transforming economies through sustainable development. The UN's High-Level Panel of Eminent Persons on the Post-2015 Development Agenda proposed five major transformations (Heinrichs 2013):

1) authorization to leave no one behind;
2) put sustainable development at the core;
3) transform economies for jobs and inclusive growth;

4) create harmony and effective, open and accountable organizations for all;
5) forge new international enterprises.

The end of extreme poverty and achieving sustainability by 2030 are the main objectives of the SDGs, reflecting the importance of poverty reduction in the topic of sustainability. Grounded in the conception of extreme poverty reduction, the goals proposed are divided into three categories (Table 20.1) (Liu et al. 2015):

1) multidimensional poverty reduction goals;
2) SDGs;
3) global partnership goals.

Table 20.1 Goal classification of the post-2015 development agenda and post-2015 sustainable development goals.

Objectives	Post-2015 Development Agenda	Post-2015 SDGs
Multidimensional poverty	• Poverty eradication • Women empowerment • Gender equality • Quality education • Lifelong learning • Healthy lives • Food security • Universal access to water and sanitation	• Poverty eradication in all its forms • End hunger, achieve food security and improved nutrition • Promote agriculture sustainability • Healthy lives and promote wellbeing • Quality education • Lifelong learning prospects • Gender equality and empower all women and girls • Universal management of water and sanitation
Sustainability	• Secure sustainable energy • Create jobs, sustainable livelihoods and equitable growth • Manage natural resources	• Access to affordable, reliable, sustainable and modern energy • Sustainable economic development • Productive employment • Build resilient infrastructure • Inequality removal • Ensure sustainable consumption and production patterns • Take action to combat climate change and its impacts • Conserve natural resources • Protect, restore and promote sustainable use of ecosystems
Governance and partnership	• Ensure good governance • Ensure stable and peaceful societies • Create a global environment to increase finance	• Promote peace • Strengthen the means of implementation • Revitalize the global partnership

20.3.5 Sustainable Consumption

Many approaches to Education for Sustainable Consumption (ESC) have developed from convention consumer information. Present consumption outlines and standard of living have been designed in various eras based on our evolution and driven by the following forces (Stanszus et al. 2017):

1) *Economic development*: This leads to improved productivity to reduce prices. It eventually leads to an increase in the income and purchasing power of an individual. Ideally, more income together with accurate knowledge leads to a more sustainable world.
2) *Technological advances*: Achievements in technology supply the world with more efficient products and technologies. They make life easier and create new visions for living conditions and lifestyles.
3) *Policy actions:* These influence lifestyle by placing rules and regulations on activities that are potentially harmful to people and nations..
4) *Socio-psychological tools*: These include personal motives and influences of the social environment. Goods and services purchased for social status, to mark the measure of success, power and happiness. Capital creates happiness to a certain level, after which other factors become the reason for being happy in life.
5) *Cultural and historical aspects*: These provide a silent code of conduct to society. To measure changes in lifestyle it is important to understand the cultural environment of a society.

Educational formats are defined as practical approaches used in the education system to enhance the learning process in the field of sustainable consumption. These practical approaches comprise assignments, exercises and other learning activities (Stanszus et al. 2017). The BiNKA (Bildung für Nachhaltigen Konsum durch Achtsamkeitstraining – German acronym for education for sustainable consumption through mindfulness) training was developed and implemented by different ESC formats. This is used to identify different educational formats that could be used for training purposes.

20.4 Potential Improvements via Education for Sustainable Development

ESD can improve people's vision, critical thinking, systematic thinking and capability to build partnerships/collaborations; details are as follows.

20.4.1 Envisioning

"Envisioning" is defined as an ability to imagine a better future. The foundation is that if we know where we want to go, we will be better able to work out how to get there. Quality education and ultimate learning are the basis for sustainable development. In the post-2015 meeting, UNESCO highlighted that education, learning and training are the universal rights to enhance empowerment, the growth of social communities and the promotion of social integrity. Education contributes to the eradication of poverty, better supremacy and involved citizens, promotion of social unity, improved health services and gender equality.

Figure 20.1 The human resources environment.

Education consultation and other governing bodies have said that education promotes comprehensive economic growth in order to build decent productive citizens. Education also plays a role in conflict prevention, peace building, disaster management and risk reduction (Yasunaga 2014). The post-2015 agenda continues to meet the goals of Agenda 2030, which include (Dubeck and Gove 2015):

1) Everyone should be given access to quality education at all levels of education. A life-long learning approach should involve primary and post-primary education, formal and informal, higher education and vocational training. Learning and training should involve all contexts of country, development challenges and priorities. The main concern is to uplift women of the society through learning and training.
2) A focus is given to quality education, its content, relevance and outcomes. This also focuses on measureable and equitable learning outcomes throughout the lifecycle. The education should also promote environment education to understand the challenges and threats to the natural ecosystem due to harmful activities. In addition it is important to provide the training and skills needed for professional development. This makes people more informed, responsible and active citizens to contribute to peaceful nations. This enhances integration and collaboration across countries to develop tranferrable technical skills. Emphasis is also given to strengthening learning environments, which means safe and healthy learning including safe buildings, clean drinking water, gender-sensitive sanitation and hygiene and, most importantly, the integration of sexual and reproductive health into the curriculum. Education based on the lives of rural populations is also given importance.
3) There is a focus on equity: the needs of abnormal and disadvantaged people, especially girls and women, to acquire literacy, numeracy, technical knowledge, vocational skills

and transferrable skills for a decent life, which should be reflected in policies and decisions made. It is important to reach out to every school child using innovative ideas, flexible and mobile interventions, and making partnerships with all non-government education bodies to ensure there is a good quality of learning and teaching in every part of the world. Data analysis should be done at specified intervals to describe the progress and achievement levels.

4) Gender equality remains a strong priority, with a focus on every level of education for girls and women in safe, supporting environments. The evidence shows that the effects of education only show development when girls complete their secondary education. Gender equality cannot be assessed through a narrow focus on education; rather the focus must be on every level simultaneously, including economic, social, political, and cultural aspects.

20.4.2 Critical Thinking

Critical thinking is defined as the ability to think clearly and logically about what to do or what to consider. It includes self-governing thinking and helps in understanding of the association between rational ideas. It is about learning to question or justify our existing belief system and to know the traditions essential for our knowledge, perspective and thoughts. It helps us to develop skills that also encourage economic, environmental, social and cultural sustainability.

The objective of critical thinking is to promote independent thinking, individual sovereignty, and consistent decisions in opinions and actions. This comprises two related dimensions: (i) the skill to raise an objective; and (ii) the outlook to organize accordingly. Critical thinking is responsible for logical and creative decisions, which may include inductive or deductive reasoning, analysis and problem solving, with the best possible approach to issues and challenges. The main aim of education is to produce informed learners who understand important ideas and the power of those ideas to contribute to society, culture and civilization (Hasslöf and Malmberg 2015).

The two objectives for education as a vehicle to support critical thinking are grounded on certain norms and, on the basis of these objectives, the ability for critical evaluation and study arises as essential for enjoying a good life (Hasslöf and Malmberg 2015). The ESD is highly dependent on the following points of critical thinking.

1) *Brains are biological units, whereas minds are created*: The curriculum in the education system is designed in to focus on mind development. The ethical requirement is needed to teach people to be autonomous centers of awareness with the major capacity to decide the outlines of their own thoughts and existence.
2) *Self-direction is required not pre-conceived roles*: Education prepares learners and not pre-conceived roles. Learners are prepared for thinking to make their way through the challenges of life independently.
3) *Forms of representation*: Education generally initiates the learner to learn different forms of knowledge representation to understand the importance of these representations in terms of critical thinking.
4) *Democratic life*: Clear thinking and thoughts, careful study and rational planning are key elements to democracy and democratic life.

20.4.3 Systemic Thinking

Systemic thinking is all about acknowledging difficulties and observing relationships and interactions while developing solutions to problems. It is required to increase the probability of lasting effects or impacts of policies and projects on adaptive sustainability. Without systemic thinking it is extremely difficult to develop the best solutions to support sustainability (Lewis et al. 2014). To facilitate systemic thinking initiatives have been taken that focus on different areas of society:

- government and public policy education;
- private education;
- youth education;
- cross-sectoral methods.

The aforementioned areas do not focus clearly on sustainability, but most of the areas focus on subdivision of systems. Thinking skills and different delivery mechanisms such as face-to-face or distance-based learning are all providing a process that can improve the ability to develop systemic understanding. Systemic thinking is considered too time-intensive or overtheoretical, not directing the action and with inadequate utility in real-world problem situations. Systemic thinking is not a theory that provides solutions to every complex problem but still has been proven to be a varied discipline and an invaluable tool to gain insights into complex problems and to provide action-oriented solutions to adaptive global challenges. Various local, national and international initiatives have been taken to implement systemic thinking in different social sectors. Systemic thinking can only be developed by integrating thinking capacity with firmly established, interconnected global initiatives (Lewis et al. 2014).

20.4.4 Building Partnerships/Collaborations

Collaboration is one of the elemental keys to unlock sustainability. It promotes negotiation and enables people to work together. No organization or company or sector of the world has the knowledge or resources to work alone. Unparalleled cooperation is required to solve the challenges in achieving sustainable development.

Collaboration is all about working together with a core objective of "a working environment based on integration where organization and people can be benefited from each other's resources and capabilities" (Tilbury and Wortman 2004). To have more a sustainable world international collaborations and partnerships are required. Diverse sectors should be open minded about functioning together, not afraid to take risks to work in partnership in new ways, and to ensure the lessons of these partnership prototypes – good or bad – are shared broadly. As the boundaries between divisions become indistinct, the interdependencies become more definite yet the complications increase; the sectors need to be improved to find mutual policies.

20.5 Global Education and Global Actions of ESD

20.5.1 Global Education

Global education is a combined set of ideas that are taught globally to enhance one's perception of the world at the global level. It is taught as a part of the curriculum around the world where education systems integrate multiple dimensions, perspectives and citizenships into lessons. It started in the 1960s and 1970s, through the initiatives of educationalists, non-government organizations and inter-governmental organizations. This global education systems works toward the implementation of an education system that is not based on subject knowledge and national culture (Standish 2014). This global education is the main factor in achieving ESD, and the ESD of any nation is dependent on the two factors that are defined below.

20.5.1.1 Brain Drain

Brain drain has been defined as the migration of skilled manpower within countries and abroad in search of better jobs and livelihood. There are many factors in brain drain and these factors can be categorized into two factors (Dinkovski and Markovska-Simoska 2014):

1) *Push factors*: The "push" factors are generally associated with the reasons of an economic nature such as low pay, political conditions, poor living standards, lack of funds, lack of work possibilities, complicated bureaucracy and the need to monitor the factors affecting economic conditions.
2) *Pull factors*: The "pull" factors are reasons related to career or research such as high salaries, high scholarships, high living standards, personal contacts and carrier advancement.

Sometimes, bad economic conditions are not the main reason to leave the motherland, but an inefficient system and corruption are the key factors in brain drain. When obedience is the criterion for advancement not competence, emigration is desirable in order to progress. With people who migrate to other lands to solve their housing problems, quality of life, security issues, etc. whether they stay permanently is decided on the basis of family relationships and contacts in other countries. Research has also discovered that brain drain has put a major development restraint on growth opportunities and has led to loss of investment as it has drained the human capital of the country that gave up large resources to nurture and raise them. The economy has lost its critical workforce that the country has financed in terms of education, and this cannot be compensated by other nations of the world. The government suffers a net loss as people decide to migrate when they start to reproduce after education completion (Dinkovski and Markovska-Simoska 2014).

Research has also revealed that human capital is placed above financial capital in importance because it is human capital that is converted to wealth. The real wealth of any country is its workforce and is not measured by money. A well-known axiom is that "money cannot teach but teachers can and money cannot treat but doctors can." Hence, from every angle human capital holds a stronger position than wealth. Brain drain affects both public and private consumption and investment. It places pressure on policies developed and on the ability to provide public services. It significantly decreases the revenue base by decreased incomes and taxes. This all affects government-provided services. Thus, brain drain results in reduced quality of life and public services. Migration is also responsible for decline in private investments and consumption as people take their savings with them when they leave the country.

20.5.1.2 Service Learning

Service learning is an educational approach that combines learning objectives with community service in order to provide a rational, advanced learning experience while meeting communal requirements. Service learning basically involves students in service projects to apply classroom learning for local organizations that can bring change in the community. The National Youth Leadership Council (NYLC) (Foong et al. 2017, p. 20) defines service learning as "a philosophy, pedagogy, and model for community development that is used as an instructional strategy to meet learning objectives and/or content standards." Jacoby (1996, p. 43) defines service learning as a "practice of experiential education in which learners involve themselves in actions that report human and community needs collected with structured prospects for consideration designed to achieve desired learning consequences."

Service learning is beyond classroom teaching and learning. It adds valuable academic skills to a student's development, builds their self-esteem and develops a sense of responsibility for decision making. For example, the Bozeman Youth Initiative, a Montana nonprofit organization that connects the youth of Bozeman to their community, started a project "From coffee to compost." Coffee grounds are rich in nitrogen and are needed to turn organic matter into compost. Thus, Bozeman's "Coffee to Compost" (C2C) program helped to save capital spent on coffee grounds to compost every year. The initiative taken was passed on to Montana State University in the Network of Environmentally Conscious Organizations (NECO).

20.5.2 Global Action Programmes on ESD

The UN DESD 2005–2014 (2005) emphasized the distribution of knowledge and education in all its forms as an essential element for achieving sustainability. In November 2014, the official follow-up to the DESD, UNESCO launched GAP for ESD with the objective of expanding the actions of ESD worldwide. Two important objectives of GAP on ESD are: (i) reorienting teaching and learning so that the whole world has the opportunity to gain the morals, skills and knowledge that allow them to contribute to sustainable development; and (ii) improving the role of education and learning in all significant agendas, programs and actions that support sustainable development. The GAP focuses on five areas and takes action for the development of these five areas: (i) progressing policy; (ii) changing learning and training atmospheres; (iii) structuring the capabilities of tutors and trainers; (iv) empowering and organizing youth; and (v) accelerating viable resolutions.

Due to its robust associations with sustainability, the GAP on ESD is responsible for an outstanding agenda for understanding the education system and types of education, training and public awareness initiatives to assist people to understand and implement elucidations for resolving the complex problems that arise due to ecological imbalance. UNESCO's work on CCE within the outlines of GAP on ESD objectives is to make education systems and education more central and visible parts of the global response to environment change; to support countries to integrate CCE into other education systems; and to support the world to achieve sustainability by switching to safer and greener activities.

Some other GAPs on ESD are described as follows.

20.5.2.1 Agenda 21

Agenda 21 is the GAP of the UN adopted to enhance sustainable development throughout the world. It was adopted in 1992 at the Earth Summit – UN Conference on Environment and Development that took place in Rio de Janeiro, Brazil. It is an action plan to achieve sustainability for the UN, other supporting organizations and many individual governments around the world so that the plan can be executed at global, national and local levels. The "21" in Agenda 21 denotes the sustainability of the twenty-first century. It addresses the critical issues faced by the global community such as protection and preservation of ecosystems, poverty elimination, hunger and health services, increasing population and illiteracy. This gives realistic solutions toward sustainability (Hurley 2009).

Agenda 21 focused on the need to eradicate poverty, which is one of the major problems in developing countries due to lack of resources and the ability to live a good-quality life. As per this agenda the developed nations have taken the responsibility to assist developing countries to achieve sustainable development. It asks government to integrate sustainability into national strategies with the help of non-governmental organizations and the public in the process. To resolve the issues local, national and international actors should work together toward international agreements with respect to the interests of all and protecting the integrity of the global environmental and developmental system.

Agenda 21 is a 350-page document, divided into 40 chapters, which are grouped into four sections (Hurley 2009):

Section 1: Social and economic dimensions. This is directed toward fighting poverty in developing nations, changing consumption patterns, promoting health, achieving a more sustainable population and making sustainable decisions.

Section 2: Conservation and management of resources for development. This section includes protection and preservation of the environment, conservation of biological diversity, pollution control, management of biotechnology and radioactive waste and preventing deforestation.

Section 3: Strengthening the role of major groups. This section defines the roles of children and youth, women, non-governmental organizations, local authorities, business and corporate sectors, employees and workers, and the role of the indigenous population, their communities and farmers of the country.

Section 4: Means of implementation. This section includes science, education, international organizations, technology and its transfer and financial mechanisms.

Changes have been made to Agenda 21 as the work was progressed worldwide in order to achieve its objectives. Later, in 1997 the UN General Assembly held a special meeting to assess the status of Agenda 21 (Rio +5). The data produced the in meeting showed an uneven development and recognized key trends including increased globalization, broader inequalities in income and worsening of the global environment. Based on these observations a new General Assembly Resolution (S-19/2) promised to work on the coming challenges toward sustainability. After 1997, The Johannesburg Plan of Implementation (Rio + 10) was agreed to at the Earth Summit (2002). The World Summit on Sustainable Development (WSSD) agreed to the UN commitment to "complete implementation of Agenda 21" with achievement of the MDGs and other international treaties and settlements.

In 2002, the first World Public Meeting on Culture (Culture 21) was held in Porto Alegre, Brazil. This meeting came up with objectives and ideas to establish guidelines to implement cultural strategies and policies. This is the same agenda for culture sustainability as Agenda 21 was for environment sustainability. Various programs were carried out at local levels to implement the ideas of this meeting.

In 2012, at the United Nation's Conference on Sustainable Development (UNCSD) the members repeated their commitment to Agenda 21 in a developed document called "The Future We Want." With the most recent meeting in 2015, Agenda 2030 agreed the UN SDGs. These work on the objectives of Agenda 21 and commit to work for sustainability based on the goals of Agenda 21. The objectives of Agenda 2030 were made to cover all the issues of people, planet, prosperity, peace and partnership (Holden et al. 2017).

For sustainable development the main authority is given to the Commission on Sustainable Development, which acts as a high-level forum and has acted as a preparative committee for assemblies and sessions on the implementation of Agenda 21. The UN division for sustainability acts as the Secretariat to the Commission on Sustainable Development and works within the context of Agenda 21 (Svanström et al. 2008). The implementation of Agenda 21 objectives by member nations varies and remains controlled. The implementation of Agenda 21 involves action at international, national, regional and local levels. Some national and local authorities implement the goals of this agenda as the programs known as "Local Agenda 21 (LA21)." For example, in the Philippines the national plan of Agenda 21 was known as "Philippines Agenda 21 (PA21)" with the same context as Agenda 21.

20.5.2.2 Global Citizenship Education

GCED is a framing model that defines the outlines to develop knowledge, skills, values and attitudes of learners to secure a more sustainable world. It recognizes the relevance of education in understanding and resolving global issues of social, economic and environmental dimensions. It also defines the role of education in developing international cooperation and promotes social transformation. GCED uses multidimensional approaches, concepts, methods and theories that are already implemented in other fields and areas of education. Its main aim is to overlap different agendas to attain a peaceful and sustainable world (Scheunpflug and Asbrand 2006).

GCED is basically composed of three core conceptual dimensions that are common to various definitions and interpretations of GECD. These core conceptual dimensions work along the outlines defined by UNESCO, which serves as the basis for defining GCED goals, learning objectives and experiences, as well as for assessment and evaluating learning. These core intangible dimensions are based on three domains of learning:

1) *Cognitive*: This is to obtain knowledge, understanding and critical thinking about global issues and the interdependency of different countries and inhabitants.
2) *Socio-emotional*: This dimension defines a sense of belonging to a common humanity, sharing beliefs and duties, compassion, harmony and respect for every culture and diversity.
3) *Behavioral*: This dimension teaches us to act efficiently and sensibly at local, national and global levels for a more peaceful and sustainable world. GCED has the same vision as ESD – both focus on empowering learners of all ages to become proactive contributors

to a more peaceful, accepting, comprehensive and sustainable world (Scheunpflug and Asbrand 2006). Both ESD and GCED focus on (Derricott 2014):

- actions, modification and revolution;
- the content and outcome of learning based on the process followed in the type of environment;
- acquiring standards and attitudes relevant for assessing global challenges;
- developing skills for cooperation, collaboration, communication and rational thinking.

GCED working along the guidelines of ESD helps learners to understand the interconnected and interdependent world in which learners live and the complexities of the global challenges. It helps to develop knowledge, skills, attitudes and values to solve the global issues responsibly and effectively (Derricott 2014). GCED responds to:

- globalization by increasing the thought of public education to global society;
- adopt the moral values of peace and human rights education;
- merge local and global issues.

GECD combines these three pedagogical fields in terms of political participation on global scale. To support and implement this initiative, various guidelines have been developed in 2013 based on series of international conferences.

20.5.2.3 UNESCO Prize

To initiate the implementation of projects and policies taken for development of education sustainability, UNESCO invited various governments and non-governmental organizations with certified connections to UNESCO to nominate three organizations or foundations working in the field of EfS or ESD (Azeiteiro et al. 2015).

The UNESCO-Japan Prize on Education for Sustainable Development gives a reward of US$150 000 to three individuals or organizations for projects favoring ESD. This award was established in 2014 by UNESCO at the initiative of the government of Japan as part of the UNESCO GAP on ESD. The selection of winners is made by an international jury at UNESCO headquarters in Paris and the first award was given in 2015.

There are certain criteria in order to be eligible for the UNESCO-Japan Prize on Education for Sustainable Development. The official criteria for assessment of individuals or organizations are: (i) transformation – use of education to support sustainability leading to individual and social change; (ii) integration – development of social, economy and environment dimensions of sustainability; and (iii) innovation – should have innovative and imaginative approach for education sustainability.

20.6 Sustainability as Human Resource Development

Since 1950, human activities have increased eightfold globally. This exponential growth has depleted the natural resource base, and has led to environment degradation and social imbalance throughout the world. These consequences of rapid economic expansion have raised questions about the use of resources with minimal wastage through human activities. Indeed, the recent focus on the sustainability agenda has been given to HR, which represents the human and social consequences of global economic sustainability. With increasing needs, viable production and consumption patterns have been the topics of civic policy discussion and scientific research throughout the last quarter of the twentieth century. Current social, political and economic trends have driven political–economic associations to respond in concrete ways to the numerous challenges under the authority of sustainability (Cohen et al. 2010).

Sustainability requires deep and systemic change in organizations to induce HRD. The historic mission of HRD uplifts the human potential to improve organizational problem solving and performance. It is important that HRD should define its own conceptual models to certify theories and practices that are associated with evolving demands and problems of the sustainability project and policies. The risk society is a sociological discourse that informs sustainability policies and projects. Its basic principle is a new form of reflective practice known as reflexivity, which has emerged in society as a response to the risks that are associated with the problems that threaten ecology and global sustainability. Reflexivity is a solution to the deep contradictions in the logical thinking and values of modern, industrial people upholding a great potential for deep organizational changes in the communities. The reflexive praxis has consequences for HRD and its role in sustainable development (Scully-Russ 2012).

There are three frameworks that are responsible for shaping the ideology of sustainable development, including environment, social and economic sustainability. Sustainability is dependent on many themes, and HRD is one of the themes affecting global development. Studies have suggested three models of HRD theories and practices that are the basis of frameworks of sustainability (Scully-Russ 2012): (i) strategic HRD; (ii) critical human resource development (CHRD); and (iii) holistic human resource development (HHRD), which are briefly described as follows.

20.6.1 Eco-Modernism and Strategic HRD

Strategic HRD is defined as a policy framework that encourages a less measurable material- and energy-intensive economy around the globe. This new framework is responsible for providing incentives to stakeholders for global development and switching to new technologies for production and consumption. These new technologies are designed in such a way so that a pattern is followed to minimize the material use and energy requirements of economic growth. It is dependent on modern beliefs in equal justice to all, which can be achieved through a free-market economy and advances in science and technology to provide solutions to problems. Scientific progress helps people to grow their own economy and promote social welfare. The strategic model is only feasible when all nations work based on similar policies, social standards and technology (Scully-Russ 2012).

This framework is a large-scale adoption, which makes its application difficult as global economic–political structures limit the deep partnership required to make this model work on a global scale. For this framework to be successful at the global level, corporate incentives are required to modify present lifestyle and consumption patterns to coordinate with the world economic activities of suppliers and consumers in the international market. The main objective of this framework is to preserve, protect, recycle and regenerate environmental resources. New global projects or policies and regulations to minimize destructive industrial activities and new investments in HRD and social welfare ensure equity and uphold a more cooperative global economic–political system. This agenda requires an additional institutional framework that can produce and influence the science, technology and knowledge needed to efficiently achieve a complex global political economy.

Strategic HRD is composed of three basic frameworks: (i) a framework of standards and rational prototypes to encourage a carbon-neutral economy and less quantifiable way of life; (ii) being centered on judgment of modernity – free-market thought and assurance in the rationality of science and technology; and (iii) finding a global political–economic agenda based on equality, in which all countries implement ecological and social standards that ensure the impartial distribution of wealth and risk throughout the world.

20.6.2 Critical HRD

CHRD works for the cause of the life world in organizations. It examines the power of relationships and how these corporate relationships regulate the corporate sector and corporate goals. It works to restore the central role of people and organizations to create a better workplace for humans. CHRD achieves its objective by encouraging the capacity for critical reflection so that people understand the deep assumptions that define their work and structure in corporate relationships. CHRD puts emphasis on the institutional, cultural–historical context of the working environment and the significance of this context in promoting relationships in the workplace. This enables removal of tensions, complexities and conflicts present in the beliefs and logics of the political economy, which enhances learning and social development. Corporate social responsibility is the key factor in achieving CHRD. Corporate social responsibility is a global network of informal groups, social activities and corporate people in the global political economy that share and work on a common project or policies to achieve HRD and sustainability (Scully-Russ 2012).

The framework of CHRD is composed of: (i) detailed analysis of the technical optimism of innovation and its free-market thought; (ii) new rights of ownership models that are responsible for development of new social institutions for the community; and (iii) converting capital from private funds to an open resource that can be used by the public to end poverty and preserve the environment.

20.6.3 Holistic HRD

HRD is a global development process based on personal and organizational opportunities that employees seek from employers and that also makes a historic change for working community.. Modernization is affecting globalization, which is changing the way people communicate and participate in the world. HHRD gives new conditions to survive as an individual and community. As per recent studies (Lee 2010), HHRD is to co-create new perceptions to understand humanity and human evolution. Studies do not give any universal definition of HHRD; rather it is defined as the behavior and emotions developed in humans with their own development. HHRD is a mutual constructive process that helps people to integrate local expressions with global ideologies to accommodate local needs, knowledge and skills at an international level. Thus, HHRD is not a debatable topic but is a co-constructed one in the context of the everyday life of humans. This will affect people's beliefs and morality, which form the basis of development of human roots (Scully-Russ 2012). The sustainable framework of HHRD is composed of: (i) a social assembly, ontology and epistemology that make symbiotic eco-consciousness a material certainty; (ii) striving to reestablish a healthy environment rich in biodiversity; and (iii) holding on to humanity's learning nature to form new, more justifiable conditions for the future growth of life.

20.7 HRD Indicators

Multinational companies and organizations have implemented tools to measure their performance to hold their position in the market and to stand out in the competitive environment of the business world. These tools are used to measure the performance of an organization and are known as HRD indicators, with an increased emphasis on HR measurement. It is said that HR measures should be impact- rather than activity-oriented, forward looking than backward looking, and should focus on the complete HR system of an organization rather than individuals (Dehli and Krogstad 2016).

Toulson and Dawe (Zamecnik 2016) recognized three difficulties in quantifying HR; i.e. lack of HR experience, accuracy and complications in measurement. There are various tools to measure HR capital. Based on research, Srimannarayana (2010) highlighted invented methods to measure HR capital:

- the multiple constituency approach;
- HR accounting system;
- 16 categories to be measured in HR;
- return of investment methodology;
- HR practices relating to the Balanced Scorecard (BSC) through productivity, people and process indicators;
- methodology of behavioral costing to measure the financial impact of HR activities;
- BSC in which strategic goals, i.e. operational objectives, are transferred into four perspectives.

All the established indicators to measure HR capital are in direct relation to HR activities. Many companies and government organizations have the following eight activities to measure HR capital:

- economic and financial indicators;
- structural indicators;
- planning and recruitment;
- compensation;
- training and development;
- performance appraisal;
- climate indicators;
- professional risk indicators.

20.8 Change in HRD for Sustainability

Companies or industries can succeed in a competitive environment long-term only if they recruit and inspire people who can respond to and outline the future challenges. These are the personalities with the capacity to create new policies and viable advantage from the opportunities created by fluctuating markets, with the aspiration to learn from trades, buyers, merchants and associates, and who possess the ability to build and encourage permanent, active and effective partnerships.

20.8.1 HR Environment

To maintain the HR environment the strategies and objectives adopted must be achieved by the HR team within defined budget, skills, experience and interests of its employees. HR strategy must respond to the broader environment of the company. The management and authorities should motivate people, and their work should be reflected in the growth of the company. The HR environment should be reflective and inspiring for the team and employees. It should help in giving sustainable returns to the investors and should address customer needs. The team should be able to identify and respond to emerging social needs. It should response positively to governmental and regulatory expectations to influence the public policy agenda. The HR team should have their own priorities that must reflect the objectives of and approaches followed by the company.

20.8.2 Strategy Planning

As a resource for strategy planning it is necessary to improve strategic orientation of staff and other resources. It is essential to understand the methodology of HR planning process and its workings. At its first step strategy planning ensures adequate employees to meet company goals, matching the right people with the best skills at the required time. To conduct strategy planning it is important to know the current position of the company in the market and where it needs to go to remain flexible (Nabi et al. 2017). It is dependent on current and future needs of the company, but there is a basic structure to follow in strategy planning:

1) Assess present HR capacity.
2) Forecast HR requirements, including demand forecasting, supply forecasting, matching demand and supply.
3) Develop talent strategies, including recruitment, selection, hiring, training and development, employee remuneration and benefits administration, performance management, employee relations.
4) Review and evaluate.

This process can be used by business leaders to plan strategic development process with the HRD to support sustainability. This can provide a resource to build a bridge between sustainable development and HR functioning. This creates opportunities for the promotion of sustainability (Nabi et al. 2017).

20.8.3 Training Materials

For the best outcome and growth, it is necessary for the employees to be provided with company training materials. This provides supporting facts or evidence to the staff. The sensible reproduction of material is acknowledged by all parties, management and staff, but the source manuscript is also referenced in all cases. The aim of training materials is to increase the knowledge and abilities of staff in performing duties and increasing their potential (Rafiei and Davari 2015). To increase the skills of the staff members, common methods are as follows:

- job training;
- education;
- learning in a similar environment;
- working things out with theoretical education;
- distance learning;
- courses outside the organization;
- introduction to the organization.

20.8.4 Staff Development Programs

Staff development programs provide organizational development services to every employee to address specific needs or concerns. These development programs increase staff efficiency, resulting in a positive impact on the growth of an organization (Rafiei and Davari 2015). Steps for staff development programs include:

- Assess educational needs;
- Set priorities;
- Develop general objectives;
- Determine resources to attain the goals;
- Develop master calendar for the year objectives;
- Maintain staff development records;
- Evaluate developmental programs.

20.9 HRD Resources for Sustainable Development

HRD is always concerned with learning procedures and enhancement of humans and ultimately of the entire world. Economic, political, social, environmental, cultural and spiritual development is required for the development of humans to achieve sustainability. Humans are the most invaluable resource of all but this is often neglected. The development of a nation always defines the development of its HR. The HRD theme of sustainability is examined by the products of development as opposed to the means of development. HRD always plays a major role in natural resource preservation, management of capital resources, transformation of technology, information management, system management and infrastructure resources. HRD can be measured in terms of indices including health, education, food consumption, income, employment, poverty, population, environment deterioration, and social and political issues. These areas are first determined under other themes in the subarea rather than at the global level as it is easy to determine the sociological, political and economic issues of HRD at lower levels (Garavan and McGuire 2010).

We are constantly challenged by the continuation of inequalities between and within nations, poverty, hunger, health, illiteracy and the degradation of the environment on which HRD depends. Despite significant development and growth in certain parts of the world, some developing nations are still fighting to attain minimal sustainability. The prime goal of HRD is to eliminate poverty as poverty is one of the elemental issues of society because it deprives people of choices and significantly reduces the quality of life. It limits the participation of humans in political and developmental processes that are associated with unemployment, less productivity, poor health, housing and security. HRD is important as without HRD in societies, the continued ecological degradation of natural resources will occur as large numbers of people will lack the skills and knowledge to become productive constituents and contributors to world progress (Garavan and McGuire 2010).

20.10 Conclusion

Since the concept of sustainability first emerged in 1987 there have been several interpretations of how to approach it. Sustainability is quantified in terms of growth indicators, which are also considered as consumption growth over time. This was accepted long ago by governing bodies to identify the best concepts and methods for supporting native, national and global organizations to achieve sustainable development. Sustainability comprises many dimensions and is most affected by the education and HR of any nation or at the global level. Sustainability is seen in an educated society as human welfare growth in the long run. Therefore, experts consider that sustainability can be defined based on education and HR. Development is associated with a better future in terms of social welfare, quality of life, learning, etc.

EfS is an educational method that aims to improve learning in every domain of society, equipping people with the beliefs and inspiration to act for sustainability in their lives. It develops the knowledge, innovative thinking, skills, values, beliefs and behavior that are essential for people to stand in line to contribute to achieving sustainability. It enables people and their societies to become engaged with the world to attain sustainability. SE is future-oriented, focusing on preservation and protection of the natural resource base and ecosystems to build a more sustainable world. It enhances growth in the areas of social, economic and ecological development and helps to understand their integration and interdependence.

HR practices have a strong influence on the sustainability of a company. Depending on the structure of the corporation it is likely to use less space and energy for progressing goals. Companies that focus on sustainable development are also favored by government and the general public. HR sets out the approaches for the implementation of more sustainable environments. Since recruitment, selection and training are integral functions of HR, HR professionals make sure that sustainability is ingrained into the ethos of new employees and through developmental programs in existing employees. As well as teaching about sustainable development, HR can conduct various developmental programs, providing a comfort zone to its employees and enhancing HRD.

Acknowledgments

We are thankful to Lalit R. Kumar for his supportive guidance in moderating earlier drafts and, in that way, improving this chapter meaningfully.

References

Apronti, P.T., Osamu, S., Otsuki, K., and Kranjac-Berisavljevic, G. (2015). Education for disaster risk reduction (DRR): linking theory with practice in Ghana's basic schools. *Sustainability* 7 (7): 9160–9186.

Azeiteiro, U.M., Bacelar-Nicolau, P., Caetano, F.J., and Caeiro, S. (2015). Education for sustainable development through e-learning in higher education: experiences from Portugal. *Journal of Cleaner Production* 106: 308–319.

Cohen, M.J., Brown, H.S., and Vergragt, P.J. (2010). Individual consumption and systemic societal transformation: introduction to the special issue. *Sustainability: Science, Practice, & Policy* 6 (2): 6–12.

Combes, B.P. (2005). The United Nations decade of education for sustainable development (2005–2014): learning to live together sustainably. *Applied Environmental Education and Communication* 4 (3): 215–219.

Dehli, K. and Krogstad, G.-A. (2016). Training, faking and recruitment strategy: a critical review of the job interview's value as a strategigic recruitment tool. Nord Universitet. https://nordopen.nord.no/nord-xmlui/bitstream/handle/11250/2409047/Krogstad.pdf?sequence=1&isAllowed=y (accessed November 2019).

Derricott, R. (2014). *Citizenship for the 21st Century: An International Perspective on Education*. Abingdon, UK: Routledge.

Dinkovski, V. and Markovska-Simoska, S. (2014) Brain drain as a function of sustainable development in the Republic of Macedonia. https://scindeks-clanci.ceon.rs/data/pdf/0350-0373/2018/0350-03731801173D.pdf (accessed November 2019).

Dubeck, M.M. and Gove, A. (2015). The early grade reading assessment (EGRA): its theoretical foundation, purpose, and limitations. *International Journal of Educational Development* 40: 315–322.

Foong, S.C., Kong, K.L., Looi, S.Y., Phua, C.S. and Wong, J.M. (2017). Assessing service-learning courses in a Malaysian higher learning institution: a study of Universiti Tunku Abdul Rahman (UTAR). Thesis. UTAR. http://eprints.utar.edu.my/2617 (accessed November 2019).

Fredericks, S.E. (2012). Justice in sustainability indicators and indexes. *International Journal of Sustainable Development & World Ecology* 19 (6): 490–499.

Garavan, T.N. and McGuire, D. (2010). Human resource development and society: human resource development's role in embedding corporate social responsibility, sustainability, and ethics in organizations. *Advances in Developing Human Resources* 12 (5): 487–507.

Hasslöf, H. and Malmberg, C. (2015). Critical thinking as room for subjectification in Education for Sustainable Development. *Environmental Education Research* 21 (2): 239–255.

Heinrichs, H. (2013). Sharing economy: a potential new pathway to sustainability. *Gaia* 22 (4): 228.

Holden, E., Linnerud, K., and Banister, D. (2017). The imperatives of sustainable development. *Sustainable Development* 25 (3): 213–226.

Hurley, J. (2009). Sustainable or status-quo: investigating sustainability assessment of residential estate development. *Proceedings of the State of Australian Cities National Conference 2009*. http://soac.fbe.unsw.edu.au/2009/PDF/Hurley%20Joe.pdf (accessed November 2019).

Jacoby, B. (1996). *Service-Learning in Higher Education: Concepts and Practices, The Jossey-Bass Higher and Adult Education Series*. Washington, DC: ERIC.

Latchem, C. (2017). *Using ICTs and Blended Learning in Transforming Technical and Vocational Education and Training*. Paris: UNESCO Publishing.

Lee, M. (2010). Shifting boundaries: the role of HRD in a changing world. *Advances in Developing Human Resources* 12 (5): 524–535.

Lewis, E., Mansfield, C., and Baudains, C. (2014). Ten tonne plan: education for sustainability from a whole systems thinking perspective. *Applied Environmental Education & Communication* 13 (2): 128–141.

Liu, Q.-Q., Yu, M., and Wang, X.-L. (2015). Poverty reduction within the framework of SDGs and Post-2015 Development Agenda. *Advances in Climate Change Research* 6 (1): 67–73.

Marope, P.T.M., Chakroun, B., and Holmes, K. (2015). *Unleashing the Potential: Transforming Technical and Vocational Education and Training*. Paris: UNESCO Publishing.

Michalos, A.C. (2017). Education, happiness and wellbeing. In: *Connecting the Quality of Life Theory to Health, Well-Being and Education*, 277–299. New York: Springer.

Nabi, M.N., Ahmed, A.A.T., and Rahman, M.S. (2017). The empirical study on human resource management practices with special reference to job satisfaction and employee turnover at Investment Corporation of Bangladesh. *Human Resource Management Research* 7 (1): 54–64.

Nichols, A., Maynard, V., Goodman, B., and Richardson, J. (2009). Health, climate change and sustainability: a systematic review and thematic analysis of the literature. *Environmental Health Insights* 3: 63.

OECD (2017). *OECD Economic Surveys: Luxembourg 2017*, 51. Paris: OECD.

Paul, P.K. and Mehera, C. (2016). Impacts of education on sustainable development: a micro study in Burdwan District of West Bengal, India. *American Journal of Educational Research* 4 (7): 551–555.

Pearce, D.W., Markandya, A., and Barbier, E. (1989). *Blueprint for a Green Economy*. London: Earthscan.

Peña-López, I. (2015). *Rethinking Education. Towards a Global Common Good?* Paris: UNESCO.

Rafiei, N. and Davari, F. (2015). The role of human resources management on enhancing the teaching skills of faculty members. *Materia Socio-medica* 27 (1): 35.

Rakodi, C. (2014). *Urban Livelihoods: A People-Centred Approach to Reducing Poverty*. Abingdon, UK: Routledge.

Scheunpflug, A. and Asbrand, B. (2006). Global education and education for sustainability. *Environmental Education Research* 12 (1): 33–46.

Scully-Russ, E. (2012). Human resource development and sustainability: beyond sustainable organizations. *Human Resource Development International* 15 (4): 399–415.

Srimannarayana, M. (2010). Status of HR measurement in India. *Vision* 14 (4): 295–307.

Standish, A. (2014). What is global education and where is it taking us? *Curriculum Journal* 25 (2): 166–186.

Stanszus, L., Fischer, D., Böhme, T. et al. (2017). Education for sustainable consumption through mindfulness training: development of a consumption-specific intervention. *Journal of Teacher Education for Sustainability* 19 (1): 5–21.

Svanström, M., Lozano-García, F.J., and Rowe, D. (2008). Learning outcomes for sustainable development in higher education. *International Journal of Sustainability in Higher Education* 9 (3): 339–351.

Tanner, T. and Allouche, J. (2011). Towards a new political economy of climate change and development. *IDS Bulletin* 42 (3): 1–14.

Tilbury, D. and Wortman, D. (2004). *Engaging People in Sustainability*. Gland, Switzerland: IUCN.

UN General Assembly (2015). Draft outcome document of the United Nations Summit for the Adoption of the Post-2015 Development Agenda. https://digitallibrary.un.org/record/800852?ln=en (accessed November 2019).

Yasunaga, M. (2014). Non-formal education as a means to meet learning needs of out-of-school children and adolescents. Background Paper prepared for Fixing the Broken Promise of Education for All: Findings from the Global Initiative on Out-of-School Children. Montreal, Canada: UNESCO Institute for Statistics.

Part III

Applications

21

Climate Change Adaptation for Sustainable Management of Water in India: Issues and Challenges

Adani Azhoni

Department of Civil Engineering, National Institute of Technology Karnataka, Surathkal, India

21.1 Introduction

Climate change is arguably the biggest challenge the world faces today (IPCC 2014a). The increase in global mean temperature will have many cascading effects on planet earth including increasing the melting of the icecaps, leading to sea level rise (Raper 2005) and displacement of people from low-lying coastal areas, in addition to sea water intrusion into freshwater aquifers. Climate change has the potential to exacerbate the distorted distribution of water both temporally and spatially (IPCC 2014a; Seneviratne et al. 2012). Moreover, glacier-fed river systems, such as those arising from the Himalayas, will undergo drastic changes in the seasonal river discharges, leading to floods and droughts in different seasons (Xu et al. 2009). The retreating glaciers (Seneviratne et al. 2012) and the changing monsoon cycle (Hassan et al. 2015; Singh 2013) in the Indian subcontinent, for example, will adversely affect freshwater availability and quality for about a billion people living in this region.

The rise in global temperatures will also trigger increased water demand (Wang et al. 2014) due to increased evapotranspiration leading to water shortages. Moreover, the precipitation patterns across the world are projected to change, with wet areas becoming even more wet and dry areas getting drier leading to floods and droughts simultaneously in different locations and different seasons (Barnett et al. 2005). In addition, this changing climate is likely to affect the availability of groundwater due to decreased recharge and increased demand (Holman 2006). Climate change also has the potential to impact water quality (Whitehead et al. 2009) and not just the availability and quantity of available water (Arnell 2004; Kundzewicz et al. 2008). It will increase the sediments during floods and increase the concentration of contaminants during the dry season (Whitehead et al. 2009). Therefore, water is the medium through which the immediate and direct impacts of climate change are going to be experienced, although it has the potential to impact almost all areas of life.

With the increasing realization of the potential impacts of climate change, adaptation is increasingly being recognized as a key priority in the climate discourse, although mitigation

Sustainability: Fundamentals and Applications, First Edition.
Edited by Rao Y. Surampalli, Tian C. Zhang, Manish Kumar Goyal, Satinder K. Brar, and R.D. Tyagi

is also not to be neglected (UNFCCC 2015). It is being emphasized that mitigation need to be complemented by adaptation, as mitigation alone is insufficient to reduce the risks due to the emissions already released and the efforts to contain greenhouse gases (GHGs) emissions are becoming more uncertain (Bosello et al. 2013; Jones et al. 2007; Mata and Budhooram 2007; Vignola et al. 2012). The recent Paris Agreement (UNFCCC 2015) by 197 countries under the leadership of the United Nations Framework Convention on Climate Change (UNFCCC) reinforces the urgency of adaptation while mitigation is also given additional thrust (Simonet and Fatorić 2015; UNFCCC 2015). It calls for strengthening mechanisms for enhanced transparency on assessing adaptation, emphasizes the need for bolstering adaptation commitments from state actors and recognizes the multilevel nature of adaptation governance (Lesnikowski et al. 2017).

The capacity of a system to conform to the effects and consequences of the changing climate and take advantage of the new opportunities, reduce its vulnerability and increase resilience are pivotal due to the impending climate change impacts (IPCC 2014b). Individuals and communities can adapt to climate change and are already adapting in certain circumstances (IPCC 2014a). However, individuals and communities can adapt only to a certain extent without the involvement of governmental as well as non-governmental institutions (Butler and Adamowski 2015). Increasingly, it is being realized that institutions, large or small, governmental or non-governmental, also need to adapt to the changing climate (Berkhout 2012; Inderberg 2011). It is now being widely recognized that public utility service institutions such as Municipal Corporations that deliver public water supplies, electricity grid companies and infrastructure development institutions also need to adapt to climate change (Fünfgeld 2015; Porter et al. 2015; Shi et al. 2015) in order to maintain the sustainability of goods and services being provided by them. Adaptation can be in various forms, such as anticipatory (Kuruppu and Willie 2014), autonomous or planned adaptation (Pittock 2009) which are further discussed in Section 21.4. Adaptation for sustainable management of water requires going beyond structural supply measures (Stakhiv 2011) and considering forecasting/warning systems, demand management and related behavioral changes (Gifford et al. 2011), economic and fiscal instruments (Tompkins et al. 2010), legislation and institutional change (Crabbé and Robin 2006; The Royal Commission on Environmental Pollution 2010).

This chapter reviews some of these challenges and issues for sustainable water management in the context of a changing climate in order to enlarge the perspectives for identifying adaptation approaches by examining adaptation initiatives from two states for sustainable water management in India. Section 21.2 highlights some key challenges of climate change for sustainable water management in India followed by a brief elaboration of the various adaptation approaches in Section 21.3, which is drawn from the literature and highlights the importance of adaptation at various levels by multiple actors (Section 21.3.5). Section 21.4 then situates the role of water management institutions in general, along with other actors and agents, in climate change adaptation, which is followed by a brief description of the water management institutions in India in Section 21.5 that sets the context. Section 21.6 bring in the context of India and highlights the recent initiatives for addressing the challenges of climate change at the national level followed by Section 21.7, which draws out two examples of how local institutions in two Himalayan states of India are trying to maintain the sustainable supply of water in the context of depleting sources.

Finally, discussions and conclusions are drawn in Section 21.8 by taking into consideration the challenges and issues facing Indian water institutions, and suggestions are made for managing water sustainably.

21.2 Climate Change Challenges to India's Water Management

The Indian subcontinent of more than a billion people is considered to be one of the places most vulnerable to changing climate (Burton and May 2004; Mall et al. 2006; Naswa and Garg 2012; Sanghi and Mendelsohn 2008). Providing adequate water supply, protecting livelihoods from floods and droughts, and tapping the rich water resources into beneficial irrigation for food security in India continues to be a challenge, and climate change potently increases these challenges. Besides the direct impacts through increased flood and drought situations leading to food security issues, climate change has the potential to increase migration and societal conflicts and other policy areas (Bates et al. 2008). Emphasizing the seriousness of the climate change threats, the then Indian Minister for Environment and Forest stressed that "no country in the world is as vulnerable, on so many dimensions, to climate change as India" (MEF 2009). Different regions in the country experience either water stress and scarcity or floods in different seasons or at different locations (MWR 2008). Climate change impacts will directly and seriously affect more than 70% of India's population who are solely dependent on agriculture and allied activities (Central Water Commission 2010).

The diverse geography from tropical forest in the east to arid desert in the west, from snow covered mountains of the north to the coastal plains of the south, and from the dry rocky plateau of the southern central to flat plains in the north of central gives rise to a wealth of diverse biological and cultural hotspots in India. The huge variation in the rainfall patterns with extreme annual rainfalls of more than 12 000 mm in the north-eastern region to less than 13 mm in the western Thar Desert (Jain et al. 2013) makes water management

Box 21.1 Prime Minister of India's Remark on the Challenge of Climate Change for India

... water, or the lack of it, could well become the limiting factor to our social and economic growth in the future. With around 18% of the world's population but only 4% of its usable fresh water, India already faces a scarcity of water, which is a vital and stressed natural resource. Climate change could further aggravate the distortions in water availability in our country. Receding glaciers would negatively impact flows in our major rivers and pose a major new threat to the welfare of millions of our people.

Dr Manmohan Singh, the then Prime Minister of India in his opening statement to the National Water Resources Council while adopting the National Water Policy 2012. Available at http://pmindia.nic.in/speech-details.php?nodeid=1266.

in India a huge challenge. To add to all these complexities is the huge variation in the seasonal rainfalls, with about 75% of annual rainfall occurring within June–September (Central Water Commission 2005) and changing monsoon patterns (Singh 2013), which makes it all the more difficult to address water problems. Climate change is likely to exacerbate all these challenges (Mall et al. 2006). India, being a developing economy, faces greater challenges because of not only greater climate change impacts but also its presumably lower adaptive capacity (Mertz et al. 2009).

21.3 Adaptation Approaches

The growing and convincing body of literature points to the need for earnest action on adaptation (IPCC 2014a). While not neglecting the ongoing efforts of mitigation, adaptation actions are urgently required (Simonet and Fatorić 2015) to complement mitigation (Füssel 2007) as the GHG emissions already in the atmosphere are likely to continue to have an impact on the global climate and the international agreements on reducing emissions remain illusory (Bhave et al. 2014a). Most of the advances in climate change adaptation research in the recent past (Preston et al. 2015) focus primarily on the impacts and vulnerability (Moser and Boykoff 2013) of ecosystems, which is needed for adaptation action. However, much less actual implementation of adaptation has been reported. Adaptation is particularly challenging for developing countries, which are struggling to provide basic infrastructures for the current needs of the population, and much of their resources are devoted toward addressing the challenges due to climate variability and the related disasters (Ford et al. 2014; Krysanova et al. 2010; Nyamwanza and Kujinga 2016), and this has to compete with other developmental priorities. Yet most studies on adaptation to date are focused on high-income countries (Berrang-Ford et al. 2011; Preston et al. 2015).

The Intergovernmental Panel on Climate Change (IPCC 2007, p. 6) defined adaptation as "adjustment in natural or human systems in response to actual or expected climatic stimuli or their effects, which moderates harm or exploits beneficial opportunities" and distinguished into anticipatory, autonomous and planned adaptation depending on how and when adaptation is initiated. These various approaches to adaptation are listed and elaborated in the next section.

21.3.1 Planned Adaptation and Autonomous Adaptation

Planned adaptation involves deliberate policy decisions based on the available information by using the present scenario to project to the future for evaluating the suitability of current and planned practices, policies and infrastructure (Füssel 2007). Deliberate alterations in the way water is managed currently and changes in the design of water infrastructure, such as including a climate change factor allowance in the calculated discharge, for example, constitutes planned adaptation for water management. However, perceptible actions in the water sector that can be considered as examples of adaptation solely aimed at adapting to a changing climate are rare or at least they are often not named so (Moser and Boykoff 2013). One of the reasons why such adaptation actions are not perceptible as climate change adaptation is that there are many other drivers affecting the decisions in making investments and the strategies. Moreover, climate change is often seen as a challenge for the future because of the focus on the current issues with the compounded complexity of the uncertainty in the projections of the likely hydrological changes (Bates et al. 2008). Planned adaptation is also used synonymously with anticipatory adaptation (Preston et al. 2013). Autonomous adaptation, on the other hand, is often triggered spontaneously by changes in the natural and human systems (Bhave et al. 2014a) and consists of "those that do not constitute a conscious response to climate stimuli, but result from changes to meet altered demands, objectives and expectations which, whilst not deliberately designed to cope with climate change, may lessen the consequences of that change" (Bates et al. 2008, p. 48).

21.3.2 Top-Down and Bottom-Up Adaptation

Adaptation approaches are also distinguished as top-down or bottom-up (Dessai and Hulme 2004) depending on how the risk and vulnerability is identified (Garcia et al. 2014). Top-down and bottom-up approaches of adaptation are also differentiated based on who initiates the adaptation agenda: national governments or local authorities/communities (Butler et al. 2015). In a top-down approach, climate risk and vulnerability is assessed based on projections from global climate models to identify the potential impacts on a particular system, such as the water sector in a particular region, based on expected scenarios to plan the adaptation strategies (Garcia et al. 2014). The vulnerability of water systems is then assessed based on the climate scenario's information (Garcia et al. 2014). The global climate models or general circulation models are then scaled down to the regional level at which adaptation is to be planned. With this approach, technical data and mathematical models drive the adaptation decisions and climate change impact assessment typically follows a mechanistic approach by forcing the hydrological data from the climate scenario (Garcia et al. 2014). Generally, national adaptation planning takes this approach, in which government institutions at the national level make the adaptation plan, based on climate change modeling and impacts projections (Butler et al. 2015), and is mostly technocratic cost–benefit driven. Top-down approaches are limited due to (i) the uncertainty in climate scenarios; (ii) smaller and non-specialized institutions may not have the capability to replicate these methodologies at their appropriate scale (Whitehead 2009); (iii) it obscures the socio-economic and political context in which adaptation has to occur (Mastrandrea et al. 2010; Preston et al. 2015); (iv) detailed local historical data required for feeding into the model may not be available (Whitehead 2009); and (v) the dynamic nature of adaptive capacity of a system means the vulnerability will vary with time (Wise et al. 2014). In addition, top-down scenario-driven adaptation approaches are criticized for ignoring the local context that shapes the development of adaptation responses (Mastrandrea et al. 2010; Preston et al. 2015).

In contrast to top-down, a bottom-up approach starts by identifying the social vulnerability based on the adaptive capacity of the system, which is dependent on the technological, economic, infrastructural and knowledge resources available within the system (Dessai and Hulme 2004). From this perspective, Dessai and Hulme (2004) suggest that systems that are able to cope with the current climate variabilities are better at adapting to future uncertain climate as well. Many studies advocate the bottom-up approach because "it takes into account important system characteristics and local capacities while framing possible adaptation options" (Garcia et al. 2014, p. 20), which can circumvent the uncertainties of climate change projections to an extent (Dessai and Hulme 2009). Bottom-up approaches give legitimacy by taking the perspectives of how stakeholders may be affected into consideration (Girard et al. 2015), although they attach insufficient importance to physical factors (Bhave et al. 2014a). Here, participatory approaches enable the identification of adaptation options based on the available local resources and the options can be evaluated by the key stakeholders. This has the advantage of weighing up the cost effectiveness of each measure while designing the adaptation strategies (Girard et al. 2015) by taking into account the local contextual challenges based on the availability of knowledge, natural and financial resources (Butler et al. 2015).

As both the bottom-up and top-down approaches have deficiencies, increasingly, recent studies (Bhave et al. 2014b; Butler et al. 2015; Girard et al. 2015; Mimura et al. 2014) advocate for a combined approach taking into consideration both the physical and sociological parameters and national and local perspectives. Examples of integrating top-down and bottom-up approaches are still limited in both the developed and developing countries (Bhave et al. 2014a; Garnier et al. 2015).

21.3.3 Direct and Indirect Adaptation

Depending on whether or not the receptor, which is the target of adaptation, is also an exposure unit, Eisenack and Stecker (2012) differentiate adaptation into "direct" and "indirect" adaptation. In direct adaptation, since the receptor is also an exposure unit, "the purpose of the action is to improve the situation of a system that is affected by the climate stimuli" (Eisenack and Stecker, 2012, p. 248). Here the receptor is the one acting to take advantage of the new situation or avert or minimize its undesirable impacts. On the other hand, when the receptor, for whom the adaptation is initiated, and the exposure unit (the actor initiating the adaptation) are separate entities, the action is aimed to enable the receptor to take necessary measures for averting the unwanted impacts of climate change and hence is called "indirect" adaptation. In this instance, it might be necessary for an actor to be provided with sufficient resources to effectively adapt. Hence, the operator agent initiates the action for the receptor (identical or unidentical with the actor) to take the desirable measures. Therefore, broadcasting information regarding climate projections is an indirect adaptation. Similarly, some actors, such as researchers and policymakers, may facilitate the adaptation action (Hinkel 2007) rather than themselves adapting. In this instance, research institutions are facilitating agencies as they generate knowledge regarding the need for adaptation and also by providing knowledge regarding adaptation options. They facilitate adaptation by enabling or providing the receptors to obtain the resources required for adaptation or creating a conducive situation for adaptation, as in the case of policymakers. Therefore, identifying and evaluating the vulnerable systems, planning adaptation strategies, disseminating and imparting knowledge and enhancing adaptive capacity, etc. would provide decision makers with suitable adaptation measures to facilitate adaptation. Eisenack and Stecker (2012) cite the example of setting up early warning systems for public information by political administration as an adaptation facilitation. In this aspect the political administration is the operator facilitating the receptor to adapt.

21.3.4 Adaptation at Different Scales

Adaptation is generally planned and occurs at multiple scales. At the international level institutions such as the UNFCCC play a leading role in climate actions – both mitigation and adaptation. At the national level, many national governments have initiated actions through the National Adaptation Programme of Actions (Bisaro et al. 2010; Hardee and Mutunga 2010; Mimura et al. 2014; Schipper et al. 2010). However, as climate change impacts and vulnerabilities are contextual, in a top-down adaptation approach, local authorities are often expected to implement the adaptation policies framed by the national institutions (Measham et al. 2011).

Adaptation takes place at multiple scales and levels of administrative governance. It involves actions to be undertaken by individuals for their own benefit or by the government institutions operating at the national to regional or provincial and local scales to protect its citizens from harmful effects (Adger et al. 2005). The implementation of top-down adaptation strategies, such as the National Action Plan on Climate Change (NAPCC) and State Action Plans in India, takes place in a hierarchical administrative structure involving different institutional bodies operating at multiple levels that interact and influence, and hence, enables or hinders the implementation of the adaptation actions at another level (Adger et al. 2005; Lyle 2015). Therefore, adaptation opportunities are expected to be shaped by the dynamics of the layers of institutions operating at different scales. Bureaucrats and officials operating at various administrative scales are strengthened or hindered by the decisions made by other actors and institutions in another sector (or department) or operating at another scale (Smajgl and Prananingtyas 2009) through information and data acquisition, sharing of natural resources or even administrative orders. The capacity of institutions to address the climate change impacts is therefore expected to be largely heightened or weakened by the deliberate commissions or omissions of other actors and agencies that are operating at a different scale. Therefore, evaluating the capacity of an institution to adapt to the changing climate will also involve understanding the networks of influences and the effectiveness of coordination between the actors and agencies operating at different administrative scales or tiers of a hierarchical governance structure (Smajgl and Prananingtyas 2009). This understanding is particularly relevant in the case of highly bureaucratic structures of governance, in which the jurisdictions and administrative mandates overlap both in terms of resource allocation and control over natural resources, and, hence, has the potential for both complementing and conflicting with the adaptation efforts.

Based on reviews of adaptation in Saint Louis, Senegal for flood risk management at the city level, Vedeld et al. (2015) suggested the concept of co-production to understand how higher level institutions can enhance the adaptive capacity of institutions operating at a lower hierarchical level of administration. On the other hand, the experiences of adaptation from implementing institutions are helpful to inform institutions operating at another scale, normally a higher administrative scale, to design more effective adaptation strategies. Administrating water in situations where the changing climate has the potential to exacerbate the already growing demands due to population growth and changing lifestyles, therefore, requires effective coordination across administrative scales in order to reduce the competing policy agendas (Vedeld et al. 2015) and also enhance the effectiveness and efficiency of sharing the resources, which may be depleting, while also multiplying the learning experiences through the implementation of the various strategies. Although studies regarding the interaction between scales are emerging, knowledge is limited regarding the feedback loops between institutions operating at different scales that influence the temporal dynamics of institutional coordination (Lyle 2015). Current knowledge regarding the flow of resources from different scales of governance through cross-scale frameworks is inadequate to perceive how socially impartial and viable acclimation to ecological change can be redesigned. Climate change affects different sectors, actors and regions differently and hence the need for adaptation varies and so does the adaptive capacity across scales, sectors and actors (Grothmann et al. 2013; Prutsch et al. 2014). Locality based studies are required to draw out contextual socio-cultural perspectives for designing effective adaptation strategies that are sustainable and implementable at the specific scale (Amaru and Chhetri 2013).

21.3.5 Multilevel Institutions and Interactions

Institutions operating at different administrative scales from national to state or provincial and block levels are required to play complementary as well as distinct roles in initiating and implementing adaptation strategies (Adger et al. 2005; Fidelman et al. 2013; Nalau et al. 2015). This is particularly so for sustainable management of water by basin level management organizations to local authorities (Bisaro et al. 2010; Finger et al. 2006; Lebel and Garden 2008; Mollinga et al. 2006; Pittock 2011; Wilby and Wood 2012) and water users in a changing climate. The degree to which the diverse stakeholders are involved in the planning and implementation of the adaptation strategies is known to shape its outcome (Juhola and Westerhoff, 2011). Azhoni et al. (2017a) supplemented other earlier research (Amundsen et al. 2010; Dannevig and Aall 2015; Vedeld et al. 2015) by assessing how the multiple institutions operating at different scales in India, from national to state and at district levels, coordinate, complement or even compete with each other in some instances, and suggest the need for understanding the coordination across scale in order to evaluate how it shapes the implementation of the adaptation strategies.

Significant gaps often exist between institutions operating at different scales both in terms of risk perceptions and adaptive capacity (Butler et al. 2015) on one hand and the availability of financial resources and knowledge expertise on the other. Knowledge regarding the impact of this variance is limited due to the scant critical attention from adaptation researchers on the key issue of the relationship between institutions operating at different levels (Wyborn 2015). Empirical knowledge is needed regarding the lack of adaptive capacity for institutions operating at one scale being circumvented by the presence at another level or the lack of adaptation plans at another level causing bottlenecks for creating adaptive capacity at another (Pahl-Wostl 2009; Wyborn 2015). Moreover, the wider socio-economic and governance structure under which the institutions operate also shapes interinstitutional interactions. Despite the growing research on understanding the multilevel nature of adaptation governance (Bizikova et al. 2015; Dannevig and Aall 2015; Lyle 2015; McGloughlin and Sweeney 2011; Schreurs 2010; Vedeld et al. 2015), the interaction of institutions across scales and how it influences the adaptive capacity at different scales are contextual and dependent on the socio-economic and governance structure of the particular region (Jörgensen et al. 2015). More empirical studies, which take the local contextual challenges into account, will, therefore, complement national level studies in understanding the critical role of national level adaptation planning on the adaptive capacity of local level institutions.

The capacity of an institution to adapt depends on various factors but critically on how effectively decision makers within those institutions can seize the opportunity by gathering the resources through identifying the relevant information and knowledge leading to the recognition of the threats and identifying options for adjusting to the new and changing scenario (Smit and Pilifosova 2003). Social networks between individuals are known to enhance adaptive capacity through an enhanced sharing of resources (Dow et al. 2013). However, interinstitutional networks function at a more complex level as institutions are made up of individuals (Pahl-Wostl 2009) with different personalities and motivations. Knowledge concerning networks among public institutions and how adaptation is actually being delivered is emerging (Arnell 2010). On the other hand, the existence of social networks does not in itself strengthen the capacity of institutions to address the changes if the network is confined only to groups of similar affinities (precluding inter-sectoral cooperation) keeping conventional wisdom perpetual and leaving out learning opportunities (Newman and Dale 2007).

21.4 Water Management Institutions and Their Role in Adaptation

Government institutions play the major role in many countries, such as India, both for climate change adaptation in general and water management in particular, and they are key in enabling society to adapt (Jogesh and Dubash 2015). Therefore, government institutions, particularly in the water sector, need to actively participate in climate change adaptation. Most of the research on actions being undertaken by institutional bodies (e.g. governmental and non-governmental as well as private and profit-making corporations) for adapting water management to climate change is focused in industrialized countries (Preston et al. 2010). Climate change adaptation studies in low-income countries are mostly focused on the societies and vulnerable communities (Ford et al. 2014). Very little knowledge exists regarding adaptation by institutional bodies, particularly by government institutions in developing countries.

Societal factors, including cultural practices, attitudes toward risks and information sharing, as well as the availability of resources and infrastructure, also affect the ability to implement climate change adaptation (Berman et al. 2012) although in a much less visible way (Helmke et al. 2012). However, government institutional bodies, having their mandate enforced by government legislations, play the chief role in adjudication and apportionment of resources and hence have the dominant role in enabling societies to adapt (Azhoni et al. 2018). They play the key role in demarcating responsibilities between actors, mediate trade-offs, and serve as authorities and facilitators of actions (Cook et al. 2010). Government agencies that administer water are thus at the very heart of how a society manages water. They regulate the provision of water for ecosystem services and goods and environmental integrity while also targeting the meeting of human needs. Hard infrastructures such as dams, canals and tanks are at their disposal and hence how they regulate the distribution and control of water resources shapes many other activities in a society. Therefore, it is important to ensure that these institutions need to adopt strategic measures to enable themselves and the society to be resilient to changing climate (Cook et al. 2010).

21.5 Indian Water Institutions

Water governance in India involves multiple institutions at multiple levels of operation. Several union government institutions (ministries) are directly involved in the development of water infrastructure and/or monitoring and collection of information related to water as well as other governmental and research institutions. While the Ministry of Water Resources along with its allied institutions, such as the Central Water Commission, Central Groundwater Board and National Institute of Hydrology and other river basin organizations, functions as the apex body at the central/union government level and lays down policy guidelines and programs for the development and regulation of the country's water resources, environmental clearances for all major hydro projects such as dams and canals for power generation (under the Ministry of Power) and irrigation are to be obtained from the Ministry of Environment and Forests. On the other hand, the erstwhile Planning Commission (replaced by National Institution for Transforming India (NITI) Aayog in 2015) plans and allocates funds for infrastructure development in the country including development of water storage systems, flood defense, and major irrigation and inland waterways. The Ministry of Earth Sciences and the Ministry of Science and Technology focus on development of technology including forecasting of weather and other related areas, while the Ministry of Environment and Forests monitors aspects related to water quality and pollution. The Union Ministry of Drinking Water and Sanitation plays the important role of overall policy planning, funding and coordination of developmental programs for water supply and sanitation services in the country although water is a state subject, according to the Constitution of India, and as such, the state government is responsible for water supply and irrigation (Iyer 1994). On the other hand, most of the rivers in India are interstate and hence conflicts over sharing of river water sources emerge (Richards and Singh 2002; Swain 1998) causing conflictual interests between the union and state government(s) or between state governments (Richards and Singh 2002). River basin management organizations have been formed by the union government to address some of these interstate challenges. However, these river basin organizations are formed based on project-specific requirements and focus primarily on the engineering structures rather than the socio-economic sustainability of the basin. Overlapping of jurisdiction for water governance between the union government and the state government institutions makes coordination between the two levels crucial. This multilayered governance of water in India, whose complexity is being intensified by a changing climate, makes it an ideal case to explore the challenges of what makes institutions adaptive. The National Water Mission, currently under implementation as one of the eight missions of the NAPCC, acknowledges the need for involvement of multiple institutions for realizing the mission. Figure 21.1, extracted from an earlier publication of the author, highlights this multiple institutional involvement.

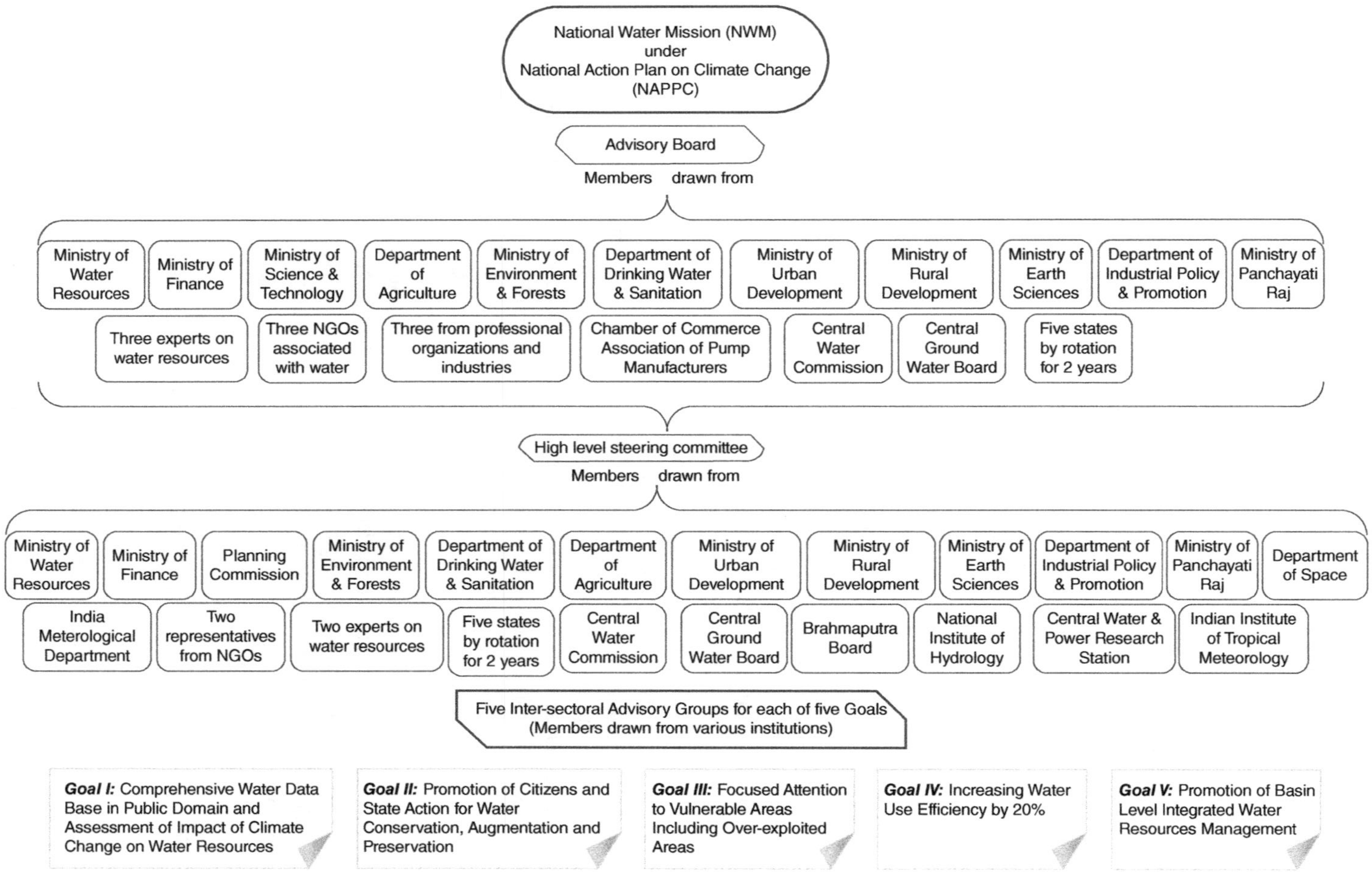

Figure 21.1 Institutions in the National Water Mission under the National Action Plan on Climate Change. Source: Azhoni et al. 2017a.

21.6 Recent Initiatives to Address Climate Change in India

Realizing the impending challenges of climate change, the Prime Minister's Council on Climate Change (PMCCC), having been instituted in 2007, came up with the NAPCC (PMCCC 2008). Recognizing that "India needs a national strategy to adapt to climate change" (PMCCC 2008, p. 1), this document identified eight National Missions among which the National Water Mission was one. A key aim of the National Water Mission is to "develop a framework to optimize water use by increasing water use efficiency by 20% through appropriate institutional mechanisms suited for effective delivery" (MWR 2011, p. 3) by taking advantage of the provisions of the National Water Policy. The expected results of the mission are to assist the country to adapt to changing climate while pursuing economic growth (PMCCC 2008). While the National Water Mission provides a broad aim, its details of implementation are still being working out via a continuous process (Byravan and Rajan 2013).

In order to enhance the adaptive capacity of Indian institutions for addressing the challenges of a changing climate, the NAPCC identified the need for bringing together key research institutions in the country through an improved network as a key strategy (PMCCC 2008). For this reason, the Government of India under the aegis of the Ministry of Environment and Forests initiated the Indian Network for Climate Change Action (INCCA) in 2007. It was a network-based program among the participating institutions with the aim to assess the drivers and implications of the changing climate and provide decision support systems and also build capacity to address the challenges of climate change in the country. The underlying assumption was that coordination among various institutions (120 in INCCA) will improve the adaptive capacity of the country through sharing of knowledge, improve decision making and planning, monitoring and collection of data and minimize conflicts of interest (MEF 2010). Among the key institutions in this network are the Ministries of Earth Sciences, Agriculture, Science and Technology, and its constituent agencies. It also includes government research institutions such as Defence Research and Development Organization (DRDO), Indian Space Research Organization (ISRO), Council of Scientific and Industrial Research (CSIR), Indian Council of Agriculture Research (ICAR) and other prominent educational institutions along with non-governmental organizations currently working in climate change and water-related issues. Likewise, one of the eight missions in NAPCC is the National Mission on Strategic Knowledge for Climate Change (NMSKCC). Among the chief assumptions for having a "Strategic Knowledge Mission" is "the current modes of data generation and information and knowledge sharing in India do not enable required integrated approaches" (MST 2010, p. 5).

21.7 Adaptation for Sustainability: Two Examples

Freshwater sources are of crucial importance for water supply, as well as for biodiversity conservation, both of which are threatened by climate change. Two of the strategies to maintain the sustainability of water, both for irrigation and domestic supply, in the Himalayan regions of the Indian state of Sikkim and Himachal Pradesh, based on the studies reported in Azhoni et al. (2017b) and Azhoni and Goyal (2018) are discussed in Section 21.7.1.

21.7.1 Investing in Perennial and Larger Sources of Water in Himachal Pradesh

The state of Himachal Pradesh in northern India is dependent on glacier-fed rivers (Jaswal et al. 2015; Rana et al. 2014) for hydropower generation, irrigation and water supply; hence the state is highly sensitive to climate change (Brenkert and Malone 2005; Mall et al. 2006). Various studies (Erlewein 2013; Jaswal et al. 2015) have indicated that the changing patterns of precipitation in Himachal Pradesh could have multiple consequences. Many of the sources for water supply are perceived to be depleting (Azhoni et al. 2017b) due to changes in land use and other factors including climate change (Rana et al. 2013).

The state government of Himachal Pradesh has initiated various actions to address the impacts of climate change (ADB 2010; Govt. of HP 2013) including the formulation of the State Action on Plan on Climate (DEST-HP 2012), which was adopted in 2012. For sustainable management of water resources in Himachal Pradesh under a changing climate change, the need for effective institutions for integrated water management, establishing of comprehensive information and data systems, catchment and agricultural planning, disaster preparedness and rural employment diversification has been suggested (ADB 2010; Govt. of HP 2013). Among the strategies for maintaining the sustainable supply of water, the state government has initiated rainwater harvesting in new government buildings. Since many of the stream sources from which the water supply is drawn are perceived to be depleting for various reasons, the state government water supply department – Irrigation and Public Health Department (IPH) – has now decided to invest in larger and perennial sources even if smaller sources are currently sufficient due to the fear that these small sources might dry up in the future (Azhoni et al. 2017b). This means, of course, the cost of the infrastructure will be higher because larger sources are generally difficult to find and further away from human settlements than the smaller sources.

21.7.2 Spring Water Rejuvenation in Sikkim

Sikkim, in northeast India, is one Himalayan state that is considered to be highly sensitive to the impacts of climate change (Jain et al. 2013). The state is vulnerable to glacier lake outburst floods (GLOFs) (Ives et al. 2010) due to the melting of snow at higher altitudes, which could also have devastating consequences for other states in the downstream. Studies have pointed out the changing seasonal discharge of the rivers originating from the state (Jeelani et al. 2012) possibly due to the changing precipitation patterns (Singh and Goyal 2016). These changing hydrological regimes will have an impact on the other regions in the downstream besides the hydropower stations that are dependent on these rivers. However, one of the key concerns of the local people in Sikkim is the perceived decrease in mountain spring water sources (Barnett et al. 2005) and changing precipitation patterns possibly due to the changing climate (Singh and Goyal 2016). Various studies have observed a change in the precipitation pattern (Singh and Goyal 2016) while other studies project the likelihood of a more erratic rainfall pattern leading to flood and drought situations (Singh and Goyal 2016). On the other hand, the projected increased temperature will mean increased crop water requirement (Biemans et al. 2013) while availability of spring water is also reduced (Jeelani et al. 2012). Such changes could have multiple and devastating cascading effects on the state's socio-cultural and economic life leading to migration and other insecurities (Goodess 2012).

The state government of Sikkim, through the State Department of Science, Technology and Climate Change have initiated some actions to understand the climate change risks in the state and have also adopted the Sikkim State Action Plan on Climate Change Plan (Government of Sikkim 2012). Among these key strategies listed by the State Action Plan are: (i) artificial recharge to revive springs by harvesting rainwater; (ii) reviving dried up hilltop lakes; (iii) increasing base flow of critical streams by rainwater harvesting; (iv) exploring the possibility of harnessing stream water for meeting household and irrigation needs; (v) exploring possibility of harnessing stream water for meeting household and irrigation needs; (vi) increasing water storage capacity by building household, community and village-level reservoirs; (vii) formation of Water Users Association to assist in regulating and conserving water; and (viii) adoption of efficient irrigation techniques such as sprinkler and drip irrigation. The Rural Management and Development Department of the state has initiated actions to improvise adaptation strategies into the developmental initiatives (Tambe et al. 2012) while risks of other natural disasters are also being assessed by other state government departments along with agencies and consultants (Azhoni and Goyal 2018; Huber and Joshi 2015).

For domestic water supply and irrigation, “spring and streams” are the sole sources of water in the state (Government of Sikkim 2012). In the recent past, it has been perceived that the spring sources (which also give rise to streams) are depleting (Tambe et al. 2012). It is being observed that during the lean season the discharge from some spring sources is decreasing. Studies have pointed out that delayed monsoon and erratic rainfall could further aggravate the situation (Singh and Goyal 2016). This has caused concern among both the local people and the state government (Azhoni and Goyal 2018). Therefore, in order to improve the infiltration of rainwater, trenches are being dug on the mountain tops, which is known locally as *dhara vikas*, with the aim of sustaining the spring water discharge. However, studies conducted by Tambe et al. (2012) show that identification of recharge locations remains a challenge. Assistance from research institutions and experts for building local capacity, and financial assistance from government institutions remain necessary for implementing successful rejuvenation of spring water.

21.8 Discussion and Conclusion

The role and effectiveness of water institutions in India is not without debate. Various researchers have highlighted the strengths and weaknesses of water institutions in India (Ananda et al. 2006; Aubriot and Prabhakar 2011; Bottrall 1992; Marothia 2003; Reddy and Reddy 2002). The National Water Policy (MWR 2002) itself emphasized the “need to reorient” the governance of water in India and restructuring of its institutions, which has been reiterated in the revised National Water Policy 2012 (MWR 2012). The lack of coordination within government departments and agencies has been a major weakness (Narain 2000, 2008). Ananda et al. (2006, p. 927) suggested the need to not only incorporate the “proven criteria of institutional design but also the changing socioeconomic, political and cultural factors.” More recently, Byravan and Rajan (2013) have identified that the fragmented authority between the central and state governments and weak institutional structures are potential barriers to effectively implementing the NAPCC. The need for "interdepartmental coordination" with "greater synergy and interaction among different organizations" has been emphasized (Byravan and Rajan 2013, p. 31).

In order to manage water sustainably in the light of new challenge being added by climate change, besides the growing demand of water due to population and economic growth, the following considerations are in place:

1) *Adaptation strategies for sustainable management of water are required to be introduced through co-benefits of development*: Adaptation actions inconsistent with developmental pathways may face greater resistance of implementation. Therefore, in circumstances of infrastructure development deficits, adaptation strategies are required to be introduced through co-benefits of development. The various challenges and issues of sustainable water management are interrelated and therefore policymakers have to approach adaptation holistically by taking into consideration the various underlying causes of adaptation implementation.
2) *Need for building knowledge and technological capacity*: Effective mainstreaming of adaptation for sustainable management of water requires knowledge capacity as well as accessibility to the existing climate information as well as development of new tools and models for risk assessment. This will require greater synergies between different government agencies and ministries, in addition to an improved coordination with research and academic institutions, and development agencies for capacity building and efficient resource allocation. This is necessary to enhance opportunities for building the capacity of local institutions for improving the sustainability of water in a changing climate requires improving both vertical and horizontal interactions across institutions.
3) *Standards and guidelines for practitioners*: Standards and guidelines for implementing sustainable water management need to be developed by incorporating the uncertainties of climate change into the design features of infrastructure. Making adaptation options, guidance and standards available to practitioners, supplemented with building greater knowledge capacities, is important in the changing scenario.

References

ADB (2010). *Climate Change Adaptation in Himachal Pradesh: Sustainable Strategies for Water Resources*. Shimla, India: Asian Development Bank.

Adger, W.N., Arnell, N.W., and Tompkins, E.L. (2005). Successful adaptation to climate change across scales. *Global Environmental Change* 15 (2): 77–86.

Amaru, S. and Chhetri, N.B. (2013). Climate adaptation: institutional response to environmental constraints, and the need for increased flexibility, participation, and integration of approaches. *Applied Geography* 39: 128–139.

Amundsen, H., Berglund, F., and Westskogh, H. (2010). Overcoming barriers to climate change adaptation – a question of multilevel governance? *Environment and Planning C: Government and Policy* 28 (2): 276–289.

Ananda, J., Crase, L., and Pagan, P.G. (2006). A preliminary assessment of water institutions in India: an institutional design perspective. *Review of Policy Research* 23 (4): 927–953.

Arnell, N.W. (2004). Climate change and global water resources: SRES emissions and socio-economic scenarios. *Global Environmental Change-Human and Policy Dimensions* 14 (1): 31–52.

Arnell, N.W. (2010). Adapting to climate change: an evolving research programme. *Climatic Change* 100 (1): 107–111.

Aubriot, O. and Prabhakar, P.I. (2011). Water institutions and the "revival" of tanks in south India: what is at stake locally? *Water Alternatives* 4 (3): 325–346.

Azhoni, A. and Goyal, M.K. (2018). Diagnosing climate change impacts and identifying adaptation strategies by involving key stakeholder organisations and farmers in Sikkim, India: challenges and opportunities. *Science of the Total Environment* 626: 468–477.

Azhoni, A., Holman, I., and Jude, S. (2017a). Adapting water management to climate change: inter-institutional networks and barriers in India. *Global Environmental Change* 44 (March): 144–157.

Azhoni, A., Holman, I.P., and Jude, S. (2017b). Contextual and interdependent causes of climate change adaptation barriers: insights from water management institutions in Himachal Pradesh, India. *Science of the Total Environment* 576 (1): 817–828.

Azhoni, A., Jude, S., and Holman, I. (2018). Adapting to climate change by water management organisations: enablers and barriers. *Journal of Hydrology* 559: 736–748.

Barnett, T.P., Adam, J.C., and Lettenmaier, D.P. (2005). Potential impacts of a warming climate on water availability in snow-dominated regions. *Nature*, Nature Publishing Group 438 (7066): 303–309.

Bates, B.C., Kundzewicz, Z.W., Wu, S., and Palutikof, J.P. (2008). *Climate Change and Water.*. Technical Paper of the Intergovernmental Panel on Climate Change. Geneva: IPCC Secretariat.

Berkhout, F. (2012). Adaptation to climate change by organizations. *Wiley Interdisciplinary Reviews: Climate Change* 3 (1): 91–106.

Berman, R., Quinn, C., and Paavola, J. (2012). The role of institutions in the transformation of coping capacity to sustainable adaptive capacity. *Environmental Development* 2 (1): 86–100.

Berrang-Ford, L., Ford, J.D., and Paterson, J. (2011). Are we adapting to climate change? *Global Environmental Change* 21 (1): 25–33.

Bhave, A.G., Mishra, A., and Raghuwanshi, N.S. (2014a). A brief review of assessment approaches that support evaluation of climate change adaptation options in the water sector. *Water Policy* 16 (5): 959–972.

Bhave, A.G., Mishra, A., and Raghuwanshi, N.S. (2014b). A combined bottom-up and top-down approach for assessment of climate change adaptation options. *Journal of Hydrology* 518: 150–161.

Biemans, H., Speelman, L.H., Ludwig, F. et al. (2013). Future water resources for food production in five South Asian river basins and potential for adaptation – a modeling study. *Science of the Total Environment* 468–469: S117–S131.

Bisaro, A., Hinkel, J., and Kranz, N. (2010). Multilevel water, biodiversity and climate adaptation governance: evaluating adaptive management in Lesotho. *Environmental Science & Policy* 13 (7): 637–647.

Bizikova, L., Parry, J.-E., Karami, J., and Echeverria, D. (2015). Review of key initiatives and approaches to adaptation planning at the national level in semi-arid areas. *Regional Environmental Change* 15 (5): 837–850.

Bosello, F., Carraro, C., and De Cian, E. (2013). Adaptation can help mitigation: an integrated approach to post-2012 climate policy. *Environment and Development Economics* 18 (3): 270–290.

Bottrall, A. (1992). Fits and misfits over time and space: technologies and institutions of water development for South Asian agriculture. *Contemporary South Asia* 1 (2): 227–247.

Brenkert, A.L. and Malone, E.L. (2005). Modeling vulnerability and resilience to climate change: a case study of India and Indian States. *Climatic Change* 72 (1–2): 57–102.

Burton, I. and May, E. (2004). The adaptation deficit in water resource management. *IDS Bulletin* 35 (3): 31–37.

Butler, C. and Adamowski, J. (2015). Empowering marginalized communities in water resources management: addressing inequitable practices in participatory model building. *Journal of Environmental Management* 153: 153–162.

Butler, J.R.A., Wise, R.M., Skewes, T.D. et al. (2015). Integrating top-down and bottom-up adaptation planning to build adaptive capacity: a structured learning approach. *Coastal Management* 43 (4): 346–364.

Byravan, S. and Rajan, S. (2013). An evaluation of India's National Action Plan on Climate Change. http://dx.edoi.org/10.2139/ssrn.2195819.

Central Water Commission (2005). *CWC Water Data Complete Book 2005*. New Delhi: Central Water Commission, Ministry of Water Resources, Govt. of India.

Central Water Commission (2010). *Water and Related Statistics*. New Delhi: Central Water Commission, Ministry of Water Resources, Govt. of India.

Cook, J., Freeman, S., Levine, E., and Hill, M. (2010). *Shifting Course : Climate Change for Water Management Institutions*. Gland, Switzerland: World Wildlife Fund (WWF).

Crabbé, P. and Robin, M. (2006). Institutional adaptation of water resource infrastructures to climate change in Eastern Ontario. *Climatic Change* 78 (1): 103–133.

Dannevig, H. and Aall, C. (2015). The regional level as boundary organization? An analysis of climate change adaptation governance in Norway. *Environmental Science & Policy* 54: 168–175.

Dessai, S. and Hulme, M. (2004). Does climate adaptation policy need probabilities? *Climate Policy* 4: 107–128.

Dessai, S. and Hulme, M. (2009). Climate prediction: a limit to adaptation? In: *Adapting to Climate Change: Thresholds, Values, Governance* (eds. W.N. Ader, I. Lorenzoni and K. O'Brien), 64–78. Cambridge, UK: Cambridge University Press.

DEST-HP (2012). *State Strategy & Action Plan on Climate Change*. Shimla, India: Department of Environment, Science & Technology. Government of Himachal Pradesh.

Dow, K., Haywood, B.K., Kettle, N.P., and Lackstrom, K. (2013). The role of ad hoc networks in supporting climate change adaptation: a case study from the Southeastern United States. *Regional Environmental Change* 13 (6): 1235–1244.

Eisenack, K. and Stecker, R. (2012). A framework for analyzing climate change adaptations as actions. *Mitigation and Adaptation Strategies for Global Change* 17 (3): 243–260.

Erlewein, A. (2013). Disappearing rivers – The limits of environmental assessment for hydropower in India. *Environmental Impact Assessment Review* 43: 135–143.

Fidelman, P.I.J., Leitch, A.M., and Nelson, D.R. (2013). Unpacking multilevel adaptation to climate change in the Great Barrier Reef, Australia. *Global Environmental Change* 23 (4): 800–812.

Finger, M., Tamiotti, L., and Allouche, J. (2006). *The Multi-Governance of Water: Four Case Studies*. Albany, NY: State University of New York Press.

Ford, J.D., Berrang-Ford, L., Bunce, A. et al. (2014). The status of climate change adaptation in Africa and Asia. *Regional Environmental Change*: 801–814.
Fünfgeld, H. (2015). Facilitating local climate change adaptation through transnational municipal networks. *Current Opinion in Environmental Sustainability* 12: 67–73.
Füssel, H.-M. (2007). Adaptation planning for climate change: concepts, assessment approaches, and key lessons. *Sustainability Science* 2 (2): 265–275.
Garcia, L.E., Mathews, J.H., Rodriguez, D.J. et al. (2014). *Beyond Downscaling: A Bottom-up Approach to Climate Adaptation for Water Management*. AGWA Report 01. Washington, DC: World Bank.
Garnier, M., Harper, D.M., Blaskovicova, L. et al. (2015). Climate change and European water bodies, a review of existing gaps and future research needs: findings of the climate water project. *Environmental Management* 56 (2): 271–285.
Gifford, R., Kormos, C., and McIntyre, A. (2011). Behavioral dimensions of climate change: drivers, responses, barriers, and interventions. *Wiley Interdisciplinary Reviews: Climate Change* 2 (6): 801–827.
Girard, C., Pulido-Velazquez, M., Rinaudo, J.-D. et al. (2015). Integrating top-down and bottom-up approaches to design global change adaptation at the river basin scale. *Global Environmental Change* 34: 132–146.
Goodess, C.M. (2012). How is the frequency, location and severity of extreme events likely to change up to 2060? *Environmental Science and Policy* 27 (Suppl. 1): S4–S14.
Government of Sikkim (2012). *The Sikkim State Action Plan on Climate Change*. Gangtok, India: Government of Sikkim.
Govt. of HP (2013). *State Water Policy 2013*. Shimla, India: Government of Himachal Pradesh.
Grothmann, T., Grecksch, K., Winges, M., and Siebenhüner, B. (2013). Assessing institutional capacities to adapt to climate change – integrating psychological dimensions in the adaptive capacity wheel. *Natural Hazards and Earth System Sciences Discussions* 1 (2): 793–828.
Hardee, K. and Mutunga, C. (2010). Strengthening the link between climate change adaptation and national development plans: lessons from the case of population in National Adaptation Programmes of Action (NAPAs). *Insectes Sociaux* 57 (2): 113–126.
Hassan, M., Du, P., Jia, S. et al. (2015). An assessment of the South Asian summer monsoon variability for present and future climatologies using a high resolution regional climate model (RegCM4.3) under the AR5 scenarios. *Atmosphere* 6 (11): 1833–1857.
Helmke, G., Levitsky, S., and Galvan, D. (2012). Politics: institutions comparative agenda. *Political Science* 2 (4): 725–740.
Hinkel, J. (2007). Adaptation problem types. Proceedings of the Amsterdam Conference on the Human Dimensions of Global Environmental Change 2007, Amsterdam.
Holman, I.P. (2006). Climate change impacts on groundwater recharge – uncertainty, shortcomings, and the way forward. *Hydrogeology Journal* 14 (5): 637–647.
Huber, A. and Joshi, D. (2015). Hydropower, anti-politics, and the opening of new political spaces in the Eastern Himalayas. *World Development* 76 (289374): 13–25.
Inderberg, T.H. (2011). Institutional constraints to adaptive capacity: adaptability to climate change in the Norwegian electricity sector. *Local Environment* 16 (4): 303–317.
IPCC (2007). *Climate Change 2007: Impacts, Adaptation and Vulnerability. Contribution of Working Group II to the Fourth Assessment Report of the Intergovernmental Panel on Climate Change* (eds. M.L. Parry, O.F. Canziani, J.P. Palutikof, et al.). Cambridge, UK: Cambridge University Press 976pp.

IPCC (2014a). Summary for policymakers. In: *Climate Change 2014: Impacts, Adaptation, and Vulnerability. Part A: Global and Sectoral Aspects*. Contribution of Working Group II to the Fifth Assessment Report of the Intergovernmental Panel on Climate Change (eds. C.B. Field, V.R. Barros, D.J. Dokken, et al.). Cambridge, UK: Cambridge University Press.

IPCC (2014b). Climate change 2014: impacts, adaptation, and vulnerability. Part A: global and sectoral aspects. In: *Contribution of Working Group II to the Fifth Assessment Report of the Intergovernmental Panel on Climate Change* (eds. C.B. Field, V.R. Barros, D.J. Dokken, et al.). Cambridge, UK: Cambridge University Press.

Ives, J.D., Shrestha, R.B., and Mool, P.K. (2010). *Formation of Glacial Lakes in the Hindu Kush-Himalayas and GLOF Risk Assessment*. Kathmandu: ICIMOD (Inernational Centre for Integrated Mountain Development).

Iyer, R.R. (1994). Indian federalism and water resources. *International Journal of Water* 10 (2): 191–202.

Jain, S.K., Kumar, V., and Saharia, M. (2013). Analysis of rainfall and temperature trends in northeast India. *International Journal of Climatology* 33 (4): 968–978.

Jaswal, A.K., Bhan, S.C., Karandikar, A.S., and Gujar, M.K. (2015). Seasonal and annual rainfall trends in Himachal Pradesh during 1951-2005 study area which lies in the Western Himalayas bounded by Jammu. *MAUSAM* 2 (April): 247–264.

Jeelani, G., Feddema, J.J., Van Der Veen, C.J., and Stearns, L. (2012). Role of snow and glacier melt in controlling river hydrology in Liddar watershed (western Himalaya) under current and future climate. *Water Resources Research* 48 (12) https://doi.org/10.1029/2011WR011590.

Jogesh, A. and Dubash, N.K. (2015). State-led experimentation or centrally-motivated replication? A study of state action plans on climate change in India. *Journal of Integrative Environmental Sciences* 12 (4): 247–266.

Jones, R.N., Dettmann, P., Park, G. et al. (2007). The relationship between adaptation and mitigation in managing climate change risks: a regional response from North Central Victoria, Australia. *Mitigation and Adaptation Strategies for Global Change* 12 (5): 685–712.

Jörgensen, K., Jogesh, A., and Mishra, A. (2015). Multi-level climate governance and the role of the subnational level. *Journal of Integrative environmental Sciences* 12 (4): 235–245.

Juhola, S. and Westerhoff, L. (2011). Challenges of adaptation to climate change across multiple scales: a case study of network governance in two European countries. *Environmental Science & Policy* 14 (3): 239–247. https://doi.org/10.1016/j.envsci.2010.12.006.

Krysanova, V., Dickens, C., Timmerman, J. et al. (2010). Cross-comparison of climate change adaptation strategies across large river basins in Europe, Africa and Asia. *Water Resources Management* 24 (14): 4121–4160.

Kundzewicz, Z.W., Mata, J.L., Arnell, W.N. et al. (2008). The implications of projected climate change for freshwater resources and their management. *Hydrological Sciences Journal*, International Association of Hydrological Sciences 53 (1): 3–10.

Kuruppu, N. and Willie, R. (2014). Barriers to reducing climate enhanced disaster risks in least developed country-small islands through anticipatory adaptation. *Weather and Climate Extremes* 7 (2015): 72–83.

Lebel, L. and Garden, P. (2008). Deliberation, negotiation and scale in the Governance of Water Resources in the Mekong Region. In: *Adaptive and Integrated Water Management Coping with Complexity and Uncertainty* (eds. C. Pahl-Wostl, P. Kabat and J. Möltgen), 205–225. Amsterdam: Springer.

Lesnikowski, A., Ford, J., Biesbroek, R. et al. (2017). What does the Paris Agreement mean for adaptation? *Climate Policy* 7 (7): 825–831. (online).

Lyle, G. (2015). Understanding the nested, multi-scale, spatial and hierarchical nature of future climate change adaptation decision making in agricultural regions: a narrative literature review. *Journal of Rural Studies* 37: 38–49.

Mall, R.K., Gupta, A., Singh, R. et al. (2006). Water resources and climate change : an Indian perspective. *Current Science* 90 (12): 1610–1626.

Marothia, D.K. (2003). Enhancing sustainable management of water resource in agriculture sector: the role of institutions. *Indian Journal of Agricultural Economics* 58 (3): 406–426.

Mastrandrea, M.D., Heller, N.E., Root, T.L., and Schneider, S.H. (2010). Bridging the gap: linking climate-impacts research with adaptation planning and management. *Climatic Change* 100 (1): 87–101.

Mata, L.J. and Budhooram, J. (2007). Complementarity between mitigation and adaptation: the water sector. *Mitigation and Adaptation Strategies for Global Change* 12 (5): 799–807.

McGloughlin, J.S. and Sweeney, J. (2011). Multi-level climate policies in Ireland. *Irish Geography* 44 (1): 137–150.

Measham, T.G., Preston, B.L., Smith, T.F. et al. (2011). Adapting to climate change through local municipal planning: barriers and challenges. *Mitigation and Adaptation Strategies for Global Change* 16 (8): 889–909.

MEF (2009). *Climate Change and India: Towards Preparation of a Comprehensive Climate Change Assessment*. New Delhi: Ministry of Environment and Forests, Government of India.

MEF (2010). *Climate Change in India: A 4X4 Assessment a Sectoral and Regional Analysis for 2030s INCCA: Indian Network for Climate Change Assessment*. New Delhi: Ministry of Environment & Forests, Government of India.

Mertz, O., Halsnæs, K., Olesen, J.E. et al. (2009). Adaptation to climate change in developing countries. *Environmental Management* 43 (5): 743–752.

Mimura, N., Pulwarty, R.S., Duc, D.M. et al. (2014). Adaptation planning and implementation. In: *Climate Change 2014: Impacts, Adaptation, and Vulnerability. Part A: Global and Sectoral Aspects*. Contribution of Working Group II to the Fifth Assessment Report of the International Panel on Climate Change (eds. C.B. Field, V.R. Barros, D.J. Dokken, et al.), 869–898. Cambridge, UK: Cambridge University Press.

Mollinga, P.P., Dixit, A., and Athukorala, K. (2006). *Integrated Water Resources management : Global Theory, Emerging Practice, and Local Needs*. London: Sage.

Moser, S. and Boykoff, M. (2013). Climate change adaptation success: the scope of the challenge. In: *Successful Adaptation to Climate Change: Linking Science and Policy in a Rapidly Changing World* (eds. S.C. Moser and M.T. Boykoff), 1–33. London and New York: Routledge.

MST (2010). *Mission Document of National Mission on Strategic Knowledge for Climate Change*. New Delhi: Ministry of Science and Technology, Government of India.

MWR (2002). *National Water Policy*. New Delhi: Ministry of Water Resources, Government of India.

MWR (2008). Preliminary consolidated report on effect of climate change on water resources. Central Water Commission, Ministry of Water Resources, Government of India, New Delhi.

MWR (2011). *National Water Mission Under National Action Plan on Climate Change. Comprehensive Mission Document Volume I*. New Delhi: Ministry of Water Resources, Government of India.

MWR (2012). *National Water Policy (2012)*. New Delhi: Ministry of Water Resources, Government of India.

Nalau, J., Preston, B.L., and Maloney, M.C. (2015). Is adaptation a local responsibility? *Environmental Science & Policy* 48: 89–98.

Narain, V. (2000). India's water crisis: the challenges of governance. *Water Policy* 2 (6): 433–444.

Narain, V. (2008). Reform in Indian canal irrigation: does technology matter? *Water International* 33 (1): 33–42.

Naswa, P. and Garg, A. (2012). Managing climate-induced risks on Indian infrastructure assets. www.environmentportal.in/files/file/Current%20Science-cc-Aug11.pdf (accessed November 2019).

Newman, L. and Dale, A. (2007). Homophily and agency: creating effective sustainable development networks. *Environment, Development and Sustainability* 9 (1): 79–90.

Nyamwanza, A.M. and Kujinga, K.K. (2016). Climate change, sustainable water management and institutional adaptation in rural sub-Saharan Africa. *Environment, Development and Sustainability* https://doi.org/10.1007/s10668-016-9762-2.

Pahl-Wostl, C. (2009). A conceptual framework for analysing adaptive capacity and multi-level learning processes in resource governance regimes. *Global Environmental Change* 19 (3): 354–365.

Pittock, J. (2009). Lessons for climate change adaptation from better management of rivers. *Climate and Development* 1 (3): 194–211.

Pittock, J. (2011). National climate change policies and sustainable water management: conflicts and synergies. *Ecology and Society* 16 (2): article 25.

PMCCC (2008). *National Action Plan on Climate Change*. New Delhi: Prime Minister's Council on Climate Change, Government of India.

Porter, J.J., Demeritt, D., and Dessai, S. (2015). The right stuff? informing adaptation to climate change in British local government. *Global Environmental Change* 35: 411–422.

Preston, B.L., Westaway, R.M., and Yuen, E.J. (2010). Climate adaptation planning in practice: an evaluation of adaptation plans from three developed nations. *Mitigation and Adaptation Strategies for Global Change* 16: 407–438.

Preston, B.L., Mustelin, J., and Maloney, M.C. (2013). Climate adaptation heuristics and the science/policy divide. *Mitigation and Adaptation Strategies for Global Change* 20 (3): 467–497.

Preston, B.L., Rickards, L., Fünfgeld, H., and Keenan, R.J. (2015). Toward reflexive climate adaptation research. *Current Opinion in Environmental Sustainability* 14: 127–135.

Prutsch, A., McCallum, S., Grothmann, T. et al. (2014). Facing the specific challenges of adaptation. In: *Climate Change Adaptation Manual Lessons Learned from European and Other Industrialised Countries* (eds. A. Prutsch, S. McCallum, T. Grothmann, et al.), 7–13. Abingdon, UK: Routledge.

Rana, R.S., Bhagat, R.M., Kalia, V. et al. (2013). Indigenous perceptions of climate change vis-a-vis mountain agricultural activities in Himachal Pradesh, India. *Indian Journal of Traditional Knowledge* 12 (4): 596–604.

Rana, R.S., Singh, S., Chander, N. et al. (2014). Impacts of changes in climate on mountain water resources of Himachal Pradesh. *MAUSAM* 2 (April): 153–160.

Raper, S.C.B. (2005). The potential for sea level rise: New estimates from glacier and ice cap area and volume distributions. *Geophysical Research Letters* 32: 10–13.

Reddy, V.R. and Reddy, P.P. (2002). Water institutions: is formalisation the answer? (A study of water user associations in Andhra Pradesh). *Indian Journal of Agricultural Economics* 57 (3): 519–534.

Richards, A. and Singh, N. (2002). Inter-state water disputes in India: institutions and policies. *International Journal of Water Resources Development* 18 (4): 611–625.

Sanghi, A. and Mendelsohn, R. (2008). The impacts of global warming on farmers in Brazil and India. *Global Environmental Change* 18 (4): 655–665.

Schipper, L., Liu, W., Krawanchid, D., and Chanthy, S. (2010). *Review of Climate Change Adaptation Methods and Tools."*. MRC Technical Paper No. 34. Vientiane: Mekong River Commission.

Schreurs, M.A. (2010). Multi-level governance and global climate change in East Asia. *Asian Economic Policy Review* 5 (1): 88–105.

Seneviratne, S.I., Nicholls, N., Easterling, D. et al. (2012). Changes in climate extremes and their impacts on the natural physical environment. In: *Managing the Risks of Extreme Events and Disasters to Advance Climate Change Adaptation* (eds. C.B. Field, V. Barros, T.F. Stocker, et al.), 109–130. Cambridge, UK: Cambridge University Press.

Shi, L., Chu, E., and Debats, J. (2015). Explaining progress in climate adaptation planning across 156 U.S. municipalities. *Journal of the American Planning Association* 81 (3): 191–202.

Simonet, G. and Fatorić, S. (2015). Does "adaptation to climate change" mean resignation or opportunity? *Regional Environmental Change* 16 https://doi.org/10.1007/s10113-015-0792-3.

Singh, C. (2013). Changing pattern of the Indian summer monsoon rainfall: an objective analysis. *Climate Dynamics* 41: 1–9.

Singh, V. and Goyal, M.K. (2016). Analysis and trends of precipitation lapse rate and extreme indices over north Sikkim eastern Himalayas under CMIP5ESM-2M RCPs experiments. *Atmospheric Research* 167: 34–60.

Smajgl, A. and Prananingtyas, S.D. (2009). Adaptation dynamics shaped by multiple tiers of governance: Climate change and deforestation in Indonesia. In: *18th World IMACS Congress and MODSIM09 International Congress on Modelling and Simulation: Interfacing Modelling and Simulation with Mathematical and Computational Sciences, Proceedings, (July)* (eds. R.S. Anderssen, R.D. Braddock and L.T.H. Newham), 3025–3031. Cairns, Australia: Modelling and Simulation Society of Australia and New Zealand and International Association for Mathematics and Computers in Simulation.

Smit, B. and Pilifosova, O. (2003). From adaptation to adaptive capacity and vulnerability reduction. *Climate Change, Adaptive Capacity and Development*: 9–28.

Stakhiv, E.Z. (2011). Pragmatic approaches for water management under climate change uncertainty. *Journal of the American Water Resources Association* 47 (6): 1183–1196.

Swain, A. (1998). Fight for the last drop: inter-state river disputes in India. *Contemporary South Asia* 7 (2): 167–180.

Tambe, S., Kharel, G., Arrawatia, M.L. et al. (2012). Reviving dying springs: climate change adaptation experiments from the Sikkim Himalaya. *Mountain Research and Development* 32 (1): 62–72.

The Royal Commission on Environmental Pollution (2010). Adapting Institutions to Climate Change. https://assets.publishing.service.gov.uk/government/uploads/system/uploads/attachment_data/file/238460/7843.pdf (accessed November 2019).

Tompkins, E.L., Adger, W.N., Boyd, E. et al. (2010). Observed adaptation to climate change: UK evidence of transition to a well-adapting society. *Global Environmental Change* 20 (4): 627–635.

UNFCCC (2015). "Paris Agreement." UNFCCC, United Nations Framework Convention on Climate Change, Paris.

Vedeld, T., Coly, A., Ndour, N.M., and Hellevik, S. (2015). Climate adaptation at what scale? Multi-level governance, resilience, and coproduction in Saint Louis, Senegal. *Natural Hazards* 32 (Suppl. 2): 173–199.

Vignola, R., Klinsky, S., Tam, J., and McDaniels, T. (2012). Public perception, knowledge and policy support for mitigation and adaption to climate change in Costa Rica: comparisons with North American and European studies. *Mitigation and Adaptation Strategies for Global Change* 18 (3): 303–323.

Wang, X., Zhang, J., Shahid, S. et al. (2014). Adaptation to climate change impacts on water demand. *Mitigation and Adaptation Strategies for Global Change* 21: 81–99.

Whitehead, J.C. (2009). Building scenarios of adaptive capacity: a case of community water systems in central Pennsylvania. Thesis. University of Pennsylvania.

Whitehead, P., Wilby, R.L., Battarbee, R.W. et al. (2009). A review of the potential impacts of climate change on surface water quality. *Hydrological Sciences Journal* (November): 37–41.

Wilby, R.L. and Wood, P.J. (2012). Introduction to adapting water management to climate change: putting our science into practice. *Area* 44 (4): 394–399.

Wise, R.M., Fazey, I., Stafford Smith, M. et al. (2014). Reconceptualising adaptation to climate change as part of pathways of change and response. *Global Environmental Change* 28: 325–336.

Wyborn, C.A. (2015). Connecting knowledge with action through coproductive capacities: adaptive governance and connectivity conservation. *Ecology and Society* 20 (1): article 11.

Xu, J., Grumbine, R.E., Shrestha, A. et al. (2009). The melting Himalayas: cascading effects of climate change on water, biodiversity, and livelihoods. *Conservation Biology* 23 (3): 520–530.

22

Sustainability of Carbon Storage and Sequestration

Gilbert Hinge[1], *Rao Y. Surampalli*[2], *and Manish Kumar Goyal*[1, 3]

[1]*Department of Civil Engineering, Indian Institute of Technology, Guwahati, India*
[2]*Global Institute for Energy, Environment, and Sustainability, Lenexa, KS, USA*
[3]*Discipline of Civil Engineering, Indian Institute of Technology, Indore, India*

22.1 Introduction

Climate change due to global warming is the major concern of the twenty-first century. The drastic increase in greenhouse gas emissions has been recognized as the major cause of global warming. Greenhouse gases, particularly carbon dioxide (CO_2), nitrous oxide (N_2O) and methane (CH_4) trap most of the reflecting longwave radiation emitted by the earth to ensure that the earth maintains the moderate temperature required to support all life processes. However, human activities since the Industrial Revolution have contributed to an increase of these gases in the atmosphere. Atmospheric CO_2 concentrations have escalated from about 280 parts per million (ppm) prior to 1750–399.5 ppm in 2013, with methane levels from 722 to 1834 parts per billion (ppb) and N_2O levels from 270 to 328 ppb (IPCC 2014). Around 72% of the total greenhouse gases emitted is CO_2, whereas 18% is CH_4 and 9% is NO_x (Baumert 2005). CO_2 emissions are therefore considered as the major cause of global warming. According to IPCC (2007), increased levels of these concentrations are predicted to:

- elevate the earth's average temperature;
- affect the patterns and intensity of precipitation;
- raise the sea level;
- deplete snow cover, ice cover as well as permafrost;
- increase the duration, intensity and frequency of various extreme events;
- raise the acidity level of the oceans;
- change the ecosystem characteristics;
- increase the risk of many diseases.

These changes will affect our agricultural supply, water supply, biological communities and even our own wellbeing. Thus, the increase in atmospheric gas concentration necessitates

Sustainability: Fundamentals and Applications, First Edition.
Edited by Rao Y. Surampalli, Tian C. Zhang, Manish Kumar Goyal, Satinder K. Brar, and R. D. Tyagi

identification of strategies to mitigate climate change. There is no single technology that will be able to address climate change but a sum of many initiatives, such as carbon sequestration, may be one of the effective ways to address climate change (Goyal and Royal 2017). The success of such initiative lies in how much we can get rid of excess CO_2 in the atmosphere.

In a push to get rid of the carbon that we release into the atmosphere, it is necessary to understand the simple choices we make every day and how they affect carbon storage. We also need to understand the ways we can sequestrate CO_2 and track carbon changes in order to understand the effectiveness of various mitigation and adaptation strategies. This chapter gives a brief understanding of various sources and sinks of carbon as well as various methods of carbon sequestration. It presents the most common methods to quantify soil carbon storage and addresses the various adaptation and mitigation strategies of carbon management for mitigating climate change.

22.2 Carbon Sources and Sinks

CO_2 and other greenhouse gases are continuously recycled on earth. Processes that release CO_2 into the atmosphere are termed as carbon sources, whereas processes that help to absorb it are termed as carbon sinks. Soil, forests, oceans, the atmosphere and fossil fuels are known as carbon reservoirs. Carbon is constantly moving between these different reservoirs, and all of them can act as either a carbon sink or source. A sink accumulates and stores more carbon than it emits, whereas a source emits more carbon than it absorbs. The total amount of CO_2 in the atmosphere at any point in time is the net balance between sources and sinks. Carbon sources can be both natural and manmade. Natural sources of CO_2 come from volcano eruptions, plant and animal respiration, decomposition of soil organic matter (SOM), etc. The largest natural sources are from the ocean release in respiration from the whole community of the marine ecosystem. Natural carbon sinks include plants, soils and oceans. In the ocean, the marine ecosystem takes up CO_2 for photosynthesis, while some of the CO_2 simply dissolves in the ocean naturally. Before the Industrial Revolution, the amount of carbon stored and transferred between these various reservoirs was relatively balanced. Forests are known to be major sinks of atmospheric CO_2; however, human activities such as deforestation and forest fires have reduced the size of these sinks, leading to more CO_2 in the atmosphere. The extraction and use of natural resources such as oil and coal are the major human carbon sources. Our daily activities such as power generation, transportation, chemical production, industrial production and even some agricultural activities are various anthropogenic activities that lead to carbon emission.

22.3 Types of Carbon Sequestration

Carbon sequestration is a term that refers to the process of capturing and storage of CO_2 (Lal 2008). Capturing can happen at the source of emission (e.g. a power plant) or naturally through a process of photosynthesis, which can be enhanced with various proper management practices. Various sequestration methods are described below.

22.3.1 Geological Sequestration

Geological sequestration includes three processes: capturing CO_2 from industries or energy-related sources (e.g. a power plant), transporting the captured CO_2 and injecting it into deep rock formations for permanent or semi-permanent storage (Bachu 2003). The CO_2 is put into the geological formation with the help of injection wells. This well is similar to an oil or water well, but, unlike the latter, it is used for injecting CO_2 rather than extracting materials (oil or water). The geological formation must possess certain features for it to be suitable for CO_2 sequestration. It should be deep enough and should contain a non-porous layer so that the injected CO_2 is not re-emitted back to the surface. The formation should also have sufficient permeability and porosity so that it can take in the injected CO_2. If the formation contains certain rocks that would react with the CO_2 and form certain compounds that will stay permanently within the subsurface, that increases the suitability of said formation for CO_2 sequestration. Suitable sites generally used for geological sequestration include depleted gas and oil fields, basalt formations or deep natural formations filled with saline water. Unlike terrestrial sequestration where carbon can go back to the atmosphere when plants die or when the soil is disturbed, geological sequestration is permanent.

22.3.2 Ocean Sequestration

The ocean represents the largest sinks for atmospheric CO_2. The ocean has been an important sink for anthropogenic CO_2 with a magnitude similar to the terrestrial sink, but with time the strength of the ocean as a carbon sink has declined. More and more CO_2 in the ocean surface has led to ocean acidification causing harm to marine ecosystems and microorganisms. In the ocean, CO_2 dissolves by reacting with carbonate ions. However, these ions are limited and, as their concentration decreases, the rate of CO_2 intake by the ocean is also declining. Ocean sequestration is, therefore, a method proposed to distribute CO_2 more evenly throughout the ocean depths so as to prevent or reduce the CO_2 impact on the ocean surface (Herzog 1998). Ocean sequestration includes two major methods: direct injections of CO_2 into the deep ocean, similar to geological sequestration, and ocean fertilization with the addition of iron to increase phytoplankton growth and photosynthesis rate. Models related to ocean sequestration have reported that by injecting CO_2 into the deep ocean, it will remain there for thousands of years, keeping it isolated from the atmosphere, and it will recycle back into the atmosphere only after the millennium scale (Herzog 1998). To date, there are still uncertainties about the feasibility of ocean sequestration; there is still a lack of clear understanding about ocean biology, and the chemical reactions that ocean sequestration may impose on the ecosystem. This is one of the ambitious schemes that involves a lot of uncertainties but may also prove to be effective to slow down climate change.

22.3.3 Terrestrial Sequestration

Terrestrial sequestration involves capturing and storing of CO_2 by plants and storage of carbon in the soil. Plants capture CO_2 from the atmosphere and use it in the process of photosynthesis. During this process, the photosynthetic cell containing chlorophyll converts the carbon from CO_2 into energy-rich organic molecules (glucose, starch) that can be stored in leaves, stem, roots, etc. Thus, plants that live for many years like trees can store carbon for a long period of time. Once the trees die, or when the trees shed their leaves or seed or when the trees drop their blossom, the plants or their materials decompose. During the process, a portion of the carbon is released back into the atmosphere, while most of the carbon is sequestrated in the soil. The stored carbon in the soil in the form of SOM is known as soil organic carbon (SOC). Batjes (1996) and Eswaran et al. (1993) reported that around 1550 Pg (1.55×10^{15} kg) of SOC can be found just in the upper 1 m of the soil, which is about three times and twice the amount of carbon stored in the aboveground biomass and atmosphere, respectively (Lal 2008).

While terrestrial sequestration is a natural process and will keep on occurring, the role of human activities to preserve the already sequestrated carbon and to bolster the carbon sequestration capacity of the land is important in order to combat global warming. For example, carbon may be sequestered along a forest cover. Since carbon is released when forest cover is converted into cropland, the owner is preserving the terrestrial carbon sequestration capacity of that particular land by preventing the switch to cropland. Terrestrial carbon sequestration can be also thought of a way to enhance and promote the carbon storage of land. An example of promoting carbon storage is through forest restoration, afforestation or forest management. Forest management includes activities that help to reduce forest fires, prevention of pest diseases, removal of diseased plants. Reforestation involves replanting of trees in areas where timber has been harvested, whereas afforestation involves the planting of trees where land has been used for other purposes such as for cropland, etc. Land use management such as practices of zero tillage, reduced tillage (West and Post 2002), use of cover crops and organic manure (Compton and Boone 2000), application of compost (Swarup 1998) and application of proper irrigation (Bordovsky et al. 1998) have all been reported to improve the terrestrial carbon sequestration capacity of the land.

Although engineering methods such as geological sequestration and ocean sequestration are promising methods to reduce CO_2 levels, this chapter focuses on the natural method, i.e. soil carbon sequestration, and on the issues associated with the sustainability of the soil carbon sink. Such a focus is necessary as soil carbon sequestration has been widely practiced, but there is an urgent need to understand how to quantify the amount of carbon stored in the various landscape and to analyze the various management systems and methods that can improve or affect its storage. Therefore, the later sections of the chapter specifically discuss the methods to quantify the soil carbon stock, the methods with which soil carbon can be modeled, and the various technologies and management practices to enhance soil carbon.

22.4 Methods for Quantification of Soil Organic Carbon Stocks

The quantity of deposited carbon in the soils relies upon various factors such as climate, soil types, vegetation, parent soil materials, topography and human activities (Dai and Huang 2006; Hinge et al. 2018a). These various factors vary spatially and temporally and affect the distribution of SOC, which hinders accurate quantification of SOC

(Wang et al. 2014; Hinge et al. 2018b). Therefore, a prior decision on the selection of appropriate sites and the number of representative samples is important in the quantification of SOC. As SOC also varies with depth, an appropriate sampling depth also needs to be considered before sampling. Methods for measuring SOC can be broadly categorized into two categories: direct methods and indirect methods (Figure 22.1), and they are described as follows.

22.4.1 Direct Methods

Direct methods are time- and cost-consuming methods, as they require sample collection (as shown in Figure 22.2), sample preparation, chemicals and technology for analysis. However, they provide accurate results compared to indirect methods. The steps involved in direct methods are discussed in the upcoming subsection.

22.4.1.1 Analytical Measurement

The amount of SOC content interpreted in term of concentration (% or $Mg\,C\,g^{-1}$ soil) can be measured easily in the laboratory with the three methods mentioned below. The value of carbon reported can be converted into carbon stock, i.e. in terms of area basis ($kg\,m^{-2}$ or $t\ ha^{-1}$) if we have information on bulk density and sampling depth. When comparing carbon of different sites or different land use, we should compare their carbon stocks rather than their carbon concentration as carbon values will vary with soils that are more or less compacted. For instance, if soil samples collected from the upper 20 cm layers of sites C and sites D have the same carbon concentration values (say 3%), and the bulk density of site C is twice ($2\,g\ cm^{-3}$) that of site D ($1\,g\ cm^{-3}$), then site C will have more carbon stock than site D. Hence it is always advisable to compare the SOC in term of carbon stock. The most popular laboratory methods that are used to calculate SOC are described below.

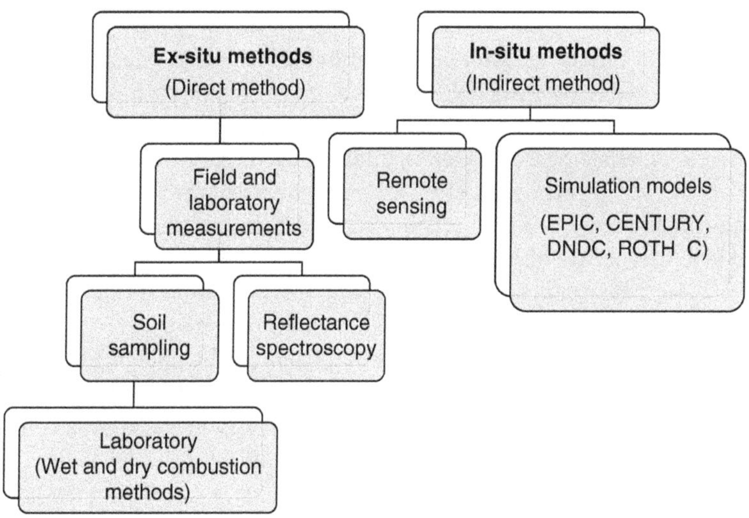

Figure 22.1 Methods for measuring soil organic carbon. *Source:* Modified from Post et al. (2001).

Figure 22.2 Soil sample collection using core cutter (left) and soil sample preparation (right).

1. *Walkley and Black method*: The Walkley and Black method (Walkley and Black 1934) is a rapid dichromate oxidation–reduction method. It works based on the following equations:

   i. $$2Cr_2O_7^{2-} + 3C^{\circ} + 16H^{+} = 4Cr^{3-} + 3CO_2 + 8H_2O \quad (22.1)$$

   ii. $$Cr_2O_7^{2-} + 6Fe^{2+} + 14H^{+} = 2Cr^{3+} + 6Fe^{3+} + 7H_2O \quad (22.2)$$

   The amount of Cr_2O_7 reduced during the above reaction is presumed to be equal to the amount of organic carbon existing in the soil sample. Excess Cr_2O_7 left over is then titrated with standard ferrous ammonium sulfate solution to discover the volume of dichromate reduced during the reaction (Chatterjee et al. 2009). The soil carbon values were finally adjusted by a correction factor of 1.32 since this method failed to give complete oxidation of soil carbon. Many studies (Sutherland 1998; Kamara et al. 2007) argue that this recovery factor is not sensible as soils arriving from diverse regions, different soil depths, with different mineralogy, are affected by different factors that will lead to variation in the recovery factor. However, Kamara et al. (2007) prove that the average value of the recovery factor in respective soil groups is not much different; thus it is possible to report the SOC value irrespective of soil types, depths, etc. using the same recovery factor. Another downside of this method is the production of hazardous chemicals after the reaction. However, despite these drawbacks, it is the most commonly used method because of its simplicity, low cost, minimum equipment requirement and availability.
2. *Loss on ignition method*: The loss on ignition method (Sutherland 1998) estimates carbon concentration based on differences in the mass of soil samples after its being ignited for a sustained period. The ranged of temperature employed in this method is from 300 to 550°C and the length of ignition period can be between 2 and 17 hours. This method assumes that organic carbon present in the soil is volatile at this temperature and duration. Thus, organic carbon is measured as the difference in mass of soil samples

before and after heating. This method is very simple and relatively inexpensive compared to Walkley and Black's method. This method requires only an oven or muffle furnace that can reach a high temperature and weighing scales, which is easily available. However, this method may overestimate the amount of organic C as combustion of soil components other than organic matter may lead to more weight loss. Carbonates present in the soils may also break down at these temperatures. Factors such as furnace type, duration and temperature of ignition may also influence the accuracy of this method.

3. *Dry combustion method*: The dry combustion method involves heating up a soil–catalyst mixture at 1000°C in a furnace in the presence of CO_2 free air or a stream of O_2, which is then followed by quantification of evolved CO_2 (Heanes 1984) An automated C analyzer is the instrument that is widely used for dry combustion. The various automated steps involved in quantification of SOC by this method are: (i) conversion of soil carbon to CO_2 in a high-temperature chamber by oxidation; (ii) separation of CO_2 from the other gases; and (iii) quantification of CO_2 concentration by mass spectrometry, thermal conductivity or infrared gas analyzing methods (Mikhailova et al. 2003). This method is accurate, quick and does not yield hazardous wastes. However, the instrument is very costly, and an increased level of training is required (Chatterjee et al. 2009).
4. *Reflectance spectroscopy*: This method utilizes the concept of differences in absorption and reflection of light by the carbon content present in the soil samples. This method was designed for remote sensing but was modified for handheld sensors on the ground. Before analysis, the instrument needs to be calibrated with known soil carbon values measured in the laboratory, particularly those obtained from combustion method for better accuracy. This method is quick and non-destructive and can measure a large number of samples in a short period of time. However, the instrument is very costly, and an increased level of training is required. Also, the presence of moisture may affect the amount of absorption and reflection of light. Hence, a prior measurement of soil moisture is required before taking any readings.

22.4.2 Indirect Methods

Indirect methods, which include remote sensing and model simulations, are generally faster and can be carried out on a large scale. However, they are less accurate compared to direct methods. Descriptions of some indirect methods are as follows.

22.4.2.1 Estimating SOC from Soil Reflectance Through Remote Sensing

SOC can be estimated indirectly from soil reflectance based on the concept that there is a correlation between reflectance in certain spectral bands and SOC (Daniel et al. 2004; Ladoni et al. 2010b). Many researchers have attempted to estimate SOC from reflectance and have come out with moderate to good results.

Soil predominantly has reflectance spectra in the 1100–2500 nm range, which comprises three distinct absorption peaks about the 1400, 1900 and 2200 nm range and a small absorption peak between 2200 and 2500 nm (Chang 2000). An increase in organic content of the soil will cause a decrease in soil reflectance. However, with soil being a complicated material, a decrease in soil reflectance may be also due to the increase of some other soil properties. Therefore, many researchers have studied and indicated the portion of the spectrum that might be best for predicting SOC. Table 22.1 shows the reflectance source, spectral ranges and different methods carried out by different researchers across the world for predicting SOC. The reported R^2 in the middle infrared, near infrared and visible region was high.

22.4.2.2 Carbon Modeling

Modeling enables us to forecast the likely changes in SOC in the future and also helps in the identification of potential areas for soil carbon sequestration. The areas with high carbon storage can be conserved and protected. It also helps to identify the potential management practices that can promote SOC storage. Many models have been used and tested in attempts to evaluate the effect of climate change on soil carbon and crop production. These models help us to compare experimental research findings across different sites, extrapolate the data and results to wider environments, and help to develop management recommendations and policies to support systems. Here, we briefly describe the four carbon models that are widely used across the world.

1. *Century model*: The model was established to simulate long-term (months to centuries) SOM, nitrogen (N), sulfur (S), phosphorus (P) and even plant growth for various plant–soil systems. It was initially established for grassland soils (Parton et al. 1987), but later it was improved and widely used for cropland and even for forest soils. The model simulates the dynamics of various ecosystems by considering climate factors, human management factors and soil properties. It works on a monthly basis, and its major variable inputs are soil texture and weather data. The Century model has been successfully applied and tested for long-term data and even plot-level experiments across various soil types and climatic conditions of the world. Lima et al. (2011) conducted a study in Brazil with the aim of evaluating the changes of SOC stock for two eucalyptus chrono-sequences using a Century model and then compared the carbon stock from the field with that simulated by the model. Their result as simulated by the Century model indicated that the carbon stock declined after 37 years of inadequately managed pasture. Also, the measured soil carbon from the field is very close to that estimated by the Century model (root mean square error = 20.9; model efficiency = 0.29). However, for a lower soil profile it was found that the model overestimated the SOC stock. Cong et al. (2014) used the Century model to simulate SOC in a mixed wheat–corn cropping system in three long-term fertilization trial sites. Their results showed that the model is capable of simulating the fertilization effect on the soil carbon dynamic (normalized root mean square error $\leq$ 15% for all treatments).
2. *DNDC model*: The denitrification–decomposition (DNDC) model was initially designed for the simulation of soil carbon and soil nitrogen in a cultivated ecosystem (Li et al. 1994). The DNDC model was developed to predict the amount of carbon sequestration in the soil, crop yield, nitrogen leaching and to trace gas emissions from the agroecosystem on a spatial scale ranging from site to region. The model works on a daily time step. In the DNDC model, the SOC comprises four pool components: plant residues, microbial biomass, active humus and passive humus. These pools are divided based on their decomposition rates, which are determined based on soil clay contents, nutrient availability, soil temperature and moisture. Several modifications of the model have been carried to improve its performance; since then various sub models such as Manure

Table 22.1 List of literature showing the prediction of soil organic carbon from remote sensing.

Study area	Reflectance source	Spectral region	Spectral range (nm)	Methods	R^2 (spectral range and organic carbon)	Authors
Australia	Infra Analyzer 500 C	NIR	1100–2500	Multiple regression	0.93	Dalal and Henry (1986)
USA	Perstorp NIR Systems 6500	VIS–NIR	400–2498	Partial least square regression	0.89	Chang (2000)
Georgia	Aerial photograph	B–G–R	—	Logarithm linear equation	0.92	Chang (2000)
U.S.A	Fourier transform spectrometer	NIR	1100–2300	Partial least square regression	0.94	Reeves et al. (2001)
Greece	Ground-based sensor	VIS–NIR	350–2500	Multiple Regression	0.80	Shepherd and Walsh (2002)
Thailand	Stellar net spectroradiometer	VIS–NIR	400–1190	Multiple regression	0.65	Daniel et al. (2004)
Australia	Varian Cary 500	NIR	700–2500	Partial least-squares regression	0.60	Viscarra Rossel et al. (2006)
Brazil	Landsat TM	G–R–NIR	500–885	Multiple linear regression	0.51	Nanni and Demattê (2006)
Germany	Hyperspectral image	VIS–NIR	350–2500	Partial least square regression	0.74	Patzold et al. (2008)
Iran	LISS iii	VIS–SWIR	520–1750	PCA, soil line, simple regression	0.75	Ladoni et al. (2010a)
India (Punjab)	ASD Field SpecPro spectro radiometer	VIS–NIR	460–2380	Partial least square regression	0.81	Srivastava et al. (2015)

NIR, near-infrared; VIS, visible, B, blue; G, green; R, red; SWIR, short-wave infrared.

DNDC, Forest DNDC, Modified DNDC for Indian rice fields (Pathak et al. 2005) and many more have been developed. It is widely applied across the world. Syp et al. (2011) conducted an experiment in Poland to estimate the effect of climate change on greenhouse gas emissions under different management practices using the DNDC model by comparing the output of the baseline climate from the model with output of future climate scenarios to get an insight into how gas emissions may change under climate change and different management practices. They reported that by applying conservation tillage the global warming potential reduces by approximately 17–19% as per the baseline scenario, 16–18% in the 2030 scenario and by 15–17% in 2050 scenarios.

3. *Roth C model*: The Roth C model is designed to calculate the turnover of SOC in non-waterlogged upper soil layer by considering the effect of parameters such as temperature, soil type, moisture content and plant cover. It operates on a monthly time step to compute the amount of total organic carbon and microbial biomass carbon for a timescale ranging from years to even centuries (Coleman and Jenkinson 2005). The model requires only a few inputs, which are easily obtainable. A new version of this model whereby the monthly time step has been replaced by daily time steps was later published. King et al. (1997) have modified the Roth C model into a much larger model for global C cycling. The model was first developed and parameterized to model the turnover of SOC of the arable upper soil layer data obtained from Rothamsted Long-Term Field Experiments – hence how it got its name. Subsequently, it was expanded to model soil carbon turnover rate in woodland and in grassland under different soils and different climates. This model was widely applied across the world. Bhattacharyya et al. (2011) conducted a study in Maharashtra India to test the performance of the Roth C model to estimates changes in total organic carbon from four different sites that have been subjected to long-term fertilizer application. Their results disclosed that Roth C could simulate changes in total organic carbon for topsoil layers. The root means square error in the four experimental sites ranged from 11.50 to 15.01, 4.70 to 11.60, 2.14 to 6.52 and 1.45 to 13.74, respectively.
4. *EPIC model*: Environmental Policy Integrated Climate (EPIC) is a simulation model designed to perform a long-term simulation of climate, soil and management relationship. It was established by the United States Department of Agriculture (USDA) modeling team to investigate the impact of soil erosion on crop productivity. The model works using daily time steps. According to Gassman et al. (2004), EPIC is vast in terms of its potential to model various biophysical processes, some of which include weather processes, irrigation, tillage effects, nutrient management, simulation of plant growth and even economic budget. The model has been redefined and extended continuously since its original development and is capable of solving many agriculture-related problems such as crop yield problems, nutrient management, nutrient cycling, irrigation studies, wind and water erosion, soil carbon sequestration studies, regional assessment, economic and environmental studies and climate change impact on crop yields (Gassman et al. 2004). Many researchers have applied EPIC across the world. Toure et al. (1995) reported that the EPIC model had high potential for analyzing the impact of climate change on crop production and soil carbon. They reported this after evaluating five crop models; namely, CERES, EPIC, Century, Sinclair and Stewart, using long-term experimental data on loam soils.

They compared the wheat yield predicted by these five models and reported that out of these five models, the EPIC model had the highest capability for evaluating the effect of climate change on soil carbon and wheat yield.

22.5 Adaptation and Mitigation Policy for Carbon Management

22.5.1 Adaptation

Adaptation to climate change means anticipating the likely effect of the current and projected climate and taking suitable measures to avoid its adverse effects or to exploit the favorable opportunities (IPCC 2007). Enhancing carbon storage in soil is one of the adaptation strategies against climate change (Lal 2008). The amount of terrestrial carbon storage relies upon numerous factors; namely soil types, vegetation types, climate, cropping systems, etc. Since human activities can greatly alter the amount of soil carbon storage (Penman et al. 2006), evaluating the impact of human activities on carbon storage is necessary for carbon management. Some of the recommended human practices that can serve as an adaptation strategy are briefly described below.

1. *Reduced tillage*: Much of the blame for the loss of carbon from the soils has been attributed to the practice of poor soil management (Reicosky 2003). Lal and Bruce (1999) reported that most of the lost carbon can be restored with appropriate management. One of the primary practices mentioned is the practice of conservation tillage such as zero tillage or reduced tillage. Several studies have reported the use of no-till systems, and how it can improve the soil carbon storage. In a 3-year experiment performed on a sandy loam of New Delhi (Saha et al. 2010), the consequences of tillage application and residue management on soil properties and soil management were tested. It was observed that zero tillage along with the addition of residue as inputs resulted in a higher SOC and microbial biomass carbon content. West and Marland (2002) conducted a study in the US and reported that carbon emissions from corn cultivation with a conventional, reduced or no-tillage farming system were 72.02, 45.27 or 23.26 kg C ha^{-1}, respectively. Van den Bygaart et al. (2003), from his study in western Canada, reported that the practice of reduced tillage improves the carbon sequestered in the soil by a value of approximate 320–150 kg ha^{-1}. However, for all that, the advantages of no-till on SOC improvement may depend on soil types and locations (Wander et al. 1998). For instance, it is reported that no-tillage does not improve SOC in lower soil layers (Luo et al. 2010). Syswerda et al. (2011) concluded that conventional tillage outperformed the no-till system when sampled to a greater depth. Due to the lack of data at a greater depth to compare no-till systems with another system, we cannot strongly conclude that the no-till system is more effective than other systems.
2. *Cover cropping*: Another promising option for soil carbon sequestration in agricultural land is the incorporation of cover crops. Cover crops are grown after the main crops to protect the bare soils and are plowed as green manure before sowing the next main crops. Besides improving soil carbon storage, cover crops can also protect the soil surface

and improve the soil water balance rate (Lal 2004). Bare and till soils are most vulnerable to water and wind erosion. The presence of cover crops can protect the soil from erosion. Standing biomass of cover crops can create a microclimate that shelters the soil from high surface temperature and thus lowers the soil water use rates compared to that of bare soils. The contribution of cover crops to increase the soil carbon content and their role in mitigating climate change has been highlighted in many studies. The result of 10 years of experimental data carried out in India (Venkatesh et al. 2013) reported that incorporation of cover crops produced a biomass of approximately 3–4.3 Mg ha^{-1} $year^{-1}$; the presence of this biomass has led to the improvement of soil carbon stocks. Olson et al. (2014) conducted a 12-year experiment to study the effects of cover crops on SOC sequestration and reported that for the same soil layer and soil types and for the same tillage treatment, the soil which had a cover crop on it was found to store more soil carbon than the one without cover crops.

3. *Nutrient management*: Nutrient management practices such as the application of chemical fertilizers, manure and retention of crop residues (straw) have been identified as the most efficient way to promote SOC accumulation and prevention of soil carbon losses. Application of fertilizers and manure in areas with nutrient deficiency will increase soil fertility and crop yield and ultimately will improve the biomass residues and root carbon inputs into the soils. Pathak et al. (2011) analyzed a rice-based long-term experiment and concluded that application of fertilizers along with manure (NPK + FYM) helps in sequestrated 0.33 MgC ha^{-1} $year^{-1}$ while the NPK treatment sequestered 0.16 MgC ha^{-1} $year^{-1}$ in comparison to the control treatment. Application of farm manure at 5–10 t ha^{-1} $year^{-1}$ combined with balanced NPK was found to improve organic carbon by 10.7% (Mandal et al. 2007). With the evolution of conservation tillage, reduced tillage along with nutrient management will certainly help to promote soil carbon sequestration. However, one must also be careful with the type of fertilizer, manure and dosage being applied as they may bring risks to the soil and plant's health.
4. *Irrigation*: Irrigation is essential for crop production as well as for increasing organic residues into the soils. Irrigation can influence the soil in many ways and can cause it to become a sink and a source to atmospheric CO_2 (Figure 22.3). An improved water supply will promote plant growth and thus lead to more inputs of carbon into the soils in the form of dead plant materials and through the roots. However, excessive irrigation will lead to higher soil moisture content, which will enhance soil microbial activity (Trost et al. 2013). This will result in a higher decomposition rate of SOC and thus lead to more carbon emissions into the atmosphere. The effect of irrigation on SOC will depend on not only the amount of soil moisture but also the interaction with various fertilizers and manure applied and with various soil microorganisms. Thus, efficient management of water along with fertilizers and manure is necessary for SOC enhancement.

22.5.2 Mitigation

The term mitigation refers to change in technologies, change in management practices and people's behavior, and the use of alternatives to reduce and prevent the emission of greenhouse gases. Producing biofuels that can replace fossil fuels, conversion of wasteland and

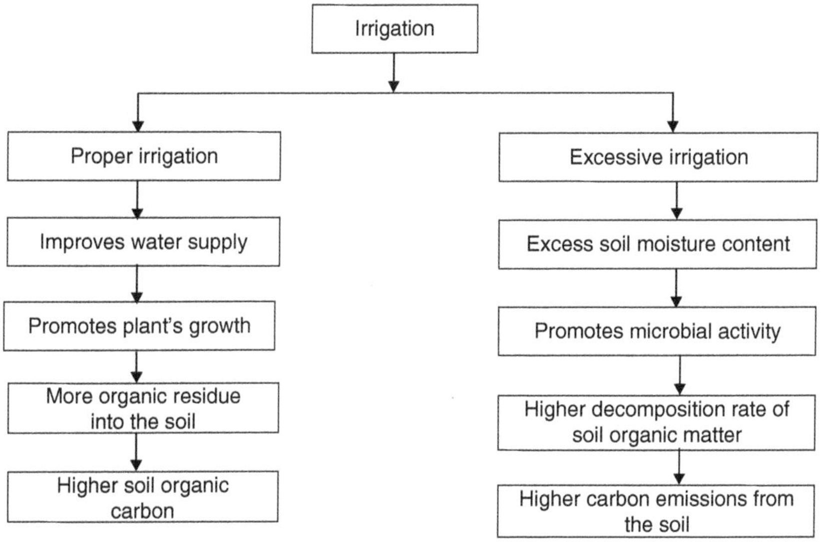

Figure 22.3 Effect of irrigation on SOC.

other land into forest land, management of water, etc. are some of the potential mitigation options. Some of these are briefly described below:

1. *Food waste recycle as biofuel*: Food waste can be recycled as a biofuel in order to reduce carbon emission from the burning of fossil fuels as well as to increase energy production. Food waste can be recycled in a number of ways. Among these, one is the conversion of yellow and other grease generated from food waste into diesel fuel energy. Restaurants are a good source to provide these greases since they produce lots of cooking oils that come either from animal fats or plant oils. Converting food waste into biofuel not only helps in reducing the amount of waste being generated but also provides a more eco-friendly fuel, which produces less than half the amount of greenhouse gases compared to diesel fuels. Another technique to convert food waste into useful energy is with the help of an anaerobic digester. This technique involves several stages in which, with the help of microorganisms, the food waste biomass is converted into methane, which can then be used to generate power for an electric generator. Pyrolysis is another versatile method that can be used to convert food waste biomass into energy. This method involves heating food waste biomass in the presence of a limited amount of oxygen where waste is converted into pyrolysis oil and other liquid waste, which can be used directly as fuel.
2. *Agroforestry*: Agroforestry represents another interesting option that can improve the carbon sequestration potential of both soils and vegetation. It refers to the method of growing crops, trees and livestock on the same land. Agroforestry has a great potential as a mitigation measure against climate change as it can sequester a huge amount of carbon with its multiple crop and tree species. In addition, various components of agroforestry provide different harvests at different times. Thus, besides helping in offsetting greenhouse gases, such practices help in reducing the risk of crop failure, provide an

alternative income to the farmer, and improve crop production and soil fertility. However, depending on the type of agroforestry system practice, it can function both as a sink and source for greenhouse gases. For example, agrosilvicultural practices are reported to be a net sink of atmospheric carbon, whereas ruminant-based silvopastoral systems are a net source of greenhouse gases. Therefore, one must adopt a proper agroforestry system that can have a positive impact on carbon storage.

3. *Bioenergy crop as an alternative energy*: Bioenergy refers to the energy acquired from living materials such as crops, animal waste or from woods. Bioenergy crops refer to any plants that can generate bioenergy. Crops that can operate as a collector of solar energy and storage of chemical energy form the basis for bioenergy crops. In fact, before the discovery of oil in 1859, agricultural and forestry crops along with their residues were the main source of energy. Depletion of oil reserves and the recognition that the increase in greenhouse gases has led to climate change have brought attention to the need to reduce dependency on fossil fuels. This, in turn, has brought attention to again promoting bioenergy and biofuel as renewable energy resources. Bioenergy crops will play a pivotal role in the reduction of greenhouse gas concentrations and can reduce the burden on fossils fuels (Karp and Shield 2008). Planting bioenergy crops on degraded land can be one of the potential ways to promote C sequestration in degraded lands. It is reported that bioenergy crops can sequester carbon in degraded land in the range of ~0.6 to 3 Mg C $ha^{-1}year^{-1}$ (Dipti and Priyanka 2013). However, even though bioenergy crops can be one of the solutions for the mitigation of greenhouse gases, the process involved is complex and contentious (Lal 2008). Though they absorb carbon dioxide from the atmosphere, they generate the same amount of carbon dioxide when burned, and, thus, bring no net balance to the amount of carbon sequestrated. They can help in reducing the amount of fuel consumption; however, they cannot entirely substitute for fossil fuels because to produce even a small amount of bioenergy a huge amount of bioenergy crops and cropland is required.
4. *Restoration of degraded land*: Restoration of degraded land by revegetation is another mitigation option to restore soil quality and carbon content of both soil and biomass in degraded land. Restoration can be of any land type such as forest, cropland, mining areas, deserts and other lands. Restoration can bring various benefits: not only can carbon can be sequestrated, but non-productive and abandoned land can be brought back into use. Degraded land has a greater potential for carbon sequestration compared to disturbed land because often these lands still contain a small amount of carbon in them (Eggleston et al. 2006). However, restoration of degraded land is not always simple, as there are always factors that restrict this potential. For every piece of degraded land, there is an underlying cause. Therefore, to restore them, first, the cause must be identified and addressed. For example, if the land is polluted with heavy metals, then those metals need to be removed before revegetation. Knowledge of vegetation and land characteristics before it became degraded is also important. Thus, because of this required intervention, extra cost is involved. If the sites are in remote areas, additional cost for transportation will be needed. Even though remediation might be expensive, it represents another option as it provides additional benefits to this land, which would otherwise remain abandoned and worthless.
5. *Other options*: Some of the other management options for mitigation of climate change include: crop rotation; intercropping; recycling of municipal waste; use of

biochar; erosion control; water conservation; reduced dependency on a single source of energy; adjustment of planting dates; biogas; and water reuse.

22.6 Conclusions

The increased concentration of atmospheric CO_2 is exacerbating the impact of global warming and environmental pollution. Carbon capture and storage in soil, plants, oceans and geological formations can be one of the initiatives to slow down the effects of global warming. This chapter has mainly focused on soil carbon sequestration, keeping in view the numerous ancillary benefits that this approach could provide, such as soil fertility and various ecosystem services. While increasing carbon in the soil is a difficult task, following some of the above-mentioned adaptation and mitigation measures may present an option to improve crop production as well as soil carbon storage to slow down climate change.

However, it must be noted that it will not be possible to overcome the problem of global warming by using only a single approach such as soil carbon sequestration or any of the engineering carbon sequestration technologies. Depending upon the availability of resources and the applicability of these methods in a particular location, these methods along with other initiatives such as reduction of fossil fuels combustions, increased use of renewable resources such as wind power, solar energy, biomass fuel, etc. must be used in combination to achieve an efficient and sustainable mitigation plan against global warming.

References

Bachu, S. (2003). Screening and ranking of sedimentary basins for sequestration of CO_2 in geological media in response to climate change. *Environmental Geology* 44 (3): 277–289.

Batjes, N.H. (1996). Total carbon and nitrogen in the soils of the world. *European Journal of Soil Science* 47 (2): 151–163.

Baumert, K. (2005). Navigating the numbers: greenhouse gas data and international climate policy. www.wri.org (accessed November 2019).

Bhattacharyya, T., Pal, D.K., Deshmukh, A.S. et al. (2011). Evaluation of RothC model using four long term fertilizer experiments in black soils, India. *Agriculture, Ecosystems and Environment* 144 (1): 222–234.

Bordovsky, D.G., Choudhary, M., and Gerard, C.J. (1998). Tillage effects on grain sorghum and wheat yields in the Texas Rolling Plains. *Agronomy Journal* 90 (5): 638–643.

Chang, C. (2000). Near-infrared reflectance spectroscopic measurement of soil properties. DPhil dissertation. Iowa State University. https://lib.dr.iastate.edu/rtd/12315 (accessed November 2019).

Chatterjee, A., Lal, R., Wielopolski, L. et al. (2009). Evaluation of different soil carbon determination methods. *Critical Reviews in Plant Sciences* 28 (3): 164–178.

Coleman, K. and Jenkinson, D.S. (2005). RothC 26.3. A model for the turnover of carbon in soil. Version Windows 2.0. (June). https://www.rothamsted.ac.uk/sites/default/files/RothC_guide_WIN.pdf (accessed November 2019).

Compton, J.E. and Boone, R.D. (2000). Long-term impacts of agriculture on soil carbon and nitrogen in New England forests. *Ecology* 81 (8): 2314–2330.

Cong, R., Wang, X., Xu, M. et al. (2014). Evaluation of the CENTURY model using long-term fertilization trials under corn-wheat cropping systems in the typical croplands of China. *PLoS One* 9 (4) https://doi.org/10.1371/journal.pone.0095142.

Dai, W. and Huang, Y. (2006). Relation of soil organic matter concentration to climate and altitude in zonal soils of China. *Catena* 65 (1): 87–94.

Dalal, R.C. and Henry, R.J. (1986). Simultaneous determination of moisture, organic carbon, and total nitrogen by near infrared reflectance spectrophotometry. *Soil Science Society of America Journal* 50 (1): 120–123.

Daniel, K.W., Tripathi, N.K., Honda, K., and Apisit, E. (2004). Analysis of VNIR (400-1100 nm) spectral signatures for estimation of soil organic matter in tropical soils of Thailand. *International Journal of Remote Sensing* 25 (3): 643–652.

Dipti and Priyanka (2013). Bioenergy crops an alternative energy. *International Journal of Environmental Engineering and Management* 4 (3): 265–272.

Eggleston, H.S., Buendia, L., Miwa, K. et al. (2006). *IPCC Guidelines for National Greenhouse Gas Inventories*, vol. 2, 48–56. Hayama, Japan: Institute for Global Environmental Strategies.

Eswaran, H., Van Den Berg, E., and Reich, P. (1993). Organic carbon in soils of the world. *Soil Science Society of America Journal* 57 (1): 192.

Gassman, P., Williams, J., Benson, V., Izaurralde, R.C., Hauck, L.M., Jones, C.A., Atwood, J.D., Kiniry, J.R. and Flowers, J.D. (2004). Historical development and applications of the EPIC and APEX models. ASAE/CSAE Annual International Meeting. Meeting Paper No. 042097, 31.

Goyal, M.K. and Royal, I. (2017). Soil carbon sequestration: an alternative option for climate change mitigation. In: *Natural Resources Management: Concepts, Methodologies, Tools, and Applications*, 188–212. Hershey, PA: IGI Global.

Heanes, D.L. (1984). Determination of total organic-C in soils by an improved chromic acid digestion and spectrophotometric procedure. *Communications in Soil Science and Plant Analysis* 15 (10): 1191–1213.

Herzog, H. (1998). Ocean sequestration of CO_2: an overview. Proceedings of the AWMA's Second International Specialty Conference, October 13–15, 1998, Washington, DC.

Hinge, G., Surampalli, R.Y., and Goyal, M.K. (2018a). Prediction of soil organic carbon stock using digital mapping approach in humid India. *Environmental Earth Sciences* 77 (5): 172.

Hinge, G., Surampalli, R.Y., and Goyal, M.K. (2018b). Regional carbon fluxes from land-use conversion and land-use management in Northeast India. *Journal of Hazardous, Toxic, and Radioactive Waste*, American Society of Civil Engineers 22 (4) 4018016.

IPCC (2007). *Climate Change 2007: The Physical Science Basis. Contribution of Working Group I to the Fourth Assessment Report of the Intergovernmental Panel on Climate Change.* Cambridge, UK: Cambridge University Press.

IPCC (2014). *Climate Change 2014: Synthesis Report Summary Chapter for Policymakers.* Geneva, IPCC.

Kamara, A., Rhodes, E.R., and Sawyerr, P.a. (2007). Dry combustion carbon, Walkley–Black carbon, and loss on ignition for aggregate size fractions on a Toposequence. *Communications in Soil Science and Plant Analysis* 38 (15–16): 2005–2012.

Karp, A. and Shield, I. (2008). Bioenergy from plants and the sustainable yield challenge. *New Phytologist* 179 (1): 15–32.

King, A.W., Post, W.M., and Wullschleger, S.D. (1997). The potential response of terrestrial carbon storage to changes in climate and atmospheric CO_2. *Climatic Change* 35 (2): 199–227.

Ladoni, M., Alavipanah, S., Bahrami, H.A., and Noroozi, A.A. (2010a). Remote sensing of soil organic carbon in semi-arid region of Iran. *Arid Land Reseach and Management* 24 (4): 271–281.

Ladoni, M., Bahrami, H.A., Alavipanah, S.K., and Norouzi, A.A. (2010b). Estimating soil organic carbon from soil reflectance: a review. *Precision Agriculture* 11 (1): 82–99.

Lal, R. (2004). Soil carbon sequestration to mitigate climate change. *Geoderma* 123: 1–22.

Lal, R. (2008). Sequestration of atmospheric CO_2 in global carbon pools. *Energy & Environmental Science* 1 (1): 86–100.

Lal, R. and Bruce, J.P. (1999). The potential of world cropland soils to sequester C and mitigate the greenhouse effect. *Environmental Science & Policy* 2 (2): 177–185.

Li, C., Frolking, S., and Harriss, R. (1994). Modeling carbon biogeochemistry in agricultural soils. *Global Biogeochemical Cycles* 8: 237–254.

Lima, A.M.N., Da Silva, I.R., Stape, J.L. et al. (2011). Modeling changes in organic carbon stocks for distinct soils in southeastern Brazil after four eucalyptus rotations using the Century model. *Revista Brasileira de Ciências do Solo* 35 (5): 833–847.

Luo, Z., Wang, E., and Sun, O.J. (2010). Can no-tillage stimulate carbon sequestration in agricultural soils? A meta-analysis of paired experiments. *Agriculture, Ecosystems and Environment* 139 (1–2): 224–231.

Mandal, B., Majumder, B., Bandyopadhyay, P.K. et al. (2007). The potential of cropping systems and soil amendments for carbon sequestration in soils under long-term experiments in subtropical India. *Global Change Biology* 13 (2): 357–369.

Mikhailova, E.A., Noble, R.R.P., and Post, C.J. (2003). Comparison of soil organic carbon recovery by Walkley-Black and dry combustion methods in the Russian Chernozein. *Communications in Soil Science and Plant Analysis* 34 (13–14): 1853–1860.

Nanni, M.R. and Demattê, J.A.M. (2006). Spectral reflectance methodology in comparison to traditional soil analysis. *Soil Science Society of America Journal* 70 (2): 393.

Olson, K., Ebelhar, S.A., and Lang, J.M. (2014). Long-term effects of cover crops on crop yields, soil organic carbon stocks and sequestration. *Open Journal of Soil Science* 4 (8): 284–292.

Parton, W., Schimel, D., Cole, C., and Ojima, D. (1987). Analysis of factors controlling soil organic matter levels in Great Plains grasslands. *Soil Science Society of America Journal* 51 (i): 1173–1179.

Pathak, H., Li, C., and Wassmann, R. (2005). Greenhouse gas emissions from Indian rice fields: calibration and upscaling using the DNDC model. *Biogeosciences Discussions* 2 (1): 77–102.

Pathak, H., Byjesh, K., Chakrabarti, B., and Aggarwal, P.K. (2011). Potential and cost of carbon sequestration in Indian agriculture: estimates from long-term field experiments. *Field Crops Research* 120 (1): 102–111.

Patzold, S., Mertens, F.M., Bornemann, L. et al. (2008). Soil heterogeneity at the field scale: a challenge for precision crop protection. *Precision Agriculture* 9 (6): 367–390.

Penman, J., Gytarsky, M., Hiraishi, T. et al. (2006). *IPCC Guidelines for National Greenhouse Gas Inventories*. Geneva: IPCC.

Post, W.M., Izaurralde, R.C., Mann, L.K., and Bliss, N. (2001). Monitoring and verifying changes of organic carbon in soil. *Climatic Change* 51 (1): 73–99.

Reeves, J.B., McCarty, G.W., and Reeves, V.B. (2001). Mid-infrared diffuse reflectance spectroscopy for the quantitative analysis of agricultural soils. *Journal of Agricultural and Food Chemistry* 49 (2): 766–772.

Reicosky, D.C. (2003). Tillage-induced CO_2 emissions and carbon sequestration: effect of secondary tillage and compaction. In: *Conservation Agriculture*, 291–300. Amsterdam: Springer.
Saha, S., Chakraborty, D., Sharma, A.R. et al. (2010). Effect of tillage and residue management on soil physical properties and crop productivity in maize (*Zea mays*)-Indian mustard (*Brassica juncea*) system. *Indian Journal of Agricultural Sciences* 80 (8): 679–685.
Shepherd, K.D. and Walsh, M.G. (2002). Development of reflectance spectral libraries for characterization of soil properties. *Soil Science Society of America Journal* 66 (3): 988.
Srivastava, R., Sarkar, D., Mukhopadhayay, S.S. et al. (2015). Development of hyperspectral model for rapid monitoring of soil organic carbon under precision farming in the Indo-Gangetic Plains of Punjab, India. *Journal of the Indian Society of Remote Sensing* 43 (4): 751–759.
Sutherland, R.A. (1998). Loss-on-ignition estimates of organic matter and relationships to organic carbon in fluvial bed sediments. *Hydrobiologia* 389: 153–167.
Swarup, A. (1998). Emerging soil fertility management issues for sustainable crop productivity in irrigated systems. Proceedings of the National Workshop on Long-term Soil Fertility Management through Integrated Plant Nutrient Supply System, April 2–4.
Syp, A., Faber, A., Kozyra, J. et al. (2011). Modeling impact of climate change and management practices on greenhouse gas emissions from arable. *The Soil* 20 (6): 1593–1602.
Syswerda, S.P., Corbin, A.T., Mokma, D.L. et al. (2011). Agricultural management and soil carbon storage in surface vs. deep layers. *Soil Science Society of America Journal* 75 (1): 92.
Toure, A., Major, D.J., and Lindwall, C.W. (1995). Comparison of five wheat simulation models in southern Alberta. *Canadian Journal of Plant Science* 75 (1): 61–68.
Trost, B., Prochnow, A., Drastig, K. et al. (2013). Irrigation, soil organic carbon and N_2O emissions. A review. *Agronomy for Sustainable Development* 33 (4): 733–749.
Van den Bygaart, A.J., Gregorich, E.G., and Angers, D.A. (2003). Influence of agricultural management on soil organic carbon: a compendium and assessment of Canadian studies. *Canadian Journal of Soil Science* 83 (4): 363–380.
Venkatesh, M.S., Hazra, K.K., Ghosh, P.K. et al. (2013). Long-term effect of pulses and nutrient management on soil carbon sequestration in Indo-Gangetic plains of India. *Canadian Journal of Soil Science* 93 (1): 127–136.
Viscarra Rossel, R.A., Walvoort, D.J.J., McBratney, A.B. et al. (2006). Visible, near infrared, mid infrared or combined diffuse reflectance spectroscopy for simultaneous assessment of various soil properties. *Geoderma* 131 (1–2): 59–75.
Walkley, A. and Black, I.A. (1934). An examination of the Degtjareff method for determining soil organic matter, and a proposed modification of the chromic acid filtration method. *Soil Science Society of America Journal* 37: 29–38.
Wander, M.M., Bidart, M.G., and Aref, S. (1998). Tillage impacts on depth distribution of total and particulate organic matter in three Illinois soils. *Soil Science and Plant Nutrition* 62: 1704–1711.
Wang, M., Su, Y., and Yang, X. (2014). Spatial distribution of soil organic carbon and its influencing factors in desert grasslands of the hexi corridor, northwest China. *PLoS One* 9 (4): 24–26.
West, T.O. and Marland, G. (2002). A synthesis of carbon sequestration, carbon emissions, and net carbon flux in agriculture: comparing tillage practices in the United States. *Agriculture, Ecosystems & Environment* 91 (1–3): 217–232.
West, T.O. and Post, W.M. (2002). Soil organic carbon sequestration rates by tillage and crop rotation: a global data analysis. *Soil Science Society of America Journal* 66: 1930–1946.

23

Environmental Degradation and Sustainability

Yong Qi[1], *Puspendu Bhunia*[2], *Tian C. Zhang*[1,2], *F. Luo*[3], *Pengzhi Lin*[4], *and Yu Chen*[4]

[1] *Civil Engineering Department, University of Nebraska-Lincoln, Omaha, NE, USA*
[2] *School of Infrastructure, Indian Institute of Technology Bhubaneswar, Odisha, India*
[3] *Huazhong University of Science and Technology, Wuhan, China*
[4] *State Key Laboratory of Hydraulics and Mountain Engineering, Sichuan University, Chengdu, China*

23.1 Introduction

Environmental degradation is the deterioration of the environment via the depletion of soil, water and air, and is defined as the reduction of the environmental capacity to meet social and ecological needs and objectives (Johnson 1997). Sustainability is a state in which the demands placed on the environment can be met without reducing its capacity to allow all people to live well, now and in the future (Blowers 2012). Historically, environmental problems were neglected for a long time, but they have gained tremendous attention since the beginning of the 1970s. While the concept and principles of sustainable development had been reorganized in US law and policy (NRC 2011), the Rio Declaration on Environment and Development (UN 1992) formally endorsed the terminology "sustainable development". Moving into the 2000s, the Johannesburg Declaration on Sustainable Development (2002) and United Nation's (UN's) Millennium Ecosystem Assessment Report (2005) have further emphasized linkages of environmental degradation to human wellbeing and sustainable development (UNEP 2018a). In 2015, the 17 Sustainable Development Goals (SDGs) were formally adopted by all 193 UN member states (UN 2018), and many SDGs were linked with environmental degradation and sustainability.

To date, environmental degradation is one of the ten threats identified by the UN (Zimmermann 2016) and is the main factor reducing sustainability. Figure 23.1 shows that environmental degradation (e.g. desertification, pollution, land degradation, misbalance of snow ablation and glacier retreat) will impact economic, environment and social issues – the three sustainability pillars (Fiksel et al. 2012), such as:

- Environmental degradation has a great impact on human health (e.g. toxic air pollutants can cause respiratory problems like pneumonia and asthma).

Sustainability: Fundamentals and Applications, First Edition.
Edited by Rao Y. Surampalli, Tian C. Zhang, Manish Kumar Goyal, Satinder K. Brar, and R. D. Tyagi

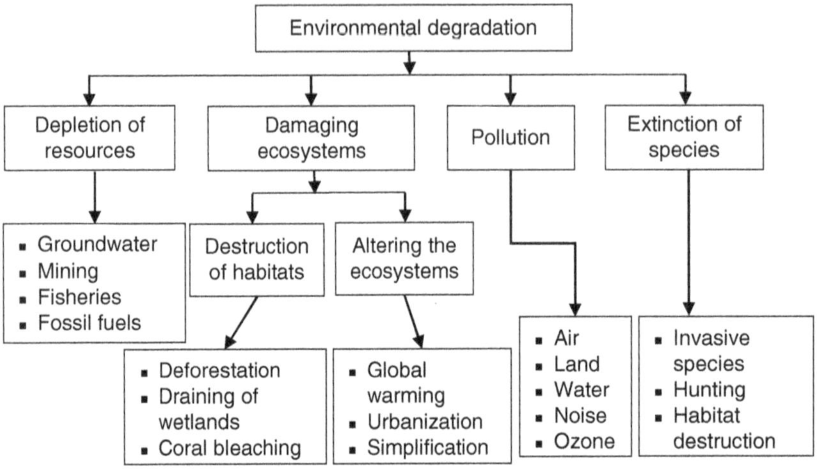

Figure 23.1 Major aspects of environmental degradation and selected impacts on environmental components, natural resources and ecosystem services. *Source:* Adapted from Donohoe (2003).

- Loss of biodiversity, which is vital to sustain the balance of the ecosystem by combating pollution, restoring nutrients, protecting water sources and stabilizing climate.
- Ozone layer depletion: the ozone layer is responsible for shielding the earth from detrimental ultraviolet rays. The presence of chlorofluorocarbons, and hydrochlorofluorocarbons in the atmosphere is causing the ozone layer to deplete.
- Losses in the tourism industry due to loss of green cover and biodiversity, and due to huge landfills, increased pollution.
- Economic impact: handling environmental degradation may require huge cost, including remediation of contaminated water and soil, restoration of green cover or river habits, cleaning up of landfills and protection of endangered species.

Although information on environmental degradation and sustainability development is readily available via the internet (e.g. UNEP 2018b; USEPA 2018), it is not easy to comprehend the subject and related issues as the information is often overwhelming – at the community, regional, national or global scale and being multidisciplinary in nature. This chapter provides a non-exhaustive overview of the subject, focusing on environmental degradation, its causes and consequences (i.e. desertification, pollution, land degradation, snow ablation and glacier retreat, dams and resettlement), their selected impacts on and the relationships with the three pillars of sustainability (e.g. the health of humans, loss of biodiversity, ozone layer depletion, environmental restoration, soil remediation and protection of endangered species), as well as an introduction to major strategies and practices for people to overcome these problems and to develop sustainable society without hurting the ecosystem and human wellbeing.

23.2 Desertification and Associated Sustainable Strategies

Nearly one-fifth of the world's land is reported to be threatened with desertification, which is a significant global ecological and environmental problem. Desertification is the conversion of a grassland or an already arid land into a desert through indiscriminate human activities magnified by droughts. Such activities include overexploitation of soil through human activities (i.e. overgrazing, repeated burning, intensive farming and stripping of vegetation for firewood) and climate change (Zhao 2011), typically losing its bodies of water as well as vegetation and wildlife. As a result of a long-term failure to balance human demand and supply of ecosystem services in drylands, desertification leads to the loss of land productivity, making the land more difficult to farm and less able to provide the necessary resources (Mao 2018).

23.2.1 Scope and Causes of Desertification

Occurring on all continents but Antarctica, desertification affects millions of people and large areas of drylands. Drylands occupy 41% of the earth's land area and are home to more than 2 billion people, nearly a third of the human population (MEA 2005). Formally, the UN definition includes all lands where the climate is classified as dry subhumid, semiarid, arid or hyperarid. Some 10–20% of drylands are already degraded, about 1–6% of the dryland people live in desertified areas, while a much larger number is under threat from further desertification. The greatest vulnerability is ascribed to sub-Saharan and central Asian drylands (MEA 2005). For example, in three key regions of Africa – the Sahel, the Horn of Africa and southeast Africa – severe droughts occur once every 30 years on average, leading to major food and health crises (Winkler 2017).

As a process, desertification combines various factors changing over time and location. The indirect factors include population pressure, socio-economic and policy factors, and international trade, while direct factors include land use patterns/practices and climate-related processes. The factors of water scarcity, intensive use of services and climate change are much greater threats in drylands compared to non-dryland areas. In general, one can call desert a natural phenomenon if it appears automatically over the natural course of a planet's lifecycle. In this case, natural conditions may be difficult to change. However, the acceleration of desertification due to human overexploitation of the environment and the ecosystems deserves much more attention. Usually, near-sighted decisions and/or practices in policy framing and social/economic development (e.g. land use patterns) often lead to soil erosion, the loss of soil organic matter and biodiversity, and habitat changes for many plants and animal species, and, thus, belong in the category of inappropriate and unsustainable activities.

23.2.2 Desertification and Sustainability

Essentially, a long-term failure to balance demand and supply of ecosystem services in drylands would lead to desertification, causing the dryland ecosystems to have increased pressure to provide services (e.g. food, forage, fuel, building materials and water) for humans and livestock (e.g. irrigation and sanitation), which would threaten future improvements in human wellbeing or even reverse gains in some regions. Therefore, desertification ranks among the greatest environmental sustainable challenges today and is a major impediment to meet basic human needs in drylands (Max 2009).

23.2.3 Response to Desertification

The UN proposed that 17 June should be the annual day to promote public awareness of and education on the issues of desertification and drought, and has implemented the UN Convention to Combat Desertification (UNCCD) in those countries (particularly in Africa) experiencing serious drought and/or desertification. Effective prevention or reversion of desertification calls for local- to global-scale management, active engagement of stakeholders and communities, and adjustment of actions according to the degree/stage of the desertification. It is important to realize that it is possible to stop the early-stage desertification process, and prevention is more cost-effective than rehabilitation. Generally, the following main prevention actions are recommended:

- Integration of land and water management to prevent soils from erosion and salinization.
- Integration of land use for grazing and farming, allowing for a more efficient cycling of nutrients in the agricultural systems.
- Shifting to alternative livelihoods (e.g. greenhouse agriculture, dryland aquaculture and tourism-related activities).
- Creating economic opportunities in dryland urban centers.

In the middle or late stages of desertification, the ecosystem services have been severely impacted, and thus, the rehabilitation and restoration approaches are necessary, which requires the implementation of a combination of policies and technologies. The selected actions to restore ecosystems in drylands include:

- Establishing seed banks.
- Reintroducing selected species.
- Planting trees.
- Enriching soil with nutrients.

To be successful, rehabilitation practices need human resources, funding, functional infrastructures, improved technologies, societal support and changed cultural perceptions (Alexandratos 2005).

23.3 Environmental Pollution and Associated Impacts

Pollution is the introduction of contaminants into the ecosystems/natural resources (i.e. air, water, soil), and, thus, causes adverse impacts to the environment and jeopardizes ecosystem sustainability. In the form of chemical substances or energy (e.g. noise, heat or light), pollution is often classified into point and non-point source, with its components being either foreign substances/energies or naturally occurring contaminants. As a worldwide issue, environment pollution is tangled with unsustainable human activities and has substantial adverse impacts on human health, causing many infectious diseases (e.g. cancer, birth defects and asthma) (McGeehin 2004; Fereidoun 2007). For example, pollution killed 9 million people in the world in 2015 (Beil 2017). The remainder of this section discusses environmental pollution and its associated impacts on sustainable development;

Section 23.6 presents strategies for sustainable development while addressing environmental pollution/degradation.

23.3.1 Air Pollution

Table 23.1 shows the main pollutants found in the air and some details on these aspects (EPHA 2009). Urban air pollution is responsible for 300 000–700 000 deaths annually and creates chronic health problems for an increasing number of people (Sharma 2008). For example, the London Smog in 1952 caused 4000 deaths and a host of other related ailments among Londoners. The causes of air pollution include natural and manmade sources.

Table 23.1 Major air pollutants: descriptions, sources and effects.

Pollutant	Description	Sources	Effects
Sulfur dioxide (SO_2)	Gaseous compound made up of sulfur and oxygen	Coal-burning power plants and industries; industrial boilers and processes; coal-burning stoves; refineries; heaters	Eye irritation; dead aquatic life; lung damage; reacts in atmosphere resulting in acidic precipitation; deteriorates buildings and statues; damages forests
Hydrocarbons (HCs)	Composed of hydrogen and carbon	Dry cleaning operations; auto paint shops; chemical plants; auto emissions; service stations; waste facilities	Asphyxiation, narcosis (i.e. depression of the central nervous system; anesthesia), cardiac arrest and aspiration (with exposure concentrations at thousands of parts per million)
Ozone (O_3)	Gaseous pollutant	Vehicle exhaust	Lung damage; eye irritation; respiratory tract problem
Particulate matter (PM)	Very small particles of soot, dust, tiny droplets of liquids or others	Diesel engines; power plants; steel industry; flour mills; windblown dust; wood stoves	Damages crops; lung damage; reduces visibility; discolors buildings and statues; eye irritation
Nitrogen oxides (NO_x)	Several gaseous compounds made up of nitrogen and oxygen	Vehicles; industrial boilers; industrial processes; power plants; commercial and residential areas; coal-burning stoves; natural gas pipelines	Lung damage; forms acid rain; damages forests, buildings and statues; forms ozone and other pollutants (smog)
Lead (Pb)	Metallic element	Vehicles burning leaded gasoline; power plants; metal refineries	Brain and kidney damage; contaminated crops and livestock; smog
Carbon monoxide (CO)	Colorless and odorless gas	Vehicles; indoor sources (kerosene- or wood-burning stoves); dry cleaners	Headaches, reduced mental alertness, death; heart damage; smog

Source: Adapted from Kumar and Omidvarborna (2018).

The manmade sources include tobacco smoke, heating, home cleaning agents, insecticide industries, automobiles, power generation, congested roads, and age and poor maintenance of vehicles. Burning of fossil fuels is one of the main sources of serious air pollution, but rapid growth in urban population, increasing industrialization, and rising demands for energy and motor vehicles are making significant contributions to increased air pollution levels (Mishra 2003).

23.3.2 Water Pollution

Resulting from many different and complicated causes, water pollution can cause environmental problems in water resource management, water/wastewater/stormwater treatment and management, and public/social services.

Currently, major issues (e.g. climate change, population explosion, rapid urbanization, improved living standards and water pollution) have made the problem of water scarcity worse. With the increasing population, water demand has been increasing, while global warming has led to an increase in evapotranspiration, which makes the existing water resources more vulnerable. Depletion of the freshwater resources on earth is a major component of environmental degradation. Only ~2.5% of all of the water on earth is freshwater, with the rest being saltwater. Approximately 69% of freshwater is frozen in ice caps located on Antarctica and Greenland, leaving only 30% of the 2.5% of freshwater available for consumption (Gleick 1996). According to the World Health Organization (WHO), ~1.1 billion people have no access to safe water, and 2.4 billion are lacking basic sanitation (EPHA 2009).

Municipal wastewater usually contains the following major constituents: (i) colloidal, suspended and floatable material; (ii) oxygen-demanding materials (e.g. organics); (iii) microorganisms (pathogens); (iv) nutrients (N and P); (v) hazardous constituents (e.g. xenobiotic compounds, heavy metals); (vi) contaminants of emerging concern; and (vii) energy-rich material (e.g. organics) (Metcalf and Eddy Aecom 2014). Water pollution could be caused by industrial discharged effluents, untreated wastewater/stormwater and other non-point source pollution (e.g. pesticides). Water pollution is the leading cause of eutrophication, ground/surface water deterioration, reduction of drinking water quality and death for humans/animals, often leads to water scarcity and affects water bodies (e.g. oceans, lakes, rivers and groundwater) via biological, chemical and geological changes (Ashraf 2010). Therefore, water pollution is of global concern and, thus, severely affects sustainability development.

23.3.3 Land Pollution and Degradation

23.3.3.1 Land Pollution

Hazardous waste, sewage, industrial activities and deforestation are the major causes (and sources) of land pollution, which affects the natural balance of the land (e.g. leading to poor plant growth, habitat loss, water pollution, soil erosion and desertification). Human activities such as mining and quarrying, sewage discharge, household garbage and industrial waste discharge may cause land pollution and degradation, and therefore negatively impact sustainability development.

As one of the major causes, improper management of solid/hazardous waste land pollution leads to land pollution. Generally, solid/hazardous wastes include municipal solid waste (MSW), industrial waste, hazardous waste, agricultural waste, medical waste, sludge, chemical waste, biological waste and e-waste. Table 23.2 shows that MSW mainly includes domestic sources (e.g. paper, glass bottles, plastic bags, kitchen waste), industrial sources (e.g. chemical residue, fly ash, metallic waste), and two other sources. Table 23.3 shows the major types of hazardous waste encountered in contaminated sites. Agricultural waste is any substance or object generated by agricultural activities (e.g. waste from a farm shop or a vegetable packing plant). Some examples of agricultural solid waste include empty pesticide containers, old silage wrap, used medicines, wormers, surplus milk, packaging waste, silage plastics, scrap metals, batteries, pesticides, agricultural wasted biomass (e.g. almond, bagasse, straw, corn stalks, cotton, peanut, rice hulls, straw and wheat straw), and agricultural residuals (e.g. pesticides and fertilizer residuals). Plastic items made from low density

Table 23.2 Sources of MSW.

Source	Activities, location of waste generated	Types of waste
Residential	Single- and multi-family dwellings/ apartments, etc.	Food waste, rubbish, ashes, special waste
Commercial and institutional	Stores, restaurants, markets, office buildings, hotels, motels, schools, print shops, auto repair shops, medical facilities and institutions	Food waste, rubbish, ashes, demolition and construction waste, special waste, occasionally hazardous waste
Open areas	Streets, alleys, parks, vacant lots, playgrounds, beaches, highways, recreational areas, etc.	Street sweepings, roadside litter, rubbish and other special waste
Treatment plants	Water/sewage/industrial wastewater treatment processes	Biosolids (sludges)

Source: Adapted from Sil and Kumar (2016).

Table 23.3 Major types of hazardous waste encountered in contaminated sites.

(1)	Petroleum hydrocarbons (~38%): (i) alphatics; (ii) alicyclics; (iii) single-ring aromatics; (iv) polycyclic aromatics; (v) heterocylic aromatics; (vi) non-chlorinated solvents; and (vii) oxygenates
(2)	Creosote for wood preservation (~28%): creosote, polynuclear aromatics, pentachlorophenol, nitrogen heterocyclic aromatics, aromatics
(3)	Solvent (~22%): (i) halogenated aliphatics and halogenated aromatics
(4)	Pesticide (~9%): (i) halogenated aliphatics; (ii) chlorinated cyclic aliphatics; (iii) halogenated aromatics; (iv) chlorinated phenoyalkanoates; (v) halogenated aniline-based (alachlor); (vi) cyclodienes; (vii) carbamates; (viii) pyrethroids; and (ix) organophosphonates
(5)	Others (~8%): (i) munition (explosives) wastes; (ii) heavy metals/inorganics; (iii) emerging contaminants; and (iv) e-waste

Source: Adapted from Zhang et al. (2017).

polyethylene (LDPE) are virtually indestructible, which creates colossal environmental hazards. Plastic is non-biodegradable; burning of plastic in garbage dumps may release highly toxic and poisonous gases (e.g. carbon monoxide, carbon dioxide, phosgene, dioxine and other poisonous chlorinated compounds). E-waste includes any items containing circuitry or electrical components with either power or battery supply (e.g. cathode ray tubes, printed circuit board, chips and other gold-plated components, keyboards, monitors and computer wires). As the global consumer demand increase, e-waste also increases continually. The major harmful materials in discarded electronics include beryllium, cadmium, mercury and lead.

23.3.3.2 Land Degradation

In the process of land degradation, the value of the biophysical environment is reduced by a combination of human activities upon the land (Figure 23.2) but excluding natural hazards as a cause. For example, up to 40% of the world's agricultural land has been reported to be seriously degraded, with the resultant side effects of land degradation (including lack of agronomic productivity and food security).

Soil erosion is the removal of topsoil by agents like water (e.g. running water, groundwater, rain, sea waves) and wind. A certain amount of erosion is natural and might be good for the ecosystem. Generally, removal of vegetative cover via overgrazing or deforestation is the major cause of soil erosion. Figure 23.2 shows that soil erosion causes soil degradation (i.e. decreased fertility and productivity) and soil degradation causes desertification, gradually leading the land to become unfit for cultivation or habitation. Also, excessive use of chemical fertilizers and pesticides causes soil acidification, which will increase salinity and alkalinity.

It is well known that climate change results in changes in patterns of wind, rain, snow, river and stream flow, and ocean circulation, etc. which cause short- and long-term effects on landscapes, presumably causing land degradation. One example of short-term effects caused by the wind can be the dust bowl, which was caused by farmers tilling over the native grasses that held the dirt in place. The long-term effects can be wind erosion. The wind wears away rocks and mountains and grinds them to dust and then picks up the dust

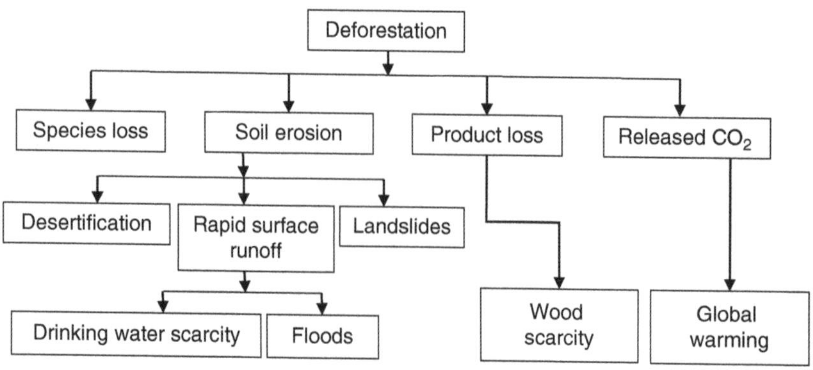

Figure 23.2 Deforestation leads to environmental degradation. *Source:* Adapted from Sharma (2008).

Table 23.5 Major federal environmental statutes enacted in the US.

The Clean Air Act (CAA) authorizes the Environmental Protection Agency to set mobile source limits, ambient air quality standards, hazardous air pollutant emission standards, standards for new pollution sources and significant deterioration requirements; and to identify areas that do not attain standards
The Clean Water Act (CWA) authorizes the regulation and enforcement of requirements that govern waste discharges into US waters, and financial assistance for wastewater treatment plant construction and improvements
The Safe Drinking Water Act (SDWA) authorizes the Environmental Protection Agency to establish primary drinking water standards, regulate underground injection disposal practices and administer a groundwater control program
The Ocean Dumping Act focuses on the regulation of the intentional disposal of materials into ocean waters and authorizes related research
The Solid Waste Disposal Act and Resource Conservation and Recovery Act (RCRA) govern the regulation of solid and hazardous waste at facilities currently operating and at those yet to be constructed
The Comprehensive Environmental Response, Compensation, and Liability Act (CERCLA) addresses the clean-up of contamination resulting from the past release of hazardous substances. CERCLA authorizes the federal government to require the parties responsible for the contamination to pay the clean-up costs. CERCLA established a Superfund Trust Fund to pay for the clean-up when the responsible parties cannot pay or cannot be found. Amendments to the Solid Waste Disposal Act included similar liability provisions for the clean-up of petroleum leaked from underground storage tanks, not covered under CERCLA, and established a Leaking Underground Storage Tank Trust Fund to clean up abandoned sites with petroleum contamination
The Emergency Planning and Community Right-to-Know Act (EPCRA) requires industrial reporting of toxic releases and encourages chemical emergency response planning
The Pollution Prevention Act (PPA) authorizes various mechanisms intended to prevent pollution by reducing the generation of pollutants at the point of origin
Toxic Substances Control Act (TSCA) requires regulation of commercial chemicals to reduce risks to human health and the environment. The most recent amendment of TSCA is the Frank R. Lautenberg Chemical Safety for the 21st Century Act passed in June 2016, which provided the Environmental Protection Agency with new authority and tools it sorely needs to better ensure the safety of chemicals in use and entering the market. These authorities and tools include a health-based safety standard, a mandate to address risks to vulnerable subpopulations, required safety reviews of both new and existing chemicals, stronger testing authority and expanded access to chemical information. Federal pesticide policies are implemented under two statutes:
The Federal Insecticide, Fungicide, and Rodenticide Act (FIFRA) governing the sale and use of pesticide products within the United States; and
The Federal Food, Drug, and Cosmetic Act (FFDCA), which limits pesticide residues on food in interstate commerce (including imports)
The National Environmental Policy Act requires federal agencies to incorporate environmental considerations in their planning and decision making through a systematic interdisciplinary approach

Source: Adapted from Surampalli et al. (2018).

sustainability pillars as well as trade-off considerations into its decision making (NRC 2011). Note that SAMP is intended to be equally applicable to all types of issues, including human health and ecological risks.

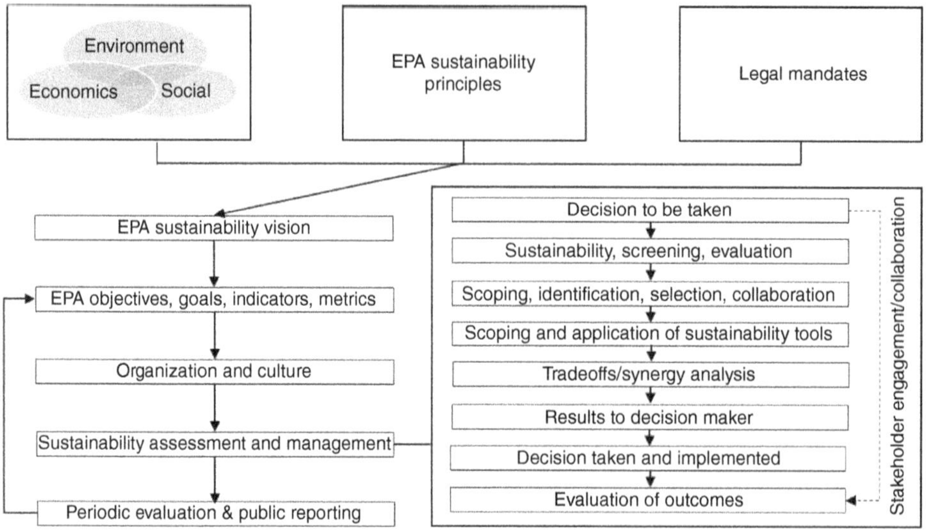

Figure 23.3 A framework for USEPA sustainability decisions. *Source:* Adapted from NRC (2011).

23.6.2 Strategies and Platforms for Holistic Planning/Solutions

Strategies and innovative technologies have been developed to holistically handle the issues based on principles of sustainability (e.g. including environmental, social and economic considerations). For example, the USEPA developed Federal Reference Methods (FRMs) for monitoring six criteria air pollutants: CO, Pb, SO_2, O_3, NO_2 and particulate matter (PM_{10} and $PM_{2.5}$) in outdoor air. FRMs are exercised by states and other monitoring organizations to attain the federal Clean Air Act's (CAA) National Ambient Air Quality Standards (NAAQS). To nurture innovation and to revolutionize new technologies, the USEPA also analyzes, examines and endorses other methods, called Federal Equivalent Methods (FEMs), which are based on different sampling and/or analyzing technologies than FRMs, but are required to provide the same decision-making quality when making NAAQS attainment determinations (Surampalli et al. 2018). The next step is to identify clear measurable targets to improve air quality, based on understanding of the health effects of the pollutants concerned and costs of emission reduction methods. For example, in the UK, the Air Quality Strategy (DEFRA 2007) sets out standards and objectives for the eight main health-threatening air pollutants: CO, Pb, SO_2, O_3, NO_2, particulate matter, benzene and 1,3-butadiene. The Strategy sets out the contribution of key sectors, such as industry, transport and local government, for achieving objectives via the best available techniques not entailing excessive cost.

Climate change and air pollution are intimately linked through emissions from common sources, even though in both science and policy these topics are often treated separately. Sustainable policy implementation for one cause can directly provide a co-benefit to another cause and vice-versa. Transportation, urban households, industries and infrastructures are key sources of GHGs and air pollution emissions in urban centers. Therefore, several adaptation and mitigations strategies have been opted by regulatory authorities to overcome the air pollution and climate change in megacities and rapidly growing urban areas. Implementation of metro and Bus Rapid Transit System (BRTS), introduction of cleaner fuel and new technology are the most viable examples of adaptation and mitigation strategies for the transportation sector, followed by various regulatory bodies to mitigate the impact of air pollution and climate change. Appropriate policy and technological interventions are key measures from a co-benefit perspective that need to be further strengthened.

To increase water sustainability, sustainable water resource management and water management are the goals, while integrated water resource management (IWRM) and development of green technologies for water management/treatment and resource recovery are the strategies for pursuing these goals. In order to implement IWRM, one should consider the sustainability principles in (i) social (e.g. clean and abundant water, safe and secure food supply, clean and stable energy supply); (ii) environmental (carbon neutral or positive, hydrologically neutral or restorative, nutrient neutral and ecologically neutral or restorative); and (iii) economic aspects (e.g. low lifecycle cost, and robust in the face of economic disruption). One should then consider all related components, including (i) integrated planning with different tools/strategies (e.g. adaptive decision making, watershed management tools, assessment tools for evaluation, guidance manuals); (ii) regulatory and programmatic change (e.g minimum standards and framing ordinances, codes and policies); (iii) management and financial (e.g. integrated decision making recognizing interconnections on different scales, financial links or market mechanisms, utilities and service providers, infrastructure development programs); and (iv) stakeholder participation and community engagement (Ojha et al. 2017).

Sustainable water management generally involves the improvement of the health of surface/groundwater systems, avoidance of overextraction of freshwater supplies, identification of alternative water resources (e.g. rainwater, stormwater, desalinated water and recycled water), and implementation of environment-related policies and educational campaigns. Development and applications of green technologies often include various innovations in water collection, treatment, distribution and drainage systems as well as enhanced approaches to management, assessment strategies, policies and regulations. After being successfully implemented, they can play crucial roles in enhancing sustainability, driving green growth and achieving a balance between economic, social and environmental factors while maintaining productivity, prosperity and efficiency (Figure 23.4).

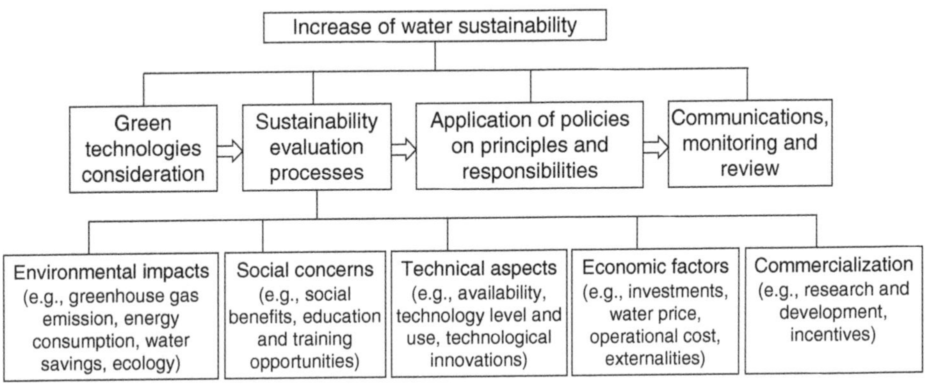

Figure 23.4 Development of green technologies for sustainable water management.

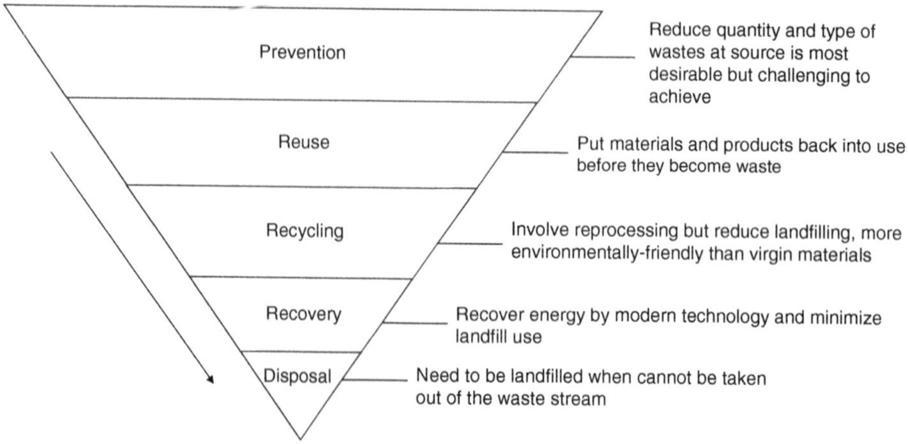

Figure 23.5 Hong Kong waste management hierarchy for sustainable use of resources (2013–2022). *Source:* Adapted from Wong et al. (2016).

For solid and hazardous waste management, the hierarchy witnessed changes recently, mainly because waste is recognized as resources nowadays. Many countries are now adopting the 4R (i.e. reduction, reuse, recycling, recovery) policies, which is the basis for the sustainable resource utilization. As shown in Figure 23.5, in Hong Kong, waste avoidance, waste reduction through producer responsibility schemes (PRSs) and waste charging are the targeted actions of the highest priority, followed by initiatives and firm actions for recycling and recovery (e.g. via waste-to-energy technologies), and treatment and disposal as the last options.

23.6.3 Addressing Cross-Cutting, Complex and Challenging Issues at Different Levels

It is critical to adapt different strategies and platforms to address the complicity of sustainable development at different levels. Innovations for sustainable development can be small- or large scale. For example, the most significant advances in sustainable biological wastewater treatment include the use of innovative technologies/methods for treatment and process monitoring, control, and management; the use of computational fluid dynamics (CFD) and computer software for process design/analysis/control; and technologies for resource recovery from wastewater and biosolids (Zhang et al. 2018). All of these technologies are at the community level or the process scale.

IWRM is a complex process and it may not always be possible to come up with win–win solutions; sometimes, compromises have to be made for the common cause. In many cases, allocation of resources between competing uses and users requires political drive. For example, quantifying water security under climate and land use change is very challenging. As shown in Figure 23.6, all the items in the figure are potentially affected by land use and/or climate change. Two particular challenges could be the impacts of land use change on river flows and incorporating climate change uncertainty into reservoir storage analysis. Water availability at catchment scales could be modeled under the influence of the two major factors (McIntyre et al. 2017).

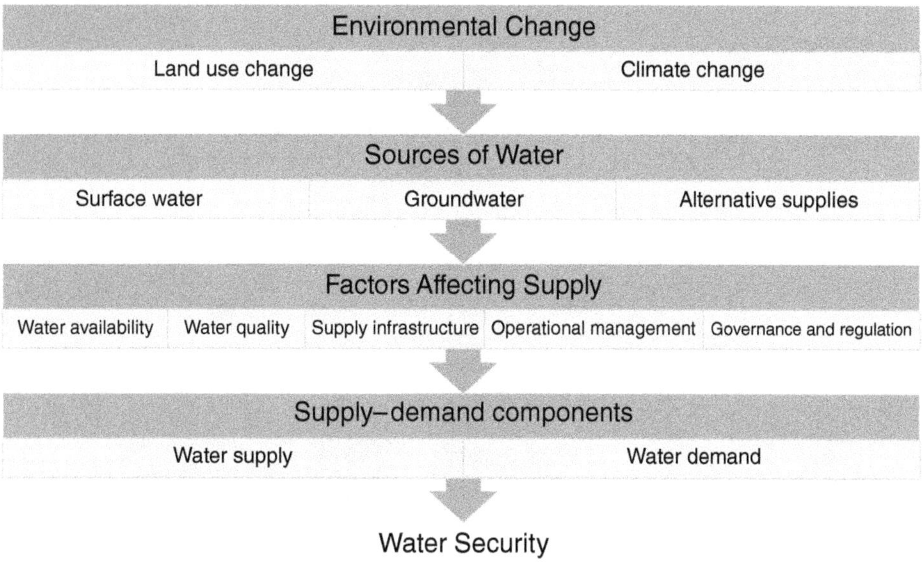

Figure 23.6 Components of water security analysis under the influence of climate and land use changes. *Source:* Adapted from McIntyre et al. (2017).

Urban sustainability depends on a supply of water in sufficient quantity and quality to meet the societal and ecosystem needs. Sustainable cities also need to be resilient to flooding, extreme weather events and climate change. Therefore, cities are not sustainable if they are managed by a disorganized set of institutions as these institutions are not be able to mobilize collective action for system organization and management. For water sustainable urban design (WSUD), the scale can be regional, and multiple factors need to be considered (e.g. water security and clean/healthy water environments, mitigating urban heat, enhancing urban biodiversity, improving water resources and protecting from flooding, improving the landscape and overall environment health). The key is to create the conditions to satisfy these factors for sustainable development and for providing sustainable ecosystem services (Figure 23.7). As one example, since 2015 the US National Science Foundation (NSF) has been supporting a research center "Urban Water Innovation Network (U-WIN)" to promote collaborations among different groups and stakeholders. To address the core challenges inherent in systems related to U-WIN and WSUD, the center is conducting research in several aspects. One task is to develop a national Urban Water Management Blueprint by creating a conceptual framework that defines the essential characteristics of sustainable urban water systems across regions. Another task is to characterize interactions and feedbacks between urban development patterns and the interconnected natural/socio-economic processes that affect the sustainability of coupled natural–human water systems. The center is targeting the identification of sustainable technological, socio-economic and management urban water solutions that reduce pressures, enhance resilience and maximize co-benefits in other sectors, and measuring their impacts and trade-offs across time and space. Finally, the center will explore and explain the systemic institutional frameworks required for successful transitions toward sustainable urban water systems in metropolitan regions across the US (NSF 2018).

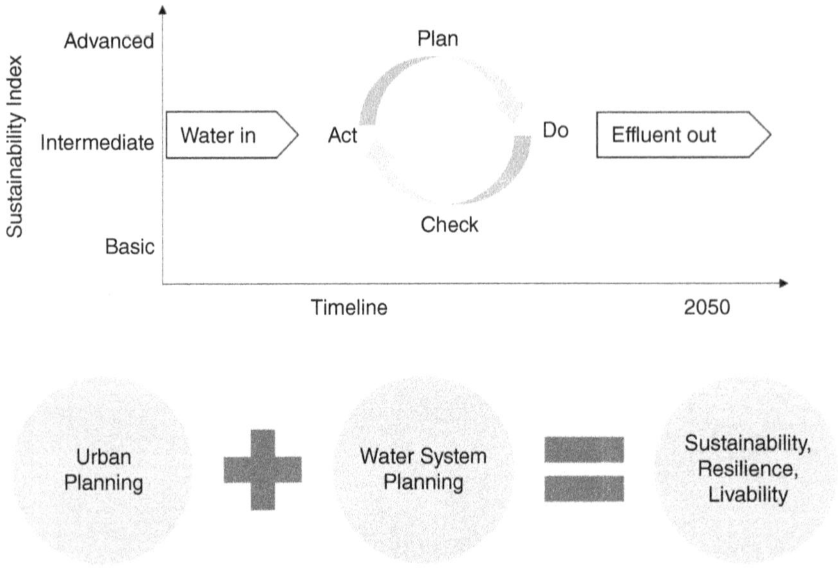

Figure 23.7 Creating conditions for water sustainable urban design. Sustainability Index is a composite index published from 1999 to 2005 that tracked 21 elements of environmental sustainability covering natural resource endowments, past and present pollution levels, environmental management efforts, contributions to protection of the global commons and a society's capacity to improve its environmental performance over time (Wikipedia 2018). *Source:* Adapted from Singh et al. (2017).

In remediation practices, the significant uncertainty requires more flexible, iterative approaches to managing uncertainty, which has forced the development of mechanisms and reevaluations, known collectively as remediation process optimization (RPO). Together with the evolution of RPO, performance-based environmental management (PBEM) and the integration of "green and sustainable remediation" (GSR) have emerged. PBEM relies on establishing and working toward performance objectives instead of managing the process only. GSR is a more holistic approach that considers all environmental effects of remedy implementation for contaminated sites and incorporates options to maximize the net environmental benefit of clean-up actions. Currently, budget constraints, policy priorities and legacy remedial systems, among other factors, are driving remedy optimization and environmental sustainability goals. Practices in RPO, PBEM and GSR would help people to achieve at lower cost and environmental protectiveness footprint (Zhang et al. 2018).

23.6.4 Engaging Publics and Stakeholders

Technological dissemination and transfer are very important. Nowadays, worldwide organizations and conferences are popular to exchange innovative ideas and strategies for solving the problems related to environmental degradation and sustainable development. Moreover, public involvement and stakeholder participation in sustainable development are extremely important, which often go hand in hand with educating the public of the needs and challenges in environmental degradation and sustainable development.

Outreach activities often are required for successful implementation of the new policies or innovative treatment facilities in a particular region or country. Currently, a collaboration among the public, private and informal sectors is evident. It is ideal to organize the environmental education and public engagement through one of these channels for effective implementation (Wong et al. 2016).

23.7 Future Trends

As shown in Figure 23.7, using the "plan → do → check → act" roadmap approach, an organization may begin its sustainability journey at the basic level, meeting a minimal level of performance, and move toward advanced sustainability goals. While the UN's SDGs are general guidelines, the roadmap can be case sensitive and affected by future trends such as (GRI 2017):

- *Climate change*: The top factors include: (i) using alternative/renewable energy; (ii) new technologies/innovation; (iii) setting targets and having a climate strategy; (iv) communicating and advocating on climate; (v) supportive governance; (vi) leadership from the top; (vii) committing and delivering real results; and (viii) reviewing products and services for climate impacts. Science-based targets (e.g. reduction of GHG emissions by food and beverage companies) are needed, and greater company and country engagement is expected following the Paris Agreement.
- *Human rights*: Key human rights issues include labor rights and issues linked to natural resources. For multi-located companies, ensuring human rights becomes a key challenge, which will be addressed by improving digital trust; implementing a framework of action (e.g. a commitment to respect basic social rights, equal opportunities, merit and non-discrimination and reporting about where the company stands on human rights and how they comply with their human rights policy); and building a diverse workforce for an innovative ecosystem.
- *Wealth inequality*: Currently, the richest 1% in the world have more wealth than the rest of the world combined. Up to now, wealth inequality has been rarely addressed directly as a mega trend or part of the wider operating context influencing business. Various challenges exist for business related to wealth inequality. Ending the era of tax havens and breaking market concentration and addressing the unequal distribution of power are imperative.
- *Data and technology*: They are part of an ever-growing field of sustainable development. Challenges include securing/obtaining sufficient sensitive and confidential data, and tools for better analyses of data. Opportunities include online reporting; embedding sustainability data into management systems; feedback loops to data providers; and better understanding the dynamics and other demands on the data across value chains (e.g. big data, networks and an internet of things).

GRI (2017) also made some key suggestions to deal with each trend. As usual, organizations should set several strategic 3-to-5-year breakthrough objectives related to sustainability implementation and performance indicators and associated metrics, and follow the approach shown in Figure 23.7 to implement them. It is certain that future expectations will be greater; targets will be cross-cutting; new tools will be developed for better data collection and analyses of present and future consequences of alternative decision options on the full range of social, environmental and economic indicators; and best practices for corporate reporting on the SDGs will be established.

23.8 Conclusions

The side effects of environmental degradation on the social, economic and environmental systems of a region or country have been proved to be significant. Environmental disasters do not recognize manmade borders, and threaten the development of human beings, especially future generations. Thus, governments, international organizations, communities and the public must work together to lower the risks associated with environmental degradation and its contributing factors, such as desertification and pollution, and to ensure that vulnerable people and regions are prepared to survive and adapt. At the same time, companies, organizations and stakeholders must also ensure that their work and activities are environmentally friendly and sustainable.

References

ACIA (2005). *Arctic Climate Impact Assessment – Scientific Report*. New York: Cambridge University Press.

Alexandratos, N. (2005). Countries with rapid population growth and resource constraints: issues of food, agriculture, and development. *Population and Development Review* 31 (2): 237–258.

Anderson, D. (2008). Proximate and distant visitors: differences in importance ratings of beneficial experiences. *Journal of Park and Recreation Administration* 26 (4): 47–65.

Ashraf, M. (2010). Effects of polluted water irrigation on environment and health of people in Jamber, District Kasur, Pakistan. *International Journal of Basic & Applied Sciences* 10 (3): 37–57.

Baran, E.J., Jantunen, T., and Chong, C.K. (2007). *Values of Inland Fisheries in the Mekong River Basin*. Phnom Penh, Cambodia: Worldfish Center.

Beck, M.C. (2012). Environmental and livelihood impacts of dams: common lessons across development gradients that challenge sustainability. *International Journal of River Basin Management* 10 (1): 73–92.

Beil, L. (2017). The list of diseases linked to air pollution is growing. ScienceNews, 192, p. 18.

Bindoff, N. (2007). Observation: oceanic climate change and sea level. In: *Climate Change 2007: The Physical Science Basis* (eds. S. Solomon, D. Qin, M. Manning, et al.), 389–422. Cambridge, UK: Cambridge University Press.

Blowers, A.B. (2012). Is sustainable development sustainable? *Journal of Integrative Environmental Sciences* 9 (1): 1–8.

Brismar, A. (2002). River systems as providers of goods and services: a basis for comparing desired and undesired effects of large dam projects. *Environmental Management* 29 (5): 598–609.

Carbognin, L.T. (2004). Eustacy and land subsidence in the Venice Lagoon at the beginning of the new millennium. *Journal of Marine Systems* 51 (1–4): 345–353.

Carey, M. (2005). Living and dying with glaciers: people's historical vulnerability to avalanches and outburst floods in Peru. *Global and Planetary Change* 47 (2–4): 122–134.

Department for Environment, Food and Rural Affairs (DEFRA) of the United Kingdom (2007). The Air Quality Strategy for England, Scotland, Wales and Northern Ireland. (Vol. 1), Ref: PB12654. https://assets.publishing.service.gov.uk/government/uploads/system/uploads/attachment_data/file/69336/pb12654-air-quality-strategy-vol1-070712.pdf (accessed 31 May 2018).

DiFrancesco, K. (2006). *Global Perspectives on Large Dams*. New Haven, CT: Yale School of Forestry and Environmental Studies Publication Series.

Donohoe, M. (2003). Causes and health consequences of environmental degradation and social injustice. *Social Science & Medicine* 56 (3): 573–587.

Dugan, P. (2010). Fish migration, dams, and loss of eco-system services in the Mekong Basin. *Ambio* 29 (4): 344–348.

EPHA (2009). Air, water pollution and health effects. European Public Health Alliance. https://epha.org/raising-the-bar-for-eu-action-on-childhood-immunisation/54 (accessed November 2019).

Fiksel, J., Eason, T., and Frederickson, H. (2012). *A Framework for Sustainability Indicators at EPA*. Cincinnati, OH: National Risk Management Research Laboratory EPA Number: EPA/600/R/12/687.

Fereidoun, H.N. (2007). The effect of long-term exposure to particulate pollution on the lung function of Teheranian and Zanjanian students. *Pakistan Journal of Physiology* 3 (2): 1–5.

Gleick, P. (1996). Water resources. In: *Encyclopedia of Climate and Weather* (ed. S.H. Schneider), 817–823. New York: Oxford University Press.

Goswami, U.P. (2018). study on climate extremes and glacial lake outburst floods. PhD dissertation. Department of Civil Engineering, Indian Institute of Technology Guwahati.

GRI (2017). future trends in sustainability reporting. https://www.globalreporting.org/resourcelibrary/GRI-CLG_Report-FutureTrends2025.pdf (accessed 1 June 2018).

Johnson, D.S.-N. (1997). Meanings of environmental terms. *Journal of Environmental Quality* 26 (3): 581–589.

Kumar, A. and Omidvarborna, H. (2018). Introduction to air pollution. In: *Handbook of Environmental Engineering* (eds. R.Y. Surampalli, T.C. Zhang, S.K. Brar, et al.), 87–92. New York: McGraw-Hill Education and American Society of Civil Engineers.

Lemke, P. (2007). Observations: changes in snow, ice and frozen ground. In: *Climate Change 2007: The Physical Science Basis* (eds. S. Solomon, D. Qin, M. Manning, et al.), 337–383. Cambridge, UK: Cambridge University Press.

Lessard, J. (2003). Effects of elevated water temperature on fish and macroinvertebrate communities below small dams. *River Research and Applications* 19 (7): 721–732.

Makinen, K. (2010). Policy considerations for greenhouse gas emissions from freshwater reservoirs. *Water Alternatives* 3 (2): 91–105.

Mao, K. (2018). The treadmill of taxation: desertification and organizational state deviance in Minqin Oasis, China. *Critical Criminology* 26 (2): 271–288.

Max, A. (2009). Inside climate news. October 20. https://insideclimatenews.org/news/20091020/desertification-threatens-food-security-and-climate (accessed November 2019).

McIntyre, N., Woldemeskel, F., Visessri, S., and Sharma, A. (2017). Quantifying surface water supplies under changing climate and land use. In: *Sustainable Water Resources Management* (eds. C.S.P. Ojha, R.Y. Surampalli, A. Bárdossy, et al.), 337–373. Reston, VA: ASCE.

McGeehin, M.A. (2004). National environmental public health tracking program: bridging the information gap. *Environmental Health Perspectives* 112 (14): 1409–1413.

MEA (2005). *Millennium Ecosystem Assessment: Ecosystems and Human Well-Being: Synthesis.* Washington, DC: Island Press.

Meixler, M.B. (2009). Predicting barrier passage and habitat suitability for migratory fish species. *Ecological Modeling* 220 (20): 2782–2791.

Metcalf & Eddy Aecom (2014). *Wastewater Engineering: Treatment and Resource Recovery*, 5e. New York: McGraw-Hill Education.

Mishra, V. (2003). Health effects of air pollution, background paper for Population–Environment Research Network (PERN) Cyberseminar. www.mnforsustain.org/climate_health_effects_of_air_pollution_mishra_pern.htm (accessed November 2019).

Moore, D.D. (2010). Special Issue: WCD+10: revisiting the large dam controversy. *Water Alternatives* 3 (2): 1–474.

National Research Council (NRC) of National Academics (2011). *Sustainability and the U.S. EPA*. Washington, DC: The National Academics Press.

National Science Foundation (NSF) (2018). SRN: Urban Water Innovation Network (U-WIN): transitioning toward sustainable urban water systems. Award Abstract #1444758. https://www.nsf.gov/awardsearch/showAward?AWD_ID=1444758&HistoricalAwards=false (accessed August 2018).

Nordli, Ø.P. (2014). Long-term temperature trends and variability on Spitsbergen: the extended Svalbard Airport temperature series, 1898–2012. *Polar Research* 33: 21349.

Ojha, C.S.P., Surampalli, R., Bardossy, A. et al. (2017). Sustainable water resources management: an introduction. In: *Sustainable Water Resources Management* (eds. C.S.P. Ojha, R.Y. Surampalli, A. Bárdossy, et al.), 1–13. Reston, VA: ASCE.

Provencher, B.S. (2008). Does small dam removal affect local property values? An empirical analysis. *Contemporary Economic Policy* 26 (2): 187–197.

Serreze, M.A. (2006). The Arctic amplification debate. *Climatic Change* 76: 241–264.

Sil, A. and Kumar, S. (2016). Waste characteristics and generation. In: *Sustainable Solid Waste Management* (eds. J.W.C. Wong, R. Surampalli, T.C. Zhang, et al.), 7–34. Reston, VA: ASCE.

Singh, R.P., Fu, D.F., and Ojha, C.S.P. (2017). Water sustainable urban designs. In: *Sustainable Water Resources Management* (eds. C.S.P. Ojha, R.Y. Surampalli, A. Bárdossy, et al.), 547–565. Reston, VA: ASCE.

Sharma, P. (2008, August 18). Effects of environmental degradation. https://saferenvironment.wordpress.com/2008/08/18/effects-of-environmental-degradation (accessed November 2019).

Surampalli, R.Y., Zhang, T.C., Kao, C.M. et al. (2018). Environmental legislation. In: *Handbook of Environmental Engineering* (eds. R.Y. Surampalli, T.C. Zhang, S.K. Brar, et al.), 3–45. New York: McGraw Hill.

United Nations (UN) (1992). Rio Declaration on Environment and Development. Report of the United Nations Conference on Environment and Development, Rio de Janeiro, June 3–14, 1992. UN publication, Sales No. E.93.I.8 and corrigenda, vols I–III. https://www.un.org/documents/ga/conf151/aconf15126-1annex1.htm (accessed May 2018).

United Nations (UN) (2002). Johannesburg Declaration on Sustainable Development, The World Summit on Sustainable Development, Johannesburg, South Africa, 26 August to 4 September 2002. www.un-documents.net/jburgdec.htm (accessed May 2018).

UN (United Nations) (2018). Sustainable Development Goals. www.undp.org/content/undp/en/home/sustainable-development-goals.html (accessed June 2018).

UNEP (2018a). Emissions Gap Report 2018. United Nations Environment Programme. http://wedocs.unep.org/bitstream/handle/20.500.11822/26895/EGR2018_FullReport_EN.pdf?sequence=1&isAllowed=y ().

UNEP (United Nations Environment Programme) (2018b). Why sustainability matters. https://www.unenvironment.org/about-un-environment/sustainability/why-sustainability-matters (accessed May 2018).

USEPA (2018). Learn about sustainability. https://www.epa.gov/sustainability/learn-about-sustainability (accessed May 2018).

UNGC (United Nations Global Compact) (2018). Roadmap for integrated sustainability. https://www.unglobalcompact.org/take-action/leadership/integrate-sustainability/roadmap (accessed June 2018).

WCD (2000). *Dams and Development – A New Framework for Decision-Making*. London: Earthscan https://www.internationalrivers.org/sites/default/files/attached-files/world_commission_on_dams_final_report.pdf (accessed November 2000).

Wikipedia (2018). Environmental sustainability index. http://en.wikipedia.org/wiki/Environmental_Sustainability_Index (accessed August 2018).

Winkler, K.G. (2017). Identifying droughts affecting agriculture in Africa based on remote sensing time series between 2000–2016: rainfall anomalies and vegetation condition in the context of ENSO. *Remote Sensing* 9: 1–27.

Wong, J.W.C., Surampalli, R., Zhang, T.C. et al. (2016). *Sustainable Solid Waste Management*. Reston, VA: ASCE.

Zhao, L.S. (2011). Long-term effects of grazing exclusion on aboveground and belowground plant species diversity in a steppe of the Loess Plateau, China. *Plant Ecology and Evolution* 144: 313–320.

Zhang, T.C., Surampalli, R.Y., Tyagi, R.D., and Benerji, S.K. (2017). Biological treatment of hazardous wastes. In: *Current Developments in Biotechnology and Bioengineering: Solid Waste management* (eds. J.W.-C. Wong, R.D. Tyagi and A. Pandey), 311–340. Amsterdam: Elsevier.

Zhang, T.C., Surampalli, R.Y., and Kao, C.M. (2018). Wastewater treatment plant design: treatment processes and recent advances. In: *Handbook of Environmental Engineering* (eds. R.Y. Surampalli, T.C. Zhang, S.K. Brar, et al.), 179–188. New York: McGraw Hill.

Zimmermann, N. (2016). Deutsche Wells (DW). November 10. www.dw.com/en/five-of-the-worlds-biggest-environmental-problems/a-35915705 (accessed November 2019).

24

Sustainability of River Water Resources Under the Influence of Climate Change

Shivam Gupta[1] *and Manish Kumar Goyal*[2]

[1] *Central Agricultural University, Imphal, India*
[2] *Discipline of Civil Engineering, Indian Institute of Technology, Indore, India*

24.1 Introduction

Several climatic events, reports and scientific studies confirm the phenomenon of climate change. The Intergovernmental Panel on Climate Change (IPCC, 2007) has reported in its assessment report that the mean temperature of the earth will rise by 0.6°C. This rise in temperature will have severe consequences for various sectors such as the environment, agriculture and water resources. Rising temperatures alter the various components of the hydrological cycle such as evapotranspiration, snowfall, snowmelt, runoff and precipitation (Bates et al. 2008). These changes in hydrological processes will affect water availability at the regional and global scale. Enhanced evapotranspiration will lead to aridity and increased water demand for agriculture; decreased precipitation will cause the frequent occurrence of droughts, and increased precipitation will lead to flood conditions.

Recent studies for the assessment of the impact of climate change on the Himalayan region also indicate several changes in the hydrology of the region. Bates et al. (2008) stated that one-sixth of the world population, who live in the river basin of snow and glacier-fed river systems, will face severe water scarcity due to change in snowmelt and snow accumulation time. This shift in snow and glacier melt will cause high flow situations in winters, whereas flow in summers and lean season will decrease significantly. Shivam Goyal and Sarma (2017) carried out a study on the change in temperature trends over the river basin of northeast India and found significant increasing trends in temperature under different emission scenarios. Freshwater availability is dependent on several factors such as runoff and groundwater recharge; threat of scarcity of freshwater availability is also a likely phenomenon in the coming future because of reduced runoff and groundwater recharge. The following sections cover the effects of climate change on different water resources along with a case study of the impact of climate change on a river basin in India and a water management approach to solve the water scarcity problem.

Sustainability: Fundamentals and Applications, First Edition.
Edited by Rao Y. Surampalli, Tian C. Zhang, Manish Kumar Goyal, Satinder K. Brar, and R. D. Tyagi

24.2 Effects of Global Warming on Observed Changes

With rapid population growth, urbanization and industrialization, the immense pressure on the available water resources is the prime concern of researchers and policymakers. There is an urgent need for a sustainable approach to water resource management in order to mitigate water scarcity problems in the coming future. Population projections for the twenty-first century along with the projections of precipitation and temperature indicate an alarming situation for available water resources and meeting the water demand. A study on the scenario of population growth and greenhouse gas emissions as well as the associated impacts on available water resources at the global scale shows that the demand for developing countries is going to increase, and in wet tropic countries the major challenge will be clean water rather than the amount of water (Vorosmarty et al. 2000). Ligtvoet et al. (2014) reported that 15% of the global population will face flooding problems because of the increased discharge in rivers due to climate change and enhanced extreme events. The World Water Development Report (WWDR) (2006) reveals that by the end of 2025 water withdrawals in the developed and developing countries will increase by 18% and 40%, respectively, which will create a 40% shortage of freshwater. The severity of water stress is likely to be intensified at the regional scale. Kiparsky and Gleick (2003) conducted a study on the water resources of California and stated that population growth and climate change tend to increase water scarcity in urban areas. A CDWR (1998) report states that water demand will exceed the available water supply by more than 2.96 billion m^3 by 2020 in California; Hidalgo et al. (2009) also stated a significant effect of climate change on the availability of water in the California region.

24.2.1 Observed Changes

- *Precipitation*: Several global-scale studies of precipitation trends indicate the spatial distribution of changes in rainfall pattern. The region between 30°N and 85°N has seen increased precipitation, whereas that between 10°S and 30°N has had decreased precipitation over the past 30–40 years. Apart from this, precipitation has significantly increased from 1900 to the 1950s, and decreased between 10°N and 30°N after about 1970 (Peterson and Vose 1997; Chen et al. 2002; Adler et al. 2003).
- *Snow and ice*: The frozen form of water on the earth's surface (cryosphere) makes up 75% of freshwater, which is also the main source of discharge in major perennial rivers originated in these regions. These river basins accommodate more than one-sixth of the total world population (Stern 2007). Recent studies on snow and glaciers in the Himalayan region imply that the glacier cover area has retreated significantly in the past few decades, which is an alarming situation for the ecosystems and population living in the vicinity. Kulkarni and Karyakarte (2014) carried out a study on the Indian Himalayan glaciers and found that 13% of the glaciated area has retreated in the last four or five decades and 19 ± 7 m glacial ice has been lost.
- *Sea level rise*: Many highly populated cities situated in coastal areas have severe risk of natural hazards from sea level rise. Increasing global mean temperature is enhancing the glacial melt rates, which is a serious threat to the cities in coastal areas. In the United States, 40% of the population lives in coastal areas, which are very much prone to intense

rainfall and urban flooding. Scientific studies reveal that the average rising rate of the global sea level has doubled from 1.7 mm/year to 3.4 mm/year in the twentieth century.

- *Evapotranspiration*: Evapotranspiration is also an important component of the hydrological cycle that determines the water availability of a region. Evapotranspiration depends on several factors, which are dependent on climate variables and local parameters such as precipitation, temperature and wind speed, etc. Global warming has increased the global mean temperature by 0.6°C in the twentieth century. This rise in temperature will alter evapotranspiration at the local and larger scale.
- *River runoff*: River runoff depends on the land cover, soil type topography and other physical factors of a watershed. Apart from the physical factors, several climatic factors also play important roles in governing river discharge. Increased precipitation and temperature play a major role in altering river flow. In mountainous regions, enhanced snow and glacier melt have increased the runoff in rivers.

24.2.2 Projected Changes

Projection of climatic variables such as precipitation and temperature gives a clear picture of change in different hydrological components. The recently released fifth assessment report of the IPCC (2012) suggests that extreme events are likely to increase in the twenty-first century with an increase in the runoff in mountainous catchment and a decrease in inland rivers. Worldwide studies on the projection of stream flow for different greenhouse gas emission scenarios indicate change in flow regimes of many large rivers. Arnell (1999) found that a decrease in water resource availability in major river basins would result from climate change and increased demand of the population. Gleick and Chalecki (1999) conducted a study on the Colorado River Basin and the Sacramento River Basin to evaluate the change in flow regimes of the river for climate change scenarios; they found that the river has gone through a major shift in its usual flow schedule because of the change in snowmelt dynamics of the region. Although the climate change study results have an element of uncertainty, the changes in river hydrology are unequivocal (Mujumdar 2011). Apart from river flow, the occurrence of several other hydrological phenomena such as droughts and floods has been studied by several researchers, and noticeable changes have occurred in the occurrence and distribution of these events.

Coundrain et al. (2005) mentioned that the increased snowmelt in glacier-fed rivers may have severe consequences with possible increased flood events and loss of life. Palmer and Raisanen (2002) found that Europe is likely to face longer wet spells during winters followed by a wet monsoon season in Asia for increased greenhouse gas emissions, which will have consequences in the form of winter flood hazards.

Milly et al. (2002) conducted a study on the 16 large river basins in the world by analyzing the changes in flood occurrences enhanced by the increased CO_2 concentration in the atmosphere; they anticipated an increase in peak flood events with a 100-year return period in the near future. Spring floods in temperate regions induced by the snowmelt contribution are likely to decline (Zhang et al. 2005). South Asian countries are also vulnerable to extreme hydrological events. Mirza et al. (2003) projected an increase of 30% in flooded areas with an increase in global temperature by 2°C. About 20% of the world's population is vulnerable to flood disasters, and this hazard will increase by the 2080s (Kleinen and

Petschel-Held 2007). Douville et al. (2002) stated that southern Europe would suffer frequent and intense drought because of a decrease in precipitation followed by increased temperatures.

24.3 Water Stress – the Supply Demand Balance

Sustainable water management is a prime concern of policymakers for rapid population growth, urbanization and industrialization. A sustainable approach to managing water resources is of paramount importance to mitigate water scarcity problems in the future. Population projections for the twenty-first century along with the projections of precipitation and temperature indicate an alarming situation for available water resources. Population growth combined with climate change impacts is creating more challenges for the water resource planners and sustainable management will largely depend upon the available hydrological, climatic and socio-economic data. Fulfilling the demands of different sectors and conflicts of interest is also a major challenge for sustainable development planning.

24.3.1 Freshwater Availability and Demand in the Future

Spatiotemporal availability of freshwater depends largely on precipitation, temperature and evapotranspiration (Chiew 2007). Freshwater availability in glacier- and snowmelt-fed river basins largely depends on the melting season of the river basin. Studies on the continental and mountainous rivers and stream flow behavior indicate that spring snowmelt peak has shifted forward, and subsequently, winter flow has increased. Because large populations reside in the river basins, which are fed by snowmelt water, an alteration in snowmelt and snowfall schedule because of climate change would pose a serious impact on resource availability (Barnett et al. 2005; Mackay 2007). Regions with projected runoff reduction in the future are likely to face problems in services such as drinking water supply, energy sources, and habitat for freshwater fauna and flora (Barnett et al. 2004; Lehner et al. 2005).

One-fourth of the total world population resides in coastal regions and these areas are undergoing immense population increase. These coastal regions have very low freshwater supply; excessive groundwater withdrawal in such coastal regions causes saltwater intrusion into the groundwater, which restricts the freshwater availability as a minor rise in sea level would cause a reduction in aquifer freshwater levels (Mackay 2007; Bobba et al. 2000). Arnell (2004) concluded that the population affected by climate change will not only depend on changes in climatic or hydrological variables but also on how management strategies are implemented. Climate projections of the major river basins, depending on the scenario and climate models in question, show that water stress is likely to increase in the 2050s (Alcamo et al. 2007).

24.3.2 Water for Hydropower and Irrigation

The majority of dams and reservoirs are constructed to fulfill the requirement of flood control, irrigation and urban water. These reservoirs and hydropower plants provide facilities to mankind; on the other hand these giant concrete structures become a major threat to ecology and the environment. With the increasing demand from different sectors such as irrigation, hydropower and drinking water supply, these reservoirs may not be able to fulfill the requirement as per their presently designed capacity. Aside from the demographic changes in the command area, climate change also poses a serious threat to the available dams and reservoirs in terms of their utility to accommodate the extreme events and increased flood events in terms of volume and frequencies. Sustainable approaches to the designing and functioning of such structures are the need of the hour.

24.4 Impacts of Climate Change on Glaciers and Mountainous Water Resources

24.4.1 Climate Change and Glaciers

Due to rise in the concentration of CO_2, the mean temperature of the earth has been increasing since the late nineteenth century. The increase in temperature has resulted in the accelerated melting of snow and glaciers throughout the globe (Haeberli and Beniston 1998). These glaciers are the source of freshwater to a large proportion of the population. For example, many Indian subcontinent rivers such as Indus, Brahmaputra and Ganga have their origin in Himalayan glaciers. These river basins are home to more than 800 million people. During the dry season, a major part of the water in these rivers comes from the melting of the glaciers (Bolch et al. 2012). Himalaya and Karakoram (H-K) glacier forms one of the largest glaciers in the world, with a total estimated area between 43 178 km^2 and 49 650 km^2 (Bolch et al. 2012). Glaciers not only play important roles in mountainous regions, but also have a significant impact on the hydrology of lowland rivers. Shrestha and Aryal (2011) conducted a study on the impact of climate change on Himalayan glaciers in the Nepal region. The study found a significant warming in the country with an average trend of 0.06°C per year, which is thought to be the main reason for the retreating glacial extent in Nepal Himalaya. Oerlemans et al. (1998) carried out a study on several glaciers and ice caps for different climate change scenarios and found different responses from individual glaciers for different scenarios. Dyurgerov and Meier (2000) explored the interaction between glaciers and their surrounding climate to study the historical climate change footprints and possible impacts of future climate change on the cold regions of the world; the study found that change in the areal extent of glaciers is time- and region-dependent with a total loss of 147 mm $year^{-1}$ of water equivalent volume during the period of 1961–1997.

24.4.2 Changes in Snowmelt Runoff

As discussed in the previous section, climate change or global warming has serious impacts on rivers originating from ountains covered with snow or glaciers. The accumulation of temporary snow during the winter and further melting in the summer form a major component of river discharge. Snowmelt is an important component of river flow, but in some cases, excessive snowmelt leads to floods in the downstream. A water balance study on the Himalayan rivers fed by snowmelt and glacier runoff reveals that 50% of the discharge in western river basins such as the Indus River is fed by snowmelt, followed by 25% for eastern catchments and less than 20% for the other rivers (Bookhagen and Burbank 2010). The study also states that changes in the occurrence and magnitude of snow- and glacier melt may have social and ecological consequences because of global warming and changes in winter rainfall patterns. Barnett et al. (2005) studied the change in water resource availability of the snow- and glacier-dominated region under the ambit of climate change and found there was a decrease in winter snowfall followed by a shift in snowmelt period. The contribution of snow and glaciers to the freshwater availability on earth shows that any change in the hydrology of these regions will have major consequences for the water availability of a large portion of the world; such problems are already diagnosed in some regions and are likely to increase if proper steps are not taken.

24.4.3 Floods in Mountainous Region

Mountains are known as the natural reservoir of the earth. Covering one-fourth of the total earth surface, mountains store ample amounts of water and divert the stored water in surrounding plains of lower elevation through the network of natural channels. However, the catchments in the higher altitudes are susceptible to intense floods. Factors contributing to the floods in mountainous catchments include high rainfall intensity, steep gradients (helping the faster accumulation of surface runoff), rocky beds with low percolation and glacial lake outburst floods (GLOFs). Weingartner et al. (2003) studied the floods in mountain areas of Switzerland. They found that the most critical zone for the high specific discharge is catchments in the range of 1000–2000 m altitude. It is well accepted that climate change is resulting in the retreat and shrinkage of glaciers throughout the world, and the Himalayas are no exception. The melting of glaciers causes formation of glacial lakes. Some of the lakes are so large that they have the potential to cause huge floods in the downstream areas if an outburst event happens. Several mountainous regions in the world have experienced many GLOF events in recent time, which have caused severe damage in terms of human lives and infrastructure (Ives et al. 2010). One such example is the Kedarnath flash floods in June 2013, which cost more than 5000 human lives and affected about 4200 villages (Durga Rao et al. 2014).

24.5 Sustainable Management of River Water Resources

Sustainable management of water resources can be defined as a system that is planned and designed to meet the demands of the various sectors for the present as well as for the future, taking into account the integrity of the ecology, environment and hydrology of the region (Loucks 2000). There are three main criteria suggested by the scientific community to assess the sustainability of water resources:

Reliability: The indicator of a water resource system's ability to meet the demand and supply needs of the river basin.

Resiliency: The indicator of a system's ability to recover from failure. It also indicates the time taken by the system to come out of a failure state.
Vulnerability: The indicator assessing the loss incurred by system failure.

A sustainable system is supposed to be highly reliable and highly resilient with low vulnerability. Any water resource system is designed taking ecological, environmental and societal factors into account; a system is affected by any change in the components of these factors. For example, a water supply system designed for a particular population will be affected by the change in population after a decade or so. Similarly, climate change is also an important factor to be considered while designing a sustainable water management system. Reliability, resilience and vulnerability are all considered to be affected by the impact of climate change, along with changes in demand and supply.

Several studies related to sustainable management of water resources have been carried out, which provide a wide framework for the adaptation of methods based on objectives and the limitations. Xu et al. (2002) applied an object-oriented system dynamics approach to evaluate the sustainability of the water resources management system in the Yellow River Basin. The system dynamics approach accounts for all the sources of available water along with all the factors that are responsible for the water supply and demand balance. An appropriate sustainable plan was adopted after measuring the sustainability index of each component of the management plan. Mujumdar (2011) suggested a participatory approach should be taken while deciding on a management plan – plan should take a bottom-up approach, which should be inclusive of all the stakeholders, rather than imposing a top-down approach. Torretta (2014) developed a decision support system for sustainable utilization of the available water resources in a major river basin of Brazil; decision making was performed by taking the water quality parameters and riverine ecosystems into account. Adaptation for future scenarios was planned based on the population projection, climate change scenarios and adaptive measures such as wastewater treatment and improved sewage systems.

24.6 Case Studies

24.6.1 Introduction

Planning of sustainable water resource development largely depends on several factors at regional scales, such as climate and demography. For sustainable planning of water resources it is important to understand the regional climatic variations and population increase rate along with the impact of climate change. Projection of climatic variables along with the changes in population provides an approximate estimation of the changes likely to happen in the region, which need to be addressed for a sustainable development program. Hydrological modeling can be used for assessment of the water resources; possible changes in the water availability and demand can be assessed by projection of climatic factors along with hydrological projections. The present case study was carried out on the tributary of the Brahmaputra River Basin to assess the changes in climate variables along with the hydrological and demographic change of the river basin, which is important for the sustainable development of water resources. The study includes the general circulation models (GCMs) for the climate projections and hydrological modeling of the river basin for assessment of the change in hydrology of the river basin. Based on the changes in demography of the basin, changes in water demand and water resource availability were assessed using hydrological modeling.

24.6.2 Study Area

The Subansiri River Basin in the northeastern states of India is a tributary of the Brahmaputra River, which originates in the Tibet Plateau (Figure 24.1). The River Basin covers parts of the Lower Subansiri district, Upper Subansiri district and Papumpare district of Arunachal Pradesh, and enters Assam state through the Lakhimpur district, where it joins the Brahmaputra River and covers partially the Dhemaji district. The Subansiri River contributes about 7.9% of the total discharge of the Brahmaputra River, observed at Pandu port (Sarma 2004).

24.6.3 Methodology

A statistical downscaling model (SDSM) is used for the downscaling of maximum temperature (T_{max}), minimum temperature (T_{min}) and precipitation. SDSM is a tool working on regression transformation equations between observed data (predictand) and GCM predictors (Wilby et al. 2002). SDSM works with two modules for model preparation, i.e. a conditional model and an unconditional model. The conditional module is used for precipitation downscaling because it establishes regression transformation functions separately for dry spells and wet spells (Wilby et al. 2002).

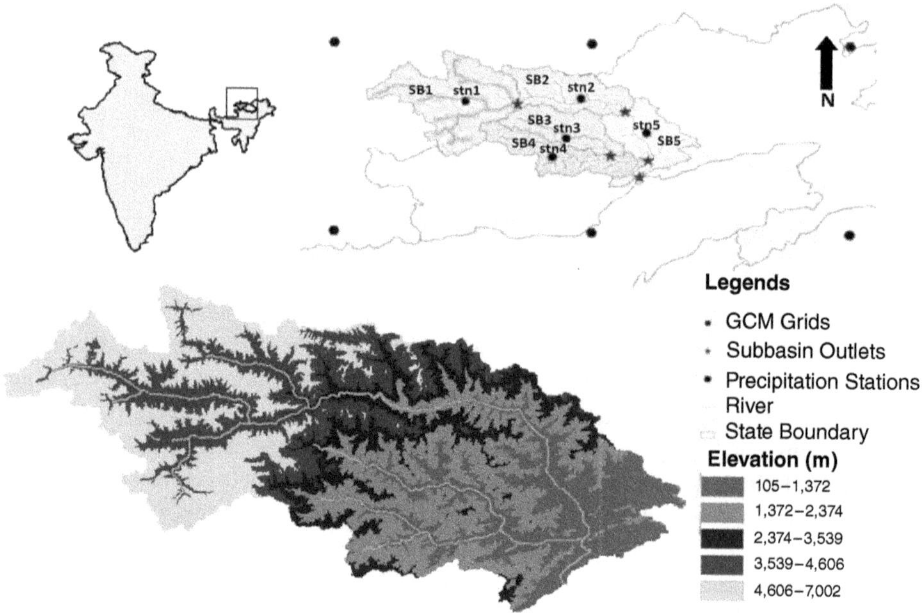

Figure 24.1 Study area map.

Hydrological modeling was performed using the soil and water assessment tool (SWAT) model. SWAT is a semi-distributed deterministic model developed initially for simulating the streamflow at ungauged sites (Arnold et al. 2012; Srinivasan et al. 2010). SWAT works on physical-based parameter inputs such as soil maps, land use/land cover maps and weather data (e.g. temperature, precipitation). SWAT works on the principle of the water balance equation, which can be written as:

$$SW_t = SW_0 + \sum_{i=1}^{t}(R_{day} - Q_{surf} - E_a - W_{seep} - Q_{gw}) \tag{24.1}$$

where SW_t denotes the soil water content in millimeters (mm) at the end of the time step; SW_0 denotes soil water content (mm) at the starting of the day i; R_{day} denotes precipitation (mm) on day i; Q_{surf} denotes surface runoff (mm) on day i; E_a denotes amount of evapotranspiration (ET) in mm on day i; W_{seep} denotes the water in the vadose zone (mm) from the soil profile on day i; and Q_{gw} denotes return flow (mm) on day i.

The top-down approach to water resource management includes decision making based on impact studies, and the decision-making process does not involve the stakeholders' priorities. Changes in water demand were calculated based on the population increase and availability of the water resources using the hydrological model. An integrated approach of hydrological modeling and a management planning model approach was adopted for the possible water resource management strategies in the river basin. Further, in-situ water conservation structures were suggested based on the geospatial analysis of the river basin for meeting the water demand in the lean season (Kusre et al. 2010; Ramakrishnan et al. 2009).

24.6.4 Results and Discussion

24.6.4.1 Change in Temperature and Precipitation

The downscaling model was prepared for precipitation and temperature projections using National Centre for Environmental Prediction (NCEP) datasets and observed precipitation and temperature data for the study area. Furthermore, a calibrated and validated precipitation and temperature downscaling model was utilized for projecting the future series of these variables. Three GCMs with three Representative Concentration Pathway (RCP) scenarios were selected in this study for future projection of precipitation, maximum temperature (T_{max}) and minimum temperature (T_{min}). In this way, a total of nine ensembles of each variable were generated to check the variability within the GCMs. Figure 24.2a shows the plot of mean annual temperature of all the nine ensembles for the period of 2011–2100 and shows an increasing pattern of a warming trend in future. Figure 24.2b shows the average of three GCMs for the respective RCP scenarios, i.e. RCP2.6, RCP6.0 and RCP8.5. Figure 24.2b indicates an increase in temperature for low to extreme emission scenarios, and this increase in temperature varies from 14°C (for RCP2.6) to 14.6°C (for RCP8.5). Figure 24.2b represents the average of three GCMs for their respective RCP scenarios and shows that the change in magnitude is increasing for all emission scenarios with magnitude increasing from low to high emission scenarios.

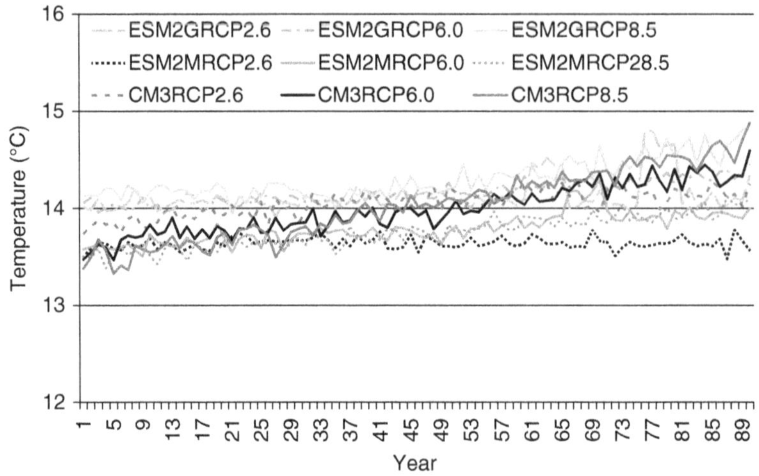

(a)

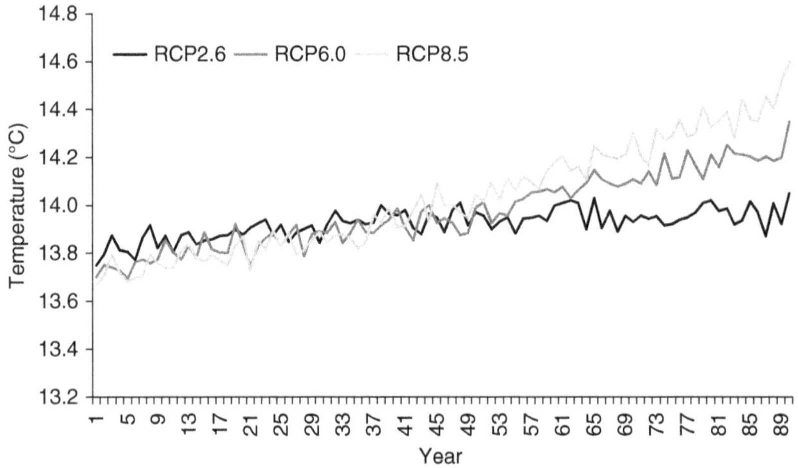

(b)

Figure 24.2 Changing pattern of maximum temperature for all general circulation model (GCM)/Representative Concentration Pathway (RCP) scenarios for 2011–2100 period. (a) All the nine ensembles and (b) average of the GCMs of respective RCPs.

Similar patterns can be seen for the T_{min} in Figure 24.3. Figure 24.3a shows the annual T_{min} patterns of all the nine ensembles for the period of 2011–2100; Figure 24.3b shows the average of three GCMs for the respective RCP scenarios. The T_{min} is likely to increase from 4.8°C for the RCP2.6 scenario to 5.6°C for RCP8.5 in 2071s. Although the T_{max} and T_{min} values show clear increasing trends, the monthly analysis of temperature for all the scenarios may provide a clearer picture of the seasonal changes in temperature. Figure 24.4 shows the month-wise changes in T_{max} as compared to the observed annual average T_{max}. Changes in magnitude are computed for a 30-year-inter-decadal time interval, i.e. 2011–2100 time period divided into three parts 2011–2040 (2011s), 2041–2070 (2041s) and 2071–2100 (2071s).

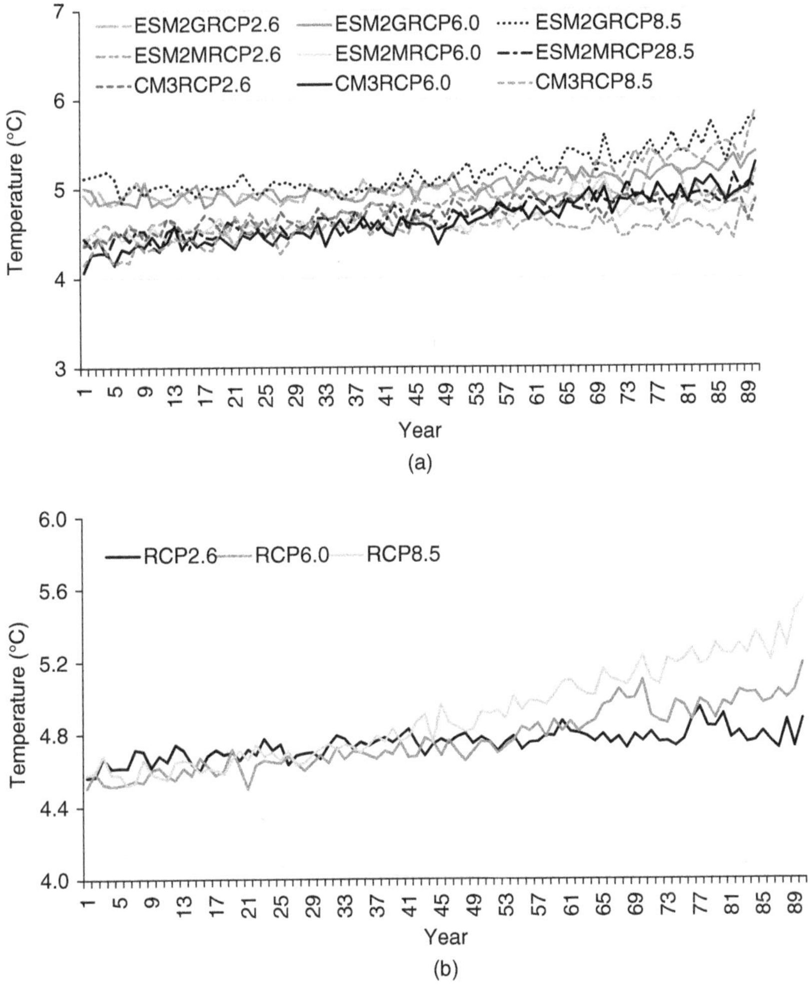

Figure 24.3 Changing pattern of minimum temperature (T_{min}) for all general circulation model (GCM)/Representative Concentration Pathway (RCP) scenarios for the 2011–2100 period. (a) For all the nine ensembles and (b) average of the GCMs of respective RCPs.

Figure 24.4 shows the changes in T_{max} for each month during the three inter-decadal time intervals, i.e. 2011s, 2041s and 2071s, as compared to the observed T_{max} for the average of three GCMs and three RCP scenarios. Figure 24.4 shows a clear increase in summer temperature along with the winter months of January and February for all the RCP scenarios. Monsoon months (July, August, September and October) show decreases in annual average T_{max}. November and December also show a gradual decrease in T_{max} from 2011s to 2071s. In January, the increase in T_{max} varies from a minimum of 0.9°C (RCP2.6, 2011s) to a maximum increase of 1.5°C (RCP8.5, 2071s). The month of February has an increase of 2.7°C at the end of the century (2071s) for the extreme emission scenario (RCP8.5). March, April and May show uniform gradual increases in T_{max} for 2011s, 2041s and 2071s for RCP scenarios. In April, the increase in T_{max} for 2011s is 0.4°C (RCP2.6) and 2.7°C for RCP8.5

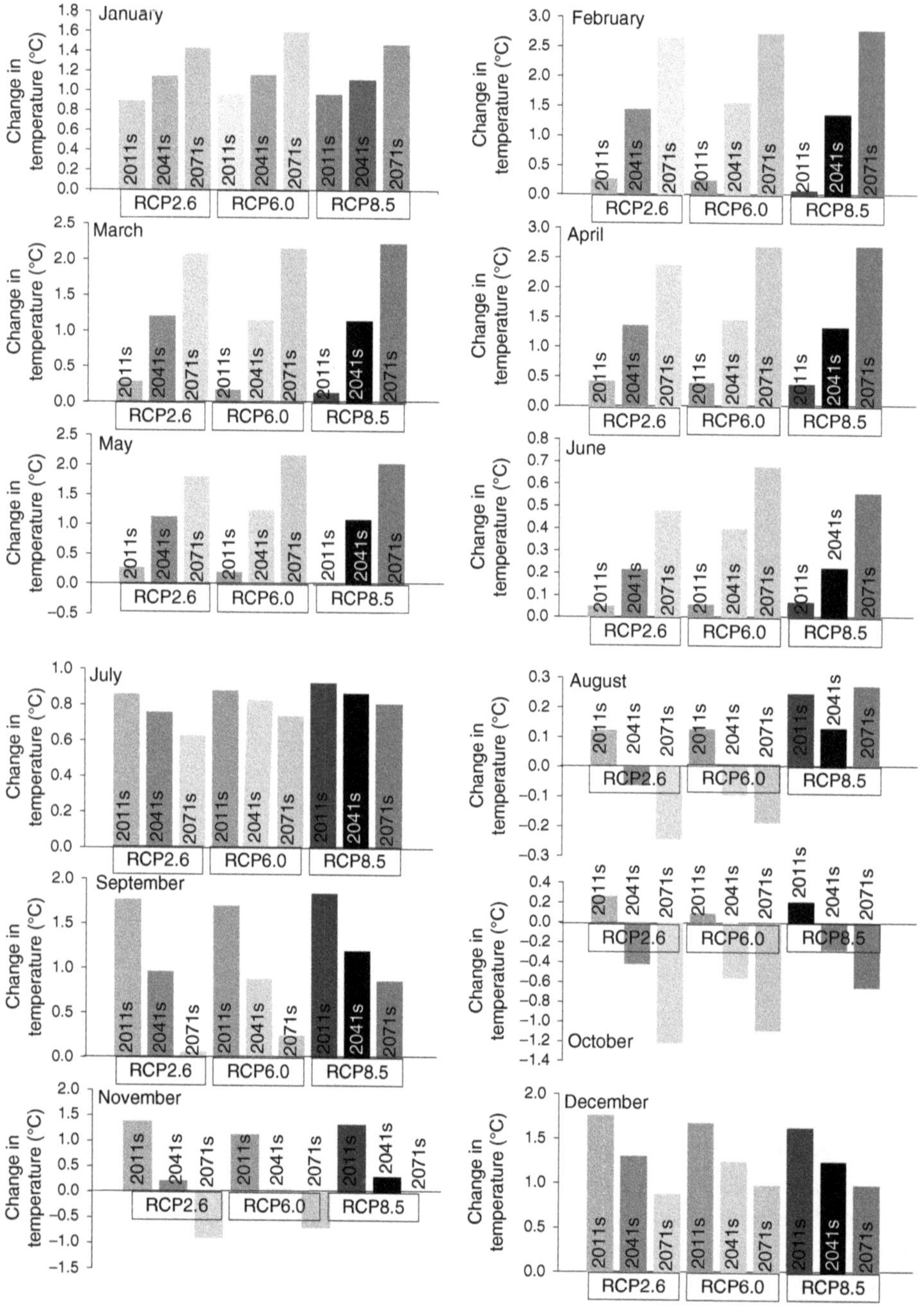

Figure 24.4 Monthly changes in maximum temperature for Representative Concentration Pathway (RCP) scenarios for different three inter-decadal time scales as compared to observed maximum temperature.

during the past 30 years (2071s). Monsoon month temperatures tend to decrease from the low emission scenario (RCP2.6) to the extreme emission scenario (RCP8.5) gradually for 2011s, 2041s and 2071s. The decrease in T_{max} of October has a magnitude of 1.1°C for RCP6.0 in 2071s.

The average annual minimum temperature (T_{min}) also shows a similar pattern to T_{max}. Figure 24.5 shows that changes in average T_{min} for all the months compared to the observed period average monthly T_{min}. The increase in T_{min} for the summer months is higher compared to the other months; the increase in T_{min} for the months of March and April ranges from the lowest 0.34°C in 2011s for RCP2.6 to the maximum of 3.1°C in 2071s for RCP8.5,

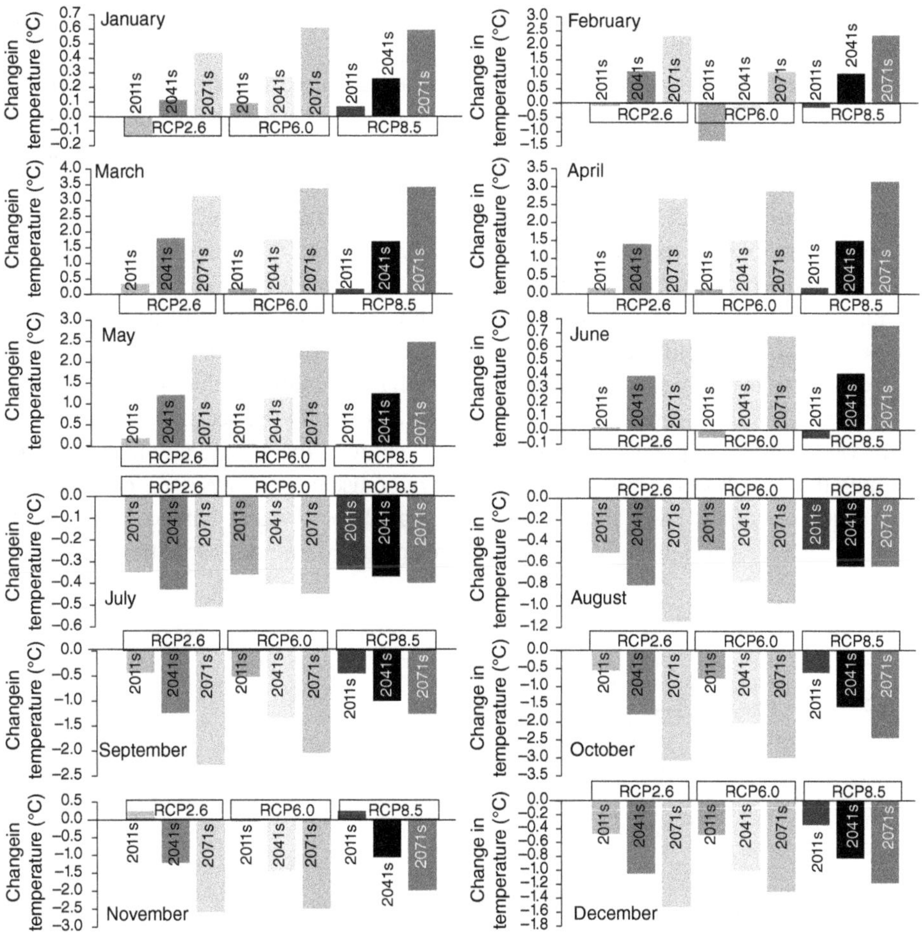

Figure 24.5 Monthly changes in minimum temperature for Representative Concentration Pathway (RCP) scenarios for different three inter-decadal time scales as compared to observed minimum temperature.

whereas the monsoon months depict a continuous decrease in the magnitude of T_{min}. The T_{min} range for October is likely to decrease up to 2.43°C for the RCP8.5 scenario in 2071s. Similar decreases in the months of November and December are observed: in 2071s the decrease in T_{min} goes up to 2.5°C in November for RCP2.6 and to 1.18°C in December for RCP8.5.

Similarly, a precipitation downscaling model was used for the projection of nine precipitation ensembles for GCM/RCP scenarios. Although there is variability found in the magnitude of change in projected precipitation series for different GCM/RCP ensembles, an increase in precipitation is evident. Projection of precipitation shows an increase in annual precipitation for all the scenarios. The average of the three GCMs for their respective RCP scenarios shows an increase in precipitation from RCP2.6 to RCP8.5. The magnitude of annual precipitation at the end of the century would increase up to 2610, 2680 and 2750 mm for RCP2.6, RCP6.0 and RCP8.5, respectively.

Downscaled precipitation and temperature data were used as primary inputs in the calibrated hydrological model for assessing the change in the amount of stream flow for different RCP scenarios. The major concern of the changing climate for a region is the availability of the water for different seasons, which can be assessed with the help of the exceedance percentage of the streamflow or the flow-duration curve. This methodology provides an idea of the probability of the dependable. Figure 24.6 shows the flow-duration curve for different GCM/RCP scenarios. Figure 24.6 provides a good insight into the dependable flow in terms of the probability; it shows that for almost all the dependable flow percentage the magnitude of stream flow is increasing; also for a particular percentage, exceedance discharge in the RCP8.5 scenario has the highest value. Figure 24.6 also indicates that the flow in the river is likely to increase for all the scenarios, and the discharge value for all the percentage exceedeance is going to increase in any RCP scenario. Although the flow-duration curve plotting of the river discharge shows a uniform increase in the discharge of the river, the subbasin-based analysis of water balance components would provide a clear pattern of changes in precipitation, evapotranspiration and water yield.

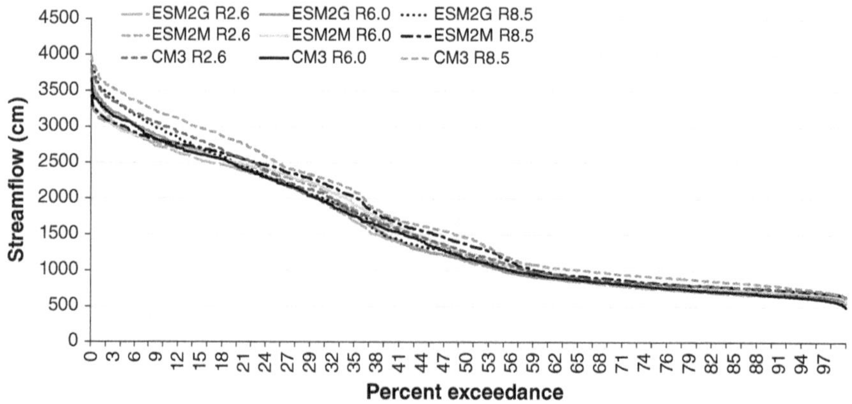

Figure 24.6 Comparison of exceedance flow for all the general circulation model (GCM)/ Representative Concentration Pathway (RCP) ensembles over the period of 2011–2100.

24.6.4.2 Change in Water Demand

Water demand was calculated for the three districts of the river basin and it was compared with the available water. The storage in the reservoir and unmet demand was calculated for the downstream districts with its corresponding projected population. Figure 24.7 shows the location of the districts that are considered as the demand sites. In this study, only population water demand was considered for estimating the water demand and availability.

Census data for the given districts were used for estimating the population growth and increase in the population between two censuses (2001–2011). The percentage growth in population was taken as the basis for estimating the annual population growth. Furthermore, the population growth was projected following the geometric increase method. Population in the downstream of the reservoir was calculated by the geometric increase method and the averages of the population growth in the last two censuses for all three districts were taken as the increase in population per year. In this case, 2% increases in population were taken as the growth rate for projection of the population up to 2050. Furthermore, water demand was calculated using the per capita water requirement of the district, and projection of the water demand was carried out based on the population projection and water availability obtained from the hydrological modeling. Figure 24.8 shows the water demand and unmet water demand for the projected years (up to 2100).

24.6.4.3 Proposed Water Management Plan

Hydrological modeling and flow-duration curve analysis of the river basin indicate that, although the annual average stream flow of the river basin is likely to increase, less precipitation in the non-monsoon period will cause a deficit in the available water resources. It was observed from the water availability calculation that the available water resources are

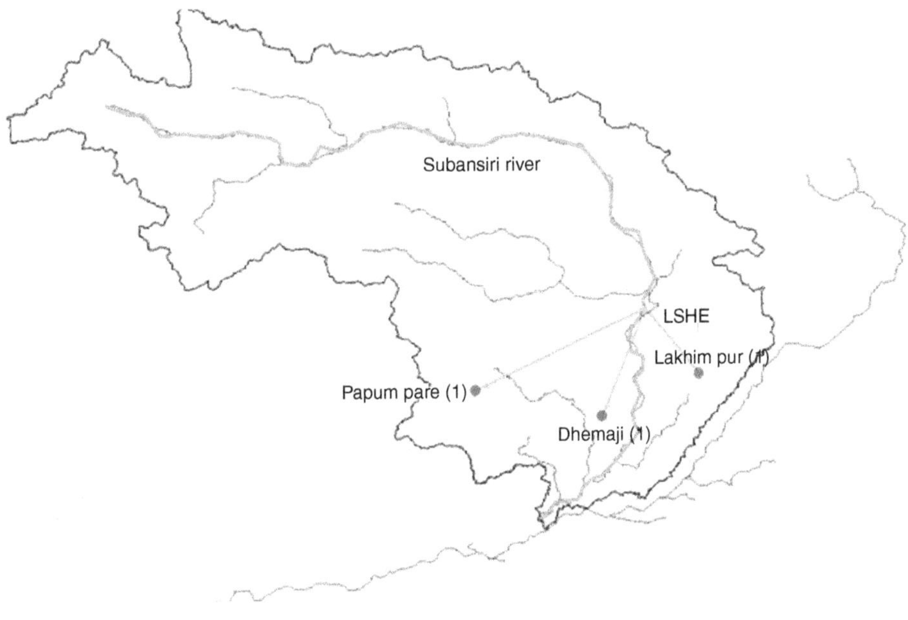

Figure 24.7 Location of demand sites in the water management model.

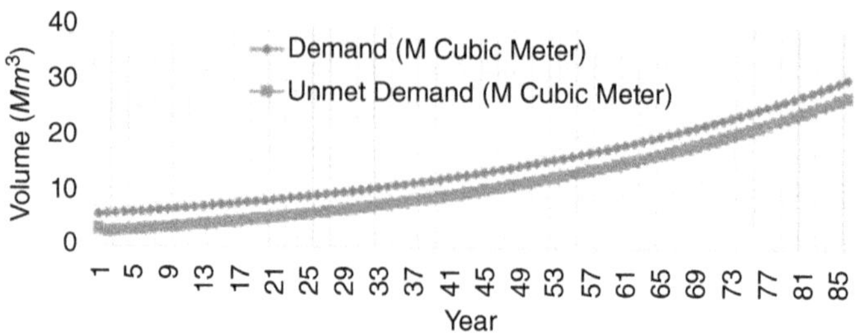

Figure 24.8 Unmet demand of command area districts for Representative Concentration Pathway (RCP) scenarios.

not enough to fulfill the water demand in the command area. In the management plan to overcome the seasonal scarcity of the command area, it is necessary to rely on local surface water resources. Since the groundwater level and utility data are not available for the region, the management plan is focused on the surface water resources only. A geospatial platform was utilized for assessing the stream density, slope and potential water retention sites of the river basin. Water availability data were taken from the hydrological model data outcome. Further, weighted overlay analysis was done to identify the potential water retention sites in the catchment. Figure 24.9 shows the proposed water storage structures in the downstream small tributaries. These proposed water storage structures would check the drainage of water to conserve it for the lean period. It was observed from the model that local water storage structure is sufficient to fulfill the water requirement in the lean period and low flow season.

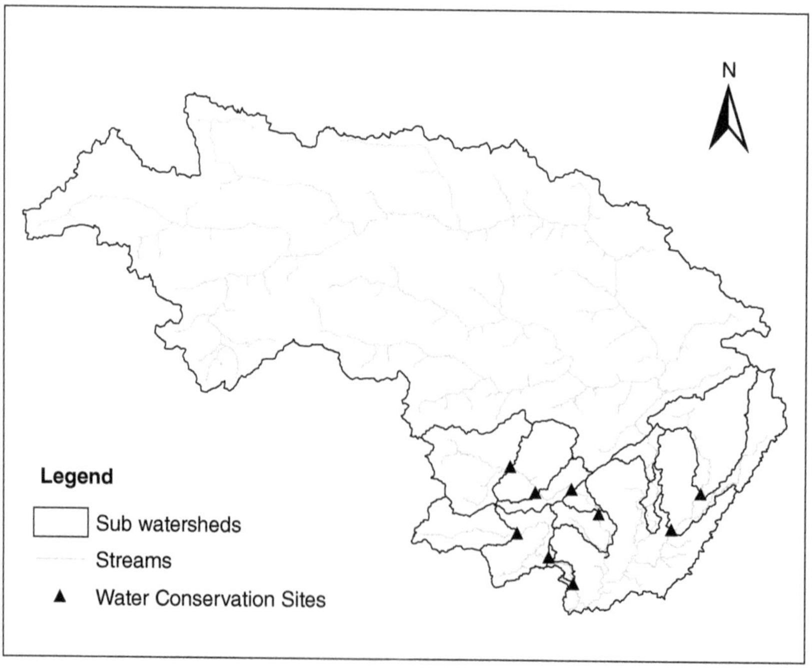

Figure 24.9 Location of proposed check-dam sites in the command area.

24.7 Conclusions

Sustainable water resource management is the need of the hour with changing global climate and increased population and industrialization. Sustainable development involves the welfare of several stakeholders participating in management planning. Conflict of interest is also a challenge for water managers while considering the interests of all the stakeholders. Thus, a basic philosophy should be followed, which is "no one wins unless everyone wins." Water resource management plans are mostly governed by regional scale factors, which brings the necessity of involvement of all the stakeholders.

The case study presented here focuses on sustainable water resource management planning under the climate change scenario by considering the demographic changes of a river basin in northeast India. The results show significant increases in annual precipitation, T_{max}, and T_{min} of the basin. Since the hydrology of any region depends largely on the climatology factors along with the physical factors of the region, the changes in precipitation and temperature subsequently affect the runoff and water yield in the river basin. Water availability and demand estimation of the river basin rshow that the water scarcity situation in the non-monsoon season needs to be addressed. This study could be helpful for policymaking, planning and mitigation strategies to curb the possible flood hazard situations for the climate change scenarios in the region.

References

Adler, R.F., Huffman, G.J., Chang, A. et al. (2003). The version-2 global precipitation climatology project (GPCP) monthly precipitation analysis (1979–present). *Journal of Hydrometeorology* 4 (6): 1147–1167.

Alcamo, J., Florke, M., and Marker, M. (2007). Future long-term changes in global water resources driven by socio-economic and climatic change. *Hydrological Sciences Journal* 52 (2): 247–275.

Arnell, N.W. (1999). The effect of climate change on hydrological regimes in Europe: a continental perspective. *Global Environmental Change* 9: 5–23.

Arnell, N.W. (2004). Climate change and global water resources: SRES scenarios and socio-economic scenarios. *Global Environmental Change* 14: 31–52.

Arnold, J.G., Moriasi, D.N., Gassman, P.W. et al. (2012). SWAT: model use, calibration, and validation. *Transactions of the ASABE* 55 (4): 1491–1508.

Barnett, T.P., Malone, R., Pennell, W. et al. (2004). The effects of climate change on water resources in the West: introduction and overview. *Climatic Change* 62: 1–11.

Barnett, T.P., Adam, J.C., and Lettenmaier, D.P. (2005). Potential impacts of a warming climate on water availability in snow dominated regions. *Nature* 438: 303–309.

Bates, B.C., Kundzewicz, Z.W., Wu, S., and Palutikof, J.P. (eds.) (2008). *Climate Change and Water*. Technical Paper of the Intergovernmental Panel on Climate Change. Geneva: IPCC Secretariat.

Bobba, A., Singh, V., Berndtsson, R., and Bengtsson, L. (2000). Numerical simulation of saltwater intrusion into Laccadive Island aquifers due to climate change. *Journal Geological Society of India* 55: 589–612.

Bolch, T., Kulkarni, A., Kaab, A. et al. (2012). The state and fate of Himalayan glaciers. *Science* 336 (6079): 310–314.

Bookhagen, B. and Burbank, D.W. (2010). Toward a complete Himalayan hydrological budget: spatiotemporal distribution of snowmelt and rainfall and their impact on river discharge. *Journal of Geophysical Research Earth Surface* 115 (3): 1–25.

California Department of Water Resources (CDWR) (1998). *California Water Plan*. Updated Bulletin 160-98. Sacramento, CA: California Department of Water Resources.

Chen, M., Xie, P., and Janowiak, J.E. (2002). Global land precipitation: a 50-yr monthly analysis based on gauge observations. *Journal of Hydrometeorology* 3: 249–266.

Chiew, F.H.S. (2007). Estimation of rainfall elasticity of streamflow in Australia. *Hydrological Sciences Journal* 51 (4): 613–625.

Coundrain, A., Francou, B., and Kundzewicz, Z.W. (2005). Glacier shrinkage in the Andes and consequences for water resources-editorial. *Hydrological Sciences Journal* 50 (6): 925–932.

Douville, H., Chauvin, F., Planton, S. et al. (2002). Sensitivity of the hydrological cycle to increasing amounts of greenhouse gases and aerosols. *Climate Dynamics* 20 (1): 45–68.

Durga Rao, K.H.V., Venkateshwar Rao, V., Dadhwal, V.K., and Diwakar, P.G. (2014). Kedarnath flash floods: a hydrological and hydraulic simulation study. *Current Sciences* 106 (4): 598–603.

Dyurgerov, M.B. and Meier, M.F. (2000). Twentieth century climate change: evidence from small glaciers. *PNAS* 97 (4): 1406–1411.

Gleick, P.H. and Chalecki, E.L. (1999). The impacts of climatic changes for water resources of the Colorado and Sacramento-San Joaquin river basins. *Journal of the American Water Resources Association* 35: 1429–1442.

Haeberli, W. and Beniston, M. (1998). Climate change and its impacts on glaciers and permafrost in the Alps. *Ambio* 27: 258–265.

Hidalgo, H.G., Das, T., Dettinger, M.D. et al. (2009). Detection and attribution of streamflow timing changes to climate change in the western United States. *Journal of Climate* 22: 3838–3855.

IPCC (2007). *Impacts, Adaptation, and Vulnerability: Working Group II Contribution to the Intergovernmental Panel on Climate Change Fourth Assessment Report, Summary for Policymakers*. Geneva: IPCC Secretariat.

IPCC (2012). Managing the risks of extreme events and disasters to advance climate change adaptation. A special report of the Intergovernmental Panel on Climate Change. Geneva: IPCC.

Ives, J.D., Shrestha, R.B., and Mool, P.K. (2010). *Formation of Glacial Lakes in the Hindu Kush-Himalayas and GLOF Risk Assessment*. Lalitpur, Nepal: ICIMOD.

Kiparsky, M. and Gleick, P.H. (2003). Climate change and California water resources: a survey and summary of the literature. California Climate Change Center, Report no. CEC-500-04-073.

Kleinen, T. and Petschel-Held, G. (2007). Integrated assessment of changes in flooding probabilities due to climate change. *Climatic Change* 81 (3): 283–312.

Kulkarni, A.V. and Karyakarte, Y. (2014). Observed changes in Himalayan glaciers. *Current Sciences*: 237–244.

Kusre, B.C., Baruah, D.C., Bordoloi, P.K., and Patra, S.C. (2010). Assessment of hydropower potential using GIS and hydrological modeling technique in Kopili River basin in Assam (India). *Applied Energy* 87 (1): 298–309.

Lehner, B., Doll, P., Alcamo, J. et al. (2005). Estimating the impact of global change on flood and drought risks in Europe: a continental integrated assessment. *Climatic Change* 75: 273–299.

Ligtvoet, W., Hilderink, H., Bouwman, A. et al. (2014). *Towards a World of Cities in 2050. An Outlook on Water-Related Challenges*. Background report to the UN-Habitat Global Report. The Hague, the Netherlands: PBL Netherlands Environmental Assessment Agency.

Loucks, D.P. (2000). Sustainable water resources management. *Water International* 25 (1): 3–10.

Mackay, A. (2007). Climate change 2007: impacts, adaptation and vulnerability. Contribution of Working Group II to the fourth assessment report of the Intergovernmental Panel on Climate Change. *Journal of Environmental Quality* 37 (6): 2407.

Milly, P.C.D., Wetherald, R.T., Dunne, K.A., and Delworth, T.L. (2002). Increasing risk of great floods in a changing climate. *Nature* 415 (6871): 514.

Mirza, M.M.Q., Warrick, R.A., and Ericksen, N.J. (2003). The implications of climate change on floods of the Ganges, Brahmaputra and Meghna rivers in Bangladesh. *Climatic Change* 57 (3): 287–318.

Mujumdar, P.P. (2011). Implications of climate change for water resources management. India Infrastructure Report, 18.

Oerlemans, J., Anderson, B., Hubbard, A. et al. (1998). Modelling the response of glaciers to climate warming. *Climate Dynamics* 14 (4): 267–274.

Palmer, T.N. and Raisanen, J. (2002). Quantifying the risk of extreme seasonal precipitation events in a changing climate. *Nature* 415: 512–514.

Peterson, T.C. and Vose, R.S. (1997). An overview of the global historical climatology network temperature database. *Bulletin of the American Meteorological Society* 78: 2837–2848.

Ramakrishnan, D., Bandyopadhyay, A., and Kusuma, K.N. (2009). SCS-CN and GIS-based approach for identifying potential water harvesting sites in the Kali Watershed, Mahi River Basin, India. *Journal of Earth System Science* 118 (4): 355–368.

Sarma, J.N. (2004). An overview of the Brahmaputra River system. In: *The Brahmaputra Basin Water Resources. Water Science and Technology Library*, vol. 47 (eds. V.P. Singh, N. Sharma and C.S.P. Ojha). Dordrecht, the Netherlands: Springer.

Shivam Goyal, M.K. and Sarma, A.K. (2017). Analysis of the change in temperature trends in Subansiri River basin for RCP scenarios using CMIP5 datasets. *Theoretical and Applied Climatology* 129 (3–4): 1175–1187.

Shrestha, A.B. and Aryal, R. (2011). Climate change in Nepal and its impact on Himalayan glaciers. *Regional Environmental Change* 11: 65–77.

Srinivasan, R., Zhang, X., and Arnold, J. (2010). SWAT ungauged: hydrological budget and crop yield predictions in the Upper Mississippi River Basin. *Transactions of the ASABE* 53 (5): 1533–1546.

Stern, N. (2007). *The Economics of Climate Change: The Stern Review*. Cambridge, UK: Cambridge University Press.

Torretta, V. (2014). The Sustainable use of water resources: a technical support for planning. A case study. *Sustainability* 6 (11): 8128–8148.

Vorosmarty, C.J., Charles, J., Green, P. et al. (2000). Global water resources: vulnerability from climate change and population growth. *Science* 289: 284–288.

Weingartner, R., Barbena, M., and Spreafico, M. (2003). Floods in mountain areas – an overview based on examples from Switzerland. *Journal of Hydrology* 282 (1–4): 10–24.

Wilby, R.L., Dawson, C.W., and Barrow, E.M. (2002). SDSM – a decision support tool for the assessment of regional climate change impacts. *Environmental Modelling and Software* 17 (2): 145–157.

WWDR (2006). World Water Development Report: Water a shared responsibility. Section 2: Changing Natural Systems. Paris: UNESCO.

Xu, Z.X., Takeuchi, K., Ishidaira, H., and Zhang, X.W. (2002). Sustainability analysis for Yellow River water resources using the system dynamics approach. *Water Resources Management* 16 (3): 239–261.

Zhang, W., Parker, K.M., Luo, Y. et al. (2005). Soil microbial responses to experimental warming and clipping in a tallgrass prairie. *Global Change Biology* 11: 266–277.

25

Sustainable Systems for Groundwater Resource Management

Manish Kumar Goyal[1], *Srinidhi Jha*[1], *and Rao Y. Surampalli*[2]

[1] *Discipline of Civil Engineering, Indian Institute of Technology, Indore, India*
[2] *Global Institute for Energy, Environment, and Sustainability, Lenexa, KS, USA*

25.1 Introduction

Groundwater is arguably one of the most vital natural resources. Out of the total water availability on planet earth, only 2.5% is present in the form of freshwater, most of which is in inaccessible to the human population. About 69% of the whole volume of freshwater is stored in the form of glaciers and ice caps, and the other 31% exists in and above the earth's crust as groundwater (30%) and surface water (1%). The total amount of groundwater is estimated to be much larger than surface water, and it is important to mention that a fraction of this water is saline. For example, subsurface water in the arid zones is often found to be unsuitable for drinking purposes because it rarely goes through renewal in thousands of years. Despite all the limitations, groundwater resources are of great value, especially in developing and underdeveloped countries as a large proportion of the population in these countries sees groundwater as a feasible source of fresh water. The distribution of surface water resources is similar to groundwater resources, but surface water is more susceptible to evaporation losses in the dry and hot seasons. Groundwater, on the other hand, is safe from such losses in comparison to surface water reservoirs. It is also more economical, more accessible and less vulnerable to pollution than surface water. Hence, on a larger scale such as public water supplies, groundwater is preferred over surface water resources. Even in economically leading countries like the United States, groundwater is the largest source of usable water. In many countries in the Middle East, Europe, Russia and North America, groundwater resources supply 80% of the water for drinking, agricultural and industrial purposes. Furthermore, with changing natural and climate dynamics, groundwater remains the most precious form of water due to its reliability against decline in quality and quantity.

From the early ages, groundwater has been extracted by human beings for their multipurpose uses. People in the past used simple practices to bring groundwater from the subsurface to ground level with little or no understanding of the environment. With the

Sustainability: Fundamentals and Applications, First Edition.
Edited by Rao Y. Surampalli, Tian C. Zhang, Manish Kumar Goyal, Satinder K. Brar, and R. D. Tyagi

advancement of technology and knowledge, such as remote sensing techniques, groundwater modeling and hydrogeology, it has become straightforward to delineate, locate and extract the groundwater resources. While this has helped people to ensure better access to freshwater, it is also true that overexploitation of subsurface water resources in many parts of the world has caused water levels to drop to dangerous depths. Poor water laws and their inefficient enforcement have led to the unrestrained withdrawal of underground water. Growing populations, urbanization, agglomeration and the ever-increasing demands of water-based industries are leading to a reckless race of groundwater excavation. Lack of waste management and skyrocketing levels of pollution have created huge problems for freshwater supplies. Moreover, it is expected that future climate conditions will influence the hydrological cycle to a great extent by the end of this century. Increasing temperatures and changing precipitation patterns will alter the occurrence and distribution of freshwater resources on earth. In such circumstances, it is necessary to understand that groundwater reserves should be used wisely keeping a sustainable framework in mind. However, this is not an easy task and requires large-scale research, planning and proper implementation of existing plans.

Sustainability is the basic need of today's world, and we must preserve this idea for the safe future of planet earth. A report by the Water Resources Institute (WRI) suggests that parts of the world such as the Mediterranean, the Middle East, western North America, China and western Asia will be facing the highest water stress by the end of 2025. With the expansion of the existing middle class in these countries, more water will be required to meet the needs of changing lifestyles. Water consumption for energy and manufacturing processes is likely to rise by 20% between 2010 and 2035, but an astounding 85% increase in drinking water consumption will be witnessed. More sophisticated plans and their implementation are required to cope with upcoming scenarios. Groundwater resources are set to play a significant role in assuring safe and feasible freshwater to a vast proportion of the population all around the globe. To ensure a better future, policymakers will have to take necessary steps as soon as possible. However, groundwater resource sustainability is itself subject to many interpretations. Groundwater sustainability is the idea of using and development of groundwater in such a way that it meets the present demands as well as those of the future. Hence, it is vital that we know the issues that contribute to groundwater scarcities, the policies that can be employed to ensure a sustainable groundwater supply, and the tools required for efficiently executing the plans and strategies. It is the right time to act and to cultivate in-depth knowledge about factors affecting groundwater quantity and quality.

The objectives of this chapter are to provide a detailed account of the qualitative and quantitative challenges to sustainable groundwater resource management and to discuss possible solutions, keeping in mind the changing natural and anthropogenic scenarios. First, we discuss the occurrence, distribution and use of groundwater resources. Then, qualitative and quantitative challenges arising for various reasons are explored. The solutions to sustainable groundwater resource management are discussed in two sections: primary and secondary. Solutions through technological advancement, water transfer, groundwater recharge, conjunctive use of groundwater and surface water, and improving groundwater efficiency are discussed in in section on primary solutions. The last section examines the utilization pf common pool resources (CPRs) regulating the use of groundwater and additional measures associated with it. Figure 25.1 shows brief details of the

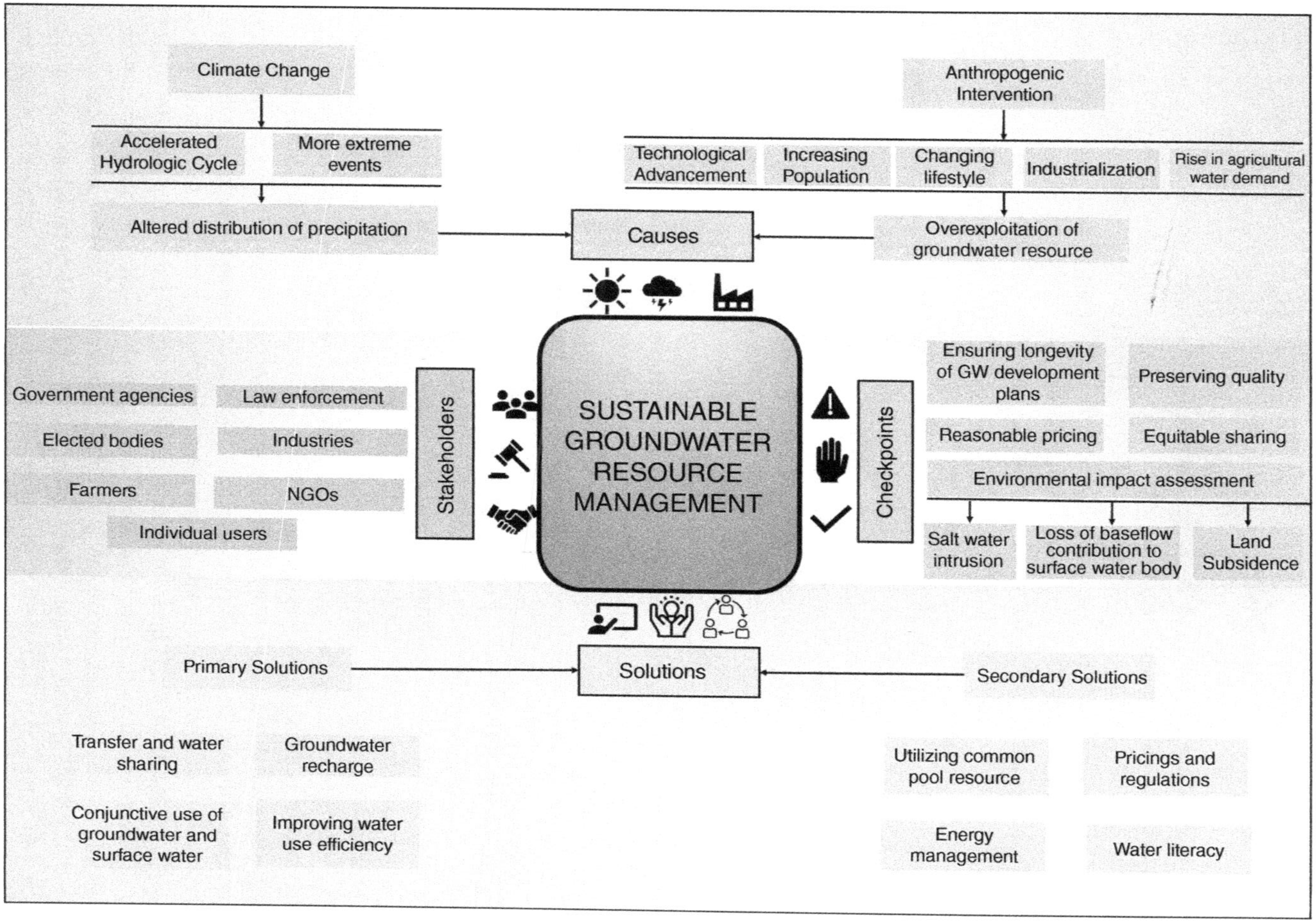

Figure 25.1 Flowchart showing important aspects of sustainable groundwater resource management.

problems, solutions, stakeholders and checkpoints involved in sustainable groundwater resource management and provides a brief outline of the contents of the chapter.

25.2 Occurrence and Distribution

Currently, available groundwater resources have originated from various sources, which are generally classified into meteoric, connate and fossils. In a normal hydrological cycle, when the groundwater is derived from rainfall and infiltration, it is defined as meteoric water. The term meteoric is used to indicate the recent contact of such water with the atmosphere. During the procedure of formation of sedimentary rocks, deposition is followed by compaction, which spills out most of the moisture available between the sediments. However, occasionally, incomplete compaction may result in retention of a small quantity of water by these rocks, which is called connate water.

During the procedure of formation of sedimentary rocks, deposition is followed by compaction, which spills out most of the moisture available between the sediments. However, occasionally, incomplete compaction may result in retention of a small quantity of water by these rocks, which is called connate water. Generally, this water is saline, and it is believed that it has been generated completely or fully from seawater. Lastly, fossil water, if fresh, may have been generated from the effect of any abrupt climate event. It can also be trapped in fossil aquifers for a considerable period of time. While most kinds of aquifers are generally recharged significantly by infiltration, fossil aquifers get a minor volume of recharge. The extraction of water from fossil aquifers causes the surrounding water table to drop at a very rapid rate, forming an area of depression. It is interesting to note that the process of pumping water from these aquifers, also known as "water mining," is sometimes blamed for the rise in sea levels (Konikow 2011). In this chapter, we will primarily deal with meteoric water.

When precipitation falls on the earth's surface, it may travel back to the atmosphere once again by evaporation and transpiration, or it may flow along the surface adding huge water volume to the runoff before finally meeting the ocean. A fraction of water volume also infiltrates down into the earth's surface through an enormous number of pore spaces. This water is held in the voids of unsaturated soil particles by the action of capillary forces. After these voids become saturated, the water travels down to lower levels because of gravity. As long as there is an adequate amount of water to continue to the state of saturation, the water will travel down before it is stopped by some impervious obstacles. If possible, this water may also flow laterally through the void spaces above these obstacles. If substantial differences in surface elevation are encountered, a spring may be generated, with water flowing out of the saturated layer. In general, when we dig a vertical hole into a saturated layer, it is expected that water will flow down into the hole or well. The rate at which water flows into the well depends upon the sufficiency of interconnecting voids. Lack of resistance to the flowing water through these interconnecting voids is referred as permeability (Todd and Mays 2005). Usually, fine-grained substances like clay or silt exhibit lower permeability, while sand and gravel have medium and high permeability. However, the case is a little different with rock fractures where permeability is a function of extent and pattern of fracture characteristics (Lamur et al. 2017). The amount of water in subsurface media is equal to the total volume of voids. The ratio of the total volume of voids to the total volume is called porosity. It is obvious that if the voids are interconnected, such volumes, which cumulatively are also known as aquifers, are likely to have high permeability. Sometimes, a significant volume of water is also held between two impervious strata. Such an aquifer is called a confined aquifer, in which water is under pressure and climbs up to a depth greater than the confining layer. These wells are popularly known as artesian wells and flowing wells if the water level rises up to the open surface.

The subsurface zones where groundwater is expected can be classified as zones of saturation and zones of aeration (Skopp 2002). The zone of aeration comprises spaces occupied partially by water and air, while, in the saturated zone, no pore spaces are left empty, and all voids are filled with water under the influence of hydrostatic pressure. This zone starts from the uppermost layer of saturation and extends deep to the point where impermeable strata are found. When this layer is not present, the water table is itself the uppermost layer of the saturated zone. This is also called the surface of atmospheric pressure and is present a little above the water table formed due to capillary action (Freeze and Cherry 1979). Here, water volume generally present in the zone of saturation is termed as groundwater.

The availability of groundwater is governed by many factors, such as climate, topography and aquifer material (Sophocleous 2002). For example, in arid regions, groundwater is available at considerable depths; however, it can be found at shallow depths in areas where the climate is known to be humid (Knutsson 1988; Lloyd 2009). In general, water seeps into streams, swamps and lakes as it travels down to lower elevations; the water table is expected to be higher at highlands and lower in the valleys.

It is evident that the occurrence of groundwater is governed by natural and human-induced factors as well, whose effects can be witnessed in changing groundwater quality and recharge environment of aquifers worldwide (Brouyère et al. 2004; Hornbeck and Keskin 2014; Kløve et al. 2014). Surface water and groundwater are often understood as separate components of total water resources, and their occurrence depends on a whole set of different factors. It should be kept in mind that they are closely associated with each other in most landscapes. Streams interact with the groundwater resources in three major ways: streams at their bed level are fed with groundwater; streams also may feed the groundwater reservoirs through the same bed level; or they can do both at different locations in the stream. For a surface water body to gain water from a groundwater source, it is necessary that surroundings of the water body have a greater water table and the opposite for the other case. However, this discharge or recharge also depends on the climate conditions and seasonal variation in the water table. Groundwater is important for surface water bodies it helps to sustain them through periods of deficient precipitation. For the sustainable development of land and water resources, more emphasis should be given to developing plans for joint management of both the resources, as withdrawal of water from one can deplete the other, or polluting the one can degrade the other. We will not go into the details of surface water occurrence here. However, it is important to mention that interaction between surface water and groundwater should be considered while talking about sustainability of groundwater resources.

The conventional hydrological cycle generally describes the main transfers of land and ocean. But for managing groundwater water resources efficiently, the cycle needs to be looked into in a more detailed way. It is easy to picture the movement of water on the land and atmosphere, but the subsurface water flow is a bit more complex to visualize.

The movement of groundwater is described in Figure 25.2. The groundwater travels to a discharge area from a recharge area along variable flow lengths. The general flow lines begin near the water table and travel through the subsurface flow systems and end at any outlet such as a stream or pumped well. The flow paths in the vicinity of streams may be of 3 to 30 m in length and may have a travel time of a few days to years, respectively. The most prominent and deepest line may have a length up to thousands of meters to miles, and time of flow varying from decades to millennia. It can be understood that a shallow groundwater table is more vulnerable to pollution from anthropogenic activities. Hence, near-surface groundwater flow paths are more important to study. The seepage line may also be affected by some small-scale geological structures in the bed of surface water bodies. Factors such as size, shape and orientation of the sediments in the river or lake bed also alter seepage patterns. If the river bed is made up of homogenous sediment type material, like sand, the inflow will be highest along the shorelines and will vary away from it in a non-linear fashion. Differences in flow can also be observed when there are several geological units that have diverse permeability values. Transpiration and combination with seasons can have adverse impacts on the recharge near the riverbank. In some places, it is also possible that groundwater flow regimes have different courses at day and night. These changes in the flow path also take place on larger time scales depending upon the precipitation and drawdown patterns during wet and dry periods of time.

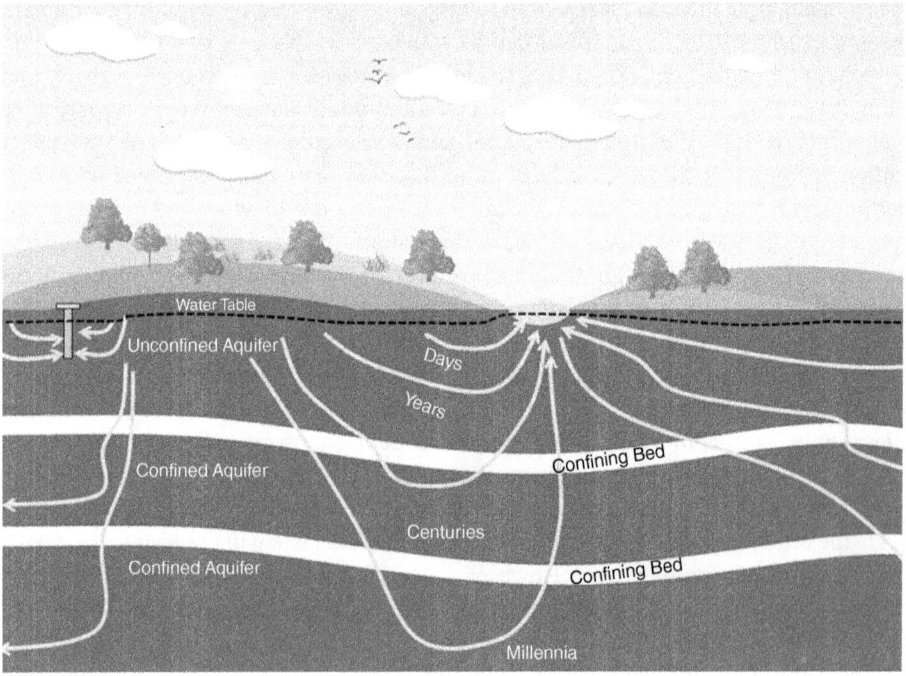

Figure 25.2 Groundwater flow system.

Apart from how groundwater occurs and flows, it is important to know the global groundwater quantity as it directly affects the equations of energy, food security and ecosystem services (Giordano and Villholth 2007; Rasul 2014; Sharma 2009). Sustainability of groundwater resources cannot be talked about without understanding its total available quantity. Along with groundwater fluxes and flow regimes, it is necessary to have an idea about groundwater storage. Proper estimation of groundwater resources can be used to understand the groundwater vulnerability toward major challenges like climate change and anthropogenic contamination. The total groundwater distribution beneath the earth's surface is hard to estimate given the dynamic nature of hydrological and climate processes. A recent estimate says that global groundwater volume in the uppermost part of the earth (<2 km) is ~22.6 million m^3 (Gleeson et al. 2016). This volume is equal to a 180-m-deep layer of water over all the land surface of earth.

There are several maps that provide global groundwater distribution on different scales. For example, the World-wide Hydrogeological Mapping and Assessment Programme (WHYMAP) initiative presents the hydrogeological characteristics of the world (Richts et al. 2011). Briefly WHYMAP divides the world's subsurface water resources into three classes: "a," "b," and "c." There are shallow unconfined to deep confined aquifers in class "a," having the common characteristics of being highly productive. Class "b" comprises highly complex hydrogeological structures, which include local volcanic or karst aquifers, deep or shallow with significant storage. In the last class, there are areas with only shallow and local aquifers like reserves in fissured rocks and alluvial settings. These different classes of aquifers are distributed in different continents in variable proportions. The class "a" aquifers are mostly available in Europe (53%) and Africa (45%). Class "b" aquifers predominantly govern the groundwater distribution in Australia (27%). Class "c" aquifers form 58% of the groundwater availability of North America (58%). These aquifers not only have different groundwater development potential, but are also different in their geological structure and hydraulic behavior.

25.3 Groundwater Usage in the World

As a freshwater source, groundwater offers a range of advantages over surface water. Since groundwater is widely available, abstraction losses during conveyance are minimized. This allows the users to extract a greater volume of water with higher productivity. Hence, a large amount of groundwater is available with limited seasonal variation and has the advantage of being replenished by precipitation. This allows groundwater resources to be more resilient to climatic disturbance and fewer regulatory facilities are required to maintain demand (Jasechko and Taylor 2015). Groundwater resources naturally are of better quality than surface water resources and require little treatment before supplying the needy. In addition, the vulnerability to human-induced pollution is also much less, especially in the case of deep groundwater aquifers. Groundwater resources are not only qualitatively and quantitatively better, but also easy to extract and use. Compared to hydraulic works, groundwater extraction works are easy to perform, and the developed infrastructure can also be easily adjusted with changing water demands in future. These advantages have caused people to use groundwater sources widely for their freshwater requirements. But later, individuals have started to consider the groundwater resource as their private property, which is not appropriate from a sustainability point of view. Because of the careless and unrestricted exploitation of freely available groundwater resources, efforts are needed to avoid its uncontrolled exploitation (Burke and Moench 2000).

The global patterns of groundwater use and the purposes of use can be understood by the quantity of its abstraction in different countries. A comparison of these data with the total available groundwater resources can give an idea about the groundwater stress generated by current utilization trends. The 10 countries in which abstraction occurred at the largest rate are India, China, the USA, Pakistan, Iran, Bangladesh, Mexico, Saudi Arabia, Indonesia and Turkey (Margat and Van der Gun 2013; Siebert et al. 2010). These 10 countries account for 72% of total groundwater abstractions. The most abstraction takes place in concentrated regions in Asia and the least in Africa.

There are many reasons for unequal groundwater distribution around the world. The rate of withdrawal depends on the size of country, economic situation, population density, climate, type of water requirement, such as industry or irrigation, and availability of other freshwater resources. Usually, countries that consume maximum per capita groundwater are located in arid zones (Brinkmann 2016). In these regions, the surface water is scarce and dependence on groundwater for irrigation purposes is greater. Some countries, such as Saudi Arabia, Malta and Denmark, use groundwater as their only source of fresh water, and groundwater in Belgium and Tunisia is 83% and 95% of their total water resources, whereas in the Netherlands, Germany and Morocco it accounts for at least 75%. (Zektser and Lorne 2004). Similarly, groundwater in most European countries comprises 80% of their drinking water and 70% of their total piped water supply (Zhu and Balke 2008). In semiarid or arid climate zones, groundwater is extensively used for irrigation. It is believed that one-third of the total land mass of the world is irrigated by groundwater sources. Of total irrigated area in the USA, Iran, Algeria and Libya, 45%, 58% and 67% is based on fresh groundwater (Eslamian and Eslamian 2017). For domestic use and drinking water supply purposes, quality groundwater is necessary. According to the EEC (European Economic Commission), most European countries' municipal drinking and domestic water supply depend upon groundwater sources. Cities such Copenhagen, Munich and Rome are fully or almost fully dependent upon groundwater, while it contributes more than 50% of the total water demand of other cities such as Amsterdam and Vienna (Zektser and Lorne 2004).

It is fair to say that domestic and drinking purposes should be given priority over industrial and other uses when water is scarce. However, due to poor regulations and lack of proper management, it has been observed that different users extract groundwater overlooking the needs of other important sectors. This overlap of different sectors is set to create problems in the future. It is also difficult to fix the limits of admissible groundwater volume withdrawal, which has two aspects: one is associated with developing the maximum possible amount of groundwater resources in terms of quality; and the other is related to overexploitation of the subsurface volume, which may have negative impact on other environment constituents. As discussed in the previous section, when understanding groundwater use, its interconnections with other elements of environment should also be considered. Full or partial renewability of groundwater resources may have resulted from the seasonal recharges because of its interconnections with the atmospheric and surface water. While considering groundwater uses and its future, the fact that close link between environment components and groundwater resources also depends upon a whole set of dynamic factors should also be considered.

Over the course of time, the importance of groundwater has changed a lot. Initially, when municipal water supplies of cities were designed, surface water bodies were considered prominently. With rising industrial pollution in the latter half of the nineteenth century, many reconstructions took place in water supply systems, which focused in on extracting freshwater from existing groundwater sources. Paris is a good example where natural springs situated at the outskirts of the city were first used as the water supply source till the early 1900s. Later, increasing consumption and pollution of these water bodies forced policymakers to reconsider their water supply plans and incorporate groundwater resources as a safer freshwater source (Zektser and Lorne 2004). As discussed groundwater has become the most prominent source of municipal water supply in most European countries as well as the USA. Similarly, groundwater remains the most important source of domestic drinking as well as industrial and agricultural water in most Asian and African countries. Safe drinking water resources are already in a stressed condition in these regions and the situation is likely to deteriorate further. In such cases, the industrial and agricultural water supply, which can manage with slightly poorer water quality, should free up its dependence on fresh groundwater resources. However, current scenarios in different countries are not encouraging enough and rampant groundwater extraction is taking place ignoring sustainable development requirements. The volume of total groundwater abstraction in different sectors is shown in Table 25.1. South Asian countries such as Bangladesh, India, Nepal and Pakistan, along with China, account for almost half of the world's total groundwater use. At the same time, they are expected to experience some really serious water deficits in upcoming years. South Asia is developing and, according to the World Water Development Report, there exists a really complex but important relationship between water resources and economic development (Connor 2015).

Table 25.1 Worldwide sectorial groundwater use.

Continent	Irrigation (km^3 $year^{-1}$)	Domestic (km^3 $year^{-1}$)	Industrial (km^3 $year^{-1}$)	Total	
				(km^3 $year^{-1}$)	%
North America	102	33	8	143	14.6
Central America and the Caribbean	4	8	3	14	1.4
South America	13	8	5	26	2.6
Europe including Russian Federation	26	33	13	72	7.3
Africa	26	13	2	41	4.1
Asia	514	111	55	680	69.3
Oceania	3	3	0	7	0.7
World	688	209	85	982	100

Source: Margat and Van der Gun (2013).

25.4 Two Main Challenges

From the supply point of view, we know that groundwater is thought of as an invisible and silent underground resource whose location, quantity and function in human and natural systems are poorly known. On the demand side, the fast growth in groundwater extraction has come to our knowledge, with a dramatic rise in consumption by millions of farmers worldwide. Most of these farmers have invested in their own resources in drilling and pumping machinery. There is no doubt that groundwater has changed millions of lives in the latter half of the twentieth century. However, now groundwater tables are dropping at very fast rates, often at a rate of more than 1 m per annum in some parts of the world (Biswas et al. 2017). Previously perennial rivers and streams, which were contributed to by the base flow of the rivers, are now at risk of drying up (Datry et al. 2014). Other impacts such as saltwater intrusion and drying of wetlands have also been observed (Van der Kamp and Hayashi 1998; Willis and Finney 1988). With deeper wells being excavated the chances of pollutants entering into the recharge system are increased. Although the importance of groundwater use is undisputed, the method of how we use it is the point of discussion. We have benefited a lot from using groundwater resources, but is it too late to realize that we have already or are very speedily approaching an irreversible state? The main aim of this section is to present the major problems associated with groundwater resources that restrict our approach to sustainable groundwater development.

25.4.1 Quantity of Groundwater

Conventionally, depletion in the water table is looked at from the perspective of safe yield, which is the measure of the rate of steady-state extraction that can be maintained for an infinite period of time without actually degrading the supply (Custodio 2002). This measure ignores certain things such as temporal and spatial variability in recharge rates. The study of groundwater within the frame of sustainable development not only requires the analysis of the concept of recharge versus withdrawal alone; it also requires the incorporation of all unwanted impacts linked to groundwater utilization and intergenerational parity, i.e. current requirements are expected to be met without compromising the requirements of forthcoming generations. There are a number of studies that have highlighted the shortcomings of using these terms, which may have resulted in potentially flawed policies. For example, from a groundwater development viewpoint, withdrawal can make storage usable for wet period recharge, weakening the flood potential and making water available in the dry season (Custodio 2002). As suggested by the ex-director of the International Water Management Institute, "The penalty of mismanagement of this valuable resource is now coming due, and it is no exaggeration to say that the results could be catastrophic for these countries [China, Pakistan, India, Mexico and most countries in North Africa/Middle East], and given their importance, for the world as a whole" (Seckler et al. 1999). The day is not far off when the rapid depletion in world aquifer systems will lead to food security issues all over the world. The two hardest hit countries will be China and India, the two largest consumers of groundwater and food. A study by the World Bank speculates that possibly one-fourth of India's food production is facing threat due to unplanned and careless groundwater management systems (Pahuja et al. 2010). Almost 50% of the wheat yield in the North China Plain areas is under threat due to the forthcoming water scarcity in the country (Brown and Halweil 1998). The study indicates that the depletion of the Hai River Basin aquifer system will directly impact China's ability to feed about one-tenth of its population.

In India, an alarming rate of decline in the water table has been registered in the hard rock aquifers of the southern peninsula region as well as in the Indo-Gangetic plains of the country since the 1990s (Singh and Singh 2002). Similar cases have been observed for African and Middle Eastern countries as well (Famiglietti 2014). The situation is not confined to the developing world: drawdown has shown alarming trends in the USA, Spain and Australia also. A very significant example is the large-scale drawdown of the Ogallala aquifer in the United States, whose groundwater reserve is observed to be vanishing at a very fast rate and has even disappeared in some parts of the region (Guru and Horne 1970). This region is known as the breadbasket of America and is responsible for supplying at least 20% of the total yearly US agricultural harvest. It is highly likely that, if the aquifer vanishes from the region, the potential to harvest food worth more than $20 billion and fiber will disappear from planet earth.

It is impossible to recover the underground reservoir, and it will take 6000 years to naturally replenish the reservoir (Little 2009). It is important to understand and keep an account of global groundwater overdraft and food security, which requires large-scale planning and modeling. For aquifers, drawdown can permanently alter the compaction ability and, hence, impair the ability to hold water and may result in irrecoverable salt water intrusion (Barlow and Reichard 2010; Minderhoud et al. 2017). Mass extraction of groundwater resources has also led to soil compaction in urban areas, which has resulted in impacts like land subsidence, which in the worst cases may damage the urban infrastructure (Chai et al. 2004). As we know from the discussions in the previous sections, groundwater and surface water have interconnections at many levels. Beyond the aquifers, drawdown may alter the relationship between ground- and surface water bodies, decreasing groundwater's capability to offer base flow to rivers, lakes and wetlands. Furthermore, it may also diminish soil moisture content levels, adversely affecting the configuration and productivity of vegetation cover, and also agriculture (Zhang and Schilling 2006). Increase in the drawdown also increases the capital required for drilling, maintaining and using wells. From an equity point of view, people have agreed to the fact that the poor, who cannot afford such updates in infrastructure, are affected the most. As the decline in the water table occurs, those who use groundwater from shallow wells suffer the most, and their supplies are cut off on the very first note as they lack access to capital to modify the wells. Moreover, during droughts, this scenario is worsened when the buffer groundwater quantity is lost.

25.4.2 Quality of Groundwater

Past discussions in the chapter have been mainly aimed at the groundwater quantity, but quality of groundwater quality is also a major issue that needs to be looked at. The quality problems can be classified into two sections: degradation of groundwater due to naturally occurring contaminants such as arsenic, iron and fluoride, and that occurring through anthropogenic influences such as mining, agricultural practices and sewage disposal. The two most important naturally occurring groundwater quantity contaminants are fluoride and arsenic. A total of 200 million inhabitants all across the globe are likely to be affected by coming into contact with arsenic-contaminated of drinking water (George et al. 2014). In Pakistan and Bangladesh alone, around 60 million and 20 million people, respectively, are susceptible to arsenic-induced diseases, also known as the biggest arsenic calamity of the world (Zhang and Schilling 2006; Guglielmi 2017). This contamination is a direct threat to the drinking water supplies of people in these developing countries. Other countries that are massively hit by the arsenic contamination of groundwater include Myanmar, Cambodia, India, Nepal, Vietnam and China. As groundwater is heavily used for irrigation purposes in this part of Asia, it is highly likely that the arsenic may enter the food chain and then cause impacts on a large scale. The mitigation of the arsenic problem is not necessarily complex, but it is expensive, which multiplies people's suffering. One possible solution is to improve the nutrition level of people as poorly nourished people are more vulnerable to arsenic poisoning. However, changing the socio-economic standards of people in the countries mentioned above is a time-consuming task and is set to affect many lives in future. Fluoride, unlike arsenic which mostly occurs in alluvial aquifers, naturally exists in the hard rock aquifers and is highly sensitive to overextraction of groundwater resources. More than normal consumption of fluoride causes devastating impacts on the human body in the form of fluorosis. It is estimated that at least 25 countries worldwide are affected by fluorosis, including countries like Jordan, Libya, Sudan, Iraq, Afghanistan, India and China. India remains the most affected country; it is estimated that more than 60 million people in India are suffering from some form of endemic fluorosis (Ramesh et al. 2016).

While on the topic of human-induced degradation, salinization due to overexploitation of groundwater is regarded as the most significant threat to aquifer sustainability. Primary salinization is basically salt water intrusion caused by aquifer drawdown. Other human-induced salt intrusion in the aquifer is caused due to salt deposition in the uppermost layers of soil, which is known as secondary salinization. An estimated 20% of the world's irrigated area in one way or another is affected due secondary salinization (Shrivastava and Kumar 2015). Among the most affected countries, China, Pakistan, India, the USA and Argentina are the most important ones as agriculture is extensively practiced on large scale. As a result of the secondary salinization, the productivity of agricultural land is reduced further impacting the upper water table. The main reason behind salinization is the excessive supply of irrigation water, which leads to a rise in the water table. This water table, when it climbs upward through the action of capillary force, carries salt particles toward the surface. The rate of deposition of soil increases during the dry season when the salt gets accumulated with increased evaporation rates. The major portion of salinity can be resolved by changing the pattern of irrigation.

In developed as well as developing countries groundwater pollution due to industrial and human waste is emerging as a major problem. As stated by the European Commission, groundwater pollution now stands as the most serious issue in the increasing water crisis of Europe (Priscoli et al. 2004). Increasing pollution in the developing countries has resulted in catastrophic situations. The pollution causing damage to the groundwater resources can be classified into two major sections, chemical and biological. While chemical pollutants are a result of industrial or agricultural waste, biological pollutants are mostly contributed by waste products of animals and human beings. The chemical pollutants affect the subsurface resources more severely and are expected to remain for a longer period of time. This allows them to disturb the deeply existing groundwater resources as well. Since biological pollutants generally survive for a day or week, they primarily affect the shallow aquifers. Further classification of pollutants can be understood in terms of point and non-point sources. The point sources are present in limited locations with lower numbers of components, such as effluent outlets and factories. The non-point sources generally originate from different expanses of areas such as agricultural waste and urban runoff.

25.5 Toward Groundwater Sustainability

For a very long period of time, groundwater has been considered as a readily available water source that is useful for human beings for many purposes. Since the latter half of the twentieth century, groundwater has played a significant role in uplifting the economic and social wellbeing of the human population in many parts of the world. With more water needed to satisfy the increasing demands, the scale of groundwater extraction has kept on enlarging. As mentioned in the previous sections, due to uncontrolled withdrawal of groundwater and changing socio-economic scenarios of the world, the overuse of this resource has caused severe threats to several global aquifer systems. Later, when people started to realize this overexploitation is impacting the natural system in a negative way, several strategies were formulated to manage these resources. Although utilization of groundwater resources have brought many positive impacts to the lives of people, its increasing demand has also caused environmental problems such as land subsidence, seawater intrusion and deterioration in water quality. It is important to mention that, with an increase in the concentration and quantities of pollutants, the possibility of a contaminant reaching the groundwater system has also increased. Earlier there were not many methods to keep track of essential components while managing groundwater resources. However, with the advancement of technology, new methods have become available for investigating, monitoring and evaluating resources more precisely. Since the dynamics of demands and supply will change drastically in future, it is the right time to take necessary steps for considering the sustainability aspects of groundwater resource management.

The idea of sustainable development was presented in the World Conservation Strategy, which considered three most important support systems for life (i.e. air, water and soil) and set out worldwide accepted principles of environment sustainability (International Union for Conservation of Nature and Natural Resources 1980). The thought was further promoted to a more significant level of international cooperation in the Brundtland Report, famously known as *Our Common Future* (Brundtland 1987). In its own words, sustainable development is "development which meets the needs of the present without compromising the ability of future generations to meet their own needs." In the context of water resource management, a plan for a water resource project is sustainable if water of ample quantity and acceptable quality at feasible rates is available to meet the present and future demands of people without causing degradation of the environment (Plate 1993). Water resources as a whole consist of systems like lakes, rivers, aquifers and wetlands. The plans that are intended to manage water resource systems must take into account their interactions with the biological, physical and chemical processes during a natural cycle. Also, developing plans and strategies for sustainable groundwater development is dynamic and continuous in nature. The socio-economic parameters will change with time. The plans that are supposed to take on the challenges opposing sustainable groundwater resource development must also be adjusted to properly achieve the target. In such cases, sometimes short-term gains may have to be ignored in order to achieve the target of a long-lasting sustainability. Overall, it is easy to see that this is not an easy task and has to take into account many parameters associated with natural and manmade interventions to groundwater resources. It requires not only more advanced research, computation and planning but also tighter cooperation between bodies at the local and global scale.

The primary aim of sustainable development of groundwater resources is to plan a strategy to manage the resources in a way that the resource is utilized and maintained over the maximum possible period of time. Most basic groundwater resource development activities start with a few pumping wells and meet the demands of the existing population of a region. It is later, when the development of the region begins with time, that the problems such as groundwater overexploitation, water quality degradation and disproportionate sharing start. There are examples of many cities which are nowadays facing water crises due to the stress on water supply plants, which have been affected by an increasingly unbalanced demand–supply equation. Hence, the focus of groundwater management plans has to be adjusted from time to time so the available resources are utilized efficiently, sustainably and equitably, profiting the economic, physical and social wellbeing of a large community.

To develop a sustainable groundwater system, a good knowledge of various natural and human-designated processes and their interactions with the atmosphere and land components is required. This knowledge, combined with technology and dedicated efforts to monitor the development of resources, can help us to efficiently formulate sustainable groundwater systems. Hence, the basic points that need to be considered while developing a sustainable groundwater framework are: considering its long-term conservation, saving it from qualitative degradation and considering human as well as environmental impacts on the interconnection of groundwater and surface water bodies. Some details about these check points are described below.

25.5.1 Conserving Groundwater for the Future

It is essential that groundwater development plans are prepared with more longevity and on a large scale. This should mean that the rate of recharge in a given period is equal to or more than the rate of extraction. When the extraction rate is greater than the recharge rate, a continuous decline in the water depth is encountered, which then needs to be carefully monitored. The lowering water depth further increases the capital required for pumping out the water, and hence, reduces the economic acceptability of such systems. When this system is utilized for agricultural purposes, it affects the production cost for farmers. In such cases, the assessment of naturally available recharge becomes primarily important in sustainable groundwater resource development. The conventional approaches at the beginning of development works can comprise studying the basic hydrological soil and water budget equations to estimate the recharge quantity. This can only be used for the initial time period, while the dynamic response of groundwater systems to the external factors should be taken into account in order to develop a long-term plan. Lately, with the availability of larger hydrogeological datasets and data obtained by closely monitoring the system, long-term plans can be designed and implemented. Since the basic aim of sustainable groundwater resources revolves around preventing water levels from reaching dangerous depths, the main limitation in formulating such plans remains in the economic aspect as well as the quantity of available groundwater resources. This limitation further multiplies when the potential deterioration of the environment caused by groundwater development is taken into account, which is discussed in the next section.

25.5.2 Preserving Groundwater Quality

Most of groundwater resource strategies are basically concerned about the quantitative aspect of management and focus on its availability to meet the same. But one of the other important aspects of developing a sustainable groundwater system is associated with its quality. This aspect becomes critically important when it is associated with the quantity. It is known that both natural and human activities can impair the quality of groundwater and the extent of deterioration is linked with the climatic and hydrogeological setting of the resource (Mileham et al. 2009). While the surface water can be easily judged for the extent of degradation in its quality, deterioration in the quality of groundwater is detected only if there is change in odor or taste, or color appears in the water that is already being consumed by people. Sometimes the degradation can also be detected by observing the immediate impacts on the users. It is often a very complex task to identify the real reasons behind such degradation because of the existence of a large number of contaminants carried by a wide range of transport processes. The quantification of the impact of deterioration in groundwater quality is also difficult. At times, it is more widely spread and hard to control. The basic theories suggest that the contamination should be restricted in such a way that the potential "causes" that could reach and react with the groundwater systems are stopped at the earliest possible point. Moreover, improved land use planning techniques can also help in preventing hazardous substances from reaching the subsurface water by keeping them away from possible water recharge sites. The structures that carry waste such as sewage industrial waste must be planned properly and possibly kept away from the potential regions of groundwater resources. It is required that no seepage from such pipes take place; if the contamination has already occurred, immediate steps must be taken to locate and stop the leak to prevent further degradation and to aid restoration of the aquifer's water quality.

25.5.3 Environmental Aspects of Groundwater Development

The foremost necessity in the progress of sustainable development for groundwater water resource is identification of the term environmental sustainability. The fluctuations in groundwater systems can have impacts on environmental systems. Moreover, by making positive changes in the environment to different components of the hydrological cycle, positive advancements in groundwater resources can be achieved. This factor becomes essentially important where the groundwater system is closely linked with surface water bodies such as rivers, springs and lakes. In some hydrogeological settings, it has been recorded that continuous lowering of the subsurface water table can cause land subsidence due to the influence of the compaction factor. The excessive groundwater extraction is then followed by droughts, floods, and loss of human lives and property. One of the most general environmental impacts of uncontrolled groundwater extraction, especially in coastal regions, is intrusion of salt into the aquifer systems. From the environmental sustainability frame of reference, the development and management of groundwater systems must be treated as a component of the integrated water management with two key objectives: supply of water for valuable uses at lowest possible cost and prevention of bad impacts on the natural environment. Groundwater management, as an important constituent of the water resource planning, is interlinked with other sectors such as industrial development, urbanization and agricultural development. The development viewpoints of these different sectors have massive influences on the sustainability component of groundwater resource development. Several efforts are being made to formulate an inclusive modeling technique for water resources at basin and subbasin level for water management considering the connections between various subcomponents of water resources and their linkage with different development areas mentioned above(Alcamo et al., 2003; Sivapalan et al. 1996). However, these efforts have also faced setbacks due to a lack of validating information and less knowledge about the interactions between individual components. Some other crucial challenges while modeling such systems are to incorporate the interactions between the biota and behavior, quality and occurrence of water.

25.6 Looking for Possible Solutions: Primary Solutions

We know that achieving groundwater sustainability is not an impossible task, but it is not possible unless we agree that major transformations are required in the evaluation, management and characterization of hydrological systems. This is not easily feasible because of the large number of stakeholders involved, the need for collaboration across authorities, and our habitually lethargic economic and political systems. Moreover, we have already stepped into the other side of the possibility, and short-term groundwater planning strategies have started to cause extensive depletion of aquifers and their contamination in many regions across the globe. Here, planning long-term objectives and reforming the earlier set objectives with time could have resulted in better sustainable groundwater use. The measures for both groundwater quality and quantity issues have to be introduced both locally and inclusively. Communities that depend on aquifers can implement strategies to attain multigenerational aims that can be revised using adaptive planning and management. For short-term objectives, formulating locally applicable goals and ideals, modifying groundwater usage and the protection of the associated environment will be crucial. Researchers will have to play an important role by formulating groundwater models that should consider the implications of management strategies and actions in accomplishing the sustainable groundwater development objectives. The role is further extended to track the usefulness of these strategies and actions on both the spatial and temporal scale. Information from modeling and monitoring gives a scientific basis for adapting management policies. It is required that local communities, government and local water suppliers should work in coordination with the researchers in deciding the long-term as well as short-term goals in order to develop a sustainable groundwater system framework.

25.6.1 Role of Technology

The technology associated with different aspects of groundwater management like assessment, abstraction, exploration, recharge, salinity and pollution control has advanced in the past. Considering the variability in the occurrence and distribution of groundwater, different technologies are required to sustainably develop the resources considering the local conditions. We know that optimizing the capital required for executing existing and new technologies can help in finding solutions to regional and local problems related to groundwater as well as improve the water use efficiency. The technological solutions to an existing problem can also give rise to new difficulties and conflicts if not supported by synchronized institutional change. In developed nations where demands surpass sustenance requirements, management of water demand becomes crucial. The necessity for demand management and conservation depends on the availability of the resource, climate perturbations and population growth. In addition to innovation in technology, education and awareness are crucial in achieving sustainable goals. Technology itself can play a very crucial role in spreading awareness and education about sustainability. One of the main challenges in deciding on the best possible technology is that it requires many folds of experimentation and large-scale efforts for its implementation. However, a more approachable method of groundwater sustainability is the path through which we reduce the magnitude of groundwater abstraction. This can be achieved by proper use of different components of water resources. Methods for introducing the concept of reuse in sustainable water planning can be characterized into two parts: substitution and regeneration. The aim of substitution shifts the uses from groundwater to rainwater and other sources, while regeneration refers to extracting potable water from recycling and wastewater reuse. In addition to broadly considering treatment and reuse of wastewater, storm water can also be considered as a potential water source. Storm water not only can be treated and used for drinking purposes but also can play a vital role in replenishing the aquifer systems of urban areas where lack of infiltration results in water level depletion.

25.6.2 Transferring and Sharing Surface Water

One of the most conventional answers to water crisis questions linked to groundwater has been the idea of importing water from other locally available water sources. For immediately fulfilling the requirement of lagging groundwater management, imported water can easily substitute for the local groundwater use, which may relieve the problems of excessive drawdown. This water can also be utilized as artificial recharge by the existing aquifers. However, the water import approach is mostly physically and economically feasible for urban areas rather than rural areas, which comprise agricultural areas with higher water demands. This model has been successfully employed in many parts of the world to provide solutions to groundwater problems. For instance, the Sardar Sarovar Dam plan in India includes water transfer schemes to the downstream regions of Gujarat state from the Narmada River from the Madhya Pradesh. This has been partially justified for the reason that it will replenish the depleting aquifers of Gujarat state. The biggest inter-basin transfer program of the world, the South-to-North Water Diversion (SNWD) scheme, has the capability of transporting 25 billion m^3 of water per annum from China's Yangtze River in the south to the more arid parts of the country in the north. The northern part of China, which comprises the Huai, Hai and Yellow River Basins, is facing an excessive shortage of water. Both urban and rural populations have extensively withdrawn groundwater, which has resulted in a drastic situation and half of the groundwater reserves in the region have now reached depths below sea level. Hence, the benefits of a water transfer network of the area, like other parts of the world, comprise balancing demands over long distance. (Barnett et al. 2015).

25.6.3 Groundwater Recharge

A second often-discussed answer is to improve water supplies by capturing locally available water and increasing groundwater recharge instead of transferring water from another place. This can be performed using a lot of techniques that are simple as well as complex in nature. In all probability, the operation of inefficient irrigation plans allowing water supplies to seep into the subsurface water table are the best methods of recharge and play a prominent role in some of Pakistan, India and other countries as well. Likewise, inefficient yield of crops whichthat grow in wet seasons such as paddy can also help in increasing the recharge. Furthermore, simpler techniques include construction of embankments (bunds) around the farms to restrict the velocity of runoff and increase the rate of infiltration. However, with time, our focus has shifted to more efficient irrigation techniques. Hence, these methods should be applied with proper consultion and care. More importantly, the conventional methods of storing water in storage tanks (ponds), lakes and wells can prove to be more efficient in increasing the recharge to the groundwater table. These traditional methods have been very efficient and traditionally have been used in many countries across the globe. It is with the rising pace of urbanization that these structures have either been demolished or encroached upon in order to search for more space for human settlement. On an individual level, one can start with rainwater harvesting and allowing the water to reach the subsurface, which can either be used for one's own purposes or to increase the recharge potential. In urban areas open spaces like parks and playgrounds in institutional and residential regions can be productively utilized for recharging the groundwater. If the scale is large enough, one can also go for a ground water dam, which is basically an underground barrier across a stream that delays the naturally flowing groundwater system and stores water below the subsurface to meet the water requirements during period of high demands. The main aim of this technique is to clog the outflowing groundwater within the subbasin and increase the storage of the underlying aquifer. Hence, the main ideas behind increasing the recharge to an aquifer system are to increase the storage to the maximum possible level. If water is saline or polluted, then it also improves the water quality through the process of dilution. Efficient planning and execution of recharge schemes can help prevent saltwater intrusion and land subsidence because of overextraction of the groundwater. The interconnection between other water resources such as rivers is also benefited, and the foremost danger, which needs to be addressed on the very first note – decline in groundwater levels – is also controlled.

25.6.4 Conjunctive Use of Groundwater and Surface Water

For irrigation purposes, usually, subsurface water is not the sole source of irrigation water and the demands are met by the combining groundwater supplies with surface water supplies. From a farmer's point of view, conjunctive utilization of resources is more effective as the groundwater can act as buffer stock to ensure the water supply during the scarcity of surface water resources at key timings. While it allows the farmers to continuously supply water to different crops with varied base and critical periods, it also lowers the economic burden by limiting the pumping costs of water from a lowered groundwater table. A continuous withdrawal from a subsurface reservoir in surplus of replenishable recharge has caused nonstop dropping of water tables. Such a situation results in serious problems, such as a rise in the pluming head for more tube wells and dug wells, and drying of shallow wells. The cure for this problem lies in providing more irrigation through surface water sources with the help of inter-basin transfers, storage reservoirs, etc. The conjunctive use framework has benefited a lot of people across the globe and has also increased the basin productivity in many regions, particularly where there were significant differences in the groundwater quality. A good example is of the Indus Valley in Pakistan, where the quality of groundwater in the upstream areas is supposed to be better than downstream areas. However, people in the upstream area use surface water supplies first, causing the farmers downstream to depend on more unsuitable and saline groundwater. A change in this pattern could benefit a large amount of population of both sides, in which the downstream users are provided with the upstream surface water and upstream users depend on the high-quality groundwater of the region. Similarly, there are many other regions in the world where adjustments like this could improve the groundwater usability of people and there exists a large scope of improvement. However, this requires coordination on many levels, and proper formulation and implementation of plans. The main constraints in the conjunctive use program include inadequate supply of surface water for recharging the groundwater basin. At times, the underground storage capacity is also insufficient, and the percolation and infiltration rates are inadequate for recharging the basin. With more land being covered by various land use patterns, areas that could be utilized as recharge zones may not be available at reasonable cost. Insufficient infrastructures on both sides is also a primary constraint in developing the conjunctive water use framework. Apart from the operational and physical constraints, there are institutional and financial constraints as well, which limit our approach toward utilizing the conjunctive use framework for development of sustainable groundwater systems.

25.6.5 Improving Water Use Efficiencies

Although recharge and water supply from different sources can easily improve groundwater supplies, water-saving technologies can increase the productivity of such supplies. A very wide range of such technologies are present in the agricultural, domestic and industrial sectors. Most of the technologies are not explicitly associated with groundwater but with total water use instead. These days, countries from both the developing and developed world are effectively using these technologies in order to improve the productivity of groundwater (Kijne et al. 2003). Similarly, productivity can be improved by shifting groundwater use from high to low groundwater-intensive products. The main question in this approach is regarding the agreement of users about the change in cropping patterns without large-scale change in policies and pricing. Trends from all around the globe suggest that users primarily respond to groundwater depletion in their own way, and initially it may be difficult for farmers to accept the changes. However, some good examples also exist to encourage this approach. Research suggests that this shift in agriculture patterns has helped in developing sustainable groundwater systems in the Indo-Gangetic plain regions of India (Ambast et al. 2006). In regions of eastern India where people generally use diesel pumps, increasing fuel prices have encouraged a large number of farmers to shift toward high-risk but high-return crops like vegetables from the traditional paddy farming. Another good story from the North China region suggests that farmers in this region are adapting to the changes in cropping patterns in response to the worsening water resource conditions accompanied by improved application of water-saving technologies (Blanke et al. 2007).

25.7 Intuitional and Organizational Reforms: Secondary Solutions

Institutional arrangements include formal regulations and laws, organizations, along with some informal norms as well. In the background of sustainable groundwater systems, we can recognize the state or national laws deciding the irrigation laws, groundwater development regulations and rulings as basic institutional norms. These arrangements, whether devised at state, national, or community or provincial levels, and whether informal or formal, define and affect the instruments to develop and manage groundwater resources. The usual instruments comprise abstraction permits, groundwater use rights, subsidies, groundwater tariffs and groundwater markets, to a certain extent. These are also known as direct instruments, provided that they are formulated to directly manage the sustainable groundwater management decisions by different stakeholders. However, there is also a set of indirect instruments that originate from other sectors and possess significant impact on groundwater use, such as agricultural produce pricing, energy pricing and trade policies. Another important component in the institutional framework for groundwater management is the organizational setting. In most countries, groundwater is generally managed by government agencies at the central and state level. There is a common thinking that the most effective answers to water scarcity questions shift from development to management as the problem intensifies. Similar arguments are also considered from the sustainable groundwater development point of view. Moreover, it is also true that this development framework of groundwater water resources will achieve its objective if proper institutional framework is present as a support. For example, technologies to improve water productivity may cause a rise in overall consumption of water use if any institutional framework is absent to restrict the use by individuals. The institutional approaches to sustainable groundwater management comprise three basic approaches: collective or community-based action or approaches, indirect approaches and instrumental approaches.

25.7.1 Utilizing Common Pool Resources

These days, the term CPRs is often talked about while discussing community-based resource management. A CPR is supposed to be a resource that exists to benefit a large number of people if used with cooperation. This benefit diminishes if people begin to pursue their own self-interest. The importance of a CPR can be abridged by its overexploitation as the supply of the resource is limited, which in turn can result in further amplification of scarcity problems. CPRs are the resources for which the incorporation of prospective consumers at one level is tough, and the use by one section can affect the quality and availability for other. Groundwater is understood, in some circumstances, as a CPR. Knowledge of the features and how different agents interact during the utilization are vital for the accomplishment and sustainability of groundwater CPRs. However, the primary aims of such initiatives are to make important resources available for a large proportion of the population. One method to improve the accessibility of groundwater resources by maximizing the possible number of users is through the implementation of technology that increases the efficiency of how people utilize the resource. Moreover, the dynamics of decision-making systems for a CPR and cooperation in its setting up and management are important. It is essential to realize that the conditions that support and work against the sustainability of such community efforts depend on both social and economic factors.

25.7.2 Regulating the Use of Groundwater: Rules and Pricing

The major difficulty in the success of community-based methods in developing sustainable groundwater management systems is that the number of consumers of a specific aquifer can be large and their geography varied, thus augmenting the expenses of partnership. Moreover, the unseen and varied environment of groundwater means that it is generally tough to keep track of use and availability, weakening the steps taken by local consumers to cooperatively set and implement norms. A frequently suggested solution by specialists is to offer centralized resources to help in data gathering and in information monitoring and development, and to enforce and set some rules or "instruments" for regulated use. These norms can function separately or form a primary agenda, within which cooperative decision making by native users can be performed.

Regulations are normally organized and postulate how consumers in general can and cannot utilize the resources. Rights are generally conferred on individuals or collectively to communities. These can be permanent or temporary and can vary from local use to even completely tradable purposes. Usufruct privileges, if recognized taking into account the aquifer environment, can prove to be useful to restrict uses to safe levels or to regulate the frequency of use so as to limit overlap between consumers. Tradable rights also permit the option of efficient profits as market practices bid water to low from high-value uses. The formation of groundwater use rules and regulations has been primarily a part of developed countries like the USA and Australia; it is encouraging that they are being formulated in the developing world too. The second major instrumental method is the variation of pricing, although it is true that deciding a commonly acceptable price is not so easy, which may create further problems in the user community.

25.7.3 Additional Measures

While discussing groundwater institutions, it is a general notion to think about sectors that are directly related to the decision-making authorities of the groundwater sector. However, the institutions, management and plans that influence groundwater range far ahead of the core sector. The two areas that are highly interlinked with the groundwater sectors are energy and agriculture. India is one of the most important examples of the possible impact of energy management and policy on groundwater resources. Within the period of 1950–1970, the use of electric and fuel-run tube wells was endorsed in India by both the external leaders and government as a way to both economic and agricultural growth. As the number of tube wells increased and the withdrawal mounted, fee collection and metering became the foremost problems and a straight tariff system was implemented partially in response. The flat tariff limited the transaction charges in fee collection, but also abridged the existing marginal incentives for agriculture-based users to utilize electricity and water efficiency. Finally, as a result, power supply for agricultural groundwater utilization has become almost freely available in India's key groundwater-using areas, causing overexploitation of groundwater and capital as well. A study by the World Bank indicates that subsidies related to the power sector in India reached around 270 billion INR (in FY 2000) whereas, India's expenditure on health was 150 billion INR and 100 billion INR on water supply and sanitation (World Bank 2002). Lately, the Indian government has started to think about restructuring and reorganization and has started to keep track of agricultural power and to resolve the problems associated with the groundwater–energy nexus. The difficulties in accomplishing instrumental and community-based approaches to groundwater management have been already discussed. The indirect approaches are not necessarily any easier to implement. India has been struggling to reform its energy sector for many years and changing the agricultural policies for any country is not easy. Still, in a broader sense, identifying the negative and positive impacts of overlapping policies in the agricultural, energy and other sectors may help to forge ahead in developing sustainable groundwater systems.

25.8 Conclusions

The underground reserves of water have many characteristics that make groundwater resources of high importance for a range of human uses, such as potable water, irrigation water supplies and a wide variety of environmental services. These advantages combined with the current growth of economically feasible pumping techniques have led to an uncontrollable rise in groundwater extraction in recent times. Groundwater is now used for almost half of all irrigation supplies and drinking water facilities. The rise in groundwater use has brought significant advantages for the world's economic development, including reducing global poverty and uplifting lives in urban areas. Lately, governance bodies that plan, manage and monitor groundwater utilization have not been able to track and regulate the changing scenarios in groundwater use. Therefore, there is a chance of extended and unfair use of resources in some areas while other regions are now suffering. Also, the aquifer quality has been observed to be declining, and the quantity in some regions has reached dangerously low levels. Groundwater is vital for the overall existence of humanity, which is why we should focus on searching for measures to develop sustainable groundwater systems. Although a range of institutional and technological solutions have been recognized and implemented to realize this objective, the free-to-use nature of groundwater resources and the economic–political scenarios in different areas of the world have restricted the success of such measures. Sustainable groundwater resource systems will need better data capturing, monitoring, accessibility and storage policies. At the same time better availability of aquifer recharge facilities and better water use efficiencies (especially in the agriculture sector) are required to improve the current scenario. Other sectors such as agriculture, energy and industry have been putting immense pressure on this vital resource. There is an urgent requirement for greater and better understanding and collaboration between individuals and institutions in the development and management of sustainable groundwater resources.

References

Alcamo, J., Döll, P., Henrichs, T. et al. (2003). Development and testing of the WaterGAP 2 global model of water use and availability. *Hydrological Sciences Journal* 48 (3): 317–337.

Ambast, S.K., Tyagi, N.K., and Raul, S.K. (2006). Management of declining groundwater in the trans Indo-Gangetic Plain (India): some options. *Agricultural Water Management* 82 (3): 279–296.

Barlow, P.M. and Reichard, E.G. (2010). Saltwater intrusion in coastal regions of North America. *Hydrogeology Journal* 18 (1): 247–260.

Barnett, J., Rogers, S., Webber, M. et al. (2015). Sustainability: transfer project cannot meet China's water needs. *Nature News* 527 (7578): 295.

Biswas, A. K., Tortajada, C., and Saklan, U. (2017). "India is facing its worst water crisis in generations." *Quartz India*, March 23, 2018.

Blanke, A., Rozelle, S., Lohmar, B. et al. (2007). Water saving technology and saving water in China. *Agricultural Water Management* 87 (2): 139–150.

Brinkmann, R. (2016). *Introduction to Sustainability*. Chichester, UK: Wiley.

Brouyère, S., Carabin, G., and Dassargues, A. (2004). Climate change impacts on groundwater resources: modelled deficits in a chalky aquifer, Geer basin, Belgium. *Hydrogeology Journal* 12 (2): 123–134.

Brown, L.R. and Halweil, B. (1998). China's water shortage could shake world food security. *World Watch* 11 (4): 10–21.

Brundtland, G.H. (1987). *Our Common Future: Report of the World Commission on Environment and Development*. Brundtland Commission, United Nations Commission. Oxford, UK: Oxford University Press.

Burke, J.J. and Moench, M.H. (2000). *Groundwater and Society: Resources, Tensions and Opportunities. Themes in Groundwater Management for the Twenty-First Century*. New York: Department of International Economic and Social Affairs, Statistical Office, United Nations.

Chai, J.-C., Shen, S.-L., Zhu, H.-H., and Zhang, X.-L. (2004). Land subsidence due to groundwater drawdown in Shanghai. *Geotechnique* 54 (2): 143–147.

Connor, R. (2015). *The United Nations World Water Development Report 2015: Water for a Sustainable World*. Paris: UNESCO Publishing.
Custodio, E. (2002). Aquifer overexploitation: what does it mean? *Hydrogeology Journal* 10 (2): 254–277.
Datry, T., Larned, S.T., and Tockner, K. (2014). Intermittent rivers: a challenge for freshwater ecology. *Bioscience* 64 (3): 229–235.
Eslamian, S. and Eslamian, F.A. (2017). *Handbook of Drought and Water Scarcity: Environmental Impacts and Analysis of Drought and Water Scarcity*. Boca Raton, FL: CRC Press.
Famiglietti, J.S. (2014). The global groundwater crisis. *Nature Climate Change* 4 (11): 945.
Freeze, R.A. and Cherry, J.A. (1979). *Groundwater*. Upper Saddle River, NJ: Prentice Hall.
George, C.M., Sima, L., Arias, M. et al. (2014). Arsenic exposure in drinking water: an unrecognized health threat in Peru. *Bulletin of the World Health Organization* 92: 565–572.
Giordano, M. and Villholth, K.G. (2007). *The Agricultural Groundwater Revolution: Opportunities and Threats to Development*. Wallingford, UK: CABI.
Gleeson, T., Befus, K.M., Jasechko, S. et al. (2016). The global volume and distribution of modern groundwater. *Nature Geoscience* 9 (2): 161.
Guglielmi, G. (2017). Arsenic in drinking water threatens up to 60 million in Pakistan. *Science* (August 23) www.sciencemag.org/news/2017/08/arsenic-drinking-water-threatens-60-million-pakistan (accessed March 15, 2018).
Guru, M. and Horne, J. (1970). *The Ogallala Aquifer*. Southampton, UK: WIT Press.
Hornbeck, R. and Keskin, P. (2014). The historically evolving impact of the Ogallala aquifer: agricultural adaptation to groundwater and drought. *American Economic Journal: Applied Economics* 6 (1): 190–219.
International Union for Conservation of Nature and Natural Resources (1980). *World Conservation Strategy: Living Resource Conservation for Sustainable Development*. Gland, Switzerland: IUCN.
Jasechko, S. and Taylor, R.G. (2015). Intensive rainfall recharges tropical groundwaters. *Environmental Research Letters* 10 (12): 124015.
Kijne, J.W., Barker, R., and Molden, D.J. (2003). *Water Productivity in Agriculture: Limits and Opportunities for Improvement*. Wallingford, UK: CABI.
Kløve, B., Ala-Aho, P., Bertrand, G. et al. (2014). Climate change impacts on groundwater and dependent ecosystems. *Journal of Hydrology* 518: 250–266.
Knutsson, G. (1988). Humid and arid zone groundwater recharge – a comparative analysis. In: *Estimation of Natural Groundwater Recharge* (ed. I. Simmers), 493–504. Amsterdam: Springer.
Konikow, L.F. (2011). Contribution of global groundwater depletion since 1900 to sea-level rise. *Geophysical Research Letters* 38 (17).
Lamur, A., Kendrick, J.E., Eggertsson, G.H. et al. (2017). The permeability of fractured rocks in pressurised volcanic and geothermal systems. *Scientific Reports* 7 (1): 6173.
Little, J.B. (2009). The Ogallala aquifer: saving a vital U.S. water source. *Scientific American* (March 24) https://www.scientificamerican.com/article/the-ogallala-aquifer (accessed March 20, 2018).
Lloyd, J.W. (2009). Groundwater in arid and semi-arid regions. In: *Groundwater – Volume I* (eds. L. Silveira and E.J. Usunoff), 284. Paris: EOLSS Publications.

Margat, J. and Van der Gun, J. (2013). *Groundwater around the World: A Geographic Synopsis*. Boca Raton, FL: CRC Press.

Mileham, L., Taylor, R.G., Todd, M. et al. (2009). The impact of climate change on groundwater recharge and runoff in a humid , equatorial catchment: sensitivity of projections to rainfall intensity The impact of climate change on groundwater recharge and runoff in a humid, equatorial catchment: sensitivity of projections to rainfall intensity. *Hydrological Sciences Journal* 54 (4): 727–738.

Minderhoud, P.S.J., Erkens, G., Pham, V.H. et al. (2017). Impacts of 25 years of groundwater extraction on subsidence in the Mekong delta, Vietnam. *Environmental Research Letters* 12 (6): 64006.

Pahuja, S., Tovey, C., Foster, S. and Garduno, H. (2010). Deep wells and prudence: towards pragmatic action for addressing groundwater overexploitation in India. Washington, DC: World Bank.

Plate, E.J. (1993). Sustainable development of water resources: a challenge to science and engineering. *Water International* 18 (2): 84–94.

Priscoli, J.D., Dooge, J.C.I., and Llamas, R. (2004). *Water and Ethics: Overview*. Paris: UNESCO.

Ramesh, M., Narasimhan, M., Krishnan, R. et al. (2016). The prevalence of dental fluorosis and its associated factors in Salem district. *Contemporary Clinical Dentistry* 7 (2): 203.

Rasul, G. (2014). Food, water, and energy security in South Asia: a nexus perspective from the Hindu Kush Himalayan region☆. *Environmental Science & Policy* 39: 35–48.

Richts, A., Struckmeier, W.F., and Zaepke, M. (2011). WHYMAP and the groundwater resources map of the world 1: 25,000,000. In: *Sustaining Groundwater Resources* (ed. J.A.A. Jones), 159–173. Amsterdam: Springer.

Seckler, D., Barker, R., and Amarasinghe, U. (1999). Water scarcity in the twenty-first century. *International Journal of Water Resources Development* 15 (1–2): 29–42.

Sharma, K.D. (2009). OPINION groundwater management for food security. *Current Science* 96 (11): 1444–1447.

Shrivastava, P. and Kumar, R. (2015). Soil salinity: a serious environmental issue and plant growth promoting bacteria as one of the tools for its alleviation. *Saudi Journal of Biological Sciences* 22 (2): 123–131.

Siebert, S., Burke, J., Faures, J.-M. et al. (2010). Groundwater use for irrigation – a global inventory. *Hydrology and Earth System Sciences* 14 (10): 1863–1880.

Singh, D.K. and Singh, A.K. (2002). Groundwater situation in India: problems and perspective. *International Journal of Water Resources Development* 18 (4): 563–580.

Sivapalan, M., Ruprecht, J.K., and Viney, N.R. (1996). Water and salt balance modelling to predict the effects of land-use changes in forested catchments. 1. Small catchment water balance model. *Hydrological Processes* 10 (3): 393–411.

Skopp, J. (2002). Vadose zone hydrology. *Journal of Environment Quality* 31 (3): 1042.

Sophocleous, M. (2002). Interactions between groundwater and surface water: the state of the science. *Hydrogeology Journal* 10 (1): 52–67.

Todd, D.K. and Mays, L.W. (2005). *Groundwater Hydrology*. Chichester, UK: Wiley.

Van der Kamp, G. and Hayashi, M. (1998). The groundwater recharge function of small wetlands in the semi-arid northern prairies. *Great Plains Research*: 39–56.

Willis, R. and Finney, B.A. (1988). Planning model for optimal control of saltwater intrusion. *Journal of Water Resources Planning and Management* 114 (2): 163–178.

World Bank (2002). *Improving the Lives of India's Farmers: How Power Sector Reforms Will Help*. Washington, DC: World Bank.

Zektser, I.S. and Lorne, E. (2004). *Groundwater Resources of the World: And their Use*, IHP Series on Groundwater. Paris: UNESCO.

Zhang, Y.-K. and Schilling, K.E. (2006). Effects of land cover on water table, soil moisture, evapotranspiration, and groundwater recharge: a field observation and analysis. *Journal of Hydrology* 319 (1–4): 328–338.

Zhu, Y. and Balke, K.-D. (2008). Groundwater protection: what can we learn from Germany? *Journal of Zhejiang University SCIENCE B* 9 (3): 227–231.

26

Sustainability and Energy Management in Facilities for Wastewater Treatment and Reuse

Joseph Sebastian, Pratik Kumar, Krishnamoorthy Hegde, Satinder K. Brar, Mausam Verma, and Rao Y. Surampalli

INRS, Université du Québec, Québec, Canada
Joseph Sebastian and Pratik Kumar have contributed equally.

26.1 Introduction

Sustainability lacks a clear definition, but it strives for economic wellbeing, environmental protection and prudent use of natural resources. The social progress of individuals, communities and the environment, in addition to economic development, is given importance in the case of sustainable development (Muga and Mihelcic 2008). It has been proven that focusing only on economic development is not sustainable. Hence, the focus has to shift toward sustainable development, which takes into consideration economic, social and environmental impacts. It is necessary to recognize that human and industrial systems need to be designed in such a way that the current use of natural resources does not lead to diminished quality of life for future generations. This approach to development is essential for fulfilling the requirements for the wellbeing of humans as well as for other organisms, especially in an urban and industrial atmosphere (Mihelcic et al. 2003).

One of the main resources that have been exploited extensively is water. Water is essential for life on this planet, but human activities, such as urbanization and industrialization, have stressed the available water resources, both surface as well as underground. It is estimated that almost 2 billion people do not have access to safe drinking water due to fecal contamination and that by the year 2025 half of world population will be living in water-stressed areas. It is forecast that the next world war will be for water. In the year 2010, the United Nations (UN) General Assembly unequivocally recognized the human right to sufficient, safe, accessible and affordable water for personal and domestic use (Bdour et al. 2009; www.who.int).

Contaminated water, particularly with fecal matter, is the main reason for many diseases, such as cholera, diarrhea and hepatitis A. In addition, the improper management of urban, industrial and agricultural wastewater has led to dangerous levels of contamination in

Sustainability: Fundamentals and Applications, First Edition.
Edited by Rao Y. Surampalli, Tian C. Zhang, Manish Kumar Goyal, Satinder K. Brar, and R. D. Tyagi

water sources. The mismanagement of the resource has necessitated the need for appropriate wastewater management to prevent polluting the available potable water sources, and the water from these management systems in itself may be used to meet some of the requirements, particular in regions with limited water resources.

Rapid growth in population, urbanization and industrialization has had an impact on the available water sources, which has resulted in wastewater treatment plants (WWTPs) becoming an integral part of the modern industrialized urban ecosystem. The establishment of a WWTP requires a considerable amount of investment. This is mainly associated with building up the infrastructure required for transportation of wastewater as well as high operational costs, which probably renders the current centralized WWTPs and associated technologies unsustainable (Bdour et al. 2009; Molinos-Senante et al. 2014). This chapter discusses sustainability and energy management in facilities for sustainable wastewater treatment and reuse (SWWTR) and the potential utilization of nutrients present in domestic and industrial wastewater for the production of chemicals as well as energy. These measures enhance the sustainability index and prevent rapid depletion of existing resources.

26.2 Sustainable Wastewater Treatment and Reuse

Currently, one of the biggest concerns is that of water shortage and the social unrest that such a situation would create, as potable water is one of the basic requirements for human wellbeing. Hence, most of the efforts have been focused toward supplying water at really low tariffs with the result that the operational, maintenance and utility costs are often not recovered. This leaves very little money for investment in the treatment of the wastewater generated, which leads to environmental and health hazards (Bdour et al. 2009).

Rapid increases in population, urbanization and industrialization have necessitated the adoption of wastewater treatment techniques that enable water recovery and its reuse. Most of the treatment systems used currently are disposal-based linear systems that do not allow for either recovery or reuse of water as well as the nutrients present in wastewater. These treatment systems are unsustainable. For example, waste stabilization ponds require a large land area, careful maintenance to prevent issues of odor and periodic removal of a high quantity of sludge and its disposal. More modern treatment techniques, such as activated sludge systems, require a large investment and have high operational costs. Hence, they are not sustainable especially for low-income communities. Therefore, planners need to take into consideration various aspects, such as sludge disposal, capital, operational cost, reliability, land requirements and environmental impact, while selecting appropriate wastewater treatment techniques for the community (Bdour et al. 2009).

The environmental and social impact of the wastewater treatment technique needs to be taken into consideration, in addition to economic issues, before selection of the most appropriate sustainable treatment technology (Popovic et al. 2013). However, wastewater treatment lacks a consensus on the definition of sustainability. Hence, evaluation of the sustainability of different techniques is essential during the decision-making process (Muga and Mihelcic 2008; Molinos-Senante et al. 2014).

26.2.1 Indicators for Sustainable Wastewater Treatment and Recovery

The indicators used for evaluating sustainable development, in general, should have/be: (i) sound scientific backing and approval of the scientific community; (ii) ease of comprehension even to non-experts; (iii) appropriate and must cover crucial aspects of sustainable development; (iv) easy to gather and update data; and (v) limited in number and fit for purpose (Muga and Mihelcic 2008). The two main approaches that are used to assess sustainability are:

1) Single-indicator approach where tools of assessment, such as the outcome of exergy analysis or lifecycle assessment (LCA), are used to study sustainability. Most often, these assessments do not take into consideration social impact and, hence, cannot provide a true assessment of sustainability (Molinos-Senante et al. 2014).
2) Multidisciplinary indicator approach, which uses different parameters for the assessment of sustainability that takes into consideration the economic, environmental and social impacts. Some of the parameters that can be assessed to evaluate sustainability and their impact are provided in Table 26.1. The selection of appropriate parameters

Table 26.1 Indicators for the evaluation of sustainability using multiple indicators.

Dimension	Indicators	Direction	Units
Economic	Investment cost	Negative	\$/GPD
	Operation and maintenance costs	Negative	\$/MGD\$/month
	User cost	Negative	
Environmental	Organic matter removal	Positive	% Removal
	Suspended solids removal	Positive	% Removal
	Nitrogen removal	Positive	% Removal
	Phosphorus removal	Positive	% Removal
	Energy consumption	Negative	kWh/MGD (kWh/m^3)
	Land area required	Negative	Acre/MGD
	Sewage sludge production	Negative	kg/m^3
	Water reusability	Positive	—
	Products recovery	Positive	—
	Reliability	Positive	—
Social	Public participation and acceptance	Positive	Qualitative measure
	Community size served	Positive	Population/MGD
	Staffing required to operate plant	Negative	Staff/MGD
	Level of education	Negative	—
	Odors	Negative	Qualitative measure
	Noise	Negative	Decibel (dB)
	Visual impact	Negative	Qualitative measure
	Complexity	Negative	Qualitative measure

MGD, million gallons per day; GPD, gallons per day.

should be made based on the requirements of the community and its geography, which will affect sustainability assessment of the individual treatment technique under consideration.

26.2.2 Multiple Indicators for Sustainability Evaluation

A brief description of the sustainability indicators that can be considered in a multi-indicator approach to studying sustainability follows:

Economic indicators: These indicators represent the costs incurred during operation of the WWTP. The choice of the appropriate treatment technique is primarily based on the economic feasibility of the technique. Economic feasibility cannot be universal and will vary based on the economic situation of the community that the treatment plant is supposed to serve. Commonly used indicators of economic feasibility are capital cost, operation and maintenance. It is vital that the lifespan of each technology be taken into consideration when an economic feasibility study is being carried out.

Environmental indicators: These indicators allude to the environmental effects of the treatment techniques being considered. This takes into consideration the effluent quality, resources consumed, emissions and the flexibility of the technology to upgrading and addition of novel processes in the future. Conventional assessment of the quality of effluent from the treatment plant involves quantifying the concentrations of organic matter, suspended solids (SS), nitrogen, phosphorus and coliforms in the effluent from the treatment plant. The regulatory requirements associated with the quality of effluent are bound to change, which will necessitate the introduction of additional processes. Moreover, the carbon footprint of the treatment techniques needs to be assessed. This will take into consideration energy consumption as well as greenhouse gas emissions and energy produced (Muga and Mihelcic 2008).

Land requirement for the setup of the treatment plant needs to be considered while assessing the environmental impact of the technique. The land requirement varies based on the technique used and hence can be a valuable tool for assessing the sustainability of the proposed treatment techniques. Most of the treatment techniques produce sludge as a byproduct and management of this is one of the most complex issues associated with wastewater treatment. The quantity of sludge produced needs to be taken into consideration while assessing sustainability. The potential to reuse the water and recover or produce products from wastewater needs to be assessed while evaluating sustainability (Muga and Mihelcic 2008). Additionally, the reliability of the process also needs to be considered. The probability of mechanical failure and its impact on the quality of effluent generated needs to be addressed while assessing sustainability and used as an indicator (Eisenberg et al. 2001).

Social indicators: These take into consideration the impact of the treatment plant and processes used in society. The societal impact of the treatment plant is often given the least priority and generally not addressed while evaluating sustainability. It is vital to include the evaluation of social impact while evaluating sustainability. Some of the indicators used for assessment are odors, noise, visual impact, public acceptance, and complexity of construction and operation. The simplicity of the process may be an important factor while determining sustainability, especially in the case of developing countries. It is important that the investment and operational costs are as minimal as possible to ensure the availability of funds for fulfilling the basic requirements of the community, such as centralized potable water supply and municipal waste management systems (Molinos-Senante et al. 2014; Muga and Mihelcic 2008).

26.3 Approaches for Sustainable Wastewater Treatment and Reuse

SWWTR can be achieved by two approaches. One is the use of appropriate sustainable technology and generation of water that can be reused. The other approach is the valorization of the wastewater generated during human and industrial activities. Another factor that has a significant impact on the sustainability of wastewater treatment is energy consumption. Wastewater treatment requires a lot of energy, especially associated with the operation of pumps. The adoption of energy-efficient treatment processes and energy recovery techniques, in the form of biogas or electricity, can improve the sustainability of WWTPs (see the detailed discussion in Section 26.4).

26.3.1 Sustainable Wastewater Treatment Technologies

Conventional mechanized techniques of wastewater treatment are not sustainable as they consume a lot of energy and need a large investment to construct, operate and maintain. Most wastewater treatments are centralized thereby requiring expensive sewer systems to transport wastewater from origin to treatment facilities. The sewers from different regions combine and increase the volume, hence requiring designing treatment plants that are able to process large volumes of water. Most often, these sewer systems are also required to handle rain and flood waters thereby requiring treatment plants to be designed to handle these increases in volumes. This further adds to the investment required. The best solution is to treat the wastewater close to the site of origin. To this effect, the decentralized system of wastewater treatment has special significance (Massoud et al. 2009).

Decentralized systems require less investment as they do not require expensive sewer systems to transport wastewater and provide the flexibility required to meet the stresses of rapid urbanization and industrialization. A decentralized system in combination with zero-discharge wastewater treatment techniques can greatly improve sustainability and allow for water and resource recovery from wastewater (Massoud et al. 2009). In addition, the transformation of traditional linear treatment systems to cyclical treatment allows for conservation of water and its reuse as well as reuse of valuable nutrient resources. It is vital that the organic components of wastewater from urban areas are recovered. The failure to do so means a huge loss of life-supporting resources and ensuing water pollution.

The shift toward zero-discharge technologies allows for the reuse of treated water. This cyclical approach, rather than a conventional linear approach, is an important measure toward sustainable water use. The fear that treated water is not safe is unwarranted as the effluent has been treated for pathogens. Moreover, they allow for reuse of the effluent for agricultural purpose and enable recycling of the valuable resource, which holds special significance for water-scarce regions (Bdour et al. 2009). Some of the sustainable zero-discharge wastewater treatment technologies are discussed below:

26.3.1.1 Lagoons and Wetlands

In this treatment, natural forces act in tandem to purify wastewater. This system utilizes a series of shallow ponds that act as stabilization lagoons. The removal of heavy metals is achieved by growing water hyacinth or duckweed and different bacteria, planktons and algae, which further purify the water. Wetland treatment technologies offers several advantages over conventional treatments as they are self-sufficient, provide ecological balance and economic viability. Moreover, this system allows for total resource recovery. It is a low-cost system, but requires a large expanse of land and is affected by climatic conditions. Hence, it may not be suitable for urban areas where the demand for flat land is high. The other disadvantages include the loss of water due to evaporation, the release of methane and issues with odor (Bdour et al. 2009).

26.3.1.2 Anaerobic Digestion

Anaerobic digestion is another sustainable wastewater treatment system that requires little land and hence is suitable for the urban environment. Anaerobic bacteria degrade organic matter to produce methane and carbon dioxide. The methane so produced can be used as an alternative energy source. A more detailed discussion about anaerobic digestion, especially up-flow anaerobic sludge blanket (UASB) reactors, is provided in Section 26.4.1. It has been observed that this technique is not able to remove nitrogen and phosphorus effectively. Hence, the effluent from this system can be used for agriculture or aquaculture purposes thereby enabling the complete resource to be recycled. Moreover, additional sustainable treatment techniques, such as sludge composting, wetland systems or stabilization ponds, may be needed for the complete removal of pathogens and in order to meet the discharge standards (Bdour et al. 2009). We will discuss in detail the process and energy inventory of anaerobic digesters and their concepts in later sections.

26.3.1.3 Soil Aquifer Treatment (SAT)

SAT is a geo-purification system. In this system, partially treated sewage effluent is used to artificially recharge the aquifers. The water is then withdrawn from the aquifer for future use. The effluent from the primary treatment undergoes additional purification while passing through soil layers before it mixes with the natural groundwater. This technique of treating wastewater enables recharge of groundwater sources, which is important for water-scarce regions. The advantages of SAT systems are that it is inexpensive to set up such a simple system of wastewater treatment and it is able to remove pathogens, biological oxygen demand (BOD) and total suspended solids (TSS) efficiently. Care should be taken to use an appropriate primary treatment technique so that algal concentration in the treated water is kept to a minimum. The presence of algal biomass can lead to clogging, which will prevent proper percolation of water through the soil layers. Also, it is important to maintain adequate distance (45–106 m) between the infiltration basin and wells to allow for adequate SAT (Rose 1999; Metcalf and Eddy et al. 2002).

26.3.1.4 Hybrid Systems

These systems of wastewater treatment use a combination of two or more of the treatment techniques to achieve complete treatment and reuse of the water. One such example is the wastewater treatment being carried out at the Omega Center for Sustainable Living (OCSL), which is a Living Building Challenge Certified (LBCC) facility. The wastewater generated at the campus is treated using a multi-stage process. Wastewater first undergoes anaerobic digestion followed by the artificial wetland. After wetland treatment, the water is pumped to aerated lagoons which are followed by sand filtration. This sand-filtered water, while safe enough for consumption, is dispersed to replenish the groundwater table and for irrigation. The hybrid system is able to treat up to 52000 gal day^{-1} of wastewater (www.treehugger.com).

The sustainable wastewater treatment techniques described above enable the reuse of treated water so that a sustainable and cyclic system for water management can be achieved. The possibility for use of the nutrient-rich treated water for irrigation also ensures the recycling of valuable resources. Moreover, further advances and research activities toward SWWTR have led to the development of novel treatment systems, such as HyVAB® (developed by Biowater Technology) (Figure 26.1), and nanofiltration.

The HyVAB® bioreactor treats wastewater in two stages in a vertical bioreactor (Figure 26.2). The wastewater enters the lower section of the bioreactor and initially undergoes anaerobic digestion. During this stage, over 75% of organic waste can be converted to methane. This is followed by a proprietary aerobic biofilm reactor that removes the remaining organic compounds and odor from the water. The main benefits of this system are that it has a small footprint, high efficiency and odorless discharges (www.biowatertechnology.com). Such advances can improve sustainability and enable both energy and water recovery.

26.3.2 Valorization of Wastewater

Water reuse and resource recovery from wastewater is the solution for polluted waterways and water shortage. Almost all nitrogen, phosphorus and micronutrient requirements can be supplied by using wastewater for irrigation. This method of irrigation holds special significance for water-scarce regions, such as North and South Africa, the Middle East, southern Europe, southwestern US, Mexico, South America, and parts of central and

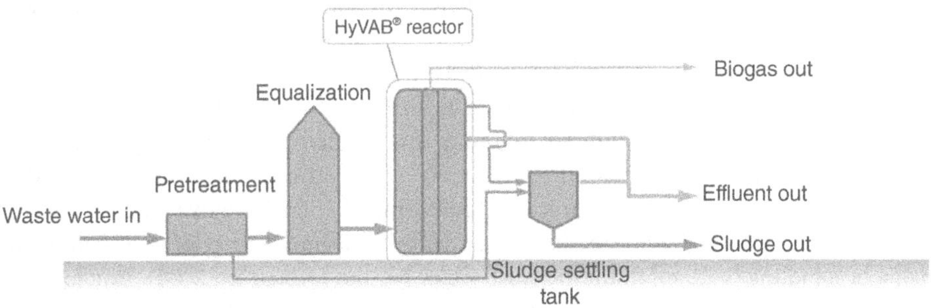

Figure 26.1 General layout of HyVAB® system (www.biowatertechnology.com).

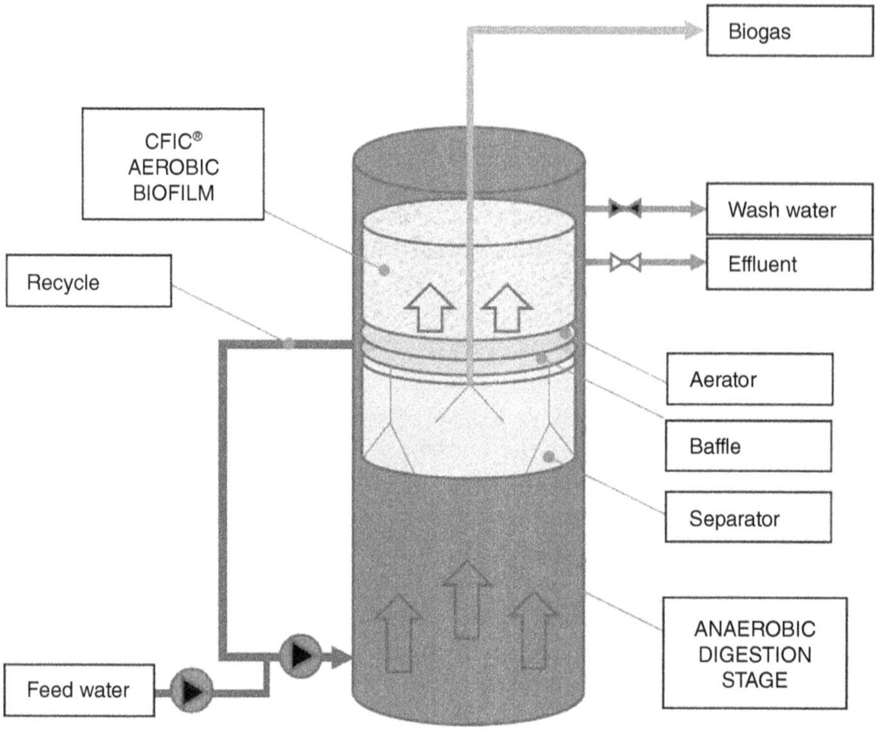

Figure 26.2 HyVAB® bioreactor (www.biowatertechnology.com).

east Asia. The objectives of removal of nutrients from sewage and producing vegetables fit for human consumption or as animal fodder can be achieved by using hydroponic systems for wastewater management (Boyden and Rababah 1996).

Integration of hydroponic cultivation to wastewater treatment was proposed as a green alternative to conventional wastewater treatment (Guterstam 1996). It has been observed that vascular plants secrete photosynthetically fixed carbon either actively or passively. The released carbon is susceptible to microbial degradation, and the roots tend to act as a stabilizing factor. Hence, it has been proposed that water hyacinth systems are more stable than activating sludge systems (Norström 2005). Similar observations may hold true for hydroponic systems of wastewater treatment where the roots of the plant provide a stable environment for microbial growth and degradation of pollutants.

One such study compared the wastewater hydroponic system with a commercial hydroponic system that utilized commercial nutrients. This study showed that the crops grown were able to remove nitrogen and phosphorous from settled primary sewage effectively. Phosphorus and nitrogen removal of 77% and 80%, respectively, were achieved (Boyden and Rababah 1996). About 90% chemical oxygen demand (COD) removal and total nitrogen reduction of 72% were achieved in another study performed at a pilot treatment plant capable of handling 0.559 m^3 domestic wastewater per day (Norström 2005). Aquaculture is another field where hydroponics has proved its feasibility and viability. The wastewater from aquaculture can be used for the culture of crops and vegetables. COD removal up to

79% was achieved in a study performed to investigate the possibility of using a hydroponic system to treat and reuse wastewater generated during aquaculture (Keeratiurai 2013). Also, a 54% higher COD removal rate was observed when lettuce was used in a hydroponic reactor versus when no plant was used. This COD removal rate depended on the type and quantity of plant used in the system (Gloger et al. 1995).

Wastewater biorefinery (WWBR) is another important technique that allows for water reuse and recovery of resources. This is very important for sustainable development and ensures careful use of one of the most important resources. WWBR deals with the simultaneous processing and treatment of wastewater to yield treated water as well as the generation of products of sufficient value, such as commodity chemicals, to make biorefining feasible. Most of the investigations of biorefining have been performed at the laboratory scale providing proof of concept for product generation from wastewater (Harrison et al. 2017). This concept of resource recovery for sustainable wastewater treatment is represented in Figure 26.3.

Generally, wastewater treatment is considered an expense due to the costs associated with construction, operation and maintenance (Harrison et al. 2017). The possibility of generating revenue by producing commodity or platform chemicals of significance can greatly improve the scope of the concept of WWBR. The aim of shifting to a bio-based economy, rather than a fossil fuel-dependent economy, has been the driving force for the research activities into the possibility of using wastewater as the feedstock for chemical production. This shift to using waste as feedstock has the potential to drastically bring down the cost of production of bio-based chemicals. Moreover, wastewater biorefineries are designed and optimized with the objectives of effective treatment of water as well as the conversion of components, especially organic compounds, to products of economic and social value. A wide variety of compounds ranging from dyes to biofuels have been investigated and proposed for production utilizing wastewater from an agro-industry and domestic sources and are represented in Table 26.2.

Industrial and domestic wastewater has the potential to act as inexpensive feedstock for the production of high-value compounds. Carbohydrate-rich wastewaters that are not suitable for human or animal consumption, such as cassava and brewery wastewater, are ideal for conversion via fermentation using pure culture or co-culture processes (Angenent et al. 2004). The main limiting factors for the widespread utilization of the alternate substrate are: (i) finding the right balance of nutrients that will promote cell growth and accumulation of product; (ii) cost efficiency; and (iii) conversion yield (Angenent et al. 2004; Nitschke and Pastore 2006). The poor conversion will most likely be the greatest barrier, and hence most efforts have been devoted to improving the amount of product formed during fermentation. Some of the strategies employed are the identification of novel microorganisms, metabolic engineering and control of carbon flux by the addition of external substrate or inhibitors (Bramucci 2000; Angenent et al. 2004; Kalyuzhnaya et al. 2015; Puyol et al. 2017; Yang et al. 2017).

In addition, process modification strategies, such as immobilization and co-culture, have shown to improve process yield. For example, immobilization of the fungal species, *Rhizopus oryzae* NRRL1526 on eggshells was able to improve fumaric acid production from 31.3 to 47.2 $g\,l^{-1}$ (Das and Brar 2014; Das et al. 2015). Co-culture of pure microbial strains is another strategy that has been shown to improve bioconversion rates. With this strategy,

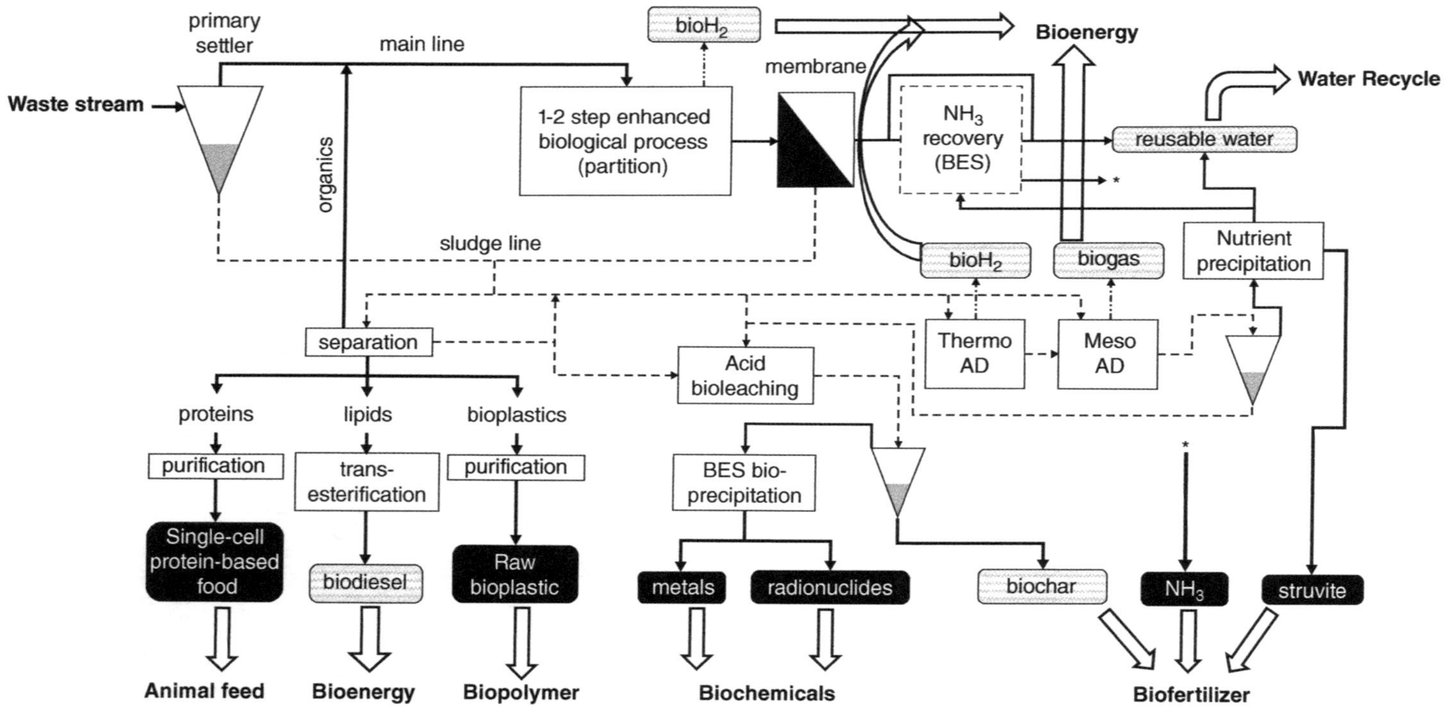

Figure 26.3 Concept of sustainable wastewater treatment and resource recovery by using different technologies. *Source:* Adapted from Puyol et al. 2017.

Table 26.2 Some wastewaters suitable for bioproduction.

Wastewater	Product	Microorganism	References
Corn steep liquor	Lipase	*Geotrichum candidum*	Burkert (2004)
Whey	Protease	*Mucor* spp.	Tubesha and Al-Delaimy (2003)
	Polyhydroxyalkanoates	Recombinant *E. coli*	Park et al. (2002), Coats et al.(2007)
Gluten-free effluent	Glycerol	*Pichia farinose*	Wang et al. (2002)
	Monascus pigment	*Monascus purpureus*	Domı'nguez-Espinosa and Webb (2003)
	Lactic acid	*L. bulgaricus*	Wang et al. (2002)
Starch-processing wastewater	Biomass protein	*Aspergillus oryzae*, *Rhizopus oligosporus* and *Rhizopus arrhizus*	Jin et al. (2010)
Salad oil	Biomass protein	Five yeasts	Zheng et al. (2001)
Milk permeate	Acetic acid	*C. thermolacticum* plus *M. thermoautotrophicus*	Collet et al. (2003)
Brewery wastewater	Fumaric acid	*R. oryzae* NRRL 1526	Das et al. (2015)
Apple residue	Citric acid	*A. niger* NRRL 567	Dhillon et al. (2012)
Domestic/industrial wastewater	Biodiesel	Algae/microalgae	Hodaifa et al. (2008), Pitmann et al. (2011), Rawat et al. (2011), Chen et al. (2015)
Palm oil processing	Butanol	*Clostridium saccharoperbutylacetonicum*	Hipolito et al. (2008)
Municipal sludge	Biopesticide	*Bacillus thuringiensis*	Yezza et al. (2006)
Mussel-processing	Nisin, Pediocin	*L. lactis*, *P. acidilactici*	Guerra and Pastrana (2002)
Rice parboiling wastewater	Carotenoid pigment	*Phaffia rhodozyma*	Rios et al. (2015)
Sochu distillery wastewater	Chitosan	*G. butleri* IF08081	Yokoi et al. (1998)
Cassava wastewater	Surfactant	*B. subtilis*	Nitschke and Pastore (2006)
Brewery wastewater	Bioflocculant	*Staphylococcus* sp./*Pseudomonas* sp.	Zhang et al. (2007)

one species performs the hydrolysis of complex nutrients and the second species converts the byproducts of the hydrolysis process into the desired products. For example, improved acetic acid production has been achieved by co-culturing bacterial species *Clostridium thermolacticum* and *Methanothermobacter thermoautotrophicus* on milk and whey permeate from the cheese industry. The introduction of hydrogenotrophic methanogen was able to improve acetic acid production by the reduction of hydrogen partial pressure (Collet et al. 2003). Also, the co-culture technique offers added advantage that it may be less susceptible to changes in the composition of industrial wastewater (Angenent et al. 2004).

26.4 Sustainable Energy Derived from Wastewater

Water and WWTPs utilize almost 3–4% (56 billion kW of electricity) of total electricity in the US. Other developed and developing countries too have demand in similar ranges. More than 40 million tons of carbon dioxide gas is emitted annually in the US itself (phys.org 2018). However, there is a huge potential for energy recovery in WWTPs. This is mainly because the wastewater system has enough organic load as compared to the raw water in the drinking water plants (DWTPs). However, gravitational energy can reduce the energy requirements in DWTPs processing natural source water intake by reducing the elevation level by around 1–1.5 m (if the source water is at elevation). Some major reforms in the existing and new models have been evaluated over a period of time for energy recovery and have succeeded in making the overall process at least 30–50% self-sustainable.

Deriving energy from the DWTP units involves much lesser potentials as compared to those achieved in WWTPs. Still, micro-hydro power installations have proven that these installations can facilitate a sustainable approach. These systems of water supply installations (though not generally a part of treatment processes) convert mechanical energy into electrical energy (100–300 kW day^{-1}). The fluid flow head and flow rate are mainly responsible for producing a high amount of convertible energy. For example, cities/towns located on elevated terrain can make use of such power generation as a back-up. This convertible energy has also been found to be sustainable. For example, in a case study in Australia, where around 330 ml day^{-1} is handled at one of the WWTPs (60 m water head), 12 GWh of electricity is generated annually, offsetting 80 000 TCO_2e (McNabola et al. 2014). These types of installations have been proven to be a great sustainability measure recovered from both water and wastewater supply (Gude 2015). However, we will focus our discussion on energy-based sustainability measures for WWTPs only.

A large amount of electrical energy (≥30%) can be saved in treating wastewater (municipal or industrial) by employing energy efficiency processes or with modifications in the existing treatment facilities (Means 2004; www.waterresearchfoundation.org). For example, a simple optimized-aeration process with improved pumping alone can contribute around a 3–6% reduction in the overall energy that is needed to carry out WWTP operations (Hoppock and Webber 2008). It has been found that most of the current WWTPs do not follow energy-efficient processes. For example, in Spain, only 10% of the WWTPs were found to be energy efficient. There could be many different variables related to this issue. These include plant size, amount of organic matter removed, aeration requirements, etc.

Energy saving is mainly measured in terms of reduced CO_2 emissions, increase in biogas production, bioelectricity generation, etc.

Wastewater treatment and energy generation go hand in hand and are mutually related. This means enhancing wastewater or water treatment can also benefit the overall energy output. However, sometimes integrated solution/approaches may not present a feasible solution and may demand a different configuration of the system to achieve sustainability. Likewise, the nature of source water plays a key role in meeting energy demand and supporting self-sustainability. For example, treating natural source water, such as rivers, lakes and ponds, by the conventional water treatment system requires approximate energy of around 0.25–1 $kWh\,m^{-3}$. On the other hand, saline water treatment requires approximately 2–20 $kWh\,m^{-3}$ on average. For wastewater treatment, it takes around 0.5–2 $kWh\,m^{-3}$ and a further 0.5–2 $k\,Wh\,m^{-3}$ of energy to make it reusable for domestic purposes.

Three major energy-based platforms in terms of bioenergy, bioelectricity and biofuel production, along with wastewater treatment, are discussed in the sections to follow. Anaerobic digestion (and its modified concepts), microbial fuel cells (MFCs) and microalgae biomass will be discussed with regard to bioenergy, bioelectricity and biofuel production, respectively, as the criteria for the sustainability measures in WWTP operation.

26.4.1 Need and Potential of Anaerobic Digesters in Sustainable Treatment of Wastewater

According to the US Department of Energy (DOE), more than 230 billion kWh of thermal energy in the form of waste hot water is discarded every year. Invoked by many climatic challenges and issues raised so often by many countries, such as greenhouse gas concern, there is a need, not only to serve wastewater treatment purposes, but also to produce some form of energy simultaneously. Surprisingly, only 10% of this energy is required to treat wastewater. Hence, it is a matter of collecting wastewater efficiently and deriving energy from it during treatment. Use of geothermal heat pumps has been shown to improve the energy utilization in WWTP. Geothermal energy has also been used successfully to operate the anaerobic digester treating agricultural wastewater, with a good amount of biogas being produced (0.25–0.66 $m^3\,kg^{-1}$ COD) (USEPA 2012). One of the main components and contributions of wastewater treatment is sludge handling, which is mainly treated by anaerobic digestion. It amounts to almost 30–40% of the economy in a WWTP. Various sludge-handling techniques are used to manage sludge efficiently, such as gasification, mechanical dewatering, thermal dewatering, incineration, etc. However, anaerobic digestion has been proved to be the most common operation in treating the volatile suspended solids (VSS) component of the sludge with simultaneous biogas production.

Energy recovery mainly comprises two parts: (i) chemical energy; and (ii) thermal energy. Chemical energy is mainly recovered from the wastewater system through a digestion process, while thermal energy recovery is still not much explored or employed as a part of the current energy recovery system. However, the concept of underground thermal energy storage is a rapidly developing energy source and could improve thermal energy recovery rates (Verstraete et al. 2009).

Chemical energy recovery from wastewater has been a key practice around the world. The high carbon source in the form of COD is mainly lost to metabolic heat. For example, a conventional activated sludge treatment plant consumes a substantial amount of energy in terms of oxygen supplied (huge aeration demand) and only a part of the energy is recovered in sludge digestion. Thus, the rise of the anaerobic digestion concept is increasing day by day to maximize the energy output thereby increasing the sustainability of the overall process. Many developing countries such as India are already processing biogas generation from wastewater sources as a potential energy bank. As much as 1227 MW of bioenergy has been generated from solid biomass against a potential of 25 700 MW (Ecogeneration 2008).

The concept of modifying conventional digesters and other techniques for the removal of COD, nitrogen and other recalcitrant compounds in WWTPs is gaining in popularity; modified concepts such as UASBs are under operation by many WWTPs. Up to 200 MJ/tear of energy in comparison to existing anaerobic systems can be saved along with potential reusable water production (Zeeman et al. 2008). A high rate of anaerobic wastewater treatment is equally potent in diversifying end-of-the-pipe treatment to resource-oriented conversion techniques (Frijns et al. 2013). Table 26.3 tabulates some of the conventional and modified anaerobic digester technologies for SWWTR.

Though a high output of energy can be derived from anaerobic treatment, still, the issues of dissolved methane in the case of diluted sewage and moderate temperature conditions persist in the environment. Methane gas remaining mainly in the effluent solution (up to 40%) contributes to the global greenhouse gas emissions later when released into the atmosphere (El-Fadel and Massoud 2001; Rao et al. 2010). However, certain hybrid systems, such as co-digestion concepts and the introduction of the supplementary treatment unit, not only hold promise to increase the treatment efficiency but also have the capability to produce bioenergy, making WWTP operations sustainable.

Another way of recovering/utilizing energy sustainably in WWTPs is the usage of the combined heat and power system (CHP) in association with an anaerobic digester. These systems work on the principle of electricity and heat energy conversion where the waste heat generation is recovered for productive use in plants. These systems can be coupled to anaerobic digestion to effectively utilize this heat energy (Dai et al. 2015). There is a mutual benefit between the working of CHP and anaerobic digestion as the biogas generated from the anaerobic digestion helps in the operation of CHP. Another possibility of a hybrid system of energy neutralization was studied by Dai et al. (2015), who evaluated the concept of integrating a membrane reactor for anaerobic digestion and nitrifitation–anammox. The process resulted in over 96% of COD removal with 223 ml methane gas/g COD production utilizing a mere 0.09 kWh m^{-3} of electrical energy. Another comparable study by Pretel et al. (2015) revealed that the anaerobic membrane bioreactor (AnMBR) outperformed the conventional combination of a primary settling stage and anaerobic digestion in terms of both performance and economically. However, treating high-sulfate wastewater was more economically optimized when the AnMBR worked with a primary settling stage and anaerobic digestion (€0.05 m^{-3}). Thus, it may depend on the kind of constituents present in the wastewater, and accordingly, the best process can be chosen for cost saving (possibly up to 45%).

Many countries have already explored the production of biogas and applied them efficiently; for example, Germany, which already has set a goal of replacing 6% of its natural gas demand with biomethane in 2020 and has a future target of 10% by 2030. However, the efficiency of biogas production from anaerobic digestion depends largely on certain factors

Table 26.3 Different wastewater facilities used for sustainable production of energy from various kinds of wastewater.

Wastewater treatment facility used	Wastewater	Energy output	Other remarks	References
Anaerobic digestion	Cattle paunch	Paunch waste contains 7.8 GJ energy	• >20% organic waste produced • Methane production 230 l CH_4 kg^{-1}	Jensen et al. (2016)
Anaerobic digestion	Plant, agricultural residues and slurries	The lower heating value of syngas was 15.7 MJ N m^{-3}, whereas pyrolysis oil exhibited a higher heating value of 23.5 MJ kg^{-1}	• Pyrolysis resulted in 8.8 wt.%, 58.4 wt.% and 32.8 wt.% of syngas, oil and char, respectively • Coupling anaerobic digestion with pyrolysis produces more electricity (42%) compared to the stand-alone anaerobic digestion (9896–14 066 kWh day^{-1})	Monlau et al. (2015)
Sludge stabilization digestion tower	Municipal wastewater plant	Electrical energy output of around 496 kWh day^{-1}	• Additionally, 145 kWh day^{-1} are generated by means of photovoltaics (PV) • Overall process was about 30% self-sufficiency	Kollmann et al. (2017)
Geothermal anaerobic digester (8 m^3)	Agro-industrial residues	Proposed process can derive 100% more biogas for value-added application consuming organic wastes	Biomass yield: cheese whey 0.66 m^3 kg^{-1} COD, olive mill 0.45 m^3 kg^{-1} COD and cattle wastewater 0.25 m^3 kg^{-1} COD	Diamantis et al. (2016)
Membrane combined process of anaerobic digestion and nitrifitation–anammox	Municipal wastewater	Stable methane yield of 1672 ml day^{-1}, corresponding to 223 ml CH_4 g^{-1} COD. The net energy consumption 0.09 kWh m^{-3}	COD removal: 96% and most CODs were converted into biogas in the anaerobic	Dai et al. (2015)
Performances of mechanical and low-temperature (<100°C) before anaerobic digestion	Municipal and IWWT	25 000 ml biogas per 20 days as compared to 20 000 ml per 20 days without treatment	Increased revenues from electricity (€/h) 35.3 49.9 53.5 for treatment at 70, 80 and 90°C, respectively	Brebbia et al. (2014)

(Continued)

Table 26.3 (Continued)

Wastewater treatment facility used	Wastewater	Energy output	Other remarks	References
Anaerobic co-digestion	Fruit and vegetable waste along with sewer sludge	8–17% increase in biogas production as compared to only sewer sludge digestion	pH stability and better reactor performance in terms of total solids handling too	Park et al. (2011)
Microbial fuel cell and anaerobic fluidized bed membrane bioreactor	Domestic wastewater	Total chemical oxygen demand (tCOD) of $210 \pm 11\ mg\,l^{-1}$, which reduced to $16 \pm 3\ mg\,l^{-1}$ TSS; from 45 ± 10 to $<1\ mg\,l^{-1}$	• The total electrical energy requirement: $0.0186\ kWh\,m^{-3}$ • Energy produced by MFCs: $0.0197\ kWh\,m^{-3}$ • Biomass yield: cheese whey $0.66\ m^3\,kg^{-1}$ COD, olive mill $0.45\ m^3\,kg^{-1}$ COD and cattle wastewater $0.25\ m^3\,kg^{-1}$ COD	Moreno-Garrido (2008)
Algae biomass	Primary effluent	>95% of N and > 96% of P removal	Positive feasibility for algal sustainable biomass treatment in urban wastewater	Selvaratnam et al. (2015)

that determine the metabolic conditions for the growth of microorganisms. There must be an optimum condition that should be met. For example, thermophilic anaerobic digestion (55–70°C) works better in terms of organic matter removal due to high reaction rate as compared to thermophilic anaerobic digestion (37°C). However, the former has a disadvantage in the production of more acids, which hampers the biogas production. This increases the amount of energy input and affects sustainability overall. On the other hand, the thermophilic anaerobic digestion process achieves relatively poor degradability and leads to a nutrient imbalance. Other factors aside from temperature include pH, C/N ratio, hydraulic retention time and organic loading rate (OLR), which affects biogas production (Dinamarca et al. 2003; Wang et al. 2014; Zielinska et al. 2013; Ekama and Wentzel 2008).

As anaerobic digestion has limitations in deriving energy in the form of biogas from the digested sludge, some recent innovations such as sludge pyrolysis have proved to be possible solutions in deriving/extracting more biogas both from the undigested and digested sludge. The byproduct in the form of biochar is environmentally resistant and holds its applicability in soil conditioning (Cao and Pawłowski 2012). Even mechanical and thermal pretreatment of wastewater before anaerobic digestion has been shown to improve biogas production. Thermally treated wastewater produced 20–30% more methane gas than the untreated samples. Another study utilized fruit and vegetable waste and sewer sludge where the performance of anaerobic digestion was improved in terms of biogas production (Park et al. 2011). An increase in 8–17% of biogas production was observed along with better stability than conventional sludge treatment of only sewer sludge indicating the possibility of co-digestion to increase the sustainability index. Thus, anaerobic digestion can be used for the sustainable treatment of wastewater in both isolated and hybrid/integrated formats.

26.4.2 Bioelectricity Generation from Microbial Fuel Cells

A MFC is an innovative fuel cell device that treats wastewater and simultaneously produces electricity due to an electrochemical reaction. Generally, dual-chamber MFCs are operated for this objective where wastewater is fed into the anode chamber (oxidation of organic food takes place due to the action of exoelectrogen bacteria), while the cathode chamber takes control of the reduction process giving out the treated water. Figure 26.4 shows dual-chamber MFCs with an electrochemical equation that occurs during the wastewater treatment. A conducting wire is connected to the anode and cathode of the cell for completing the circuit promoting charge (electrons) to flow and thereby producing electricity.

MFCs have been successfully and intensively studied around the world to achieve sustainable wastewater treatment. For air-cathode MFCs, no aeration is required and hence the treatment process is highly economical. It was found that around 0.025–0.075 $kWh\,kg^{-1}$-COD is consumed as compared to more than 0.1 $kWh\,kg^{-1}$-COD of electrical energy production, while the activated sludge process consumes around 0.6 $kWh\,kg^{-1}$-COD energy with no production of energy in any form (Li et al. 2013). On the other hand, the cost of establishing MFCs for the state-of-the-art practice for the treatment of domestic wastewater was evaluated to be around \$3/kg COD as compared to <\$0.5/kg COD for the conventional wastewater treatment. However, a low-cost treatment can be effected by changing to

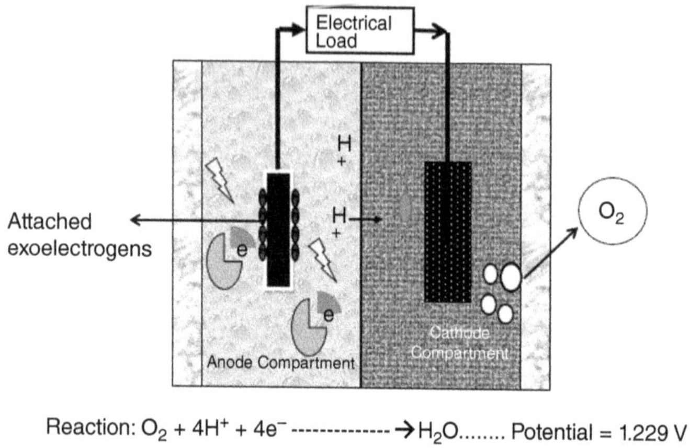

Figure 26.4 Dual-chamber MFC for wastewater treatment and simultaneous production of energy.

material such as clayware instead of plexiglass or glass. The research group of Ghangrekar et al. (Kharagpur, India) evaluated the performance of MFC with a low-cost material such as clayware using distinct clay compositions and found the performance of MFCs was very similar to, if not better than the expensive materials. The cost per unit treatment of wastewater using clayware as compared to the expensive glassware or other such fabricated reactor material has significantly reduced the capital cost over the years and provided a more sustainable approach to wastewater treatment. An added advantage of MFCs is that they comparatively produce less sludge (0.1 g VSS/g COD) as compared to 0.4–0.8 g VSS/g COD in an activated sludge system. This directly relates to the sustainability measurement and reduced energy footprint as sludge management itself amounts to around 30–40% of the WWTP economy. Moreover, electricity generation is a surplus benefit (Zhang et al. 2013). Table 26.4 shows various wastewater treatment studies with different kinds of MFCs.

Considering the sustainability criteria, MFCs can be upscaled and placed in WWTPs to enhance the electrical output and economize the treatment process, killing two birds with one stone. For example, an MFC-centered hybrid approach can be used in a WWTP, where high organic load wastewater can be first fed to the anaerobic reactor for pre-fermentation where biogas production occurs. Then, the favorable wastewater composition can be fed into the MFC system for effective utilization of organic material and simultaneous production of electricity, thereby maintaining the net positive energy balance for the WWTP operations. Anaerobic treatment using UASB before feeding wastewater into MFCs can incur biogas production in large amounts, which can be converted into electrical energy (as much as 28% electrical energy can be recovered in this way) in conjugation with a CHP system. This energy can be utilized in maintaining heat energy, which is required to run the reactor (USEPA 2007). In the case of wastewater with a high nutrient value, effluent can be first pre-treated for the struvite precipitation and then can be fed into the MFC system for energy exploration. Even though MFCs are not versatile enough to remove residual nutrients, the possibility of employing algal treatment for nutrient removal can be explored, where further energy output can be gained, increasing effluent quality in terms of environmental guidelines and the overall sustainability and energy footprint.

Table 26.4 Distinct kinds of microbial fuel cells (MFCs) for sustainable treatment of various wastewaters.

MFC type	Kind of wastewater treated	Power generation/ current generation	Other information/Remarks	Organic loading rate (OLR) studied, or degradation rate achieved	References
Open-air-cathode microbial fuel cell	Composite vegetable waste	Higher power output (57.38 $mW\,m^{-2}$)	Chemical oxygen demand (COD) removal (62.86%), carbohydrates removal (79.84%) and turbidity removal (55.12%)	2.08, 1.39 and 0.70 kg COD m^3 day^{-1} (OLR)	Venkata Mohan et al. (2010a,b)
Solid-phase MFC	Canteen based WW	Power density: 170.81 $mW\,m^{-2}$	COD removal of 64.83%	OLR1, 1.01 kg COD $m^{-3}\,day^{-1}$; OLR2, 1.74 kg COD m^{-3} day^{-1}; OLR3, 2.61 kg COD m^{-3}- day^{-1}	Goud et al. (2011)
Dual-chambered, salt bridge MFC	Food-processing wastewater	Power density: 123.8 $mW\,m^{-2}$, current density 54.3 $mA\,m^{-2}$, power yield of 110 $mW\,kg^{-1}$	COD removal of 98.9%	7760 $mg\,l^{-1}\,day^{-1}$	Sangeetha and Muthukumar (2011)
Two-compartment MFC reactor	Food-processing wastewater	Power density: 230 $mW\,m^{-2}$	COD, BOD_5 (biochemical oxygen demand), NH_3, P, (total suspended solids) TSS, VSS (volatile suspended solids), SO_4 and alkalinity were 86, 79, 73, 18, 68, 62, 30 and 58%, respectively	Studied OLR: 0.364 g COD l^{-1} day^{-1} 0.182 g COD $l^{-1}\,day^{-1}$	Mansoorian et al. (2013)
Single chambered MFC Further P_d was increased by 158% to 528 $mW\,m^{-2}$ by adding 200 mM phosphate buffer	Beer brewery wastewater	Power density: 205 $mW\,m^{-2}$	COD removal and coulombic efficiency (CE) of 87%, and 10%, respectively	Feed COD of 2240 $mg\,l^{-1}\,day^{-1}$	Feng et al. (2008)

(Continued)

Table 26.4 (Continued)

MFC type	Kind of wastewater treated	Power generation/ current generation	Other information/Remarks	Organic loading rate (OLR) studied, or degradation rate achieved	References
MFC, open-air-cathode	Real field dairy wastewater	Power density: $1.10\,W\,m^{-3}$	COD removal of 95.49%, protein removal (78.07%), carbohydrates (91.98%) and turbidity (99.02%)	$4.44\,kg\,COD\,m^{-3}$, COD, 95.49%; (digestion rate), $1.059\,kg\,COD\,m^{-3}\,day^{-1}$	Venkata Mohan et al. (2010a,b)
Catalyst-less and mediator-less membrane MFC	Dairy industry wastewater	Power density: $621.13\,mW\,m^{-2}$	CE and COD removal of 37.16%, 90.46%, respectively	$53.22\,kg\,COD\,m^{-3}\,day^{-1}$	Mansoorian et al. (2016)
Earthen pot MFC	Rice mill wastewater	Power density: $2.3\,W\,m^{-3}$	(COD) removal efficiencies of 96.5%	Around $1100–1125\,mg\,l^{-1}\,day^{-1}$	Behera et al. (2010)
Up-flow tubular microbial fuel cells	Animal carcass wastewater	Power density: $2.19\,W\,m^{-3}$	and nitrate removal: 50.66% and 79.76%	$>24\,g\,l^{-1}\,day^{-1}$	Li et al. (2013a,b)
Dual-chambered MFCs	Pharmaceutical industrial effluent	Power density: $205.61\,mW\,m^{-2}$	NA	OLR: $7.98\,kg\,COD\,l^{-1}$	Velvizhi and Venkata Mohan (2012)
Mediator-less MFCs	Paper mill effluent	Power density: $24\,mW\,m^{-2}$	$29.4 \pm 2.4\%$ Total chemical oxygen demand (tCOD) removal	$0.112\,kg\,COD\,m^{-3}\,day$ (degradation rate)	Kassongo (2011)
Single-chamber biocathode-microbial fuel cell	Real dye wastewater	Power density: $8\,W\,m^{-3}$	COD removal, almost 80%	$>500\,mg\,l^{-1}\,day^{-1}$ (feed rate)	Kalathil et al. (2012)
Two-chamber MFC	Human feces wastewater	Power density: $70.8\,mW\,m^{-2}$	COD removal and ammonia removal, >75% and 44%, respectively	$>4\,g$ COD(starch) $day^{-1}\,l^{-1}$ (Feed rate)	Fangzhou et al. (2011)

The ability of bacterial inoculum to oxidize the substrate results in the electron transfer process from the anode region toward the cathode. Upon reaching the cathode chamber/side these electrons are reduced when combined with hydronium ions and oxygen molecules (which acts as a terminal electron acceptor) in the catholyte. This reduction process is shown in the form of the chemical equation in Figure 26.4. The production of bioelectricity in MFCs is directly linked with the kinds of substrates present in the wastewater/kinds of wastewater under treatment. For example, under similar MFC design and reactor volume conditions, maximum power production values with respect to anode surface area for acetate wastewater, swine wastewater and domestic wastewater were reported to be 506, 261 and 146 $mW\,m^{-2}$, respectively (Liu et al. 2005; Min et al. 2005).

Ren et al. (2014) evaluated the performance of two-stage MFC-AFMBR (microbial fuel cells and anaerobic fluidized membrane bioreactor). It was found that the total energy production and total energy required to operate the dual system were equal (i.e. electrical energy required: 0.0186 $kWh\,m^{-3}$ as compared to 0.0197 $kWh\,m^{-3}$ of energy produced). Other such combinations such as MFC-MBR and MFC-CW (MFC-constructing wetlands) have already demonstrated successful bioelectricity production at a small scale (15.73 $mW\,m^{-2}$ for MFC-CW and 14.5 $W\,m^{-3}$ for MFC-MBR) along with wastewater treatment. This suggests that MFCs can prove to be a proper sustainable approach that can be undertaken with different kinds of wastewater undergoing treatment.

However, during large-scale MFC operations, a higher cathode surface area is required. Practically, it may be difficult to engineer such reactors. Even non-air-cathode MFCs require aeration, which becomes energetically expensive to operate. Other problems such as high power loss and voltage reversal damage the bioanode in the long run, which incurs high replacement costs. A microbial electrolysis cell (MEC) resolves the above issues and involves generation of useful compounds in the form of hydrogen (quite pure) and other power consumers. Unlike MFCs, which involve a spontaneous reaction, MECs enhances the substrate removal rates by outcompeting the aerobic heterotrophs, at the same applied potential (between electrodes) (Ren et al. 2014). By achieving this, the reactor size can be substantially reduced along with a decrease in the start-up period (Kumar et al. 2017). A recent example of a hydrogen-fueled train launched in Germany perfectly depicts the application of these fuel cells working in the direction of a sustainable future. Other countries including Canada, France, the Netherlands, Norway and Italy have already planned to bring it on track by 2025.

26.4.3 Sustainability Approach in WWTPs Related to Algal Biomass

The use of microalgae has been very popular in WWTPs for the treatment of nutrients. Cultivation of microalgae in wastewater for biofuel production utilizing the biomass produced is a sustainable approach practiced in some WWTPs (Delrue et al. 2012). The advantage of using algae is that they produce double the biomass compared to bacteria and almost 20% more energy per unit nutrient uptake. Also, the added benefit of incorporating both internal energy from wastewater and the solar energy captured through photosynthesis fuels the bioenergy production and promotes the sustainable approach. The traditional activated sludge process employed in DWTP consumes 1 kWh of electrical energy producing 0.45 kg biomass/kg BOD removal (resulting in 1 kg of fossil-CO_2 emission). In contrast,

1 kg BOD removed through photosynthetic oxygenation in the algal system requires no energy inputs. Moreover, the algal biomass growth is anaerobically digested to produce more biogas (than anaerobically digested biomass formed from the conventional aeration active sludge process), which produces 1 kWh of electrical power (Moreno-Garrido 2008). However, to enhance the sustainable approach using microalgae in WWTP, more optimization of mass culture is needed. The problem with treating high-nutrient wastewater such as swine wastewater, agricultural wastewater, etc. is that it limits the treatment efficiency of various operational units and requires more input energy, making WWTP less sustainable and costlier to manage. Also, some WWTPs fail to meet the guidelines for nitrate and phosphate levels. This is when microalgae wastewater treatment comes in handy as these nutrients are effectively utilized in the formation of algal biomass, producing biofuels and simultaneously carrying out treatment. This dual working of microalgae for phytoremediation and biomass production for sustainable biofuel production is certainly a novel option for the future.

Considering the economics of biofuel production (in the form of bioalcohol), the efficiency can be enhanced significantly using a biorefinery production strategy by feeding in all the biomass raw materials that help wastewater to be converted into useful products. This has proven a good way to minimize environmental impact and maximize economic benefits (Briens et al. 2008). Algal biomass can be digested/burned to produce fuel gas or can be biochemically fermented to produce bioalcohol, biodiesel or bio-oil. Bioethanol from algal biomass can be derived mainly by employing two methods: (i) a biochemical process, i.e. fermentation; ot (ii) a thermo-chemical process (gasification). Since microalgae are rich in nutrients, carbohydrates and protein, they can be used as a potential carbon source in the fermentation process. The spent biomass after fermentation is sent to the anaerobic digester for methane production (along with CO_2 gas) thus effectively utilizing the algal biomass that is formed from WWTPs in the tertiary treatment units. However, the process needs optimization and is yet to be commercialized on a larger scale (Harun et al. 2010). Biomethane production involves unused biomass at the end of the process, and this biomass can be used to make biofertilizers. However, the production of biogas followed by biopesticide production later depends on the organic loading that is under treatment and thus a lot also depends on the kind of wastewater treated.

High-rate algal pond (HRAP) systems have been put to use (for some decades) for the above purpose, where nutrients are effectively removed. These treatment methods require less land area, shorter generation time and lead to high biofuel production. However, the drawbacks of the difficulty in harvesting them can be overcome by proper kinetics estimation under suspended form (Moreno-Garrido 2008). These algal ponds are designed in a raceway shape including a large paddle wheel to create a channel velocity and the algal cells are kept under suspension for biomass growth by periodically exposing them to light. The most sustainable velocity that is kept for operating such ponds lies between 10 and 30 $cm\,s^{-1}$ (Richmond and Grobbelaar 1986). HRAP systems are by far the most cost-effective treatment option available for overall liquid waste management along with capturing solar energy. The advanced integrated ponds developed by Oswald LLC consumed almost 0–0.57 $kWh\,kg^{-1}$-BOD as compared to 0.80–6.41 $kWh\,kg^{-1}$-BOD for the conventional mechanical aerated pond (Oswald 1995).

Some studies have also formulated approaches of using a algal biomass treatment technique as a post-primary treatment process in WWTPs (Passos et al. 2017). The concept is that primary sludge obtained from primary treatment units in the WWTP is co-digested with the harvested microbial algal biomass (MAB) to produce biogas and bioenergy. Pilot-scale studies were conducted among four different scenarios for HRAP systems followed by: (i) co-digestion with thermal pretreatment of MAB and heat generation from boilers; (ii) anaerobic co-digestion of MAB and primary sludge; (iii) co-digestion with thermal pretreatment of MAB followed by generation from biogas in a CHP unit; and (iv) co-digestion and heat generation from biogas in a boiler. It was found that the net energy balance (energy output – energy input for operation) for scenarios ii and iii was always positive while scenario iii was the best. On the other hand, when produced biogas was only used to produce heat (i and iv), the heat requirement was duly met all year round. Thus, algal biomass production due to wastewater treatment has enough potential to produce biofuel post-digestion and could prove to be a long-run sustainable treatment approach in the future.

26.5 Conclusions

Depletion of fossil fuels, environmental pollution and overexploitation of resources such as water have led to the emphasis on sustainable development. The acute water shortage across the world has forced researchers and policymakers to look for alternative treatment systems. Conventional treatment systems are linear in nature with effluent from treatment plants ending up in the nearby water source. These systems are not sustainable. The alternative treatment strategies discussed in the previous sections tend to utilize the inherent cyclical nature of these alternative systems, allowing for recycling of water and valuable resources, therefore being sustainable. It is vital that current systems are retrofitted with alternative systems that allow for the recycling of water. Since conventional systems of treatment consume a lot of energy, the primary target may be to produce energy for operation in-house by using systems such as UASB or MFCs. This will lead to the conventional systems being less dependent on external energy sources, which will reduce the operational costs considerably. Hence, the sustainable approach will help in reducing the carbon footprint of the treatment system.

Wastewater treatment is essential to prevent contamination of water sources as well as for general public health. The selection and management of treatment strategies should be specific to the site and take into consideration social, environmental and economic conditions of the area. The most appropriate technology should be affordable, environmentally sustainable and socially acceptable. Moreover, the treatment system adopted should be flexible enough to meet the stresses of the future mainly brought about by population growth and concomitant urbanization. Energy management of the employed process in the form of modified anaerobic digestion, MFCs and algal biomass could be good sustainable treatment approaches in WWTPs. However, their scale-up studies need proper optimization before they can be employed with full confidence to treat high volumes of wastewater.

References

Angenent, L.T., Karim, K., Al-Dahhan, M.H. et al. (2004). Production of bioenergy and biochemicals from industrial and agricultural wastewater. *Trends in Biotechnology* 22 (9): 477–485.

Bdour, A.N., Hamdi, M.R., and Tarawneh, Z. (2009). Perspectives on sustainable wastewater treatment technologies and reuse options in the urban areas of the Mediterranean region. *Desalination* 237 (1–3): 162–174.

Behera, M., Jana, P.S., More, T.T., and Ghangrekar, M.M. (2010). Rice mill wastewater treatment in microbial fuel cells fabricated using proton exchange membrane and earthen pot at different pH. *Bioelectrochemistry* 79 (2): 228–233.

Boyden, B.H. and Rababah, A.A. (1996). Recycling nutrients from municipal wastewater. *Desalination* 106 (1–3): 241–246.

Bramucci, M. (2000). Industrial wastewater bioreactors: sources of novel microorganisms for biotechnology. *Trends in Biotechnology* 18 (12): 501–505.

Brebbia, C.A., Magaril, E.R., and Khodorovsky, M.Y. (2014). *Energy Production and Management in the 21st Century: The Quest for Sustainable Energy*. Southampton, UK: WIT Press.

Briens, C., Piskorz, J., and Berruti, F. (2008). Biomass valorization for fuel and chemicals production - a review. *International Journal of Chemical Reactor Engineering* 6 (1) https://doi.org/10.2202/1542-6580.1674.

Burkert, J. (2004). Optimization of extracellular lipase production by *Geotrichum* sp. using factorial design. *Bioresource Technology* 91 (1): 77–84.

Cao, Y. and Pawłowski, A. (2012). Sewage sludge-to-energy approaches based on anaerobic digestion and pyrolysis: brief overview and energy efficiency assessment. *Renewable and Sustainable Energy Reviews* 16 (3): 1657–1665.

Chen, G., Zhao, L., and Qi, Y. (2015). Enhancing the productivity of microalgae cultivated in wastewater toward biofuel production: a critical review. *Applied Energy* 137: 282–291.

Coats, E.R., Loge, F.J., Wolcott, M.P. et al. (2007). Synthesis of polyhydroxyalkanoates in municipal wastewater treatment. *Water Environment Research* 79 (12): 2396–2403.

Collet, C., Schwitzguebel, J.P., and Peringer, P. (2003). Improvement of acetate production from lactose by growing Clostridium thermolacticum in mixed batch culture. *Journal of Applied Microbiology* 95 (4): 824–831.

Dai, W., Xu, X., Liu, B., and Yang, F. (2015). Toward energy-neutral wastewater treatment: a membrane combined process of anaerobic digestion and nitritation–anammox for biogas recovery and nitrogen removal. *Chemical Engineering Journal* 279: 725–734.

Das, R.K. and Brar, S.K. (2014). Enhanced fumaric acid production from brewery wastewater and insight into the morphology of *Rhizopus oryzae* 1526. *Applied Biochemistry and Biotechnology* 172 (6): 2974–2988.

Das, R.K., Brar, S.K., and Verma, M. (2015). Enhanced fumaric acid production from brewery wastewater by immobilization technique. *Journal of Chemical Technology & Biotechnology* 90 (8): 1473–1479.

Delrue, F., Setier, P.A., Sahut, C. et al. (2012). An economic, sustainability, and energetic model of biodiesel production from microalgae. *Bioresource Technology* 111: 191–200. https://doi.org/10.1016/j.biortech.2012.02.020.

Dhillon, G.S., Brar, S.K., and Verma, M. (2012). Biotechnological potential of industrial wastes for economical citric acid bioproduction by *Aspergillus niger* through submerged fermentation. *International Journal of Food Science & Technology* 47 (3): 542–548.

Diamantis, V., Tataki, V., Eftaxias, A. et al. (2016). Geothermal energy valorisation for enhanced biogas production from agro-industrial residues. *Environmental Processes* 3 (S1): 81–90.

Dinamarca, S., Aroca, G., Chamy, R., and Guerrero, L. (2003). The influence of pH in the hydrolytic stage of anaerobic digestion of the organic fraction of urban solid waste. *Water Science and Technology* 48 (6): 249–254.

Domínguez-Espinosa, R.M. (2003). Submerged fermentation in wheat substrates for production of *Monascus* pigments. *World Journal of Microbiology and Biotechnology* 19 (3): 329–336.

EcoGeneration (2008). Powering Sydney Waters carbon neutral future. News article. July–August 2008.

Eisenberg, D., Soller, J., Sakaji, R., and Olivieri, A. (2001). A methodology to evaluate water and wastewater treatment plant reliability. *Water Science and Technology* 43 (10): 91–99. https://doi.org/10.2166/wst.2001.0589.

Ekama, G.A. and Wentzel, M.C. (2008). Organic material removal. In: *Biological Wastewater Treatment: Principles, Modelling and Design* (eds. M. Henze, M. van Loosdrecht, G. Ekama and D. Brdjanovic), 53. London: IWA Publishing.

El-Fadel, M. and Massoud, M. (2001). Methane emissions from wastewater management. *Environmental Pollution* 114 (2): 177–185.

Fangzhou, D., Zhenglong, L., Shaoqiang, Y. et al. (2011). Electricity generation directly using human feces wastewater for life support system. *Acta Astronautica* 68 (9–10): 1537–1547.

Feng, Y., Wang, X., Logan, B.E., and Lee, H. (2008). Brewery wastewater treatment using air-cathode microbial fuel cells. *Applied Microbiology and Biotechnology* 78 (5): 873–880.

Frijns, J., Hofman, J., and Nederlof, M. (2013). The potential of (waste)water as energy carrier. *Energy Conversion and Management* 65: 357–363.

Gloger, K.C., Rakocy, J.E., Conter, J.B. et al. (1995). Contribution of lettuce to wastewater treatment capacity of raft hydroponics in a closed recirculating fish culture system. In: *Aquacultural Engineering and Waste Management*. Proceedings from the Aquaculture Expo VIII and Aquaculture in the Mid-Atlantic Conference (ed. M.B. Timmons), 272–300. Washington DC: World Aquaculture Society.

Goud, R.K., Babu, P.S., and Mohan, S.V. (2011). Canteen based composite food waste as potential anodic fuel for bioelectricity generation in single chambered microbial fuel cell (MFC): bio-electrochemical evaluation under increasing substrate loading condition. *International Journal of Hydrogen Energy* 36 (10): 6210–6218.

Gude, V.G. (2015). Energy and water autarky of wastewater treatment and power generation systems. *Renewable and Sustainable Energy Reviews* 45: 52–68.

Guerra, N.P. and Pastrana, L. (2002). Nisin and pediocin production on mussel-processing waste supplemented with glucose and five nitrogen sources. *Letters in Applied Microbiology* 34 (2): 114–118.

Guterstam, B. (1996). Demonstrating ecological engineering for wastewater treatment in a Nordic climate using aquaculture principles in a greenhouse mesocosm. *Ecological Engineering* 6 (1–3): 73–97.

Harrison, S.L. et al. (2017). Toward wastewater biorefineries: integrated bioreactor and process design for combined water treatment and resource productivity. Report to the Water Research Commission, South Africa.

Harun, R., Singh, M., Forde, G.M., and Danquah, M.K. (2010). Bioprocess engineering of microalgae to produce a variety of consumer products. *Renewable and Sustainable Energy Reviews* 14 (3): 1037–1047.

Hipolito, C.N., Crabbe, E., Badillo, C.M. et al. (2008). Bioconversion of industrial wastewater from palm oil processing to butanol by *Clostridium saccharoperbutylacetonicum* N1-4 (ATCC 13564). *Journal of Cleaner Production* 16 (5): 632–638.

Hodaifa, G., Martínez, M.E., and Sánchez, S. (2008). Use of industrial wastewater from olive-oil extraction for biomass production of *Scenedesmus obliquus. Bioresource Technology* 99 (5): 1111–1117.

Hoppock, D.C. and Webber, M.E. (2008). Energy needs and opportunities at POTWs in the United States. In Proceedings of the American Society of Mechanical Engineers (ASME) 2nd International Conference on Energy Sustainability, Jacksonville, Florida, USA, August 10–14.

Jensen, P.D., Mehta, C.M., Carney, C., and Batstone, D.J. (2016). Recovery of energy and nutrient resources from cattle paunch waste using temperature phased anaerobic digestion. *Waste Management* 51: 72–80.

Jin, B., Yu, Q., van Leeuwen, J.H., and Hung, Y.-T. (2010). An integrated biotechnological process for fungal biomass protein production and wastewater reclamation. In: *Handbook of Environmental Bioengineering* (eds. L. Wang, J.H. Tay, S. Tay and Y.T. Hung), 699–721. Totowa, NJ: Humana Press.

Kalathil, S., Lee, J., and Cho, M.H. (2012). Efficient decolorization of real dye wastewater and bioelectricity generation using a novel single chamber biocathode-microbial fuel cell. *Bioresource Technology* 119: 22–27.

Kalyuzhnaya, M.G., Puri, A.W., and Lidstrom, M.E. (2015). Metabolic engineering in methanotrophic bacteria. *Metabolic Engineering* 29: 142–152.

Kassongo, J. (2011). Evaluation of full-strength paper mill effluent for electricity generation in mediator-less microbial fuel cells. *African Journal of Biotechnology* 10 (69).

Keeratiurai, P. (2013). Efficiency of wastewater treatment with hydroponics. *ARPN Journal of Agricultural and Biological Science* 8 (12): 800–805.

Kollmann, R., Neugebauer, G., Kretschmer, F. et al. (2017). Renewable energy from wastewater – practical aspects of integrating a wastewater treatment plant into local energy supply concepts. *Journal of Cleaner Production* 155: 119–129.

Kumar, G., Bakonyi, P., Zhen, G. et al. (2017). Microbial electrochemical systems for sustainable biohydrogen production: surveying the experiences from a start-up viewpoint. *Renewable and Sustainable Energy Reviews* 70: 589–597. https://doi.org/10.1016/j.rser.2016.11.107.

Li, X., Zhu, N., Wang, Y. et al. (2013a). Animal carcass wastewater treatment and bioelectricity generation in up-flow tubular microbial fuel cells: effects of HRT and non-precious metallic catalyst. *Bioresource Technology* 128: 454–460.

Li, W.-W., Yu, H.-Q., and He, Z. (2013b). Towards sustainable wastewater treatment by using microbial fuel cells-centered technologies. *Energy & Environmental Science* 7 (3): 911–924.

Liu, H., Cheng, S.A., and Logan, B.E. (2005). Production of electricity from acetate or butyrate using a single-chamber microbial fuel cell. *Environmental Science & Technology* 39: 658–662.

Mansoorian, H.J., Mahvi, A.H., Jafari, A.J. et al. (2013). Bioelectricity generation using two chamber microbial fuel cell treating wastewater from food processing. *Enzyme and Microbial Technology* 52 (6–7): 352–357.

Mansoorian, H.J., Mahvi, A.H., Jafari, A.J., and Khanjani, N. (2016). Evaluation of dairy industry wastewater treatment and simultaneous bioelectricity generation in a catalyst-less and mediator-less membrane microbial fuel cell. *Journal of Saudi Chemical Society* 20 (1): 88–100.

Massoud, M.A., Tarhini, A., and Nasr, J.A. (2009). Decentralized approaches to wastewater treatment and management: applicability in developing countries. *Journal of Environmental Management* 90 (1): 652–659.

McNabola, A., Coughlan, P., Corcoran, L. et al. (2014). Energy recovery in the water industry using micro-hydropower: an opportunity to improve sustainability. *Water Policy* 16 (1): 168.

Means, E.G. (2004). *Water and Wastewater Industry Energy Efficiency: A Research Roadmap*, 28. Denver, CO: Water Research Foundation www.waterresearchfoundation.org/research/topicsandprojects/execSum/2923.aspx (accessed February 11, 2018.

Metcalf & Eddy, Tchobanoglous, G., Burton, F., and Stensel, H.D. (2002). *Wastewater Engineering: Treatment and Reuse*, 4e. New York: McGraw Hill.

Mihelcic, J.R., Crittenden, J.C., Small, M.J. et al. (2003). Sustainability science and engineering: the emergence of a new metadiscipline. *Environmental Science & Technology* 37 (23): 5314–5324.

Min, B., Kim, J.R., Oh, S. et al. (2005). Electricity generation from swine wastewater using microbial fuel cells. *Water Research* 39: 4961–4968.

Molinos-Senante, M., Gómez, T., Garrido-Baserba, M. et al. (2014). Assessing the sustainability of small wastewater treatment systems: a composite indicator approach. *Science of the Total Environment* 497–498: 607–617.

Monlau, F., Sambusiti, C., Antoniou, N. et al. (2015). A new concept for enhancing energy recovery from agricultural residues by coupling anaerobic digestion and pyrolysis process. *Applied Energy* 148: 32–38.

Moreno-Garrido, I. (2008). Microalgae immobilization: current techniques and uses. *Bioresource Technology* 99 (10): 3949–3964.

Muga, H.E. and Mihelcic, J.R. (2008). Sustainability of wastewater treatment technologies. *Journal of Environmental Management* 88 (3): 437–447.

Nitschke, M. and Pastore, G.M. (2006). Production and properties of a surfactant obtained from *Bacillus subtilis* grown on cassava wastewater. *Bioresource Technology* 97 (2): 336–341.

Norström, A. (2005). Treatment of domestic wastewater using microbiological processes and hydroponics in Sweden. Doctoral thesis. Department of Biotechnology, Royal Institute of Technology, Stockholm, Sweden.

Oswald, W.J. (1995). Ponds in the twenty-first century. *Water Science and Technology* 31 (12) https://doi.org/10.1016/0273-1223(95)00487-8.

Park, N.D., Thring, R.W., Garton, R.P. et al. (2011). Increased biogas production in a wastewater treatment plant by anaerobic co-digestion of fruit and vegetable waste and sewer sludge – a full scale study. *Water Science & Technology* 64 (9): 1851.

Park, S.J., Park, J.P., and Lee, S.Y. (2002). Production of poly(3-hydroxybutyrate) from whey by fed-batch culture of recombinant Escherichia coli in a pilot-scale fermenter. *Biotechnology Letters* 24 (3): 185–189.

Passos, F., Gutiérrez, R., Uggetti, E. et al. (2017). Towards energy neutral microalgae-based wastewater treatment plants. *Algal Research* 28: 235–243.

phys.org (2018). Geothermal heat pumps reduce energy costs at wastewater plants. November 3, 2010, University of Illinois at Chicago, phys.org. https://phys.org/news/2010-11-geothermal-energy-wastewater.html (accessed November, 2019).

Pittman, J.K., Dean, A.P., and Osundeko, O. (2011). The potential of sustainable algal biofuel production using wastewater resources. *Bioresource Technology* 102 (1): 17–25.

Popovic, T., Kraslawski, A., and Avramenko, Y. (2013). Applicability of sustainability indicators to wastewater treatment processes. *Computer Aided Process Engineering* 32: 931–936.

Pretel, R., Shoener, B.D., Ferrer, J., and Guest, J.S. (2015). Navigating environmental, economic, and technological trade-offs in the design and operation of submerged anaerobic membrane bioreactors (AnMBRs). *Water Research* 87: 531–541. https://doi.org/10.1016/j.watres.2015.07.002.

Puyol, D., Batstone, D.J., Hülsen, T. et al. (2017). Resource recovery from wastewater by biological technologies: opportunities, challenges, and prospects. *Frontiers in Microbiology* 7: 1–23.

Rao, P.V., Baral, S.S., Dey, R., and Mutnuri, S. (2010). Biogas generation potential by anaerobic digestion for sustainable energy development in India. *Renewable and Sustainable Energy Reviews* 14 (7): 2086–2094.

Rawat, I., Ranjith Kumar, R., Mutanda, T., and Bux, F. (2011). Dual role of microalgae: Phycoremediation of domestic wastewater and biomass production for sustainable biofuels production. *Applied Energy* 88 (10): 3411–3424.

Ren, L., Zhang, X., He, W., and Logan, B.E. (2014). High current densities enable exoelectrogens to outcompete aerobic heterotrophs for substrate. *Biotechnology and Bioengineering* 111 (11): 2163–2169. https://doi.org/10.1002/bit.25290.

Richmond, A. and Grobbelaar, J.U. (1986). Factors affecting the output rate of *Spirulina platensis* with reference to mass cultivation. *Biomass* 10 (4): 253–264. https://doi.org/10.1016/0144-4565(86)90002-8.

Rios, D.A.d.S., Borba, T.d.M.d., Kalil, S.J., and Burkert, J.F.d.M. (2015). Rice parboiling wastewater in the maximization of carotenoids bioproduction by *Phaffia rhodozyma*. *Ciência e Agrotecnologia* 39 (4): 401–410.

Rose, G. D. (1999). Community-based technologies for domestic wastewater. Treatment and reuse: options for urban agriculture. N.C. Division of Pollution Prevention and Environmental Assistance, CFP Report Series, Report 27.

Sangeetha, T. and Muthukumar, M. (2011). Catholyte performance as an influencing factor on electricity production in a dual-chambered microbial fuel cell employing food processing wastewater. *Energy Sources, Part A: Recovery, Utilization, and Environmental Effects* 33 (16): 1514–1522.

Selvaratnam, T., Pegallapati, A., Montelya, F. et al. (2015). Feasibility of algal systems for sustainable wastewater treatment. *Renewable Energy* 82: 71–76.

Tubesha, Z.A. and Al-Delaimy, K.S. (2003). Rennin-like milk coagulant enzyme produced by a local isolate of *Mucor*. *International Journal of Dairy Technology* 56 (4): 237–241.

USEPA (2007). Opportunities for and benefits of combined heat and power at wastewater treatment facilities. Washington, DC: USEPA.

USEPA (2012). Evaluation of combined heat and Power technologies for wastewater treatment facilities. EPA 832-R-10-006. Washington, DC: USEPA.

Velvizhi, G. and Venkata Mohan, S. (2012). Electrogenic activity and electron losses under increasing organic load of recalcitrant pharmaceutical wastewater. *International Journal of Hydrogen Energy* 37 (7): 5969–5978.

Venkata Mohan, S., Mohanakrishna, G., Velvizhi, G. et al. (2010a). Bio-catalyzed electrochemical treatment of real field dairy wastewater with simultaneous power generation. *Biochemical Engineering Journal* 51 (1–2): 32–39.

Venkata Mohan, S., Mohanakrishna, G., and Sarma, P.N. (2010b). Composite vegetable waste as renewable resource for bioelectricity generation through non-catalyzed open-air cathode microbial fuel cell. *Bioresource Technology* 101 (3): 970–976.

Verstraete, W., Van de Caveye, P., and Diamantis, V. (2009). Maximum use of resources present in domestic "used water". *Bioresource Technology* 100 (23): 5537–5545.

Wang, R., Domínguez-Espinosa, R.M., Leonard, K. et al. (2002). The application of a generic feedstock from wheat for microbial fermentations. *Biotechnology Progress* 18 (5): 1033–1038.

Wang, X., Lu, X., Li, F., and Yang, G. (2014). Effects of temperature and carbon–nitrogen (C/N) ratio on the performance of anaerobic co-digestion of dairy manure, chicken manure and rice straw: focusing on ammonia inhibition. *PLoS One* 9: e97265.

Yang, D., Cho, J.S., Choi, K.R. et al. (2017). Systems metabolic engineering as an enabling technology in accomplishing sustainable development goals. *Microbial Biotechnology* 10 (5): 1254–1258.

Yezza, A., Tyagi, R.D., Valéro, J.R., and Surampalli, R.Y. (2006). Bioconversion of industrial wastewater and wastewater sludge into *Bacillus thuringiensis* based biopesticides in pilot fermentor. *Bioresource Technology* 97 (15): 1850–1857.

Yokoi, H., Aratake, T., Nishio, S. et al. (1998). Chitosan production from shochu distillery wastewater by funguses. *Journal of Fermentation and Bioengineering* 85 (2): 246–249.

Zeeman, G., Kujawa, K., de Mes, T. et al. (2008). Anaerobic treatment as a core technology for energy, nutrients and water recovery from source-separated domestic waste(water). *Water Science & Technology* 57 (8): 1207.

Zhang, Z.-Q., Lin, B., Xia, S.-Q. et al. (2007). Production and application of a novel bioflocculant by multiple-microorganism consortia using brewery wastewater as carbon source. *Journal of Environmental Sciences* 19 (6): 667–673.

Zhang, F., Ge, Z., Grimaud, J. et al. (2013). Long-term performance of liter-scale microbial fuel cells treating primary effluent installed in a municipal wastewater treatment facility. *Environmental Science & Technology* 47 (9): 4941–4948.

Zheng, S., Yang, M., Lv, W., and Liu, F. (2001). Study on sludge expansion during treatment of salad oil manufacturing wastewater by yeast. *Environmental Technology* 22 (5): 533–542.

Zielinska, M., Cydzik-Kwiatkowska, A., Zielinski, M., and Debowski, M. (2013). Impact of temperature, microwave radiation and organic loading rate on methanogenic community and biogas production during fermentation of dairy wastewater. *Bioresource Technology* 129: 308–314.

27

Energy Needs for Sustainable Buildings and Transportation

Rehan Khan, Suchit Deshmukh, and Ritunesh Kumar

Discipline of Mechanical Engineering, Indian Institute of Technology, Indore, India

27.1 Introduction

Energy is the key ingredient for the process of development of human civilization. All life on the planet earth is made possible by solar energy captured through plants and passed through ecosystems. The rise of human civilization is marked by innovation and development in utilizing and acquiring diverse sources of energy. The human population is increasing at a rapid pace, and it has achieved the present level of 7.2 billion (Pimentel and Burgess 2018). The rate of our development and its furtherance are unarguably dependent on how efficiently we can harvest and utilize energy obtained from different resources. Different forms of energy like electrical, thermal etc. are required for applications like buildings, transport, industries and agriculture. The energy demand has increased tremendously since the mid twentieth century because of urbanization and industrialization. According to Figure 27.1, energy consumption has increased sharply from 1990 to 2012 by approximately 114% with a mean increase of 11.87% in every year. The energy consumption increase rate was the highest (~34% increase) in 2000–2010 owning to the industrialization boom (Leahy et al. 2013). The total energy is mainly utilized in three sectors: buildings (residential and commercial), transportation and industry. Globally the energy utilized in the transportation and building sector was around 33% and 35% of the total energy intake in the year 2010 (Nejat et al. 2015). With the global increase in population, the demand for energy will increase gradually and so too the CO_2 emissions. The increasing demand for energy in the future can be met through sustainable means. Sustainability is defined as patterns of economic, environmental and social progress that meet the needs of the present day without reducing the capacity to meet future needs (Randolph and Masters 2008). The chapter gives a comprehensive review of the sustainable and renewable energy development and utilization in the two major sectors: buildings and transportation, as these two sectors cover the maximum primary energy consumption.

Sustainability: Fundamentals and Applications, First Edition.
Edited by Rao Y. Surampalli, Tian C. Zhang, Manish Kumar Goyal, Satinder K. Brar, and R. D. Tyagi

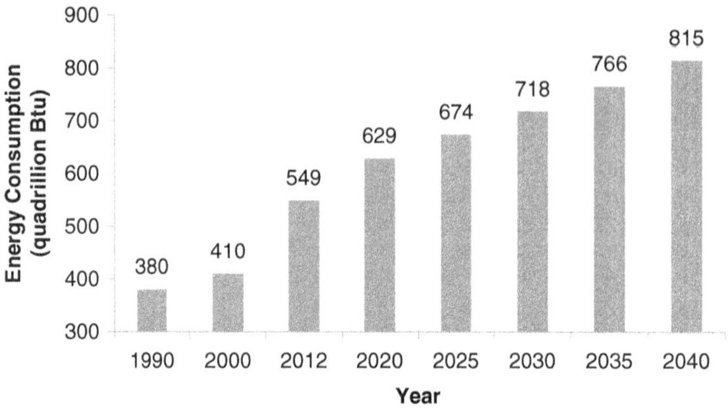

Figure 27.1 World energy consumption, 1990–2040 (Leahy et al. 2013).

27.2 Building and Energy

Building energy represents an indispensable part of the energy consumed globally. The energy intake for heating, cooling and ventilation accounts for approximately 15–20% of the energy consumed in buildings (Choudhury et al. 2010). Buildings are the largest energy-consuming sector in the world and account for an equal amount of CO_2 emissions (IEA 2013). With an expected increase in population, improvement in economy and standard of living, energy intake in the building sector is set to surge sharply, placing an additional burden on energy sector. Modern buildings consist of numerous advanced and sophisticated systems. These systems include structure, plumbing, lighting and heating ventilation and air conditioning (HVAC) systems. Air quality is one of the important parameters for maintaining productivity, health and comfort of occupants. The function of the HVAC system is to supply cool air during summer and warm air during winter. Development and promotion of low-energy buildings and zero-energy buildings is one of the emerging energy policies in many countries (Vanaga et al. 2018). The pursuit of sustainability has become the mainstream building design objective.

27.2.1 Solar Energy for Buildings

The largest demand for building energy is residential space heating and cooling followed by lighting. The design and orientation of building significantly affects the solar heat gain during summer and winter conditions. Use of window overhangs and window coatings that admit natural daylight and reject unwanted thermal gains are important techniques to minimize the cooling load and heating load.

27.2.2 Solar Angle to Help Design Overhangs

The sun rises in the east, reaches maximum height during the day time (solar noon) and sets somewhere in the west, and also the sun's height in the sky changes with latitude and with seasons. The length of overhang should be such that it allows the winter sun to enter our

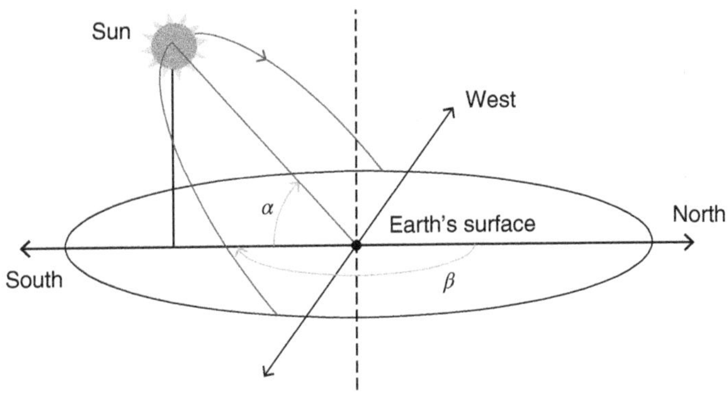

Figure 27.2 Solar angles: azimuth angle and zenith angle (Ferdaus et al. 2014).

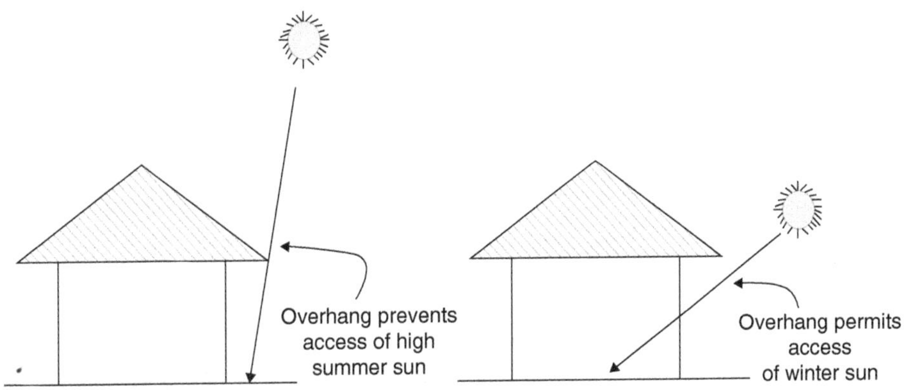

Figure 27.3 Overhangs.

home but blocks the summer sun's hot rays. The sun's height is at an angle called altitude (the angle of the sun height measured from the earth). Figure 27.2 shows the altitude and zenith angle of the sun (Ferdaus et al. 2014). The sun reaches its highest point in the sky around noon on the summer solstice (June 21, the longest day of the year) and lowest maximum height at noon on the winter solstice the shortest day of the year (December 21). To calculate the summer/winter sun altitude on the summer/winter solstice for any location the following equation is used: sun's altitude at noon on summer/winter solstice $= 90 \pm 23.5 - (\text{latitude})$. Figure 27.3 depicts that solar gains through south-facing windows are easy to control using overhangs, but east- and west-facing windows can cause overheating in the summer.

27.2.3 The Use of the Sun Path Chart

The sun path chart is a graph that shows the path of the sun as it moves across the sky during a typical day in each month. Figure 27.4 shows the sun path chart of the city of Indore. A sun path chart is essential for any site when designing building structures for light access, passive solar gain, and reducing glare and overheating. The abscissa is the azimuth or the direction that the sun is in from the observer. The ordinate is the sun's elevation or altitude.

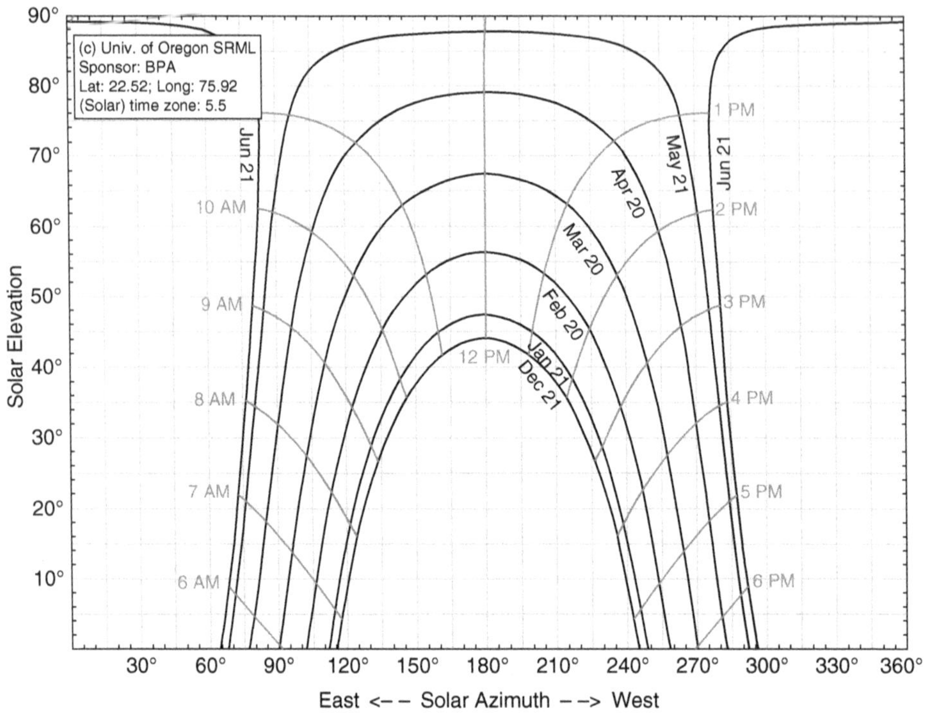

Figure 27.4 Sun path chart.

Each sun path chart is created for a specific location. To create a sun path chart for any location, the website http://solardat.uoregon.edu/SunChartProgram.html of the University of Oregon is especially handy. Sun path charts are helpful for doing quick site analysis to determine obstructions such as trees or building shadows in the proposed location.

27.2.4 Solar Radiation

Sun path and sun angle are useful, but in most cases it is essential to know the amount of solar radiation falling on a given location. The sun radiation intercepted by the earth varies according to the time, season and place. Solar radiation allows us to estimate the available solar energy for generating electricity, heating buildings and domestic water. Moreover, it is important to calculate the extra cooling load required during summer as much of the sunlight enters from the windows. Solar radiation (insolation) data have been compiled by the National Renewable Energy Laboratory in the US and are readily available on the internet.

27.2.5 Importance of Building Orientation

Wong and Fan (2013) suggested that it is vital to correctly orient a building so that it can receive a large amount of solar energy. Building location and orientation is a basic step to ensure that the building utilizes the available solar energy to the optimum. With rising

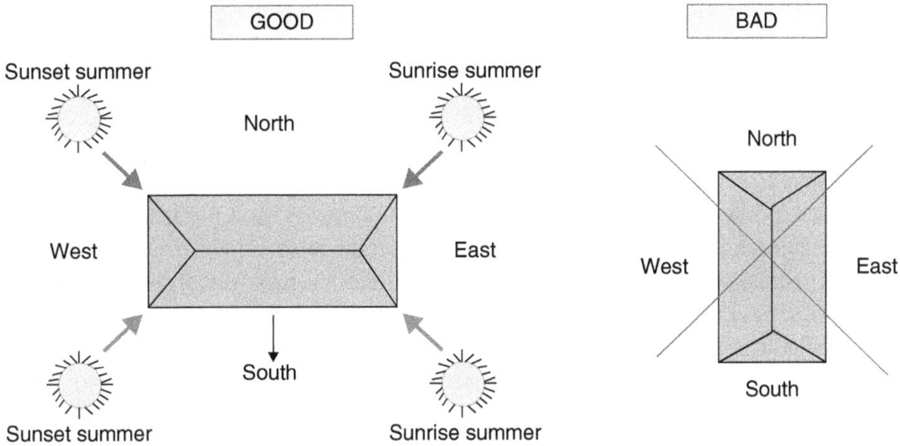

Figure 27.5 Building orientation along east–west direction (Randolph and Masters 2008).

energy costs it is becoming increasingly important for builders to orient buildings to capitalize on the sun's free energy. Properly oriented buildings will maximize the daylight through the building structure thereby minimizing the use of artificial light. Buildings that maximize daylight are ideal for implementing passive solar collection techniques to minimize electricity and enhance occupant comfort. A correct methodology can also reduce overheating and sun glare when sunlight is excessive. The relative position of the sun is a significant factor for thermal heat gain in buildings, which makes correct orientation of buildings a basis step in passive solar design. Figure 27.5 illustrates the importance of building orientation (Randolph and Masters 2008). A well-designed building makes the most of the available natural light without compromising energy efficiency. In summer it keeps cool and during winter it uses the sun's free energy to make it warm. The design and position of windows are critical to well-designed buildings. The fact that sun is higher in the sky during summer and lower in winter allows us to design and construct buildings accordingly. A building with its long axis along the east–west direction helps maximize solar gains in the winter while minimizing morning and afternoon solar exposure in the summer. Plants along the east and west sides can help control summer overheating without affecting winter solar gains. Abanda and Byers (2016) studied the impact of building orientation on energy consumption through various tools of Building Information Modeling (BIM). A correctly designed and oriented building can save a substantial amount of energy throughout its lifecycle. It was reported that a total annual energy cost saving of €878 for a period of 30 years with the best possible orientation of a building could be obtained.

27.2.6 Passive Solar Heating

Passive solar heating of building includes properly designed overhangs and reflective coated windows to provide comfort during the summer and cooling seasons. Passive solar heating is simple, cheap, reliable and affordable. It is simply encouraging the use of solar heat to pass through the window to gain the required heat. Passive solar heating maximizes the sun to the living area by aligning the length of the building along the east–west direction. This minimizes the need to heat rooms in winter and supplies more natural light. In summer when the sun is high in the sky, by designing the home with windows facing in the north direction and with proper overhangs, the home is covered with shade throughout the day. The sun does not stream into the north-facing windows, which keeps the room cool during summer. During winter the sun is at a lower height, meaning that it can penetrate through the window creating thermal gains and reducing the need for artificial heating. Once the design and orientation are right, it's important to shade it correctly so that in winter heat gain is obtained from the sun and in summer it is protected from the sun's hot rays. The basic design principles of passive solar heating are simple: (i) try to orient the building along an east–west axis to control solar gains; (ii) provide south-facing glazing systems to admit solar energy; (iii) provide well-designed overhangs to minimize natural daylight during summer; and (iv) provide adequate thermal mass to absorb excess solar energy in daytime.

27.2.6.1 Controlling Solar Gains with Better Windows

The design and selection of window glass should be such that it transmits the maximum light while blocking the thermal heat. Some of the solar radiation will bring visible light that helps to reduce the use of artificial light. The traditional approach in controlling solar heat gain in commercial buildings was to use either tinted or reflective glass. Tinted glass blocks the solar radiation by absorbing it making it difficult to sit next to it as it radiates heat; on the other hand, reflective windows eliminate the heating problem but unfortunately block most of the visible light wavelengths making it necessary to use artificial light. The most modern and sustainable approach for controlling solar heat gain is the use of smart windows. Several new windows made of different glass materials have been investigated by researchers (Alawadhi 2012; Liang et al. 2018). Liang et al. (2018) analyzed the performance of five different thermochromic windows for different climatic conditions of China. Thermochromic materials are temperature-sensitive materials that have the ability to regulate sunlight and thermal heat gain. When the temperature of the thermochromic materials becomes greater than its transition temperature, less solar radiation will enter the building. The authors reported that all different types of thermochromic windows could achieve both energy conservation and daylight. Alawadhi (2012) investigated the use of phase change material (PCM) in aluminum rolling window shutters through the finite element method. The motivation for using the PCM was to absorb solar radiation to minimize the solar heat gain. The performance of the PCM shutter was compared with aluminum foam shutters and it was found that the PCM shutter reduces heat gain by 23.29%.

27.2.6.2 Thermal Mass

Thermal mass is the ability of building material to absorb and store heat energy and radiate it when required (like a thermal battery). Heavy material has high thermal mass, which can be quite beneficial during a hot, sunny day. For example, concrete slab and masonry (high density material) walls absorb heat during hot sunny days and later on release it slowly by night, keeping the house comfortable. In winter time, thermal mass absorbs heat from the sunlight and releases it at night, keeping the home warm. It is particularly useful where there is a large difference between night and day ambient temperatures. Another promising approach to thermal mass is PCM where energy is stored as the phase changes. The state of the material changes from solid to liquid as it absorbs heat, and transforms back to liquid to give back the heat. PCM like hydrated calcium chloride absorbs 186 kJ kg^{-1} when it melts at 27°C. It can store around 18 600 kJ kg^{-1} when it changes phase at a density of 1500 kg m^{-3} (Randolph and Masters 2008).

27.2.7 Climate Change and Buildings

In the past decade, there has been a sharp rise in CO_2 emissions due to various human activities. In the period between 2002 and 2011, CO_2 emissions rose by 54% because of human activities as compared to 1990 (Mazo et al. 2012). The increase in CO_2 emissions has created many hazardous situations like global warming, rises in sea level, increase in levels of ultraviolet radiation and change in rainfall patterns (Cellura et al. 2018). The energy consumption and carbon emissions of buildings are substantial. Therefore, efficient design of buildings will play a critical part in achieving low carbon emissions. According to the European Union (EU), 36% of CO_2 emissions are related to the building sector, which was addressed in the EU Directive on the energy performance of buildings. The lifetime of a building is approximately 50–100 years, which coincides with the timeline over which the climate is predicted to change; hence, in the future there is a need to develop more resilient and ecofriendly buildings.

27.2.7.1 Green Buildings

The amount of energy used for maintaining buildings is immense. Buildings use a large fraction of energy in heating, cooling, ventilation and lighting. Compared to green buildings conventional buildings are not as efficient in reducing the required amount of energy, while contemporary green buildings are superior both in carbon saving and cost saving. Green building uses 21% less energy than conventional building (Lin et al. 2016). The concept of green buildings has become popular worldwide. The harmful effect of construction on environment significantly promotes the concept of green buildings (Ding et al. 2018). These green buildings create less harm and do better, which creates a sustainable environment for all. The US Environment Protection Agency (EPA) says "Green building is the practice of creating structures and using processes that are environmentally responsible and resource-efficient throughout a building's lifecycle: from siting to design, construction, operation, maintenance, renovation, and deconstruction." This practice expands and complements the classical building design concerns of economy, utility, durability and comfort. Green buildings are also known as sustainable or high-performance buildings. The features that can make a building green are: (i) efficient and economic use of energy, water and other resources; (ii) use of renewable energy such as solar energy and geothermal energy; (iii) comfortable and healthy indoor air quality; (iv) use of smart and non-toxic materials; (v) reduction in pollution, waste management and recycling; (vi) environment and ecological consideration in design, operation and construction of buildings; and (vii) sustainable design that can adapt with changes in the environment.

Green buildings are thermally more comfortable than conventional buildings. Hong et al. (2009) found that conventional buildings over 15 years old are half as comfortable as green buildings that are less than 15 years old. Building environment and indoor air quality affect the health of occupants. The primary cause of harmful ailments like asthma is indoor air quality and building environment (Singh et al. 2010). According to Abbaszadeh et al. (2006), green buildings have better indoor air quality compared to conventional buildings. More light obtained in green buildings increases the productivity and eliminates the need for artificial light. A study was conducted by Freed (2006) on how much sunlight enters the student-based classroom. The study concluded that student exam scores were 15% higher with classrooms lit by natural sunlight. The need for green buildings is felt globally more than ever; carbon emissions, increasing energy cost and human health are at stake. A strong environmental policy is needed to create good environments and to promote the rapid development of green buildings.

27.2.7.2 Advantages of Green Building

The advantages of green building include: (i) environmental benefits (e.g. reduction in energy consumption; reduce wastage of water and other resources; improved air quality; environmental and ecological protection); (ii) economic benefits (e.g. reduction in operating cost of systems; improvement in occupant productivity); (iii) social benefits (e.g. improvement in quality of life; better health and comfort of occupants).

27.2.7.3 Building Energy Ratings (BERs)

BER measures the energy performance of a building and is based on assessment of the energy use for space heating and cooling, ventilation and lighting for standard occupancy. The rating is expressed in the form of performance bands from A to G, with "A" being the most energy efficient and "G" being the least energy efficient.

Certain measures can be taken to improve BER and reduce annual energy bills. Some simple measures are: (i) insulating hot water tanks and pipes; (ii) installing energy-saving electrical equipment like compact fluorescent light bulbs; (iii) increasing insulation in walls and floor; (iv) replacing old inefficient systems with improved efficient systems (boilers); and (v) using renewable energy heating systems.

27.2.8 Heating Ventilation and Air Conditioning Systems

Although solar passive gains and improvement in structural design of buildings provide solar heating during winter and alleviate thermal gain during summer, still some kind of heating and cooling mechanism is needed to maintain the required comfort level. A major portion of the energy for cooling/heating load in many countries is expended in dehumidifying/humidifying the ventilation and infiltration air. Traditionally, the cooling and heating requirement is fulfilled by a vapor compression system. The fundamental components of a vapor compression system are condenser, evaporator, expansion valve and compressor. These systems are based on a closed-cycle system, where a fixed amount of working fluid known as refrigerant continuously flows through each component. A large amount of energy is consumed through the vapor compression system to achieve the required comfort condition. The energy intake for the heating and cooling system can be reduced effectively by integrating these systems with solar energy and geothermal energy.

27.2.9 Solar Collectors

Solar radiation can be utilized effectively to meet a portion of the hot water demand for domestic water heating. A solar thermal collector is a device that absorbs solar radiation and converts it into a more usable and restorable form. The solar collector transfers sun energy into thermal energy that can be used to heat water.

27.2.9.1 Flat-Plate Solar Collectors

A simple flat-plate solar collector is the heart of most hot water systems. The collector consists of a black absorber plate with a circulating tube attached at the back of the plate. The tube and the plate are kept in a casing and covered with glass to allow the ingress of sunlight. A pump or sometimes natural gravity causes the water to circulate from the storage tank to the collector. Solar flat-plate collectors tend to be more cost effective due to their simple design. Solar flat-plate collector may come with many possible options. They may have single glazing, double glazing or no glazing; or selective surfaces to absorb maximum solar radiation.

27.2.9.2 Evacuated-Tube Solar Collector

This new and quite exciting type of solar-evacuated collector consists of an absorber plate placed inside a cylindrical glass tube filled with a vacuum. The vacuum eliminates the conductive and convective loss from plate to glass, greatly improving the efficiency of the collector. To transfer thermal energy to circulating water these systems use a heat pipe filled with fluid. At higher temperatures the fluid evaporates and moves to the header position where heat is transferred to the water.

27.2.10 Solar Heat Pump System

The basic work of a solar collector is to absorb solar radiation in the form of heat. It comprises fluid running inside a pipe attached to the absorber plate. The working fluid absorbs the energy, which is transferred to a heat pump through an indirect intermediate heat exchanger. The heat pump transfers energy from a cooler region to a warmer region. In a heat pump cycle, the heating effect is obtained as the working fluid is cooled to a liquid state in condenser. Figure 27.6 shows the working of a solar-assisted heat pump. A Solar-assisted heat pump is the integration of a vapor compression heat pump and a solar thermal collector. The heat pump and solar collector are connected through an intermediate heat exchanger to transfer solar energy. The heat pump includes elements such as a condenser, expansion valve and evaporator (intermediate heat exchanger). The solar system comprises a solar panel, pump and intermediate heat exchanger. The working fluid in the solar system, e.g. water or glycol, feeds the solar energy to the refrigerant (R-134a) flowing in the evaporator of the heat pump. After evaporation, the refrigerant rejects heat during condensation in the condenser. The evaporator operates at higher temperatures, improving the coefficient of performance of the system. The advantage of combining a solar system with a heat pump is to improve the performance of the subsystem, which eventually increases the performance of the solar heat pump system as a whole. The system works in the presence of solar radiation; hence, in its absence, an auxiliary heat exchanger can be used.

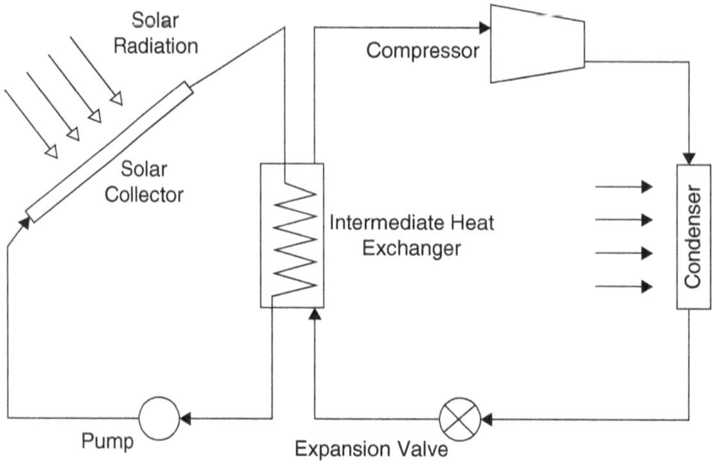

Figure 27.6 Solar heat pump.

27.2.11 Solar Domestic Water Heating

A solar water heater is a device used to heat water by trapping the infrared solar radiation. A number of options for solar heating system are available. The difference is how the water is transferred from solar collector to the storage tank. Figure 27.7 shows passive and active solar water heating systems. Thermosiphoning (passive) systems eliminate the use of pump and controller by placing the storage water tank above the solar collector. Cold water enters the collector while the hot water rises unassisted from the solar collector to the tank. The tank might be placed on the roof with the collector (closed-couple system) or within the roof cavity above the collector (gravity feed system). The system may consist of a simple flat-plate collector or an evacuated-tube collector. The evacuated-tube solar collector enjoys the advantage of receiving solar radiation orthogonally for most the day and, combined with their vacuum insulation, means better performance. The flat-plate collector, particularly used in a closed-couple system, is more affordable than evacuated-tube collectors. However, the prices of evacuated-tube collectors have decreased in recent years. In the active system, a small pump is used to move water from solar collector to tank. The location of the tank can be on the ground or within the roof. A passive system is space saving and has no moving parts, while an active system uses a pump with flexible tank position. During winter the solar collector needs to be protected when the temperature dips below the freezing point of water. The freezing water expands and can easily break the solar collector with potentially harmful consequences. Non-toxic antifreeze such as propylene glycol can be used to prevent freezing of water in winter conditions (Randolph and Masters 2008).

27.2.12 Geothermal Heating Systems

Geothermal energy means the energy available under the earth's surface. Geothermal energy is the thermal heat from the earth's crust. It is a clean and sustainable source of

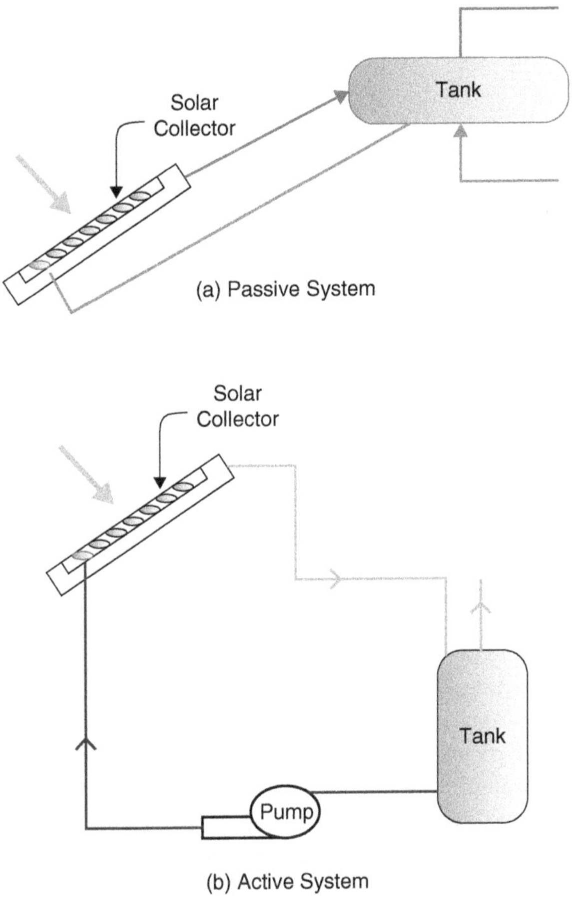

Figure 27.7 Passive and active system.

renewable energy, and it is constantly available. Heat from the earth's crust warms up the available underground fluid. In some places when the water becomes hot enough it can break through the earth's surface as steam or hot water. The atmosphere of the earth keeps on changing with respect to the weather conditions, but deep inside the layers of earth, the temperature is almost constant. Geothermal energy utilization in building will mitigate carbon emissions and will also save energy. Thus, geothermal energy can be effectively utilized to meet the future energy needs of the heating and cooling of buildings. One of the promising applications of geothermal energy is the earth-air heat exchanger. Figure 27.8 shows the schematic of an earth-air heat exchanger (ground-coupled heat exchanger). In this technology, a pipe is installed inside the earth up to certain depth. Air flows inside the pipe taking away the thermal energy from the earth, which is subsequently utilized for cooling and heating depending on the need. Depending on the requirement the fluid can be either air or water. In the case of air, it can be directly supplied to the conditioned space. The performance of the heat exchanger depends on the properties of soil (De Paepe and Janssens 2003).

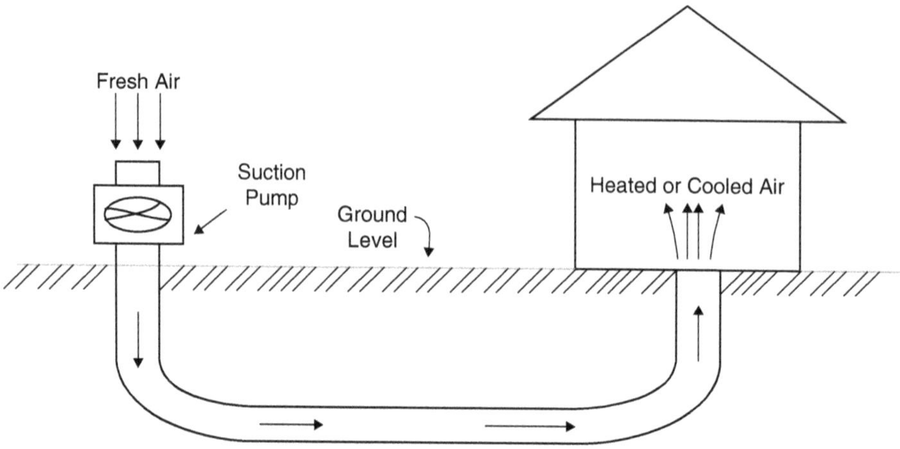

Figure 27.8 Geothermal heat exchanger.

27.3 Energy for Transportation

For sustainable development, the transport sector plays a vital role from the viewpoint of energy security as well as emission concerns. A lot of investment is being made in this sector as energy demand will rise at a faster pace because of the increasing need for transportation in the developing world. The transport sector energy demand is expected to increase to 3400 million tons of oil equivalent (mtoe); more than 80% of this energy need will be satisfied by oil as a primary source. It can be seen from Figure 27.9 that the share of oil was highest (31%) in terms of global energy demand compared to other sources (International Energy Agency 2016). It will reduce to almost 27% in 2040 but still maintaining its highest consumption status. The other renewables will see the highest relative growth in future energy scenarios. The demand for oil will rise to 103.5 mb day^{-1} in 2040 from 90.6 mb day^{-1} in 2014, out of which, around 55% was used for the transport sector in 2014 (Figure 27.10) (International Energy Agency 2016). This grows to 58% in a future scenario, with 60.4 mb day^{-1} of oil required for transport alone. Most of the increase in energy demand of the transport sector is due to road transport, though other modes of transport like rail or aviation continue to grow at a steady pace.

Crude oil is readily available fuel. There are well-established facilities for its extraction and processing. However, there is a rising concern about its availability in the near future considering its non-renewable nature, environmental and health problems due to the emissions. In addition to this, countries like India have to rely on a select few crude oil-exporting countries that control the per barrel rate of crude oil. There is a lot of debate going on regarding the future of the transportation energy supply. This brings us to the idea of sustainable development of the transportation sector. Sustainability analysis is necessary to assess the ability of different fuels to satisfy current and expected energy needs without compromising the ecological balance. This assessment should consider the environmental, economic, social and technological aspects. Efforts are being made to find new and renewable alternative fuels that not only can contribute to the demand–supply equation but also

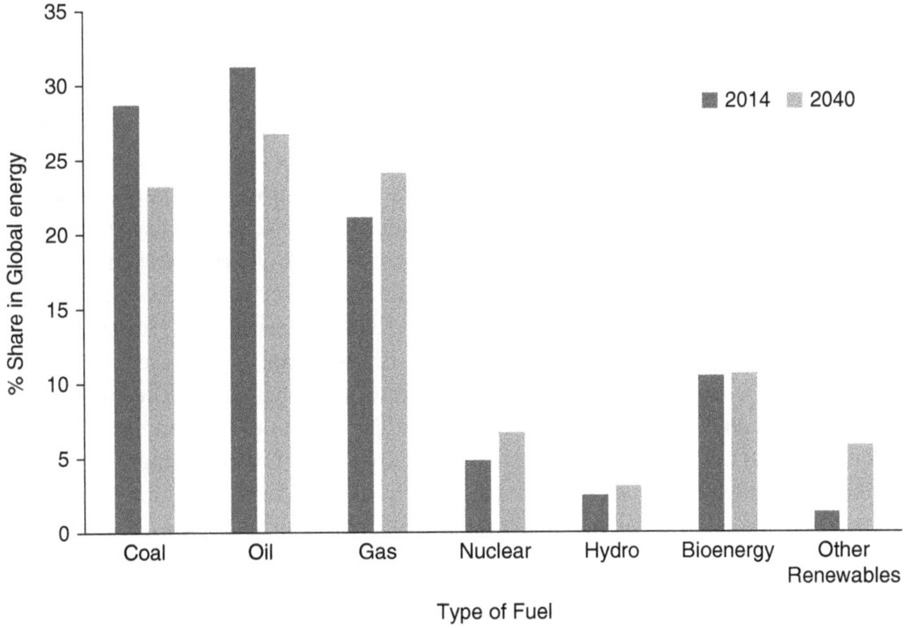

Figure 27.9 Percentage demand for different sources of energy.

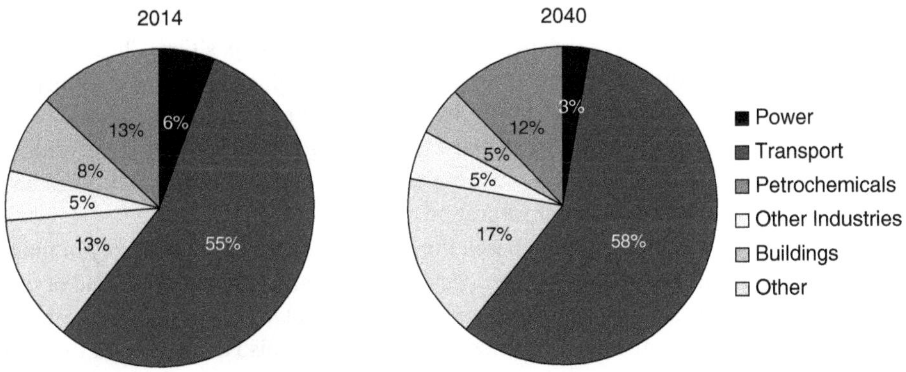

Figure 27.10 Use of oil in different sectors.

will be sustainable in nature. The government policies in this regard are crucial factors in the support of biofuel programs in countries. The demand for biofuels is expected to grow to 4 million barrels of oil equivalent (mboe)/day in 2040 – almost three times the current use. Keeping this in mind, the following paragraphs discuss different alternative sources of fuels for the transport sector considering mainly road transport. Battery-operated vehicles have also been covered briefly as they are being considered as the next big thing in the automobile industry. The pros and cons of them will help in comparisons of the future of the transport sector.

27.3.1 Biodiesel

Biodiesel is a mono alkyl ester of long chain fatty acids obtained from edible and non-edible oils as well as animal fats. It is an alternative fuel to diesel owing to similar thermo-physical properties. It is a clean, renewable fuel with less carbon monoxide, unburned hydrocarbon and smoke, net carbon dioxide and sulfur emissions (Atabani et al. 2012). It is not entirely new technology as Rudolf Diesel himself tested his engine with vegetable oil. However, cheap and readily available petroleum products became favorable, and the use of vegetable oil as a fuel remained just an idea. The current increase in crude oil prices and fear for their long-term availability has brought back the concept of use of oils as a fuel. As the oil has high viscosity compared to diesel fuel, a reaction called transesterification is used to convert oil into biodiesel. It is a reaction of triglyceride with methanol in the presence of a catalyst (acid or base), which form biodiesel and glycerol as a byproduct. The use of acid (H_2SO_4) or base (NaOH) catalyst mostly depends on the content of free fatty acids (FFAs) in the oil or animal fat. The reaction is carried out at around 60°C with continuous stirring. The conversion efficiency of oil to biodiesel is fairly high when the reaction is properly optimized for various parameters, namely ratio of alcohol to oil, catalyst used, time, temperature and mixing intensity (Meher et al. 2006).

As already stated, chemically, biodiesel is a mixture of different fatty acid methyl esters. The fatty acids can be divided broadly into saturated and unsaturated fatty acids based on whether they contain a double bond or not. The fatty acid chain length in biodiesel also varies from 12 to 24 carbon atoms. Physical properties like viscosity, density heating value, etc. will be dissimilar for different fatty acids. Saturated and unsaturated fatty acids have both favorable and unfavorable properties (Knothe 2005). Therefore, biodiesels have varying thermo-physical properties that depend on their fatty acid composition. These properties determine how well the fuel can be utilized in the diesel engine in terms of combustion as well as emission parameters. For example, biodiesel obtained from palm oil has a large percentage of palmitic acid (C16 : 0) so the saturated fatty acid content is very high. Therefore, palm biodiesel has better cetane number, density and oxidation stability but has inferior viscosity because of the higher saturation content.

Despite the various advantages of biodiesel, the source of oil for biodiesel is key in making the fuel viable as an alternative to diesel. Table 27.1 presents the potential yield of different sources of biodiesel. The biodiesel source should not only provide biodiesel in abundance so that the growing need for energy is satisfied, but also the quality of the produced biodiesel should be in accordance with the established norms like ASTM D675 or EN14214.

Initial biodiesel sources are called first generation sources of biodiesel. They are mainly edible oil plants like soybean, sunflower, etc. Since they pose a problem of food vs fuel security, it was later decided to switch to second-generation biodiesel sources like Jatropha, *Pongamia pinnata*, etc. They have the advantage that they are non-edible oil sources and can grow on waste land, therefore do not compete with food crops. They also produce biodiesel with a good mix of saturated and unsaturated fatty acids making them favorable for most of the properties. However, researchers have reported that second-generation biodiesel sources do not have the potential to satisfy the demand–supply equation of diesel vehicles. Recently, microalgae have been regarded as a third-generation source of biodiesel with vast potential to replace fossil fuel consumption

Table 27.1 Estimated oil yield from different feedstocks (Atabani et al. 2012).

Feedstocks	Oil content (%)	Oil yield (l (ha year^{-1})$^{-1}$)
First generation		
Sunflower	25–35	952
Soybean	15–20	446
Palm	30–60	5950
Second generation		
Jatropha	35–40	1892
Jojoba	45–50	1818
Calophyllum inophyllum	65	4648
3rd generation		
Microalgae	30 (low oil content)	58 700

because of the high growth rate and high oil-yielding potential. Figure 27.11 shows the main stages of biodiesel production through microalgae. Table 27.2 presents the oil content of different species of microalgae. The photosynthetic ability of microalgae is very high, consuming CO_2 at a very high rate. Thus, they help in reducing atmospheric CO_2 emitted from industries and power plants. They can be grown in water in tanks or specially designed photo bioreactors. Simultaneously, they can be grown in wastewater for phycoremediation. Such a beautiful scenario of microalgae biodiesel is hindered by some challenges in the efficient biodiesel production process. Even though the growth rate is high, harvesting of the biomass from water is difficult. The harvested biomass needs to be dried before oil extraction, which itself needs innovative techniques as the conventional expeller press oil extraction technique will not work in the case of dried microalgae powder. The advanced techniques that are applied for microalgae

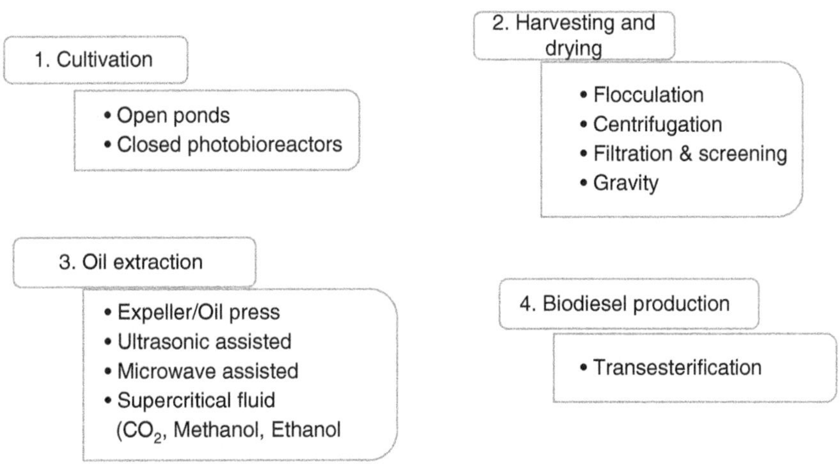

Figure 27.11 Microalgae biodiesel production process.

Table 27.2 Oil content in different microalgae species (Illman et al. 2000; Mata et al. 2010; Rodolfi et al. 2009; Yoo et al. 2010).

Microalgae species	Oil content (%)
Botryococcus braunii UTEX 572	20.8
Chaetoceros calcitrans CS 178	39.8
Chlorella vulgaris CCAP 211/11B	5.4–14.9
Chlorella emersonii CCAP 211/11N	8.1–49.9
Chlorella sorokiniana IAM-212	44.7
Isochrysis sp. (T-ISO) CS 177	37.7
Nannochloropsis CS 246	49.7
Scenedesmus sp. KCTC AG20831	20.7
Ankistrodesmus sp.	24.0–31.0
Chlorococcum sp.	19.3
Dunaliella salina	6–25
Pavlova salina	30.9

oil extraction include the use of ultrasonication, microwaves, supercritical fluid like CO_2 and ionic liquids. These processes, along with harvesting, require additional energy, which can make the net energy output of the microalgae biodiesel production process negative. Many companies are doing research and many of them are making claims about the high yield potential of microalgae biodiesel and that they can produce it economically. But almost all of them are still not able to commercialize the product to a satisfactory level. The microalgae biodiesel production process will be economical only when it is combined with other applications such as wastewater treatment, CO_2 sequestration from flue gases and production of other valuable products such as pigments.

From the sustainability point of view, many lifecycle analysis studies have been carried out to determine the net energy output of the biodiesel production process. Lardon et al. (2009) studied the microalgae biodiesel production process with normal fertilizer and nitrogen stress conditions. They also considered dry and wet oil extraction methods. They underlined that microalgae biodiesel seems promising but only if we reduce fertilizer cost and energy required for the oil extraction process. Out of the total energy consumed in the biodiesel production process in their work, about 70–90% was utilized for the oil extraction process alone. Dutta et al. (2016) also carried out a lifecycle analysis study of microalgae biodiesel and stressed the idea of reducing cost by using wastewater for microalgae growth and production of value-added co-products along with biodiesel. Oil extraction and recovery of solvent also play a crucial role in economics as well as in greenhouse gas emissions. A lot of technological advancement is needed to produce microalgae biodiesel at a large scale economically, which should also address the environmental concerns. Already stiff competition has arisen in the form of battery-operated vehicles wherein government policies and investment from many automobile companies are overshadowing biodiesel's viability.

27.3.2 Ethanol

Ethanol or bioethanol is obtained from biomass by fermentation of sugar. Its applications include use as a solvent, as an intermediate material for production of chemicals and use in the beverage industry. It is also used as a raw material in the cosmetic and plastic industries. Notably, it is one of the chemicals used in very large volumes. It is a clean-burning chemical with a high octane rating owing to the presence of oxygen. As it can be obtained from biomass, bioethanol is a renewable source of energy. Its easy miscibility with gasoline fuel makes it a suitable candidate for use as a blend in gasoline-powered vehicles. Blending 10% ethanol in gasoline will increase its octane rating by three (Sakthivel et al. 2018). This will also offset the reduced energy output because of the lower heating value of ethanol. This will make a considerable saving in fuel and allow the gasoline engine to operate at a higher compression ratio (CR) just like its diesel counterpart. The improved combustion due to ethanol addition results in lower emissions; hence, the blend is suitable for recent stringent emission norms. In current engines, up to a 20% ethanol blend in gasoline does not require any modifications. Complete replacement with ethanol as a fuel in gasoline engines will require some changes to the engines. Rather than modifying the existing engines, it is advised to specially design engines to run on ethanol fuel. Interestingly, ethanol can also be used in diesel engines, and many researchers are exploiting the use of diesel–biodiesel–ethanol blends in engines for possible reduction in emissions (Zhu et al. 2011). The miscibility of ethanol in biodiesel makes the above blend easy to form, and it remains stable for a sufficient amount of time depending on the concentration of ethanol and biodiesel. The low viscosity and low vapor pressure of ethanol lowers the overall viscosity and vapor pressure of the blend compared to that of a diesel–biodiesel blend. Therefore, it improves the fuel spray and atomization characteristics and helps in complete combustion and reducing emissions.

In India, in order to reduce the strain on foreign currency reserves and to reduce the pollution levels of highly populated cities, the government has recognized the potential of ethanol as an alternative fuel at the start of this century and accepted an ethanol-blended petrol program since 2003. Brazil adopted an ethanol fuel policy a long while back and has achieved phenomenal statistics since then. It's a leading country in terms of the use of ethanol as a fuel, with the primary source being the sugarcane industry. The USA is also a top manufacturer of ethanol; together these two countries produce 86% of total ethanol production in the world (Sakthivel et al. 2018). The total capacity of India's ethanol production is 2.3 billion l as of 2015 according to the Indian sugar mills association, lagging behind the top producing countries by a large margin. Out of this, only around 30% is used for gasoline blending. With this statistic, the country's aim of reducing oil imports by use of alternative fuels seems a distant dream.

The technology to produce ethanol is not new as it has been known for thousands of years. Ethanol is produced mainly by fermentation of sugar. Therefore, it requires a resource that has sugar, which includes material like sugar cane. Apart from this, a material like starch or cellulose can also be used for ethanol production by converting it first into sugar and then fermenting it into ethanol. This includes feed stocks like corn or grains. This requires an additional step called hydrolysis. Even though the fermentation is a simple and well-established process, use of ethanol as a fuel produced by this process is ill-advised because the raw material is costly and also it will give rise to the food-versus-fuel issue. An integrated approach where, along with ethanol, other products are also produced is a key to make the process beneficial. The fermentation requires specific microorganisms and the reaction slightly differs based on what sugar molecule is used. One mole of glucose gives two moles of ethanol and two moles of CO_2. The reaction of xylose is not straightforward and can proceed in different pathways. One mole of xylose can give two moles of ethanol, one mole of CO_2 and one mole of water. Other possible pathways are: three moles of xylose produce five moles of ethanol and five moles of CO_2; or three moles of xylose can also produce four moles of ethanol and seven moles of CO_2. Any material containing cellulose can be used for the above process. The efficiency of the production of ethanol of the above processes ranges from 40–60% on mass basis, but it also produces CO_2, whose disposal can lead to further problems as it is a greenhouse gas. The other feedstock that can be used for the above process is waste from agriculture and industries that contain sufficient amounts of cellulose or starch. The amount of such waste produced in the world is tremendous.

The other method of ethanol production uses lignocellulose material. It is a cheap source as it consists of material mostly obtained from agricultural residue and waste, forest residue and waste, industrial and municipal waste, but the conversion to sugar is complex because of the structure of lignocellulose. It consists of three components, namely crystalline cellulose, hemicellulose and lignin. The first two components, which are used for ethanol production, are attached to lignin by different bonds. This makes the structure complex and difficult to depolymerize. The lignocellulose biomass is first processed for preparing slurry by the addition of water. Drying or grinding is done if necessary. The slurry is pretreated by biological, chemical or physical means to break the complex structure to separate cellulose, hemicellulose and lignin so that the former two are available for hydrolysis. This step is expensive but essential otherwise the complex structure of lignocellulose does not permit hydrolysis easily. The hydrolysis breaks the cellulose and hemicellulose into sugars and then the production of ethanol proceeds by fermentation. Compared to the use of material like sugarcane or corn, use of lignocellulosic biomass does not produce the fuel-vs-food issue, and separate land requirement is also not necessary. The potential of lignocellulosic biomass is high for bioethanol production with an estimated value of 442 billion l every year including forest residue (Zabed et al. 2016). Still, due to lack of technological advances, use of lignocellulosic biomass for ethanol production needs further research. The production of ethanol can be summarized in Figure 27.12.

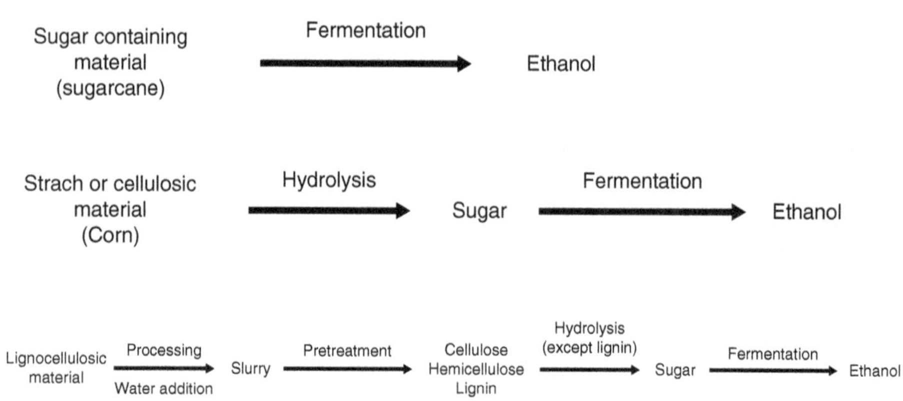

Figure 27.12 Ethanol production pathways from different feedstocks.

To make ethanol a viable alternative fuel, the energy requirement of the production process is a key factor. Lifecycle analysis studies done in the past propose ambiguous results with some expecting net energy output of the process to be positive while others are suggesting some additional steps for the process to yield positive net energy. The cost required for transport of raw material is also important as reducing it would drastically reduce the production cost of ethanol. More studies with the complete scenario considered could help solve these problems.

27.3.3 Electric Vehicles

Electric vehicles (EVs) are an alternative to traditional gasoline- or diesel-powered vehicles. A significant and straightforward advantage of EVs is an emission-free and noiseless drive as compared to traditional vehicles. It has very few moving parts, does not require frequent oil change and lubrication. This reduces the maintenance cost drastically.

The EVs are of two types, namely all-EVs and plug-in hybrid vehicles. All-EVs use only electricity as a fuel source with a battery and electric drive as two main components. The plug-in hybrid EVs use both electricity and conventional fossil fuels. Electricity can be used for the required range and when the battery is about to be fully discharged, international combustion engine mode can be switched on and the vehicle now runs as a conventional vehicle. In both these types, the battery needs to be charged by connecting it to external power source or grid. The charging time depends on what type of charger and battery the owner of the car is using. The current technological advancement in the area of battery technology is driving the vehicles to go all electric instead of hybrid vehicles. Tesla motor is doing pioneering work in this field with many of its all-electric models currently running successfully in most areas of the world. To fulfill our future energy demands for road transport, EVs are claimed to be the only possible solution by many experts and researchers. This is not only because of the development in electricity generation capacity of many countries, but also because of the mere simplicity of the EVs. After the initial hiccups, the EV industry is now taking a big leap. The power required for EVs is in the form of electrical energy. Currently, in India, most electricity generation is done by using coal. However, even if we consider that energy for EVs is coming from coal thermal power plants, the types of emissions produced are still different; their place and timing is beneficial to us considering the pollution of vehicles in the middle of the city. Environmental concerns are forcing the use of cleaner unconventional sources of energy for the generation of electricity. These sources include solar, wind, nuclear and hydropower. In a scenario in which these cleaner energy sources are used for electricity and this electricity is used for charging of EVs, the emissions of the process are near zero.

Including the maintenance cost, the operational cost of EVs is actually less than that of internal combustion engine-powered vehicles. Also, the better torque speed characteristics of the electric motors make them more responsive to drive. However, there are still some glitches that need to be addressed. The current capacity to produce batteries in the world is not sufficient to replace all the conventional vehicles. The Giga factory of Tesla may help solve this issue. The extraction of raw material used for the production of lithium ion batteries produces certain issues like land and policy conflicts, environmental problems and local labor issues. Hybrid EVs seem to be a promising alternative for now until we are ready to completely shift to fully EVs.

27.4 Conclusions

Energy security is vital for both the building and transport sector as fossil fuels do not fall under the category of renewable and sustainable fuels. The aforementioned sources of energy (biofuels, batteries, geothermal and solar energy) are slowly making their way into the energy demand for building and transport. However, there are some challenges and hurdles to overcome. The world is skeptical about whether biofuels and solar energy can actually meet our energy demands at affordable rates. The concept of green building is promising but at a fairly large capital cost. In the case of third-generation biofuels like microalgae, the economic analysis of its production stages illustrates discouraging statistics. The space requirement for solar equipment installation and biofuel production is tremendous. Therefore, we cannot afford to lose too much land to producing biofuels and renewable energy. Certain concrete steps, both research- and government policy-driven may help solve our energy problems by use of biofuels and renewable energy. EVs also have their share of disadvantages such as the battery weight is quite high for high-powered vehicles, current lack of charging infrastructure, battery production capacity and need of thermal management as batteries tend to overheat. The use of solar collectors and passive solar heating are efficient to alleviate the electricity consumption. Overall, sustainable development of the building and transport sectors is possible but there are still many miles to cover before we reach our desired goals.

References

Abanda, F.H. and Byers, L. (2016). An investigation of the impact of building orientation on energy consumption in a domestic building using emerging BIM (building information modelling). *Energy* 97: 517–527.

Abbaszadeh, S., Zagreus, L., Lehrer, D. and Huizenga, C. (2006). Occupant satisfaction with indoor environmental quality in green buildings. Proceedings of Healthy Buildings 2006, Lisbon, 3, pp. 365–370.

Alawadhi, E.M. (2012). Using phase change materials in window shutter to reduce the solar heat gain. *Energy and Buildings* 47: 421–429.

Atabani, A.E., Silitonga, A.S., Badruddin, I.A. et al. (2012). A comprehensive review on biodiesel as an alternative energy resource and its characteristics. *Renewable and Sustainable Energy Reviews* 16 (4): 2070–2093.

Cellura, M., Guarino, F., Longo, S., and Tumminia, G. (2018). Climate change and the building sector: modelling and energy implications to an office building in southern Europe. *Energy for Sustainable Development* 45: 46–65.

Choudhury, B., Chatterjee, P.K., and Sarkar, J.P. (2010). Review paper on solar-powered air-conditioning through adsorption route. *Renewable and Sustainable Energy Reviews* 14 (8): 2189–2195.

De Paepe, M. and Janssens, A. (2003). Thermo-hydraulic design of earth-air heat exchangers. *Energy and Buildings* 35: 389–397.

Ding, Z., Fan, Z., Tam, V.W.Y. et al. (2018). Green building evaluation system implementation. *Building and Environment* 133: 32–40.

Dutta, S., Neto, F., and Coelho, M.C. (2016). Microalgae biofuels: a comparative study on techno-economic analysis and life-cycle assessment. *Algal Research* 20: 44–52.

Ferdaus, R.A., Mohammed, M.A., Rahman, S. et al. (2014). Energy efficient hybrid dual axis solar tracking system. *Journal of Renewable Energy* 2014 Art. No. 629717.

Freed. E.C. (2006). As the green architect: why should I care about green building anyway? https://blogs.umass.edu/natsci397a-eross (accessed November 2019).

Hong, S.H., Gilbertson, J., Oreszczyn, T. et al. (2009). A field study of thermal comfort in low-income dwellings in England before and after energy efficient refurbishment. *Building and Environment* 44 (6): 1228–1236.

IEA (2013). Transition to sustainable buildings. Paris: IEA. https://www.iea.org/publications/freepublications/publication/Building2013_free.pdf (accessed November 2019).

IEA (2016). World energy outlook (2016). Paris: IEA.

Illman, A.M., Scragg, A.H., and Shales, S.W. (2000). Increase in Chlorella strains calorific values when grown in low nitrogen medium. *Enzyme and Microbial Technology* 27 (8): 631–635.

Knothe, G. (2005). Dependence of biodiesel fuel properties on the structure of fatty acid alkyl esters. *Fuel Processing Technology* 86 (10): 1059–1070.

Lardon, L., Helias, A., Sialve, B. et al. (2009). Life-cycle assessment of biodiesel production from microalgae. *Environmental Science & Technology* 43: 6475–6481.

Leahy, M., Barden, J. L., Murphy, B. T., Slater-Thompson, N. and Peterson, D. (2013). International energy outlook 2013. Washington, DC: EIA. https://www.eia.gov/outlooks/ieo/pdf/0484(2013).pdf (accessed November 2019).

Liang, R., Sun, Y., Aburas, M. et al. (2018). Evaluation of the thermal and optical performance of thermochromic windows for office buildings in China. *Energy and Buildings* 176: 216–231.

Lin, B., Liu, Y., Wang, Z. et al. (2016). Measured energy use and indoor environment quality in green office buildings in China. *Energy and Buildings* 129: 9–18.

Mata, T.M., Martins, A.A., and Caetano, N.S. (2010). Microalgae for biodiesel production and other applications: a review. *Renewable and Sustainable Energy Reviews* 14 (1): 217–232.

Mazo, J., Delgado, M., Marin, J.M., and Zalba, B. (2012). Modeling a radiant floor system with phase change material (PCM) integrated into a building simulation tool: analysis of a case study of a floor heating system coupled to a heat pump. *Energy and Buildings* 47: 458–466.

Meher, L.C., Dharmagadda, V.S.S., and Naik, S.N. (2006). Optimization of alkali-catalyzed transesterification of *Pongamia pinnata* oil for production of biodiesel. *Bioresource Technology* 97 (12): 1392–1397.

Nejat, P., Jomehzadeh, F., Taheri, M.M. et al. (2015). A global review of energy consumption, CO_2 emissions and policy in the residential sector (with an overview of the top ten CO_2 emitting countries). *Renewable and Sustainable Energy Reviews* 43: 843–862.

Pimentel, D. and Burgess, M. (2018). *World Human Population Problems. Encyclopedia of the Anthropocene*. Amsterdam: Elsevier Inc.

Randolph, J. and Masters, G. (2008). *Energy for Sustainability: Technology, Planning, Policy*. Washington, DC: Island Press.

Rodolfi, L., Zittelli, G.C., Bassi, N. et al. (2009). Microalgae for oil: strain selection, induction of lipid synthesis and outdoor mass cultivation in a low-cost photobioreactor. *Biotechnology and Bioengineering* 102 (1): 100–112.

Sakthivel, P., Subramanian, K.A., and Mathai, R. (2018). Indian scenario of ethanol fuel and its utilization in automotive transportation sector. *Resources, Conservation and Recycling* 132: 102–120.

Singh, A., Syal, M., Grady, S.C., and Korkmaz, S. (2010). Effects of green buildings on employee health and productivity. *American Journal of Public Health* 100 (9): 1665–1668.

Vanaga, R., Blumberga, A., Freimanis, R. et al. (2018). Solar facade module for nearly zero energy building. *Energy* 157: 1025–1034.

Wong, K. and Fan, Q. (2013). Building information modelling (BIM) for sustainable building design. *Facilities* 31 (3/4): 138–157.

Yoo, C., Jun, S.Y., Lee, J.Y. et al. (2010). Selection of microalgae for lipid production under high levels carbon dioxide. *Bioresource Technology* 101: 71–74.

Zabed, H., Sahu, J.N., Boyce, A.N., and Faruq, G. (2016). Fuel ethanol production from lignocellulosic biomass: an overview on feedstocks and technological approaches. *Renewable and Sustainable Energy Reviews* 66: 751–774.

Zhu, L., Cheung, C.S., Zhang, W.G., and Huang, Z. (2011). Combustion, performance and emission characteristics of a DI diesel engine fueled with ethanol-biodiesel blends. *Fuel* 90 (5): 1743–1750.

28

Remote Sensing and GIS Applications in Sustainability

Manish Kumar Goyal[1], Ashutosh Sharma[2], and Rao Y. Surampalli[3]

[1]*Discipline of Civil Engineering, Indian Institute of Technology, Indore, India*
[2]*Department of Civil Engineering, Indian Institute of Technology, Guwahati, India*
[3]*Global Institute for Energy, Environment, and Sustainability, Lenexa, KS, USA*

28.1 Introduction

The overexploitation of available natural resources for fulfilling the increasing demand for food, fuel and fiber of the ceaselessly increasing population has resulted in severe environmental degradation. Rapid industrialization and infrastructure development along with deforestation and the release of an enormous pool of greenhouse gases (GHGs) in different parts of the world have resulted in increasing the GHG concentration in the earth's atmosphere causing global warming and climate change (IPCC 2007, 2014). It has become extremely important to cease the ongoing overexploitation of natural resources and focus on sustainability. Sustainable use of natural resources focuses on an optimal utilization of limited available natural resources such that it does not have any harmful impact on the environment and we leave enough natural resources for future generations to fulfill their needs (Brundtland 1987). It is all about maintaining a balance between conservation and production through recognition and monitoring of critical areas (Rao 2000). This chapter presents a brief introduction to remote sensing and geographic information system (GIS) and their applications in sustainability. The role of remote sensing and GIS in earth observation (EO) and data acquisition, and sustainable resource management are discussed. The chapter concludes with a case study on the application of remote sensing and GIS for the assessment of ecosystem resilience. Each section of this chapter contains many references to related studies around the world, which allow interested readers to pursue specialist topics further if they wish.

28.1.1 Remote Sensing

Remote sensing is the science or process of monitoring and detection of different features on the earth's surface and their geo-bio-physical characteristics without making actual physical contact while using electromagnetic radiation (EMR) as the medium of interaction. In contrast to on-site observation, there is no physical contact with the objects in

Sustainability: Fundamentals and Applications, First Edition.
Edited by Rao Y. Surampalli, Tian C. Zhang, Manish Kumar Goyal, Satinder K. Brar, and R. D. Tyagi

remote sensing. It has been widely used in different fields such as geology, ecology, geography, hydrology, land surveying and natural resource management (Chebud et al. 2012; Cracknell and Hayes 2007; Prakash 2000; Xie et al. 2008). It also plays an important role in military, intelligence, commercial, economic, planning and humanitarian applications (Cracknell and Hayes 2007). Over the past few decades, remote sensing has seen a phenomenal advancement due to the significant increase in the number of EO satellites, the improvement in tools and processing methods, and the availability of huge amounts of data for new applications (Prakash 2000). Remote sensing generally utilizes an artificial satellite or an aircraft-based sensor to detect features on the earth's surface. Through the analysis of the characteristics of the propagated signals (i.e. frequency and wavelength of EMR), different objects on the surface of the earth or in the atmosphere or in the oceans can be studied. Though remote sensing achieved tremendous success after it started using artificial satellites, the use of remote sensing to gather information dates from long before the launch of the first artificial satellite (Cracknell and Hayes 2007). Apart from satellites, there are many ways to collect data, such as aerial photography using balloons or aircraft. Remote sensing includes the recording of reflected or emitted energy (i.e. EMR) and then processing and analysis of the data. Based on the source of EMR, remote sensing is classified into "active" and "passive." "Active" and "passive" remote sensing have different applications. Passive remote sensing is used in most of the remote sensing systems used these days. It uses solar energy as the source of EMR. Passive remote sensors include spectrometers, radiometers, spectroradiometers and imaging radiometers. Active remote sensing systems have an inbuilt source of energy (EMR) to illuminate the object or scene they observe. These systems include lidar (light detection and ranging), Scatterometers, radar (radio detection and ranging) and laser altimeters. The applications of remote sensing in EO and data acquisition are discussed in Section 28.2.

28.1.2 Geographic Information System (GIS)

GIS refers to a computer system that is designed for acquiring, storing, querying, studying, managing and visualizing spatial/geographic/geospatial data (Chang 2017). Geospatial data comprises both the geographic location and the characteristics of spatial features. In addition to geospatial data, GIS has other components, which include hardware, software, people and organizations. GIS is used by people from different sectors such as individuals, different organizations, research institutes, researchers, health societies, companies and government organizations at all levels. GIS has different applications, such as remote sensing applications, natural hazards assessment, natural resource management, land-use or forest planning, hydrology, waste management, wildlife management and riparian zone regulation. In addition, GIS is also used in transport applications, military operations, public health and emergency planning. Due to the geospatial nature of the input data in water resources studies, GIS has been widely utilized for hydrological modeling (DeVantier and Feldman 1993). GIS offers a digital geospatial depiction of watershed features using digital elevation models (DEMs), land use and land cover (LULC) and soil maps. In the case of surface water, groundwater, wastewater or storm water systems, most of the information in water resources planning and management comes with geographic attributes, which makes use of GIS unavoidable (Shamsi 2005). According to the American Water Works

Association (AWWA), about 90% of the water-related works in the USA utilized different GIS technologies in 2000 (Shamsi 2005). The practice of GIS has been growing and the number of GIS users has increased significantly. Advantages of GIS as discussed by Shamsi (2005) include:

- GIS technology is faster compared to conventional tools and, therefore, it helps in increasing productivity and efficiency. This results in saving both time and money. GIS applications are time- and cost-efficient compared to conventional methods (EPA 2000).
- The main strength of GIS lies in its capability to assimilate different types of data with a geographic component. It provides integrated solutions for different areas.
- As GIS can handle different types of data in conjunction with geographic information, it can be effectively used as a decision-making tool (Kadiyala et al. 2015; Malczewski 1996; Pereira and Duckstein 1993).
- GIS helps in better communication and cooperation among various stakeholders and better project management (Sui and Goodchild 2001).

Shamsi (2005) has discussed various applications of GIS in water resource systems. Sui and Goodchild (2001) presented GIS as media and indicated that GIS has progressively become a communication medium for providing certain information to the public. Likewise, there are numerous applications and advantages of GIS in different fields.

28.1.3 Sustainability

The concept of sustainability dates back to 1713, when the word *nachhaltigkeit* (the German term for sustainability) was first used by a German scientist, Hans Carl von Carlowitz, working in the field of forestry (von Carlowitz 1713; Wilderer 2007). In forestry, the concept of sustainability has been an accepted principle since the eighteenth century (Wiersum 1995). Sustainability as a policy concept started with the Brundtland Report of 1987 (Brundtland 1987). It states sustainable development is (Brundtland 1987): "development that meets the needs of the present without compromising the ability of future generations to meet their own needs." After the release of the Brundtland Report ("Our Common Future"), the principle of sustainable development has achieved international acceptance for policymaking. Ever since, the concept of sustainability has been widely used in almost all fields of science and technology (S&T), including socio-ecological systems (Ostrom 2009; Shrivastava 1995), agriculture (Tilman et al. 2002), hydrology and water resource management (Ajami et al. 2008; Bobba and Bengtsson 1997; Loucks 2000), business and marketing (Belz and Peattie 2009) and tourism (Mowforth and Munt 2015).

Sustainability is the quantitative and qualitative continuity in the utilization of resources (Rao 2000). Sustainability could be defined as (Gallopín 1996):

$$V\left(O_{t+1}\right) \geq V\left(O_t\right) \tag{28.1}$$

where V is a value function of the system's outputs and t is the time. The role of sustainability is not only in global issues such as global warming, deforestation, natural resource management and ozone layer degradation, but it is also applicable to local or regional issues such as sustainment of the eco- and socio-eco-systems or a combination of these (Graymore et al. 2008; Rao 2000). It is essential to study sustainability at finer spatial scales (i.e. regional or local) to accomplish our goal of global sustainability (Graymore et al. 2008). Nidumolu et al. (2009) clearly indicated that there is no alternative to sustainable development. It has been realized long ago that S&T has an important role in sustainability to ensure that future generations are able to fulfill their needs (Malaska and Kaivo-oja 1996). The contribution of S&T to sustainability has increased over the time with the involvement of several researchers, practitioners, scientific academies and development organizations from around the world (Clark and Dickson 2003). In the research and development, various efforts have been made to promote sustainability, which extend over different fields and ways such as ecosystem resilience, industrial ecology and energy systems (Clark and Dickson 2003; Holling 1973).

28.1.4 Role of Remote Sensing and GIS in Sustainability

Both remote sensing and GIS are widely used tools for natural resource management (Dahdouh-Guebas 2002). These are generally integrated to study a variety of natural resources and their characteristics (Navalgund et al. 2007; Patil et al. 2014; Rajitha et al. 2007; Saraf and Choudhury 1998; Wilkinson 1996). Major advancements have been accomplished in integrating remote sensing and GIS (Wilkinson 1996). Advanced techniques such as machine learning, neural networks and knowledge-based systems have revolutionized the extraction and analysis of geographical information from the remotely sensed satellite imagery (Wilkinson 1996). Both remote sensing and GIS complement each other in the following ways (Wilkinson 1996):

- Remote sensing is used for acquiring information for GIS. For example, satellite imageries from different multispectral sensors (such as Landsat) are widely used for LULC mapping in a GIS environment.
- GIS datasets are generally used as supportive information to ameliorate the quality of remote sensing products.
- Both remote sensing and GIS datasets can easily be coupled for environmental modeling and analysis.

Therefore, remote sensing and GIS have a close relationship. Their integration provides information on the spatial variation, extent, and potential and limitations of natural resources, which is a prerequisite for planning the strategy for sustainable development (Rao 2000). They have been used together for generating information on ecosystem functioning (Kasischke et al. 1997; Tang et al. 2017), soil characterization (Mu et al. 2007b), surface water (Bastiaanssen et al. 2000; Mu et al. 2009), groundwater (Minor et al. 1994), and LULC and forest cover mapping (Hansen and Loveland 2012; Roy and Giriraj 2008; Sudhakar Reddy et al. 2016). Rao (2000) discussed the different aspects of the role of remote sensing and GIS in sustainable development. Remote sensing enables the continuous monitoring of a large area at regular intervals with spaceborne sensors in a time- and cost-efficient way. In addition, it also enables us to obtain information for remote locations, which are otherwise inaccessible. It has been found to be effective in keeping track of snow cover, glacier extent and runoff simulation in high elevated Himalayan regions (Immerzeel et al. 2009). Furthermore, GIS provides a supportive environment for the integration of spatial and attribute data, which is further used for a variety of applications. Most hydrological or ecological models are developed in a GIS framework. It is used for planning and management of natural resources and decision making (Malczewski 1996). Campagna (2006) has discussed the role of GIS in sustainable development, which includes sustainable development planning, natural and cultural heritage, urban planning, society and environment and public participation. Remote sensing and GIS technologies offer effective and reliable tools for the analysis, management, problem solving, planning, policymaking and decision making needed to pursue the goal of sustainability.

28.2 Remote Sensing for Earth Observation (EO) and Data Acquisition

EO refers to the collection of information about the earth's biological, chemical and physical systems. It is used for continuous monitoring of the natural environment and resources. With the increasing number of EO satellites and advancement in the technology, EO has become increasingly sophisticated. Many countries now have their own EO satellites. More than 320 EO satellites have been launched globally since 1962 (Huadong 2013), covering the land, atmosphere and oceans. In its early days, EO was mainly focused on defense applications, but it has now expanded to various fields. EO now plays a crucial role in different fields such as meteorology, natural resource management, natural disasters, the health sector and defense. The first land monitoring sensor, Landsat, was launched in 1972, which has produced more than 2 million images so far (Kansakar and Hossain 2016). The United States Geological Survey (USGS) publicly released the Landsat archive in 2008, which makes it the largest compilation of remote sensing imagery. Landsat has been helping researchers in developing countries in solving regional problems using remote sensing tools. With the launch of Landsat-4 in 1982 began the current Thematic Mapper (TM) class, which continued with Landsat-5 in 1984 and Landsat-7 ETM+ (Enhanced Thematic Mapper Plus) in 1999 (Markham et al. 2004). Both Landsat-5 and Landsat-7 are operational and still providing data.

The focus of today's EO is not limited to mere curiosity or pure research, but to solving the most threatening problems. It includes problems related to water, food security, natural disasters, climate change, human health and habitat endangerment. EO has a dramatic impact on human society due to its capability of minimizing negative impacts and improving social and economic wellbeing. Applications of EO in different fields are presented here along with a review of relevant past studies.

28.2.1 Land Use and Land Cover (LULC) Mapping

Satellite images are widely used to prepare LULC maps, which are effective tools for planning, management, and utilization of land resources. These maps also provide information on human-caused alterations to the earth's surface and land cover. LULC maps are crucial to identifying the causes and consequences of climate change for hydrological systems, carbon dynamics, sequestration, ecosystem productivity, biodiversity, urbanization and population. Studies in the past have claimed that humans have altered 39–50% of the earth's land surface, and the concentration of atmospheric CO_2 has increased by about 30%

from the commencement of the Industrial Revolution (Vitousek et al. 1997). It should be noted that global measurements of LULC changes or land transformation are challenging or nearly impossible. Remote sensing-aided EO plays inevitable roles in such studies. Landsat remains the most consistent and reliable source of satellite images for LULC mapping, and has a four-decade collection of images. Sterling and Ducharne (2008) created a large dataset of global land cover change through the synthesis of satellite imagery, ecological modeling and country surveys. The study further showed that anthropogenic land cover sums about 40% of earth's surface, and nearly 92% of natural grasslands have been converted for human use, which mainly includes conversions to grazing land and cropland. Xiuwan (2002) used remote sensing and GIS for analyzing the land cover change from multitemporal satellite data (Landsat 4 TM), which was studied with regard to the impacts on sustainable development. Tian et al. (2014) used district and state-level historical records along with remote sensing datasets (Resourcesat-1) to produce a LULC dataset for the period 1880–2010 in India. The study found a huge decrease in forest cover (from 89 million ha to 63 million ha). All these studies were based on remote sensing datasets.

28.2.2 Carbon Biomass and Vegetation Productivity Assessment

Vegetation or forest biomass is a major carbon sink and therefore has a major role to play in climate change mitigation (Melillo et al. 1993). Estimation of biomass is important for carbon sequestration, as it is believed that about half of the forest dry biomass is carbon (Vicharnakorn et al. 2014).

Some of the resources that are used for carbon biomass assessment include Shuttle Radar Topography Mission (SRTM), Landsat imagery, EOS (Earth Observing System), MODIS (Moderate Resolution Imaging Spectroradiometer), etc. (Kansakar and Hossain 2016). Recently, EO has also been used in the continuous monitoring of global vegetation productivity. Running et al. (2004) presented estimates for annual net primary production (NPP) and gross primary production (GPP) over 109 782 756 km^2 of vegetated land surface using MODIS. Vicca et al. (2016) used the Enhanced Vegetation Index (EVI) to estimate the effect of extreme droughts on GPP based on a remote sensing dataset. Zhang et al. (2017a,b) developed a global GPP dataset for 2000–2016 using eddy covariance data, MODIS and NCEP Reanalysis II climate data. MODIS products have been widely used for vegetation productivity assessment and are improved from time to time (Zhang et al. 2017a,b; Zhao et al. 2005). Before MODIS, the Advanced Very High-Resolution Radiometer (AVHRR) was used for estimating vegetation productivity utilizing the Normalized Difference Vegetation Index (NDVI) (Myneni and Los 1995).

28.2.3 Hydrology and Water Resource Management

Water security is a major challenge for water resource managers in both developed and developing countries. Remote sensing-based EO is used in river basin management, hydrologic parameter observations and drought monitoring. In hydrologic monitoring and modeling, the applications of remote sensing are in LULC and catchment characteristics, estimation of soil moisture, detection of surface and subsurface water, estimation of the depth of water bodies, detection and estimation of snow cover and glacial lakes, estimation of water quality parameters such as suspended sediments, temperature and oil spills, real-time flood monitoring, etc. (Schultz and Engman 1991). Multitemporal remote sensing data is effective in studying the drought conditions over a region (Tucker and Choudhury 1987). Droughts generally spread over a larger spatial extent compared to other hydroclimatic disturbances and, hence, are difficult to monitor through on-site observation, especially in developing nations where the network of measuring stations is scarce. Moderate- or high-resolution satellite data are helpful for monitoring and detection of drought (Kogan 1995). Remote sensing is also used for the supervision of water quality parameters such as the presence of suspended sediment or chlorophyll in water, and its temperature. Different types of sensors (optical or thermal) are placed on aircraft, boats or satellites for spatiotemporal monitoring or assessment of water quality parameters to formulate management policies to improve the water quality (Ritchie et al. 2003). Hyperspectral remote sensing has been shown to be effective for the assessment of water quality of estuarine and coastal waters (Brando and Dekker 2003).

28.2.4 Agriculture and Food Security

Remote sensing observations provide information about vegetation growth, greenness, crop water requirement and crop yields. Remotely sensed information has demonstrated its applicability in understanding the soil–vegetation–atmosphere–transfer process, which improves our knowledge of the relationship between crop growth and water management (Bastiaanssen et al. 2000). Data provided by satellites related to environment parameters at a global scale, including precipitation, ocean productivity, vegetation health, temperature and soil moisture conditions, are used to better understand the impact of climate on agricultural productivity (Brown 2015). Remote sensing-based vegetation indices have been developed over the years to detect vegetation growth characteristics such as NDVI, EVI and leaf area index (LAI) (Fang et al. 2013; Revadekar et al. 2012). The relationship of these indices to climatic factors strengthens our knowledge of climate–soil–vegetation interactions (Kumar and Rao 2013). Many new crop-specific or region-specific vegetation indices are developed using MODIS or hyperspectral datasets (NASA 2016). Droughts are closely linked with the reduction in food production or agricultural drought (Zhang et al. 2017a,b).

28.2.5 Disaster Management

During periods of disaster management, remote sensing plays a vital role in developing warnings and mitigation strategies. Obtaining ground observation and real-time information during such calamities is challenging. Satellite EO has an exceptional potential to assist in providing ground information. Pre and post-disaster monitoring of disasters like landslides, floods and earthquake are carried out using remote sensing and GIS. Advanced techniques in conjunction with continuous images from satellites enable real-time monitoring of flood inundation, cyclones, volcanic eruptions and forest fires. Advancement in the remote sensing of atmospheric circulation has enabled us to accurately predict the trajectory of a cyclone in advance. Development of glacial lakes due to landslides or tectonic activities can also be monitored using real-time satellite observations, which help in the prediction of glacial lake outburst floods (GLOFs). Tralli et al. (2005) summarized the application of satellite remote sensing in different disasters such as floods, earthquakes, landslides, volcanos and coastal inundation.

28.2.6 Oceans and Atmosphere

Oceans cover a vast area of earth and play a significant role in maintaining life on earth. Information about the oceans is crucial to understand different oceanic and atmospheric processes. Different remote sensing techniques such as optical, thermal and microwave remote sensing are used to collect information related to chemical, physical and biological characteristics of the water (Rajeesh and Dwarakish 2015). The parameters that are quantified using remote sensing include color, temperature, chlorophyll, bathymetry, suspended particulates, surface winds, surface roughness, surface winds and wave height. In addition, remote sensing is also used to identify the Potential Fishing Zone (PFZ), which helps in catching fish on a larger scale (Devi et al. 2015). Remote sensing is also used to monitor air quality parameters including aerosols, tropospheric ozone, tropospheric nitrogen dioxide, carbon monoxide and sulfur dioxide (Martin 2008). Global CO_2 observations used in climate change studies are taken using satellite data. For example, Gupta et al. (2006) used MODIS observations of aerosol optical thickness (AOT) retrievals along with observed records of the mass concentration of $PM_{2.5}$ to measure particulate matter and air quality over global cities including Hong Kong, Delhi, New York, Switzerland and Sydney.

28.3 Sustainable Resource Management Using Remote Sensing and GIS

As discussed in the previous sections, remote sensing and GIS have become important tools for the collection of information about different earth systems, analysis and modeling of the acquired information, and decision making. In this section, a review of applications of integrated remote sensing and GIS in sustainable resource management is presented.

Burrough (1986) described the principles of GIS for applications in environmental and natural resource inventory and analysis. Goodchild et al. (1992) discussed the issues faced during the early days of the integration of GIS and spatial data analysis (SDA). The discussion focused on three main topics: (i) the advantages of integration of GIS and SDA; (ii) identification of a specific area of SDA to be integrated with GIS and (iii) how the integration should be carried out. Wilkinson (1996) reviewed the different issues in the integration of GIS and remote sensing. The author highlighted the use of knowledge-based systems and neural networks in improving GIS and remote sensing systems. Dahdouh-Guebas (2002) used remote sensing for the sustainable management of tropical coastal ecosystems. The study showed that remote sensing could be integrated with other scientific tools over the long term to both look back at and predict the deprivation of coastal ecosystems and to take preventive measures at an early stage. Monitoring of the earth's surface has a multidisciplinary character and allows a wide range of issues to be solved (Napryushkin and Vertinskaya 2006). Napryushkin and Vertinskaya (2006) discussed remote sensing-based thematic mapping methods for the collection of ecosystem data, such as surface waters, vegetation canopy, soils and anthropogenic landscape components. Geodemographics and lifestyle data are scarcely used for planning and resource management applications, but they have valuable information (See and Gibson 2006). See and Gibson (2006) utilized these datasets in conjunction with GIS for sustainable development. Rashed (2006) presented the concept of sustainable hazard mitigation and proposed an empirical GIS and remote sensing-based methodology for understanding spatiotemporal patterns of vulnerability in urban areas. Mohammad et al. (2007) used remote sensing and GIS for assessment of the land use in Kuala Selangor, Malaysia for sustainable natural resources. The maps of vegetation density, soil suitability and erosion risk were overlaid using GIS. Henderson and Lewis (2008) reviewed the application of radar in the detection of wetlands and discussed the diversity in radar applications in terms of wavelength, incident angle, polarization, environment and analysis techniques. They concluded that radar technology helped in the study and protected the fragile wetland environments better than multispectral sensors, through its ability to collect data during the day as well as nighttime, and under all weather conditions. Dewan and Yamaguchi (2009) assessed the LULC alterations and urban growth in Greater Dhaka, Bangladesh, over the period of 1975–2003 using remote sensing data and socio-economic records to promote sustainable urbanization. They reported that hasty growth in urbanization through the clearance of vegetation and the infilling of low-lying areas had resulted in environmental degradation. Chowdary et al. (2009) pointed out that integrated sustainable water resource development requires different types of interrelated information. Remote sensing can provide valuable and up-to-date information about natural resources, whereas GIS is capable of integration and analysis of different formats of remote sensing and conventional information. They adopted the Integrated Mission for Sustainable Development (IMSD) for generating explicit watershed development plans for Mayurakshi watershed. Duro (2012) used various advanced tools from the field like geomatics, remote sensing and GIS to make an analytical framework for the sustainable development of agricultural environments. McLaughlin and Kinzelbach (2015) related global food security to sustainable resource management. They highlighted that, though on a global scale there is not much shortage of water and land for agriculture, local shortages pose a serious threat to food security. Increased agricultural production will have adverse environmental impacts such as unsustainable depletion of water and soil resources. Jacobs et al. (2016) evaluated and compared the role of scientific information and participatory governance in decision making in four basins spread across Brazil, Mexico, Thailand and the USA for sustainable water resource management. The study suggests that the water resource management institutions in every basin have evolved in the past few decades. The evolution is from centralized state- or national-level water management structures to a basin-scale decision structure involving water users.

28.4 Case Study

28.4.1 Introduction

This case study presents the analysis of the resilience of terrestrial ecosystems to hydroclimatic disturbances in northeast India. The resilience concept focuses on the ability of an ecosystem to withstand disturbances and maintain its functioning under disrupted environments (Holling 1973). It can also be presented as the ability to sustain in the face of alteration and to endure dynamic environments (Folke 2016). It indicates the ability of a system to acclimatize or even change to a new equilibrium state (regime) under the changing environment. Many researchers see the resilience concept as a subclass of sustainability science (Anderies et al. 2013; Walker and Salt 2006). Marchese et al. (2018) highlighted that the concepts of resilience and sustainability have many similarities, and there is no clear difference in the meaning and purpose of both. The concept of sustainability takes into account environmental, social and economic considerations, whereas resilience describes the system's ability to prepare for disturbances, withstand its impacts and to recover after the disturbance. The main similarity is that both approaches deal with the persistence of a system to sustain its normal functionality over time in response to disturbances (Marchese et al. 2018).

28.4.2 Study Area and Data

28.4.2.1 Study Area

The study area was chosen as the seven northeastern states of India (Figure 28.1). The states include Arunachal Pradesh (AR), Assam (AS), Manipur (MN), Meghalaya (MG), Mizoram (MZ), Nagaland (NG) and Tripura (TR). Figure 28.1 shows the LULC of northeast India. AR, NG, MN and MZ have vast forest cover of more than 80% of the geographical area of the states (Figure 28.2a). MZ has the largest forest cover – over 89% of the area. Assam (AS) has only 21% forest cover, whereas the state has 56% cropland. There is a substantial difference between the states in terms of precipitation (Figure 28.2b). The mean annual precipitation ranges from 1614 mm in MN to 2954 mm in MG. Figure 28.2c shows the spatial variation of the mean annual precipitation.

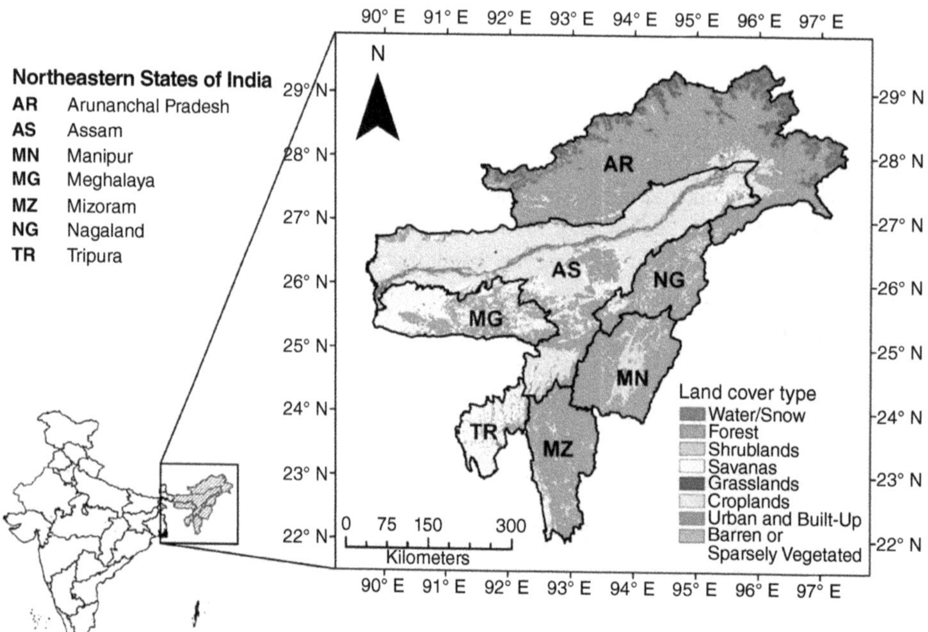

Figure 28.1 Seven northeastern states of India considered for this study. The map shows the land use and land cover (LULC) of the study area.

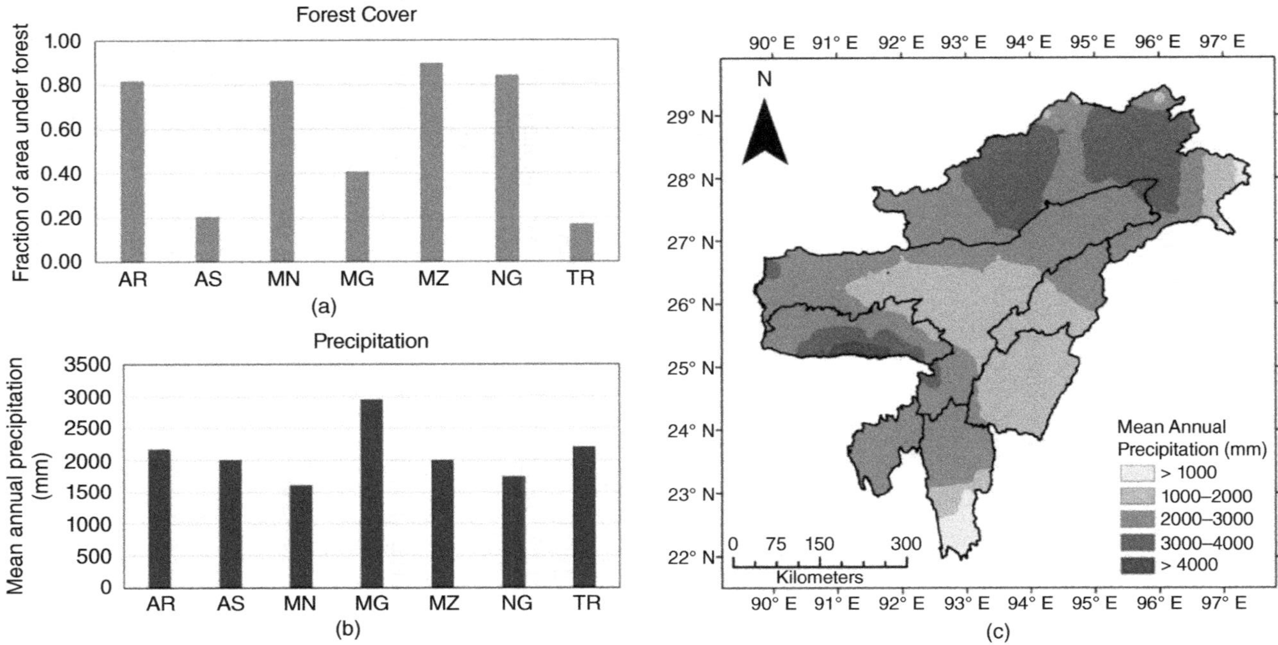

Figure 28.2 (a) Forest cover, (b) mean annual precipitation and (c) spatial distribution of precipitation in seven northeastern states.

28.4.2.2 Data

Due to the unavailability of ground measurements, we used the datasets developed by the University of Montana (UMT) based Numerical Terradynamic Simulation Group (NTSG). These datasets were prepared using MODIS satellite from the NASA EOS program. The datasets used in this study include: (i) NPP from the MOD17A3 product; and (ii) evapotranspiration (ET) from the MOD16A3 product. These products were used at the annual timescale and are available at the NTSG website (www.ntsg.umt.edu). Both datasets were available at 1 km spatial resolution from 2000 to 2014. The NPP dataset is prepared based on the MOD17 algorithm (Running et al. 2004; Zhao et al. 2005), whereas the ET dataset uses the modified version of the MOD16 algorithm (Mu et al. 2007a, 2011, 2013). For precipitation, we used a gridded rainfall dataset (IMD4) obtained from the Indian Meteorological Department (IMD) (Pai et al. 2014). The dataset is available at high resolution ($0.25° \times 0.25°$) on a daily basis. We used LULC classification from the USGC Land Cover Institute (Broxton et al. 2014).

28.4.3 Methods

In this study, we assessed the resilience of the ecosystem to hydroclimatic disturbance. A disturbance is defined as a short-term variation in environmental conditions that produces a significant change in the ecosystem functioning or behavior. Here, we consider a hydroclimatic disturbance in terms of a water-limited period or drought. To quantify the impact of this disturbance on ecosystem functioning, ecosystem water use efficiency (WUE) was chosen as the indicator (Ponce Campos et al. 2013; Sharma and Goyal 2018a). WUE indicates the connection between carbon and water cycles. It is generally quantified as the ratio of the rate of carbon uptake by the terrestrial ecosystem and the corresponding water loss (Ponce Campos et al. 2013; Song et al. 2017; Tang et al. 2014). We calculated WUE as the ratio of NPP and ET. We compared the WUE during the dry period with the mean WUE to analyze the impact of water-limited conditions on the terrestrial ecosystem's productivity. During the dry conditions, the ecosystem shows its resilience by enhancing the WUE (Ponce Campos et al. 2013). Increased WUE indicates the vegetation is using the water more efficiently and, hence, can survive. More details of the method can be found in Sharma and Goyal (2018a,b). The analysis was performed at the state level.

28.4.4 Results and Discussions

Due to the large variation in LULC and precipitation in northeast India, both NPP and ET showed significant spatial variation (Figure 28.3). NPP ranges between 0 and 1872 $gC\,m^{-2}$. ET ranges between 185 and 1629 mm. Forest-dominated areas had higher NPP and ET. The state average value of mean NPP for AR, AS, MN, MG, MZ, NG and TR were 905, 995, 1078, 1196, 949, 1378, and 838 respectively. NG had the highest NPP. There was no significant trend found in NPP for any of the states. The spatial pattern of WUE is shown in Figure 28.4. The plains of Brahmaputra in AS had a higher WUE. These regions had moderate NPP and lower ET, which resulted in higher WUE (=NPP/ET). The state average values of mean WUE for AR, AS, MN, MG, MZ, NG and TR were 0.962, 1.262, 0.957, 1.056, 0.753, 1.231 and 0.698 $gC\,m^{-2}\,mm^{-1}$ (Figure 28.5). AS had the highest WUE.

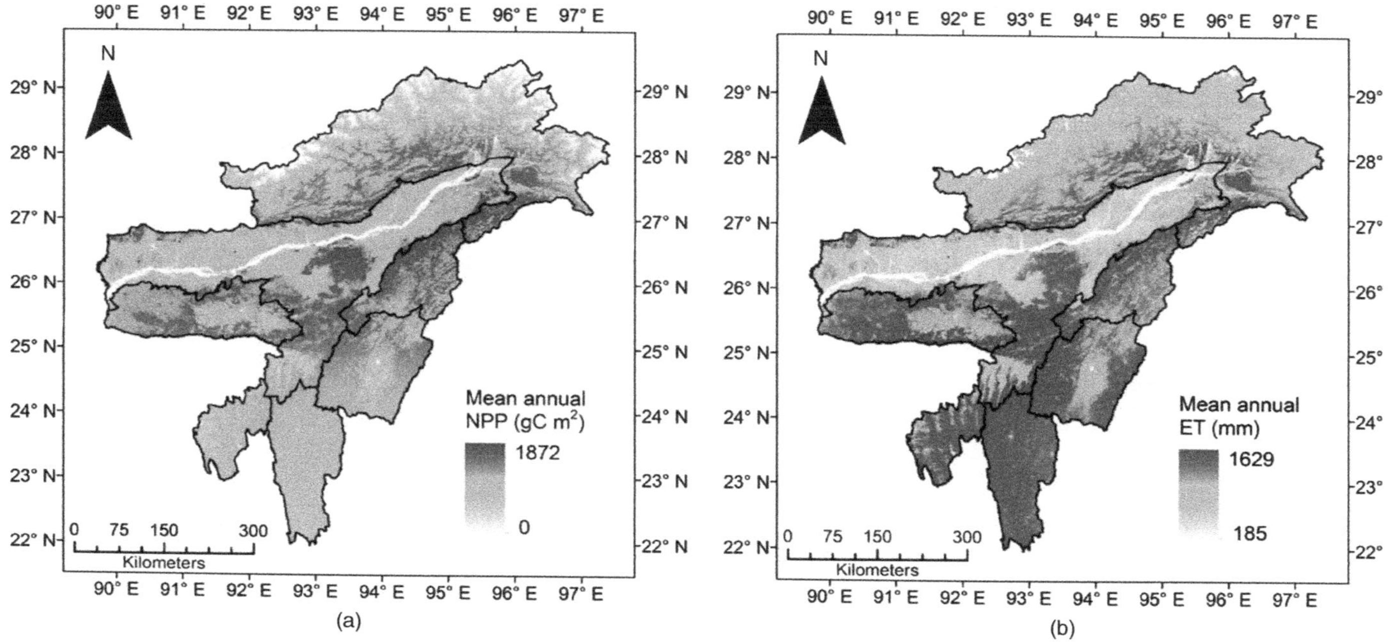

Figure 28.3 Spatial variation of (a) mean annual net primary production and (b) mean annual evapotranspiration over the period 2000–2014.

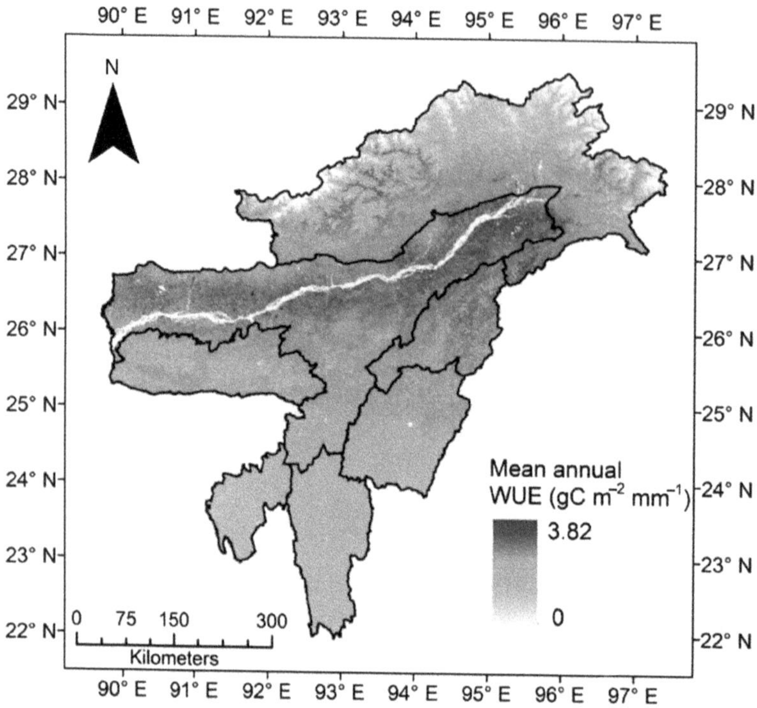

Figure 28.4 Mean annual water use efficiency over the period 2000–2014.

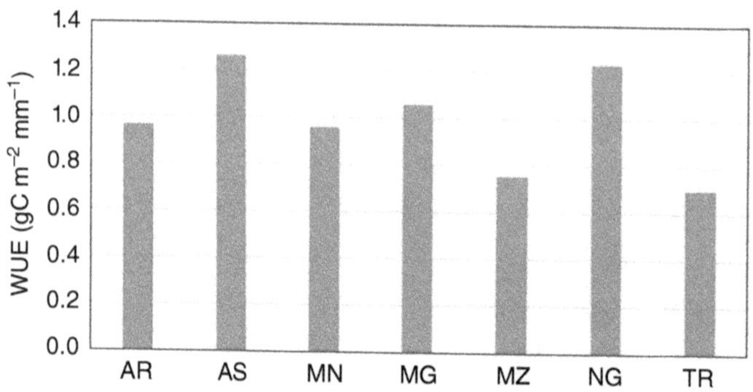

Figure 28.5 State average mean values of water use efficiency for the period 2000–2014.

Table 28.1 shows the ecosystem resilience analysis of the seven states in northeast India. The analysis shows that out of seven states, three (AR, MN and NG) were found to be resilient. These three states have forest cover of 82%, 82% and 84%, respectively, for AR, MN and NG. In contradiction of this, MZ had the largest forest cover (89%) but was found to be non-resilient. It should be noted that we used MODIS LULC data for this study, which categorizes forest into five different classes. To simplify the analysis, we combined all five forest classes into one. There are differences in the resilience characteristic of different forest types. AS and TR have less forest cover, and both were found to be non-resilient. The scarcity of forest in these states could be the reason for these non-resilient characteristics.

Table 28.1 Resilience analysis of seven states in northeast India.

State	Mean precipitation (mm)	Driest year precipitation (mm)	Mean water use efficiency (WUE) (gCm^{-2} mm^{-1})	Driest year WUE (gCm^{-2} mm^{-1})	Resilience
AR	2177	1416	0.96	0.99	Yes
AS	2011	1665	1.26	1.24	No
MN	1615	1161	0.96	1.01	Yes
MG	2954	1492	1.06	1.05	No
MZ	2002	1627	0.75	0.73	No
NG	1753	1432	1.23	1.24	Yes
TR	2212	1803	0.70	0.62	No

Arunachal Pradesh (AR), Assam (AS), Manipur (MN), Meghalaya (MG), Mizoram (MZ), Nagaland (NG) and Tripura (TR).

The resilience analysis of different land cover types showed that forests had a higher chance of being resilient (Sharma and Goyal 2018a). In addition to land cover types, climate also plays a vital role in the ecosystem's behavior. Northeast India mainly has "Cwa" and "Cwb" classes of Koppen–Geiger climate classification. The resilience analysis for different climate types revealed that "Cwb" is more resilient compared to "Cwa." All three resilient states have a significant presence of "Cwb" climate, whereas, in MZ, the presence of "Cwb" is much less. This could be another reason why MZ is not resilient despite the large area of forest cover.

28.4.5 Summary and Conclusions from Case Study

This case study presented the ecosystem resilience analysis of seven states in northeast India. The resilience concept is linked to sustainability in many ways. Resilience is defined as the ability of a system to absorb threats and sustain its functioning under adverse conditions. We assessed ecosystem resilience in terms of the ability of the terrestrial ecosystem to sustain its primary production under water-limited conditions (i.e. droughts). Ecosystem WUE was used as an indicator of ecosystem functioning and response to disturbance, which was calculated as the ratio of NPP to ET. A significant variation in WUE was found in northeast India: the highest was for AS and the least for MZ. Ecosystem resilience analysis revealed that three states, AR, MN and NG, were resilient. These states have large areas of forest cover.

28.5 Future Perspectives

So far, remote sensing and GIS have been a revolution in the field of natural resource management and sustainability. The presence of remote sensing and GIS can be seen in various fields. EO is improving day by day with the launch of new sensors and the advancement of technologies. Despite the enormous technological advancements, there are some applications where remote sensing and GIS have not been explored. According to National Oceanic and Atmospheric Administration (NOAA), "the future of remote sensing lies in making numerous types of accurate, current, and high-resolution remotely sensed data and derived geospatial information products readily available for every area of interest" (NOAA 2017). NOAA identifies the following challenges for remote sensing for the coming decades:

- Improvement in the resolution (spatial and temporal), accuracy and accessibility of remote sensing data.
- Development of early warning and rapid response systems for better management and preparedness for natural and human-induced disaster.
- Improvement in the distribution of remote sensing data for users.
- Development of technologies for more efficient and cost-effective collection of data.
- Improvement in software for quick and easy processing of data.

In addition to the improvement of remote sensing and GIS technologies, ongoing advances in the field of computer science have a key role to play in sustainability through advanced processing and analysis tools. Nowadays, various machine learning algorithms are widely used in remote sensing for processing and analysis of airborne- and satellite-derived imagery (Belgiu and Drăgu 2016; Mountrakis et al. 2011; Rodriguez-Galiano et al. 2012). The scarcity of data has been a challenge in remote sensing studies. The ability of machine learning algorithms to generalize and to perform well with limited data has proved advantageous in remote sensing. In addition, machine learning techniques are found to be more accurate than traditional methods of classification, detection and prediction. Use of deep learning in remote sensing and GIS should be explored.

28.6 Conclusions

This chapter presented different aspects of the role of remote sensing and GIS in sustainability. The integration of remote sensing and GIS provides information on the spatial variation, extent, and potential and limitations of natural resources, which is a prerequisite for planning the strategy for sustainable development. Remote sensing has improved EO and data acquisition. It enables the continuous monitoring of large areas at regular intervals using spaceborne sensors in a time- and cost-effective way. In addition, it also enables information to be obtained for remote locations, which are otherwise inaccessible. In addition to EO and data acquisition, remote sensing and GIS are crucial for sustainable development planning, natural and cultural heritage, society and environment, and public participation. These technologies offer effective and reliable tools for the analysis, management, problem solving, planning, policymaking and decision making needed to pursue the goal of sustainability.

References

Ajami, N.K., Hornberger, G.M., and Sunding, D.L. (2008). Sustainable water resource management under hydrological uncertainty. *Water Resources Research* 44 (11): 1–10.

Anderies, J.M., Folke, C., Walker, B., and Ostrom, E. (2013). Aligning key concepts for global change policy: robustness, resilience, and sustainability. *Ecology and Society* 18 (2): article 8. http://dx.doi.org/10.5751/ES-05178-180208.

Bastiaanssen, W.G.M., Molden, D.J., and Makin, I.W. (2000). Remote sensing for irrigated agriculture: examples from research and possible applications. *Agricultural Water Management* 46 (2): 137–155.

Belgiu, M. and Drăgu, L. (2016). Random forest in remote sensing: a review of applications and future directions. *ISPRS Journal of Photogrammetry and Remote Sensing* 114: 24–31.

Belz, F.-M. and Peattie, K.J. (2009). *Sustainability Marketing: A Global Perspective*. Chichester, UK: Wiley.

Bobba, A.G. and Bengtsson, L. (1997). Sustainable development of water resources in India. *Environmental Management* 21 (3): 367–393.

Brando, V.E. and Dekker, A.G. (2003). Satellite hyperspectral remote sensing for estimating estuarine and coastal water quality. *IEEE Transactions on Geoscience and Remote Sensing* 41 (6): 1378–1387.

Brown, M.E. (2015). Satellite remote sensing in agriculture and food security assessment. *Procedia Environmental Sciences* 29 (Agri): 307.

Broxton, P.D., Zeng, X., Sulla-Menashe, D., and Troch, P.A. (2014). A global land cover climatology using MODIS data. *Journal of Applied Meteorology and Climatology* 53 (6): 1593–1605.

Brundtland, G.H. (1987). *Our Common Future: Report of the World Commission on Environment and Development*. Brundtland Commission, United Nations Commission. Oxford, UK: Oxford University Press.

Burrough, P.A. (1986). *Principles of Geographical Information Systems for Land Resources Assessment*. Oxford, UK: Clarendon Press.

Campagna, M. (2006). *GIS for Sustainable Development*. Boca Raton, FL: CRC Press, Taylor and Francis Group.

Chang, K.-T. (2017). Geographic information system. In: *International Encyclopedia of Geography: People, the Earth, Environment and Technology* (eds. D. Richardson, N. Castree, M.F. Goodchild, et al.), 1–9. Chichester, UK: Wiley.

Chebud, Y., Naja, G.M., Rivero, R.G., and Melesse, A.M. (2012). Water quality monitoring using remote sensing and an artificial neural network. *Water, Air, & Soil Pollution* 223 (8): 4875–4887.

Chowdary, V.M., Ramakrishnan, D., Srivastava, Y.K. et al. (2009). Integrated water resource development plan for sustainable management of mayurakshi watershed, India using remote sensing and GIS. *Water Resources Management* 23 (8): 1581–1602.

Clark, W.C. and Dickson, N.M. (2003). Sustainability science: the emerging research program. *Proceedings of the National Academy of Sciences United States of America* 100 (14): 8059–8061.

Cracknell, A.P. and Hayes, L. (2007). *Introduction of Remote Sensing*. Boca Raton, FL: CRC Press, Taylor and Francis Group.

Dahdouh-Guebas, F. (2002). The use of remote sensing and GIS in the sustainable management of tropical coastal ecosystems. *Environment, Development and Sustainability* 4: 93–112.

DeVantier, B.A. and Feldman, A.D. (1993). Review of GIS applications in hydrologic modeling. *Journal of Water Resources Planning and Management* 119 (2): 246–261.

Devi, G.K., Ganasri, B.P., and Dwarakish, G.S. (2015). Applications of remote sensing in satellite oceanography: a review. *Aquatic Procedia* 4 (Icwrcoe): 579–584.

Dewan, A.M. and Yamaguchi, Y. (2009). Land use and land cover change in greater Dhaka, Bangladesh: using remote sensing to promote sustainable urbanization. *Applied Geography* 29 (3): 390–401.

Duro, D.C. (2012). *Remote Sensing and GIS in Support of Sustainable Agricultural Development*. Sakatoon, Canada: School of Environment and Sustainability, University of Saskatchewan.

EPA (2000). *Environmental Planning for Communities – a Guide to the Environmental Visioning Process Utilizing a Geographic Information System (GIS). EPA/625/R- 98/003*. Cincinnati, OH: EPA.

Fang, H., Jiang, C., Li, W. et al. (2013). Characterization and intercomparison of global moderate resolution leaf area index (LAI) products: analysis of climatologies and theoretical uncertainties. *Journal of Geophysical Research: Biogeosciences* 118 (2): 529–548.

Folke, C. (2016). Resilience (republished). *Ecology and Society* 21 (4): 44.

Gallopín, G.C. (1996). Environmental and sustainability indicators and the concept of situational indicators. A systems approach. *Environmental Modeling & Assessment* 1 (3): 101–117.

Goodchild, M.F., Haining, R., and Wise, S. (1992). Integrating GIS and spatial data analysis: problems and possibilities. *International Journal of Geographical Information Systems* 6 (5): 407–423.

Graymore, M.L.M., Sipe, N.G., and Rickson, R.E. (2008). Regional sustainability: how useful are current tools of sustainability assessment at the regional scale? *Ecological Economics* 67 (3): 362–372.

Gupta, P., Christopher, S.A., Wang, J. et al. (2006). Satellite remote sensing of particulate matter and air quality assessment over global cities. *Atmospheric Environment* 40 (30): 5880–5892.

Hansen, M.C. and Loveland, T.R. (2012). A review of large area monitoring of land cover change using Landsat data. *Remote Sensing of Environment* 122: 66–74.

Henderson, F.M. and Lewis, A.J. (2008). Radar detection of wetland ecosystems: a review. *International Journal of Remote Sensing* 29 (20): 5809–5835.

Holling, C.S. (1973). Resilience and stability of ecological systems. *Annual Review of Ecology and Systematics* 4: 1–23.

Huadong, G. (2013). Earth observations in China and the world: history and development in 50 years. *Bulletin of the Chinese Academy of Sciences* 27 (2): 96–98.

Immerzeel, W.W., Droogers, P., de Jong, S.M., and Bierkens, M.F.P. (2009). Large-scale monitoring of snow cover and runoff simulation in Himalayan river basins using remote sensing. *Remote Sensing of Environment* 113 (1): 40–49.

IPCC (2007). *Climate Change 2007: Synthesis Report*. Contribution of working group I, II and III to the fourth assessment report of the Intergovernmental Panel on Climate Change. Geneva, Switzerland: IPCC.

IPCC (2014). *Climate Change 2014: Synthesis Report*. Contribution of Working Groups I, II and III to the Fifth Assessment Report of the Intergovernmental Panel on Climate Change. Geneva, Switzerland: IPCC.

Jacobs, K., Lebel, L., Buizer, J. et al. (2016). Linking knowledge with action in the pursuit of sustainable water-resources management. *Proceedings of the National Academy of Sciences United States of America* 113 (17): 4591–4596.

Kadiyala, M.D.M., Nedumaran, S., Singh, P.S.C. et al. (2015). An integrated crop model and GIS decision support system for assisting agronomic decision making under climate change. *Science of the Total Environment* 521–522: 123–134.

Kansakar, P. and Hossain, F. (2016). A review of applications of satellite earth observation data for global societal benefit and stewardship of planet earth. *Space Policy* 36: 46–54.

Kasischke, E.S., Melack, J.M., and Craig Dobson, M. (1997). The use of imaging radars for ecological applications – a review. *Remote Sensing of Environment* 59 (2): 141–156.

Kogan, F.N. (1995). Application of vegetation index and brightness temperature for drought detection. *Advances in Space Research* 15 (11): 91–100.

Kumar, T.L. and Rao, K.K. (2013). Studies on spatial pattern of NDVI over India and its relationship with rainfall, air temperature, soil moisture adequacy and ENSO. *Geofizika* 30: 1–19.

Loucks, D.P. (2000). Sustainable water resources management. *International Water Resources Association* 25 (1): 3–10.

Malaska, P. and Kaivo-oja, J. (1996). Science and technology for sustainable development. *GAIA – Ecological Perspectives for Science and Society* 5: 302–309.

Malczewski, J. (1996). A GIS-based approach to multiple criteria group decision-making. *International Journal of Geographical Information Systems* 10 (8): 955–971.

Marchese, D., Reynolds, E., Bates, M.E. et al. (2018). Resilience and sustainability: similarities and differences in environmental management applications. *Science of the Total Environment* 613–614: 1275–1283.

Markham, B.L., Storey, J.C., Williams, D.L., and Irons, J.R. (2004). Landsat sensor performance: history and current status. *IEEE Transactions on Geoscience and Remote Sensing* 42 (12): 2691–2694.

Martin, R.V. (2008). Satellite remote sensing of surface air quality. *Atmospheric Environment* 42 (34): 7823–7843.

McLaughlin, D. and Kinzelbach, W. (2015). Food security and sustainable resource management. *Water Resource Research* 51: 4966–4985.

Melillo, J.M., McGuire, A.D., Kicklighter, D.W. et al. (1993). Global climate change and terrestrial net primary production. *Nature* 363: 234–240.

Minor, T.B., Carter, J.A., Chesley, M.M. et al. (1994). *The Use of GIS and Remote Sensing in Groundwater Exploration for Developing Countries*. Fort Belvoir, VA: Army Topographic Engineering Center.

Mohammad, N.A., Sa, S.M., and Akhir, J.M. (2007). Land use evaluation for Kuala Selangor, Malaysia using remote sensing and GIS technologies. *Malaysia Journal of Society and Space* 3: 1–18.

Mountrakis, G., Im, J., and Ogole, C. (2011). Support vector machines in remote sensing: a review. *ISPRS Journal of Photogrammetry and Remote Sensing* 66 (3): 247–259.

Mowforth, M. and Munt, I. (2015). *Tourism and Sustainability: Development, Globalisation and New Tourism in the Third World*. Abingdon, UK: Routledge.

Mu, Q., Heinsch, F.A., Zhao, M., and Running, S.W. (2007a). Development of a global evapotranspiration algorithm based on MODIS and global meteorology data. *Remote Sensing of Environment* 111: 519–536.

Mu, Q., Zhao, M., Heinsch, F.A. et al. (2007b). Evaluating water stress controls on primary production in biogeochemical and remote sensing based models. *Journal of Geophysical Research: Biogeosciences* 112 (1): 1–13.

Mu, Q., Jones, L.A., Kimball, J.S. et al. (2009). Satellite assessment of land surface evapotranspiration for the pan-Arctic domain. *Water Resources Research* 45 (9): 1–20.

Mu, Q., Zhao, M., and Running, S.W. (2011). Improvements to a MODIS global terrestrial evapotranspiration algorithm. *Remote Sensing of Environment* 115 (8): 1781–1800.

Mu, Q., Zhao, M., and Running, S.W. (2013). *MODIS Global Terrestrial Evapotranspiration (ET) Product (MOD16A2/A3) – ATBD Collection 5*. NASA Headquarters, Missoula, MT.

Myneni, R.B. and Los, S.O. (1995). Potential gross primary productivity of terrestrial vegetation from 1982–1990. *Geophysical Research Letters* 22 (19): 2617–2620.

Napryushkin, A.A. and Vertinskaya, E.V. (2006). Advanced remote sensing techniques for ecosystem data collection. In: *GIS for Sustainable Development* (ed. M. Campagna), 107–122. Boca Raton, FL: CRC Press, Taylor & Francis Group.

NASA (2016). *Vegetation Indices*. Washington, DC: NASA Goddard Space Flight Center.

Navalgund, R.R., Jayaraman, V., and Roy, P.S. (2007). Remote sensing applications: an overview. *Current Science* 93 (12): 1747–1766.

Nidumolu, R., Prahalad, C.K., and Rangaswami, M.R. (2009). Why sustainability is now the key driver of innovation. *Harvard Business Review* 87 (9): 57–64.

NOAA (2017). *Bringing the Big Picture into Focus: The Future of Remote Sensing at NOAA*. Silver Spring, MD: National Oceanic and Atmospheric Administration.

Ostrom, E. (2009). A general framework for analyzing sustainability of social-ecological systems. *Science* 325 (July): 419–423.

Pai, D.S., Sridhar, L., Rajeevan, M. et al. (2014). Development of a new high spatial resolution ($0.25^\circ \times 0.25^\circ$) long period (1901–2010) daily gridded rainfall data set over India and its comparison with existing data sets over the region. *Mausam* 65 (1): 1–18.

Patil, R.J., Sharma, S.K., and Tignath, S. (2014). Remote sensing and GIS based soil erosion assessment from an agricultural watershed. *Arabian Journal of Geosciences* 8 (9): 6967–6984.

Pereira, J.M.C. and Duckstein, L. (1993). A multiple criteria decision-making approach to GIS- based land suitability evaluation. *International Journal of Geographical Information Systems* 7 (5): 407–424.

Ponce Campos, G.E., Moran, M.S., Huete, A. et al. (2013). Ecosystem resilience despite large-scale altered hydroclimatic conditions. *Nature* 494 (7437): 349–352.

Prakash, A. (2000). *Thermal Remote Sensing: Concepts, Issues and Applications*, 239–243. Amsterdam: International Archives of Photogrammetry and Remote Sensing.

Rajeesh, R. and Dwarakish, G.S. (2015). Satellite oceanography – a review. *Aquatic Procedia* 4 (Icwrcoe): 165–172.

Rajitha, K., Mukherjee, C.K., and Vinu Chandran, R. (2007). Applications of remote sensing and GIS for sustainable management of shrimp culture in India. *Aquacultural Engineering* 36 (1): 1–17.

Rao, D.P. (2000). Role of remote sensing and geographic information system in sustainable development. *International Archives of Photogrammetry and Remote Sensing* XXXIII (B7): 1231–1251.

Rashed, T. (2006). Sustainable hazards mitigation. In: *GIS for Sustainable Development* (ed. M. Campagna), 288–305. Boca Raton, Florida: CRC Press, Taylor & Francis Group.

Revadekar, J.V., Tiwari, Y.K., and Kumar, K.R. (2012). Impact of climate variability on NDVI over the Indian region during 1981–2010. *International Journal of Remote Sensing* 33 (22): 7132–7150.

Ritchie, J.C.J., Zimba, P.P.V., and Everitt, J.H.J. (2003). Remote sensing techniques to assess water quality. *Photogrammetric Engineering & Remote Sensing* 69 (6): 695–704.

Rodriguez-Galiano, V.F., Ghimire, B., Rogan, J. et al. (2012). An assessment of the effectiveness of a random forest classifier for land-cover classification. *ISPRS Journal of Photogrammetry and Remote Sensing*, International Society for Photogrammetry and Remote Sensing, Inc. (ISPRS) 67 (1): 93–104.

Roy, P.S. and Giriraj, A. (2008). Land use and land cover analysis in Indian context. *Journal of Applied Sciences* 8 (8): 1346–1353.

Running, S.W., Nemani, R.R., Heinsch, F.A. et al. (2004). A continuous satellite-derived measure of global terrestrial primary production. *Bioscience* 54 (6): 547.

Saraf, A.K. and Choudhury, P.R. (1998). Integrated remote sensing and gis for groundwater exploration and identification of artificial recharge sites. *International Journal of Remote Sensing* 19 (10): 1825–1841.

Schultz, G.A. and Engman, E.T. (1991). *Remote Sensing in Hydrology and Water Management.* New York: Springer.

See, L. and Gibson, P. (2006). Using geodemographics and GIS for sustainable development. In: *GIS for Sustainable Development* (ed. M. Campagna), 211–221. Boca Raton, Florida: CRC Press, Taylor & Francis Group.

Shamsi, U.M. (2005). *GIS Applications for Water, Wastewater, and Stormwater Systems.* Boca Raton, FL: CRC Press, Taylor and Francis Group.

Sharma, A. and Goyal, M.K. (2018a). Assessment of ecosystem resilience to hydroclimatic disturbances in India. *Global Change Biology* 24 (2): e432–e441.

Sharma, A. and Goyal, M.K. (2018b). District-level assessment of the ecohydrological resilience to hydroclimatic disturbances and its controlling factors in India. *Journal of Hydrology* 564: 1048–1057.

Shrivastava, P. (1995). The role of corporations in achieving ecological sustainability. *The Academy of Management Review* 20 (4): 936–960.

Song, Q.-H., Fei, X.-H., Zhang, Y.-P. et al. (2017). Water use efficiency in a primary subtropical evergreen forest in Southwest China. *Scientific Reports* 7: 43031.

Sterling, S. and Ducharne, A. (2008). Comprehensive data set of global land cover change for land surface model applications. *Global Biogeochemical Cycles* 22 (3): GB3017.

Sudhakar Reddy, C., Jha, C.S., Dadhwal, V.K. et al. (2016). Quantification and monitoring of deforestation in India over eight decades (1930–2013). *Biodiversity and Conservation* 25 (1): 93–116.

Sui, D.Z. and Goodchild, M.F. (2001). GIS as media? *International Journal of Geographical Information Science* 15 (5): 387–390.

Tang, X., Li, H., Desai, A.R. et al. (2014). How is water-use efficiency of terrestrial ecosystems distributed and changing on earth? *Scientific Reports* 4: 7483.

Tang, X., Ma, M., Ding, Z. et al. (2017). Remotely monitoring ecosystem water use efficiency of grassland and cropland in China's arid and semi-arid regions with MODIS data. *Remote Sensing* 9 (6): 616.

Tian, H., Banger, K., Bo, T., and Dadhwal, V.K. (2014). History of land use in India during 1880-2010: large-scale land transformations reconstructed from satellite data and historical archives. *Global and Planetary Change* 121: 78–88.

Tilman, D., Cassman, K.G., Matson, P.A. et al. (2002). Agricultural sustainability and intensive production practices. *Nature* 418 (6898): 671–677.

Tralli, D.M., Blom, R.G., Zlotnicki, V. et al. (2005). Satellite remote sensing of earthquake, volcano, flood, landslide and coastal inundation hazards. *ISPRS Journal of Photogrammetry and Remote Sensing* 59 (4): 185–198.

Tucker, C.J. and Choudhury, B.J. (1987). Satellite remote sensing of drought conditions. *Remote Sensing of Environment* 23 (2): 243–251.

Vicca, S., Balzarolo, M., Filella, I. et al. (2016). Remotely-sensed detection of effects of extreme droughts on gross primary production. *Scientific Reports* 6 (March): 28269.

Vicharnakorn, P., Shrestha, R.P., Nagai, M. et al. (2014). Carbon stock assessment using remote sensing and forest inventory data in Savannakhet, Lao PDR. *Remote Sensing* 6 (6): 5452–5479.

Vitousek, P.M., Mooney, H.a., Lubchenco, J., and Melillo, J.M. (1997). Human domination of earth's ecosystems. *Science* 277 (5325): 494–499.

von Carlowitz, H.C. (1713). *Sylvicultura Oeconomica*. Reprint published 2000. Freiberg, Germany: Bergakademie.

Walker, B. and Salt, D. (2006). *Resilience Thinking: Sustaining Ecosystems and People in a Changing World*. Washington, DC: Island Press.

Wiersum, K.F. (1995). 200 years of sustainability in forestry: lessons from history. *Environmental Management* 19 (3): 321–329.

Wilderer, P.A. (2007). Sustainable water resource management: the science behind the scene. *Sustainability Science* 2 (1): 1–4.

Wilkinson, G. (1996). A review of current issues in the integration of GIS and remote sensing data. *International Journal of Geographical Information Systems* 10 (1): 85–101.

Xie, Y., Sha, Z., and Yu, M. (2008). Remote sensing imagery in vegetation mapping: a review. *Journal of Plant Ecology* 1 (1): 9–23.

Xiuwan, C. (2002). Using remote sensing and GIS to analyse land cover change and its impacts on regional sustainable development. *International Journal of Remote Sensing* 23 (1): 107–124.

Zhang, X., Obringer, R., Wei, C. et al. (2017a). Droughts in India from 1981 to 2013 and implications to wheat production. *Scientific Reports* 7: 44552.

Zhang, Y., Xiao, X., Wu, X. et al. (2017b). A global moderate resolution dataset of gross primary production of vegetation for 2000–2016. *Scientific Data* 4 (170165).

Zhao, M., Heinsch, F.A., Nemani, R.R., and Running, S.W. (2005). Improvements of the MODIS terrestrial gross and net primary production global data set. *Remote Sensing of Environment* 95 (2): 164–176.

29

Artificial Intelligence and Computational Sustainability

S.K. Ram, and R.D. Tyagi

INRS, Université du Québec, Québec, Canada

29.1 Introduction

Global estimates suggest that by 2050 the global population will reach a whopping 10 billion human habitants. The dramatic exhaustion of our natural resources at an alarming rate is slowly pushing our planet and the livelihood of all its inhabitants to extinction. The World Commission on Environment and Development in its report, *Our Common Future* (Keeble 1988) first introduced the concept of sustainable development as development that fulfills the needs of the present and at the same time doesn't compromise the needs of the next generations.

Essential issues in designing policies for development will require complex decision making, which needs to be supported by real-life data and facts. These policies will decide the future course of the pattern of consumption of our natural resources. Making such decisions in an optimized manner requires quantified facts acquired from actual data about the ecosystem, natural resources, history of use and abuse, and their consequences. In other words, we are talking about a huge mass of data, which directs us to computational sciences and thereby pushes us to the concept of artificial intelligence (AI) and computational sustainability.

As defined by Russell and Norvig (2003), artificial intelligence (better known by its acronym, AI) is the intelligence displayed by machines or software, and the field of computer science that studies these machines and software with intelligence. AI researchers have often termed it as "the study and design of intelligent agents," where the term intelligent agent refers to a system that recognizes its environment and reacts so that it can maximize its chances of success for any given set of objectives. Eric Horvitz (Microsoft Research) states that "AI is not really a single thing – it is a set of rich sub-disciplines and methods, vision, perception, speech and dialogue, robotics and so on. We have to consider all these different disciplines and methods in seeking true solutions in delivering value to human beings and organizations" (Dourish 2006). AI is probably the most revolutionizing technology that has ever been created, and it could radically transform our future. Over the

Sustainability: Fundamentals and Applications, First Edition.
Edited by Rao Y. Surampalli, Tian C. Zhang, Manish Kumar Goyal, Satinder K. Brar, and R. D. Tyagi

past few decades, AI has developed and evolved dramatically, transforming the world on social, economic and political fronts. AI with its unique abilities to handle a high volume of data (information) has left behind many classically followed approaches to problem solving. AI has been handy in high-bulk data organization, pattern recognition, optimization and recommending solutions. AI is useful in every aspect of life, and everyone can benefit from the technology including single users, organizations, government policymakers and even global international stakeholders. AI has affected humanity in many ways. AI has been implemented in the areas of health, law, weaponry, automation, finance, trade, education, environment preservation, governance and policymaking, public safety, employment, entertainment and, in general, is improving the quality of life and wellbeing. According to a spokesperson in the United Nations, (UN's) Global Summit for AI for Good 2017, AI conforms with the UN's 17 SDGs (Sustainable Development Goals) and if used in a judicious way can yield miraculous results.

There is no doubt that AI is a powerful tool, but it must be used very cautiously. Skeptics believe that AI can be used and misused at the same time. It can have various negative impacts on society. For example, antisocial elements can harness the potential of AI to develop devastating instruments of mass destruction, which can negatively impact global peace and social wellbeing.

This chapter explores the power and potential of AI toward the sustainable development of society and examines the debate of whether AI is a boon or doom.

29.2 Basic Elements of AI Implementation

The term "intelligence" refers to the ability to acquire knowledge, process it and implement it wherever and whenever necessary (Cohen and Feigenbaum 2014). There exists no precise definition of AI: it varies from one researcher to another and even within the domain of its application, but the fundamental core element of AI remains the same. The basic idea of AI is to develop intelligent machines that can learn from existing data about their environment (system or surroundings in the physical world) and use this knowledge to react, are stimulated in response to stimuli or simulate appropriately targeted responses. The ability to learn behaviors from the surroundings remains the defining element for a software or hardware to be called intelligent and fall within the ambit of AI. For example, an arithmetic calculator (machine) is indeed fast and accurate and almost never makes mistakes, but it is not AI. Calculators have a fixed set of information and rules already fed in, and cannot learn progressively. However, different speech recognition systems developed by many enterprises these days are AI as the model learns the owner's voice and recognizes it using intelligent software tools embedded in the system.

When compared with the human brain, no system can match the amount of intelligence it carries because of its abilities to reason, decide, understand and generate language (communicate), perceive and respond to sensory inputs, feel empathy and sympathy. AI always follows its primary learning mechanism. AI is the science and set of computational technologies that are inspired by the functional mechanisms of the human brain. The typical jargon of AI involves artificial neural networks (ANN) and deep learning mechanisms. An ANN is a mathematical modeling tool that consists of three operational layers of neurons (nodes) (Majumder 2015). In modeling terms, these neurons are computational mathematical equations and limiting conditions (system boundaries). The first layer is an input layer that receives the independent input parameters from the system (surroundings). The second layer (also called the hidden layer) is the combination of different functional operations (operators combined with logic). This layer is composed of a considerable number of operations (neurons) that need to be performed to arrive at the final output results (outcome, dependent variables). The types of operations in hidden layer functions (or operations) depend upon the system being handled and interactions among the independent parameters and their mutual sensitivity. Finally, the results are transformed to a meaningful outcome and transmitted to the output layer (the third layer). The mathematical model can receive a very high volume of data (depending upon the efficient design of the model) and predict outcomes. In order to have a definite outcome, the model needs to be pre-trained with some known empirical outcomes. The larger the training dataset available, the more robust the model prediction and design are. The model uses various learning algorithms (supervised or unsupervised) to learn the system using the training dataset to apply this knowledge to any unknown test data. This learning is often associated with the term deep learning, which reflects the intensity of datasets. With advancement in computational abilities, larger data volumes can be processed, and thereby more powerful AI tools can be generated. AI tools like ANN can use this learning ability of models to explore new behaviors, organize and recognize data patterns, and predict (or command) behaviors for any systems. For example, in the automation and robotics industry AI is often implemented to train robots and study their behaviors in response to changes in their milieu via cognitive analytics.

Research in AI is widely divided depending upon the niche areas of its application. The primary areas of AI research are large-scale machine learning, natural language processing, reinforcement learning, robotics, computer vision, deep learning, crowdsourcing, collaborative systems, human computation, algorithmic game theory and computational social choice, neuromorphic computing and development of the internet of things (IoT) (Alharbi et al. 2018).

29.3 Concept of Computational Sustainability

The concept of computational sustainability is the study of computational methods that can contribute to the sustainable management of the earth's ecosystems and natural resources (Sullivan et al. 2014). The complete overall process is a series of individual steps that should be performed in tandem in the context of any given problem/system. The essential key elements of the computational sustainability studies involve: (i) data acquisition; (ii) data interpretation; (iii) model fitting; (iv) solution optimization; and (v) solution execution.

The problem that is to be addressed by the concept of computational sustainability could relate to weather forecasting, agriculture, business design, information and telecommunication, healthcare education or any field where a colossal data computation is involved and whose solution can yield sustainable solutions or policies. For example, stationing the position of weather stations to maximize the coverage at the lowest cost for accurate weather forecasting in poor countries to help farmers with their agricultural practices can solve the problem of food security, which is considered as one of the SDGs (Sullivan et al. 2014).

29.3.1 Data Acquisition

It is very important to collect a complete universal set of data but due to limited reach and technological development, data acquisition is the first hurdle. There are some remote areas of the world that may not have access to advanced gadgets or even cellphones to contribute to the digital footprints of information. African countries, for example, lack the necessary infrastructure for meteorological predictions. Many of the weather stations are either non-functional or obsolete in terms of the advanced technologies required (and available) to be on a par with rest of the world. Therefore, data available to policymakers in developing countries are scarce and noisy (Quinn et al. 2014). The only ways to acquire accurate data are via cost-incurring surveys and interviews. AI can provide more networked solutions to data collection by using personal handheld mobile devices (cellphones) and crowdsourcing the data. Although these crowdsourced data are powerful tools for data acquisition, there are accompanying issues with these methods. There is always the issue of maintaining the privacy of the contributing elements in the data pool and avoiding a data breach. Robust algorithms need to be designed that can handle such a vast amount of data. This is more of a computational problem that needs further research. Representability of the extracted data is another concern. By virtue of the anonymity of the contributors, the analysts have no idea about what faction of the society is contributing the data, and whether it represents the complete demography evenly. Further analysis at the urban or national scale requires different access mechanisms and outreach programs. Ultimately, comprehensive visualization of the data is necessary so that policymakers who may not be data scientists can infer valuable information from the pile of analytics.

29.3.2 Data Interpretation

Data interpretation is the core computing part of the problem in this series of computational sustainability studies. It is imperative to treat the abundant amount of data in a logical manner to draw comprehensive conclusions from the data heaps. In different ecological studies, satellite images or images taken by hovering drones are the feed data, and these images need to be analyzed by computer vision methods (Lytle et al. 2010) in the form of the millions and sometimes trillions of pixels they generate to make significant conclusions.

29.3.3 Model Fitting

In almost all ecological systems, the available data are in discrete forms spatially and temporally distributed. The data need to be integrated together to be fed into the model for a continuous data feed. Once the data are integrated and ready for treatment, a suitable model should be chosen to fit and simulate the data. Researchers have developed many case-specific models that can accommodate the nature of the variables and their data ranges. For example, for species determination, species distribution models are available, whereas for studying seasonal patterns, migration and dispersal models are used.

29.3.4 Solution Optimization

Once a given suitable model is available, researchers simulate different scenarios in the model in order to compute optimal scenarios (policies) for managing ecosystems. In this step, models should be able to incorporate the uncertainty factors that may occur in the future. Various computational tools are available for the optimization of models, e.g. MDP (Markov decision problems); POMDPs (partially observable MDPs); and point-based solvers (Krause et al. 2008).

29.3.5 Solution Execution

After finding an optimized solution, policymakers must enforce the solution for a probationary period and observe the current state. The feedback from the observations should be fed back into the models to update them or reoptimize them if necessary. Once the global optimum has been achieved with respect to all the limitations, the policy should be executed over a more extended phase.

29.4 Computational Sustainability and Ecological Preservation

Various researchers have addressed the problem of ecological preservation using computational tools. Researchers have focused on various issues related to preserving biodiversity and maintaining the biosphere. Lai et al. (2011) addressed the issue of solving the problem of interlinking existing bioreserves so that species can cross-breed with each other (Lai et al. 2011). These spaces are proposed to be connected by a corridor. The problem is to design the safest, most economical and efficient corridors to avoid cross-interference between humans and wildlife. Golovin (2011) studied the similar issues and integrated the constraints of limited budget, utilities and costs into the scenario for developing corridors and acquiring bioreserves; they modeled the behavior of species and their mobility rate to predict whether the species would inhabit within the limits of the reserve or wander in insecure regions (Golovin et al. 2011). These studies used machine learning to develop algorithms and synthesize models for simulating large-scale wildlife, and for resource modeling and planning (Dietterich et al. 2012).

During policy design, it is of high interest to know the services rendered by the ecosystem to human beneficiaries. Such an evaluation will help in prioritizing the ecological preservation plans. This kind of evaluation has been made possible by AI. Villa (2009) developed ARIES (Artificial Intelligence for Ecosystem Services), which is a methodology and web-based application to evaluate the benefits of a given ecosystem for the wellbeing of humans (Villa et al. 2009). Such an application gives a quantitative picture of the advantages of a given ecosystem that can be derived, enabling policymakers to make a sound decision.

Designing a large-scale sensor network for data collection is also a computational problem, particularly when deciding the spatial distributions of sensors. NSF (National Science Foundation) supports many initiatives for autonomous networks (Hinckley et al. 2016) like

National Ecological Observatory Network (NEON). Data collection in many cases has been democratized, and contribution is open to the public. Such a mechanism is called citizen science or citizen data. This is one of the crowdsourced formats for large-scale data collection. Galaxy Zoo is one such initiative, which is open to public access, where the general public can contribute data and their inferences (Dissanayake et al. 2017). eBird is another initiative that engages the general public in observing birds using standard protocols (Sullivan et al. 2014). AI can also solve the problem of extensive exploitation of marine life by pisciculture. To maintain ecological balance, fish catching activities should be regulated. Limits should be imposed on different species of fish. Such policies require concrete knowledge of biodiversity in the oceans. Hybrid dynamic optimization models can be applied in these scenarios to help in forming regulations for fishing activities. A rich mapping of discrete information of grasslands and water resources by using combinations of graphic images and computer vision can provide a database to model migratory patterns for pastoralists, who need to migrate their herds for food and water. Similar databases generated by AI can help in providing humanitarian aid in the case of catastrophes, famines and natural disasters.

To meet the sustainable goal of energy security, policy planners often struggle to choose among the various renewable energy resources available. Renewable energy sources can vary in technology, feasibility, availability and their potential to cater to the specific needs in question. However, in planning any energy initiative at the policy level, many variables play crucial roles in decentralized production, localized or short-term cost, distributed and long-term benefits, stakeholders available and associated risk (Polatidis et al. 2015). Many factors like social, technical, economic, institutional, social barriers, etc. need to be considered. The multidimensional nature of energy planning requires multi-objective optimization tools (or Multiple Objective Decision Support Systems, MODSSs). Such types of computational problem can be modeled and solved using various AI tools (El-Swaify and Yakowitz 1997). AI is already in extensive use to develop different intelligent tools for solving environmental problems. A few examples include: OASIS (Operations Assistant and Simulated Intelligent System); STORMCAST (weather prediction tool); MEXSES (environmental impact assessment for water resource development projects); FRAME (air pollution model selection tool); MEDEX (Mediterranean meteorology database tool for weather forecast); DCHEM (Distributed Chemical emergencies manager, applicable in the field of chemical accidents); CHARADE (combining human assessment and reasoning aid for decision making in environmental emergencies, tools for defining strategies in environmental emergencies); DAI-DEPUR (multi-level supervisory agent-based architecture for wastewater treatment plants operation); and INFORMS-R8 (integrated forest resource management system) (Cortés et al. 2000).

29.5 Potential Applications of AI in Various Sectors

In recent years, the roots of AI have reached the farthest corners of technology and have influenced nearly every technologically enhanced sector that has been using computers, electronics and automation. Using the rejuvenating powers of computational sciences and our enhanced abilities to handle big data, AI is now applicable to almost every walk of life making the lives of human more convenient and taking measures to sustain the natural ecological balance of the planet. AI is widely applicable to many different sectors. The prominent fields of application are discussed in this section.

29.5.1 Healthcare

Health is one of the primary concerns that the UN expressed in designing the SDGs. This aspect of life cannot be overlooked irrespective of the economic class of the country. Both developed and developing nations require a secure and easily accessible healthcare system. In developed countries, best healthcare practices are already available, but the developing countries are still struggling to reach average standards of healthcare facilities (Quinn et al. 2014). AI and computational sustainability studies can radically improve the current situation of healthcare in developing countries.

Many feats have been achieved by researchers in healthcare using the power of AI. Medical diagnosis has been made much easier by computer-aided tools for the interpretation of medical images. A typical use of such a tool is in the detection of tumors. On similar lines, the Watson Project initiated by the IBM's oncology division has been using AI to assist doctors in cancer-related cases. Many companion robots are a new concept in households especially for assisting disabled or elderly people in day-to-day life (Yorita and Kubota 2011). The PARO robot is an AI robot that looks like a baby seal, and is an official treatment tool of the UK's National Health Service, which can be used for dementia and Alzheimer's patients (McGlynn et al. 2014). Medical histories of a patient and other patient's medical histories (electronic health records, EHR) with the same disease can be integrated together using AI to design personalized treatment plans for the patients to take care of their specific needs like allergies and response rates.

Researchers have been successful in developing point of care (POC) devices, which are small mobile diagnostic devices that can carry out routine diagnostics for general diseases that were once epidemic and claimed hundreds of lives in developing nations. The POC devices are often equipped with high-end advanced technologies for detection of disease-causing agents (Gomaa et al. 2016). Unlike classical methods of manual optical visualization of blood samples, which are limited by the observational skills of a diagnostic professional, these POC devices are easy to use and accurate in their results. Real-time rapid diagnosis is possible in the most remote places using these POC devices.

Disease control and surveillance are among the most important applications of AI and computational tools. Using AI, disease surveillance systems can be generated to update real-time disease occurrence data. The disease occurrence and associated symptom data can be used by AI tools to predict early the possibilities of any disease outbreak and curb the potential epidemic health threats. For example, in 2011 Punjab Intelligent Disease Surveillance System (IDSS) released a mobile hotline application where users could report dengue-like symptoms (Quinn et al. 2014). IDSS used data from their server and reported confirmed data from hospitals to accurately predict the disease outbreak location at an early point. IDSS uses complex algorithms to combine information from their server, the news and results from hospitals to estimate the scale of the problem and plan adequate measures to control the situation. Similar computational tools are Flubreaks, LUMS and Google Flu Trends (Klembczyk et al. 2016; Wiwanitkit 2015). These tools can give disease occurrence density maps over a given space–time frame, which can help doctors to prepare for outbreak control situations.

Developing countries often lack the analytical abilities to handle big data to profit from the precious and valuable knowledge generated by high-tech data acquisition systems integrated into mobile healthcare and surveillance mechanisms. During a disease outbreak governments often call for closedown of schools and businesses to decrease human mobility and the spread rate of disease. This closedown has a severely adverse economic impact on the society not to mention the day-to-day discomfort people experience. Human mobility data are computed from an emerging region only by census data, which are taken every 5–10 years. AI can provide solutions to capture real-time updated quantitative data of human mobility using current networked device systems. Such knowledge is crucial in deciding the extent and active regions of closedown to have the least economic loss. AlertImpact is one such computational tool developed by researchers, which focuses on approaches in modeling the evolution of epidemics under different policy scenarios (Frias-Martinez et al. 2012). The tool utilizes cellphone networks to locate and acquire human (subjects) mobility data and social patterns of populations. The call records are registered with events of regular phone usage (call-dial, call-receive, SMS, MMS, etc.) Using the mobility data, the software tool simulates and predicts the virus propagation using the susceptible–exposed–infectious–recovered (SEIR) model.

29.5.2 Food Security and Agriculture

Many developing countries economies thrive on primary sector economic activities like agriculture. The agricultural activities in these countries are fragile and heavily dependent upon the climate and its natural resources. A considerable loss is suffered by farmers due to various crop diseases, which require regular monitoring. Immediate measures are needed to make sure farmers can furnish the required food products. AI can help in tracking and monitoring diseases in staple crops. Simple mobile applications can be used to transmit real-time images of infected crops, and their visual analysis by computer vision tools can give a farmer an immediate diagnosis and preventive measures to control the disease. The database of crop disease occurrence can be used to identify patterns and suggest measures to be undertaken to prevent such events in future or to pinpoint the real cause of the disease. Such a system has already been deployed in Uganda to monitor viral diseases in cassava (Owomugisha et al. 2014; Quinn et al. 2014). A comprehensive large-scale monitoring can also be performed using satellite images of croplands to hint at the state of health of the crops. AI tools can provide valuable weather forecasting, water resource availability, a soil nutrient database, the history of crop growth and a yield database, which can equally help farmers and policymakers to address the issues related to food production. Predictive models can use rigorous computational techniques to analyze the significant amount of data obtained from satellite images to calculate food yield and use the current demographic data amassed via advanced data acquisition techniques to compare food supply and demand with household-level precision. The food supply and demand information can facilitate predictive information of any upcoming food shortage for an economy. Satellite images can also be used to draw historical maps of land use patterns. Due to rapid urbanization, many agrarian economies are losing their arable lands to new cities or townships. This may cause a lack of agricultural land and pose threats to the food security of the nation; therefore, such information is invaluable to policymakers to understand and regulate land use patterns.

29.5.3 Transportation

Transportation is often called the lifeline of an economy and is largely responsible for all forms of inland trade, the daily commute by inhabitants of the country and the general mobility of various goods and services. AI can radically transform transport methods and the way they are perceived in the current age. Driverless vehicles have been a very promising hit in the transport domain using AI. Using AI, it is possible to use 3D maps of cities and corridors to navigate through the city using camera, sound, lasers global positioning systems (GPS) and radio sensors. Driverless vehicles will make roads safer and more convenient for disabled people thereby making them eligible for jobs that require them to move around. In general, driverless vehicles are going to make human lives better as the time they spend in driving can be spent elsewhere to improve their wellbeing or productivity. AI is already in place to design intelligent parking systems, lane departure systems, blind spot monitoring and lane changing systems.

Commuting matrices are a characterization of the transition of the population between two different geographic locations indicating the origin and the destination. Cellphones can be used to retrieve the mobility data to generate such commuting matrices. Quinn et al. (2014) have developed a new technique to generate commuting matrices by using call data records obtained from cellphones (Quinn et al. 2014). Experimental results and validation data were found to be very accurate and can be used instead of classical commuting matrices computed via data generated from expensive censuses and surveys. Traffic congestion information obtained (crowdsourced) by user's mobile phones is precious and can be used by transport and urban planning engineers. These data are close to real and invalidate assumptions that all vehicles are driven in the fixed lane. CCTV footage can be used for various traffic signals to give true estimates of congestion and can help in regulating traffic lights and in designing congestion collapse strategies.

Using AI-based search methods, Sachenbacher (2011) developed a tool to find the optimal route of navigation between source and destination for electric vehicles (Sachenbacher et al. 2011). The tool gives you the optimal solution to achieve different targets like minimal energy usage or minimum commute time. The model makes dynamic energy estimates based on changing payload weights and air conditioning. In a similar study, Quin et al. (2014) planned commutes to minimize commute time. The model accounts for the data about traffic lights and estimates the time spent at each traffic light. On similar lines, Au (2008) proposed a smart multi-agent intersection system that reduces the uncertainty about time spent at traffic lights (Au et al. 2008). With this intelligent traffic system, smart vehicles requested passage and were guided through as requests are honored. This system reduces wait time at lights and balances traffic flows.

29.5.4 Public Safety and Security

AI has already been contributing to safeguarding our day-to-day lives – from air traffic controls to regular traffic light logistics. Security cameras are an integrated part of security tools used by many safety agencies to collect data every hour. These data can be used to detect anomalies that can hint at possible crimes or catastrophes. The most successful use of AI has been in detecting white-collar crimes like credit card fraud, internet fraud and cybercrimes. Intelligent software tools can be used to detect spam, spyware and ransomware to protect public assets and valuable data. These security measures are of high importance in the case of high-stakes situations like border security, weaponry warehouse security, intelligence against terrorism and nuclear safety programs. AI tools can help solve civil crimes and prevent them to a great extent. COMPStat 2.0 is an intelligent software tool used by the New York Police Department for its predictive policing abilities to predict possible crimes (Molinari 2016). AI can also be used to develop intelligent simulations for training law enforcement agents. The European Union (EU) through its Horizon 2020 program supports such initiatives as LawTrain (Bitan et al. 2017). Highly intelligent tools can be developed to scan billions of tweets and social media to detect possible security threats. AI tools can be used to infer deception and criminal behavior, which can be useful for international security agents like coastguards, borders, immigration security, airport security, etc.

29.5.5 Human Resources – Employment and Workplace

The employment sector is the sector where the effect of AI has been most debated. There are various schools of thought as to whether AI is a boon or a curse to the human population as far as jobs are concerned. Prior to AI, due to a revolution in the digital world, the advent of IC (integrated circuits) and other technologies the job sector was profoundly affected. The effects were uneven as they helped a particular sector of the industry while negatively affecting the other. Digital technology most affected the skilled middle-level personnel who perform mundane routine tasks, while the lowest and highest skilled jobs were least affected. Similar effects are anticipated from AI. Optimistic economists believe that, while AI might replace existing jobs, new jobs will be created that require new skills to be learned. Although predictions of the new job skills that will be needed are ambiguous, the skills that will be displaced are visible. AI can engage in mass production, which previously required humans, thereby lowering the prices of goods. The jobs lost are more salient to people than the diffusive benefits of price lowering by AI. Active measures are required to make sure the changes in the economics of society take place slowly and gradually, giving people time to adjust accordingly.

Another critical application of AI is in keeping tabs on human resources. The census data of any country is valuable information that reflects the socio-economic state of the region. Conducting a census is sometimes difficult due to the low accessibility of remote areas, people's lack of interest in participating in long, time-consuming interviews and so on. Quinn (2014) developed CenCell, a computational technology-based census tool. It consists of an array of supervised (and unsupervised) techniques to calculate the socio-economic level of a region (Quinn et al. 2014). The tools use average consumption, mobility and social network pattern data obtained from their call records. The tool results have been tested and validated with millions of cellphone records in urban population, and the results match with National Systems of Information (Soto et al. 2011). Likewise, a system called the Aadhar Card system has been executed in India, which can help maintain digital records of all the citizens in the country so that policy decisions can be made with more reliable data (Krishna et al. 2014).

29.5.6 Education

AI will change the methods and content of educational pedagogy, implying that in future we might be learning to work with many automated machines in tandem rather than just learning the physics and mathematics behind them (Hilty and Aebischer, 2015). Quality education will always require human engagement, but many academic activities can be performed using AI based tools like interactive remote presentations. Schools were earlier limited by space and infrastructure, but AI can completely negate the whole concept of a school classroom, and a class size of thousands can be imagined using virtual environments. Teaching robots have been one of the big leaps in education. Ozobot is an intelligent robot that teaches kids to code and reason (Prato, 2017). It can also be programmed to dance and play with color-coded patterns. Cubelets helps children with logical thinking (Correll et al. 2014). PLEO is an interactive robot that teaches and helps kids to learn biology (Fernández-Baena et al. 2015).

The tutoring system has been part of an age-old tradition. Intelligent Tutoring Systems (ITSs) have been developed by initiatives like the Why-2 Atlas, which helps in solving problems in physics. Software like Duolingo gives you personalized experience for language learning at your own pace. An ITS, SHERLOCK is now available that teaches air force technicians to diagnose electrical problems in planes (Lesgold 1988). AI brought with it a technical boom of online learning platforms often termed as MOOC (Massive Open Online Courses) (Hew and Cheung, 2014). Educational Testing Services and Pearson (educational organizations) are working together to develop assessment tools to grade essays in standardized tests like GMAT, SAT, TOEFL and GRE. Many recent MOOC platforms that are in vogue are EDX, Coursera, Udemy and UDacity, which use neuro-linguistic programming, machine learning and crowdsourcing methods for grading and evaluations. However, the progressive transformation of education is not as rapid as expected, which can be attributed to many challenging factors like lack of resources and data-establishing technologies. In the next two decades, it is possible to see AI-powered technologies assisting teachers in academia. Virtual realities and augmented realities will be used to improve visualization of knowledge. AI is going to blur the line of difference between classroom teaching and individual self-learning.

29.5.7 Home and Services

AI is having an impact on household chores, replacing many daily duties with the help of automation equipment. Robots already exist in homes to perform many different scheduled activities. Advances in robotics and AI will further change the way and standard of living in the near future. Home robots are now existing realities. These robots are equipped with expertise in navigation and daily handling of household objects. Recently the addition of VSLAM (simultaneous visual location and mapping) – an AI-based technology – has enabled robots to make a 3D map of the complete house, which further helps them in navigation and locomotive actions (Eade et al. 2016). In the early days, robots were considered to be machines that can only perform scheduled activities, but with the abilities of machine learning, these robots can learn from humans and their environment, making them more interactive, responsive and accurate in their actions.

AI at the community level also has some critical applications. Many researchers have used machine learning and AI to understand anthropological settings. Mady (2011) studied heating ventilation and air conditioning (HVAC) controllers to reduce the energy usage by buildings (Mady et al. 2011). HVAC utilizes the Markov model to integrate building occupancy with different elements (Wei et al. 2015). The model also factors in the comfort of the occupant in relation to model temperature and desired temperatures. Unlike the model-based approach, Kolter and Ferreira (2011) follow the regression approach to predict a building's total energy usage. This method utilizes data related to the building like monthly electricity and gas usage, building location and layout, and compares the results with various regression methods (Kolter and Ferreira 2011). This can help building managers to point out use or abuse of energy by buildings.

29.6 AI for UN Sustainable Development Goals

Many economists and entrepreneurs believe that AI is a powerful accelerator of the 17 UN SDGs. At a global summit: "AI for Good" in 2017, Microsoft CEO Satya Nadella called AI the ultimate breakthrough technology. AI has penetrated all the domains, continuously helping each of them to a better and sustainable future. Table 29.1 lists how broadly AI is contributing to individual UN SDGs.

29.7 Challenges Associated with AI

Dietterich (2012) very accurately points out that the primary challenge that machine learning-based solutions face is the acquisition of data (Dietterich et al. 2012). Many of the sustainability-threatening events (issues) are large-scale ranging from countries and continents to global. The mobility of the problem in context is another issue that poses great challenges to data collection. Steady objects like buildings and farmland are relatively less dynamic in their changes, while human populations during disease control and migratory birds in sanctuaries are very mobile. The methods and mechanisms of data collection vary and are specific to each problem. In most of the cases when the scale of the problem is sizable, a small set of data collected by a handful of scientists can't represent the entire true state of the problem. Therefore, data can be crowdsourced from the general public. For example, in natural ecosystem studies on population dynamics of birds, the data collected by the general public have a high degree of variation for the same given observation. Once the data are collected, quality control of the data is another big hurdle to manage as the amount of data now available is easily outnumbering the data scientists available to inspect them for quality control manually. Automated measures are required to refine and screen the input for high-quality data for further processing; otherwise, computational resources will often face breakdowns. Even high-quality crowdsourced data can experience observer bias as the expertise of the observation may vary with the observer. Machine learning methods (algorithms) are needed to accommodate these data quality and bias issues. For modeling a dynamically changing problem set, the ideal solution will require collecting data from each individual element. For example, in population dynamics problems in

Table 29.1 Contributions of artificial intelligence to United Nations' Sustainable Development Goals (Ericsson.com 2015).

UN Sustainable Development Goal (SDG)	Artificial intelligence contribution
SDG 1: Poverty	Satellite mapping to locate poverty, data analysis to plan resource allocation
SDG 2: Zero hunger	Increase agriculture productivity using predictive analysis of weather, land usage, soil quality. Lowering crop diseases and prediction of food requirements and inefficiencies
SDG 3: Health and wellbeing	Preventive healthcare initiatives, rapid diagnostics analysis, digital health records, disease control in outbreak situation and point of care analytics
SDG 4: Quality education	Virtual classrooms, interactive robots as teaching assistants, online courses and evaluation, high-volume classrooms online and a higher number of graduates. Facilitate education for countries low on budget
SDG 5: Gender equality	By digital census, AI helps in identifying gender biases and help in correcting them by accounting for them in policies
SDG 6: Clean water	Internet of things and sensors feeding into AI of everything can predict sanitation and consumption patterns for improved water provisioning
SDG 7: Clean and affordable energy	Green energy, optimized usage and efficiencies by AI real-time analysis
SDG 8: Work and economic growth	AI will improve the work environment, lessen the burden on humans, increase productivity and restore work–life balance for many workers. Catalyzed economic growth in association with automation
SDG 9: Industry innovation and infrastructure	Hybrid manufacturing with integration of AI, internet of things sensors and 4D printing
SDG 10: Reduce inequality	Sensors and bionics improve the lives of people with disabilities giving them equal opportunities in society
SDG 11: Sustainable cities and communities	Digital database of smart cities, smart devices, crowdsourced data from mobile data and wearable devices give rise to sustainable communities
SDG 12: Responsible consumption and production	AI helps in analyzing consumption patterns of resources making people liable for their usage, eliminating or minimization of wastage. Vast improvements in yield and resource efficiency
SDG 13: Climate actions	Climate change analysis by real-time data, improved accuracy and easily accessible history of data. Prediction of disasters and management
SDG 14: Aquatic life	Marine life migrations, regulation of fishing activities, population-level monitoring, warning and policies
SDG 15: Life on land	Pattern recognition, migration of land animals, diversity studies warning of endangered species and aiding their preservation
SDG 16: Peace, justice and strong legal institutions	Reduce discrimination, digital identities of citizens, reduce corruption, responsive, intelligent services and cyber threats with enhanced securities
SDG 17: Collaborations for goals	Multidisciplinary collaboration for safe and ethical use of AI, collaborations of researchers, government and business enterprises are essential for developing a code of conduct for AI development

(Continued)

natural ecosystems like bird sanctuaries, the population is changing due to the high mobility of each individual element (birds and animals). Machine learning methods should accommodate rapidly changing datasets to model the population accurately.

Skeptics have believed that AI can also be a threat to global peace if not dealt with responsibly (Müller and Bostrom 2016). Usage of AI comes along with many ethical challenges. AI is not governed or regulated by any international body. Thus, AI's usage and its limitations are not defined, and it can be used for any cause (e.g. social or antisocial). If used by antisocial elements, it can threaten the very concept of growth development and sustainability. Thus, it is imperative that all the advanced knowledge and technology developed via AI is always in safe hands. While we develop AI-integrated technologies manual overrides should be available as safety precautions in case of unforeseen situations. When AI-integrated technology is in use and if anything goes wrong, who can be held accountable? It is something of a gray area as to who is accountable when a remote-controlled killer bot, during a war, commits a war crime. Equally, who is responsible for a wrong diagnosis made by an AI?

AI applications require manipulation of a massive amount of data, which might involve personal medical records or social media profiles (Rajbanshi et al. 2017). There are many cybersecurity issues with significant data analysis and processing. Breaches of privacy and unethical use of data can limit the usage of AI. It is essential to revisit the laws for the tasks that were earlier carried out exclusively by humans. AI-based humanized robots who look like humans must pass the test of uncanny valley (Hamet and Tremblay 2017). The "uncanny valley" explains that the more a robot looks like humans, the more creepy or fear-instilling it becomes for people. Changing public perceptions of humanized robots equipped with speech recognition, sensors and locomotive abilities remains a challenge.

The earlier electronic digital revolution was mild, but AI is going to lead to a substantial displacement of humans at work (MacCrory et al. 2014). A few optimistic visionaries disagree with the concept of machines replacing men by arguing that the economy will shift its equilibrium, and new replacement jobs will be created in the automation industry. New technologies will expand the range of products and skilled people needed to make them so employment will not be stopped. It is okay even if no skilled labor is required as AI-based machines will replace these positions. The economy will find other ways to make humans consumers. Arguably, despite the hyped fear in history, computers have not yet replaced humans on a net calculation basis (Autor et al. 2003). Jobs that can't be codified or termed as a "task", which require complex rationality, judgment, creativity, persuasion and decision making rather than merely following protocols need to be carried out by a human. Jobs that require interactions will never be replaced by automation like visual and language recognition. Friedrich (2016) polarized jobs as high skilled and low skilled. In managerial and technical jobs, people are benefited by computerization while low-skill jobs have suffered. Low-profile jobs where computerization has no contribution have remained as the payment for intense labor remains the same. Some skillful people (e.g. priests, food-tasters) can't be replaced by automated machines, and they will remain immune to advances in computational technology.

At the inception of an AI-based technology, it is going to be expensive. Only the highly resourced fraction of society will have access to such technologies. Similarly, the advanced technologies will be available to developed countries to improve their standards of living.

Thus, AI can create massive class differences, making the rich richer and the poor further impoverished. To facilitate equality in society, it is crucial to democratize technology, making it available to everyone equally (Brynjolfsson et al. 2018; Helbing et al. 2019). This normalization may require a massive amount of economic resources, and it is still ambiguous as who is going to be responsible for the aftermath of applications of technology. Therefore, the redistribution and maintenance of the socio-political system will be another big challenge. This poses a significant challenge to policymakers. What should a policymaker do? Should society opt for AI technology when it is technically feasible? How do we debate this? What is the platform (international or national)? Who are the stakeholders to contribute to the debate (Horvitz 2014)? Should the matter or legislative reforms be open to democratic discussion? There are many such questions that we need to figure out before AI can be accepted.

On the technical end, AI is a black box to users, and its technical details are not available to anyone other than the creators of the technology. Any subsequent application of the technology will require the collaboration of the technology creators, which necessarily requires a dedicated engagement of the researchers across disciplines together dissolving their international identities. However, in the present international political situation, it may not be that easy. Unregulated use of AI can endanger human rights. It may give supreme powers to authoritarian leaders reinforcing prejudice and discrimination (Horvitz 2014). Given that AI learns from historical databases, it will continue to learn all the social biases and discrimination. For example, historically, the number of arrests in the US of the black ethnic group has been higher than Native Americans and Caucasians. AI will innocuously learn these biases as a fact and integrate them into its analysis, giving misleading policy suggestions.

29.8 AI and Its Socio-Economic Impact

Classically, economic growth is seen only an increase (or improve in efficiency) in capital and labor. AI is not just a productivity enhancer but a new factor in growth and productivity, which can transform economies because of its speed, accuracy and potential to replace humans. For example, a review of 1000 reports, which could take 6 months for three men, can be done in a week with the aid of AI. AI could bring about a complete paradigm shift in the economy. According to an Accenture report (Purdy and Daugherty 2016), if AI is factored in to economic evaluation, a significant increase in gross value addition (GVA) will be observed by 2035 in the USA (as shown in Figure 29.1).

AI, along with its integration into automation technologies, can provide cheaper solutions for warehouse-handling problems. Fetch Robotics made 3D depth sensors and uses AI technology to facilitate warehouse handling efficiently. Language recognition abilities and personal assistance, Google Home, Apple's SIRI, Microsoft Cortana and Amazon Alexa further enhance the communication between machines and humans, allowing better control of machines. Such a channel of communication provides new learning avenues for machines as well. Amelia (an intelligent agent developed by IP Soft) has remarkable abilities to learn from its environment. RELAY, a robot developed by Savioke, is already an application for delivering utilities in luxury hotels (Purdy and Daugherty 2016).

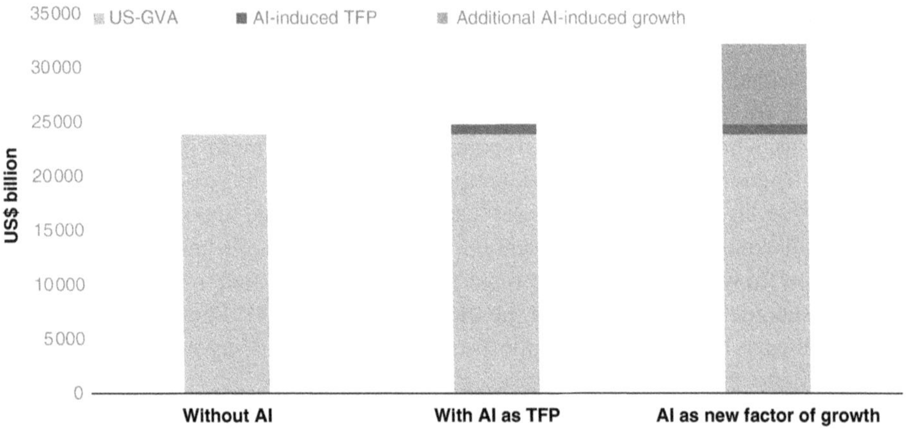

Figure 29.1 Impact of artificial intelligence as a factor of production on US economy. GVA, gross value addition; TFP, total factor productivity.

AI has also played a role in improving the manufacturing industry. Fanuc Intelligent Edge Link and Drive (FIELD) is a machine learning tool used by the manufacturing industry that derives data from various sections of the manufacturing process to analyze and optimize them. AI, as discussed earlier, will also lead to the creation of new job opportunities. When a driverless vehicle is made, many collaborations are generated. Parts manufacturing for different sensors required will shoot up ;leading to more job opportunities. In this way, the innovations will push the economy further. Driverless vehicles change the way of life for drivers as they can spend more quality time with family, and these intelligently guided systems will result in fewer road accidents. Furthermore, economic reports generated by Accenture and Frontier economics projects indicate that significant improvement in labor productivity will be observed in many developing countries (as shown in Figure 29.2).

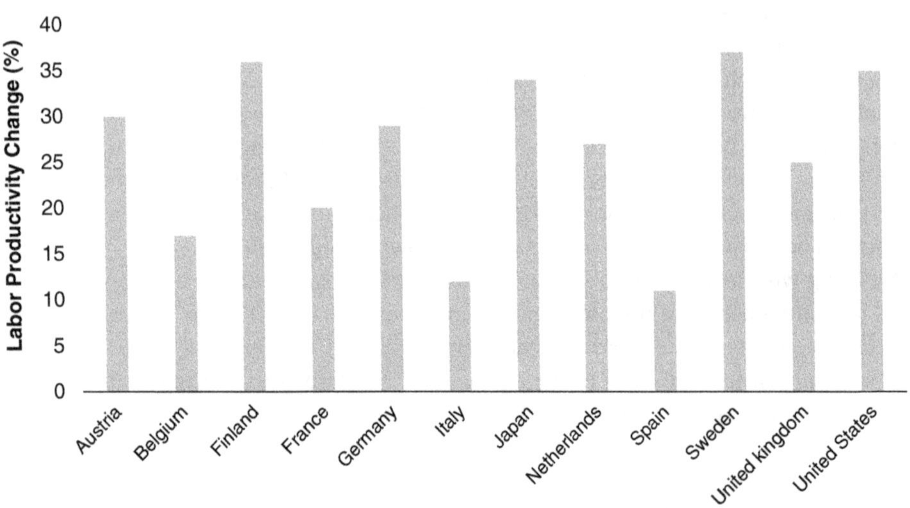

Figure 29.2 Percentage increase in labor productivity from baseline comparison if AI is implemented in policies for developing countries in 2035.

AI can compensate for the limitations in technical expertise in developing countries where technical education (or even primary education) is not always available. There are not many doctors and engineers present in developing countries to cater to the disproportionate amount of people who need their technical expertise. AI can complement and compensate for the lack of technical expertise in such countries, making the job easier. For example, automation of diagnosis and preliminary medical advice can decrease the piling-up ratio of patients to doctor. AI speech recognition and language cognition tools can bridge the language gap between the technical experts (or doctors) and the locals inhabiting remote areas of the country. For example, AI can break down the language barrier for a doctor providing healthcare in remote areas of the country where people are illiterate and they do not speak the same language.

Developing countries often lack various resources like sanitation, telecom stations, etc. The very core of this problem is engineering, which falls within the context of resource allocation optimization. AI can help in allocating the resources to maximize the outcomes. Resource allocation is preceded by the problem of having the necessary intelligence on the state and extent of the problem. A suitable data acquisition mechanism can provide the real state of the issues, and appropriate measures can be taken. This same approach can be illustrated in other real-world problems like agriculture, health, transportation and public policy.

29.9 AI for Developing Countries: A Case Study of Zimbabwe

It is quite apparent that AI has an enormous impact on developing nations with significant opportunities to raise the standard of living. Mupfiga and Fohlo (2015) reported a case study conducted in Zimbabwe about the applications of AI to enhance governance, the economy and life in general for developing countries. Zimbabwe's initiative to accommodate and embed AI in their policies is via a program called Zimasset (Zimbabwe Agenda for Sustainable Socio-Economic Transformation). The initiative is aiming toward sustainable development and social equity. The main sectors where the overt application of AI is available are discussed in the following sections.

29.9.1 Food Security

To cater to the food requirements of the country and ensure the resolution of the food security issue, the initiative proposes a small AI-enabled automated manufacturing unit process called islands of automation. A simple train of these islands of automation in tandem can result in an overall system of automation for production. Each unit operation is equipped with AI-based technologies and can perform individual modules of the tasks in manufacturing. These automated assemblies have been already enforced by wholly automated manufacturing line-ups.

AI is used extensively for its accurate and robust weather forecasting abilities. ANN have been designed and trained in local weather data for Zimbabwe. These models can give predictions for not only weather but also water network data, soil type data and output yield predictions. Surveillance using AI tools can furnish data regarding the economic health of farmers. The government will use these data to design strategies to help them in the complete process chain of food production. Land usage models like MAgPIE, e-Agriculture, by WSIS (World Summit on Information Society) and data obtained from similar economies can further help farmers to maximize their economic return without compromising food security. The entire manufacturing process is now automated, controlled and maintained by a computer system in car manufacture, machine tool production, computer chip production and almost every high-tech process. Automated machines equipped with sensors and AI abilities can perform dangerous tasks like handling hazardous radioactive materials and mining. Overall AI can bring down the cost of production for poor countries that lack economic resources and rely on foreign investments.

29.9.2 Poverty and Social Services

To achieve the goals of poverty eradication, implementation of an e-Government system, with the digitalization of the complete governance system working in parallel with AI and intelligent decision support systems, is recommended. This will allow complete transparency, and democratization of the government policies, rules and activities are making the ruling government more responsible and accountable to the general public. Once all the resources are allocated without corruption and in an utterly transparent manner, poverty eradication policies will be useful in their actions.

In the healthcare sector, AI systems can be used to organize bed schedules, make staff rotations and provide medical information. ANN can be used in various clinical diagnostic and decision support systems, such as in concept processing technology in electronic medical records software. Many of the tasks in a hospital require visual observation of test results; such routine tasks can easily be performed by AI. Such systems scan digital images, like computed tomography, and pinpoint abnormalities to flag any possible occurrence of disease. Such computer-aided visualization systems are beneficial in detection of tumors.

29.9.3 Infrastructure and Utilities

In Zimbabwe, the mining sector is the primary source of economy. It remains the pillar for economic growth by value addition and economic activities. The mining sector has always faced the challenge of limitations in energy/transport facilities, low international mineral prices and a shortage of utilities. Mining precious minerals is a very cost-intensive process with meager yields. High-throughput X-ray equipped technology plants are the future of the diamond processing industry. These plants can find and recover diamonds more easily, efficiently and safely while spending less. This type of technology can be trained to extract Type II rare diamonds, which are abundantly present in Zimbabwe. The project requires procurement and installation of high-throughput Bourevestnik (BV) for the concentration and sorting of diamond-containing ore at Marange-Chiyadzwa. Similarly, gold mining can be enhanced by AI application.

29.9.4 Tourism

Tourism is another source of economy for Zimbabwe. The tourism sector can be further enhanced by the application of AI in maintaining and managing tourism-related issues. Tourism essentials include the ability to perform internet bookings, checking the weather, flights, hotel reservations, outdoor excursions and advanced VISA payments. AI-based agents or assistants can be facilitated to perform most of these tasks. AI avatars online can help tourist to perform most of the planning and scheduling, giving them opportunities to make a well-informed decision. Tourists always are concerned about security. AI can be used to strengthen security and improve communications. AI with multiple language abilities can help locals to communicate better with tourists.

29.10 Conclusions and Future Prospects

AI is now no longer just science fiction but is becoming a day-to-day reality. Classical methods of data analysis and solution finding were linear, but the world is expanding in terms of data in an exponential manner. To cater for computational and analytical needs, AI is the most competent solution. The powers and potential of AI are yet to be explored entirely but with caution. The makers and users of AI need to find the fine line between use and abuse of AI. Expositions and summits are required to hold international dialogues between stakeholders of technology from different sectors and have a responsible exchange of ideas and expertise. AI can mine multiple databases and detect patterns to give solutions. It can help in making personal health choices and improving business, and much more. The possibilities are endless, which started to be apparent in the recent past. With advances in quantum computing and computer logic, AI will grow further stronger. Although one shouldn't fear the fantasy of machines ruling the world (the theme of much science fiction), the way it's going to impact humanity is entirely our responsibility.

Although many summits and forums are already happening, more such discussions on the ethics and responsible use of AI should take place. AI is developing very rapidly, but the discourse on its responsible use is prolonged due to the absence of liabilities. New guidelines should be made, and an international ethical code of conduct should be enforced by international agencies for regulation of AI. Nations and significant stakeholders should have incentives to sign memoranda like those on nuclear weapons (non-proliferation treaties) to make responsible use of AI. With full bridged communication between different stakeholders, unified legislation on the use of big data is expected in future for developing safe AI-equipped technologies.

References

Alharbi, O.M., Khattab, R.A., and Ali, I. (2018). Health and environmental effects of persistent organic pollutants. *Journal of Molecular Liquids* 263: 442–453.

Au, T.-C., Kuter, U., and Nau, D. (2008). Planning for interactions among autonomous agents. In: *Proceedings of the International Workshop on Programming Multi-Agent Systems*, 1–23. New York: Springer.

Autor, D.H., Levy, F., and Murnane, R.J. (2003). The skill content of recent technological change: an empirical exploration. *The Quarterly Journal of Economics* 118 (4): 1279–1333.

Bitan, M., Nahari, G., Nisin, Z., Roth, A. and Kraus, S. (2017). Psychologically based virtual-suspect for interrogative interview training. Proceedings of the Thirty-First AAAI Conference on Artificial Intelligence, San Francisco, CA.

Brynjolfsson, E., Rock, D., and Syverson, C. (2018). Artificial intelligence and the modern productivity paradox: a clash of expectations and statistics. In: *The Economics of Artificial Intelligence: An Agenda* (eds. A. Agrawal, J. Gans and A. Goldfarb), 23–57. Chicago, IL: University of Chicago Press.

Cohen, P.R. and Feigenbaum, E.A. (2014). *The Handbook of Artificial Intelligence.* Oxford: Butterworth-Heinemann.

Correll, N., Wailes, C., and Slaby, S. (2014). A one-hour curriculum to engage middle school students in robotics and computer science using cubelets. In: *Distributed Autonomous Robotic Systems* (eds. R. Gross, A. Kolling, S. Berman, et al.), 165–176. New York: Springer.

Cortés, U., Sànchez-Marrè, M., Ceccaroni, L. et al. (2000). Artificial intelligence and environmental decision support systems. *Applied Intelligence* 13 (1): 77–91.

Dietterich, T.G., Dereszynski, E.W., Hutchinson, R.A. and Sheldon, D.R. (2012). Machine learning for computational sustainability. Proceedings of the 2012 International Green Computing Conference. San José, CA, June 2–4.

Dissanayake, R., Stevenson, M., Allavena, R., and Henning, J. (2017). The value of citizen science data for passive surveillance of wildlife. In: *Proceedings of the 3rd International Animal Health Surveillance Conference*, 53–55. Frederiksberg, Denmark: New Zealand Veterinary Association.

Dourish, P. (2006). *Ubiquitous Computing: 8th International Conference, Orange County, Proceedings*, September 17–21. Berlin: Springer Science & Business Media.

Eade, E., Munich, M.E. and Fong, P. (2016). Systems and methods for VSLAM optimization. Google Patents.

El-Swaify, S. and Yakowitz, D. (1997). *Multiple Objective Decision Making for Land, Water, and Environmental Management.* Boca Raton, FL: Lewis Publishers.

Ericsson.com (2015). *How Information and Communications Technology can Achieve the Sustainable Development Goals.* New York: Ericsson in collaboration with Earth Institute at Columbia University https://www.ericsson.com/res/docs/2015/ict-and-sdginterim-report.pdf (accessed October 2018).

Fernández-Baena, A., Boldú, R., Albo-Canals, J. and Miralles, D. (2015). Interaction between Vleo and Pleo, a virtual social character and a social robot. RO-MAN 2015 – The 24th International Symposium on Robot and Human Interactive Communication, Kobe, Japan August 31 – September 4, 694–699.

Friedrich, S. (2016). Automation and artificial intelligence – what could it mean for sustainable development? Brief for GSDR, United Nations Department of Economic and Social Affairs, https://sustainabledevelopment.un.org/content/documents/968825_Soltau_Automation%20and%20artificial%20intelligence%20-%20what%20could%20it%20mean%20for%20sustainable%20development.pdf (accessed November 2019).

Frias-Martinez, V., Rubio, A. and Frias-Martinez, E. (2012). Measuring the impact of epidemic alerts on human mobility using cell-phone network data. Proceedings of the Second Workshop on Pervasive Urban Applications@ Pervasive, MIT, Boston, MA.

Golovin, D., Krause, A., Gardner, B., Converse, S.J. and Morey, S. (2011). Dynamic resource allocation in conservation planning. Proceedings of the Twenty-fifth AAAI Conference on Artificial Intelligence. San Francisco, CA. https://www.aaai.org/Press/Proceedings/aaai11.php (accessed November 2019).

Gomaa, D., Shinn, J.O., Dowd, J.R., and Branson, R.D. (2016). Point of care testing: a comparison of two devices in the intensive care unit. *Respiratory Care* 61 (10): OF18.

Hamet, P. and Tremblay, J. (2017). Artificial intelligence in medicine. *Metabolism* 69: S36–S40.

Helbing, D., Frey, B.S., Gigerenzer, G. et al. (2019). Will democracy survive big data and artificial intelligence? In: *Towards Digital Enlightenment* (ed. D. Helbing), 73–98. New York: Springer.

Hew, K.F. and Cheung, W.S. (2014). Students' and instructors' use of massive open online courses (MOOCs): motivations and challenges. *Educational Research Review* 12: 45–58.

Hilty, L.M. and Aebischer, B. (2015). ICT for sustainability: an emerging research field. In: *ICT Innovations for Sustainability* (eds. L.M. Hilty and B. Aebischer), 3–36. New York: Springer.

Hinckley, E.L.S., Bonan, G.B., Bowen, G.J. et al. (2016). The soil and plant biogeochemistry sampling design for The National Ecological Observatory Network. *Ecosphere* 7 (3): e01234.

Horvitz, E. (2014). Reflections and framing, one hundred year study on artificial intelligence. Oxford University Research Paper. Microsoft.com, California, USA.

Keeble, B.R. (1988). The Brundtland Report: "Our Common Future". *Medicine and War* 4 (1): 17–25.

Klembczyk, J.J., Jalalpour, M., Levin, S. et al. (2016). Google flu trends spatial variability validated against emergency department influenza-related visits. *Journal of Medical Internet Research* 18 (6): e175.

Kolter, J.Z., and Ferreira Jr, J. (2011). A large-scale study on predicting and contextualizing building energy usage. Proceedings of the Twenty-fifth AAAI Conference on Artificial Intelligence. San Francisco, CA. https://www.aaai.org/Press/Proceedings/aaai11.php (accessed November 2019)

Krause, A., Singh, A., and Guestrin, C. (2008). Near-optimal sensor placements in Gaussian processes: theory, efficient algorithms and empirical studies. *Journal of Machine Learning Research* 9 (Feb): 235–284.

Krishna, D., Hemalatha, T., Shankar, G.D.M. et al. (2014). Aadhar based electronic voting system and providing authentication. *International Journal of Engineering and Advanced Technology*: 2250–3676.

Lai, K.J., Gomes, C.P., Schwartz, M.K., McKelvey, K.S., Calkin, D.E. and Montgomery, C.A. (2011). The Steiner multigraph problem: wildlife corridor design for multiple species. Proceedings of the Twenty-fifth AAAI Conference on Artificial Intelligence. San Francisco, CA. https://www.aaai.org/Press/Proceedings/aaai11.php (accessed November 2019).

Lesgold, A. (1988). *SHERLOCK: A Coached Practice Environment for an Electronics Troubleshooting Job*. Philadelphia, PA: University of Pennsylvania.

Lytle, D.A., Martínez-Muñoz, G., Zhang, W. et al. (2010). Automated processing and identification of benthic invertebrate samples. *Journal of the North American Benthological Society* 29 (3): 867–874.

MacCrory, F., Westerman, G., Alhammadi, Y. and Brynjolfsson, E. (2014). Racing with and against the machine: changes in occupational skill composition in an era of rapid technological advance. Thirty Fifth International Conference on Information Systems, Auckland.

Mady, A.E.-D., Provan, G.M., Ryan, C. and Brown, K.N. (2011). Stochastic model predictive controller for the integration of building use and temperature regulation. Proceedings of the Twenty-Fifth AAAI Conference on Artificial Intelligence. San Francisco, CA. https://www.aaai.org/Press/Proceedings/aaai11.php (accessed November 2019).

Majumder, M. (2015). Artificial neural network. In: *Impact of Urbanization on Water Shortage in Face of Climatic Aberrations* (ed. M. Majumder), 49–54. New York: Springer.

McGlynn, S., Snook, B., Kemple, S., Mitzner, T.L. and Rogers, W.A. (2014). Therapeutic robots for older adults: investigating the potential of paro. Proceedings of the 2014 ACM/IEEE International Conference on Human–Robot Interaction, ACM, New York, 246–247.

Molinari, M.C. (2016). *Implementing Compstat Principles into Critical Infrastructure Protection and Improvement.* Monterey, CA: Naval Postgraduate School Monterey.

Müller, V.C. and Bostrom, N. (2016). Future progress in artificial intelligence: a survey of expert opinion. In: *Fundamental Issues of Artificial Intelligence* (ed. V.C. Müller), 555–572. New York: Springer.

Mupfiga, P.S. and Fohlo, S. (2015). The role of artificial intelligence and expert systems in the implementation of Zimasset. *The International Journal of Engineering and Science* 2.67: 72–78.

Owomugisha, G., Quinn, J.A., Mwebaze, E. and Lwasa, J. (2014). Automated vision-based diagnosis of banana bacterial wilt disease and black sigatoka disease. Proceedings of the International Conference on the Use of Mobile ICT in Africa, Citeseer, Stellenbosch, South Africa.

Polatidis, H., Haralambidou, K., and Haralambopoulos, D. (2015). Multi-criteria decision analysis for geothermal energy: a comparison between the ELECTRE III and the PROMETHEE II methods. *Energy Sources, Part B: Economics, Planning, and Policy* 10 (3): 241–249.

Prato, S.C. (2017). Beyond the computer age: a best practices intro for implementing library coding programs. *Children & Libraries* 15 (1): 19.

Purdy, M. and Daugherty, P. (2016). *Why Artificial Intelligence is the Future of Growth.* Canada: Accenture.

Quinn, J., Frias-Martinez, V., and Subramanian, L. (2014). Computational sustainability and artificial intelligence in the developing world. *AI Magazine* 35 (3): 36.

Rajbanshi, A., Bhimrajka, S., and Raina, C. (2017). Artificial intelligence in cyber security. *International Journal of Scientific Research in Computer Science, Engineering and Information Technology* 2 (3).

Russell, S.J. and Norvig, P. (2003). *Artificial Intelligence: A Modern Approach.* Kuala Lumpur, Malaysia: Pearson Education Limited.

Sachenbacher, M., Leucker, M., Artmeier, A. and Haselmayr, J. (2011). Efficient energy-optimal routing for electric vehicles. Proceedings of the Twenty-Fifth AAAI Conference on Artificial Intelligence. San Francisco, CA. https://www.aaai.org/Press/Proceedings/aaai11.php (accessed November 2019), 1402–1407.

Soto, V., Frias-Martinez, V., Virseda, J., and Frias-Martinez, E. (2011). Prediction of socioeconomic levels using cell phone records. In: *Proceedings of the International Conference on User Modeling, Adaptation, and Personalization*, 377–388. New York: Springer.

Sullivan, B.L., Aycrigg, J.L., Barry, J.H. et al. (2014). The eBird enterprise: an integrated approach to development and application of citizen science. *Biological Conservation* 169: 31–40.

Villa, F., Ceroni, M., Bagstad, K., Johnson, G. and Krivov, S. (2009). ARIES (Artificial Intelligence for Ecosystem Services): a new tool for ecosystem services assessment, planning, and valuation. Proceedings of the 11th Annual BIOECON Conference on Economic Instruments to Enhance the Conservation and Sustainable Use of Biodiversity, Conference Proceedings, Venice, Italy.

Wei, X., Kusiak, A., Li, M. et al. (2015). Multi-objective optimization of the HVAC (heating, ventilation, and air conditioning) system performance. *Energy* 83: 294–306.

Wiwanitkit, V. (2015). Google Trends (GT) related to influenza. *Cad Saude Publica* 31: 1334–1335.

Yorita, A. and Kubota, N. (2011). Cognitive development in partner robots for information support to elderly people. *IEEE Transactions on Autonomous Mental Development* 3 (1): 64–73.

Index

Sustainability: Fundamentals and Applications, First Edition.
Edited by Rao Y. Surampalli, Tian C. Zhang, Manish Kumar Goyal, Satinder K. Brar, and R. D. Tyagi

www.ingramcontent.com/pod-product-compliance
Lightning Source LLC
Chambersburg PA
CBHW081211180926
51613CB00003B/210
9781119433965